W9-CSM-505

Fourth Edition

Linear
Algebra

Stephen H. Friedberg
Arnold J. Insel
Lawrence E. Spence

Illinois State University

Prentice
Hall

Pearson Education International

Acquisitions Editor: *George Lobell*
Editor in Chief: *Sally Yagan*
Production Editor: *Lynn Savino Wendel*
Vice President/Director of Production and Manufacturing: *David W. Riccardi*
Senior Managing Editor: *Linda Mihatov Behrens*
Assistant Managing Editor: *Bayani DeLeon*
Executive Managing Editor: *Kathleen Schiaparelli*
Manufacturing Buyer: *Michael Bell*
Manufacturing Manager: *Trudy Pisciotti*
Editorial Assistant: *Jennifer Brady*
Marketing Manager: *Halee Dinsey*
Marketing Assistant: *Rachel Beckman*
Art Director: *Jayne Conte*
Cover Designer: *Bruce Kenselaar*
Cover Photo Credits: *Anni Albers, Wandbehang We 791 (Orange), 1926/64. Dreifachgewebe: Baumwolle und Kunstseide, schwarz, weiß, Orange 175 × 118 cm. Foto: Gunter Lepkowski, Berlin. Bauhaus-Archiv, Berlin, Inv. Nr. 1575. Lit.: Das Bauhaus webt, Berlin 1998, Nr. 38.*

This edition may be sold only
in those countries to which it is consigned
by Pearson Education International.
It is not to be re-exported and is
not for sale in the U.S.A., Mexico,
or Canada.

Printed in the United States of America
10 9 8 7 6 5 4 3 2

ISBN 0-13-120266-9

Pearson Education, Ltd.
Pearson Education Australia Pty. Limited
Pearson Education Singapore, Pte., Ltd
Pearson Education North Asia Ltd
Pearson Education Canada, Ltd.
Pearson Educacion de Mexico, S.A. de C.V.
Pearson Education -- Japan
Pearson Education Malaysia, Pte. Ltd
Pearson Education, Upper Saddle River, New Jersey

To our families:
Ruth Ann, Rachel, Jessica, and Jeremy
Barbara, Thomas, and Sara
Linda, Stephen, and Alison

Contents

---

*Sections denoted by an asterisk are optional.

7 Canonical Forms 482

Appendices 549

Answers to Selected Exercises 571

Index 589

Preface

The language and concepts of matrix theory and, more generally, of linear algebra have come into widespread usage in the social and natural sciences, computer science, and statistics. In addition, linear algebra continues to be of great importance in modern treatments of geometry and analysis.

The primary purpose of this fourth edition of *Linear Algebra* is to present a careful treatment of the principal topics of linear algebra and to illustrate the power of the subject through a variety of applications. Our major thrust emphasizes the symbiotic relationship between linear transformations and matrices. However, where appropriate, theorems are stated in the more general infinite-dimensional case. For example, this theory is applied to finding solutions to a homogeneous linear differential equation and the best approximation by a trigonometric polynomial to a continuous function.

Although the only formal prerequisite for this book is a one-year course in calculus, it requires the mathematical sophistication of typical junior and senior mathematics majors. This book is especially suited for a second course in linear algebra that emphasizes abstract vector spaces, although it can be used in a first course with a strong theoretical emphasis.

The book is organized to permit a number of different courses (ranging from three to eight semester hours in length) to be taught from it. The core material (vector spaces, linear transformations and matrices, systems of linear equations, determinants, diagonalization, and inner product spaces) is found in Chapters 1 through 5 and Sections 6.1 through 6.5. Chapters 6 and 7, on inner product spaces and canonical forms, are completely independent and may be studied in either order. In addition, throughout the book are applications to such areas as differential equations, economics, geometry, and physics. These applications are not central to the mathematical development, however, and may be excluded at the discretion of the instructor.

We have attempted to make it possible for many of the important topics of linear algebra to be covered in a one-semester course. This goal has led us to develop the major topics with fewer preliminaries than in a traditional approach. (Our treatment of the Jordan canonical form, for instance, does not require any theory of polynomials.) The resulting economy permits us to cover the core material of the book (omitting many of the optional sections and a detailed discussion of determinants) in a one-semester four-hour course for students who have had some prior exposure to linear algebra.

Chapter 1 of the book presents the basic theory of vector spaces: subspaces, linear combinations, linear dependence and independence, bases, and dimension. The chapter concludes with an optional section in which we prove that every infinite-dimensional vector space has a basis.

Linear transformations and their relationship to matrices are the subject of Chapter 2. We discuss the null space and range of a linear transformation, matrix representations of a linear transformation, isomorphisms, and change of coordinates. Optional sections on dual spaces and homogeneous linear differential equations end the chapter.

The application of vector space theory and linear transformations to systems of linear equations is found in Chapter 3. We have chosen to defer this important subject so that it can be presented as a consequence of the preceding material. This approach allows the familiar topic of linear systems to illuminate the abstract theory and permits us to avoid messy matrix computations in the presentation of Chapters 1 and 2. There are occasional examples in these chapters, however, where we solve systems of linear equations. (Of course, these examples are not a part of the theoretical development.) The necessary background is contained in Section 1.4.

Determinants, the subject of Chapter 4, are of much less importance than they once were. In a short course (less than one year), we prefer to treat determinants lightly so that more time may be devoted to the material in Chapters 5 through 7. Consequently we have presented two alternatives in Chapter 4—a complete development of the theory (Sections 4.1 through 4.3) and a summary of important facts that are needed for the remaining chapters (Section 4.4). Optional Section 4.5 presents an axiomatic development of the determinant.

Chapter 5 discusses eigenvalues, eigenvectors, and diagonalization. One of the most important applications of this material occurs in computing matrix limits. We have therefore included an optional section on matrix limits and Markov chains in this chapter even though the most general statement of some of the results requires a knowledge of the Jordan canonical form. Section 5.4 contains material on invariant subspaces and the Cayley–Hamilton theorem.

Inner product spaces are the subject of Chapter 6. The basic mathematical theory (inner products; the Gram–Schmidt process; orthogonal complements; the adjoint of an operator; normal, self-adjoint, orthogonal and unitary operators; orthogonal projections; and the spectral theorem) is contained in Sections 6.1 through 6.6. Sections 6.7 through 6.11 contain diverse applications of the rich inner product space structure.

Canonical forms are treated in Chapter 7. Sections 7.1 and 7.2 develop the Jordan canonical form, Section 7.3 presents the minimal polynomial, and Section 7.4 discusses the rational canonical form.

There are five appendices. The first four, which discuss sets, functions, fields, and complex numbers, respectively, are intended to review basic ideas used throughout the book. Appendix E on polynomials is used primarily in Chapters 5 and 7, especially in Section 7.4. We prefer to cite particular results from the appendices as needed rather than to discuss the appendices independently.

The following diagram illustrates the dependencies among the various chapters.

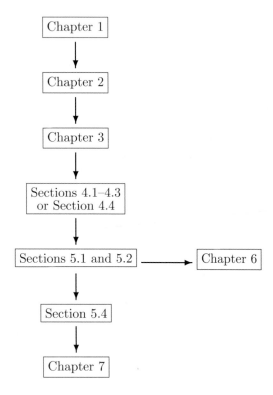

One final word is required about our notation. Sections and subsections labeled with an asterisk (∗) are optional and may be omitted as the instructor sees fit. An exercise accompanied by the dagger symbol (†) is not optional, however—we use this symbol to identify an exercise that is cited in some later section that is not optional.

DIFFERENCES BETWEEN THE THIRD AND FOURTH EDITIONS

The principal content change of this fourth edition is the inclusion of a new section (Section 6.7) discussing the singular value decomposition and the pseudoinverse of a matrix or a linear transformation between finite-dimensional inner product spaces. Our approach is to treat this material as a generalization of our characterization of normal and self-adjoint operators.

The organization of the text is essentially the same as in the third edition. Nevertheless, this edition contains many significant local changes that improve the book. Section 5.1 (Eigenvalues and Eigenvectors) has been stream-lined, and some material previously in Section 5.1 has been moved to Section 2.5 (The Change of Coordinate Matrix). Further improvements include revised proofs of some theorems, additional examples, new exercises, and literally hundreds of minor editorial changes.

We are especially indebted to Jane M. Day (San Jose State University) for her extensive and detailed comments on the fourth edition manuscript. Additional comments were provided by the following reviewers of the fourth edition manuscript: Thomas Banchoff (Brown University), Christopher Heil (Georgia Institute of Technology), and Thomas Shemanske (Dartmouth College).

To find the latest information about this book, consult our web site on the World Wide Web. We encourage comments, which can be sent to us by e-mail or ordinary post. Our web site and e-mail addresses are listed below.

web site: http://www.math.ilstu.edu/linalg

e-mail: linalg@math.ilstu.edu

Stephen H. Friedberg
Arnold J. Insel
Lawrence E. Spence

1

Vector Spaces

1.1   INTRODUCTION

Many familiar physical notions, such as forces, velocities,[1] and accelerations, involve both a magnitude (the amount of the force, velocity, or acceleration) and a direction. Any such entity involving both magnitude and direction is called a "vector." A vector is represented by an arrow whose length denotes the magnitude of the vector and whose direction represents the direction of the vector. In most physical situations involving vectors, only the magnitude and direction of the vector are significant; consequently, we regard vectors with the same magnitude and direction as being equal irrespective of their positions. In this section the geometry of vectors is discussed. This geometry is derived from physical experiments that test the manner in which two vectors interact.

Familiar situations suggest that when two like physical quantities act simultaneously at a point, the magnitude of their effect need not equal the sum of the magnitudes of the original quantities. For example, a swimmer swimming upstream at the rate of 2 miles per hour against a current of 1 mile per hour does not progress at the rate of 3 miles per hour. For in this instance the motions of the swimmer and current oppose each other, and the rate of progress of the swimmer is only 1 mile per hour upstream. If, however, the

---

[1]The word *velocity* is being used here in its scientific sense—as an entity having both magnitude and direction. The magnitude of a velocity (without regard for the direction of motion) is called its **speed**.

swimmer is moving downstream (with the current), then his or her rate of progress is 3 miles per hour downstream.

Experiments show that if two like quantities act together, their effect is predictable. In this case, the vectors used to represent these quantities can be combined to form a resultant vector that represents the combined effects of the original quantities. This resultant vector is called the *sum* of the original vectors, and the rule for their combination is called the *parallelogram law*. (See Figure 1.1.)

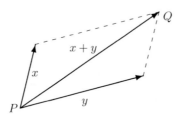

Figure 1.1

Parallelogram Law for Vector Addition.   *The sum of two vectors* x *and* y *that act at the same point* P *is the vector beginning at* P *that is represented by the diagonal of parallelogram having* x *and* y *as adjacent sides.*

Since opposite sides of a parallelogram are parallel and of equal length, the endpoint Q of the arrow representing $x + y$ can also be obtained by allowing x to act at P and then allowing y to act at the endpoint of x. Similarly, the endpoint of the vector $x + y$ can be obtained by first permitting y to act at P and then allowing x to act at the endpoint of y. Thus two vectors x and y that both act at the point P may be added "tail-to-head"; that is, either x or y may be applied at P and a vector having the same magnitude and direction as the other may be applied to the endpoint of the first. If this is done, the endpoint of the second vector is the endpoint of $x + y$.

The addition of vectors can be described algebraically with the use of analytic geometry. In the plane containing x and y, introduce a coordinate system with P at the origin. Let (a_1, a_2) denote the endpoint of x and (b_1, b_2) denote the endpoint of y. Then as Figure 1.2(a) shows, the endpoint Q of $x+y$ is $(a_1 + b_1, a_2 + b_2)$. Henceforth, when a reference is made to the coordinates of the endpoint of a vector, the vector should be assumed to emanate from the origin. Moreover, since a vector beginning at the origin is completely determined by its endpoint, we sometimes refer to *the point* x rather than *the endpoint of the vector* x if x is a vector emanating from the origin.

Besides the operation of vector addition, there is another natural operation that can be performed on vectors—the length of a vector may be magnified

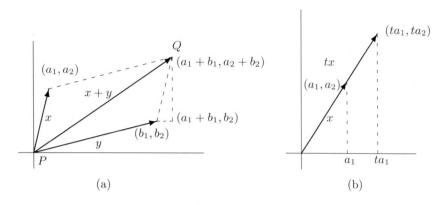

Figure 1.2

or contracted. This operation, called *scalar multiplication*, consists of multiplying the vector by a real number. If the vector x is represented by an arrow, then for any real number t, the vector tx is represented by an arrow in the same direction if $t \geq 0$ and in the opposite direction if $t < 0$. The length of the arrow tx is $|t|$ times the length of the arrow x. Two nonzero vectors x and y are called **parallel** if $y = tx$ for some nonzero real number t. (Thus nonzero vectors having the same or opposite directions are parallel.)

To describe scalar multiplication algebraically, again introduce a coordinate system into a plane containing the vector x so that x emanates from the origin. If the endpoint of x has coordinates (a_1, a_2), then the coordinates of the endpoint of tx are easily seen to be (ta_1, ta_2). (See Figure 1.2(b).)

The algebraic descriptions of vector addition and scalar multiplication for vectors in a plane yield the following properties:

1. For all vectors x and y, $x + y = y + x$.
2. For all vectors x, y, and z, $(x + y) + z = x + (y + z)$.
3. There exists a vector denoted 0 such that $x + 0 = x$ for each vector x.
4. For each vector x, there is a vector y such that $x + y = 0$.
5. For each vector x, $1x = x$.
6. For each pair of real numbers a and b and each vector x, $(ab)x = a(bx)$.
7. For each real number a and each pair of vectors x and y, $a(x + y) = ax + ay$.
8. For each pair of real numbers a and b and each vector x, $(a + b)x = ax + bx$.

Arguments similar to the preceding ones show that these eight properties, as well as the geometric interpretations of vector addition and scalar multiplication, are true also for vectors acting in space rather than in a plane. These results can be used to write equations of lines and planes in space.

Consider first the equation of a line in space that passes through two distinct points A and B. Let O denote the origin of a coordinate system in space, and let u and v denote the vectors that begin at O and end at A and B, respectively. If w denotes the vector beginning at A and ending at B, then "tail-to-head" addition shows that $u + w = v$, and hence $w = v - u$, where $-u$ denotes the vector $(-1)u$. (See Figure 1.3, in which the quadrilateral $OABC$ is a parallelogram.) Since a scalar multiple of w is parallel to w but possibly of a different length than w, any point on the line joining A and B may be obtained as the endpoint of a vector that begins at A and has the form tw for some real number t. Conversely, the endpoint of every vector of the form tw that begins at A lies on the line joining A and B. Thus an equation of the line through A and B is $x = u + tw = u + t(v - u)$, where t is a real number and x denotes an arbitrary point on the line. Notice also that the endpoint C of the vector $v - u$ in Figure 1.3 has coordinates equal to the difference of the coordinates of B and A.

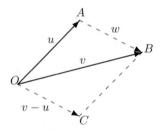

Figure 1.3

Example 1

Let A and B be points having coordinates $(-2, 0, 1)$ and $(4, 5, 3)$, respectively. The endpoint C of the vector emanating from the origin and having the same direction as the vector beginning at A and terminating at B has coordinates $(4, 5, 3) - (-2, 0, 1) = (6, 5, 2)$. Hence the equation of the line through A and B is

$$x = (-2, 0, 1) + t(6, 5, 2). \quad \blacklozenge$$

Now let A, B, and C denote any three noncollinear points in space. These points determine a unique plane, and its equation can be found by use of our previous observations about vectors. Let u and v denote vectors beginning at A and ending at B and C, respectively. Observe that any point in the plane containing A, B, and C is the endpoint S of a vector x beginning at A and having the form $su + tv$ for some real numbers s and t. The endpoint of su is the point of intersection of the line through A and B with the line through S

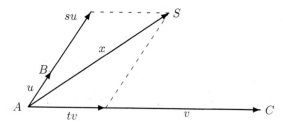

Figure 1.4

parallel to the line through A and C. (See Figure 1.4.) A similar procedure locates the endpoint of tv. Moreover, for any real numbers s and t, the vector $su + tv$ lies in the plane containing A, B, and C. It follows that an equation of the plane containing A, B, and C is

$$x = A + su + tv,$$

where s and t are arbitrary real numbers and x denotes an arbitrary point in the plane.

Example 2

Let A, B, and C be the points having coordinates $(1,0,2)$, $(-3,-2,4)$, and $(1,8,-5)$, respectively. The endpoint of the vector emanating from the origin and having the same length and direction as the vector beginning at A and terminating at B is

$$(-3,-2,4) - (1,0,2) = (-4,-2,2).$$

Similarly, the endpoint of a vector emanating from the origin and having the same length and direction as the vector beginning at A and terminating at C is $(1,8,-5) - (1,0,2) = (0,8,-7)$. Hence the equation of the plane containing the three given points is

$$x = (1,0,2) + s(-4,-2,2) + t(0,8,-7). \quad \blacklozenge$$

Any mathematical structure possessing the eight properties on page 3 is called a *vector space*. In the next section we formally define a vector space and consider many examples of vector spaces other than the ones mentioned above.

EXERCISES

1. Determine whether the vectors emanating from the origin and terminating at the following pairs of points are parallel.

    (a) $(3, 1, 2)$ and $(6, 4, 2)$
    (b) $(-3, 1, 7)$ and $(9, -3, -21)$
    (c) $(5, -6, 7)$ and $(-5, 6, -7)$
    (d) $(2, 0, -5)$ and $(5, 0, -2)$

2. Find the equations of the lines through the following pairs of points in space.

    (a) $(3, -2, 4)$ and $(-5, 7, 1)$
    (b) $(2, 4, 0)$ and $(-3, -6, 0)$
    (c) $(3, 7, 2)$ and $(3, 7, -8)$
    (d) $(-2, -1, 5)$ and $(3, 9, 7)$

3. Find the equations of the planes containing the following points in space.

    (a) $(2, -5, -1)$, $(0, 4, 6)$, and $(-3, 7, 1)$
    (b) $(3, -6, 7)$, $(-2, 0, -4)$, and $(5, -9, -2)$
    (c) $(-8, 2, 0)$, $(1, 3, 0)$, and $(6, -5, 0)$
    (d) $(1, 1, 1)$, $(5, 5, 5)$, and $(-6, 4, 2)$

4. What are the coordinates of the vector 0 in the Euclidean plane that satisfies property 3 on page 3? Justify your answer.

5. Prove that if the vector x emanates from the origin of the Euclidean plane and terminates at the point with coordinates (a_1, a_2), then the vector tx that emanates from the origin terminates at the point with coordinates (ta_1, ta_2).

6. Show that the midpoint of the line segment joining the points (a, b) and (c, d) is $((a + c)/2, (b + d)/2)$.

7. Prove that the diagonals of a parallelogram bisect each other.

1.2   VECTOR SPACES

In Section 1.1, we saw that with the natural definitions of vector addition and scalar multiplication, the vectors in a plane satisfy the eight properties listed on page 3. Many other familiar algebraic systems also permit definitions of addition and scalar multiplication that satisfy the same eight properties. In this section, we introduce some of these systems, but first we formally define this type of algebraic structure.

    Definitions.   *A **vector space** (or **linear space**) V over a field[2] F consists of a set on which two operations (called **addition** and **scalar multiplication**, respectively) are defined so that for each pair of elements x, y,*

---

[2]Fields are discussed in Appendix C.

in V there is a unique element $x + y$ in V, and for each element a in F and each element x in V there is a unique element ax in V, such that the following conditions hold.

(VS 1) For all x, y in V, $x + y = y + x$ (commutativity of addition).

(VS 2) For all x, y, z in V, $(x + y) + z = x + (y + z)$ (associativity of addition).

(VS 3) There exists an element in V denoted by 0 such that $x + 0 = x$ for each x in V.

(VS 4) For each element x in V there exists an element y in V such that $x + y = 0$.

(VS 5) For each element x in V, $1x = x$.

(VS 6) For each pair of elements a, b in F and each element x in V, $(ab)x = a(bx)$.

(VS 7) For each element a in F and each pair of elements x, y in V, $a(x + y) = ax + ay$.

(VS 8) For each pair of elements a, b in F and each element x in V, $(a + b)x = ax + bx$.

The elements $x + y$ and ax are called the **sum** of x and y and the **product** of a and x, respectively.

The elements of the field F are called **scalars** and the elements of the vector space V are called **vectors**. The reader should not confuse this use of the word "vector" with the physical entity discussed in Section 1.1: the word "vector" is now being used to describe any element of a vector space.

A vector space is frequently discussed in the text without explicitly mentioning its field of scalars. The reader is cautioned to remember, however, that *every vector space is regarded as a vector space over a given field, which is denoted by F.* Occasionally we restrict our attention to the fields of real and complex numbers, which are denoted R and C, respectively.

Observe that (VS 2) permits us to unambiguously define the addition of any finite number of vectors (without the use of parentheses).

In the remainder of this section we introduce several important examples of vector spaces that are studied throughout this text. Observe that in describing a vector space, it is necessary to specify not only the vectors but also the operations of addition and scalar multiplication.

An object of the form $(a_1, a_2, \ldots, a_n)$, where the entries $a_1, a_2, \ldots, a_n$ are elements of a field F, is called an **n-tuple** with entries from F. The elements

$a_1, a_2, \ldots, a_n$ are called the **entries** or **components** of the n-tuple. Two n-tuples $(a_1, a_2, \ldots, a_n)$ and $(b_1, b_2, \ldots, b_n)$ with entries from a field F are called **equal** if $a_i = b_i$ for $i = 1, 2, \ldots, n$.

Example 1

The set of all n-tuples with entries from a field F is denoted by F^n. This set is a vector space over F with the operations of coordinatewise addition and scalar multiplication; that is, if $u = (a_1, a_2, \ldots, a_n) \in \mathsf{F}^n$, $v = (b_1, b_2 \ldots, b_n) \in \mathsf{F}^n$, and $c \in F$, then

$$u + v = (a_1 + b_1, a_2 + b_2, \ldots, a_n + b_n) \quad \text{and} \quad cu = (ca_1, ca_2, \ldots, ca_n).$$

Thus R^3 is a vector space over R. In this vector space,

$$(3, -2, 0) + (-1, 1, 4) = (2, -1, 4) \quad \text{and} \quad -5(1, -2, 0) = (-5, 10, 0).$$

Similarly, C^2 is a vector space over C. In this vector space,

$$(1 + i, 2) + (2 - 3i, 4i) = (3 - 2i, 2 + 4i) \quad \text{and} \quad i(1 + i, 2) = (-1 + i, 2i).$$

Vectors in F^n may be written as **column vectors**

$$\begin{pmatrix} a_1 \\ a_2 \\ \vdots \\ a_n \end{pmatrix}$$

rather than as **row vectors** $(a_1, a_2, \ldots, a_n)$. Since a 1-tuple whose only entry is from F can be regarded as an element of F, we usually write F rather than F^1 for the vector space of 1-tuples with entry from F.   ◆

An $m \times n$ **matrix** with entries from a field F is a rectangular array of the form

$$\begin{pmatrix} a_{11} & a_{12} & \cdots & a_{1n} \\ a_{21} & a_{22} & \cdots & a_{2n} \\ \vdots & \vdots & & \vdots \\ a_{m1} & a_{m2} & \cdots & a_{mn} \end{pmatrix},$$

where each entry a_{ij} ($1 \leq i \leq m$, $1 \leq j \leq n$) is an element of F. We call the entries a_{ij} with $i = j$ the **diagonal entries** of the matrix. The entries $a_{i1}, a_{i2}, \ldots, a_{in}$ compose the *i*th **row** of the matrix, and the entries $a_{1j}, a_{2j}, \ldots, a_{mj}$ compose the *j*th **column** of the matrix. The rows of the preceding matrix are regarded as vectors in F^n, and the columns are regarded as vectors in F^m. The $m \times n$ matrix in which each entry equals zero is called the **zero matrix** and is denoted by O.

In this book, we denote matrices by capital italic letters (e.g., A, B, and C), and we denote the entry of a matrix A that lies in row i and column j by A_{ij}. In addition, if the number of rows and columns of a matrix are equal, the matrix is called **square**.

Two $m \times n$ matrices A and B are called **equal** if all their corresponding entries are equal, that is, if $A_{ij} = B_{ij}$ for $1 \leq i \leq m$ and $1 \leq j \leq n$.

Example 2

The set of all $m \times n$ matrices with entries from a field F is a vector space, which we denote by $\mathsf{M}_{m \times n}(F)$, with the following operations of **matrix addition** and **scalar multiplication**: For $A, B \in \mathsf{M}_{m \times n}(F)$ and $c \in F$,

$$(A + B)_{ij} = A_{ij} + B_{ij} \quad \text{and} \quad (cA)_{ij} = cA_{ij}$$

for $1 \leq i \leq m$ and $1 \leq j \leq n$. For instance,

$$\begin{pmatrix} 2 & 0 & -1 \\ 1 & -3 & 4 \end{pmatrix} + \begin{pmatrix} -5 & -2 & 6 \\ 3 & 4 & -1 \end{pmatrix} = \begin{pmatrix} -3 & -2 & 5 \\ 4 & 1 & 3 \end{pmatrix}$$

and

$$-3 \begin{pmatrix} 1 & 0 & -2 \\ -3 & 2 & 3 \end{pmatrix} = \begin{pmatrix} -3 & 0 & 6 \\ 9 & -6 & -9 \end{pmatrix}.$$

in $\mathsf{M}_{2 \times 3}(R)$.   ◆

Example 3

Let S be any nonempty set and F be any field, and let $\mathcal{F}(S, F)$ denote the set of all functions from S to F. Two functions f and g in $\mathcal{F}(S, F)$ are called **equal** if $f(s) = g(s)$ for each $s \in S$. The set $\mathcal{F}(S, F)$ is a vector space with the operations of addition and scalar multiplication defined for $f, g \in \mathcal{F}(S, F)$ and $c \in F$ by

$$(f + g)(s) = f(s) + g(s) \quad \text{and} \quad (cf)(s) = c[f(s)]$$

for each $s \in S$. Note that these are the familiar operations of addition and scalar multiplication for functions used in algebra and calculus.   ◆

A **polynomial** with coefficients from a field F is an expression of the form

$$f(x) = a_n x^n + a_{n-1} x^{n-1} + \cdots + a_1 x + a_0,$$

where n is a nonnegative integer and each a_k, called the **coefficient** of x^k, is in F. If $f(x) = 0$, that is, if $a_n = a_{n-1} = \cdots = a_0 = 0$, then $f(x)$ is called the **zero polynomial** and, for convenience, its degree is defined to be -1;

otherwise, the **degree** of a polynomial is defined to be the largest exponent of x that appears in the representation

$$f(x) = a_n x^n + a_{n-1} x^{n-1} + \cdots + a_1 x + a_0$$

with a nonzero coefficient. Note that the polynomials of degree zero may be written in the form $f(x) = c$ for some nonzero scalar c. Two polynomials,

$$f(x) = a_n x^n + a_{n-1} x^{n-1} + \cdots + a_1 x + a_0$$

and

$$g(x) = b_m x^m + b_{m-1} x^{m-1} + \cdots + b_1 x + b_0,$$

are called **equal** if $m = n$ and $a_i = b_i$ for $i = 0, 1, \ldots, n$.

When F is a field containing infinitely many scalars, we usually regard a polynomial with coefficients from F as a function from F into F. (See page 569.) In this case, the value of the function

$$f(x) = a_n x^n + a_{n-1} x^{n-1} + \cdots + a_1 x + a_0$$

at $c \in F$ is the scalar

$$f(c) = a_n c^n + a_{n-1} c^{n-1} + \cdots + a_1 c + a_0.$$

Here either of the notations f or $f(x)$ is used for the polynomial function

$$f(x) = a_n x^n + a_{n-1} x^{n-1} + \cdots + a_1 x + a_0.$$

Example 4

Let

$$f(x) = a_n x^n + a_{n-1} x^{n-1} + \cdots + a_1 x + a_0$$

and

$$g(x) = b_m x^m + b_{m-1} x^{m-1} + \cdots + b_1 x + b_0$$

be polynomials with coefficients from a field F. Suppose that $m \leq n$, and define $b_{m+1} = b_{m+2} = \cdots = b_n = 0$. Then $g(x)$ can be written as

$$g(x) = b_n x^n + b_{n-1} x^{n-1} + \cdots + b_1 x + b_0.$$

Define

$$f(x) + g(x) = (a_n + b_n)x^n + (a_{n-1} + b_{n-1})x^{n-1} + \cdots + (a_1 + b_1)x + (a_0 + b_0)$$

and for any $c \in F$, define

$$cf(x) = ca_n x^n + ca_{n-1} x^{n-1} + \cdots + ca_1 x + ca_0.$$

With these operations of addition and scalar multiplication, the set of all polynomials with coefficients from F is a vector space, which we denote by $P(F)$.   ◆

We will see in Exercise 23 of Section 2.4 that the vector space defined in the next example is essentially the same as $\mathsf{P}(F)$.

Example 5

Let F be any field. A **sequence** in F is a function σ from the positive integers into F. In this book, the sequence σ such that $\sigma(n) = a_n$ for $n = 1, 2, \ldots$ is denoted $\{a_n\}$. Let V consist of all sequences $\{a_n\}$ in F that have only a finite number of nonzero terms a_n. If $\{a_n\}$ and $\{b_n\}$ are in V and $t \in F$, define

$$\{a_n\} + \{b_n\} = \{a_n + b_n\} \quad \text{and} \quad t\{a_n\} = \{ta_n\}.$$

With these operations V is a vector space.   ◆

Our next two examples contain sets on which addition and scalar multiplication are defined, but which are *not* vector spaces.

Example 6

Let $S = \{(a_1, a_2) : a_1, a_2 \in R\}$. For $(a_1, a_2), (b_1, b_2) \in S$ and $c \in R$, define

$$(a_1, a_2) + (b_1, b_2) = (a_1 + b_1, a_2 - b_2) \quad \text{and} \quad c(a_1, a_2) = (ca_1, ca_2).$$

Since (VS 1), (VS 2), and (VS 8) fail to hold, S is not a vector space with these operations.   ◆

Example 7

Let S be as in Example 6. For $(a_1, a_2), (b_1, b_2) \in S$ and $c \in R$, define

$$(a_1, a_2) + (b_1, b_2) = (a_1 + b_1, 0) \quad \text{and} \quad c(a_1, a_2) = (ca_1, 0).$$

Then S is not a vector space with these operations because (VS 3) (hence (VS 4)) and (VS 5) fail.   ◆

We conclude this section with a few of the elementary consequences of the definition of a vector space.

Theorem 1.1 (Cancellation Law for Vector Addition).   *If x, y, and z are vectors in a vector space V such that $x + z = y + z$, then $x = y$.*

Proof. There exists a vector v in V such that $z + v = 0$ (VS 4). Thus

$$x = x + 0 = x + (z + v) = (x + z) + v$$
$$= (y + z) + v = y + (z + v) = y + 0 = y$$

by (VS 2) and (VS 3).   ∎

Corollary 1.   *The vector 0 described in (VS 3) is unique.*

Proof. Exercise.   ∎

Corollary 2. *The vector y described in (VS 4) is unique.*

Proof. Exercise.   ∎

The vector 0 in (VS 3) is called the **zero vector** of V, and the vector y in (VS 4) (that is, the unique vector such that $x + y = 0$) is called the **additive inverse** of x and is denoted by $-x$.

The next result contains some of the elementary properties of scalar multiplication.

Theorem 1.2. *In any vector space* V, *the following statements are true:*
(a) $0x = 0$ *for each* $x \in$ V.
(b) $(-a)x = -(ax) = a(-x)$ *for each* $a \in F$ *and each* $x \in$ V.
(c) $a0 = 0$ *for each* $a \in F$.

Proof. (a) By (VS 8), (VS 3), and (VS 1), it follows that

$$0x + 0x = (0 + 0)x = 0x = 0x + 0 = 0 + 0x.$$

Hence $0x = 0$ by Theorem 1.1.

(b) The vector $-(ax)$ is the unique element of V such that $ax + [-(ax)] = 0$. Thus if $ax + (-a)x = 0$, Corollary 2 to Theorem 1.1 implies that $(-a)x = -(ax)$. But by (VS 8),

$$ax + (-a)x = [a + (-a)]x = 0x = 0$$

by (a). Consequently $(-a)x = -(ax)$. In particular, $(-1)x = -x$. So, by (VS 6),

$$a(-x) = a[(-1)x] = [a(-1)]x = (-a)x.$$

The proof of (c) is similar to the proof of (a).   ∎

EXERCISES

1. Label the following statements as true or false.
    (a) Every vector space contains a zero vector.
    (b) A vector space may have more than one zero vector.
    (c) In any vector space, $ax = bx$ implies that $a = b$.
    (d) In any vector space, $ax = ay$ implies that $x = y$.
    (e) A vector in F^n may be regarded as a matrix in $\mathsf{M}_{n \times 1}(F)$.
    (f) An $m \times n$ matrix has m columns and n rows.
    (g) In $\mathsf{P}(F)$, only polynomials of the same degree may be added.
    (h) If f and g are polynomials of degree n, then $f + g$ is a polynomial of degree n.
    (i) If f is a polynomial of degree n and c is a nonzero scalar, then cf is a polynomial of degree n.

(j) A nonzero scalar of F may be considered to be a polynomial in $P(F)$ having degree zero.

(k) Two functions in $\mathcal{F}(S, F)$ are equal if and only if they have the same value at each element of S.

2. Write the zero vector of $M_{3\times 4}(F)$.

3. If

$$M = \begin{pmatrix} 1 & 2 & 3 \\ 4 & 5 & 6 \end{pmatrix},$$

what are M_{13}, M_{21}, and M_{22}?

4. Perform the indicated operations.

(a) $\begin{pmatrix} 2 & 5 & -3 \\ 1 & 0 & 7 \end{pmatrix} + \begin{pmatrix} 4 & -2 & 5 \\ -5 & 3 & 2 \end{pmatrix}$

(b) $\begin{pmatrix} -6 & 4 \\ 3 & -2 \\ 1 & 8 \end{pmatrix} + \begin{pmatrix} 7 & -5 \\ 0 & -3 \\ 2 & 0 \end{pmatrix}$

(c) $4 \begin{pmatrix} 2 & 5 & -3 \\ 1 & 0 & 7 \end{pmatrix}$

(d) $-5 \begin{pmatrix} -6 & 4 \\ 3 & -2 \\ 1 & 8 \end{pmatrix}$

(e) $(2x^4 - 7x^3 + 4x + 3) + (8x^3 + 2x^2 - 6x + 7)$
(f) $(-3x^3 + 7x^2 + 8x - 6) + (2x^3 - 8x + 10)$
(g) $5(2x^7 - 6x^4 + 8x^2 - 3x)$
(h) $3(x^5 - 2x^3 + 4x + 2)$

Exercises 5 and 6 show why the definitions of matrix addition and scalar multiplication (as defined in Example 2) are the appropriate ones.

5. Richard Gard ("Effects of Beaver on Trout in Sagehen Creek, California," *J. Wildlife Management*, **25**, 221-242) reports the following number of trout having crossed beaver dams in Sagehen Creek.

Upstream Crossings

	Fall	Spring	Summer
Brook trout	8	3	1
Rainbow trout	3	0	0
Brown trout	3	0	0

Downstream Crossings

	Fall	Spring	Summer
Brook trout	9	1	4
Rainbow trout	3	0	0
Brown trout	1	1	0

Record the upstream and downstream crossings in two 3×3 matrices, and verify that the sum of these matrices gives the total number of crossings (both upstream and downstream) categorized by trout species and season.

6. At the end of May, a furniture store had the following inventory.

	Early American	Spanish	Mediterranean	Danish
Living room suites	4	2	1	3
Bedroom suites	5	1	1	4
Dining room suites	3	1	2	6

Record these data as a 3×4 matrix M. To prepare for its June sale, the store decided to double its inventory on each of the items listed in the preceding table. Assuming that none of the present stock is sold until the additional furniture arrives, verify that the inventory on hand after the order is filled is described by the matrix $2M$. If the inventory at the end of June is described by the matrix

$$A = \begin{pmatrix} 5 & 3 & 1 & 2 \\ 6 & 2 & 1 & 5 \\ 1 & 0 & 3 & 3 \end{pmatrix},$$

interpret $2M - A$. How many suites were sold during the June sale?

7. Let $S = \{0, 1\}$ and $F = R$. In $\mathcal{F}(S, R)$, show that $f = g$ and $f + g = h$, where $f(t) = 2t + 1$, $g(t) = 1 + 4t - 2t^2$, and $h(t) = 5^t + 1$.

8. In any vector space V, show that $(a + b)(x + y) = ax + ay + bx + by$ for any $x, y \in$ V and any $a, b \in F$.

9. Prove Corollaries 1 and 2 of Theorem 1.1 and Theorem 1.2(c).

10. Let V denote the set of all differentiable real-valued functions defined on the real line. Prove that V is a vector space with the operations of addition and scalar multiplication defined in Example 3.

11. Let $V = \{0\}$ consist of a single vector 0 and define $0 + 0 = 0$ and $c0 = 0$ for each scalar c in F. Prove that V is a vector space over F. (V is called the **zero vector space**.)

12. A real-valued function f defined on the real line is called an **even function** if $f(-t) = f(t)$ for each real number t. Prove that the set of even functions defined on the real line with the operations of addition and scalar multiplication defined in Example 3 is a vector space.

13. Let V denote the set of ordered pairs of real numbers. If (a_1, a_2) and (b_1, b_2) are elements of V and $c \in R$, define

$$(a_1, a_2) + (b_1, b_2) = (a_1 + b_1, a_2 b_2) \quad \text{and} \quad c(a_1, a_2) = (ca_1, a_2).$$

Is V a vector space over R with these operations? Justify your answer.

14. Let $V = \{(a_1, a_2, \ldots, a_n): a_i \in C \text{ for } i = 1, 2, \ldots n\}$; so V is a vector space over C by Example 1. Is V a vector space over the field of real numbers with the operations of coordinatewise addition and multiplication?

15. Let $V = \{(a_1, a_2, \ldots, a_n): a_i \in R \text{ for } i = 1, 2, \ldots n\}$; so V is a vector space over R by Example 1. Is V a vector space over the field of complex numbers with the operations of coordinatewise addition and multiplication?

16. Let V denote the set of all $m \times n$ matrices with real entries; so V is a vector space over R by Example 2. Let F be the field of rational numbers. Is V a vector space over F with the usual definitions of matrix addition and scalar multiplication?

17. Let $V = \{(a_1, a_2): a_1, a_2 \in F\}$, where F is a field. Define addition of elements of V coordinatewise, and for $c \in F$ and $(a_1, a_2) \in V$, define

$$c(a_1, a_2) = (a_1, 0).$$

Is V a vector space over F with these operations? Justify your answer.

18. Let $V = \{(a_1, a_2): a_1, a_2 \in R\}$. For $(a_1, a_2), (b_1, b_2) \in V$ and $c \in R$, define

$$(a_1, a_2) + (b_1, b_2) = (a_1 + 2b_1, a_2 + 3b_2) \quad \text{and} \quad c(a_1, a_2) = (ca_1, ca_2).$$

Is V a vector space over R with these operations? Justify your answer.

19. Let $V = \{(a_1, a_2): a_1, a_2 \in R\}$. Define addition of elements of V coordinatewise, and for (a_1, a_2) in V and $c \in R$, define

$$c(a_1, a_2) = \begin{cases} (0,0) & \text{if } c = 0 \\ \left(ca_1, \dfrac{a_2}{c}\right) & \text{if } c \neq 0. \end{cases}$$

Is V a vector space over R with these operations? Justify your answer.

20. Let V be the set of sequences $\{a_n\}$ of real numbers. (See Example 5 for the definition of a sequence.) For $\{a_n\}, \{b_n\} \in V$ and any real number t, define

$$\{a_n\} + \{b_n\} = \{a_n + b_n\} \quad \text{and} \quad t\{a_n\} = \{ta_n\}.$$

Prove that, with these operations, V is a vector space over R.

21. Let V and W be vector spaces over a field F. Let

$$Z = \{(v, w): v \in V \text{ and } w \in W\}.$$

Prove that Z is a vector space over F with the operations

$$(v_1, w_1) + (v_2, w_2) = (v_1 + v_2, w_1 + w_2) \quad \text{and} \quad c(v_1, w_1) = (cv_1, cw_1).$$

22. How many matrices are there in the vector space $M_{m \times n}(Z_2)$? (See Appendix C.)

1.3   SUBSPACES

In the study of any algebraic structure, it is of interest to examine subsets that possess the same structure as the set under consideration. The appropriate notion of substructure for vector spaces is introduced in this section.

Definition.   *A subset* W *of a vector space* V *over a field* F *is called a* **subspace** *of* V *if* W *is a vector space over* F *with the operations of addition and scalar multiplication defined on* V.

In any vector space V, note that V and $\{0\}$ are subspaces. The latter is called the **zero subspace** of V.

Fortunately it is not necessary to verify all of the vector space properties to prove that a subset is a subspace. Because properties (VS 1), (VS 2), (VS 5), (VS 6), (VS 7), and (VS 8) hold for all vectors in the vector space, these properties automatically hold for the vectors in any subset. Thus a subset W of a vector space V is a subspace of V if and only if the following four properties hold.

1. $x + y \in W$ whenever $x \in W$ and $y \in W$. (W is **closed under addition.**)
2. $cx \in W$ whenever $c \in F$ and $x \in W$. (W is **closed under scalar multiplication.**)
3. W has a zero vector.
4. Each vector in W has an additive inverse in W.

The next theorem shows that the zero vector of W must be the same as the zero vector of V and that property 4 is redundant.

Theorem 1.3. *Let V be a vector space and W a subset of V. Then W is a subspace of V if and only if the following three conditions hold for the operations defined in V.*

(a) $0 \in W$.
(b) $x + y \in W$ *whenever* $x \in W$ *and* $y \in W$.
(c) $cx \in W$ *whenever* $c \in F$ *and* $x \in W$.

Proof. If W is a subspace of V, then W is a vector space with the operations of addition and scalar multiplication defined on V. Hence conditions (b) and (c) hold, and there exists a vector $0' \in W$ such that $x + 0' = x$ for each $x \in W$. But also $x + 0 = x$, and thus $0' = 0$ by Theorem 1.1 (p. 11). So condition (a) holds.

Conversely, if conditions (a), (b), and (c) hold, the discussion preceding this theorem shows that W is a subspace of V if the additive inverse of each vector in W lies in W. But if $x \in W$, then $(-1)x \in W$ by condition (c), and $-x = (-1)x$ by Theorem 1.2 (p. 12). Hence W is a subspace of V.   ∎

The preceding theorem provides a simple method for determining whether or not a given subset of a vector space is a subspace. Normally, it is this result that is used to prove that a subset is, in fact, a subspace.

The **transpose** A^t of an $m \times n$ matrix A is the $n \times m$ matrix obtained from A by interchanging the rows with the columns; that is, $(A^t)_{ij} = A_{ji}$. For example,

$$\begin{pmatrix} 1 & -2 & 3 \\ 0 & 5 & -1 \end{pmatrix}^t = \begin{pmatrix} 1 & 0 \\ -2 & 5 \\ 3 & -1 \end{pmatrix} \quad \text{and} \quad \begin{pmatrix} 1 & 2 \\ 2 & 3 \end{pmatrix}^t = \begin{pmatrix} 1 & 2 \\ 2 & 3 \end{pmatrix}.$$

A **symmetric matrix** is a matrix A such that $A^t = A$. For example, the 2×2 matrix displayed above is a symmetric matrix. Clearly, a symmetric matrix must be square. The set W of all symmetric matrices in $M_{n \times n}(F)$ is a subspace of $M_{n \times n}(F)$ since the conditions of Theorem 1.3 hold:

1. The zero matrix is equal to its transpose and hence belongs to W.

It is easily proved that for any matrices A and B and any scalars a and b, $(aA + bB)^t = aA^t + bB^t$. (See Exercise 3.) Using this fact, we show that the set of symmetric matrices is closed under addition and scalar multiplication.

2. If $A \in W$ and $B \in W$, then $A^t = A$ and $B^t = B$. Thus $(A + B)^t = A^t + B^t = A + B$, so that $A + B \in W$.

3. If $A \in W$, then $A^t = A$. So for any $a \in F$, we have $(aA)^t = aA^t = aA$. Thus $aA \in W$.

The examples that follow provide further illustrations of the concept of a subspace. The first three are particularly important.

Example 1

Let n be a nonnegative integer, and let $P_n(F)$ consist of all polynomials in $P(F)$ having degree less than or equal to n. Since the zero polynomial has degree -1, it is in $P_n(F)$. Moreover, the sum of two polynomials with degrees less than or equal to n is another polynomial of degree less than or equal to n, and the product of a scalar and a polynomial of degree less than or equal to n is a polynomial of degree less than or equal to n. So $P_n(F)$ is closed under addition and scalar multiplication. It therefore follows from Theorem 1.3 that $P_n(F)$ is a subspace of $P(F)$.   ◆

Example 2

Let $C(R)$ denote the set of all continuous real-valued functions defined on R. Clearly $C(R)$ is a subset of the vector space $\mathcal{F}(R, R)$ defined in Example 3 of Section 1.2. We claim that $C(R)$ is a subspace of $\mathcal{F}(R, R)$. First note that the zero of $\mathcal{F}(R, R)$ is the constant function defined by $f(t) = 0$ for all $t \in R$. Since constant functions are continuous, we have $f \in C(R)$. Moreover, the sum of two continuous functions is continuous, and the product of a real number and a continuous function is continuous. So $C(R)$ is closed under addition and scalar multiplication and hence is a subspace of $\mathcal{F}(R, R)$ by Theorem 1.3.   ◆

Example 3

An $n \times n$ matrix M is called a **diagonal matrix** if $M_{ij} = 0$ whenever $i \neq j$, that is, if all its nondiagonal entries are zero. Clearly the zero matrix is a diagonal matrix because all of its entries are 0. Moreover, if A and B are diagonal $n \times n$ matrices, then whenever $i \neq j$,

$$(A + B)_{ij} = A_{ij} + B_{ij} = 0 + 0 = 0 \quad \text{and} \quad (cA)_{ij} = cA_{ij} = c0 = 0$$

for any scalar c. Hence $A + B$ and cA are diagonal matrices for any scalar c. Therefore the set of diagonal matrices is a subspace of $M_{n \times n}(F)$ by Theorem 1.3.   ◆

Example 4

The **trace** of an $n \times n$ matrix M, denoted $\text{tr}(M)$, is the sum of the diagonal entries of M; that is,

$$\text{tr}(M) = M_{11} + M_{22} + \cdots + M_{nn}.$$

It follows from Exercise 6 that the set of $n \times n$ matrices having trace equal to zero is a subspace of $\mathsf{M}_{n \times n}(F)$.   ◆

Example 5

The set of matrices in $\mathsf{M}_{m \times n}(R)$ having nonnegative entries is not a subspace of $\mathsf{M}_{m \times n}(R)$ because it is not closed under scalar multiplication (by negative scalars).   ◆

The next theorem shows how to form a new subspace from other subspaces.

Theorem 1.4. *Any intersection of subspaces of a vector space* V *is a subspace of* V.

Proof. Let $\mathcal{C}$ be a collection of subspaces of V, and let W denote the intersection of the subspaces in $\mathcal{C}$. Since every subspace contains the zero vector, $0 \in$ W. Let $a \in F$ and $x, y \in$ W. Then x and y are contained in each subspace in $\mathcal{C}$. Because each subspace in $\mathcal{C}$ is closed under addition and scalar multiplication, it follows that $x + y$ and ax are contained in each subspace in $\mathcal{C}$. Hence $x + y$ and ax are also contained in W, so that W is a subspace of V by Theorem 1.3. ∎

Having shown that the intersection of subspaces of a vector space V is a subspace of V, it is natural to consider whether or not the union of subspaces of V is a subspace of V. It is easily seen that the union of subspaces must contain the zero vector and be closed under scalar multiplication, but in general the union of subspaces of V need not be closed under addition. In fact, it can be readily shown that the union of two subspaces of V is a subspace of V if and only if one of the subspaces contains the other. (See Exercise 19.) There is, however, a natural way to combine two subspaces W_1 and W_2 to obtain a subspace that contains both W_1 and W_2. As we already have suggested, the key to finding such a subspace is to assure that it must be closed under addition. This idea is explored in Exercise 23.

EXERCISES

1. Label the following statements as true or false.
    (a) If V is a vector space and W is a subset of V that is a vector space, then W is a subspace of V.
    (b) The empty set is a subspace of every vector space.
    (c) If V is a vector space other than the zero vector space, then V contains a subspace W such that $W \neq$ V.
    (d) The intersection of any two subsets of V is a subspace of V.

(e) An $n \times n$ diagonal matrix can never have more than n nonzero entries.

(f) The trace of a square matrix is the product of its diagonal entries.

(g) Let W be the xy-plane in R^3; that is, $W = \{(a_1, a_2, 0) : a_1, a_2 \in R\}$. Then $W = R^2$.

2. Determine the transpose of each of the matrices that follow. In addition, if the matrix is square, compute its trace.

(a) $\begin{pmatrix} -4 & 2 \\ 5 & -1 \end{pmatrix}$

(b) $\begin{pmatrix} 0 & 8 & -6 \\ 3 & 4 & 7 \end{pmatrix}$

(c) $\begin{pmatrix} -3 & 9 \\ 0 & -2 \\ 6 & 1 \end{pmatrix}$

(d) $\begin{pmatrix} 10 & 0 & -8 \\ 2 & -4 & 3 \\ -5 & 7 & 6 \end{pmatrix}$

(e) $\begin{pmatrix} 1 & -1 & 3 & 5 \end{pmatrix}$

(f) $\begin{pmatrix} -2 & 5 & 1 & 4 \\ 7 & 0 & 1 & -6 \end{pmatrix}$

(g) $\begin{pmatrix} 5 \\ 6 \\ 7 \end{pmatrix}$

(h) $\begin{pmatrix} -4 & 0 & 6 \\ 0 & 1 & -3 \\ 6 & -3 & 5 \end{pmatrix}$

3. Prove that $(aA + bB)^t = aA^t + bB^t$ for any $A, B \in M_{m \times n}(F)$ and any $a, b \in F$.

4. Prove that $(A^t)^t = A$ for each $A \in M_{m \times n}(F)$.

5. Prove that $A + A^t$ is symmetric for any square matrix A.

6. Prove that $\text{tr}(aA + bB) = a \, \text{tr}(A) + b \, \text{tr}(B)$ for any $A, B \in M_{n \times n}(F)$.

7. Prove that diagonal matrices are symmetric matrices.

8. Determine whether the following sets are subspaces of R^3 under the operations of addition and scalar multiplication defined on R^3. Justify your answers.

(a) $W_1 = \{(a_1, a_2, a_3) \in R^3 : a_1 = 3a_2 \text{ and } a_3 = -a_2\}$
(b) $W_2 = \{(a_1, a_2, a_3) \in R^3 : a_1 = a_3 + 2\}$
(c) $W_3 = \{(a_1, a_2, a_3) \in R^3 : 2a_1 - 7a_2 + a_3 = 0\}$
(d) $W_4 = \{(a_1, a_2, a_3) \in R^3 : a_1 - 4a_2 - a_3 = 0\}$
(e) $W_5 = \{(a_1, a_2, a_3) \in R^3 : a_1 + 2a_2 - 3a_3 = 1\}$
(f) $W_6 = \{(a_1, a_2, a_3) \in R^3 : 5a_1^2 - 3a_2^2 + 6a_3^2 = 0\}$

9. Let W_1, W_3, and W_4 be as in Exercise 8. Describe $W_1 \cap W_3$, $W_1 \cap W_4$, and $W_3 \cap W_4$, and observe that each is a subspace of R^3.

10. Prove that $W_1 = \{(a_1, a_2, \ldots, a_n) \in \mathsf{F}^n : a_1 + a_2 + \cdots + a_n = 0\}$ is a subspace of F^n, but $W_2 = \{(a_1, a_2, \ldots, a_n) \in \mathsf{F}^n : a_1 + a_2 + \cdots + a_n = 1\}$ is not.

11. Is the set $W = \{f(x) \in \mathsf{P}(F) : f(x) = 0 \text{ or } f(x) \text{ has degree } n\}$ a subspace of $\mathsf{P}(F)$ if $n \geq 1$? Justify your answer.

12. An $m \times n$ matrix A is called **upper triangular** if all entries lying below the diagonal entries are zero, that is, if $A_{ij} = 0$ whenever $i > j$. Prove that the upper triangular matrices form a subspace of $\mathsf{M}_{m \times n}(F)$.

13. Let S be a nonempty set and F a field. Prove that for any $s_0 \in S$, $\{f \in \mathcal{F}(S, F) : f(s_0) = 0\}$, is a subspace of $\mathcal{F}(S, F)$.

14. Let S be a nonempty set and F a field. Let $\mathcal{C}(S, F)$ denote the set of all functions $f \in \mathcal{F}(S, F)$ such that $f(s) = 0$ for all but a finite number of elements of S. Prove that $\mathcal{C}(S, F)$ is a subspace of $\mathcal{F}(S, F)$.

15. Is the set of all differentiable real-valued functions defined on R a subspace of $\mathsf{C}(R)$? Justify your answer.

16. Let $\mathsf{C}^n(R)$ denote the set of all real-valued functions defined on the real line that have a continuous nth derivative. Prove that $\mathsf{C}^n(R)$ is a subspace of $\mathcal{F}(R, R)$.

17. Prove that a subset W of a vector space V is a subspace of V if and only if $W \neq \varnothing$, and, whenever $a \in F$ and $x, y \in W$, then $ax \in W$ and $x + y \in W$.

18. Prove that a subset W of a vector space V is a subspace of V if and only if $0 \in W$ and $ax + y \in W$ whenever $a \in F$ and $x, y \in W$.

19. Let W_1 and W_2 be subspaces of a vector space V. Prove that $W_1 \cup W_2$ is a subspace of V if and only if $W_1 \subseteq W_2$ or $W_2 \subseteq W_1$.

20.[†] Prove that if W is a subspace of a vector space V and $w_1, w_2, \ldots, w_n$ are in W, then $a_1 w_1 + a_2 w_2 + \cdots + a_n w_n \in W$ for any scalars $a_1, a_2, \ldots, a_n$.

21. Show that the set of convergent sequences $\{a_n\}$ (i.e., those for which $\lim_{n \to \infty} a_n$ exists) is a subspace of the vector space V in Exercise 20 of Section 1.2.

22. Let F_1 and F_2 be fields. A function $g \in \mathcal{F}(F_1, F_2)$ is called an **even function** if $g(-t) = g(t)$ for each $t \in F_1$ and is called an **odd function** if $g(-t) = -g(t)$ for each $t \in F_1$. Prove that the set of all even functions in $\mathcal{F}(F_1, F_2)$ and the set of all odd functions in $\mathcal{F}(F_1, F_2)$ are subspaces of $\mathcal{F}(F_1, F_2)$.

[†]A dagger means that this exercise is essential for a later section.

The following definitions are used in Exercises 23–30.

Definition. *If S_1 and S_2 are nonempty subsets of a vector space V, then the* **sum** *of S_1 and S_2, denoted $S_1 + S_2$, is the set $\{x + y \colon x \in S_1 \text{ and } y \in S_2\}$.*

Definition. *A vector space V is called the* **direct sum** *of W_1 and W_2 if W_1 and W_2 are subspaces of V such that $W_1 \cap W_2 = \{0\}$ and $W_1 + W_2 = V$. We denote that V is the direct sum of W_1 and W_2 by writing $V = W_1 \oplus W_2$.*

23. Let W_1 and W_2 be subspaces of a vector space V.

(a) Prove that $W_1 + W_2$ is a subspace of V that contains both W_1 and W_2.

(b) Prove that any subspace of V that contains both W_1 and W_2 must also contain $W_1 + W_2$.

24. Show that F^n is the direct sum of the subspaces

$$W_1 = \{(a_1, a_2, \ldots, a_n) \in F^n \colon a_n = 0\}$$

and

$$W_2 = \{(a_1, a_2, \ldots, a_n) \in F^n \colon a_1 = a_2 = \cdots = a_{n-1} = 0\}.$$

25. Let W_1 denote the set of all polynomials $f(x)$ in $P(F)$ such that in the representation

$$f(x) = a_n x^n + a_{n-1} x^{n-1} + \cdots + a_1 x + a_0,$$

we have $a_i = 0$ whenever i is even. Likewise let W_2 denote the set of all polynomials $g(x)$ in $P(F)$ such that in the representation

$$g(x) = b_m x^m + b_{m-1} x^{m-1} + \cdots + b_1 x + b_0,$$

we have $b_i = 0$ whenever i is odd. Prove that $P(F) = W_1 \oplus W_2$.

26. In $M_{m \times n}(F)$ define $W_1 = \{A \in M_{m \times n}(F) \colon A_{ij} = 0 \text{ whenever } i > j\}$ and $W_2 = \{A \in M_{m \times n}(F) \colon A_{ij} = 0 \text{ whenever } i \leq j\}$. ($W_1$ is the set of all upper triangular matrices defined in Exercise 12.) Show that $M_{m \times n}(F) = W_1 \oplus W_2$.

27. Let V denote the vector space consisting of all upper triangular $n \times n$ matrices (as defined in Exercise 12), and let W_1 denote the subspace of V consisting of all diagonal matrices. Show that $V = W_1 \oplus W_2$, where $W_2 = \{A \in V \colon A_{ij} = 0 \text{ whenever } i \geq j\}$.

28. A matrix M is called **skew-symmetric** if $M^t = -M$. Clearly, a skew-symmetric matrix is square. Let F be a field. Prove that the set W_1 of all skew-symmetric $n \times n$ matrices with entries from F is a subspace of $\mathsf{M}_{n \times n}(F)$. Now assume that F is not of characteristic 2 (see Appendix C), and let W_2 be the subspace of $\mathsf{M}_{n \times n}(F)$ consisting of all symmetric $n \times n$ matrices. Prove that $\mathsf{M}_{n \times n}(F) = \mathsf{W}_1 \oplus \mathsf{W}_2$.

29. Let F be a field that is not of characteristic 2. Define

$$\mathsf{W}_1 = \{A \in \mathsf{M}_{n \times n}(F) \colon A_{ij} = 0 \text{ whenever } i \leq j\}$$

and W_2 to be the set of all symmetric $n \times n$ matrices with entries from F. Both W_1 and W_2 are subspaces of $\mathsf{M}_{n \times n}(F)$. Prove that $\mathsf{M}_{n \times n}(F) = \mathsf{W}_1 \oplus \mathsf{W}_2$. Compare this exercise with Exercise 28.

30. Let W_1 and W_2 be subspaces of a vector space V. Prove that V is the direct sum of W_1 and W_2 if and only if each vector in V can be *uniquely* written as $x_1 + x_2$, where $x_1 \in \mathsf{W}_1$ and $x_2 \in \mathsf{W}_2$.

31. Let W be a subspace of a vector space V over a field F. For any $v \in \mathsf{V}$ the set $\{v\} + \mathsf{W} = \{v + w \colon w \in \mathsf{W}\}$ is called the **coset** of W **containing** v. It is customary to denote this coset by $v + \mathsf{W}$ rather than $\{v\} + \mathsf{W}$.

(a) Prove that $v + \mathsf{W}$ is a subspace of V if and only if $v \in \mathsf{W}$.
(b) Prove that $v_1 + \mathsf{W} = v_2 + \mathsf{W}$ if and only if $v_1 - v_2 \in \mathsf{W}$.

Addition and scalar multiplication by scalars of F can be defined in the collection $S = \{v + \mathsf{W} \colon v \in \mathsf{V}\}$ of all cosets of W as follows:

$$(v_1 + \mathsf{W}) + (v_2 + \mathsf{W}) = (v_1 + v_2) + \mathsf{W}$$

for all $v_1, v_2 \in \mathsf{V}$ and

$$a(v + \mathsf{W}) = av + \mathsf{W}$$

for all $v \in \mathsf{V}$ and $a \in F$.

(c) Prove that the preceding operations are well defined; that is, show that if $v_1 + \mathsf{W} = v_1' + \mathsf{W}$ and $v_2 + \mathsf{W} = v_2' + \mathsf{W}$, then

$$(v_1 + \mathsf{W}) + (v_2 + \mathsf{W}) = (v_1' + \mathsf{W}) + (v_2' + \mathsf{W})$$

and

$$a(v_1 + \mathsf{W}) = a(v_1' + \mathsf{W})$$

for all $a \in F$.

(d) Prove that the set S is a vector space with the operations defined in (c). This vector space is called the **quotient space of** V **modulo** W and is denoted by V/W.

1.4   LINEAR COMBINATIONS AND SYSTEMS OF LINEAR EQUATIONS

In Section 1.1, it was shown that the equation of the plane through three noncollinear points A, B, and C in space is $x = A + su + tv$, where u and v denote the vectors beginning at A and ending at B and C, respectively, and s and t denote arbitrary real numbers. An important special case occurs when A is the origin. In this case, the equation of the plane simplifies to $x = su + tv$, and the set of all points in this plane is a subspace of R^3. (This is proved as Theorem 1.5.) Expressions of the form $su + tv$, where s and t are scalars and u and v are vectors, play a central role in the theory of vector spaces. The appropriate generalization of such expressions is presented in the following definitions.

Definitions. *Let* V *be a vector space and* S *a nonempty subset of* V. *A vector* $v \in \mathsf{V}$ *is called a **linear combination** of vectors of* S *if there exist a finite number of vectors* $u_1, u_2, \ldots, u_n$ *in* S *and scalars* $a_1, a_2, \ldots, a_n$ *in* F *such that* $v = a_1 u_1 + a_2 u_2 + \cdots + a_n u_n$. *In this case we also say that* v *is a linear combination of* $u_1, u_2, \ldots, u_n$ *and call* $a_1, a_2, \ldots, a_n$ *the **coefficients** of the linear combination.*

Observe that in any vector space V, $0v = 0$ for each $v \in \mathsf{V}$. Thus the zero vector is a linear combination of any nonempty subset of V.

Example 1

TABLE 1.1 Vitamin Content of 100 Grams of Certain Foods

	A (units)	B$_1$ (mg)	B$_2$ (mg)	Niacin (mg)	C (mg)
Apple butter	0	0.01	0.02	0.2	2
Raw, unpared apples (freshly harvested)	90	0.03	0.02	0.1	4
Chocolate-coated candy with coconut center	0	0.02	0.07	0.2	0
Clams (meat only)	100	0.10	0.18	1.3	10
Cupcake from mix (dry form)	0	0.05	0.06	0.3	0
Cooked farina (unenriched)	(0)[a]	0.01	0.01	0.1	(0)
Jams and preserves	10	0.01	0.03	0.2	2
Coconut custard pie (baked from mix)	0	0.02	0.02	0.4	0
Raw brown rice	(0)	0.34	0.05	4.7	(0)
Soy sauce	0	0.02	0.25	0.4	0
Cooked spaghetti (unenriched)	0	0.01	0.01	0.3	0
Raw wild rice	(0)	0.45	0.63	6.2	(0)

Source: Bernice K. Watt and Annabel L. Merrill, *Composition of Foods* (Agriculture Handbook Number 8), Consumer and Food Economics Research Division, U.S. Department of Agriculture, Washington, D.C., 1963.

[a]Zeros in parentheses indicate that the amount of a vitamin present is either none or too small to measure.

Table 1.1 shows the vitamin content of 100 grams of 12 foods with respect to vitamins $\mathbf{A}$, $\mathbf{B}_1$ (thiamine), $\mathbf{B}_2$ (riboflavin), niacin, and $\mathbf{C}$ (ascorbic acid).

The vitamin content of 100 grams of each food can be recorded as a column vector in R^5—for example, the vitamin vector for apple butter is

$$\begin{pmatrix} 0.00 \\ 0.01 \\ 0.02 \\ 0.20 \\ 2.00 \end{pmatrix}.$$

Considering the vitamin vectors for cupcake, coconut custard pie, raw brown rice, soy sauce, and wild rice, we see that

$$\begin{pmatrix} 0.00 \\ 0.05 \\ 0.06 \\ 0.30 \\ 0.00 \end{pmatrix} + \begin{pmatrix} 0.00 \\ 0.02 \\ 0.02 \\ 0.40 \\ 0.00 \end{pmatrix} + \begin{pmatrix} 0.00 \\ 0.34 \\ 0.05 \\ 4.70 \\ 0.00 \end{pmatrix} + 2 \begin{pmatrix} 0.00 \\ 0.02 \\ 0.25 \\ 0.40 \\ 0.00 \end{pmatrix} = \begin{pmatrix} 0.00 \\ 0.45 \\ 0.63 \\ 6.20 \\ 0.00 \end{pmatrix}.$$

Thus the vitamin vector for wild rice is a linear combination of the vitamin vectors for cupcake, coconut custard pie, raw brown rice, and soy sauce. So 100 grams of cupcake, 100 grams of coconut custard pie, 100 grams of raw brown rice, and 200 grams of soy sauce provide exactly the same amounts of the five vitamins as 100 grams of raw wild rice. Similarly, since

$$2\begin{pmatrix} 0.00 \\ 0.01 \\ 0.02 \\ 0.20 \\ 2.00 \end{pmatrix} + \begin{pmatrix} 90.00 \\ 0.03 \\ 0.02 \\ 0.10 \\ 4.00 \end{pmatrix} + \begin{pmatrix} 0.00 \\ 0.02 \\ 0.07 \\ 0.20 \\ 0.00 \end{pmatrix} + \begin{pmatrix} 0.00 \\ 0.01 \\ 0.01 \\ 0.10 \\ 0.00 \end{pmatrix} + \begin{pmatrix} 10.00 \\ 0.01 \\ 0.03 \\ 0.20 \\ 2.00 \end{pmatrix} + \begin{pmatrix} 0.00 \\ 0.01 \\ 0.01 \\ 0.30 \\ 0.00 \end{pmatrix} = \begin{pmatrix} 100.00 \\ 0.10 \\ 0.18 \\ 1.30 \\ 10.00 \end{pmatrix},$$

200 grams of apple butter, 100 grams of apples, 100 grams of chocolate candy, 100 grams of farina, 100 grams of jam, and 100 grams of spaghetti provide exactly the same amounts of the five vitamins as 100 grams of clams. ◆

Throughout Chapters 1 and 2 we encounter many different situations in which it is necessary to determine whether or not a vector can be expressed as a linear combination of other vectors, and if so, how. This question often reduces to the problem of solving a system of linear equations. In Chapter 3, we discuss a general method for using matrices to solve any system of linear equations. For now, we illustrate how to solve a system of linear equations by showing how to determine if the vector $(2, 6, 8)$ can be expressed as a linear combination of

$$u_1 = (1, 2, 1), \quad u_2 = (-2, -4, -2), \quad u_3 = (0, 2, 3),$$

$$u_4 = (2, 0, -3), \quad \text{and} \quad u_5 = (-3, 8, 16).$$

Thus we must determine if there are scalars $a_1, a_2, a_3, a_4,$ and a_5 such that

$$
\begin{aligned}
(2, 6, 8) &= a_1 u_1 + a_2 u_2 + a_3 u_3 + a_4 u_4 + a_5 u_5 \\
&= a_1(1, 2, 1) + a_2(-2, -4, -2) + a_3(0, 2, 3) \\
&\qquad\qquad\qquad\qquad + a_4(2, 0, -3) + a_5(-3, 8, 16) \\
&= (a_1 - 2a_2 + 2a_4 - 3a_5, 2a_1 - 4a_2 + 2a_3 + 8a_5, \\
&\qquad\qquad\qquad\qquad a_1 - 2a_2 + 3a_3 - 3a_4 + 16a_5).
\end{aligned}
$$

Hence $(2, 6, 8)$ can be expressed as a linear combination of $u_1, u_2, u_3, u_4,$ and u_5 if and only if there is a 5-tuple of scalars $(a_1, a_2, a_3, a_4, a_5)$ satisfying the system of linear equations

$$
\begin{aligned}
a_1 - 2a_2 \qquad\quad + 2a_4 - \;\; 3a_5 &= 2 \\
2a_1 - 4a_2 + 2a_3 \qquad\quad + \;\; 8a_5 &= 6 \\
a_1 - 2a_2 + 3a_3 - 3a_4 + 16a_5 &= 8,
\end{aligned}
\tag{1}
$$

which is obtained by equating the corresponding coordinates in the preceding equation.

To solve system (1), we replace it by another system with the same solutions, but which is easier to solve. The procedure to be used expresses some of the unknowns in terms of others by eliminating certain unknowns from all the equations except one. To begin, we eliminate a_1 from every equation except the first by adding -2 times the first equation to the second and -1 times the first equation to the third. The result is the following new system:

$$
\begin{aligned}
a_1 - 2a_2 \qquad\quad + 2a_4 - \;\; 3a_5 &= 2 \\
2a_3 - 4a_4 + 14a_5 &= 2 \\
3a_3 - 5a_4 + 19a_5 &= 6.
\end{aligned}
\tag{2}
$$

In this case, it happened that while eliminating a_1 from every equation except the first, we also eliminated a_2 from every equation except the first. This need not happen in general. We now want to make the coefficient of a_3 in the second equation equal to 1, and then eliminate a_3 from the third equation. To do this, we first multiply the second equation by $\frac{1}{2}$, which produces

$$
\begin{aligned}
a_1 - 2a_2 \qquad\quad + 2a_4 - \;\; 3a_5 &= 2 \\
a_3 - 2a_4 + \;\; 7a_5 &= 1 \\
3a_3 - 5a_4 + 19a_5 &= 6.
\end{aligned}
$$

Next we add -3 times the second equation to the third, obtaining

$$
\begin{aligned}
a_1 - 2a_2 \qquad\quad + 2a_4 - 3a_5 &= 2 \\
a_3 - 2a_4 + 7a_5 &= 1 \\
a_4 - 2a_5 &= 3.
\end{aligned}
\tag{3}
$$

We continue by eliminating a_4 from every equation of (3) except the third. This yields

$$
\begin{aligned}
a_1 - 2a_2 \quad\quad + \quad a_5 &= -4 \\
a_3 \quad + 3a_5 &= 7 \\
a_4 - 2a_5 &= 3.
\end{aligned}
\tag{4}
$$

System (4) is a system of the desired form: It is easy to solve for the first unknown present in each of the equations (a_1, a_3, and a_4) in terms of the other unknowns (a_2 and a_5). Rewriting system (4) in this form, we find that

$$
\begin{aligned}
a_1 &= 2a_2 - \quad a_5 - 4 \\
a_3 &= \quad\quad - 3a_5 + 7 \\
a_4 &= \quad\quad 2a_5 + 3.
\end{aligned}
$$

Thus for any choice of scalars a_2 and a_5, a vector of the form

$$(a_1, a_2, a_3, a_4, a_5) = (2a_2 - a_5 - 4, a_2, -3a_5 + 7, 2a_5 + 3, a_5)$$

is a solution to system (1). In particular, the vector $(-4, 0, 7, 3, 0)$ obtained by setting $a_2 = 0$ and $a_5 = 0$ is a solution to (1). Therefore

$$(2, 6, 8) = -4u_1 + 0u_2 + 7u_3 + 3u_4 + 0u_5,$$

so that $(2, 6, 8)$ is a linear combination of u_1, u_2, u_3, u_4, and u_5.

The procedure just illustrated uses three types of operations to simplify the original system:

1. interchanging the order of any two equations in the system;
2. multiplying any equation in the system by a *nonzero* constant;
3. adding a constant multiple of any equation to another equation in the system.

In Section 3.4, we prove that these operations do not change the set of solutions to the original system. Note that we employed these operations to obtain a system of equations that had the following properties:

1. The first nonzero coefficient in each equation is one.
2. If an unknown is the first unknown with a nonzero coefficient in some equation, then that unknown occurs with a zero coefficient in each of the other equations.
3. The first unknown with a nonzero coefficient in any equation has a larger subscript than the first unknown with a nonzero coefficient in any preceding equation.

To help clarify the meaning of these properties, note that none of the following systems meets these requirements.

$$\begin{aligned} x_1 + 3x_2 \quad + \quad x_4 &= 7 \\ 2x_3 - 5x_4 &= -1 \end{aligned} \tag{5}$$

$$\begin{aligned} x_1 - 2x_2 + 3x_3 \quad + \quad x_5 &= -5 \\ x_3 \quad - 2x_5 &= 9 \\ x_4 + 3x_5 &= 6 \end{aligned} \tag{6}$$

$$\begin{aligned} x_1 \quad - 2x_3 \quad + \quad x_5 &= 1 \\ x_4 - 6x_5 &= 0 \\ x_2 + 5x_3 \quad - 3x_5 &= 2. \end{aligned} \tag{7}$$

Specifically, system (5) does not satisfy property 1 because the first nonzero coefficient in the second equation is 2; system (6) does not satisfy property 2 because x_3, the first unknown with a nonzero coefficient in the second equation, occurs with a nonzero coefficient in the first equation; and system (7) does not satisfy property 3 because x_2, the first unknown with a nonzero coefficient in the third equation, does not have a larger subscript than x_4, the first unknown with a nonzero coefficient in the second equation.

Once a system with properties 1, 2, and 3 has been obtained, it is easy to solve for some of the unknowns in terms of the others (as in the preceding example). *If, however, in the course of using operations 1, 2, and 3 a system containing an equation of the form $0 = c$, where c is nonzero, is obtained, then the original system has no solutions.* (See Example 2.)

We return to the study of systems of linear equations in Chapter 3. We discuss there the theoretical basis for this method of solving systems of linear equations and further simplify the procedure by use of matrices.

Example 2

We claim that

$$2x^3 - 2x^2 + 12x - 6$$

is a linear combination of

$$x^3 - 2x^2 - 5x - 3 \quad \text{and} \quad 3x^3 - 5x^2 - 4x - 9$$

in $P_3(R)$, but that

$$3x^3 - 2x^2 + 7x + 8$$

is not. In the first case we wish to find scalars a and b such that

$$2x^3 - 2x^2 + 12x - 6 = a(x^3 - 2x^2 - 5x - 3) + b(3x^3 - 5x^2 - 4x - 9)$$

$$= (a + 3b)x^3 + (-2a - 5b)x^2 + (-5a - 4b)x + (-3a - 9b).$$

Thus we are led to the following system of linear equations:

$$\begin{aligned} a + 3b &= 2 \\ -2a - 5b &= -2 \\ -5a - 4b &= 12 \\ -3a - 9b &= -6. \end{aligned}$$

Adding appropriate multiples of the first equation to the others in order to eliminate a, we find that

$$\begin{aligned} a + 3b &= 2 \\ b &= 2 \\ 11b &= 22 \\ 0b &= 0. \end{aligned}$$

Now adding the appropriate multiples of the second equation to the others yields

$$\begin{aligned} a &= -4 \\ b &= 2 \\ 0 &= 0 \\ 0 &= 0. \end{aligned}$$

Hence

$$2x^3 - 2x^2 + 12x - 6 = -4(x^3 - 2x^2 - 5x - 3) + 2(3x^3 - 5x^2 - 4x - 9).$$

In the second case, we wish to show that there are no scalars a and b for which

$$3x^3 - 2x^2 + 7x + 8 = a(x^3 - 2x^2 - 5x - 3) + b(3x^3 - 5x^2 - 4x - 9).$$

Using the preceding technique, we obtain a system of linear equations

$$\begin{aligned} a + 3b &= 3 \\ -2a - 5b &= -2 \\ -5a - 4b &= 7 \\ -3a - 9b &= 8. \end{aligned} \tag{8}$$

Eliminating a as before yields

$$\begin{aligned} a + 3b &= 3 \\ b &= 4 \\ 11b &= 22 \\ 0 &= 17. \end{aligned}$$

But the presence of the inconsistent equation $0 = 17$ indicates that (8) has no solutions. Hence $3x^3 - 2x^2 + 7x + 8$ is not a linear combination of $x^3 - 2x^2 - 5x - 3$ and $3x^3 - 5x^2 - 4x - 9$.   ◆

Throughout this book, we form the set of all linear combinations of some set of vectors. We now name such a set of linear combinations.

Definition. *Let S be a nonempty subset of a vector space* V. *The* **span** *of S, denoted span(S), is the set consisting of all linear combinations of the vectors in S. For convenience, we define* span$(\varnothing) = \{0\}$.

In R^3, for instance, the span of the set $\{(1,0,0),(0,1,0)\}$ consists of all vectors in R^3 that have the form $a(1,0,0) + b(0,1,0) = (a,b,0)$ for some scalars a and b. Thus the span of $\{(1,0,0),(0,1,0)\}$ contains all the points in the xy-plane. In this case, the span of the set is a subspace of R^3. This fact is true in general.

Theorem 1.5. *The span of any subset S of a vector space* V *is a subspace of* V. *Moreover, any subspace of* V *that contains S must also contain the span of S.*

Proof. This result is immediate if $S = \varnothing$ because span$(\varnothing) = \{0\}$, which is a subspace that is contained in any subspace of V.

If $S \neq \varnothing$, then S contains a vector z. So $0z = 0$ is in span(S). Let $x, y \in$ span(S). Then there exist vectors $u_1, u_2, \ldots, u_m, v_1, v_2, \ldots, v_n$ in S and scalars $a_1, a_2, \ldots, a_m, b_1, b_2, \ldots, b_n$ such that

$$x = a_1u_1 + a_2u_2 + \cdots + a_mu_m \quad \text{and} \quad y = b_1v_1 + b_2v_2 + \cdots + b_nv_n.$$

Then

$$x + y = a_1u_1 + a_2u_2 + \cdots + a_mu_m + b_1v_1 + b_2v_2 + \cdots + b_nv_n$$

and, for any scalar c,

$$cx = (ca_1)u_1 + (ca_2)u_2 + \cdots + (ca_m)u_m$$

are clearly linear combinations of the vectors in S; so $x + y$ and cx are in span(S). Thus span(S) is a subspace of V.

Now let W denote any subspace of V that contains S. If $w \in$ span(S), then w has the form $w = c_1w_1 + c_2w_2 + \cdots + c_kw_k$ for some vectors $w_1, w_2, \ldots, w_k$ in S and some scalars $c_1, c_2, \ldots, c_k$. Since $S \subseteq$ W, we have $w_1, w_2, \ldots, w_k \in$ W. Therefore $w = c_1w_1 + c_2w_2 + \cdots + c_kw_k$ is in W by Exercise 20 of Section 1.3. Because w, an arbitrary vector in span(S), belongs to W, it follows that span$(S) \subseteq$ W. ∎

Definition. *A subset S of a vector space* V **generates** *(or* **spans***)* V *if* span$(S) = $ V. *In this case, we also say that the vectors of S generate (or span)* V.

Example 3

The vectors $(1,1,0), (1,0,1)$, and $(0,1,1)$ generate R^3 since an arbitrary vector (a_1, a_2, a_3) in R^3 is a linear combination of the three given vectors; in fact, the scalars r, s, and t for which

$$r(1,1,0) + s(1,0,1) + t(0,1,1) = (a_1, a_2, a_3)$$

are

$$r = \frac{1}{2}(a_1 + a_2 - a_3), \ \ s = \frac{1}{2}(a_1 - a_2 + a_3), \ \ \text{and} \ \ t = \frac{1}{2}(-a_1 + a_2 + a_3). \ \blacklozenge$$

Example 4

The polynomials $x^2 + 3x - 2$, $2x^2 + 5x - 3$, and $-x^2 - 4x + 4$ generate $\mathsf{P}_2(R)$ since each of the three given polynomials belongs to $\mathsf{P}_2(R)$ and each polynomial $ax^2 + bx + c$ in $\mathsf{P}_2(R)$ is a linear combination of these three, namely,

$$(-8a + 5b + 3c)(x^2 + 3x - 2) + (4a - 2b - c)(2x^2 + 5x - 3)$$

$$+(-a + b + c)(-x^2 - 4x + 4) = ax^2 + bx + c. \qquad \blacklozenge$$

Example 5

The matrices

$$\begin{pmatrix} 1 & 1 \\ 1 & 0 \end{pmatrix}, \ \ \begin{pmatrix} 1 & 1 \\ 0 & 1 \end{pmatrix}, \ \ \begin{pmatrix} 1 & 0 \\ 1 & 1 \end{pmatrix}, \ \ \text{and} \ \ \begin{pmatrix} 0 & 1 \\ 1 & 1 \end{pmatrix}$$

generate $\mathsf{M}_{2\times 2}(R)$ since an arbitrary matrix A in $\mathsf{M}_{2\times 2}(R)$ can be expressed as a linear combination of the four given matrices as follows:

$$\begin{pmatrix} a_{11} & a_{12} \\ a_{21} & a_{22} \end{pmatrix} = (\frac{1}{3}a_{11} + \frac{1}{3}a_{12} + \frac{1}{3}a_{21} - \frac{2}{3}a_{22}) \begin{pmatrix} 1 & 1 \\ 1 & 0 \end{pmatrix}$$

$$+ \ (\frac{1}{3}a_{11} + \frac{1}{3}a_{12} - \frac{2}{3}a_{21} + \frac{1}{3}a_{22}) \begin{pmatrix} 1 & 1 \\ 0 & 1 \end{pmatrix}$$

$$+ \ (\frac{1}{3}a_{11} - \frac{2}{3}a_{12} + \frac{1}{3}a_{21} + \frac{1}{3}a_{22}) \begin{pmatrix} 1 & 0 \\ 1 & 1 \end{pmatrix}$$

$$+ \ (-\frac{2}{3}a_{11} + \frac{1}{3}a_{12} + \frac{1}{3}a_{21} + \frac{1}{3}a_{22}) \begin{pmatrix} 0 & 1 \\ 1 & 1 \end{pmatrix}.$$

On the other hand, the matrices

$$\begin{pmatrix} 1 & 0 \\ 0 & 1 \end{pmatrix}, \ \ \begin{pmatrix} 1 & 1 \\ 0 & 1 \end{pmatrix}, \ \ \text{and} \ \ \begin{pmatrix} 1 & 0 \\ 1 & 1 \end{pmatrix}$$

do not generate $M_{2\times2}(R)$ because each of these matrices has equal diagonal entries. So any linear combination of these matrices has equal diagonal entries. Hence not every 2×2 matrix is a linear combination of these three matrices.   ◆

At the beginning of this section we noted that the equation of a plane through three noncollinear points in space, one of which is the origin, is of the form $x = su + tv$, where $u, v \in R^3$ and s and t are scalars. Thus $x \in R^3$ is a linear combination of $u, v \in R^3$ if and only if x lies in the plane containing u and v. (See Figure 1.5.)

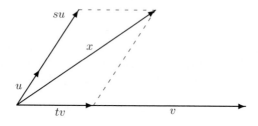

Figure 1.5

Usually there are many different subsets that generate a subspace W. (See Exercise 13.) It is natural to seek a subset of W that generates W and is as small as possible. In the next section we explore the circumstances under which a vector can be removed from a generating set to obtain a smaller generating set.

EXERCISES

1. Label the following statements as true or false.

    (a) The zero vector is a linear combination of any nonempty set of vectors.
    (b) The span of ∅ is ∅.
    (c) If S is a subset of a vector space V, then span(S) equals the intersection of all subspaces of V that contain S.
    (d) In solving a system of linear equations, it is permissible to multiply an equation by any constant.
    (e) In solving a system of linear equations, it is permissible to add any multiple of one equation to another.
    (f) Every system of linear equations has a solution.

2. Solve the following systems of linear equations by the method introduced in this section.

   (a) $\begin{aligned} 2x_1 - 2x_2 - 3x_3 \quad\quad &= -2 \\ 3x_1 - 3x_2 - 2x_3 + 5x_4 &= 7 \\ x_1 - x_2 - 2x_3 - x_4 &= -3 \end{aligned}$

   (b) $\begin{aligned} 3x_1 - 7x_2 + 4x_3 &= 10 \\ x_1 - 2x_2 + x_3 &= 3 \\ 2x_1 - x_2 - 2x_3 &= 6 \end{aligned}$

   (c) $\begin{aligned} x_1 + 2x_2 - x_3 + x_4 &= 5 \\ x_1 + 4x_2 - 3x_3 - 3x_4 &= 6 \\ 2x_1 + 3x_2 - x_3 + 4x_4 &= 8 \end{aligned}$

   (d) $\begin{aligned} x_1 + 2x_2 + 2x_3 \quad\quad &= 2 \\ x_1 \quad\quad + 8x_3 + 5x_4 &= -6 \\ x_1 + x_2 + 5x_3 + 5x_4 &= 3 \end{aligned}$

   (e) $\begin{aligned} x_1 + 2x_2 - 4x_3 - x_4 + x_5 &= 7 \\ -x_1 \quad\quad + 10x_3 - 3x_4 - 4x_5 &= -16 \\ 2x_1 + 5x_2 - 5x_3 - 4x_4 - x_5 &= 2 \\ 4x_1 + 11x_2 - 7x_3 - 10x_4 - 2x_5 &= 7 \end{aligned}$

   (f) $\begin{aligned} x_1 + 2x_2 + 6x_3 &= -1 \\ 2x_1 + x_2 + x_3 &= 8 \\ 3x_1 + x_2 - x_3 &= 15 \\ x_1 + 3x_2 + 10x_3 &= -5 \end{aligned}$

3. For each of the following lists of vectors in R^3, determine whether the
   first vector can be expressed as a linear combination of the other two.

   (a) $(-2, 0, 3), (1, 3, 0), (2, 4, -1)$
   (b) $(1, 2, -3), (-3, 2, 1), (2, -1, -1)$
   (c) $(3, 4, 1), (1, -2, 1), (-2, -1, 1)$
   (d) $(2, -1, 0), (1, 2, -3), (1, -3, 2)$
   (e) $(5, 1, -5), (1, -2, -3), (-2, 3, -4)$
   (f) $(-2, 2, 2), (1, 2, -1), (-3, -3, 3)$

4. For each list of polynomials in $P_3(R)$, determine whether the first polynomial can be expressed as a linear combination of the other two.

   (a) $x^3 - 3x + 5, x^3 + 2x^2 - x + 1, x^3 + 3x^2 - 1$
   (b) $4x^3 + 2x^2 - 6, x^3 - 2x^2 + 4x + 1, 3x^3 - 6x^2 + x + 4$
   (c) $-2x^3 - 11x^2 + 3x + 2, x^3 - 2x^2 + 3x - 1, 2x^3 + x^2 + 3x - 2$
   (d) $x^3 + x^2 + 2x + 13, 2x^3 - 3x^2 + 4x + 1, x^3 - x^2 + 2x + 3$
   (e) $x^3 - 8x^2 + 4x, x^3 - 2x^2 + 3x - 1, x^3 - 2x + 3$
   (f) $6x^3 - 3x^2 + x + 2, x^3 - x^2 + 2x + 3, 2x^3 - 3x + 1$

5. In each part, determine whether the given vector is in the span of S.
   (a) $(2, -1, 1)$, $S = \{(1, 0, 2), (-1, 1, 1)\}$
   (b) $(-1, 2, 1)$, $S = \{(1, 0, 2), (-1, 1, 1)\}$
   (c) $(-1, 1, 1, 2)$, $S = \{(1, 0, 1, -1), (0, 1, 1, 1)\}$
   (d) $(2, -1, 1, -3)$, $S = \{(1, 0, 1, -1), (0, 1, 1, 1)\}$
   (e) $-x^3 + 2x^2 + 3x + 3$, $S = \{x^3 + x^2 + x + 1, x^2 + x + 1, x + 1\}$
   (f) $2x^3 - x^2 + x + 3$, $S = \{x^3 + x^2 + x + 1, x^2 + x + 1, x + 1\}$

   (g) $\begin{pmatrix} 1 & 2 \\ -3 & 4 \end{pmatrix}$, $S = \left\{ \begin{pmatrix} 1 & 0 \\ -1 & 0 \end{pmatrix}, \begin{pmatrix} 0 & 1 \\ 0 & 1 \end{pmatrix}, \begin{pmatrix} 1 & 1 \\ 0 & 0 \end{pmatrix} \right\}$

   (h) $\begin{pmatrix} 1 & 0 \\ 0 & 1 \end{pmatrix}$, $S = \left\{ \begin{pmatrix} 1 & 0 \\ -1 & 0 \end{pmatrix}, \begin{pmatrix} 0 & 1 \\ 0 & 1 \end{pmatrix}, \begin{pmatrix} 1 & 1 \\ 0 & 0 \end{pmatrix} \right\}$

6. Show that the vectors $(1, 1, 0)$, $(1, 0, 1)$, and $(0, 1, 1)$ generate F^3.

7. In F^n, let e_j denote the vector whose jth coordinate is 1 and whose other coordinates are 0. Prove that $\{e_1, e_2, \ldots, e_n\}$ generates F^n.

8. Show that $\mathsf{P}_n(F)$ is generated by $\{1, x, \ldots, x^n\}$.

9. Show that the matrices

$$\begin{pmatrix} 1 & 0 \\ 0 & 0 \end{pmatrix}, \quad \begin{pmatrix} 0 & 1 \\ 0 & 0 \end{pmatrix}, \quad \begin{pmatrix} 0 & 0 \\ 1 & 0 \end{pmatrix}, \quad \text{and} \quad \begin{pmatrix} 0 & 0 \\ 0 & 1 \end{pmatrix}$$

generate $\mathsf{M}_{2\times 2}(F)$.

10. Show that if

$$M_1 = \begin{pmatrix} 1 & 0 \\ 0 & 0 \end{pmatrix}, \quad M_2 = \begin{pmatrix} 0 & 0 \\ 0 & 1 \end{pmatrix}, \quad \text{and} \quad M_3 = \begin{pmatrix} 0 & 1 \\ 1 & 0 \end{pmatrix},$$

then the span of $\{M_1, M_2, M_3\}$ is the set of all symmetric 2×2 matrices.

11.[†] Prove that $\text{span}(\{x\}) = \{ax : a \in F\}$ for any vector x in a vector space. Interpret this result geometrically in R^3.

12. Show that a subset W of a vector space V is a subspace of V if and only if $\text{span}(W) = W$.

13.[†] Show that if S_1 and S_2 are subsets of a vector space V such that $S_1 \subseteq S_2$, then $\text{span}(S_1) \subseteq \text{span}(S_2)$. In particular, if $S_1 \subseteq S_2$ and $\text{span}(S_1) = V$, deduce that $\text{span}(S_2) = V$.

14. Show that if S_1 and S_2 are arbitrary subsets of a vector space V, then $\text{span}(S_1 \cup S_2) = \text{span}(S_1) + \text{span}(S_2)$. (The sum of two subsets is defined in the exercises of Section 1.3.)

15. Let S_1 and S_2 be subsets of a vector space V. Prove that $\text{span}(S_1 \cap S_2) \subseteq$ $\text{span}(S_1) \cap \text{span}(S_2)$. Give an example in which $\text{span}(S_1 \cap S_2)$ and $\text{span}(S_1) \cap \text{span}(S_2)$ are equal and one in which they are unequal.

16. Let V be a vector space and S a subset of V with the property that whenever $v_1, v_2, \ldots, v_n \in S$ and $a_1 v_1 + a_2 v_2 + \cdots + a_n v_n = 0$, then $a_1 = a_2 = \cdots = a_n = 0$. Prove that every vector in the span of S can be *uniquely* written as a linear combination of vectors of S.

17. Let W be a subspace of a vector space V. Under what conditions are there only a finite number of distinct subsets S of W such that S generates W?

1.5   LINEAR DEPENDENCE AND LINEAR INDEPENDENCE

Suppose that V is a vector space over an infinite field and that W is a subspace of V. Unless W is the zero subspace, W is an infinite set. It is desirable to find a "small" finite subset S that generates W because we can then describe each vector in W as a linear combination of the finite number of vectors in S. Indeed, the smaller that S is, the fewer computations that are required to represent vectors in W. Consider, for example, the subspace W of R^3 generated by $S = \{u_1, u_2, u_3, u_4\}$, where $u_1 = (2, -1, 4)$, $u_2 = (1, -1, 3)$, $u_3 = (1, 1, -1)$, and $u_4 = (1, -2, -1)$. Let us attempt to find a proper subset of S that also generates W. The search for this subset is related to the question of whether or not some vector in S is a linear combination of the other vectors in S. Now u_4 is a linear combination of the other vectors in S if and only if there are scalars a_1, a_2, and a_3 such that

$$u_4 = a_1 u_1 + a_2 u_2 + a_3 u_3,$$

that is, if and only if there are scalars a_1, a_2, and a_3 satisfying

$$(1, -2, -1) = (2a_1 + a_2 + a_3, -a_1 - a_2 + a_3, 4a_1 + 3a_2 - a_3).$$

Thus u_4 is a linear combination of u_1, u_2, and u_3 if and only if the system of linear equations

$$\begin{array}{rcrcrcr} 2a_1 & + & a_2 & + & a_3 & = & 1 \\ -a_1 & - & a_2 & + & a_3 & = & -2 \\ 4a_1 & + & 3a_2 & - & a_3 & = & -1 \end{array}$$

has a solution. The reader should verify that no such solution exists. This does not, however, answer our question of whether *some* vector in S is a linear combination of the other vectors in S. It can be shown, in fact, that u_3 is a linear combination of u_1, u_2, and u_4, namely, $u_3 = 2u_1 - 3u_2 + 0u_4$.

In the preceding example, checking that some vector in S is a linear combination of the other vectors in S could require that we solve several different systems of linear equations before we determine which, if any, of u_1, u_2, u_3, and u_4 is a linear combination of the others. By formulating our question differently, we can save ourselves some work. Note that since $u_3 = 2u_1 - 3u_2 + 0u_4$, we have

$$-2u_1 + 3u_2 + u_3 - 0u_4 = 0.$$

That is, because some vector in S is a linear combination of the others, the zero vector can be expressed as a linear combination of the vectors in S using coefficients that are not all zero. The converse of this statement is also true: If the zero vector can be written as a linear combination of the vectors in S in which not all the coefficients are zero, then some vector in S is a linear combination of the others. For instance, in the example above, the equation $-2u_1 + 3u_2 + u_3 - 0u_4 = 0$ can be solved for any vector having a nonzero coefficient; so u_1, u_2, or u_3 (but not u_4) can be written as a linear combination of the other three vectors. Thus, rather than asking whether some vector in S is a linear combination of the other vectors in S, it is more efficient to ask whether the zero vector can be expressed as a linear combination of the vectors in S with coefficients that are not all zero. This observation leads us to the following definition.

Definition. *A subset S of a vector space* V *is called **linearly dependent** if there exist a finite number of distinct vectors $u_1, u_2, \ldots, u_n$ in S and scalars $a_1, a_2, \ldots, a_n$, not all zero, such that*

$$a_1u_1 + a_2u_2 + \cdots + a_nu_n = 0.$$

In this case we also say that the vectors of S are linearly dependent.

For any vectors $u_1, u_2, \ldots, u_n$, we have $a_1u_1 + a_2u_2 + \cdots + a_nu_n = 0$ if $a_1 = a_2 = \cdots = a_n = 0$. We call this the **trivial representation** of 0 as a linear combination of $u_1, u_2, \ldots, u_n$. Thus, for a set to be linearly dependent, there must exist a nontrivial representation of 0 as a linear combination of vectors in the set. Consequently, any subset of a vector space that contains the zero vector is linearly dependent, because $0 = 1 \cdot 0$ is a nontrivial representation of 0 as a linear combination of vectors in the set.

Example 1

Consider the set

$$S = \{(1, 3, -4, 2), (2, 2, -4, 0), (1, -3, 2, -4), (-1, 0, 1, 0)\}$$

in R^4. We show that S is linearly dependent and then express one of the vectors in S as a linear combination of the other vectors in S. To show that

S is linearly dependent, we must find scalars a_1, a_2, a_3, and a_4, not all zero, such that

$$a_1(1, 3, -4, 2) + a_2(2, 2, -4, 0) + a_3(1, -3, 2, -4) + a_4(-1, 0, 1, 0) = 0.$$

Finding such scalars amounts to finding a nonzero solution to the system of linear equations

$$\begin{aligned}
a_1 + 2a_2 + a_3 - a_4 &= 0 \\
3a_1 + 2a_2 - 3a_3 &= 0 \\
-4a_1 - 4a_2 + 2a_3 + a_4 &= 0 \\
2a_1 \qquad\quad - 4a_3 &= 0.
\end{aligned}$$

One such solution is $a_1 = 4$, $a_2 = -3$, $a_3 = 2$, and $a_4 = 0$. Thus S is a linearly dependent subset of R^4, and

$$4(1, 3, -4, 2) - 3(2, 2, -4, 0) + 2(1, -3, 2, -4) + 0(-1, 0, 1, 0) = 0. \quad \blacklozenge$$

Example 2

In $\mathsf{M}_{2\times 3}(R)$, the set

$$\left\{ \begin{pmatrix} 1 & -3 & 2 \\ -4 & 0 & 5 \end{pmatrix}, \begin{pmatrix} -3 & 7 & 4 \\ 6 & -2 & -7 \end{pmatrix}, \begin{pmatrix} -2 & 3 & 11 \\ -1 & -3 & 2 \end{pmatrix} \right\}$$

is linearly dependent because

$$5\begin{pmatrix} 1 & -3 & 2 \\ -4 & 0 & 5 \end{pmatrix} + 3\begin{pmatrix} -3 & 7 & 4 \\ 6 & -2 & -7 \end{pmatrix} - 2\begin{pmatrix} -2 & 3 & 11 \\ -1 & -3 & 2 \end{pmatrix} = \begin{pmatrix} 0 & 0 & 0 \\ 0 & 0 & 0 \end{pmatrix}. \blacklozenge$$

Definition. *A subset S of a vector space that is not linearly dependent is called **linearly independent**. As before, we also say that the vectors of S are linearly independent.*

The following facts about linearly independent sets are true in any vector space.

1. The empty set is linearly independent, for linearly dependent sets must be nonempty.
2. A set consisting of a single nonzero vector is linearly independent. For if $\{u\}$ is linearly dependent, then $au = 0$ for some nonzero scalar a. Thus

$$u = a^{-1}(au) = a^{-1}0 = 0.$$

3. A set is linearly independent if and only if the only representations of 0 as linear combinations of its vectors are trivial representations.

The condition in item 3 provides a useful method for determining whether a finite set is linearly independent. This technique is illustrated in the examples that follow.

Example 3

To prove that the set

$$S = \{(1,0,0,-1),(0,1,0,-1),(0,0,1,-1),(0,0,0,1)\}$$

is linearly independent, we must show that the only linear combination of vectors in S that equals the zero vector is the one in which all the coefficients are zero. Suppose that a_1, a_2, a_3, and a_4 are scalars such that

$$a_1(1,0,0,-1) + a_2(0,1,0,-1) + a_3(0,0,1,-1) + a_4(0,0,0,1) = (0,0,0,0).$$

Equating the corresponding coordinates of the vectors on the left and the right sides of this equation, we obtain the following system of linear equations.

$$\begin{aligned}
a_1 &= 0 \\
a_2 &= 0 \\
a_3 &= 0 \\
-a_1 - a_2 - a_3 + a_4 &= 0
\end{aligned}$$

Clearly the only solution to this system is $a_1 = a_2 = a_3 = a_4 = 0$, and so S is linearly independent.   ◆

Example 4

For $k = 0, 1, \ldots, n$ let $p_k(x) = x^k + x^{k+1} + \cdots + x^n$. The set

$$\{p_0(x), p_1(x), \ldots, p_n(x)\}$$

is linearly independent in $\mathsf{P}_n(F)$. For if

$$a_0 p_0(x) + a_1 p_1(x) + \cdots + a_n p_n(x) = 0$$

for some scalars $a_0, a_1, \ldots, a_n$, then

$$a_0 + (a_0 + a_1)x + (a_0 + a_1 + a_2)x^2 + \cdots + (a_0 + a_1 + \cdots + a_n)x^n = 0.$$

By equating the coefficients of x^k on both sides of this equation for $k = 1, 2, \ldots, n$, we obtain

$$\begin{aligned}
a_0 &= 0 \\
a_0 + a_1 &= 0 \\
a_0 + a_1 + a_2 &= 0 \\
&\vdots \\
a_0 + a_1 + a_2 + \cdots + a_n &= 0.
\end{aligned}$$

Clearly the only solution to this system of linear equations is $a_0 = a_1 = \cdots = a_n = 0$.   ◆

The following important results are immediate consequences of the definitions of linear dependence and linear independence.

Theorem 1.6. *Let* V *be a vector space, and let* $S_1 \subseteq S_2 \subseteq \mathsf{V}$. *If* S_1 *is linearly dependent, then* S_2 *is linearly dependent.*

Proof. Exercise.   ∎

Corollary. *Let* V *be a vector space, and let* $S_1 \subseteq S_2 \subseteq \mathsf{V}$. *If* S_2 *is linearly independent, then* S_1 *is linearly independent.*

Proof. Exercise.   ∎

Earlier in this section, we remarked that the issue of whether S is the smallest generating set for its span is related to the question of whether some vector in S is a linear combination of the other vectors in S. Thus the issue of whether S is the smallest generating set for its span is related to the question of whether S is linearly dependent. To see why, consider the subset $S = \{u_1, u_2, u_3, u_4\}$ of R^3, where $u_1 = (2, -1, 4)$, $u_2 = (1, -1, 3)$, $u_3 = (1, 1, -1)$, and $u_4 = (1, -2, -1)$. We have previously noted that S is linearly dependent; in fact,

$$-2u_1 + 3u_2 + u_3 - 0u_4 = 0.$$

This equation implies that u_3 (or alternatively, u_1 or u_2) is a linear combination of the other vectors in S. For example, $u_3 = 2u_1 - 3u_2 + 0u_4$. Therefore every linear combination $a_1 u_1 + a_2 u_2 + a_3 u_3 + a_4 u_4$ of vectors in S can be written as a linear combination of u_1, u_2, and u_4:

$$a_1 u_1 + a_2 u_2 + a_3 u_3 + a_4 u_4 = a_1 u_1 + a_2 u_2 + a_3(2u_1 - 3u_2 + 0u_4) + a_4 u_4$$
$$= (a_1 + 2a_3)u_1 + (a_2 - 3a_3)u_2 + a_4 u_4.$$

Thus the subset $S' = \{u_1, u_2, u_4\}$ of S has the same span as S!

More generally, suppose that S is any linearly dependent set containing two or more vectors. Then some vector $v \in S$ can be written as a linear combination of the other vectors in S, and the subset obtained by removing v from S has the same span as S. It follows that *if no proper subset of S generates the span of S, then S must be linearly independent.* Another way to view the preceding statement is given in Theorem 1.7.

Theorem 1.7. *Let* S *be a linearly independent subset of a vector space* V, *and let* v *be a vector in* V *that is not in* S. *Then* $S \cup \{v\}$ *is linearly dependent if and only if* $v \in \text{span}(S)$.

Proof. If $S \cup \{v\}$ is linearly dependent, then there are vectors $u_1, u_2, \ldots, u_n$ in $S \cup \{v\}$ such that $a_1 u_1 + a_2 u_2 + \cdots + a_n u_n = 0$ for some nonzero scalars $a_1, a_2, \ldots, a_n$. Because S is linearly independent, one of the u_i's, say u_1, equals v. Thus $a_1 v + a_2 u_2 + \cdots + a_n u_n = 0$, and so

$$v = a_1^{-1}(-a_2 u_2 - \cdots - a_n u_n) = -(a_1^{-1} a_2)u_2 - \cdots - (a_1^{-1} a_n)u_n.$$

Since v is a linear combination of $u_2, \ldots, u_n$, which are in S, we have $v \in \text{span}(S)$.

Conversely, let $v \in \text{span}(S)$. Then there exist vectors $v_1, v_2, \ldots, v_m$ in S and scalars $b_1, b_2, \ldots, b_m$ such that $v = b_1 v_1 + b_2 v_2 + \cdots + b_m v_m$. Hence

$$0 = b_1 v_1 + b_2 v_2 + \cdots + b_m v_m + (-1)v.$$

Since $v \neq v_i$ for $i = 1, 2, \ldots, m$, the coefficient of v in this linear combination is nonzero, and so the set $\{v_1, v_2, \ldots, v_m, v\}$ is linearly dependent. Therefore $S \cup \{v\}$ is linearly dependent by Theorem 1.6. ∎

Linearly independent generating sets are investigated in detail in Section 1.6.

EXERCISES

1. Label the following statements as true or false.

   (a) If S is a linearly dependent set, then each vector in S is a linear combination of other vectors in S.

   (b) Any set containing the zero vector is linearly dependent.

   (c) The empty set is linearly dependent.

   (d) Subsets of linearly dependent sets are linearly dependent.

   (e) Subsets of linearly independent sets are linearly independent.

   (f) If $a_1 x_1 + a_2 x_2 + \cdots + a_n x_n = 0$ and $x_1, x_2, \ldots, x_n$ are linearly independent, then all the scalars a_i are zero.

2.[3] Determine whether the following sets are linearly dependent or linearly independent.

   (a) $\left\{ \begin{pmatrix} 1 & -3 \\ -2 & 4 \end{pmatrix}, \begin{pmatrix} -2 & 6 \\ 4 & -8 \end{pmatrix} \right\}$ in $\mathsf{M}_{2 \times 2}(R)$

   (b) $\left\{ \begin{pmatrix} 1 & -2 \\ -1 & 4 \end{pmatrix}, \begin{pmatrix} -1 & 1 \\ 2 & -4 \end{pmatrix} \right\}$ in $\mathsf{M}_{2 \times 2}(R)$

   (c) $\{x^3 + 2x^2, -x^2 + 3x + 1, x^3 - x^2 + 2x - 1\}$ in $\mathsf{P}_3(R)$

---

[3]The computations in Exercise 2(g), (h), (i), and (j) are tedious unless technology is used.

(d) $\{x^3 - x, 2x^2 + 4, -2x^3 + 3x^2 + 2x + 6\}$ in $P_3(R)$

(e) $\{(1, -1, 2), (1, -2, 1), (1, 1, 4)\}$ in R^3

(f) $\{(1, -1, 2), (2, 0, 1), (-1, 2, -1)\}$ in R^3

(g) $\left\{ \begin{pmatrix} 1 & 0 \\ -2 & 1 \end{pmatrix}, \begin{pmatrix} 0 & -1 \\ 1 & 1 \end{pmatrix}, \begin{pmatrix} -1 & 2 \\ 1 & 0 \end{pmatrix}, \begin{pmatrix} 2 & 1 \\ -4 & 4 \end{pmatrix} \right\}$ in $M_{2\times2}(R)$

(h) $\left\{ \begin{pmatrix} 1 & 0 \\ -2 & 1 \end{pmatrix}, \begin{pmatrix} 0 & -1 \\ 1 & 1 \end{pmatrix}, \begin{pmatrix} -1 & 2 \\ 1 & 0 \end{pmatrix}, \begin{pmatrix} 2 & 1 \\ 2 & -2 \end{pmatrix} \right\}$ in $M_{2\times2}(R)$

(i) $\{x^4 - x^3 + 5x^2 - 8x + 6, -x^4 + x^3 - 5x^2 + 5x - 3,$
    $x^4 + 3x^2 - 3x + 5, 2x^4 + 3x^3 + 4x^2 - x + 1, x^3 - x + 2\}$ in $P_4(R)$

(j) $\{x^4 - x^3 + 5x^2 - 8x + 6, -x^4 + x^3 - 5x^2 + 5x - 3,$
    $x^4 + 3x^2 - 3x + 5, 2x^4 + x^3 + 4x^2 + 8x\}$ in $P_4(R)$

3. In $M_{2\times3}(F)$, prove that the set

$$\left\{ \begin{pmatrix} 1 & 1 \\ 0 & 0 \\ 0 & 0 \end{pmatrix}, \begin{pmatrix} 0 & 0 \\ 1 & 1 \\ 0 & 0 \end{pmatrix}, \begin{pmatrix} 0 & 0 \\ 0 & 0 \\ 1 & 1 \end{pmatrix}, \begin{pmatrix} 1 & 0 \\ 1 & 0 \\ 1 & 0 \end{pmatrix}, \begin{pmatrix} 0 & 1 \\ 0 & 1 \\ 0 & 1 \end{pmatrix} \right\}$$

is linearly dependent.

4. In F^n, let e_j denote the vector whose jth coordinate is 1 and whose other coordinates are 0. Prove that $\{e_1, e_2, \cdots, e_n\}$ is linearly independent.

5. Show that the set $\{1, x, x^2, \ldots, x^n\}$ is linearly independent in $P_n(F)$.

6. In $M_{m\times n}(F)$, let E^{ij} denote the matrix whose only nonzero entry is 1 in the ith row and jth column. Prove that $\{E^{ij} : 1 \le i \le m,\ 1 \le j \le n\}$ is linearly independent.

7. Recall from Example 3 in Section 1.3 that the set of diagonal matrices in $M_{2\times2}(F)$ is a subspace. Find a linearly independent set that generates this subspace.

8. Let $S = \{(1, 1, 0), (1, 0, 1), (0, 1, 1)\}$ be a subset of the vector space F^3.

    (a) Prove that if $F = R$, then S is linearly independent.

    (b) Prove that if F has characteristic 2, then S is linearly dependent.

9.† Let u and v be distinct vectors in a vector space V. Show that $\{u, v\}$ is linearly dependent if and only if u or v is a multiple of the other.

10. Give an example of three linearly dependent vectors in R^3 such that none of the three is a multiple of another.

11. Let $S = \{u_1, u_2, \ldots, u_n\}$ be a linearly independent subset of a vector space V over the field Z_2. How many vectors are there in span(S)? Justify your answer.

12. Prove Theorem 1.6 and its corollary.

13. Let V be a vector space over a field of characteristic not equal to two.

    (a) Let u and v be distinct vectors in V. Prove that $\{u, v\}$ is linearly independent if and only if $\{u + v, u - v\}$ is linearly independent.

    (b) Let u, v, and w be distinct vectors in V. Prove that $\{u, v, w\}$ is linearly independent if and only if $\{u + v, u + w, v + w\}$ is linearly independent.

14. Prove that a set S is linearly dependent if and only if $S = \{0\}$ or there exist distinct vectors $v, u_1, u_2, \ldots, u_n$ in S such that v is a linear combination of $u_1, u_2, \ldots, u_n$.

15. Let $S = \{u_1, u_2, \ldots, u_n\}$ be a finite set of vectors. Prove that S is linearly dependent if and only if $u_1 = 0$ or $u_{k+1} \in \text{span}(\{u_1, u_2, \ldots, u_k\})$ for some k $(1 \le k < n)$.

16. Prove that a set S of vectors is linearly independent if and only if each finite subset of S is linearly independent.

17. Let M be a square upper triangular matrix (as defined in Exercise 12 of Section 1.3) with nonzero diagonal entries. Prove that the columns of M are linearly independent.

18. Let S be a set of nonzero polynomials in P(F) such that no two have the same degree. Prove that S is linearly independent.

19. Prove that if $\{A_1, A_2, \ldots, A_k\}$ is a linearly independent subset of $\mathsf{M}_{n \times n}(F)$, then $\{A_1^t, A_2^t, \ldots, A_k^t\}$ is also linearly independent.

20. Let $f, g, \in \mathcal{F}(R, R)$ be the functions defined by $f(t) = e^{rt}$ and $g(t) = e^{st}$, where $r \ne s$. Prove that f and g are linearly independent in $\mathcal{F}(R, R)$.

1.6   BASES AND DIMENSION

We saw in Section 1.5 that if S is a generating set for a subspace W and no proper subset of S is a generating set for W, then S must be linearly independent. A linearly independent generating set for W possesses a very useful property—every vector in W can be expressed in one and only one way as a linear combination of the vectors in the set. (This property is proved below in Theorem 1.8.) It is this property that makes linearly independent generating sets the building blocks of vector spaces.

Definition. *A **basis** β for a vector space V is a linearly independent subset of V that generates V. If β is a basis for V, we also say that the vectors of β form a basis for V.*

Example 1

Recalling that span($\varnothing$) = $\{0\}$ and $\varnothing$ is linearly independent, we see that $\varnothing$ is a basis for the zero vector space.   ◆

Example 2

In F^n, let $e_1 = (1,0,0,\ldots,0), e_2 = (0,1,0,\ldots,0),\ldots,e_n = (0,0,\ldots,0,1)$; $\{e_1,e_2,\ldots,e_n\}$ is readily seen to be a basis for F^n and is called the **standard basis** for F^n.   ◆

Example 3

In $M_{m\times n}(F)$, let E^{ij} denote the matrix whose only nonzero entry is a 1 in the ith row and jth column. Then $\{E^{ij} : 1 \le i \le m, 1 \le j \le n\}$ is a basis for $M_{m\times n}(F)$.   ◆

Example 4

In $P_n(F)$ the set $\{1, x, x^2, \ldots, x^n\}$ is a basis. We call this basis the **standard basis** for $P_n(F)$.   ◆

Example 5

In $P(F)$ the set $\{1, x, x^2, \ldots\}$ is a basis.   ◆

Observe that Example 5 shows that a basis need not be finite. In fact, later in this section it is shown that no basis for $P(F)$ can be finite. Hence not every vector space has a finite basis.

The next theorem, which is used frequently in Chapter 2, establishes the most significant property of a basis.

Theorem 1.8. *Let V be a vector space and $\beta = \{u_1, u_2, \ldots, u_n\}$ be a subset of V. Then β is a basis for V if and only if each $v \in V$ can be uniquely expressed as a linear combination of vectors of β, that is, can be expressed in the form*

$$v = a_1u_1 + a_2u_2 + \cdots + a_nu_n$$

for unique scalars $a_1, a_2, \ldots, a_n$.

Proof. Let β be a basis for V. If $v \in V$, then $v \in$ span(β) because span(β) = V. Thus v is a linear combination of the vectors of β. Suppose that

$$v = a_1u_1 + a_2u_2 + \cdots + a_nu_n \quad \text{and} \quad v = b_1u_1 + b_2u_2 + \cdots + b_nu_n$$

are two such representations of v. Subtracting the second equation from the first gives

$$0 = (a_1 - b_1)u_1 + (a_2 - b_2)u_2 + \cdots + (a_n - b_n)u_n.$$

Since β is linearly independent, it follows that $a_1 - b_1 = a_2 - b_2 = \cdots = a_n - b_n = 0$. Hence $a_1 = b_1, a_2 = b_2, \cdots, a_n = b_n$, and so v is uniquely expressible as a linear combination of the vectors of β.

The proof of the converse is an exercise. ∎

Theorem 1.8 shows that if the vectors $u_1, u_2, \ldots, u_n$ form a basis for a vector space V, then every vector in V can be uniquely expressed in the form

$$v = a_1 u_1 + a_2 u_2 + \cdots + a_n u_n$$

for appropriately chosen scalars $a_1, a_2, \ldots, a_n$. Thus v determines a unique n-tuple of scalars $(a_1, a_2, \ldots, a_n)$ and, conversely, each n-tuple of scalars determines a unique vector $v \in \mathsf{V}$ by using the entries of the n-tuple as the coefficients of a linear combination of $u_1, u_2, \ldots, u_n$. This fact suggests that V is like the vector space F^n, where n is the number of vectors in the basis for V. We see in Section 2.4 that this is indeed the case.

In this book, we are primarily interested in vector spaces having finite bases. Theorem 1.9 identifies a large class of vector spaces of this type.

Theorem 1.9. *If a vector space V is generated by a finite set S, then some subset of S is a basis for V. Hence V has a finite basis.*

Proof. If $S = \varnothing$ or $S = \{0\}$, then $\mathsf{V} = \{0\}$ and $\varnothing$ is a subset of S that is a basis for V. Otherwise S contains a nonzero vector u_1. By item 2 on page 37, $\{u_1\}$ is a linearly independent set. Continue, if possible, choosing vectors $u_2, \ldots, u_k$ in S such that $\{u_1, u_2, \ldots, u_k\}$ is linearly independent. Since S is a finite set, we must eventually reach a stage at which $\beta = \{u_1, u_2, \ldots, u_k\}$ is a linearly independent subset of S, but adjoining to β any vector in S not in β produces a linearly dependent set. We claim that β is a basis for V. Because β is linearly independent by construction, it suffices to show that β spans V. By Theorem 1.5 (p. 30) we need to show that $S \subseteq \mathrm{span}(\beta)$. Let $v \in S$. If $v \in \beta$, then clearly $v \in \mathrm{span}(\beta)$. Otherwise, if $v \notin \beta$, then the preceding construction shows that $\beta \cup \{v\}$ is linearly dependent. So $v \in \mathrm{span}(\beta)$ by Theorem 1.7 (p. 39). Thus $S \subseteq \mathrm{span}(\beta)$. ∎

Because of the method by which the basis β was obtained in the proof of Theorem 1.9, this theorem is often remembered as saying that *a finite spanning set for V can be reduced to a basis for V*. This method is illustrated in the next example.

Example 6

Let

$$S = \{(2, -3, 5), (8, -12, 20), (1, 0, -2), (0, 2, -1), (7, 2, 0)\}.$$

It can be shown that S generates R^3. We can select a basis for R^3 that is a subset of S by the technique used in proving Theorem 1.9. To start, select any nonzero vector in S, say $(2, -3, 5)$, to be a vector in the basis. Since $4(2, -3, 5) = (8, -12, 20)$, the set $\{(2, 3, -5), (8, -12, 20)\}$ is linearly dependent by Exercise 9 of Section 1.5. Hence we do not include $(8, -12, 20)$ in our basis. On the other hand, $(1, 0, -2)$ is not a multiple of $(2, -3, 5)$ and vice versa, so that the set $\{(2, -3, 5), (1, 0, -2)\}$ is linearly independent. Thus we include $(1, 0, -2)$ as part of our basis.

Now we consider the set $\{(2, -3, 5), (1, 0, -2), (0, 2, -1)\}$ obtained by adjoining another vector in S to the two vectors that we have already included in our basis. As before, we include $(0, 2, -1)$ in our basis or exclude it from the basis according to whether $\{(2, -3, 5), (1, 0, -2), (0, 2, -1)\}$ is linearly independent or linearly dependent. An easy calculation shows that this set is linearly independent, and so we include $(0, 2, -1)$ in our basis. In a similar fashion the final vector in S is included or excluded from our basis according to whether the set

$$\{(2, -3, 5), (1, 0, -2), (0, 2, -1), (7, 2, 0)\}$$

is linearly independent or linearly dependent. Because

$$2(2, -3, 5) + 3(1, 0, -2) + 4(0, 2, -1) - (7, 2, 0) = (0, 0, 0),$$

we exclude $(7, 2, 0)$ from our basis. We conclude that

$$\{(2, -3, 5), (1, 0, -2), (0, 2, -1)\}$$

is a subset of S that is a basis for R^3.   ♦

The corollaries of the following theorem are perhaps the most significant results in Chapter 1.

Theorem 1.10 (Replacement Theorem).   *Let* V *be a vector space that is generated by a set* G *containing exactly* n *vectors, and let* L *be a linearly independent subset of* V *containing exactly* m *vectors. Then* $m \le n$ *and there exists a subset* H *of* G *containing exactly* $n - m$ *vectors such that* $L \cup H$ *generates* V.

Proof. The proof is by mathematical induction on m. The induction begins with $m = 0$; for in this case $L = \varnothing$, and so taking $H = G$ gives the desired result.

Now suppose that the theorem is true for some integer $m \geq 0$. We prove that the theorem is true for $m + 1$. Let $L = \{v_1, v_2, \ldots, v_{m+1}\}$ be a linearly independent subset of V consisting of $m + 1$ vectors. By the corollary to Theorem 1.6 (p. 39), $\{v_1, v_2, \ldots, v_m\}$ is linearly independent, and so we may apply the induction hypothesis to conclude that $m \leq n$ and that there is a subset $\{u_1, u_2, \ldots, u_{n-m}\}$ of G such that $\{v_1, v_2, \ldots, v_m\} \cup \{u_1, u_2, \ldots, u_{n-m}\}$ generates V. Thus there exist scalars $a_1, a_2, \ldots, a_m, b_1, b_2, \ldots, b_{n-m}$ such that

$$a_1 v_1 + a_2 v_2 + \cdots + a_m v_m + b_1 u_1 + b_2 u_2 + \cdots + b_{n-m} u_{n-m} = v_{m+1}. \qquad (9)$$

Note that $n - m > 0$, lest v_{m+1} be a linear combination of $v_1, v_2, \ldots, v_m$, which by Theorem 1.7 (p. 39) contradicts the assumption that L is linearly independent. Hence $n > m$; that is, $n \geq m + 1$. Moreover, some b_i, say b_1, is nonzero, for otherwise we obtain the same contradiction. Solving (9) for u_1 gives

$$u_1 = (-b_1^{-1} a_1) v_1 + (-b_1^{-1} a_2) v_2 + \cdots + (-b_1^{-1} a_m) v_m + (b_1^{-1}) v_{m+1}$$
$$+ (-b_1^{-1} b_2) u_2 + \cdots + (-b_1^{-1} b_{n-m}) u_{n-m}.$$

Let $H = \{u_2, \ldots, u_{n-m}\}$. Then $u_1 \in \text{span}(L \cup H)$, and because $v_1, v_2, \ldots, v_m$, $u_2, \ldots, u_{n-m}$ are clearly in $\text{span}(L \cup H)$, it follows that

$$\{v_1, v_2, \ldots, v_m, u_1, u_2, \ldots, u_{n-m}\} \subseteq \text{span}(L \cup H).$$

Because $\{v_1, v_2, \ldots, v_m, u_1, u_2, \ldots, u_{n-m}\}$ generates V, Theorem 1.5 (p. 30) implies that $\text{span}(L \cup H) = \mathsf{V}$. Since H is a subset of G that contains $(n - m) - 1 = n - (m + 1)$ vectors, the theorem is true for $m + 1$. This completes the induction. ∎

Corollary 1. *Let* V *be a vector space having a finite basis. Then every basis for* V *contains the same number of vectors.*

Proof. Suppose that β is a finite basis for V that contains exactly n vectors, and let γ be any other basis for V. If γ contains more than n vectors, then we can select a subset S of γ containing exactly $n + 1$ vectors. Since S is linearly independent and β generates V, the replacement theorem implies that $n + 1 \leq n$, a contradiction. Therefore γ is finite, and the number m of vectors in γ satisfies $m \leq n$. Reversing the roles of β and γ and arguing as above, we obtain $n \leq m$. Hence $m = n$. ∎

If a vector space has a finite basis, Corollary 1 asserts that the number of vectors in *any* basis for V is an intrinsic property of V. This fact makes possible the following important definitions.

Definitions. *A vector space is called* **finite-dimensional** *if it has a basis consisting of a finite number of vectors. The unique number of vectors* *in each basis for* V *is called the* **dimension** *of* V *and is denoted by* $\dim(V)$. *A vector space that is not finite-dimensional is called* **infinite-dimensional**.

The following results are consequences of Examples 1 through 4.

Example 7

The vector space $\{0\}$ has dimension zero.   ◆

Example 8

The vector space F^n has dimension n.   ◆

Example 9

The vector space $M_{m \times n}(F)$ has dimension mn.   ◆

Example 10

The vector space $P_n(F)$ has dimension $n + 1$.   ◆

The following examples show that the dimension of a vector space depends on its field of scalars.

Example 11

Over the field of complex numbers, the vector space of complex numbers has dimension 1. (A basis is $\{1\}$.)   ◆

Example 12

Over the field of real numbers, the vector space of complex numbers has dimension 2. (A basis is $\{1, i\}$.)   ◆

In the terminology of dimension, the first conclusion in the replacement theorem states that if V is a finite-dimensional vector space, then no linearly independent subset of V can contain more than $\dim(V)$ vectors. From this fact it follows that the vector space $P(F)$ is infinite-dimensional because it has an infinite linearly independent set, namely $\{1, x, x^2, \ldots\}$. This set is, in fact, a basis for $P(F)$. Yet nothing that we have proved in this section guarantees an infinite-dimensional vector space must have a basis. In Section 1.7 it is shown, however, that *every vector space has a basis.*

Just as no linearly independent subset of a finite-dimensional vector space V can contain more than $\dim(V)$ vectors, a corresponding statement can be made about the size of a generating set.

Corollary 2. *Let* V *be a vector space with dimension* n.
(a) *Any finite generating set for* V *contains at least* n *vectors, and a generating set for* V *that contains exactly* n *vectors is a basis for* V.

(b) *Any linearly independent subset of* V *that contains exactly* n *vectors is a basis for* V.

(c) *Every linearly independent subset of* V *can be extended to a basis for* V.

Proof. Let β be a basis for V.

(a) Let G be a finite generating set for V. By Theorem 1.9 some subset H of G is a basis for V. Corollary 1 implies that H contains exactly n vectors. Since a subset of G contains n vectors, G must contain at least n vectors. Moreover, if G contains exactly n vectors, then we must have $H = G$, so that G is a basis for V.

(b) Let L be a linearly independent subset of V containing exactly n vectors. It follows from the replacement theorem that there is a subset H of β containing $n - n = 0$ vectors such that $L \cup H$ generates V. Thus $H = \varnothing$, and L generates V. Since L is also linearly independent, L is a basis for V.

(c) If L is a linearly independent subset of V containing m vectors, then the replacement theorem asserts that there is a subset H of β containing exactly $n - m$ vectors such that $L \cup H$ generates V. Now $L \cup H$ contains at most n vectors; therefore (a) implies that $L \cup H$ contains exactly n vectors and that $L \cup H$ is a basis for V. ∎

Example 13

It follows from Example 4 of Section 1.4 and (a) of Corollary 2 that

$$\{x^2 + 3x - 2, 2x^2 + 5x - 3, -x^2 - 4x + 4\}$$

is a basis for $P_2(R)$. ◆

Example 14

It follows from Example 5 of Section 1.4 and (a) of Corollary 2 that

$$\left\{ \begin{pmatrix} 1 & 1 \\ 1 & 0 \end{pmatrix}, \begin{pmatrix} 1 & 1 \\ 0 & 1 \end{pmatrix}, \begin{pmatrix} 1 & 0 \\ 1 & 1 \end{pmatrix}, \begin{pmatrix} 0 & 1 \\ 1 & 1 \end{pmatrix} \right\}$$

is a basis for $M_{2 \times 2}(R)$. ◆

Example 15

It follows from Example 3 of Section 1.5 and (b) of Corollary 2 that

$$\{(1, 0, 0, -1), (0, 1, 0, -1), (0, 0, 1, -1), (0, 0, 0, 1)\}$$

is a basis for R^4. ◆

Example 16

For $k = 0, 1, \ldots, n$, let $p_k(x) = x^k + x^{k+1} + \cdots + x^n$. It follows from Example 4 of Section 1.5 and (b) of Corollary 2 that

$$\{p_0(x), p_1(x), \ldots, p_n(x)\}$$

is a basis for $\mathsf{P}_n(F)$.   ◆

A procedure for reducing a generating set to a basis was illustrated in Example 6. In Section 3.4, when we have learned more about solving systems of linear equations, we will discover a simpler method for reducing a generating set to a basis. This procedure also can be used to extend a linearly independent set to a basis, as (c) of Corollary 2 asserts is possible.

An Overview of Dimension and Its Consequences

Theorem 1.9 as well as the replacement theorem and its corollaries contain a wealth of information about the relationships among linearly independent sets, bases, and generating sets. For this reason, we summarize here the main results of this section in order to put them into better perspective.

A basis for a vector space V is a linearly independent subset of V that generates V. If V has a finite basis, then every basis for V contains the same number of vectors. This number is called the dimension of V, and V is said to be finite-dimensional. Thus if the dimension of V is n, every basis for V contains exactly n vectors. Moreover, every linearly independent subset of V contains *no more* than n vectors and can be extended to a basis for V by including appropriately chosen vectors. Also, each generating set for V contains *at least* n vectors and can be reduced to a basis for V by excluding appropriately chosen vectors. The Venn diagram in Figure 1.6 depicts these relationships.

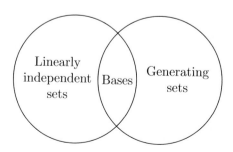

Figure 1.6

The Dimension of Subspaces

Our next result relates the dimension of a subspace to the dimension of the vector space that contains it.

Theorem 1.11. *Let* W *be a subspace of a finite-dimensional vector space* V. *Then* W *is finite-dimensional and* $\dim(W) \leq \dim(V)$. *Moreover, if* $\dim(W) = \dim(V)$, *then* V = W.

Proof. Let $\dim(V) = n$. If $W = \{0\}$, then W is finite-dimensional and $\dim(W) = 0 \leq n$. Otherwise, W contains a nonzero vector x_1; so $\{x_1\}$ is a linearly independent set. Continue choosing vectors, $x_1, x_2, \ldots, x_k$ in W such that $\{x_1, x_2, \ldots, x_k\}$ is linearly independent. Since no linearly independent subset of V can contain more than n vectors, this process must stop at a stage where $k \leq n$ and $\{x_1, x_2, \ldots, x_k\}$ is linearly independent but adjoining any other vector from W produces a linearly dependent set. Theorem 1.7 (p. 39) implies that $\{x_1, x_2, \ldots, x_k\}$ generates W, and hence it is a basis for W. Therefore $\dim(W) = k \leq n$.

If $\dim(W) = n$, then a basis for W is a linearly independent subset of V containing n vectors. But Corollary 2 of the replacement theorem implies that this basis for W is also a basis for V; so W = V. ∎

Example 17

Let

$$W = \{(a_1, a_2, a_3, a_4, a_5) \in \mathsf{F}^5 : a_1 + a_3 + a_5 = 0, a_2 = a_4\}.$$

It is easily shown that W is a subspace of F^5 having

$$\{(-1, 0, 1, 0, 0), (-1, 0, 0, 0, 1), (0, 1, 0, 1, 0)\}$$

as a basis. Thus $\dim(W) = 3$. ◆

Example 18

The set of diagonal $n \times n$ matrices is a subspace W of $\mathsf{M}_{n \times n}(F)$ (see Example 3 of Section 1.3). A basis for W is

$$\{E^{11}, E^{22}, \ldots, E^{nn}\},$$

where E^{ij} is the matrix in which the only nonzero entry is a 1 in the ith row and jth column. Thus $\dim(W) = n$. ◆

Example 19

We saw in Section 1.3 that the set of symmetric $n \times n$ matrices is a subspace W of $\mathsf{M}_{n \times n}(F)$. A basis for W is

$$\{A^{ij} : 1 \leq i \leq j \leq n\},$$

where A^{ij} is the $n \times n$ matrix having 1 in the ith row and jth column, 1 in the jth row and ith column, and 0 elsewhere. It follows that

$$\dim(\mathsf{W}) = n + (n-1) + \cdots + 1 = \frac{1}{2}n(n+1). \quad \blacklozenge$$

Corollary. *If* W *is a subspace of a finite-dimensional vector space* V, *then any basis for* W *can be extended to a basis for* V.

Proof. Let S be a basis for W. Because S is a linearly independent subset of V, Corollary 2 of the replacement theorem guarantees that S can be extended to a basis for V. ∎

Example 20

The set of all polynomials of the form

$$a_{18}x^{18} + a_{16}x^{16} + \cdots + a_2 x^2 + a_0,$$

where $a_{18}, a_{16}, \ldots, a_2, a_0 \in F$, is a subspace W of $\mathsf{P}_{18}(F)$. A basis for W is $\{1, x^2, \ldots, x^{16}, x^{18}\}$, which is a subset of the standard basis for $\mathsf{P}_{18}(F)$. ♦

We can apply Theorem 1.11 to determine the subspaces of R^2 and R^3. Since R^2 has dimension 2, subspaces of R^2 can be of dimensions 0, 1, or 2 only. The only subspaces of dimension 0 or 2 are $\{0\}$ and R^2, respectively. Any subspace of R^2 having dimension 1 consists of all scalar multiples of some nonzero vector in R^2 (Exercise 11 of Section 1.4).

If a point of R^2 is identified in the natural way with a point in the Euclidean plane, then it is possible to describe the subspaces of R^2 geometrically: A subspace of R^2 having dimension 0 consists of the origin of the Euclidean plane, a subspace of R^2 with dimension 1 consists of a line through the origin, and a subspace of R^2 having dimension 2 is the entire Euclidean plane.

Similarly, the subspaces of R^3 must have dimensions 0, 1, 2, or 3. Interpreting these possibilities geometrically, we see that a subspace of dimension zero must be the origin of Euclidean 3-space, a subspace of dimension 1 is a line through the origin, a subspace of dimension 2 is a plane through the origin, and a subspace of dimension 3 is Euclidean 3-space itself.

The Lagrange Interpolation Formula

Corollary 2 of the replacement theorem can be applied to obtain a useful formula. Let $c_0, c_1, \ldots, c_n$ be distinct scalars in an infinite field F. The polynomials $f_0(x), f_1(x), \ldots, f_n(x)$ defined by

$$f_i(x) = \frac{(x - c_0) \cdots (x - c_{i-1})(x - c_{i+1}) \cdots (x - c_n)}{(c_i - c_0) \cdots (c_i - c_{i-1})(c_i - c_{i+1}) \cdots (c_i - c_n)} = \prod_{\substack{k=0 \\ k \neq i}}^{n} \frac{x - c_k}{c_i - c_k}.$$

are called the **Lagrange polynomials** (associated with $c_0, c_1, \ldots, c_n$). Note that each $f_i(x)$ is a polynomial of degree n and hence is in $\mathsf{P}_n(F)$. By regarding $f_i(x)$ as a polynomial function $f_i \colon F \to F$, we see that

$$f_i(c_j) = \begin{cases} 0 & \text{if } i \neq j \\ 1 & \text{if } i = j. \end{cases} \tag{10}$$

This property of Lagrange polynomials can be used to show that $\beta = \{f_0, f_1, \ldots, f_n\}$ is a linearly independent subset of $\mathsf{P}_n(F)$. Suppose that

$$\sum_{i=0}^{n} a_i f_i = 0 \quad \text{for some scalars } a_0, a_1, \ldots, a_n,$$

where 0 denotes the zero function. Then

$$\sum_{i=0}^{n} a_i f_i(c_j) = 0 \quad \text{for } j = 0, 1, \ldots, n.$$

But also

$$\sum_{i=0}^{n} a_i f_i(c_j) = a_j$$

by (10). Hence $a_j = 0$ for $j = 0, 1, \ldots, n$; so β is linearly independent. Since the dimension of $\mathsf{P}_n(F)$ is $n+1$, it follows from Corollary 2 of the replacement theorem that β is a basis for $\mathsf{P}_n(F)$.

Because β is a basis for $\mathsf{P}_n(F)$, every polynomial function g in $\mathsf{P}_n(F)$ is a linear combination of polynomial functions of β, say,

$$g = \sum_{i=0}^{n} b_i f_i.$$

It follows that

$$g(c_j) = \sum_{i=0}^{n} b_i f_i(c_j) = b_j;$$

so

$$g = \sum_{i=0}^{n} g(c_i) f_i$$

is the unique representation of g as a linear combination of elements of β. This representation is called the **Lagrange interpolation formula**. Notice that the preceding argument shows that if $b_0, b_1, \ldots, b_n$ are any $n+1$ scalars in F (not necessarily distinct), then the polynomial function

$$g = \sum_{i=0}^{n} b_i f_i$$

is the unique polynomial in $P_n(F)$ such that $g(c_j) = b_j$. Thus we have found the unique polynomial of degree not exceeding n that has specified values b_j at given points c_j in its domain ($j = 0, 1, \ldots, n$). For example, let us construct the real polynomial g of degree at most 2 whose graph contains the points $(1, 8), (2, 5)$, and $(3, -4)$. (Thus, in the notation above, $c_0 = 1$, $c_1 = 2$, $c_2 = 3$, $b_0 = 8$, $b_1 = 5$, and $b_2 = -4$.) The Lagrange polynomials associated with c_0, c_1, and c_2 are

$$f_0(x) = \frac{(x-2)(x-3)}{(1-2)(1-3)} = \frac{1}{2}(x^2 - 5x + 6),$$

$$f_1(x) = \frac{(x-1)(x-3)}{(2-1)(2-3)} = -1(x^2 - 4x + 3),$$

and

$$f_2(x) = \frac{(x-1)(x-2)}{(3-1)(3-2)} = \frac{1}{2}(x^2 - 3x + 2).$$

Hence the desired polynomial is

$$g(x) = \sum_{i=0}^{2} b_i f_i(x) = 8f_0(x) + 5f_1(x) - 4f_2(x)$$
$$= 4(x^2 - 5x + 6) - 5(x^2 - 4x + 3) - 2(x^2 - 3x + 2)$$
$$= -3x^2 + 6x + 5.$$

An important consequence of the Lagrange interpolation formula is the following result: If $f \in P_n(F)$ and $f(c_i) = 0$ for $n+1$ distinct scalars $c_0, c_1, \ldots, c_n$ in F, then f is the zero function.

EXERCISES

1. Label the following statements as true or false.
    (a) The zero vector space has no basis.
    (b) Every vector space that is generated by a finite set has a basis.
    (c) Every vector space has a finite basis.
    (d) A vector space cannot have more than one basis.

(e) If a vector space has a finite basis, then the number of vectors in every basis is the same.

(f) The dimension of $P_n(F)$ is n.

(g) The dimension of $M_{m \times n}(F)$ is $m + n$.

(h) Suppose that V is a finite-dimensional vector space, that S_1 is a linearly independent subset of V, and that S_2 is a subset of V that generates V. Then S_1 cannot contain more vectors than S_2.

(i) If S generates the vector space V, then every vector in V can be written as a linear combination of vectors in S in only one way.

(j) Every subspace of a finite-dimensional space is finite-dimensional.

(k) If V is a vector space having dimension n, then V has exactly one subspace with dimension 0 and exactly one subspace with dimension n.

(l) If V is a vector space having dimension n, and if S is a subset of V with n vectors, then S is linearly independent if and only if S spans V.

2. Determine which of the following sets are bases for R^3.

(a) $\{(1, 0, -1), (2, 5, 1), (0, -4, 3)\}$

(b) $\{(2, -4, 1), (0, 3, -1), (6, 0, -1)\}$

(c) $\{(1, 2, -1), (1, 0, 2), (2, 1, 1)\}$

(d) $\{(-1, 3, 1), (2, -4, -3), (-3, 8, 2)\}$

(e) $\{(1, -3, -2), (-3, 1, 3), (-2, -10, -2)\}$

3. Determine which of the following sets are bases for $P_2(R)$.

(a) $\{-1 - x + 2x^2, 2 + x - 2x^2, 1 - 2x + 4x^2\}$

(b) $\{1 + 2x + x^2, 3 + x^2, x + x^2\}$

(c) $\{1 - 2x - 2x^2, -2 + 3x - x^2, 1 - x + 6x^2\}$

(d) $\{-1 + 2x + 4x^2, 3 - 4x - 10x^2, -2 - 5x - 6x^2\}$

(e) $\{1 + 2x - x^2, 4 - 2x + x^2, -1 + 18x - 9x^2\}$

4. Do the polynomials $x^3 - 2x^2 + 1$, $4x^2 - x + 3$, and $3x - 2$ generate $P_3(R)$? Justify your answer.

5. Is $\{(1, 4, -6), (1, 5, 8), (2, 1, 1), (0, 1, 0)\}$ a linearly independent subset of R^3? Justify your answer.

6. Give three different bases for F^2 and for $M_{2 \times 2}(F)$.

7. The vectors $u_1 = (2, -3, 1)$, $u_2 = (1, 4, -2)$, $u_3 = (-8, 12, -4)$, $u_4 = (1, 37, -17)$, and $u_5 = (-3, -5, 8)$ generate R^3. Find a subset of the set $\{u_1, u_2, u_3, u_4, u_5\}$ that is a basis for R^3.

8. Let W denote the subspace of R^5 consisting of all the vectors having coordinates that sum to zero. The vectors

$$
\begin{aligned}
u_1 &= (2, -3, 4, -5, 2), & u_2 &= (-6, 9, -12, 15, -6), \\
u_3 &= (3, -2, 7, -9, 1), & u_4 &= (2, -8, 2, -2, 6), \\
u_5 &= (-1, 1, 2, 1, -3), & u_6 &= (0, -3, -18, 9, 12), \\
u_7 &= (1, 0, -2, 3, -2), & u_8 &= (2, -1, 1, -9, 7)
\end{aligned}
$$

generate W. Find a subset of the set $\{u_1, u_2, \ldots, u_8\}$ that is a basis for W.

9. The vectors $u_1 = (1, 1, 1, 1)$, $u_2 = (0, 1, 1, 1)$, $u_3 = (0, 0, 1, 1)$, and $u_4 = (0, 0, 0, 1)$ form a basis for F^4. Find the unique representation of an arbitrary vector (a_1, a_2, a_3, a_4) in F^4 as a linear combination of u_1, u_2, u_3, and u_4.

10. In each part, use the Lagrange interpolation formula to construct the polynomial of smallest degree whose graph contains the following points.

   (a) $(-2, -6)$, $(-1, 5)$, $(1, 3)$
   (b) $(-4, 24)$, $(1, 9)$, $(3, 3)$
   (c) $(-2, 3)$, $(-1, -6)$, $(1, 0)$, $(3, -2)$
   (d) $(-3, -30)$, $(-2, 7)$, $(0, 15)$, $(1, 10)$

11. Let u and v be distinct vectors of a vector space V. Show that if $\{u, v\}$ is a basis for V and a and b are nonzero scalars, then both $\{u + v, au\}$ and $\{au, bv\}$ are also bases for V.

12. Let u, v, and w be distinct vectors of a vector space V. Show that if $\{u, v, w\}$ is a basis for V, then $\{u+v+w, v+w, w\}$ is also a basis for V.

13. The set of solutions to the system of linear equations

$$
\begin{aligned}
x_1 - 2x_2 + x_3 &= 0 \\
2x_1 - 3x_2 + x_3 &= 0
\end{aligned}
$$

is a subspace of R^3. Find a basis for this subspace.

14. Find bases for the following subspaces of F^5:

$$W_1 = \{(a_1, a_2, a_3, a_4, a_5) \in F^5 : a_1 - a_3 - a_4 = 0\}$$

and

$$W_2 = \{(a_1, a_2, a_3, a_4, a_5) \in F^5 : a_2 = a_3 = a_4 \text{ and } a_1 + a_5 = 0\}.$$

What are the dimensions of W_1 and W_2?

15. The set of all $n \times n$ matrices having trace equal to zero is a subspace W of $M_{n \times n}(F)$ (see Example 4 of Section 1.3). Find a basis for W. What is the dimension of W?

16. The set of all upper triangular $n \times n$ matrices is a subspace W of $M_{n \times n}(F)$ (see Exercise 12 of Section 1.3). Find a basis for W. What is the dimension of W?

17. The set of all skew-symmetric $n \times n$ matrices is a subspace W of $M_{n \times n}(F)$ (see Exercise 28 of Section 1.3). Find a basis for W. What is the dimension of W?

18. Find a basis for the vector space in Example 5 of Section 1.2. Justify your answer.

19. Complete the proof of Theorem 1.8.

20.† Let V be a vector space having dimension n, and let S be a subset of V that generates V.

(a) Prove that there is a subset of S that is a basis for V. (Be careful not to assume that S is finite.)
(b) Prove that S contains at least n vectors.

21. Prove that a vector space is infinite-dimensional if and only if it contains an infinite linearly independent subset.

22. Let W_1 and W_2 be subspaces of a finite-dimensional vector space V. Determine necessary and sufficient conditions on W_1 and W_2 so that $\dim(W_1 \cap W_2) = \dim(W_1)$.

23. Let $v_1, v_2, \ldots, v_k, v$ be vectors in a vector space V, and define $W_1 = \text{span}(\{v_1, v_2, \ldots, v_k\})$, and $W_2 = \text{span}(\{v_1, v_2, \ldots, v_k, v\})$.

(a) Find necessary and sufficient conditions on v such that $\dim(W_1) = \dim(W_2)$.
(b) State and prove a relationship involving $\dim(W_1)$ and $\dim(W_2)$ in the case that $\dim(W_1) \neq \dim(W_2)$.

24. Let $f(x)$ be a polynomial of degree n in $P_n(R)$. Prove that for any $g(x) \in P_n(R)$ there exist scalars $c_0, c_1, \ldots, c_n$ such that

$$g(x) = c_0 f(x) + c_1 f'(x) + c_2 f''(x) + \cdots + c_n f^{(n)}(x),$$

where $f^{(n)}(x)$ denotes the nth derivative of $f(x)$.

25. Let V, W, and Z be as in Exercise 21 of Section 1.2. If V and W are vector spaces over F of dimensions m and n, determine the dimension of Z.

26. For a fixed $a \in R$, determine the dimension of the subspace of $\mathsf{P}_n(R)$ defined by $\{f \in \mathsf{P}_n(R)\colon f(a) = 0\}$.

27. Let W_1 and W_2 be the subspaces of $\mathsf{P}(F)$ defined in Exercise 25 in Section 1.3. Determine the dimensions of the subspaces $\mathsf{W}_1 \cap \mathsf{P}_n(F)$ and $\mathsf{W}_2 \cap \mathsf{P}_n(F)$.

28. Let V be a finite-dimensional vector space over C with dimension n. Prove that if V is now regarded as a vector space over R, then $\dim \mathsf{V} = 2n$. (See Examples 11 and 12.)

Exercises 29–34 require knowledge of the sum and direct sum of subspaces, as defined in the exercises of Section 1.3.

29. (a) Prove that if W_1 and W_2 are finite-dimensional subspaces of a vector space V, then the subspace $\mathsf{W}_1 + \mathsf{W}_2$ is finite-dimensional, and $\dim(\mathsf{W}_1 + \mathsf{W}_2) = \dim(\mathsf{W}_1) + \dim(\mathsf{W}_2) - \dim(\mathsf{W}_1 \cap \mathsf{W}_2)$. *Hint:* Start with a basis $\{u_1, u_2, \ldots, u_k\}$ for $\mathsf{W}_1 \cap \mathsf{W}_2$ and extend this set to a basis $\{u_1, u_2, \ldots, u_k, v_1, v_2, \ldots v_m\}$ for W_1 and to a basis $\{u_1, u_2, \ldots, u_k, w_1, w_2, \ldots w_p\}$ for W_2.
(b) Let W_1 and W_2 be finite-dimensional subspaces of a vector space V, and let $\mathsf{V} = \mathsf{W}_1 + \mathsf{W}_2$. Deduce that V is the direct sum of W_1 and W_2 if and only if $\dim(\mathsf{V}) = \dim(\mathsf{W}_1) + \dim(\mathsf{W}_2)$.

30. Let

$$V = \mathsf{M}_{2\times 2}(F), \quad \mathsf{W}_1 = \left\{ \begin{pmatrix} a & b \\ c & a \end{pmatrix} \in V\colon a, b, c \in F \right\},$$

and

$$\mathsf{W}_2 = \left\{ \begin{pmatrix} 0 & a \\ -a & b \end{pmatrix} \in V\colon a, b \in F \right\}.$$

Prove that W_1 and W_2 are subspaces of V, and find the dimensions of W_1, W_2, $\mathsf{W}_1 + \mathsf{W}_2$, and $\mathsf{W}_1 \cap \mathsf{W}_2$.

31. Let W_1 and W_2 be subspaces of a vector space V having dimensions m and n, respectively, where $m \geq n$.

(a) Prove that $\dim(\mathsf{W}_1 \cap \mathsf{W}_2) \leq n$.
(b) Prove that $\dim(\mathsf{W}_1 + \mathsf{W}_2) \leq m + n$.

32. (a) Find an example of subspaces W_1 and W_2 of R^3 with dimensions m and n, where $m > n > 0$, such that $\dim(\mathsf{W}_1 \cap \mathsf{W}_2) = n$.
(b) Find an example of subspaces W_1 and W_2 of R^3 with dimensions m and n, where $m > n > 0$, such that $\dim(\mathsf{W}_1 + \mathsf{W}_2) = m + n$.

    (c) Find an example of subspaces W_1 and W_2 of R^3 with dimensions m and n, where $m \geq n$, such that both $\dim(W_1 \cap W_2) < n$ and $\dim(W_1 + W_2) < m + n$.

33. (a) Let W_1 and W_2 be subspaces of a vector space V such that $V = W_1 \oplus W_2$. If β_1 and β_2 are bases for W_1 and W_2, respectively, show that $\beta_1 \cap \beta_2 = \varnothing$ and $\beta_1 \cup \beta_2$ is a basis for V.

    (b) Conversely, let β_1 and β_2 be disjoint bases for subspaces W_1 and W_2, respectively, of a vector space V. Prove that if $\beta_1 \cup \beta_2$ is a basis for V, then $V = W_1 \oplus W_2$.

34. (a) Prove that if W_1 is any subspace of a finite-dimensional vector space V, then there exists a subspace W_2 of V such that $V = W_1 \oplus W_2$.

    (b) Let $V = R^2$ and $W_1 = \{(a_1, 0) : a_1 \in R\}$. Give examples of two different subspaces W_2 and W_2' such that $V = W_1 \oplus W_2$ and $V = W_1 \oplus W_2'$.

The following exercise requires familiarity with Exercise 31 of Section 1.3.

35. Let W be a subspace of a finite-dimensional vector space V, and consider the basis $\{u_1, u_2, \ldots, u_k\}$ for W. Let $\{u_1, u_2, \ldots, u_k, u_{k+1}, \ldots, u_n\}$ be an extension of this basis to a basis for V.

    (a) Prove that $\{u_{k+1} + W, u_{k+2} + W, \ldots, u_n + W\}$ is a basis for V/W.

    (b) Derive a formula relating $\dim(V)$, $\dim(W)$, and $\dim(V/W)$.

1.7*   MAXIMAL LINEARLY INDEPENDENT SUBSETS

In this section, several significant results from Section 1.6 are extended to infinite-dimensional vector spaces. Our principal goal here is to prove that every vector space has a basis. This result is important in the study of infinite-dimensional vector spaces because it is often difficult to construct an explicit basis for such a space. Consider, for example, the vector space of real numbers over the field of rational numbers. There is no obvious way to construct a basis for this space, and yet it follows from the results of this section that such a basis does exist.

    The difficulty that arises in extending the theorems of the preceding section to infinite-dimensional vector spaces is that the principle of mathematical induction, which played a crucial role in many of the proofs of Section 1.6, is no longer adequate. Instead, a more general result called the *maximal principle* is needed. Before stating this principle, we need to introduce some terminology.

    Definition. *Let $\mathcal{F}$ be a family of sets. A member M of $\mathcal{F}$ is called* **maximal** *(with respect to set inclusion) if M is contained in no member of $\mathcal{F}$ other than M itself.*

Example 1

Let $\mathcal{F}$ be the family of all subsets of a nonempty set S. (This family $\mathcal{F}$ is called the **power set** of S.) The set S is easily seen to be a maximal element of $\mathcal{F}$. ◆

Example 2

Let S and T be disjoint nonempty sets, and let $\mathcal{F}$ be the union of their power sets. Then S and T are both maximal elements of $\mathcal{F}$. ◆

Example 3

Let $\mathcal{F}$ be the family of all finite subsets of an infinite set S. Then $\mathcal{F}$ has no maximal element. For if M is any member of $\mathcal{F}$ and s is any element of S that is not in M, then $M\cup\{s\}$ is a member of $\mathcal{F}$ that contains M as a proper subset. ◆

Definition. *A collection of sets $\mathcal{C}$ is called a **chain** (or **nest** or **tower**) if for each pair of sets A and B in $\mathcal{C}$, either $A\subseteq B$ or $B\subseteq A$.*

Example 4

For each positive integer n let $A_n=\{1,2,\ldots,n\}$. Then the collection of sets $\mathcal{C}=\{A_n\colon n=1,2,3,\ldots\}$ is a chain. In fact, $A_m\subseteq A_n$ if and only if $m\le n$. ◆

With this terminology we can now state the maximal principle.

Maximal Principle.[4] *Let $\mathcal{F}$ be a family of sets. If, for each chain $\mathcal{C}\subseteq\mathcal{F}$, there exists a member of $\mathcal{F}$ that contains each member of $\mathcal{C}$, then $\mathcal{F}$ contains a maximal member.*

Because the maximal principle guarantees the existence of maximal elements in a family of sets satisfying the hypothesis above, it is useful to reformulate the definition of a basis in terms of a maximal property. In Theorem 1.12, we show that this is possible; in fact, the concept defined next is equivalent to a basis.

Definition. *Let S be a subset of a vector space V. A **maximal linearly independent subset** of S is a subset B of S satisfying both of the following conditions.*
(a) *B is linearly independent.*
(b) *The only linearly independent subset of S that contains B is B itself.*

---

[4]The *Maximal Principle* is logically equivalent to the *Axiom of Choice,* which is an assumption in most axiomatic developments of set theory. For a treatment of set theory using the Maximal Principle, see John L. Kelley, *General Topology,* Graduate Texts in Mathematics Series, Vol. 27, Springer-Verlag, 1991.

Example 5

Example 2 of Section 1.4 shows that

$$\{x^3 - 2x^2 - 5x - 3, 3x^3 - 5x^2 - 4x - 9\}$$

is a maximal linearly independent subset of

$$S = \{2x^3 - 2x^2 + 12x - 6, x^3 - 2x^2 - 5x - 3, 3x^3 - 5x^2 - 4x - 9\}$$

in $P_2(R)$. In this case, however, any subset of S consisting of two polynomials is easily shown to be a maximal linearly independent subset of S. Thus maximal linearly independent subsets of a set need not be unique.   ◆

A basis β for a vector space V is a maximal linearly independent subset of V, because

1. β is linearly independent by definition.
2. If $v \in V$ and $v \notin \beta$, then $\beta \cup \{v\}$ is linearly dependent by Theorem 1.7 (p. 39) because span(β) = V.

Our next result shows that the converse of this statement is also true.

Theorem 1.12. *Let V be a vector space and S a subset that generates V. If β is a maximal linearly independent subset of S, then β is a basis for V.*

Proof. Let β be a maximal linearly independent subset of S. Because β is linearly independent, it suffices to prove that β generates V. We claim that $S \subseteq$ span(β), for otherwise there exists a $v \in S$ such that $v \notin$ span(β). Since Theorem 1.7 (p. 39) implies that $\beta \cup \{v\}$ is linearly independent, we have contradicted the maximality of β. Therefore $S \subseteq$ span(β). Because span(S) = V, it follows from Theorem 1.5 (p. 30) that span(β) = V.   ∎

Thus a subset of a vector space is a basis if and only if it is a maximal linearly independent subset of the vector space. Therefore we can accomplish our goal of proving that every vector space has a basis by showing that every vector space contains a maximal linearly independent subset. This result follows immediately from the next theorem.

Theorem 1.13. *Let S be a linearly independent subset of a vector space V. There exists a maximal linearly independent subset of V that contains S.*

Proof. Let $\mathcal{F}$ denote the family of all linearly independent subsets of V that contain S. In order to show that $\mathcal{F}$ contains a maximal element, we must show that if $\mathcal{C}$ is a chain in $\mathcal{F}$, then there exists a member U of $\mathcal{F}$ that contains each member of $\mathcal{C}$. We claim that U, the union of the members of $\mathcal{C}$, is the desired set. Clearly U contains each member of $\mathcal{C}$, and so it suffices to prove that $U \in \mathcal{F}$ (i.e., that U is a linearly independent subset of V that contains S). Because each member of $\mathcal{C}$ is a subset of V containing S, we have $S \subseteq U \subseteq$ V. Thus we need only prove that U is linearly independent. Let $u_1, u_2, \ldots, u_n$ be in U and $a_1, a_2, \ldots, a_n$ be scalars such that $a_1u_1 + a_2u_2 + \cdots + a_nu_n = 0$. Because $u_i \in U$ for $i = 1, 2, \ldots, n$, there exists a set A_i in $\mathcal{C}$ such that $u_i \in A_i$. But since $\mathcal{C}$ is a chain, one of these sets, say A_k, contains all the others. Thus $u_i \in A_k$ for $i = 1, 2, \ldots, n$. However, A_k is a linearly independent set; so $a_1u_1 + a_2u_2 + \cdots + a_nu_n = 0$ implies that $a_1 = a_2 = \cdots = a_n = 0$. It follows that U is linearly independent.

The maximal principle implies that $\mathcal{F}$ has a maximal element. This element is easily seen to be a maximal linearly independent subset of V that contains S. ∎

Corollary. *Every vector space has a basis.*

It can be shown, analogously to Corollary 1 of the replacement theorem (p. 46), that every basis for an infinite-dimensional vector space has the same *cardinality*. (Sets have the same cardinality if there is a one-to-one and onto mapping between them.) (See, for example, N. Jacobson, *Lectures in Abstract Algebra*, vol. 2, Linear Algebra, D. Van Nostrand Company, New York, 1953, p. 240.)

Exercises 4-7 extend other results from Section 1.6 to infinite-dimensional vector spaces.

EXERCISES

1. Label the following statements as true or false.
    (a) Every family of sets contains a maximal element.
    (b) Every chain contains a maximal element.
    (c) If a family of sets has a maximal element, then that maximal element is unique.
    (d) If a chain of sets has a maximal element, then that maximal element is unique.
    (e) A basis for a vector space is a maximal linearly independent subset of that vector space.
    (f) A maximal linearly independent subset of a vector space is a basis for that vector space.

2. Show that the set of convergent sequences is an infinite-dimensional subspace of the vector space of all sequences of real numbers. (See Exercise 21 in Section 1.3.)

3. Let V be the set of real numbers regarded as a vector space over the field of rational numbers. Prove that V is infinite-dimensional. *Hint:*

Use the fact that π is transcendental, that is, π is not a zero of any polynomial with rational coefficients.

4. Let W be a subspace of a (not necessarily finite-dimensional) vector space V. Prove that any basis for W is a subset of a basis for V.

5. Prove the following infinite-dimensional version of Theorem 1.8 (p. 43): Let β be a subset of an infinite-dimensional vector space V. Then β is a basis for V if and only if for each nonzero vector v in V, there exist unique vectors $u_1, u_2, \ldots, u_n$ in β and unique nonzero scalars $c_1, c_2, \ldots, c_n$ such that $v = c_1 u_1 + c_2 u_2 + \cdots + c_n u_n$.

6. Prove the following generalization of Theorem 1.9 (p. 44): Let S_1 and S_2 be subsets of a vector space V such that $S_1 \subseteq S_2$. If S_1 is linearly independent and S_2 generates V, then there exists a basis β for V such that $S_1 \subseteq \beta \subseteq S_2$. *Hint:* Apply the maximal principle to the family of all linearly independent subsets of S_2 that contain S_1, and proceed as in the proof of Theorem 1.13.

7. Prove the following generalization of the replacement theorem. Let β be a basis for a vector space V, and let S be a linearly independent subset of V. There exists a subset S_1 of β such that $S \cup S_1$ is a basis for V.

INDEX OF DEFINITIONS FOR CHAPTER 1

2

Linear Transformations and Matrices

In Chapter 1, we developed the theory of abstract vector spaces in considerable detail. It is now natural to consider those functions defined on vector spaces that in some sense "preserve" the structure. These special functions are called *linear transformations*, and they abound in both pure and applied mathematics. In calculus, the operations of differentiation and integration provide us with two of the most important examples of linear transformations (see Examples 6 and 7 of Section 2.1). These two examples allow us to reformulate many of the problems in differential and integral equations in terms of linear transformations on particular vector spaces (see Sections 2.7 and 5.2).

In geometry, rotations, reflections, and projections (see Examples 2, 3, and 4 of Section 2.1) provide us with another class of linear transformations. Later we use these transformations to study rigid motions in R^n (Section 6.10).

In the remaining chapters, we see further examples of linear transformations occurring in both the physical and the social sciences. Throughout this chapter, we assume that all vector spaces are over a common field F.

2.1   LINEAR TRANSFORMATIONS, NULL SPACES, AND RANGES

In this section, we consider a number of examples of linear transformations. Many of these transformations are studied in more detail in later sections. Recall that a function T with domain V and codomain W is denoted by

64

T: V → W. (See Appendix B.)

Definition.   *Let* V *and* W *be vector spaces (over* F*). We call a function* T: V → W *a* **linear transformation from** V **to** W *if, for all* $x, y \in$ V *and* $c \in F$, *we have*
   (a) T$(x + y) = $ T$(x) + $ T(y) *and*
   (b) T$(cx) = c$T(x).

If the underlying field F is the field of rational numbers, then (a) implies (b) (see Exercise 37), but, in general (a) and (b) are logically independent. See Exercises 38 and 39.

We often simply call T **linear**. The reader should verify the following properties of a function T: V → W. (See Exercise 7.)

1. If T is linear, then T$(0) = 0$.
2. T is linear if and only if T$(cx + y) = c$T$(x) + T(y)$ for all $x, y \in$ V and $c \in F$.
3. If T is linear, then T$(x - y) = T(x) - T(y)$ for all $x, y \in$ V.
4. T is linear if and only if, for $x_1, x_2, \dots, x_n \in$ V and $a_1, a_2, \dots, a_n \in F$, we have

$$\mathsf{T}\left(\sum_{i=1}^{n} a_i x_i\right) = \sum_{i=1}^{n} a_i \mathsf{T}(x_i).$$

We generally use property 2 to prove that a given transformation is linear.

Example 1

Define

$$\mathsf{T}: \mathsf{R}^2 \to \mathsf{R}^2 \text{ by } \mathsf{T}(a_1, a_2) = (2a_1 + a_2, a_1).$$

To show that T is linear, let $c \in R$ and $x, y \in \mathsf{R}^2$, where $x = (b_1, b_2)$ and $y = (d_1, d_2)$. Since

$$cx + y = (cb_1 + d_1, cb_2 + d_2),$$

we have

$$\mathsf{T}(cx + y) = (2(cb_1 + d_1) + cb_2 + d_2, cb_1 + d_1).$$

Also

$$\begin{aligned}
c\mathsf{T}(x) + \mathsf{T}(y) &= c(2b_1 + b_2, b_1) + (2d_1 + d_2, d_1) \\
&= (2cb_1 + cb_2 + 2d_1 + d_2, cb_1 + d_1) \\
&= (2(cb_1 + d_1) + cb_2 + d_2, cb_1 + d_1).
\end{aligned}$$

So T is linear.   ♦

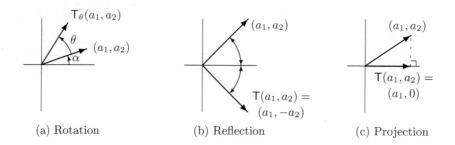

(a) Rotation   (b) Reflection   (c) Projection

Figure 2.1

As we will see in Chapter 6, the applications of linear algebra to geometry are wide and varied. The main reason for this is that most of the important geometrical transformations are linear. Three particular transformations that we now consider are rotation, reflection, and projection. We leave the proofs of linearity to the reader.

Example 2

For any angle θ, define $\mathsf{T}_\theta \colon \mathsf{R}^2 \to \mathsf{R}^2$ by the rule: $\mathsf{T}_\theta(a_1, a_2)$ is the vector obtained by rotating (a_1, a_2) counterclockwise by θ if $(a_1, a_2) \neq (0, 0)$, and $\mathsf{T}_\theta(0, 0) = (0, 0)$. Then $\mathsf{T}_\theta \colon \mathsf{R}^2 \to \mathsf{R}^2$ is a linear transformation that is called the **rotation by** θ.

We determine an explicit formula for T_θ. Fix a nonzero vector $(a_1, a_2) \in \mathsf{R}^2$. Let α be the angle that (a_1, a_2) makes with the positive x-axis (see Figure 2.1(a)), and let $r = \sqrt{a_1^2 + a_2^2}$. Then $a_1 = r \cos \alpha$ and $a_2 = r \sin \alpha$. Also, $\mathsf{T}_\theta(a_1, a_2)$ has length r and makes an angle $\alpha + \theta$ with the positive x-axis. It follows that

$$
\begin{aligned}
\mathsf{T}_\theta(a_1, a_2) &= (r \cos(\alpha + \theta), r \sin(\alpha + \theta)) \\
&= (r \cos \alpha \cos \theta - r \sin \alpha \sin \theta, r \cos \alpha \sin \theta + r \sin \alpha \cos \theta) \\
&= (a_1 \cos \theta - a_2 \sin \theta, a_1 \sin \theta + a_2 \cos \theta).
\end{aligned}
$$

Finally, observe that this same formula is valid for $(a_1, a_2) = (0, 0)$.

It is now easy to show, as in Example 1, that T_θ is linear.   ◆

Example 3

Define $\mathsf{T} \colon \mathsf{R}^2 \to \mathsf{R}^2$ by $\mathsf{T}(a_1, a_2) = (a_1, -a_2)$. T is called the **reflection about the x -axis**. (See Figure 2.1(b).)   ◆

Example 4

Define $\mathsf{T} \colon \mathsf{R}^2 \to \mathsf{R}^2$ by $\mathsf{T}(a_1, a_2) = (a_1, 0)$. T is called the **projection on the x-axis**. (See Figure 2.1(c).)   ◆

We now look at some additional examples of linear transformations.

Example 5

Define $\mathsf{T}\colon \mathsf{M}_{m \times n}(F) \to \mathsf{M}_{n \times m}(F)$ by $\mathsf{T}(A) = A^t$, where A^t is the transpose of A, defined in Section 1.3. Then T is a linear transformation by Exercise 3 of Section 1.3.   ◆

Example 6

Define $\mathsf{T}\colon \mathsf{P}_n(R) \to \mathsf{P}_{n-1}(R)$ by $\mathsf{T}(f(x)) = f'(x)$, where $f'(x)$ denotes the derivative of $f(x)$. To show that T is linear, let $g(x), h(x) \in \mathsf{P}_n(R)$ and $a \in R$. Now

$$\mathsf{T}(ag(x) + h(x)) = (ag(x) + h(x))' = ag'(x) + h'(x) = a\mathsf{T}(g(x)) + \mathsf{T}(h(x)).$$

So by property 2 above, T is linear.   ◆

Example 7

Let $\mathsf{V} = \mathsf{C}(R)$, the vector space of continuous real-valued functions on R. Let $a, b \in R$, $a < b$. Define $\mathsf{T}\colon \mathsf{V} \to R$ by

$$\mathsf{T}(f) = \int_a^b f(t)\, dt$$

for all $f \in \mathsf{V}$. Then T is a linear transformation because the definite integral of a linear combination of functions is the same as the linear combination of the definite integrals of the functions.   ◆

Two very important examples of linear transformations that appear frequently in the remainder of the book, and therefore deserve their own notation, are the identity and zero transformations.

For vector spaces V and W (over F), we define the **identity transformation** $\mathsf{I}_\mathsf{V}\colon \mathsf{V} \to \mathsf{V}$ by $\mathsf{I}_\mathsf{V}(x) = x$ for all $x \in \mathsf{V}$ and the **zero transformation** $\mathsf{T}_0\colon \mathsf{V} \to \mathsf{W}$ by $\mathsf{T}_0(x) = 0$ for all $x \in \mathsf{V}$. It is clear that both of these transformations are linear. We often write I instead of I_V.

We now turn our attention to two very important sets associated with linear transformations: the *range* and *null space*. The determination of these sets allows us to examine more closely the intrinsic properties of a linear transformation.

Definitions. *Let V and W be vector spaces, and let $\mathsf{T}\colon \mathsf{V} \to \mathsf{W}$ be linear. We define the **null space** (or **kernel**) $\mathsf{N}(\mathsf{T})$ of T to be the set of all vectors x in V such that $\mathsf{T}(x) = 0$; that is, $\mathsf{N}(\mathsf{T}) = \{x \in \mathsf{V}\colon \mathsf{T}(x) = 0\}$.*

*We define the **range** (or **image**) $\mathsf{R}(\mathsf{T})$ of T to be the subset of W consisting of all images (under T) of vectors in V; that is, $\mathsf{R}(\mathsf{T}) = \{\mathsf{T}(x)\colon x \in \mathsf{V}\}$.*

Example 8

Let V and W be vector spaces, and let I: V → V and T_0: V → W be the identity and zero transformations, respectively. Then $N(I) = \{0\}$, $R(I) = V$, $N(T_0) = V$, and $R(T_0) = \{0\}$. ◆

Example 9

Let T: $R^3 \to R^2$ be the linear transformation defined by

$$T(a_1, a_2, a_3) = (a_1 - a_2, 2a_3).$$

It is left as an exercise to verify that

$$N(T) = \{(a, a, 0): a \in R\} \quad \text{and} \quad R(T) = R^2. \quad ◆$$

In Examples 8 and 9, we see that the range and null space of each of the linear transformations is a subspace. The next result shows that this is true in general.

Theorem 2.1. *Let* V *and* W *be vector spaces and* T: V → W *be linear. Then* N(T) *and* R(T) *are subspaces of* V *and* W, *respectively.*

Proof. To clarify the notation, we use the symbols 0_V and 0_W to denote the zero vectors of V and W, respectively.

Since $T(0_V) = 0_W$, we have that $0_V \in N(T)$. Let $x, y \in N(T)$ and $c \in F$. Then $T(x+y) = T(x) + T(y) = 0_W + 0_W = 0_W$, and $T(cx) = cT(x) = c0_W = 0_W$. Hence $x + y \in N(T)$ and $cx \in N(T)$, so that $N(T)$ is a subspace of V.

Because $T(0_V) = 0_W$, we have that $0_W \in R(T)$. Now let $x, y \in R(T)$ and $c \in F$. Then there exist v and w in V such that $T(v) = x$ and $T(w) = y$. So $T(v+w) = T(v) + T(w) = x + y$, and $T(cv) = cT(v) = cx$. Thus $x + y \in R(T)$ and $cx \in R(T)$, so $R(T)$ is a subspace of W. ∎

The next theorem provides a method for finding a spanning set for the range of a linear transformation. With this accomplished, a basis for the range is easy to discover using the technique of Example 6 of Section 1.6.

Theorem 2.2. *Let* V *and* W *be vector spaces, and let* T: V → W *be linear. If* $\beta = \{v_1, v_2, \ldots, v_n\}$ *is a basis for* V, *then*

$$R(T) = \mathrm{span}(T(\beta)) = \mathrm{span}(\{T(v_1), T(v_2), \ldots, T(v_n)\}).$$

Proof. Clearly $T(v_i) \in R(T)$ for each i. Because $R(T)$ is a subspace, $R(T)$ contains $\mathrm{span}(\{T(v_1), T(v_2), \ldots, T(v_n)\}) = \mathrm{span}(T(\beta))$ by Theorem 1.5 (p. 30).

Now suppose that $w \in R(T)$. Then $w = T(v)$ for some $v \in V$. Because β is a basis for V, we have

$$v = \sum_{i=1}^{n} a_i v_i \quad \text{for some } a_1, a_2, \ldots, a_n \in F.$$

Since T is linear, it follows that

$$w = T(v) = \sum_{i=1}^{n} a_i T(v_i) \in \text{span}(T(\beta)).$$

So $R(T)$ is contained in $\text{span}(T(\beta))$.   ∎

It should be noted that Theorem 2.2 is true if β is infinite, that is, $R(T) = \text{span}(\{T(v) : v \in \beta\})$. (See Exercise 33.)

The next example illustrates the usefulness of Theorem 2.2.

Example 10

Define the linear transformation $T \colon P_2(R) \to M_{2\times 2}(R)$ by

$$T(f(x)) = \begin{pmatrix} f(1) - f(2) & 0 \\ 0 & f(0) \end{pmatrix}.$$

Since $\beta = \{1, x, x^2\}$ is a basis for $P_2(R)$, we have

$$R(T) = \text{span}(T(\beta)) = \text{span}(\{T(1), T(x), T(x^2)\})$$

$$= \text{span}\left(\left\{\begin{pmatrix} 0 & 0 \\ 0 & 1 \end{pmatrix}, \begin{pmatrix} -1 & 0 \\ 0 & 0 \end{pmatrix}, \begin{pmatrix} -3 & 0 \\ 0 & 0 \end{pmatrix}\right\}\right)$$

$$= \text{span}\left(\left\{\begin{pmatrix} 0 & 0 \\ 0 & 1 \end{pmatrix}, \begin{pmatrix} -1 & 0 \\ 0 & 0 \end{pmatrix}\right\}\right).$$

Thus we have found a basis for $R(T)$, and so $\dim(R(T)) = 2$.   ◆

As in Chapter 1, we measure the "size" of a subspace by its dimension. The null space and range are so important that we attach special names to their respective dimensions.

Definitions.   *Let V and W be vector spaces, and let $T \colon V \to W$ be linear. If N(T) and R(T) are finite-dimensional, then we define the **nullity** of T, denoted* nullity(T), *and the **rank** of T, denoted* rank(T), *to be the dimensions of N(T) and R(T), respectively.*

Reflecting on the action of a linear transformation, we see intuitively that the larger the nullity, the smaller the rank. In other words, the more vectors that are carried into 0, the smaller the range. The same heuristic reasoning tells us that the larger the rank, the smaller the nullity. This balance between rank and nullity is made precise in the next theorem, appropriately called the *dimension theorem.*

Theorem 2.3 (Dimension Theorem). *Let* V *and* W *be vector spaces, and let* $T: V \rightarrow W$ *be linear. If* V *is finite-dimensional, then*

$$\text{nullity}(T) + \text{rank}(T) = \dim(V).$$

Proof. Suppose that $\dim(V) = n$, $\dim(N(T)) = k$, and $\{v_1, v_2, \ldots, v_k\}$ is a basis for $N(T)$. By the corollary to Theorem 1.11 (p. 51), we may extend $\{v_1, v_2, \ldots, v_k\}$ to a basis $\beta = \{v_1, v_2, \ldots, v_n\}$ for V. We claim that $S = \{T(v_{k+1}), T(v_{k+2}), \ldots, T(v_n)\}$ is a basis for $R(T)$.

First we prove that S generates $R(T)$. Using Theorem 2.2 and the fact that $T(v_i) = 0$ for $1 \le i \le k$, we have

$$R(T) = \text{span}(\{T(v_1), T(v_2), \ldots, T(v_n)\}$$
$$= \text{span}(\{T(v_{k+1}), T(v_{k+2}), \ldots, T(v_n)\}) = \text{span}(S).$$

Now we prove that S is linearly independent. Suppose that

$$\sum_{i=k+1}^{n} b_i T(v_i) = 0 \quad \text{for } b_{k+1}, b_{k+2}, \ldots, b_n \in F.$$

Using the fact that T is linear, we have

$$T\left(\sum_{i=k+1}^{n} b_i v_i\right) = 0.$$

So

$$\sum_{i=k+1}^{n} b_i v_i \in N(T).$$

Hence there exist $c_1, c_2, \ldots, c_k \in F$ such that

$$\sum_{i=k+1}^{n} b_i v_i = \sum_{i=1}^{k} c_i v_i \quad \text{or} \quad \sum_{i=1}^{k} (-c_i) v_i + \sum_{i=k+1}^{n} b_i v_i = 0.$$

Since β is a basis for V, we have $b_i = 0$ for all i. Hence S is linearly independent. Notice that this argument also shows that $T(v_{k+1}), T(v_{k+2}), \ldots, T(v_n)$ are distinct; therefore $\text{rank}(T) = n - k$. ∎

If we apply the dimension theorem to the linear transformation T in Example 9, we have that $\text{nullity}(T) + 2 = 3$, so $\text{nullity}(T) = 1$.

The reader should review the concepts of "one-to-one" and "onto" presented in Appendix B. Interestingly, for a linear transformation, both of these concepts are intimately connected to the rank and nullity of the transformation. This is demonstrated in the next two theorems.

Theorem 2.4. *Let* V *and* W *be vector spaces, and let* $T: V \to W$ *be linear. Then* T *is one-to-one if and only if* $N(T) = \{0\}$.

Proof. Suppose that T is one-to-one and $x \in N(T)$. Then $T(x) = 0 = T(0)$. Since T is one-to-one, we have $x = 0$. Hence $N(T) = \{0\}$.

Now assume that $N(T) = \{0\}$, and suppose that $T(x) = T(y)$. Then $0 = T(x) - T(y) = T(x - y)$ by property 3 on page 65. Therefore $x - y \in N(T) = \{0\}$. So $x - y = 0$, or $x = y$. This means that T is one-to-one. ∎

The reader should observe that Theorem 2.4 allows us to conclude that the transformation defined in Example 9 is not one-to-one.

Surprisingly, the conditions of one-to-one and onto are equivalent in an important special case.

Theorem 2.5. *Let* V *and* W *be vector spaces of equal (finite) dimension, and let* $T: V \to W$ *be linear. Then the following are equivalent.*
(a) T *is one-to-one.*
(b) T *is onto.*
(c) $\mathrm{rank}(T) = \dim(V)$.

Proof. From the dimension theorem, we have

$$\mathrm{nullity}(T) + \mathrm{rank}(T) = \dim(V).$$

Now, with the use of Theorem 2.4, we have that T is one-to-one if and only if $N(T) = \{0\}$, if and only if $\mathrm{nullity}(T) = 0$, if and only if $\mathrm{rank}(T) = \dim(V)$, if and only if $\mathrm{rank}(T) = \dim(W)$, and if and only if $\dim(R(T)) = \dim(W)$. By Theorem 1.11 (p. 50), this equality is equivalent to $R(T) = W$, the definition of T being onto. ∎

We note that if V is not finite-dimensional and $T: V \to V$ is linear, then it does *not* follow that one-to-one and onto are equivalent. (See Exercises 15, 16, and 21.)

The linearity of T in Theorems 2.4 and 2.5 is essential, for it is easy to construct examples of functions from R into R that are not one-to-one, but are onto, and vice versa.

The next two examples make use of the preceding theorems in determining whether a given linear transformation is one-to-one or onto.

Example 11

Let $T: P_2(R) \to P_3(R)$ be the linear transformation defined by

$$T(f(x)) = 2f'(x) + \int_0^x 3f(t)\,dt.$$

Now

$$R(T) = \operatorname{span}(\{T(1), T(x), T(x^2)\}) = \operatorname{span}(\{3x, 2 + \frac{3}{2}x^2, 4x + x^3\}).$$

Since $\{3x, 2 + \frac{3}{2}x^2, 4x + x^3\}$ is linearly independent, $\operatorname{rank}(T) = 3$. Since $\dim(P_3(R)) = 4$, T is not onto. From the dimension theorem, $\operatorname{nullity}(T) + 3 = 3$. So $\operatorname{nullity}(T) = 0$, and therefore, $N(T) = \{0\}$. We conclude from Theorem 2.4 that T is one-to-one. ♦

Example 12

Let $T\colon F^2 \to F^2$ be the linear transformation defined by

$$T(a_1, a_2) = (a_1 + a_2, a_1).$$

It is easy to see that $N(T) = \{0\}$; so T is one-to-one. Hence Theorem 2.5 tells us that T must be onto. ♦

In Exercise 14, it is stated that if T is linear and one-to-one, then a subset S is linearly independent if and only if $T(S)$ is linearly independent. Example 13 illustrates the use of this result.

Example 13

Let $T\colon P_2(R) \to R^3$ be the linear transformation defined by

$$T(a_0 + a_1 x + a_2 x^2) = (a_0, a_1, a_2).$$

Clearly T is linear and one-to-one. Let $S = \{2 - x + 3x^2, x + x^2, 1 - 2x^2\}$. Then S is linearly independent in $P_2(R)$ because

$$T(S) = \{(2, -1, 3), (0, 1, 1), (1, 0, -2)\}$$

is linearly independent in R^3. ♦

In Example 13, we transferred a property from the vector space of polynomials to a property in the vector space of 3-tuples. This technique is exploited more fully later.

One of the most important properties of a linear transformation is that it is completely determined by its action on a basis. This result, which follows from the next theorem and corollary, is used frequently throughout the book.

Theorem 2.6. *Let* V *and* W *be vector spaces over* F, *and suppose that* $\{v_1, v_2, \ldots, v_n\}$ *is a basis for* V. *For* $w_1, w_2, \ldots, w_n$ *in* W, *there exists exactly one linear transformation* $T\colon V \to W$ *such that* $T(v_i) = w_i$ *for* $i = 1, 2, \ldots, n$.

Proof. Let $x \in V$. Then

$$x = \sum_{i=1}^{n} a_i v_i,$$

where $a_1 a_2, \ldots, a_n$ are unique scalars. Define

$$\mathsf{T} \colon V \to W \quad \text{by} \quad \mathsf{T}(x) = \sum_{i=1}^{n} a_i w_i.$$

(a) T is linear: Suppose that $u, v \in V$ and $d \in F$. Then we may write

$$u = \sum_{i=1}^{n} b_i v_i \quad \text{and} \quad v = \sum_{i=1}^{n} c_i v_i$$

for some scalars $b_1, b_2, \ldots, b_n, c_1, c_2, \ldots, c_n$. Thus

$$du + v = \sum_{i=1}^{n} (db_i + c_i) v_i.$$

So

$$\mathsf{T}(du + v) = \sum_{i=1}^{n} (db_i + c_i) w_i = d \sum_{i=1}^{n} b_i w_i + \sum_{i=1}^{n} c_i w_i = d\mathsf{T}(u) + \mathsf{T}(v).$$

(b) Clearly

$$\mathsf{T}(v_i) = w_i \quad \text{for } i = 1, 2, \ldots, n.$$

(c) T is unique: Suppose that $\mathsf{U} \colon V \to W$ is linear and $\mathsf{U}(v_i) = w_i$ for $i = 1, 2, \ldots, n$. Then for $x \in V$ with

$$x = \sum_{i=1}^{n} a_i v_i,$$

we have

$$\mathsf{U}(x) = \sum_{i=1}^{n} a_i \mathsf{U}(v_i) = \sum_{i=1}^{n} a_i w_i = \mathsf{T}(x).$$

Hence $\mathsf{U} = \mathsf{T}$.   ∎

Corollary. *Let V and W be vector spaces, and suppose that V has a finite basis $\{v_1, v_2, \ldots, v_n\}$. If $\mathsf{U}, \mathsf{T} \colon V \to W$ are linear and $\mathsf{U}(v_i) = \mathsf{T}(v_i)$ for $i = 1, 2, \ldots, n$, then $\mathsf{U} = \mathsf{T}$.*

Example 14

Let $T: R^2 \to R^2$ be the linear transformation defined by

$$T(a_1, a_2) = (2a_2 - a_1, 3a_1),$$

and suppose that $U: R^2 \to R^2$ is linear. If we know that $U(1, 2) = (3, 3)$ and $U(1, 1) = (1, 3)$, then $U = T$. This follows from the corollary and from the fact that $\{(1, 2), (1, 1)\}$ is a basis for R^2.   ◆

EXERCISES

1. Label the following statements as true or false. In each part, V and W are finite-dimensional vector spaces (over F), and T is a function from V to W.

   (a) If T is linear, then T preserves sums and scalar products.
   (b) If $T(x + y) = T(x) + T(y)$, then T is linear.
   (c) T is one-to-one if and only if the only vector x such that $T(x) = 0$ is $x = 0$.
   (d) If T is linear, then $T(0_V) = 0_W$.
   (e) If T is linear, then $\text{nullity}(T) + \text{rank}(T) = \dim(W)$.
   (f) If T is linear, then T carries linearly independent subsets of V onto linearly independent subsets of W.
   (g) If $T, U: V \to W$ are both linear and agree on a basis for V, then $T = U$.
   (h) Given $x_1, x_2 \in V$ and $y_1, y_2 \in W$, there exists a linear transformation $T: V \to W$ such that $T(x_1) = y_1$ and $T(x_2) = y_2$.

For Exercises 2 through 6, prove that T is a linear transformation, and find bases for both $N(T)$ and $R(T)$. Then compute the nullity and rank of T, and verify the dimension theorem. Finally, use the appropriate theorems in this section to determine whether T is one-to-one or onto.

2. $T: R^3 \to R^2$ defined by $T(a_1, a_2, a_3) = (a_1 - a_2, 2a_3)$.

3. $T: R^2 \to R^3$ defined by $T(a_1, a_2) = (a_1 + a_2, 0, 2a_1 - a_2)$.

4. $T: M_{2 \times 3}(F) \to M_{2 \times 2}(F)$ defined by

$$T \begin{pmatrix} a_{11} & a_{12} & a_{13} \\ a_{21} & a_{22} & a_{23} \end{pmatrix} = \begin{pmatrix} 2a_{11} - a_{12} & a_{13} + 2a_{12} \\ 0 & 0 \end{pmatrix}.$$

5. $T: P_2(R) \to P_3(R)$ defined by $T(f(x)) = xf(x) + f'(x)$.

6. $\mathsf{T}\colon \mathsf{M}_{n\times n}(F) \to F$ defined by $\mathsf{T}(A) = \mathrm{tr}(A)$. Recall (Example 4, Section 1.3) that

$$\mathrm{tr}(A) = \sum_{i=1}^{n} A_{ii}.$$

7. Prove properties 1, 2, 3, and 4 on page 65.

8. Prove that the transformations in Examples 2 and 3 are linear.

9. In this exercise, $\mathsf{T}\colon \mathsf{R}^2 \to \mathsf{R}^2$ is a function. For each of the following parts, state why T is *not* linear.

    (a) $\mathsf{T}(a_1, a_2) = (1, a_2)$
    (b) $\mathsf{T}(a_1, a_2) = (a_1, a_1^2)$
    (c) $\mathsf{T}(a_1, a_2) = (\sin a_1, 0)$
    (d) $\mathsf{T}(a_1, a_2) = (|a_1|, a_2)$
    (e) $\mathsf{T}(a_1, a_2) = (a_1 + 1, a_2)$

10. Suppose that $\mathsf{T}\colon \mathsf{R}^2 \to \mathsf{R}^2$ is linear, $\mathsf{T}(1,0) = (1,4)$, and $\mathsf{T}(1,1) = (2,5)$. What is $\mathsf{T}(2,3)$? Is T one-to-one?

11. Prove that there exists a linear transformation $\mathsf{T}\colon \mathsf{R}^2 \to \mathsf{R}^3$ such that $\mathsf{T}(1,1) = (1,0,2)$ and $\mathsf{T}(2,3) = (1,-1,4)$. What is $\mathsf{T}(8,11)$?

12. Is there a linear transformation $\mathsf{T}\colon \mathsf{R}^3 \to \mathsf{R}^2$ such that $\mathsf{T}(1,0,3) = (1,1)$ and $\mathsf{T}(-2,0,-6) = (2,1)$?

13. Let V and W be vector spaces, let $\mathsf{T}\colon V \to W$ be linear, and let $\{w_1, w_2, \dots, w_k\}$ be a linearly independent subset of $\mathsf{R}(\mathsf{T})$. Prove that if $S = \{v_1, v_2, \dots, v_k\}$ is chosen so that $\mathsf{T}(v_i) = w_i$ for $i = 1, 2, \dots, k$, then S is linearly independent.

14. Let V and W be vector spaces and $\mathsf{T}\colon V \to W$ be linear.

    (a) Prove that T is one-to-one if and only if T carries linearly independent subsets of V onto linearly independent subsets of W.
    (b) Suppose that T is one-to-one and that S is a subset of V. Prove that S is linearly independent if and only if $\mathsf{T}(S)$ is linearly independent.
    (c) Suppose $\beta = \{v_1, v_2, \dots, v_n\}$ is a basis for V and T is one-to-one and onto. Prove that $\mathsf{T}(\beta) = \{\mathsf{T}(v_1), \mathsf{T}(v_2), \dots, \mathsf{T}(v_n)\}$ is a basis for W.

15. Recall the definition of $P(R)$ on page 10. Define

$$\mathsf{T}\colon P(R) \to P(R) \quad \text{by} \quad \mathsf{T}(f(x)) = \int_0^x f(t)\, dt.$$

Prove that T linear and one-to-one, but not onto.

16. Let $T: P(R) \to P(R)$ be defined by $T(f(x)) = f'(x)$. Recall that T is linear. Prove that T is onto, but not one-to-one.

17. Let V and W be finite-dimensional vector spaces and $T: V \to W$ be linear.

  (a) Prove that if $\dim(V) < \dim(W)$, then T cannot be onto.
  (b) Prove that if $\dim(V) > \dim(W)$, then T cannot be one-to-one.

18. Give an example of a linear transformation $T: R^2 \to R^2$ such that $N(T) = R(T)$.

19. Give an example of distinct linear transformations T and U such that $N(T) = N(U)$ and $R(T) = R(U)$.

20. Let V and W be vector spaces with subspaces V_1 and W_1, respectively. If $T: V \to W$ is linear, prove that $T(V_1)$ is a subspace of W and that $\{x \in V: T(x) \in W_1\}$ is a subspace of V.

21. Let V be the vector space of sequences described in Example 5 of Section 1.2. Define the functions $T, U: V \to V$ by

$$T(a_1, a_2, \ldots) = (a_2, a_3, \ldots) \quad \text{and} \quad U(a_1, a_2, \ldots) = (0, a_1, a_2, \ldots).$$

T and U are called the **left shift** and **right shift** operators on V, respectively.

  (a) Prove that T and U are linear.
  (b) Prove that T is onto, but not one-to-one.
  (c) Prove that U is one-to-one, but not onto.

22. Let $T: R^3 \to R$ be linear. Show that there exist scalars a, b, and c such that $T(x, y, z) = ax + by + cz$ for all $(x, y, z) \in R^3$. Can you generalize this result for $T: F^n \to F$? State and prove an analogous result for $T: F^n \to F^m$.

23. Let $T: R^3 \to R$ be linear. Describe geometrically the possibilities for the null space of T. *Hint:* Use Exercise 22.

The following definition is used in Exercises 24–27 and in Exercise 30.

  Definition. *Let V be a vector space and W_1 and W_2 be subspaces of V such that $V = W_1 \oplus W_2$. (Recall the definition of direct sum given in the exercises of Section 1.3.) A function $T: V \to V$ is called the* **projection on** W_1 **along** W_2 *if, for $x = x_1 + x_2$ with $x_1 \in W_1$ and $x_2 \in W_2$, we have $T(x) = x_1$.*

24. Let $T: R^2 \to R^2$. Include figures for each of the following parts.

(a) Find a formula for $T(a, b)$, where T represents the projection on the y-axis along the x-axis.

(b) Find a formula for $T(a, b)$, where T represents the projection on the y-axis along the line $L = \{(s, s): s \in R\}$.

25. Let $T: R^3 \to R^3$.

(a) If $T(a, b, c) = (a, b, 0)$, show that T is the projection on the xy-plane along the z-axis.

(b) Find a formula for $T(a, b, c)$, where T represents the projection on the z-axis along the xy-plane.

(c) If $T(a, b, c) = (a - c, b, 0)$, show that T is the projection on the xy-plane along the line $L = \{(a, 0, a): a \in R\}$.

26. Using the notation in the definition above, assume that $T: V \to V$ is the projection on W_1 along W_2.

(a) Prove that T is linear and $W_1 = \{x \in V: T(x) = x\}$.

(b) Prove that $W_1 = R(T)$ and $W_2 = N(T)$.

(c) Describe T if $W_1 = V$.

(d) Describe T if W_1 is the zero subspace.

27. Suppose that W is a subspace of a finite-dimensional vector space V.

(a) Prove that there exists a subspace W' and a function $T: V \to V$ such that T is a projection on W along W'.

(b) Give an example of a subspace W of a vector space V such that there are *two* projections on W along two (distinct) subspaces.

The following definitions are used in Exercises 28–32.

Definitions. Let V be a vector space, and let $T: V \to V$ be linear. A subspace W of V is said to be T-*invariant* if $T(x) \in W$ for every $x \in W$, that is, $T(W) \subseteq W$. If W is T-invariant, we define the **restriction of T on W** to be the function $T_W: W \to W$ defined by $T_W(x) = T(x)$ for all $x \in W$.

Exercises 28-32 assume that W is a subspace of a vector space V and that $T: V \to V$ is linear. *Warning*: Do not assume that W is T-invariant or that T is a projection unless explicitly stated.

28. Prove that the subspaces $\{0\}$, V, R(T), and N(T) are all T-invariant.

29. If W is T-invariant, prove that T_W is linear.

30. Suppose that T is the projection on W along some subspace W'. Prove that W is T-invariant and that $T_W = I_W$.

31. Suppose that $V = R(T) \oplus W$ and W is T-invariant. (Recall the definition of *direct sum* given in the exercises of Section 1.3.)

(a) Prove that $W \subseteq N(T)$.
(b) Show that if V is finite-dimensional, then $W = N(T)$.
(c) Show by example that the conclusion of (b) is not necessarily true if V is not finite-dimensional.

32. Suppose that W is T-invariant. Prove that $N(T_W) = N(T) \cap W$ and $R(T_W) = T(W)$.

33. Prove Theorem 2.2 for the case that β is infinite, that is, $R(T) = \text{span}(\{T(v) : v \in \beta\})$.

34. Prove the following generalization of Theorem 2.6: Let V and W be vector spaces over a common field, and let β be a basis for V. Then for any function $f : \beta \to W$ there exists exactly one linear transformation $T : V \to W$ such that $T(x) = f(x)$ for all $x \in \beta$.

Exercises 35 and 36 assume the definition of *direct sum* given in the exercises of Section 1.3.

35. Let V be a finite-dimensional vector space and $T : V \to V$ be linear.

(a) Suppose that $V = R(T) + N(T)$. Prove that $V = R(T) \oplus N(T)$.
(b) Suppose that $R(T) \cap N(T) = \{0\}$. Prove that $V = R(T) \oplus N(T)$.

Be careful to say in each part where finite-dimensionality is used.

36. Let V and T be as defined in Exercise 21.

(a) Prove that $V = R(T) + N(T)$, but V is not a direct sum of these two spaces. Thus the result of Exercise 35(a) above cannot be proved without assuming that V is finite-dimensional.
(b) Find a linear operator T_1 on V such that $R(T_1) \cap N(T_1) = \{0\}$ but V is not a direct sum of $R(T_1)$ and $N(T_1)$. Conclude that V being finite-dimensional is also essential in Exercise 35(b).

37. A function $T : V \to W$ between vector spaces V and W is called **additive** if $T(x + y) = T(x) + T(y)$ for all $x, y \in V$. Prove that if V and W are vector spaces over the field of rational numbers, then any additive function from V into W is a linear transformation.

38. Let $T : C \to C$ be the function defined by $T(z) = \bar{z}$. Prove that T is additive (as defined in Exercise 37) but not linear.

39. Prove that there is an additive function $T : R \to R$ (as defined in Exercise 37) that is not linear. *Hint:* Let V be the set of real numbers regarded as a vector space over the field of rational numbers. By the corollary to Theorem 1.13 (p. 60), V has a basis β. Let x and y be two distinct vectors in β, and define $f : \beta \to V$ by $f(x) = y$, $f(y) = x$, and $f(z) = z$ otherwise. By Exercise 34, there exists a linear transformation

$\mathsf{T}\colon \mathsf{V} \to \mathsf{V}$ such that $\mathsf{T}(u) = f(u)$ for all $u \in \beta$. Then T is additive, but for $c = y/x$, $\mathsf{T}(cx) \neq c\mathsf{T}(x)$.

The following exercise requires familiarity with the definition of *quotient space* given in Exercise 31 of Section 1.3.

40. Let V be a vector space and W be a subspace of V. Define the mapping $\eta\colon \mathsf{V} \to \mathsf{V}/\mathsf{W}$ by $\eta(v) = v + \mathsf{W}$ for $v \in \mathsf{V}$.

    (a) Prove that η is a linear transformation from V onto V/W and that $\mathsf{N}(\eta) = \mathsf{W}$.

    (b) Suppose that V is finite-dimensional. Use (a) and the dimension theorem to derive a formula relating $\dim(\mathsf{V})$, $\dim(\mathsf{W})$, and $\dim(\mathsf{V}/\mathsf{W})$.

    (c) Read the proof of the dimension theorem. Compare the method of solving (b) with the method of deriving the same result as outlined in Exercise 35 of Section 1.6.

2.2   THE MATRIX REPRESENTATION OF A LINEAR TRANSFORMATION

Until now, we have studied linear transformations by examining their ranges and null spaces. In this section, we embark on one of the most useful approaches to the analysis of a linear transformation on a finite-dimensional vector space: the representation of a linear transformation by a matrix. In fact, we develop a one-to-one correspondence between matrices and linear transformations that allows us to utilize properties of one to study properties of the other.

We first need the concept of an *ordered basis* for a vector space.

Definition. *Let V be a finite-dimensional vector space. An **ordered basis** for V is a basis for V endowed with a specific order; that is, an ordered basis for V is a finite sequence of linearly independent vectors in V that generates V.*

Example 1

In F^3, $\beta = \{e_1, e_2, e_3\}$ can be considered an ordered basis. Also $\gamma = \{e_2, e_1, e_3\}$ is an ordered basis, but $\beta \neq \gamma$ as ordered bases.   ◆

For the vector space F^n, we call $\{e_1, e_2, \ldots, e_n\}$ the **standard ordered basis** for F^n. Similarly, for the vector space $\mathsf{P}_n(F)$, we call $\{1, x, \ldots, x^n\}$ the **standard ordered basis** for $\mathsf{P}_n(F)$.

Now that we have the concept of ordered basis, we can identify abstract vectors in an n-dimensional vector space with n-tuples. This identification is provided through the use of *coordinate vectors*, as introduced next.

Definition.   Let $\beta = \{u_1, u_2, \ldots, u_n\}$ be an ordered basis for a finite-dimensional vector space V. For $x \in \mathsf{V}$, let $a_1, a_2, \ldots, a_n$ be the unique scalars such that

$$x = \sum_{i=1}^{n} a_i u_i.$$

We define the **coordinate vector of** x **relative to** β, denoted $[x]_\beta$, by

$$[x]_\beta = \begin{pmatrix} a_1 \\ a_2 \\ \vdots \\ a_n \end{pmatrix}.$$

Notice that $[u_i]_\beta = e_i$ in the preceding definition. It is left as an exercise to show that the correspondence $x \to [x]_\beta$ provides us with a linear transformation from V to F^n. We study this transformation in Section 2.4 in more detail.

Example 2

Let $\mathsf{V} = \mathsf{P}_2(R)$, and let $\beta = \{1, x, x^2\}$ be the standard ordered basis for V. If $f(x) = 4 + 6x - 7x^2$, then

$$[f]_\beta = \begin{pmatrix} 4 \\ 6 \\ -7 \end{pmatrix}. \quad \blacklozenge$$

Let us now proceed with the promised matrix representation of a linear transformation. Suppose that V and W are finite-dimensional vector spaces with ordered bases $\beta = \{v_1, v_2, \ldots, v_n\}$ and $\gamma = \{w_1, w_2, \ldots, w_m\}$, respectively. Let $\mathsf{T} \colon \mathsf{V} \to \mathsf{W}$ be linear. Then for each j, $1 \leq j \leq n$, there exist unique scalars $a_{ij} \in F$, $1 \leq i \leq m$, such that

$$\mathsf{T}(v_j) = \sum_{i=1}^{m} a_{ij} w_i \quad \text{for } 1 \leq j \leq n.$$

Definition.   Using the notation above, we call the $m \times n$ matrix A defined by $A_{ij} = a_{ij}$ the **matrix representation of** T **in the ordered bases** β **and** γ and write $A = [\mathsf{T}]_\beta^\gamma$. If $\mathsf{V} = \mathsf{W}$ and $\beta = \gamma$, then we write $A = [\mathsf{T}]_\beta$.

Notice that the jth column of A is simply $[\mathsf{T}(v_j)]_\gamma$. Also observe that if $\mathsf{U} \colon \mathsf{V} \to \mathsf{W}$ is a linear transformation such that $[\mathsf{U}]_\beta^\gamma = [\mathsf{T}]_\beta^\gamma$, then $\mathsf{U} = \mathsf{T}$ by the corollary to Theorem 2.6 (p. 73).

We illustrate the computation of $[\mathsf{T}]_\beta^\gamma$ in the next several examples.

Example 3

Let $T\colon R^2 \to R^3$ be the linear transformation defined by

$$T(a_1, a_2) = (a_1 + 3a_2, 0, 2a_1 - 4a_2).$$

Let β and γ be the standard ordered bases for R^2 and R^3, respectively. Now

$$T(1,0) = (1,0,2) = 1e_1 + 0e_2 + 2e_3$$

and

$$T(0,1) = (3,0,-4) = 3e_1 + 0e_2 - 4e_3.$$

Hence

$$[T]_\beta^\gamma = \begin{pmatrix} 1 & 3 \\ 0 & 0 \\ 2 & -4 \end{pmatrix}.$$

If we let $\gamma' = \{e_3, e_2, e_1\}$, then

$$[T]_\beta^{\gamma'} = \begin{pmatrix} 2 & -4 \\ 0 & 0 \\ 1 & 3 \end{pmatrix}. \quad \blacklozenge$$

Example 4

Let $T\colon P_3(R) \to P_2(R)$ be the linear transformation defined by $T(f(x)) = f'(x)$. Let β and γ be the standard ordered bases for $P_3(R)$ and $P_2(R)$, respectively. Then

$$T(1) = 0\cdot 1 + 0\cdot x + 0\cdot x^2$$
$$T(x) = 1\cdot 1 + 0\cdot x + 0\cdot x^2$$
$$T(x^2) = 0\cdot 1 + 2\cdot x + 0\cdot x^2$$
$$T(x^3) = 0\cdot 1 + 0\cdot x + 3\cdot x^2.$$

So

$$[T]_\beta^\gamma = \begin{pmatrix} 0 & 1 & 0 & 0 \\ 0 & 0 & 2 & 0 \\ 0 & 0 & 0 & 3 \end{pmatrix}.$$

Note that when $T(x^j)$ is written as a linear combination of the vectors of γ, its coefficients give the entries of column $j+1$ of $[T]_\beta^\gamma$. $\blacklozenge$

Now that we have defined a procedure for associating matrices with linear transformations, we show in Theorem 2.8 that this association "preserves" addition and scalar multiplication. To make this more explicit, we need some preliminary discussion about the addition and scalar multiplication of linear transformations.

Definition. *Let* $\mathsf{T}, \mathsf{U}: \mathsf{V} \to \mathsf{W}$ *be arbitrary functions, where* V *and* W *are vector spaces over* F, *and let* $a \in F$. *We define* $\mathsf{T} + \mathsf{U}: \mathsf{V} \to \mathsf{W}$ *by* $(\mathsf{T} + \mathsf{U})(x) = \mathsf{T}(x) + \mathsf{U}(x)$ *for all* $x \in \mathsf{V}$, *and* $a\mathsf{T}: \mathsf{V} \to \mathsf{W}$ *by* $(a\mathsf{T})(x) = a\mathsf{T}(x)$ *for all* $x \in \mathsf{V}$.

Of course, these are just the usual definitions of addition and scalar multiplication of functions. We are fortunate, however, to have the result that both sums and scalar multiples of linear transformations are also linear.

Theorem 2.7. *Let* V *and* W *be vector spaces over a field* F, *and let* $\mathsf{T}, \mathsf{U}: \mathsf{V} \to \mathsf{W}$ *be linear.*
  (a) *For all* $a \in F$, $a\mathsf{T} + \mathsf{U}$ *is linear.*
  (b) *Using the operations of addition and scalar multiplication in the preceding definition, the collection of all linear transformations from* V *to* W *is a vector space over* F.

Proof. (a) Let $x, y \in \mathsf{V}$ and $c \in F$. Then

$$
\begin{aligned}
(a\mathsf{T} + \mathsf{U})(cx + y) &= a\mathsf{T}(cx + y) + \mathsf{U}(cx + y) \\
&= a[\mathsf{T}(cx + y)] + c\mathsf{U}(x) + \mathsf{U}(y) \\
&= a[c\mathsf{T}(x) + \mathsf{T}(y)] + c\mathsf{U}(x) + \mathsf{U}(y) \\
&= ac\mathsf{T}(x) + c\mathsf{U}(x) + a\mathsf{T}(y) + \mathsf{U}(y) \\
&= c(a\mathsf{T} + \mathsf{U})(x) + (a\mathsf{T} + \mathsf{U})(y).
\end{aligned}
$$

So $a\mathsf{T} + \mathsf{U}$ is linear.

  (b) Noting that T_0, the zero transformation, plays the role of the zero vector, it is easy to verify that the axioms of a vector space are satisfied, and hence that the collection of all linear transformations from V into W is a vector space over F. ∎

Definitions.   *Let* V *and* W *be vector spaces over* F. *We denote the vector space of all linear transformations from* V *into* W *by* $\mathcal{L}(\mathsf{V}, \mathsf{W})$. *In the case that* $\mathsf{V} = \mathsf{W}$, *we write* $\mathcal{L}(\mathsf{V})$ *instead of* $\mathcal{L}(\mathsf{V}, \mathsf{W})$.

In Section 2.4, we see a complete identification of $\mathcal{L}(\mathsf{V}, \mathsf{W})$ with the vector space $\mathsf{M}_{m \times n}(F)$, where n and m are the dimensions of V and W, respectively. This identification is easily established by the use of the next theorem.

Theorem 2.8. *Let* V *and* W *be finite-dimensional vector spaces with ordered bases* β *and* γ, *respectively, and let* $\mathsf{T}, \mathsf{U}: \mathsf{V} \to \mathsf{W}$ *be linear transformations. Then*

(a) $[T + U]_\beta^\gamma = [T]_\beta^\gamma + [U]_\beta^\gamma$ and

(b) $[aT]_\beta^\gamma = a[T]_\beta^\gamma$ for all scalars a.

Proof. Let $\beta = \{v_1, v_2, \ldots, v_n\}$ and $\gamma = \{w_1, w_2, \ldots, w_m\}$. There exist unique scalars a_{ij} and b_{ij} $(1 \le i \le m, 1 \le j \le n)$ such that

$$T(v_j) = \sum_{i=1}^m a_{ij} w_i \quad \text{and} \quad U(v_j) = \sum_{i=1}^m b_{ij} w_i \quad \text{for } 1 \le j \le n.$$

Hence

$$(T + U)(v_j) = \sum_{i=1}^m (a_{ij} + b_{ij}) w_i.$$

Thus

$$([T + U]_\beta^\gamma)_{ij} = a_{ij} + b_{ij} = ([T]_\beta^\gamma + [U]_\beta^\gamma)_{ij}.$$

So (a) is proved, and the proof of (b) is similar. ∎

Example 5

Let $T: R^2 \to R^3$ and $U: R^2 \to R^3$ be the linear transformations respectively defined by

$$T(a_1, a_2) = (a_1 + 3a_2, 0, 2a_1 - 4a_2) \text{ and } U(a_1, a_2) = (a_1 - a_2, 2a_1, 3a_1 + 2a_2).$$

Let β and γ be the standard ordered bases of R^2 and R^3, respectively. Then

$$[T]_\beta^\gamma = \begin{pmatrix} 1 & 3 \\ 0 & 0 \\ 2 & -4 \end{pmatrix},$$

(as computed in Example 3), and

$$[U]_\beta^\gamma = \begin{pmatrix} 1 & -1 \\ 2 & 0 \\ 3 & 2 \end{pmatrix}.$$

If we compute $T + U$ using the preceding definitions, we obtain

$$(T + U)(a_1, a_2) = (2a_1 + 2a_2, 2a_1, 5a_1 - 2a_2).$$

So

$$[T + U]_\beta^\gamma = \begin{pmatrix} 2 & 2 \\ 2 & 0 \\ 5 & -2 \end{pmatrix},$$

which is simply $[T]_\beta^\gamma + [U]_\beta^\gamma$, illustrating Theorem 2.8.   ◆

EXERCISES

1. Label the following statements as true or false. Assume that V and W are finite-dimensional vector spaces with ordered bases β and γ, respectively, and $T, U: V \to W$ are linear transformations.

   (a) For any scalar a, $aT + U$ is a linear transformation from V to W.
   (b) $[T]_\beta^\gamma = [U]_\beta^\gamma$ implies that $T = U$.
   (c) If $m = \dim(V)$ and $n = \dim(W)$, then $[T]_\beta^\gamma$ is an $m \times n$ matrix.
   (d) $[T + U]_\beta^\gamma = [T]_\beta^\gamma + [U]_\beta^\gamma$.
   (e) $\mathcal{L}(V, W)$ is a vector space.
   (f) $\mathcal{L}(V, W) = \mathcal{L}(W, V)$.

2. Let β and γ be the standard ordered bases for R^n and R^m, respectively. For each linear transformation $T: R^n \to R^m$, compute $[T]_\beta^\gamma$.

   (a) $T: R^2 \to R^3$ defined by $T(a_1, a_2) = (2a_1 - a_2, 3a_1 + 4a_2, a_1)$.
   (b) $T: R^3 \to R^2$ defined by $T(a_1, a_2, a_3) = (2a_1 + 3a_2 - a_3, a_1 + a_3)$.
   (c) $T: R^3 \to R$ defined by $T(a_1, a_2, a_3) = 2a_1 + a_2 - 3a_3$.
   (d) $T: R^3 \to R^3$ defined by

   $$T(a_1, a_2, a_3) = (2a_2 + a_3, -a_1 + 4a_2 + 5a_3, a_1 + a_3).$$

   (e) $T: R^n \to R^n$ defined by $T(a_1, a_2, \dots, a_n) = (a_1, a_1, \dots, a_1)$.
   (f) $T: R^n \to R^n$ defined by $T(a_1, a_2, \dots, a_n) = (a_n, a_{n-1}, \dots, a_1)$.
   (g) $T: R^n \to R$ defined by $T(a_1, a_2, \dots, a_n) = a_1 + a_n$.

3. Let $T: R^2 \to R^3$ be defined by $T(a_1, a_2) = (a_1 - a_2, a_1, 2a_1 + a_2)$. Let β be the standard ordered basis for R^2 and $\gamma = \{(1, 1, 0), (0, 1, 1), (2, 2, 3)\}$. Compute $[T]_\beta^\gamma$. If $\alpha = \{(1, 2), (2, 3)\}$, compute $[T]_\alpha^\gamma$.

4. Define

   $$T: M_{2 \times 2}(R) \to P_2(R) \quad \text{by} \quad T\begin{pmatrix} a & b \\ c & d \end{pmatrix} = (a + b) + (2d)x + bx^2.$$

   Let

   $$\beta = \left\{ \begin{pmatrix} 1 & 0 \\ 0 & 0 \end{pmatrix}, \begin{pmatrix} 0 & 1 \\ 0 & 0 \end{pmatrix}, \begin{pmatrix} 0 & 0 \\ 1 & 0 \end{pmatrix}, \begin{pmatrix} 0 & 0 \\ 0 & 1 \end{pmatrix} \right\} \quad \text{and} \quad \gamma = \{1, x, x^2\}.$$

   Compute $[T]_\beta^\gamma$.

5. Let

   $$\alpha = \left\{ \begin{pmatrix} 1 & 0 \\ 0 & 0 \end{pmatrix}, \begin{pmatrix} 0 & 1 \\ 0 & 0 \end{pmatrix}, \begin{pmatrix} 0 & 0 \\ 1 & 0 \end{pmatrix}, \begin{pmatrix} 0 & 0 \\ 0 & 1 \end{pmatrix} \right\},$$
   $$\beta = \{1, x, x^2\},$$

   and

   $$\gamma = \{1\}.$$

(a) Define $T\colon M_{2\times 2}(F) \to M_{2\times 2}(F)$ by $T(A) = A^t$. Compute $[T]_\alpha$.

(b) Define

$$T\colon P_2(R) \to M_{2\times 2}(R) \quad \text{by} \quad T(f(x)) = \begin{pmatrix} f'(0) & 2f(1) \\ 0 & f''(3) \end{pmatrix},$$

where $'$ denotes differentiation. Compute $[T]_\beta^\alpha$.

(c) Define $T\colon M_{2\times 2}(F) \to F$ by $T(A) = \text{tr}(A)$. Compute $[T]_\alpha^\gamma$.

(d) Define $T\colon P_2(R) \to R$ by $T(f(x)) = f(2)$. Compute $[T]_\beta^\gamma$.

(e) If

$$A = \begin{pmatrix} 1 & -2 \\ 0 & 4 \end{pmatrix},$$

compute $[A]_\alpha$.

(f) If $f(x) = 3 - 6x + x^2$, compute $[f(x)]_\beta$.

(g) For $a \in F$, compute $[a]_\gamma$.

6. Complete the proof of part (b) of Theorem 2.7.

7. Prove part (b) of Theorem 2.8.

8.[†] Let V be an n-dimensional vector space with an ordered basis β. Define $T\colon V \to F^n$ by $T(x) = [x]_\beta$. Prove that T is linear.

9. Let V be the vector space of complex numbers over the field R. Define $T\colon V \to V$ by $T(z) = \bar{z}$, where $\bar{z}$ is the complex conjugate of z. Prove that T is linear, and compute $[T]_\beta$, where $\beta = \{1, i\}$. (Recall by Exercise 38 of Section 2.1 that T is not linear if V is regarded as a vector space over the field C.)

10. Let V be a vector space with the ordered basis $\beta = \{v_1, v_2, \ldots, v_n\}$. Define $v_0 = 0$. By Theorem 2.6 (p. 72), there exists a linear transformation $T\colon V \to V$ such that $T(v_j) = v_j + v_{j-1}$ for $j = 1, 2, \ldots, n$. Compute $[T]_\beta$.

11. Let V be an n-dimensional vector space, and let $T\colon V \to V$ be a linear transformation. Suppose that W is a T-invariant subspace of V (see the exercises of Section 2.1) having dimension k. Show that there is a basis β for V such that $[T]_\beta$ has the form

$$\begin{pmatrix} A & B \\ O & C \end{pmatrix},$$

where A is a $k \times k$ matrix and O is the $(n-k) \times k$ zero matrix.

12. Let V be a finite-dimensional vector space and T be the projection on W along W', where W and W' are subspaces of V. (See the definition in the exercises of Section 2.1 on page 76.) Find an ordered basis β for V such that $[T]_\beta$ is a diagonal matrix.

13. Let V and W be vector spaces, and let T and U be nonzero linear transformations from V into W. If $R(T) \cap R(U) = \{0\}$, prove that $\{T, U\}$ is a linearly independent subset of $\mathcal{L}(V, W)$.

14. Let $V = P(R)$, and for $j \geq 1$ define $T_j(f(x)) = f^{(j)}(x)$, where $f^{(j)}(x)$ is the jth derivative of $f(x)$. Prove that the set $\{T_1, T_2, \ldots, T_n\}$ is a linearly independent subset of $\mathcal{L}(V)$ for any positive integer n.

15. Let V and W be vector spaces, and let S be a subset of V. Define $S^0 = \{T \in \mathcal{L}(V, W) : T(x) = 0 \text{ for all } x \in S\}$. Prove the following statements.

    (a) S^0 is a subspace of $\mathcal{L}(V, W)$.
    (b) If S_1 and S_2 are subsets of V and $S_1 \subseteq S_2$, then $S_2^0 \subseteq S_1^0$.
    (c) If V_1 and V_2 are subspaces of V, then $(V_1 + V_2)^0 = V_1^0 \cap V_2^0$.

16. Let V and W be vector spaces such that $\dim(V) = \dim(W)$, and let $T: V \to W$ be linear. Show that there exist ordered bases β and γ for V and W, respectively, such that $[T]_\beta^\gamma$ is a diagonal matrix.

2.3   COMPOSITION OF LINEAR TRANSFORMATIONS AND MATRIX MULTIPLICATION

In Section 2.2, we learned how to associate a matrix with a linear transformation in such a way that both sums and scalar multiples of matrices are associated with the corresponding sums and scalar multiples of the transformations. The question now arises as to how the matrix representation of a composite of linear transformations is related to the matrix representation of each of the associated linear transformations. The attempt to answer this question leads to a definition of matrix multiplication. We use the more convenient notation of UT rather than $U \circ T$ for the composite of linear transformations U and T. (See Appendix B.)

Our first result shows that the composite of linear transformations is linear.

Theorem 2.9. *Let V, W, and Z be vector spaces over the same field F, and let $T: V \to W$ and $U: W \to Z$ be linear. Then $UT: V \to Z$ is linear.*

Proof. Let $x, y \in V$ and $a \in F$. Then

$$UT(ax + y) = U(T(ax + y)) = U(aT(x) + T(y))$$
$$= aU(T(x)) + U(T(y)) = a(UT)(x) + UT(y). \quad \blacksquare$$

The following theorem lists some of the properties of the composition of linear transformations.

Theorem 2.10. *Let* V *be a vector space. Let* $T, U_1, U_2 \in \mathcal{L}(V)$. *Then*
(a) $T(U_1 + U_2) = TU_1 + TU_2$ *and* $(U_1 + U_2)T = U_1T + U_2T$
(b) $T(U_1U_2) = (TU_1)U_2$
(c) $TI = IT = T$
(d) $a(U_1U_2) = (aU_1)U_2 = U_1(aU_2)$ *for all scalars* a.

Proof. Exercise. ∎

A more general result holds for linear transformations that have domains unequal to their codomains. (See Exercise 8.)

Let $T\colon V \to W$ and $U\colon W \to Z$ be linear transformations, and let $A = [U]_\beta^\gamma$ and $B = [T]_\alpha^\beta$, where $\alpha = \{v_1, v_2, \ldots, v_n\}$, $\beta = \{w_1, w_2, \ldots, w_m\}$, and $\gamma = \{z_1, z_2, \ldots, z_p\}$ are ordered bases for V, W, and Z, respectively. We would like to define the product AB of two matrices so that $AB = [UT]_\alpha^\gamma$. Consider the matrix $[UT]_\alpha^\gamma$. For $1 \le j \le n$, we have

$$(UT)(v_j) = U(T(v_j)) = U\left(\sum_{k=1}^{m} B_{kj}w_k\right) = \sum_{k=1}^{m} B_{kj}U(w_k)$$

$$= \sum_{k=1}^{m} B_{kj}\left(\sum_{i=1}^{p} A_{ik}z_i\right) = \sum_{i=1}^{p}\left(\sum_{k=1}^{m} A_{ik}B_{kj}\right)z_i$$

$$= \sum_{i=1}^{p} C_{ij}z_i,$$

where

$$C_{ij} = \sum_{k=1}^{m} A_{ik}B_{kj}.$$

This computation motivates the following definition of matrix multiplication.

Definition.   Let A be an $m \times n$ matrix and B be an $n \times p$ matrix. We define the **product** of A and B, denoted AB, to be the $m \times p$ matrix such that

$$(AB)_{ij} = \sum_{k=1}^{n} A_{ik}B_{kj} \quad \text{for } 1 \le i \le m, \quad 1 \le j \le p.$$

Note that $(AB)_{ij}$ is the sum of products of corresponding entries from the ith row of A and the jth column of B. Some interesting applications of this definition are presented at the end of this section.

The reader should observe that in order for the product AB to be defined, there are restrictions regarding the relative sizes of A and B. The following mnemonic device is helpful: "$(m \times n) \cdot (n \times p) = (m \times p)$"; that is, in order for the product AB to be defined, the two "inner" dimensions must be equal, and the two "outer" dimensions yield the size of the product.

Example 1

We have

$$\begin{pmatrix} 1 & 2 & 1 \\ 0 & 4 & -1 \end{pmatrix} \begin{pmatrix} 4 \\ 2 \\ 5 \end{pmatrix} = \begin{pmatrix} 1 \cdot 4 + 2 \cdot 2 + 1 \cdot 5 \\ 0 \cdot 4 + 4 \cdot 2 + (-1) \cdot 5 \end{pmatrix} = \begin{pmatrix} 13 \\ 3 \end{pmatrix}.$$

Notice again the symbolic relationship $(2 \times 3) \cdot (3 \times 1) = 2 \times 1$.   ◆

As in the case with composition of functions, we have that matrix multiplication is not commutative. Consider the following two products:

$$\begin{pmatrix} 1 & 1 \\ 0 & 0 \end{pmatrix} \begin{pmatrix} 0 & 1 \\ 1 & 0 \end{pmatrix} = \begin{pmatrix} 1 & 1 \\ 0 & 0 \end{pmatrix} \quad \text{and} \quad \begin{pmatrix} 0 & 1 \\ 1 & 0 \end{pmatrix} \begin{pmatrix} 1 & 1 \\ 0 & 0 \end{pmatrix} = \begin{pmatrix} 0 & 0 \\ 1 & 1 \end{pmatrix}.$$

Hence we see that even if both of the matrix products AB and BA are defined, it need not be true that $AB = BA$.

Recalling the definition of the transpose of a matrix from Section 1.3, we show that if A is an $m \times n$ matrix and B is an $n \times p$ matrix, then $(AB)^t = B^t A^t$. Since

$$(AB)^t_{ij} = (AB)_{ji} = \sum_{k=1}^{n} A_{jk} B_{ki}$$

and

$$(B^t A^t)_{ij} = \sum_{k=1}^{n} (B^t)_{ik} (A^t)_{kj} = \sum_{k=1}^{n} B_{ki} A_{jk},$$

we are finished. Therefore the transpose of a product is the product of the transposes *in the opposite order*.

The next theorem is an immediate consequence of our definition of matrix multiplication.

Theorem 2.11. *Let* V, W, *and* Z *be finite-dimensional vector spaces with ordered bases* α, β, *and* γ, *respectively. Let* T: V → W *and* U: W → Z *be linear transformations. Then*

$$[UT]^\gamma_\alpha = [U]^\gamma_\beta [T]^\beta_\alpha.$$

Corollary. *Let* V *be a finite-dimensional vector space with an ordered basis* β. *Let* T, U $\in \mathcal{L}(V)$. *Then* $[UT]_\beta = [U]_\beta [T]_\beta$.

We illustrate Theorem 2.11 in the next example.

Example 2

Let U: $P_3(R) \to P_2(R)$ and T: $P_2(R) \to P_3(R)$ be the linear transformations respectively defined by

$$U(f(x)) = f'(x) \qquad \text{and} \qquad T(f(x)) = \int_0^x f(t)\, dt.$$

Let α and β be the standard ordered bases of $P_3(R)$ and $P_2(R)$, respectively. From calculus, it follows that UT = I, the identity transformation on $P_2(R)$. To illustrate Theorem 2.11, observe that

$$[UT]_\beta = [U]_\alpha^\beta [T]_\beta^\alpha = \begin{pmatrix} 0 & 1 & 0 & 0 \\ 0 & 0 & 2 & 0 \\ 0 & 0 & 0 & 3 \end{pmatrix} \begin{pmatrix} 0 & 0 & 0 \\ 1 & 0 & 0 \\ 0 & \frac{1}{2} & 0 \\ 0 & 0 & \frac{1}{3} \end{pmatrix} = \begin{pmatrix} 1 & 0 & 0 \\ 0 & 1 & 0 \\ 0 & 0 & 1 \end{pmatrix} = [I]_\beta. \quad \blacklozenge$$

The preceding 3×3 diagonal matrix is called an *identity matrix* and is defined next, along with a very useful notation, the *Kronecker delta*.

Definitions. *We define the* **Kronecker delta** δ_{ij} *by* $\delta_{ij} = 1$ *if* $i = j$ *and* $\delta_{ij} = 0$ *if* $i \neq j$. *The* $n \times n$ **identity matrix** I_n *is defined by* $(I_n)_{ij} = \delta_{ij}$.

Thus, for example,

$$I_1 = (1), \quad I_2 = \begin{pmatrix} 1 & 0 \\ 0 & 1 \end{pmatrix}, \quad \text{and} \quad I_3 = \begin{pmatrix} 1 & 0 & 0 \\ 0 & 1 & 0 \\ 0 & 0 & 1 \end{pmatrix}.$$

The next theorem provides analogs of (a), (c), and (d) of Theorem 2.10. Theorem 2.10(b) has its analog in Theorem 2.16. Observe also that part (c) of the next theorem illustrates that the identity matrix acts as a multiplicative identity in $M_{n \times n}(F)$. When the context is clear, we sometimes omit the subscript n from I_n.

Theorem 2.12. *Let* A *be an* $m \times n$ *matrix,* B *and* C *be* $n \times p$ *matrices, and* D *and* E *be* $q \times m$ *matrices. Then*
(a) $A(B + C) = AB + AC$ *and* $(D + E)A = DA + EA$.
(b) $a(AB) = (aA)B = A(aB)$ *for any scalar* a.
(c) $I_m A = A = A I_n$.
(d) *If* V *is an* n-*dimensional vector space with an ordered basis* β, *then* $[I_V]_\beta = I_n$.

Proof. We prove the first half of (a) and (c) and leave the remaining proofs as an exercise. (See Exercise 5.)

(a) We have

$$[A(B+C)]_{ij} = \sum_{k=1}^{n} A_{ik}(B+C)_{kj} = \sum_{k=1}^{n} A_{ik}(B_{kj} + C_{kj})$$

$$= \sum_{k=1}^{n}(A_{ik}B_{kj} + A_{ik}C_{kj}) = \sum_{k=1}^{n} A_{ik}B_{kj} + \sum_{k=1}^{n} A_{ik}C_{kj}$$

$$= (AB)_{ij} + (AC)_{ij} = [AB + AC]_{ij}.$$

So $A(B+C) = AB + AC$.

(c) We have

$$(I_m A)_{ij} = \sum_{k=1}^{m}(I_m)_{ik} A_{kj} = \sum_{k=1}^{m} \delta_{ik} A_{kj} = A_{ij}.$$ ∎

Corollary. Let A be an $m \times n$ matrix, $B_1, B_2, \ldots, B_k$ be $n \times p$ matrices, $C_1, C_2, \ldots, C_k$ be $q \times m$ matrices, and $a_1, a_2, \ldots, a_k$ be scalars. Then

$$A\left(\sum_{i=1}^{k} a_i B_i\right) = \sum_{i=1}^{k} a_i AB_i$$

and

$$\left(\sum_{i=1}^{k} a_i C_i\right) A = \sum_{i=1}^{k} a_i C_i A.$$

Proof. Exercise. ∎

For an $n \times n$ matrix A, we define $A^1 = A$, $A^2 = AA$, $A^3 = A^2 A$, and, in general, $A^k = A^{k-1} A$ for $k = 2, 3, \ldots$. We define $A^0 = I_n$.

With this notation, we see that if

$$A = \begin{pmatrix} 0 & 0 \\ 1 & 0 \end{pmatrix},$$

then $A^2 = O$ (the zero matrix) even though $A \neq O$. Thus the cancellation property for multiplication in fields is not valid for matrices. To see why, assume that the cancellation law is valid. Then, from $A \cdot A = A^2 = O = A \cdot O$, we would conclude that $A = O$, which is false.

Theorem 2.13. Let A be an $m \times n$ matrix and B be an $n \times p$ matrix. For each j $(1 \leq j \leq p)$ let u_j and v_j denote the jth columns of AB and B, respectively. Then

(a) $u_j = Av_j$

(b) $v_j = Be_j$, where e_j is the jth standard vector of F^p.

Proof. (a) We have

$$u_j = \begin{pmatrix} (AB)_{1j} \\ (AB)_{2j} \\ \vdots \\ (AB)_{mj} \end{pmatrix} = \begin{pmatrix} \displaystyle\sum_{k=1}^{n} A_{1k}B_{kj} \\ \displaystyle\sum_{k=1}^{n} A_{2k}B_{kj} \\ \vdots \\ \displaystyle\sum_{k=1}^{n} A_{mk}B_{kj} \end{pmatrix} = A \begin{pmatrix} B_{1j} \\ B_{2j} \\ \vdots \\ B_{nj} \end{pmatrix} = Av_j.$$

Hence (a) is proved. The proof of (b) is left as an exercise. (See Exercise 6.) ∎

It follows (see Exercise 14) from Theorem 2.13 that column j of AB is a linear combination of the columns of A with the coefficients in the linear combination being the entries of column j of B. An analogous result holds for rows; that is, row i of AB is a linear combination of the rows of B with the coefficients in the linear combination being the entries of row i of A.

The next result justifies much of our past work. It utilizes both the matrix representation of a linear transformation and matrix multiplication in order to evaluate the transformation at any given vector.

Theorem 2.14. *Let* V *and* W *be finite-dimensional vector spaces having ordered bases* β *and* γ, *respectively, and let* $\mathsf{T}\colon \mathsf{V} \to \mathsf{W}$ *be linear. Then, for each* $u \in \mathsf{V}$, *we have*

$$[\mathsf{T}(u)]_\gamma = [\mathsf{T}]_\beta^\gamma [u]_\beta.$$

Proof. Fix $u \in \mathsf{V}$, and define the linear transformations $f\colon F \to \mathsf{V}$ by $f(a) = au$ and $g\colon F \to \mathsf{W}$ by $g(a) = a\mathsf{T}(u)$ for all $a \in F$. Let $\alpha = \{1\}$ be the standard ordered basis for F. Notice that $g = \mathsf{T}f$. Identifying column vectors as matrices and using Theorem 2.11, we obtain

$$[\mathsf{T}(u)]_\gamma = [g(1)]_\gamma = [g]_\alpha^\gamma = [\mathsf{T}f]_\alpha^\gamma = [\mathsf{T}]_\beta^\gamma [f]_\alpha^\beta = [\mathsf{T}]_\beta^\gamma [f(1)]_\beta = [\mathsf{T}]_\beta^\gamma [u]_\beta. \quad ∎$$

Example 3

Let $\mathsf{T}\colon \mathsf{P}_3(R) \to \mathsf{P}_2(R)$ be the linear transformation defined by $\mathsf{T}(f(x)) = f'(x)$, and let β and γ be the standard ordered bases for $\mathsf{P}_3(R)$ and $\mathsf{P}_2(R)$, respectively. If $A = [\mathsf{T}]_\beta^\gamma$, then, from Example 4 of Section 2.2, we have

$$A = \begin{pmatrix} 0 & 1 & 0 & 0 \\ 0 & 0 & 2 & 0 \\ 0 & 0 & 0 & 3 \end{pmatrix}.$$

We illustrate Theorem 2.14 by verifying that $[\mathsf{T}(p(x))]_\gamma = [\mathsf{T}]_\beta^\gamma[p(x)]_\beta$, where $p(x) \in \mathsf{P}_3(R)$ is the polynomial $p(x) = 2 - 4x + x^2 + 3x^3$. Let $q(x) = \mathsf{T}(p(x))$; then $q(x) = p'(x) = -4 + 2x + 9x^2$. Hence

$$[\mathsf{T}(p(x))]_\gamma = [q(x)]_\gamma = \begin{pmatrix} -4 \\ 2 \\ 9 \end{pmatrix},$$

but also

$$[\mathsf{T}]_\beta^\gamma[p(x)]_\beta = A[p(x)]_\beta = \begin{pmatrix} 0 & 1 & 0 & 0 \\ 0 & 0 & 2 & 0 \\ 0 & 0 & 0 & 3 \end{pmatrix} \begin{pmatrix} 2 \\ -4 \\ 1 \\ 3 \end{pmatrix} = \begin{pmatrix} -4 \\ 2 \\ 9 \end{pmatrix}. \quad \blacklozenge$$

We complete this section with the introduction of the *left-multiplication transformation* L_A, where A is an $m \times n$ matrix. This transformation is probably the most important tool for transferring properties about transformations to analogous properties about matrices and vice versa. For example, we use it to prove that matrix multiplication is associative.

Definition.   Let A be an $m \times n$ matrix with entries from a field F. We denote by L_A the mapping $\mathsf{L}_A\colon F^n \to F^m$ defined by $\mathsf{L}_A(x) = Ax$ (the matrix product of A and x) for each column vector $x \in F^n$. We call L_A a **left-multiplication transformation.**

Example 4

Let

$$A = \begin{pmatrix} 1 & 2 & 1 \\ 0 & 1 & 2 \end{pmatrix}.$$

Then $A \in \mathsf{M}_{2\times3}(R)$ and $\mathsf{L}_A\colon R^3 \to R^2$. If

$$x = \begin{pmatrix} 1 \\ 3 \\ -1 \end{pmatrix},$$

then

$$\mathsf{L}_A(x) = Ax = \begin{pmatrix} 1 & 2 & 1 \\ 0 & 1 & 2 \end{pmatrix} \begin{pmatrix} 1 \\ 3 \\ -1 \end{pmatrix} = \begin{pmatrix} 6 \\ 1 \end{pmatrix}. \quad \blacklozenge$$

We see in the next theorem that not only is L_A linear, but, in fact, it has a great many other useful properties. These properties are all quite natural and so are easy to remember.

Theorem 2.15. *Let A be an $m \times n$ matrix with entries from F. Then the left-multiplication transformation $\mathsf{L}_A\colon F^n \to F^m$ is linear. Furthermore, if B is any other $m \times n$ matrix (with entries from F) and β and γ are the standard ordered bases for F^n and F^m, respectively, then we have the following properties.*

(a) $[\mathsf{L}_A]_\beta^\gamma = A$.
(b) $\mathsf{L}_A = \mathsf{L}_B$ *if and only if* $A = B$.
(c) $\mathsf{L}_{A+B} = \mathsf{L}_A + \mathsf{L}_B$ *and* $\mathsf{L}_{aA} = a\mathsf{L}_A$ *for all* $a \in F$.
(d) *If* $\mathsf{T}\colon F^n \to F^m$ *is linear, then there exists a unique $m \times n$ matrix C such that* $\mathsf{T} = \mathsf{L}_C$. *In fact,* $C = [\mathsf{T}]_\beta^\gamma$.
(e) *If* E *is an* $n \times p$ *matrix, then* $\mathsf{L}_{AE} = \mathsf{L}_A \mathsf{L}_E$.
(f) *If* $m = n$, *then* $\mathsf{L}_{I_n} = \mathsf{I}_{F^n}$.

Proof. The fact that L_A is linear follows immediately from Theorem 2.12.

(a) The jth column of $[\mathsf{L}_A]_\beta^\gamma$ is equal to $\mathsf{L}_A(e_j)$. However $\mathsf{L}_A(e_j) = Ae_j$, which is also the jth column of A by Theorem 2.13(b). So $[\mathsf{L}_A]_\beta^\gamma = A$.

(b) If $\mathsf{L}_A = \mathsf{L}_B$, then we may use (a) to write $A = [\mathsf{L}_A]_\beta^\gamma = [\mathsf{L}_B]_\beta^\gamma = B$. Hence $A = B$. The proof of the converse is trivial.

(c) The proof is left as an exercise. (See Exercise 7.)

(d) Let $C = [\mathsf{T}]_\beta^\gamma$. By Theorem 2.14, we have $[\mathsf{T}(x)]_\gamma = [\mathsf{T}]_\beta^\gamma [x]_\beta$, or $\mathsf{T}(x) = Cx = \mathsf{L}_C(x)$ for all $x \in F^n$. So $\mathsf{T} = \mathsf{L}_C$. The uniqueness of C follows from (b).

(e) For any j ($1 \leq j \leq p$), we may apply Theorem 2.13 several times to note that $(AE)e_j$ is the jth column of AE and that the jth column of AE is also equal to $A(Ee_j)$. So $(AE)e_j = A(Ee_j)$. Thus

$$\mathsf{L}_{AE}(e_j) = (AE)e_j = A(Ee_j) = \mathsf{L}_A(Ee_j) = \mathsf{L}_A(\mathsf{L}_E(e_j)).$$

Hence $\mathsf{L}_{AE} = \mathsf{L}_A \mathsf{L}_E$ by the corollary to Theorem 2.6 (p. 73).

(f) The proof is left as an exercise. (See Exercise 7.) ∎

We now use left-multiplication transformations to establish the associativity of matrix multiplication.

Theorem 2.16. *Let $A, B,$ and C be matrices such that $A(BC)$ is defined. Then $(AB)C$ is also defined and $A(BC) = (AB)C$; that is, matrix multiplication is associative.*

Proof. It is left to the reader to show that $(AB)C$ is defined. Using (e) of Theorem 2.15 and the associativity of functional composition (see Appendix B), we have

$$\mathsf{L}_{A(BC)} = \mathsf{L}_A \mathsf{L}_{BC} = \mathsf{L}_A(\mathsf{L}_B \mathsf{L}_C) = (\mathsf{L}_A \mathsf{L}_B)\mathsf{L}_C = \mathsf{L}_{AB}\mathsf{L}_C = \mathsf{L}_{(AB)C}.$$

So from (b) of Theorem 2.15, it follows that $A(BC) = (AB)C$. ∎

Needless to say, this theorem could be proved directly from the definition of matrix multiplication (see Exercise 18). The proof above, however, provides a prototype of many of the arguments that utilize the relationships between linear transformations and matrices.

Applications

A large and varied collection of interesting applications arises in connection with special matrices called *incidence matrices*. An **incidence matrix** is a square matrix in which all the entries are either zero or one and, for convenience, all the diagonal entries are zero. If we have a relationship on a set of n objects that we denote by $1, 2, \ldots, n$, then we define the associated incidence matrix A by $A_{ij} = 1$ if i is related to j, and $A_{ij} = 0$ otherwise.

To make things concrete, suppose that we have four people, each of whom owns a communication device. If the relationship on this group is "can transmit to," then $A_{ij} = 1$ if i can send a message to j, and $A_{ij} = 0$ otherwise. Suppose that

$$A = \begin{pmatrix} 0 & 1 & 0 & 0 \\ 1 & 0 & 0 & 1 \\ 0 & 1 & 0 & 1 \\ 1 & 1 & 0 & 0 \end{pmatrix}.$$

Then since $A_{34} = 1$ and $A_{14} = 0$, we see that person 3 can send to 4 but 1 cannot send to 4.

We obtain an interesting interpretation of the entries of A^2. Consider, for instance,

$$(A^2)_{31} = A_{31}A_{11} + A_{32}A_{21} + A_{33}A_{31} + A_{34}A_{41}.$$

Note that any term $A_{3k}A_{k1}$ equals 1 if and only if both A_{3k} and A_{k1} equal 1, that is, if and only if 3 can send to k and k can send to 1. Thus $(A^2)_{31}$ gives the number of ways in which 3 can send to 1 in two *stages* (or in one *relay*). Since

$$A^2 = \begin{pmatrix} 1 & 0 & 0 & 1 \\ 1 & 2 & 0 & 0 \\ 2 & 1 & 0 & 1 \\ 1 & 1 & 0 & 1 \end{pmatrix},$$

we see that there are two ways 3 can send to 1 in two stages. In general, $(A + A^2 + \cdots + A^m)_{ij}$ is the number of ways in which i can send to j in at most m stages.

A maximal collection of three or more people with the property that any two can send to each other is called a **clique**. The problem of determining cliques is difficult, but there is a simple method for determining if someone belongs to a clique. If we define a new matrix B by $B_{ij} = 1$ if i and j can send to each other, and $B_{ij} = 0$ otherwise, then it can be shown (see Exercise 19) that person i belongs to a clique if and only if $(B^3)_{ii} > 0$. For example, suppose that the incidence matrix associated with some relationship is

$$A = \begin{pmatrix} 0 & 1 & 0 & 1 \\ 1 & 0 & 1 & 0 \\ 1 & 1 & 0 & 1 \\ 1 & 1 & 1 & 0 \end{pmatrix}.$$

To determine which people belong to cliques, we form the matrix B, described earlier, and compute B^3. In this case,

$$B = \begin{pmatrix} 0 & 1 & 0 & 1 \\ 1 & 0 & 1 & 0 \\ 0 & 1 & 0 & 1 \\ 1 & 0 & 1 & 0 \end{pmatrix} \quad \text{and} \quad B^3 = \begin{pmatrix} 0 & 4 & 0 & 4 \\ 4 & 0 & 4 & 0 \\ 0 & 4 & 0 & 4 \\ 4 & 0 & 4 & 0 \end{pmatrix}.$$

Since all the diagonal entries of B^3 are zero, we conclude that there are no cliques in this relationship.

Our final example of the use of incidence matrices is concerned with the concept of *dominance*. A relation among a group of people is called a **dominance relation** if the associated incidence matrix A has the property that for all distinct pairs i and j, $A_{ij} = 1$ if and only if $A_{ji} = 0$, that is, given any two people, exactly one of them *dominates* (or, using the terminology of our first example, can send a message to) the other. Since A is an incidence matrix, $A_{ii} = 0$ for all i. For such a relation, it can be shown (see Exercise 21) that the matrix $A + A^2$ has a row [column] in which each entry is positive except for the diagonal entry. In other words, there is at least one person who dominates [is dominated by] all others in one or two stages. In fact, it can be shown that any person who dominates [is dominated by] the greatest number of people in the first stage has this property. Consider, for example, the matrix

$$A = \begin{pmatrix} 0 & 1 & 0 & 1 & 0 \\ 0 & 0 & 1 & 0 & 0 \\ 1 & 0 & 0 & 1 & 0 \\ 0 & 1 & 0 & 0 & 1 \\ 1 & 1 & 1 & 0 & 0 \end{pmatrix}.$$

The reader should verify that this matrix corresponds to a dominance relation. Now

$$A + A^2 = \begin{pmatrix} 0 & 2 & 1 & 1 & 1 \\ 1 & 0 & 1 & 1 & 0 \\ 1 & 2 & 0 & 2 & 1 \\ 1 & 2 & 2 & 0 & 1 \\ 2 & 2 & 2 & 2 & 0 \end{pmatrix}.$$

Thus persons 1, 3, 4, and 5 dominate (can send messages to) all the others in at most two stages, while persons 1, 2, 3, and 4 are dominated by (can receive messages from) all the others in at most two stages.

EXERCISES

1. Label the following statements as true or false. In each part, V, W, and Z denote vector spaces with ordered (finite) bases α, β, and γ, respectively; $T: V \to W$ and $U: W \to Z$ denote linear transformations; and A and B denote matrices.

   (a) $[UT]_\alpha^\gamma = [T]_\alpha^\beta [U]_\beta^\gamma$.
   (b) $[T(v)]_\beta = [T]_\alpha^\beta [v]_\alpha$ for all $v \in V$.
   (c) $[U(w)]_\beta = [U]_\alpha^\beta [w]_\beta$ for all $w \in W$.
   (d) $[I_V]_\alpha = I$.
   (e) $[T^2]_\alpha^\beta = ([T]_\alpha^\beta)^2$.
   (f) $A^2 = I$ implies that $A = I$ or $A = -I$.
   (g) $T = L_A$ for some matrix A.
   (h) $A^2 = O$ implies that $A = O$, where O denotes the zero matrix.
   (i) $L_{A+B} = L_A + L_B$.
   (j) If A is square and $A_{ij} = \delta_{ij}$ for all i and j, then $A = I$.

2. (a) Let

$$A = \begin{pmatrix} 1 & 3 \\ 2 & -1 \end{pmatrix}, \quad B = \begin{pmatrix} 1 & 0 & -3 \\ 4 & 1 & 2 \end{pmatrix},$$

$$C = \begin{pmatrix} 1 & 1 & 4 \\ -1 & -2 & 0 \end{pmatrix}, \quad \text{and} \quad D = \begin{pmatrix} 2 \\ -2 \\ 3 \end{pmatrix}.$$

   Compute $A(2B + 3C), (AB)D$, and $A(BD)$.

   (b) Let

$$A = \begin{pmatrix} 2 & 5 \\ -3 & 1 \\ 4 & 2 \end{pmatrix}, \quad B = \begin{pmatrix} 3 & -2 & 0 \\ 1 & -1 & 4 \\ 5 & 5 & 3 \end{pmatrix}, \quad \text{and} \quad C = \begin{pmatrix} 4 & 0 & 3 \end{pmatrix}.$$

   Compute A^t, $A^t B$, BC^t, CB, and CA.

3. Let $g(x) = 3 + x$. Let $T: P_2(R) \to P_2(R)$ and $U: P_2(R) \to R^3$ be the linear transformations respectively defined by

$$T(f(x)) = f'(x)g(x) + 2f(x) \quad \text{and} \quad U(a + bx + cx^2) = (a + b, c, a - b).$$

   Let β and γ be the standard ordered bases of $P_2(R)$ and R^3, respectively.

(a) Compute $[U]_\beta^\gamma, [T]_\beta$, and $[UT]_\beta^\gamma$ directly. Then use Theorem 2.11 to verify your result.

(b) Let $h(x) = 3 - 2x + x^2$. Compute $[h(x)]_\beta$ and $[U(h(x))]_\gamma$. Then use $[U]_\beta^\gamma$ from (a) and Theorem 2.14 to verify your result.

4. For each of the following parts, let T be the linear transformation defined in the corresponding part of Exercise 5 of Section 2.2. Use Theorem 2.14 to compute the following vectors:

(a) $[T(A)]_\alpha$, where $A = \begin{pmatrix} 1 & 4 \\ -1 & 6 \end{pmatrix}$.

(b) $[T(f(x))]_\alpha$, where $f(x) = 4 - 6x + 3x^2$.

(c) $[T(A)]_\gamma$, where $A = \begin{pmatrix} 1 & 3 \\ 2 & 4 \end{pmatrix}$.

(d) $[T(f(x))]_\gamma$, where $f(x) = 6 - x + 2x^2$.

5. Complete the proof of Theorem 2.12 and its corollary.

6. Prove (b) of Theorem 2.13.

7. Prove (c) and (f) of Theorem 2.15.

8. Prove Theorem 2.10. Now state and prove a more general result involving linear transformations with domains unequal to their codomains.

9. Find linear transformations $U, T: F^2 \to F^2$ such that $UT = T_0$ (the zero transformation) but $TU \neq T_0$. Use your answer to find matrices A and B such that $AB = O$ but $BA \neq O$.

10. Let A be an $n \times n$ matrix. Prove that A is a diagonal matrix if and only if $A_{ij} = \delta_{ij} A_{ij}$ for all i and j.

11. Let V be a vector space, and let $T: V \to V$ be linear. Prove that $T^2 = T_0$ if and only if $R(T) \subseteq N(T)$.

12. Let V, W, and Z be vector spaces, and let $T: V \to W$ and $U: W \to Z$ be linear.

(a) Prove that if UT is one-to-one, then T is one-to-one. Must U also be one-to-one?

(b) Prove that if UT is onto, then U is onto. Must T also be onto?

(c) Prove that if U and T are one-to-one and onto, then UT is also.

13. Let A and B be $n \times n$ matrices. Recall that the trace of A is defined by

$$\text{tr}(A) = \sum_{i=1}^{n} A_{ii}.$$

Prove that $\text{tr}(AB) = \text{tr}(BA)$ and $\text{tr}(A) = \text{tr}(A^t)$.

14. Assume the notation in Theorem 2.13.

    (a) Suppose that z is a (column) vector in F^p. Use Theorem 2.13(b) to prove that Bz is a linear combination of the columns of B. In particular, if $z = (a_1, a_2, \ldots, a_p)^t$, then show that

$$Bz = \sum_{j=1}^{p} a_j v_j.$$

    (b) Extend (a) to prove that column j of AB is a linear combination of the columns of A with the coefficients in the linear combination being the entries of column j of B.

    (c) For any row vector $w \in \mathsf{F}^m$, prove that wA is a linear combination of the rows of A with the coefficients in the linear combination being the coordinates of w. *Hint:* Use properties of the transpose operation applied to (a).

    (d) Prove the analogous result to (b) about rows: Row i of AB is a linear combination of the rows of B with the coefficients in the linear combination being the entries of row i of A.

15.† Let M and A be matrices for which the product matrix MA is defined. If the jth column of A is a linear combination of a set of columns of A, prove that the jth column of MA is a linear combination of the corresponding columns of MA with the same corresponding coefficients.

16. Let V be a finite-dimensional vector space, and let $\mathsf{T}\colon \mathsf{V} \to \mathsf{V}$ be linear.

    (a) If $\text{rank}(\mathsf{T}) = \text{rank}(\mathsf{T}^2)$, prove that $\mathsf{R}(\mathsf{T}) \cap \mathsf{N}(\mathsf{T}) = \{0\}$. Deduce that $\mathsf{V} = \mathsf{R}(\mathsf{T}) \oplus \mathsf{N}(\mathsf{T})$ (see the exercises of Section 1.3).

    (b) Prove that $\mathsf{V} = \mathsf{R}(\mathsf{T}^k) \oplus \mathsf{N}(\mathsf{T}^k)$ for some positive integer k.

17. Let V be a vector space. Determine all linear transformations $\mathsf{T}\colon \mathsf{V} \to \mathsf{V}$ such that $\mathsf{T} = \mathsf{T}^2$. *Hint:* Note that $x = \mathsf{T}(x) + (x - \mathsf{T}(x))$ for every x in V, and show that $\mathsf{V} = \{y\colon \mathsf{T}(y) = y\} \oplus \mathsf{N}(\mathsf{T})$ (see the exercises of Section 1.3).

18. Using only the definition of matrix multiplication, prove that multiplication of matrices is associative.

19. For an incidence matrix A with related matrix B defined by $B_{ij} = 1$ if i is related to j and j is related to i, and $B_{ij} = 0$ otherwise, prove that i belongs to a clique if and only if $(B^3)_{ii} > 0$.

20. Use Exercise 19 to determine the cliques in the relations corresponding to the following incidence matrices.

(a) $\begin{pmatrix} 0 & 1 & 0 & 1 \\ 1 & 0 & 0 & 0 \\ 0 & 1 & 0 & 1 \\ 1 & 0 & 1 & 0 \end{pmatrix}$ (b) $\begin{pmatrix} 0 & 0 & 1 & 1 \\ 1 & 0 & 0 & 1 \\ 1 & 0 & 0 & 1 \\ 1 & 0 & 1 & 0 \end{pmatrix}$

21. Let A be an incidence matrix that is associated with a dominance relation. Prove that the matrix $A + A^2$ has a row [column] in which each entry is positive except for the diagonal entry.

22. Prove that the matrix

$$A = \begin{pmatrix} 0 & 1 & 0 \\ 0 & 0 & 1 \\ 1 & 0 & 0 \end{pmatrix}$$

corresponds to a dominance relation. Use Exercise 21 to determine which persons dominate [are dominated by] each of the others within two stages.

23. Let A be an $n \times n$ incidence matrix that corresponds to a dominance relation. Determine the number of nonzero entries of A.

2.4   INVERTIBILITY AND ISOMORPHISMS

The concept of invertibility is introduced quite early in the study of functions. Fortunately, many of the intrinsic properties of functions are shared by their inverses. For example, in calculus we learn that the properties of being continuous or differentiable are generally retained by the inverse functions. We see in this section (Theorem 2.17) that the inverse of a linear transformation is also linear. This result greatly aids us in the study of *inverses* of matrices. As one might expect from Section 2.3, the inverse of the left-multiplication transformation L_A (when it exists) can be used to determine properties of the inverse of the matrix A.

In the remainder of this section, we apply many of the results about invertibility to the concept of *isomorphism*. We will see that finite-dimensional vector spaces (over F) of equal dimension may be identified. These ideas will be made precise shortly.

The facts about inverse functions presented in Appendix B are, of course, true for linear transformations. Nevertheless, we repeat some of the definitions for use in this section.

Definition. *Let* V *and* W *be vector spaces, and let* $\mathsf{T}\colon \mathsf{V} \to \mathsf{W}$ *be linear. A function* $\mathsf{U}\colon \mathsf{W} \to \mathsf{V}$ *is said to be an* **inverse** *of* T *if* $\mathsf{TU} = \mathsf{I}_\mathsf{W}$ *and* $\mathsf{UT} = \mathsf{I}_\mathsf{V}$. *If* T *has an inverse, then* T *is said to be* **invertible**. *As noted in Appendix B, if* T *is invertible, then the inverse of* T *is unique and is denoted by* T^{-1}.

The following facts hold for invertible functions T and U.

1. $(TU)^{-1} = U^{-1}T^{-1}$.
2. $(T^{-1})^{-1} = T$; in particular, T^{-1} is invertible.

We often use the fact that a function is invertible if and only if it is both one-to-one and onto. We can therefore restate Theorem 2.5 as follows.

3. Let $T: V \to W$ be a linear transformation, where V and W are finite-dimensional spaces of equal dimension. Then T is invertible if and only if $\text{rank}(T) = \dim(V)$.

Example 1

Let $T: P_1(R) \to R^2$ be the linear transformation defined by $T(a + bx) = (a, a+b)$. The reader can verify directly that $T^{-1}: R^2 \to P_1(R)$ is defined by $T^{-1}(c, d) = c + (d - c)x$. Observe that T^{-1} is also linear. As Theorem 2.17 demonstrates, this is true in general. ◆

Theorem 2.17. *Let* V *and* W *be vector spaces, and let* $T: V \to W$ *be linear and invertible. Then* $T^{-1}: W \to V$ *is linear.*

Proof. Let $y_1, y_2 \in W$ and $c \in F$. Since T is onto and one-to-one, there exist unique vectors x_1 and x_2 such that $T(x_1) = y_1$ and $T(x_2) = y_2$. Thus $x_1 = T^{-1}(y_1)$ and $x_2 = T^{-1}(y_2)$; so

$$T^{-1}(cy_1 + y_2) = T^{-1}[cT(x_1) + T(x_2)] = T^{-1}[T(cx_1 + x_2)]$$
$$= cx_1 + x_2 = cT^{-1}(y_1) + T^{-1}(y_2). \quad \blacksquare$$

It now follows immediately from Theorem 2.5 (p. 71) that if T is a linear transformation between vector spaces of equal (finite) dimension, then the conditions of being invertible, one-to-one, and onto are all equivalent.

We are now ready to define the inverse of a matrix. The reader should note the analogy with the inverse of a linear transformation.

Definition. *Let* A *be an* $n \times n$ *matrix. Then* A *is* **invertible** *if there exists an* $n \times n$ *matrix* B *such that* $AB = BA = I$.

If A is invertible, then the matrix B such that $AB = BA = I$ is unique. (If C were another such matrix, then $C = CI = C(AB) = (CA)B = IB = B$.) The matrix B is called the **inverse** of A and is denoted by A^{-1}.

Example 2

The reader should verify that the inverse of

$$\begin{pmatrix} 5 & 7 \\ 2 & 3 \end{pmatrix} \quad \text{is} \quad \begin{pmatrix} 3 & -7 \\ -2 & 5 \end{pmatrix}. \quad ◆$$

In Section 3.2, we learn a technique for computing the inverse of a matrix. At this point, we develop a number of results that relate the inverses of matrices to the inverses of linear transformations.

Lemma. *Let* T *be an invertible linear transformation from* V *to* W. *Then* V *is finite-dimensional if and only if* W *is finite-dimensional. In this case,* $\dim(V) = \dim(W)$.

Proof. Suppose that V is finite-dimensional. Let $\beta = \{x_1, x_2, \ldots, x_n\}$ be a basis for V. By Theorem 2.2 (p. 68), $T(\beta)$ spans $R(T) = W$; hence W is finite-dimensional by Theorem 1.9 (p. 44). Conversely, if W is finite-dimensional, then so is V by a similar argument, using T^{-1}.

Now suppose that V and W are finite-dimensional. Because T is one-to-one and onto, we have

$$\text{nullity}(T) = 0 \quad \text{and} \quad \text{rank}(T) = \dim(R(T)) = \dim(W).$$

So by the dimension theorem (p. 70), it follows that $\dim(V) = \dim(W)$. ∎

Theorem 2.18. *Let* V *and* W *be finite-dimensional vector spaces with ordered bases* β *and* γ, *respectively. Let* $T: V \to W$ *be linear. Then* T *is invertible if and only if* $[T]_\beta^\gamma$ *is invertible. Furthermore,* $[T^{-1}]_\gamma^\beta = ([T]_\beta^\gamma)^{-1}$.

Proof. Suppose that T is invertible. By the lemma, we have $\dim(V) = \dim(W)$. Let $n = \dim(V)$. So $[T]_\beta^\gamma$ is an $n \times n$ matrix. Now $T^{-1}: W \to V$ satisfies $TT^{-1} = I_W$ and $T^{-1}T = I_V$. Thus

$$I_n = [I_V]_\beta = [T^{-1}T]_\beta = [T^{-1}]_\gamma^\beta [T]_\beta^\gamma.$$

Similarly, $[T]_\beta^\gamma [T^{-1}]_\gamma^\beta = I_n$. So $[T]_\beta^\gamma$ is invertible and $\left([T]_\beta^\gamma\right)^{-1} = [T^{-1}]_\gamma^\beta$.

Now suppose that $A = [T]_\beta^\gamma$ is invertible. Then there exists an $n \times n$ matrix B such that $AB = BA = I_n$. By Theorem 2.6 (p. 72), there exists $U \in \mathcal{L}(W, V)$ such that

$$U(w_j) = \sum_{i=1}^n B_{ij} v_i \quad \text{for } j = 1, 2, \ldots, n,$$

where $\gamma = \{w_1, w_2, \ldots, w_n\}$ and $\beta = \{v_1, v_2, \ldots, v_n\}$. It follows that $[U]_\gamma^\beta = B$. To show that $U = T^{-1}$, observe that

$$[UT]_\beta = [U]_\gamma^\beta [T]_\beta^\gamma = BA = I_n = [I_V]_\beta$$

by Theorem 2.11 (p. 88). So $UT = I_V$, and similarly, $TU = I_W$. ∎

Example 3

Let β and γ be the standard ordered bases of $\mathsf{P}_1(R)$ and R^2, respectively. For T as in Example 1, we have

$$[\mathsf{T}]_\beta^\gamma = \begin{pmatrix} 1 & 0 \\ 1 & 1 \end{pmatrix} \quad \text{and} \quad [\mathsf{T}^{-1}]_\gamma^\beta = \begin{pmatrix} 1 & 0 \\ -1 & 1 \end{pmatrix}.$$

It can be verified by matrix multiplication that each matrix is the inverse of the other. ◆

Corollary 1. *Let* V *be a finite-dimensional vector space with an ordered basis* β, *and let* $\mathsf{T}: \mathsf{V} \to \mathsf{V}$ *be linear. Then* T *is invertible if and only if* $[\mathsf{T}]_\beta$ *is invertible. Furthermore,* $[\mathsf{T}^{-1}]_\beta = ([\mathsf{T}]_\beta)^{-1}$.

Proof. Exercise.   ∎

Corollary 2. *Let* A *be an* $n \times n$ *matrix. Then* A *is invertible if and only if* L_A *is invertible. Furthermore,* $(\mathsf{L}_A)^{-1} = \mathsf{L}_{A^{-1}}$.

Proof. Exercise.   ∎

The notion of invertibility may be used to formalize what may already have been observed by the reader, that is, that certain vector spaces strongly resemble one another except for the form of their vectors. For example, in the case of $\mathsf{M}_{2 \times 2}(F)$ and F^4, if we associate to each matrix

$$\begin{pmatrix} a & b \\ c & d \end{pmatrix}$$

the 4-tuple (a, b, c, d), we see that sums and scalar products associate in a similar manner; that is, in terms of the vector space structure, these two vector spaces may be considered identical or *isomorphic*.

Definitions. *Let* V *and* W *be vector spaces. We say that* V *is* **isomorphic** *to* W *if there exists a linear transformation* $\mathsf{T}: \mathsf{V} \to \mathsf{W}$ *that is invertible. Such a linear transformation is called an* **isomorphism** *from* V *onto* W.

We leave as an exercise (see Exercise 13) the proof that "is isomorphic to" is an equivalence relation. (See Appendix A.) So we need only say that V and W are isomorphic.

Example 4

Define $\mathsf{T}: F^2 \to \mathsf{P}_1(F)$ by $\mathsf{T}(a_1, a_2) = a_1 + a_2 x$. It is easily checked that T is an isomorphism; so F^2 is isomorphic to $\mathsf{P}_1(F)$. ◆

Example 5

Define

$$T \colon P_3(R) \to M_{2 \times 2}(R) \quad \text{by } T(f) = \begin{pmatrix} f(1) & f(2) \\ f(3) & f(4) \end{pmatrix}.$$

It is easily verified that T is linear. By use of the Lagrange interpolation formula in Section 1.6, it can be shown (compare with Exercise 22) that $T(f) = O$ only when f is the zero polynomial. Thus T is one-to-one (see Exercise 11). Moreover, because $\dim(P_3(R)) = \dim(M_{2 \times 2}(R))$, it follows that T is invertible by Theorem 2.5 (p. 71). We conclude that $P_3(R)$ is isomorphic to $M_{2 \times 2}(R)$. ◆

In each of Examples 4 and 5, the reader may have observed that isomorphic vector spaces have equal dimensions. As the next theorem shows, this is no coincidence.

Theorem 2.19. *Let* V *and* W *be finite-dimensional vector spaces (over the same field). Then* V *is isomorphic to* W *if and only if* $\dim(V) = \dim(W)$.

Proof. Suppose that V is isomorphic to W and that $T \colon V \to W$ is an isomorphism from V to W. By the lemma preceding Theorem 2.18, we have that $\dim(V) = \dim(W)$.

Now suppose that $\dim(V) = \dim(W)$, and let $\beta = \{v_1, v_2, \ldots, v_n\}$ and $\gamma = \{w_1, w_2, \ldots, w_n\}$ be bases for V and W, respectively. By Theorem 2.6 (p. 72), there exists $T \colon V \to W$ such that T is linear and $T(v_i) = w_i$ for $i = 1, 2, \ldots, n$. Using Theorem 2.2 (p. 68), we have

$$R(T) = \text{span}(T(\beta)) = \text{span}(\gamma) = W.$$

So T is onto. From Theorem 2.5 (p. 71), we have that T is also one-to-one. Hence T is an isomorphism. ∎

By the lemma to Theorem 2.18, if V and W are isomorphic, then either both of V and W are finite-dimensional or both are infinite-dimensional.

Corollary. *Let* V *be a vector space over* F. *Then* V *is isomorphic to* F^n *if and only if* $\dim(V) = n$.

Up to this point, we have associated linear transformations with their matrix representations. We are now in a position to prove that, as a vector space, the collection of all linear transformations between two given vector spaces may be identified with the appropriate vector space of $m \times n$ matrices.

Theorem 2.20. *Let* V *and* W *be finite-dimensional vector spaces over* F *of dimensions* n *and* m, *respectively, and let* β *and* γ *be ordered bases for* V *and* W, *respectively. Then the function* $\Phi \colon \mathcal{L}(V, W) \to M_{m \times n}(F)$, *defined by* $\Phi(T) = [T]_\beta^\gamma$ *for* $T \in \mathcal{L}(V, W)$, *is an isomorphism.*

Proof. By Theorem 2.8 (p. 82), Φ is linear. Hence we must show that Φ is one-to-one and onto. This is accomplished if we show that for every $m \times n$ matrix A, there exists a unique linear transformation $\mathsf{T} \colon \mathsf{V} \to \mathsf{W}$ such that $\Phi(\mathsf{T}) = A$. Let $\beta = \{v_1, v_2, \dots, v_n\}$, $\gamma = \{w_1, w_2, \dots, w_m\}$, and let A be a given $m \times n$ matrix. By Theorem 2.6 (p. 72), there exists a unique linear transformation $\mathsf{T} \colon \mathsf{V} \to \mathsf{W}$ such that

$$\mathsf{T}(v_j) = \sum_{i=1}^{m} A_{ij} w_i \quad \text{for } 1 \le j \le n.$$

But this means that $[\mathsf{T}]_\beta^\gamma = A$, or $\Phi(\mathsf{T}) = A$. Thus Φ is an isomorphism. ∎

Corollary. *Let* V *and* W *be finite-dimensional vector spaces of dimensions* n *and* m, *respectively. Then* $\mathcal{L}(\mathsf{V}, \mathsf{W})$ *is finite-dimensional of dimension* mn.

Proof. The proof follows from Theorems 2.20 and 2.19 and the fact that $\dim(\mathsf{M}_{m \times n}(F)) = mn$. ∎

We conclude this section with a result that allows us to see more clearly the relationship between linear transformations defined on abstract finite-dimensional vector spaces and linear transformations from F^n to F^m.

We begin by naming the transformation $x \to [x]_\beta$ introduced in Section 2.2.

Definition. *Let* β *be an ordered basis for an* n-*dimensional vector space* V *over the field* F. *The* **standard representation of** V **with respect to** β *is the function* $\phi_\beta \colon \mathsf{V} \to \mathsf{F}^n$ *defined by* $\phi_\beta(x) = [x]_\beta$ *for each* $x \in \mathsf{V}$.

Example 6

Let $\beta = \{(1,0), (0,1)\}$ and $\gamma = \{(1,2), (3,4)\}$. It is easily observed that β and γ are ordered bases for R^2. For $x = (1, -2)$, we have

$$\phi_\beta(x) = [x]_\beta = \begin{pmatrix} 1 \\ -2 \end{pmatrix} \quad \text{and} \quad \phi_\gamma(x) = [x]_\gamma = \begin{pmatrix} -5 \\ 2 \end{pmatrix}. \quad \blacklozenge$$

We observed earlier that ϕ_β is a linear transformation. The next theorem tells us much more.

Theorem 2.21. *For any finite-dimensional vector space* V *with ordered basis* β, ϕ_β *is an isomorphism.*

Proof. Exercise. ∎

This theorem provides us with an alternate proof that an n-dimensional vector space is isomorphic to F^n (see the corollary to Theorem 2.19).

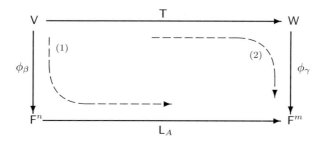

Figure 2.2

Let V and W be vector spaces of dimension n and m, respectively, and let $T: V \to W$ be a linear transformation. Define $A = [T]_\beta^\gamma$, where β and γ are arbitrary ordered bases of V and W, respectively. We are now able to use ϕ_β and ϕ_γ to study the relationship between the linear transformations T and $L_A: F^n \to F^m$.

Let us first consider Figure 2.2. Notice that there are two composites of linear transformations that map V into F^m:

1. Map V into F^n with ϕ_β and follow this transformation with L_A; this yields the composite $L_A\phi_\beta$.
2. Map V into W with T and follow it by ϕ_γ to obtain the composite $\phi_\gamma T$.

These two composites are depicted by the dashed arrows in the diagram. By a simple reformulation of Theorem 2.14 (p. 91), we may conclude that

$$L_A\phi_\beta = \phi_\gamma T;$$

that is, the diagram "commutes." Heuristically, this relationship indicates that after V and W are identified with F^n and F^m via ϕ_β and ϕ_γ, respectively, we may "identify" T with L_A. This diagram allows us to transfer operations on abstract vector spaces to ones on F^n and F^m.

Example 7

Recall the linear transformation $T: P_3(R) \to P_2(R)$ defined in Example 4 of Section 2.2 ($T(f(x)) = f'(x)$). Let β and γ be the standard ordered bases for $P_3(R)$ and $P_2(R)$, respectively, and let $\phi_\beta: P_3(R) \to R^4$ and $\phi_\gamma: P_2(R) \to R^3$ be the corresponding standard representations of $P_3(R)$ and $P_2(R)$. If $A = [T]_\beta^\gamma$, then

$$A = \begin{pmatrix} 0 & 1 & 0 & 0 \\ 0 & 0 & 2 & 0 \\ 0 & 0 & 0 & 3 \end{pmatrix}.$$

Consider the polynomial $p(x) = 2+x-3x^2+5x^3$. We show that $L_A\phi_\beta(p(x)) = \phi_\gamma\mathsf{T}(p(x))$. Now

$$L_A\phi_\beta(p(x)) = \begin{pmatrix} 0 & 1 & 0 & 0 \\ 0 & 0 & 2 & 0 \\ 0 & 0 & 0 & 3 \end{pmatrix} \begin{pmatrix} 2 \\ 1 \\ -3 \\ 5 \end{pmatrix} = \begin{pmatrix} 1 \\ -6 \\ 15 \end{pmatrix}.$$

But since $\mathsf{T}(p(x)) = p'(x) = 1 - 6x + 15x^2$, we have

$$\phi_\gamma\mathsf{T}(p(x)) = \begin{pmatrix} 1 \\ -6 \\ 15 \end{pmatrix}.$$

So $L_A\phi_\beta(p(x)) = \phi_\gamma\mathsf{T}(p(x))$.   ◆

Try repeating Example 7 with different polynomials $p(x)$.

EXERCISES

1. Label the following statements as true or false. In each part, V and W are vector spaces with ordered (finite) bases α and β, respectively, $\mathsf{T}\colon \mathsf{V} \to \mathsf{W}$ is linear, and A and B are matrices.

(a) $\left([\mathsf{T}]_\alpha^\beta\right)^{-1} = [\mathsf{T}^{-1}]_\alpha^\beta$.
(b) T is invertible if and only if T is one-to-one and onto.
(c) $\mathsf{T} = L_A$, where $A = [\mathsf{T}]_\alpha^\beta$.
(d) $\mathsf{M}_{2\times 3}(F)$ is isomorphic to F^5.
(e) $\mathsf{P}_n(F)$ is isomorphic to $\mathsf{P}_m(F)$ if and only if $n = m$.
(f) $AB = I$ implies that A and B are invertible.
(g) If A is invertible, then $(A^{-1})^{-1} = A$.
(h) A is invertible if and only if L_A is invertible.
(i) A must be square in order to possess an inverse.

2. For each of the following linear transformations T, determine whether T is invertible and justify your answer.

(a) $\mathsf{T}\colon \mathsf{R}^2 \to \mathsf{R}^3$ defined by $\mathsf{T}(a_1, a_2) = (a_1 - 2a_2, a_2, 3a_1 + 4a_2)$.
(b) $\mathsf{T}\colon \mathsf{R}^2 \to \mathsf{R}^3$ defined by $\mathsf{T}(a_1, a_2) = (3a_1 - a_2, a_2, 4a_1)$.
(c) $\mathsf{T}\colon \mathsf{R}^3 \to \mathsf{R}^3$ defined by $\mathsf{T}(a_1, a_2, a_3) = (3a_1 - 2a_3, a_2, 3a_1 + 4a_2)$.
(d) $\mathsf{T}\colon \mathsf{P}_3(R) \to \mathsf{P}_2(R)$ defined by $\mathsf{T}(p(x)) = p'(x)$.
(e) $\mathsf{T}\colon \mathsf{M}_{2\times 2}(R) \to \mathsf{P}_2(R)$ defined by $\mathsf{T}\begin{pmatrix} a & b \\ c & d \end{pmatrix} = a + 2bx + (c+d)x^2$.
(f) $\mathsf{T}\colon \mathsf{M}_{2\times 2}(R) \to \mathsf{M}_{2\times 2}(R)$ defined by $\mathsf{T}\begin{pmatrix} a & b \\ c & d \end{pmatrix} = \begin{pmatrix} a+b & a \\ c & c+d \end{pmatrix}$.

3. Which of the following pairs of vector spaces are isomorphic? Justify your answers.

    (a) F^3 and $\mathsf{P}_3(F)$.
    (b) F^4 and $\mathsf{P}_3(F)$.
    (c) $\mathsf{M}_{2\times 2}(R)$ and $\mathsf{P}_3(R)$.
    (d) $\mathsf{V} = \{A \in \mathsf{M}_{2\times 2}(R) \colon \operatorname{tr}(A) = 0\}$ and R^4.

4.[†] Let A and B be $n \times n$ invertible matrices. Prove that AB is invertible and $(AB)^{-1} = B^{-1}A^{-1}$.

5.[†] Let A be invertible. Prove that A^t is invertible and $(A^t)^{-1} = (A^{-1})^t$.

6. Prove that if A is invertible and $AB = O$, then $B = O$.

7. Let A be an $n \times n$ matrix.

    (a) Suppose that $A^2 = O$. Prove that A is not invertible.
    (b) Suppose that $AB = O$ for some nonzero $n \times n$ matrix B. Could A be invertible? Explain.

8. Prove Corollaries 1 and 2 of Theorem 2.18.

9. Let A and B be $n \times n$ matrices such that AB is invertible. Prove that A and B are invertible. Give an example to show that arbitrary matrices A and B need not be invertible if AB is invertible.

10.[†] Let A and B be $n \times n$ matrices such that $AB = I_n$.

    (a) Use Exercise 9 to conclude that A and B are invertible.
    (b) Prove $A = B^{-1}$ (and hence $B = A^{-1}$). (We are, in effect, saying that for square matrices, a "one-sided" inverse is a "two-sided" inverse.)
    (c) State and prove analogous results for linear transformations defined on finite-dimensional vector spaces.

11. Verify that the transformation in Example 5 is one-to-one.

12. Prove Theorem 2.21.

13. Let $\sim$ mean "is isomorphic to." Prove that $\sim$ is an equivalence relation on the class of vector spaces over F.

14. Let

$$\mathsf{V} = \left\{ \begin{pmatrix} a & a+b \\ 0 & c \end{pmatrix} \colon a, b, c \in F \right\}.$$

Construct an isomorphism from V to F^3.

15. Let V and W be finite-dimensional vector spaces, and let $T: V \rightarrow W$ be a linear transformation. Suppose that β is a basis for V. Prove that T is an isomorphism if and only if $T(\beta)$ is a basis for W.

16. Let B be an $n \times n$ invertible matrix. Define $\Phi: M_{n\times n}(F) \rightarrow M_{n\times n}(F)$ by $\Phi(A) = B^{-1}AB$. Prove that Φ is an isomorphism.

17.[†] Let V and W be finite-dimensional vector spaces and $T: V \rightarrow W$ be an isomorphism. Let V_0 be a subspace of V.

(a) Prove that $T(V_0)$ is a subspace of W.
(b) Prove that $\dim(V_0) = \dim(T(V_0))$.

18. Repeat Example 7 with the polynomial $p(x) = 1 + x + 2x^2 + x^3$.

19. In Example 5 of Section 2.1, the mapping $T: M_{2\times 2}(R) \rightarrow M_{2\times 2}(R)$ defined by $T(M) = M^t$ for each $M \in M_{2\times 2}(R)$ is a linear transformation. Let $\beta = \{E^{11}, E^{12}, E^{21}, E^{22}\}$, which is a basis for $M_{2\times 2}(R)$, as noted in Example 3 of Section 1.6.

(a) Compute $[T]_\beta$.
(b) Verify that $L_A \phi_\beta(M) = \phi_\beta T(M)$ for $A = [T]_\beta$ and

$$M = \begin{pmatrix} 1 & 2 \\ 3 & 4 \end{pmatrix}.$$

20.[†] Let $T: V \rightarrow W$ be a linear transformation from an n-dimensional vector space V to an m-dimensional vector space W. Let β and γ be ordered bases for V and W, respectively. Prove that $\text{rank}(T) = \text{rank}(L_A)$ and that $\text{nullity}(T) = \text{nullity}(L_A)$, where $A = [T]_\beta^\gamma$. *Hint:* Apply Exercise 17 to Figure 2.2.

21. Let V and W be finite-dimensional vector spaces with ordered bases $\beta = \{v_1, v_2, \ldots, v_n\}$ and $\gamma = \{w_1, w_2, \ldots, w_m\}$, respectively. By Theorem 2.6 (p. 72), there exist linear transformations $T_{ij}: V \rightarrow W$ such that

$$T_{ij}(v_k) = \begin{cases} w_i & \text{if } k = j \\ 0 & \text{if } k \neq j. \end{cases}$$

First prove that $\{T_{ij}: 1 \leq i \leq m, \ 1 \leq j \leq n\}$ is a basis for $\mathcal{L}(V, W)$. Then let M^{ij} be the $m \times n$ matrix with 1 in the ith row and jth column and 0 elsewhere, and prove that $[T_{ij}]_\beta^\gamma = M^{ij}$. Again by Theorem 2.6, there exists a linear transformation $\Phi: \mathcal{L}(V, W) \rightarrow M_{m\times n}(F)$ such that $\Phi(T_{ij}) = M^{ij}$. Prove that Φ is an isomorphism.

22. Let $c_0, c_1, \ldots, c_n$ be distinct scalars from an infinite field F. Define $\mathsf{T} \colon \mathsf{P}_n(F) \to F^{n+1}$ by $\mathsf{T}(f) = (f(c_0), f(c_1), \ldots, f(c_n))$. Prove that T is an isomorphism. *Hint:* Use the Lagrange polynomials associated with $c_0, c_1, \ldots, c_n$.

23. Let V denote the vector space defined in Example 5 of Section 1.2, and let $\mathsf{W} = \mathsf{P}(F)$. Define

$$\mathsf{T} \colon \mathsf{V} \to \mathsf{W} \quad \text{by} \quad \mathsf{T}(\sigma) = \sum_{i=0}^{n} \sigma(i) x^i,$$

where n is the largest integer such that $\sigma(n) \neq 0$. Prove that T is an isomorphism.

The following exercise requires familiarity with the concept of *quotient space* defined in Exercise 31 of Section 1.3 and with Exercise 40 of Section 2.1.

24. Let $\mathsf{T} \colon \mathsf{V} \to \mathsf{Z}$ be a linear transformation of a vector space V onto a vector space Z. Define the mapping

$$\overline{\mathsf{T}} \colon \mathsf{V}/\mathsf{N}(\mathsf{T}) \to \mathsf{Z} \quad \text{by} \quad \overline{\mathsf{T}}(v + \mathsf{N}(\mathsf{T})) = \mathsf{T}(v)$$

for any coset $v + \mathsf{N}(\mathsf{T})$ in $\mathsf{V}/\mathsf{N}(\mathsf{T})$.

(a) Prove that $\overline{\mathsf{T}}$ is well-defined; that is, prove that if $v + \mathsf{N}(\mathsf{T}) = v' + \mathsf{N}(\mathsf{T})$, then $\mathsf{T}(v) = \mathsf{T}(v')$.
(b) Prove that $\overline{\mathsf{T}}$ is linear.
(c) Prove that $\overline{\mathsf{T}}$ is an isomorphism.
(d) Prove that the diagram shown in Figure 2.3 commutes; that is, prove that $\mathsf{T} = \overline{\mathsf{T}}\eta$.

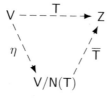

Figure 2.3

25. Let V be a nonzero vector space over a field F, and suppose that S is a basis for V. (By the corollary to Theorem 1.13 (p. 60) in Section 1.7, every vector space has a basis). Let $\mathcal{C}(S, F)$ denote the vector space of all functions $f \in \mathcal{F}(S, F)$ such that $f(s) = 0$ for all but a finite number of vectors in S. (See Exercise 14 of Section 1.3.) Let $\Psi : C(S, F) \to \mathsf{V}$ be defined by $\Psi(f) = 0$ if f is the zero function, and

$$\Psi(f) = \sum_{s \in S, f(s) \neq 0} f(s)s,$$

otherwise. Prove that Ψ is an isomorphism. Thus every nonzero vector space can be viewed as a space of functions.

2.5   THE CHANGE OF COORDINATE MATRIX

In many areas of mathematics, a change of variable is used to simplify the appearance of an expression. For example, in calculus an antiderivative of $2xe^{x^2}$ can be found by making the change of variable $u = x^2$. The resulting expression is of such a simple form that an antiderivative is easily recognized:

$$\int 2xe^{x^2} \, dx = \int e^u \, du = e^u + c = e^{x^2} + c.$$

Similarly, in geometry the change of variable

$$x = \frac{2}{\sqrt{5}}x' - \frac{1}{\sqrt{5}}y'$$

$$y = \frac{1}{\sqrt{5}}x' + \frac{2}{\sqrt{5}}y'$$

can be used to transform the equation $2x^2 - 4xy + 5y^2 = 1$ into the simpler equation $(x')^2 + 6(y')^2 = 1$, in which form it is easily seen to be the equation of an ellipse. (See Figure 2.4.) We see how this change of variable is determined in Section 6.5. Geometrically, the change of variable

$$\begin{pmatrix} x \\ y \end{pmatrix} \to \begin{pmatrix} x' \\ y' \end{pmatrix}$$

is a change in the way that the position of a point P in the plane is described. This is done by introducing a new frame of reference, an $x'y'$-coordinate system with coordinate axes rotated from the original xy-coordinate axes. In this case, the new coordinate axes are chosen to lie in the direction of the axes of the ellipse. The unit vectors along the x'-axis and the y'-axis form an ordered basis

$$\beta' = \left\{ \frac{1}{\sqrt{5}} \begin{pmatrix} 2 \\ 1 \end{pmatrix}, \frac{1}{\sqrt{5}} \begin{pmatrix} -1 \\ 2 \end{pmatrix} \right\}$$

for R^2, and the change of variable is actually a change from $[P]_\beta = \begin{pmatrix} x \\ y \end{pmatrix}$, the coordinate vector of P relative to the standard ordered basis $\beta = \{e_1, e_2\}$, to

$[P]_{\beta'} = \begin{pmatrix} x' \\ y' \end{pmatrix}$, the coordinate vector of P relative to the new rotated basis β'.

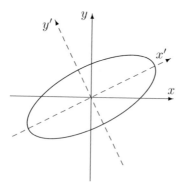

Figure 2.4

A natural question arises: How can a coordinate vector relative to one basis be changed into a coordinate vector relative to the other? Notice that the system of equations relating the new and old coordinates can be represented by the matrix equation

$$\begin{pmatrix} x \\ y \end{pmatrix} = \frac{1}{\sqrt{5}} \begin{pmatrix} 2 & -1 \\ 1 & 2 \end{pmatrix} \begin{pmatrix} x' \\ y' \end{pmatrix}.$$

Notice also that the matrix

$$Q = \frac{1}{\sqrt{5}} \begin{pmatrix} 2 & -1 \\ 1 & 2 \end{pmatrix}$$

equals $[\mathsf{I}]_{\beta'}^{\beta}$, where I denotes the identity transformation on R^2. Thus $[v]_\beta = Q[v]_{\beta'}$ for all $v \in \mathsf{R}^2$. A similar result is true in general.

Theorem 2.22. *Let β and β' be two ordered bases for a finite-dimensional vector space V, and let $Q = [\mathsf{I}]_{\beta'}^{\beta}$. Then*
(a) *Q is invertible.*
(b) *For any $v \in \mathsf{V}$, $[v]_\beta = Q[v]_{\beta'}$.*

Proof. (a) Since I_V is invertible, Q is invertible by Theorem 2.18 (p. 101).
(b) For any $v \in \mathsf{V}$,

$$[v]_\beta = [\mathsf{I}_\mathsf{V}(v)]_\beta = [\mathsf{I}_\mathsf{V}]_{\beta'}^{\beta}[v]_{\beta'} = Q[v]_{\beta'}$$

by Theorem 2.14 (p. 91).   ∎

The matrix $Q = [\mathsf{I}_\mathsf{V}]_{\beta'}^\beta$ defined in Theorem 2.22 is called a **change of coordinate matrix**. Because of part (b) of the theorem, we say that Q **changes β'-coordinates into β-coordinates**. Observe that if $\beta = \{x_1, x_2, \dots, x_n\}$ and $\beta' = \{x_1', x_2', \dots, x_n'\}$, then

$$x_j' = \sum_{i=1}^n Q_{ij} x_i$$

for $j = 1, 2, \dots, n$; that is, the jth column of Q is $[x_j']_\beta$.

Notice that if Q changes β'-coordinates into β-coordinates, then Q^{-1} changes β-coordinates into β'-coordinates. (See Exercise 11.)

Example 1

In R^2, let $\beta = \{(1, 1), (1, -1)\}$ and $\beta' = \{(2, 4), (3, 1)\}$. Since

$$(2, 4) = 3(1, 1) - 1(1, -1) \quad \text{and} \quad (3, 1) = 2(1, 1) + 1(1, -1),$$

the matrix that changes β'-coordinates into β-coordinates is

$$Q = \begin{pmatrix} 3 & 2 \\ -1 & 1 \end{pmatrix}.$$

Thus, for instance,

$$[(2, 4)]_\beta = Q[(2, 4)]_{\beta'} = Q \begin{pmatrix} 1 \\ 0 \end{pmatrix} = \begin{pmatrix} 3 \\ -1 \end{pmatrix}. \quad \blacklozenge$$

For the remainder of this section, we consider only linear transformations that map a vector space V into itself. Such a linear transformation is called a **linear operator** on V. Suppose now that T is a linear operator on a finite-dimensional vector space V and that β and β' are ordered bases for V. Then V can be represented by the matrices $[\mathsf{T}]_\beta$ and $[\mathsf{T}]_{\beta'}$. What is the relationship between these matrices? The next theorem provides a simple answer using a change of coordinate matrix.

Theorem 2.23. *Let* T *be a linear operator on a finite-dimensional vector space* V, *and let* β *and* β' *be ordered bases for* V. *Suppose that* Q *is the change of coordinate matrix that changes* β'-*coordinates into* β-*coordinates. Then*

$$[\mathsf{T}]_{\beta'} = Q^{-1}[\mathsf{T}]_\beta Q.$$

Proof. Let I be the identity transformation on V. Then $\mathsf{T} = \mathsf{IT} = \mathsf{TI}$; hence, by Theorem 2.11 (p. 88),

$$Q[\mathsf{T}]_{\beta'} = [\mathsf{I}]_{\beta'}^\beta [\mathsf{T}]_{\beta'}^{\beta'} = [\mathsf{IT}]_{\beta'}^\beta = [\mathsf{TI}]_{\beta'}^\beta = [\mathsf{T}]_\beta^\beta [\mathsf{I}]_{\beta'}^\beta = [\mathsf{T}]_\beta Q.$$

Therefore $[\mathsf{T}]_{\beta'} = Q^{-1}[\mathsf{T}]_\beta Q.$ $\blacksquare$

Example 2

Let T be the linear operator on R^2 defined by

$$T\begin{pmatrix} a \\ b \end{pmatrix} = \begin{pmatrix} 3a - b \\ a + 3b \end{pmatrix},$$

and let β and β' be the ordered bases in Example 1. The reader should verify that

$$[T]_\beta = \begin{pmatrix} 3 & 1 \\ -1 & 3 \end{pmatrix}.$$

In Example 1, we saw that the change of coordinate matrix that changes β'-coordinates into β-coordinates is

$$Q = \begin{pmatrix} 3 & 2 \\ -1 & 1 \end{pmatrix},$$

and it is easily verified that

$$Q^{-1} = \frac{1}{5}\begin{pmatrix} 1 & -2 \\ 1 & 3 \end{pmatrix}.$$

Hence, by Theorem 2.23,

$$[T]_{\beta'} = Q^{-1}[T]_\beta Q = \begin{pmatrix} 4 & 1 \\ -2 & 2 \end{pmatrix}.$$

To show that this is the correct matrix, we can verify that the image under T of each vector of β' is the linear combination of the vectors of β' with the entries of the corresponding column as its coefficients. For example, the image of the second vector in β' is

$$T\begin{pmatrix} 3 \\ 1 \end{pmatrix} = \begin{pmatrix} 8 \\ 6 \end{pmatrix} = 1\begin{pmatrix} 2 \\ 4 \end{pmatrix} + 2\begin{pmatrix} 3 \\ 1 \end{pmatrix}.$$

Notice that the coefficients of the linear combination are the entries of the second column of $[T]_{\beta'}$.   ◆

It is often useful to apply Theorem 2.23 to compute $[T]_\beta$, as the next example shows.

Example 3

Recall the reflection about the x-axis in Example 3 of Section 2.1. The rule $(x, y) \rightarrow (x, -y)$ is easy to obtain. We now derive the less obvious rule for the reflection T about the line $y = 2x$. (See Figure 2.5.) We wish to find an expression for $T(a, b)$ for any (a, b) in R^2. Since T is linear, it is completely

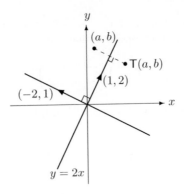

Figure 2.5

determined by its values on a basis for R^2. Clearly, $T(1,2) = (1,2)$ and $T(-2,1) = -(-2,1) = (2,-1)$. Therefore if we let

$$\beta' = \left\{ \begin{pmatrix} 1 \\ 2 \end{pmatrix}, \begin{pmatrix} -2 \\ 1 \end{pmatrix} \right\},$$

then β' is an ordered basis for R^2 and

$$[T]_{\beta'} = \begin{pmatrix} 1 & 0 \\ 0 & -1 \end{pmatrix}.$$

Let β be the standard ordered basis for R^2, and let Q be the matrix that changes β'-coordinates into β-coordinates. Then

$$Q = \begin{pmatrix} 1 & -2 \\ 2 & 1 \end{pmatrix}$$

and $Q^{-1}[T]_\beta Q = [T]_{\beta'}$. We can solve this equation for $[T]_\beta$ to obtain that $[T]_\beta = Q[T]_{\beta'} Q^{-1}$. Because

$$Q^{-1} = \frac{1}{5} \begin{pmatrix} 1 & 2 \\ -2 & 1 \end{pmatrix},$$

the reader can verify that

$$[T]_\beta = \frac{1}{5} \begin{pmatrix} -3 & 4 \\ 4 & 3 \end{pmatrix}.$$

Since β is the standard ordered basis, it follows that T is left-multiplication by $[T]_\beta$. Thus for any (a,b) in R^2, we have

$$T \begin{pmatrix} a \\ b \end{pmatrix} = \frac{1}{5} \begin{pmatrix} -3 & 4 \\ 4 & 3 \end{pmatrix} \begin{pmatrix} a \\ b \end{pmatrix} = \frac{1}{5} \begin{pmatrix} -3a + 4b \\ 4a + 3b \end{pmatrix}. \quad \blacklozenge$$

A useful special case of Theorem 2.23 is contained in the next corollary, whose proof is left as an exercise.

Corollary. *Let $A \in \mathsf{M}_{n\times n}(F)$, and let γ be an ordered basis for F^n. Then $[\mathsf{L}_A]_\gamma = Q^{-1}AQ$, where Q is the $n \times n$ matrix whose jth column is the jth vector of γ.*

Example 4

Let

$$A = \begin{pmatrix} 2 & 1 & 0 \\ 1 & 1 & 3 \\ 0 & -1 & 0 \end{pmatrix},$$

and let

$$\gamma = \left\{ \begin{pmatrix} -1 \\ 0 \\ 0 \end{pmatrix}, \begin{pmatrix} 2 \\ 1 \\ 0 \end{pmatrix}, \begin{pmatrix} 1 \\ 1 \\ 1 \end{pmatrix} \right\},$$

which is an ordered basis for R^3. Let Q be the 3×3 matrix whose jth column is the jth vector of γ. Then

$$Q = \begin{pmatrix} -1 & 2 & 1 \\ 0 & 1 & 1 \\ 0 & 0 & 1 \end{pmatrix} \quad \text{and} \quad Q^{-1} = \begin{pmatrix} -1 & 2 & -1 \\ 0 & 1 & -1 \\ 0 & 0 & 1 \end{pmatrix}.$$

So by the preceding corollary,

$$[\mathsf{L}_A]_\gamma = Q^{-1}AQ = \begin{pmatrix} 0 & 2 & 8 \\ -1 & 4 & 6 \\ 0 & -1 & -1 \end{pmatrix}. \quad \blacklozenge$$

The relationship between the matrices $[\mathsf{T}]_{\beta'}$ and $[\mathsf{T}]_\beta$ in Theorem 2.23 will be the subject of further study in Chapters 5, 6, and 7. At this time, however, we introduce the name for this relationship.

Definition. *Let A and B be matrices in $\mathsf{M}_{n\times n}(F)$. We say that B is* **similar** *to A if there exists an invertible matrix Q such that $B = Q^{-1}AQ$.*

Observe that the relation of similarity is an equivalence relation (see Exercise 9). So we need only say that A and B are similar.

Notice also that in this terminology Theorem 2.23 can be stated as follows: If T is a linear operator on a finite-dimensional vector space V, and if β and β' are any ordered bases for V, then $[\mathsf{T}]_{\beta'}$ is similar to $[\mathsf{T}]_\beta$.

Theorem 2.23 can be generalized to allow $\mathsf{T}\colon \mathsf{V} \to \mathsf{W}$, where V is distinct from W. In this case, we can change bases in V as well as in W (see Exercise 8).

EXERCISES

1. Label the following statements as true or false.

   (a) Suppose that $\beta = \{x_1, x_2, \ldots, x_n\}$ and $\beta' = \{x'_1, x'_2, \ldots, x'_n\}$ are ordered bases for a vector space and Q is the change of coordinate matrix that changes β'-coordinates into β-coordinates. Then the jth column of Q is $[x_j]_{\beta'}$.

   (b) Every change of coordinate matrix is invertible.

   (c) Let T be a linear operator on a finite-dimensional vector space V, let β and β' be ordered bases for V, and let Q be the change of coordinate matrix that changes β'-coordinates into β-coordinates. Then $[\mathsf{T}]_\beta = Q[\mathsf{T}]_{\beta'}Q^{-1}$.

   (d) The matrices $A, B \in \mathsf{M}_{n \times n}(F)$ are called similar if $B = Q^t AQ$ for some $Q \in \mathsf{M}_{n \times n}(F)$.

   (e) Let T be a linear operator on a finite-dimensional vector space V. Then for any ordered bases β and γ for V, $[\mathsf{T}]_\beta$ is similar to $[\mathsf{T}]_\gamma$.

2. For each of the following pairs of ordered bases β and β' for R^2, find the change of coordinate matrix that changes β'-coordinates into β-coordinates.

   (a) $\beta = \{e_1, e_2\}$ and $\beta' = \{(a_1, a_2), (b_1, b_2)\}$
   (b) $\beta = \{(-1, 3), (2, -1)\}$ and $\beta' = \{(0, 10), (5, 0)\}$
   (c) $\beta = \{(2, 5), (-1, -3)\}$ and $\beta' = \{e_1, e_2\}$
   (d) $\beta = \{(-4, 3), (2, -1)\}$ and $\beta' = \{(2, 1), (-4, 1)\}$

3. For each of the following pairs of ordered bases β and β' for $\mathsf{P}_2(R)$, find the change of coordinate matrix that changes β'-coordinates into β-coordinates.

   (a) $\beta = \{x^2, x, 1\}$ and
       $\beta' = \{a_2 x^2 + a_1 x + a_0, b_2 x^2 + b_1 x + b_0, c_2 x^2 + c_1 x + c_0\}$
   (b) $\beta = \{1, x, x^2\}$ and
       $\beta' = \{a_2 x^2 + a_1 x + a_0, b_2 x^2 + b_1 x + b_0, c_2 x^2 + c_1 x + c_0\}$
   (c) $\beta = \{2x^2 - x, 3x^2 + 1, x^2\}$ and $\beta' = \{1, x, x^2\}$
   (d) $\beta = \{x^2 - x + 1, x + 1, x^2 + 1\}$ and
       $\beta' = \{x^2 + x + 4, 4x^2 - 3x + 2, 2x^2 + 3\}$
   (e) $\beta = \{x^2 - x, x^2 + 1, x - 1\}$ and
       $\beta' = \{5x^2 - 2x - 3, -2x^2 + 5x + 5, 2x^2 - x - 3\}$
   (f) $\beta = \{2x^2 - x + 1, x^2 + 3x - 2, -x^2 + 2x + 1\}$ and
       $\beta' = \{9x - 9, x^2 + 21x - 2, 3x^2 + 5x + 2\}$

4. Let T be the linear operator on R^2 defined by

$$\mathsf{T}\begin{pmatrix} a \\ b \end{pmatrix} = \begin{pmatrix} 2a + b \\ a - 3b \end{pmatrix},$$

let β be the standard ordered basis for R^2, and let

$$\beta' = \left\{ \begin{pmatrix} 1 \\ 1 \end{pmatrix}, \begin{pmatrix} 1 \\ 2 \end{pmatrix} \right\}.$$

Use Theorem 2.23 and the fact that

$$\begin{pmatrix} 1 & 1 \\ 1 & 2 \end{pmatrix}^{-1} = \begin{pmatrix} 2 & -1 \\ -1 & 1 \end{pmatrix}$$

to find $[T]_{\beta'}$.

5. Let T be the linear operator on $P_1(R)$ defined by $T(p(x)) = p'(x)$, the derivative of $p(x)$. Let $\beta = \{1, x\}$ and $\beta' = \{1 + x, 1 - x\}$. Use Theorem 2.23 and the fact that

$$\begin{pmatrix} 1 & 1 \\ 1 & -1 \end{pmatrix}^{-1} = \begin{pmatrix} \frac{1}{2} & \frac{1}{2} \\ \frac{1}{2} & -\frac{1}{2} \end{pmatrix}$$

to find $[T]_{\beta'}$.

6. For each matrix A and ordered basis β, find $[L_A]_\beta$. Also, find an invertible matrix Q such that $[L_A]_\beta = Q^{-1}AQ$.

(a) $A = \begin{pmatrix} 1 & 3 \\ 1 & 1 \end{pmatrix}$ and $\beta = \left\{ \begin{pmatrix} 1 \\ 1 \end{pmatrix}, \begin{pmatrix} 1 \\ 2 \end{pmatrix} \right\}$

(b) $A = \begin{pmatrix} 1 & 2 \\ 2 & 1 \end{pmatrix}$ and $\beta = \left\{ \begin{pmatrix} 1 \\ 1 \end{pmatrix}, \begin{pmatrix} 1 \\ -1 \end{pmatrix} \right\}$

(c) $A = \begin{pmatrix} 1 & 1 & -1 \\ 2 & 0 & 1 \\ 1 & 1 & 0 \end{pmatrix}$ and $\beta = \left\{ \begin{pmatrix} 1 \\ 1 \\ 1 \end{pmatrix}, \begin{pmatrix} 1 \\ 0 \\ 1 \end{pmatrix}, \begin{pmatrix} 1 \\ 1 \\ 2 \end{pmatrix} \right\}$

(d) $A = \begin{pmatrix} 13 & 1 & 4 \\ 1 & 13 & 4 \\ 4 & 4 & 10 \end{pmatrix}$ and $\beta = \left\{ \begin{pmatrix} 1 \\ 1 \\ -2 \end{pmatrix}, \begin{pmatrix} 1 \\ -1 \\ 0 \end{pmatrix}, \begin{pmatrix} 1 \\ 1 \\ 1 \end{pmatrix} \right\}$

7. In R^2, let L be the line $y = mx$, where $m \neq 0$. Find an expression for $T(x, y)$, where

(a) T is the reflection of R^2 about L.
(b) T is the projection on L along the line perpendicular to L. (See the definition of *projection* in the exercises of Section 2.1.)

8. Prove the following generalization of Theorem 2.23. Let $T: V \rightarrow W$ be a linear transformation from a finite-dimensional vector space V to a finite-dimensional vector space W. Let β and β' be ordered bases for

V, and let γ and γ' be ordered bases for W. Then $[\mathsf{T}]_{\beta'}^{\gamma'} = P^{-1}[\mathsf{T}]_\beta^\gamma Q$, where Q is the matrix that changes β'-coordinates into β-coordinates and P is the matrix that changes γ'-coordinates into γ-coordinates.

9. Prove that "is similar to" is an equivalence relation on $\mathsf{M}_{n\times n}(F)$.

10. Prove that if A and B are similar $n \times n$ matrices, then $\mathrm{tr}(A) = \mathrm{tr}(B)$. *Hint:* Use Exercise 13 of Section 2.3.

11. Let V be a finite-dimensional vector space with ordered bases α, β, and γ.

   (a) Prove that if Q and R are the change of coordinate matrices that change α-coordinates into β-coordinates and β-coordinates into γ-coordinates, respectively, then RQ is the change of coordinate matrix that changes α-coordinates into γ-coordinates.

   (b) Prove that if Q changes α-coordinates into β-coordinates, then Q^{-1} changes β-coordinates into α-coordinates.

12. Prove the corollary to Theorem 2.23.

13.† Let V be a finite-dimensional vector space over a field F, and let $\beta = \{x_1, x_2, \ldots, x_n\}$ be an ordered basis for V. Let Q be an $n \times n$ invertible matrix with entries from F. Define

$$x'_j = \sum_{i=1}^n Q_{ij} x_i \quad \text{for } 1 \le j \le n,$$

and set $\beta' = \{x'_1, x'_2, \ldots, x'_n\}$. Prove that β' is a basis for V and hence that Q is the change of coordinate matrix changing β'-coordinates into β-coordinates.

14. Prove the converse of Exercise 8: If A and B are each $m \times n$ matrices with entries from a field F, and if there exist invertible $m \times m$ and $n \times n$ matrices P and Q, respectively, such that $B = P^{-1}AQ$, then there exist an n-dimensional vector space V and an m-dimensional vector space W (both over F), ordered bases β and β' for V and γ and γ' for W, and a linear transformation $\mathsf{T} \colon \mathsf{V} \to \mathsf{W}$ such that

$$A = [\mathsf{T}]_\beta^\gamma \quad \text{and} \quad B = [\mathsf{T}]_{\beta'}^{\gamma'}.$$

Hints: Let $\mathsf{V} = F^n$, $\mathsf{W} = F^m$, $\mathsf{T} = \mathsf{L}_A$, and β and γ be the standard ordered bases for F^n and F^m, respectively. Now apply the results of Exercise 13 to obtain ordered bases β' and γ' from β and γ via Q and P, respectively.

2.6* DUAL SPACES

In this section, we are concerned exclusively with linear transformations from a vector space V into its field of scalars F, which is itself a vector space of dimension 1 over F. Such a linear transformation is called a **linear functional** on V. We generally use the letters $\mathsf{f}, \mathsf{g}, \mathsf{h}, \ldots$ to denote linear functionals. As we see in Example 1, the definite integral provides us with one of the most important examples of a linear functional in mathematics.

Example 1

Let V be the vector space of continuous real-valued functions on the interval $[0, 2\pi]$. Fix a function $g \in \mathsf{V}$. The function $\mathsf{h} \colon \mathsf{V} \to R$ defined by

$$\mathsf{h}(x) = \frac{1}{2\pi} \int_0^{2\pi} x(t) g(t) \, dt$$

is a linear functional on V. In the cases that $g(t)$ equals $\sin nt$ or $\cos nt$, $\mathsf{h}(x)$ is often called the **nth Fourier coefficient of** x.   ♦

Example 2

Let $\mathsf{V} = \mathsf{M}_{n \times n}(F)$, and define $\mathsf{f} \colon \mathsf{V} \to F$ by $\mathsf{f}(A) = \mathrm{tr}(A)$, the trace of A. By Exercise 6 of Section 1.3, we have that f is a linear functional.   ♦

Example 3

Let V be a finite-dimensional vector space, and let $\beta = \{x_1, x_2, \ldots, x_n\}$ be an ordered basis for V. For each $i = 1, 2, \ldots, n$, define $\mathsf{f}_i(x) = a_i$, where

$$[x]_\beta = \begin{pmatrix} a_1 \\ a_2 \\ \vdots \\ a_n \end{pmatrix}$$

is the coordinate vector of x relative to β. Then f_i is a linear functional on V called the **ith coordinate function with respect to the basis** β. Note that $\mathsf{f}_i(x_j) = \delta_{ij}$, where δ_{ij} is the Kronecker delta. These linear functionals play an important role in the theory of dual spaces (see Theorem 2.24).   ♦

   Definition. *For a vector space V over F, we define the **dual space** of V to be the vector space $\mathcal{L}(\mathsf{V}, F)$, denoted by V^*.*

   Thus V^* is the vector space consisting of all linear functionals on V with the operations of addition and scalar multiplication as defined in Section 2.2. Note that if V is finite-dimensional, then by the corollary to Theorem 2.20 (p. 104)

$$\dim(\mathsf{V}^*) = \dim(\mathcal{L}(\mathsf{V}, F)) = \dim(\mathsf{V}) \cdot \dim(F) = \dim(\mathsf{V}).$$

Hence by Theorem 2.19 (p. 103), V and V* are isomorphic. We also define the **double dual** V** of V to be the dual of V*. In Theorem 2.26, we show, in fact, that there is a natural identification of V and V** in the case that V is finite-dimensional.

Theorem 2.24. *Suppose that* V *is a finite-dimensional vector space with the ordered basis* $\beta = \{x_1, x_2, \ldots, x_n\}$. *Let* f_i $(1 \leq i \leq n)$ *be the ith coordinate function with respect to* β *as just defined, and let* $\beta^* = \{f_1, f_2, \ldots, f_n\}$. *Then* β^* *is an ordered basis for* V*, *and, for any* $f \in$ V*, *we have*

$$f = \sum_{i=1}^{n} f(x_i)f_i.$$

Proof. Let $f \in$ V*. Since $\dim(V^*) = n$, we need only show that

$$f = \sum_{i=1}^{n} f(x_i)f_i,$$

from which it follows that β^* generates V*, and hence is a basis by Corollary 2(a) to the replacement theorem (p. 47). Let

$$g = \sum_{i=1}^{n} f(x_i)f_i.$$

For $1 \leq j \leq n$, we have

$$g(x_j) = \left(\sum_{i=1}^{n} f(x_i)f_i \right)(x_j) = \sum_{i=1}^{n} f(x_i)f_i(x_j)$$

$$= \sum_{i=1}^{n} f(x_i)\delta_{ij} = f(x_j).$$

Therefore $f = g$ by the corollary to Theorem 2.6 (p. 72). ∎

Definition. *Using the notation of Theorem 2.24, we call the ordered basis* $\beta^* = \{f_1, f_2, \ldots, f_n\}$ *of* V* *that satisfies* $f_i(x_j) = \delta_{ij}$ $(1 \leq i, j \leq n)$ *the* **dual basis** *of* β.

Example 4

Let $\beta = \{(2, 1), (3, 1)\}$ be an ordered basis for R^2. Suppose that the dual basis of β is given by $\beta^* = \{f_1, f_2\}$. To explicitly determine a formula for f_1, we need to consider the equations

$$1 = f_1(2, 1) = f_1(2e_1 + e_2) = 2f_1(e_1) + f_1(e_2)$$
$$0 = f_1(3, 1) = f_1(3e_1 + e_2) = 3f_1(e_1) + f_1(e_2).$$

Solving these equations, we obtain $f_1(e_1) = -1$ and $f_1(e_2) = 3$; that is, $f_1(x, y) = -x + 3y$. Similarly, it can be shown that $f_2(x, y) = x - 2y$. ◆

We now assume that V and W are finite-dimensional vector spaces over F with ordered bases β and γ, respectively. In Section 2.4, we proved that there is a one-to-one correspondence between linear transformations $T: V \rightarrow W$ and $m \times n$ matrices (over F) via the correspondence $T \leftrightarrow [T]_\beta^\gamma$. For a matrix of the form $A = [T]_\beta^\gamma$, the question arises as to whether or not there exists a linear transformation U associated with T in some natural way such that U may be represented in some basis as A^t. Of course, if $m \neq n$, it would be impossible for U to be a linear transformation from V into W. We now answer this question by applying what we have already learned about dual spaces.

Theorem 2.25. *Let V and W be finite-dimensional vector spaces over F with ordered bases β and γ, respectively. For any linear transformation $T: V \rightarrow W$, the mapping $T^t: W^* \rightarrow V^*$ defined by $T^t(g) = gT$ for all $g \in W^*$ is a linear transformation with the property that $[T^t]_{\gamma^*}^{\beta^*} = ([T]_\beta^\gamma)^t$.*

Proof. For $g \in W^*$, it is clear that $T^t(g) = gT$ is a linear functional on V and hence is in V^*. Thus T^t maps W^* into V^*. We leave the proof that T^t is linear to the reader.

To complete the proof, let $\beta = \{x_1, x_2, \ldots, x_n\}$ and $\gamma = \{y_1, y_2, \ldots, y_m\}$ with dual bases $\beta^* = \{f_1, f_2, \ldots, f_n\}$ and $\gamma^* = \{g_1, g_2, \ldots, g_m\}$, respectively. For convenience, let $A = [T]_\beta^\gamma$. To find the jth column of $[T^t]_{\gamma^*}^{\beta^*}$, we begin by expressing $T^t(g_j)$ as a linear combination of the vectors of β^*. By Theorem 2.24, we have

$$T^t(g_j) = g_j T = \sum_{s=1}^n (g_j T)(x_s) f_s.$$

So the row i, column j entry of $[T^t]_{\gamma^*}^{\beta^*}$ is

$$(g_j T)(x_i) = g_j(T(x_i)) = g_j \left(\sum_{k=1}^m A_{ki} y_k \right)$$

$$= \sum_{k=1}^m A_{ki} g_j(y_k) = \sum_{k=1}^m A_{ki} \delta_{jk} = A_{ji}.$$

Hence $[T^t]_{\gamma^*}^{\beta^*} = A^t$.   ∎

The linear transformation T^t defined in Theorem 2.25 is called the **transpose** of T. It is clear that T^t is the unique linear transformation U such that $[U]_{\gamma^*}^{\beta^*} = ([T]_\beta^\gamma)^t$.

We illustrate Theorem 2.25 with the next example.

Example 5

Define $T: P_1(R) \to R^2$ by $T(p(x)) = (p(0), p(2))$. Let β and γ be the standard ordered bases for $P_1(R)$ and R^2, respectively. Clearly,

$$[T]_\beta^\gamma = \begin{pmatrix} 1 & 0 \\ 1 & 2 \end{pmatrix}.$$

We compute $[T^t]_{\gamma^*}^{\beta^*}$ directly from the definition. Let $\beta^* = \{f_1, f_2\}$ and $\gamma^* = \{g_1, g_2\}$. Suppose that $[T^t]_{\gamma^*}^{\beta^*} = \begin{pmatrix} a & b \\ c & d \end{pmatrix}$. Then $T^t(g_1) = af_1 + cf_2$. So

$$T^t(g_1)(1) = (af_1 + cf_2)(1) = af_1(1) + cf_2(1) = a(1) + c(0) = a.$$

But also

$$(T^t(g_1))(1) = g_1(T(1)) = g_1(1, 1) = 1.$$

So $a = 1$. Using similar computations, we obtain that $c = 0$, $b = 1$, and $d = 2$. Hence a direct computation yields

$$[T^t]_{\gamma^*}^{\beta^*} = \begin{pmatrix} 1 & 1 \\ 0 & 2 \end{pmatrix} = \left([T]_\beta^\gamma\right)^t,$$

as predicted by Theorem 2.25. ◆

We now concern ourselves with demonstrating that any finite-dimensional vector space V can be identified in a natural way with its double dual V^{**}. There is, in fact, an isomorphism between V and V^{**} that does not depend on any choice of bases for the two vector spaces.

For a vector $x \in V$, we define $\hat{x}: V^* \to F$ by $\hat{x}(f) = f(x)$ for every $f \in V^*$. It is easy to verify that $\hat{x}$ is a linear functional on V^*, so $\hat{x} \in V^{**}$. The correspondence $x \leftrightarrow \hat{x}$ allows us to define the desired isomorphism between V and V^{**}.

Lemma. Let V be a finite-dimensional vector space, and let $x \in V$. If $\hat{x}(f) = 0$ for all $f \in V^*$, then $x = 0$.

Proof. Let $x \neq 0$. We show that there exists $f \in V^*$ such that $\hat{x}(f) \neq 0$. Choose an ordered basis $\beta = \{x_1, x_2, \ldots, x_n\}$ for V such that $x_1 = x$. Let $\{f_1, f_2, \ldots, f_n\}$ be the dual basis of β. Then $f_1(x_1) = 1 \neq 0$. Let $f = f_1$. ∎

Theorem 2.26. Let V be a finite-dimensional vector space, and define $\psi: V \to V^{**}$ by $\psi(x) = \hat{x}$. Then ψ is an isomorphism.

Proof. (a) ψ is linear: Let $x, y \in V$ and $c \in F$. For $f \in V^*$, we have

$$\psi(cx + y)(f) = f(cx + y) = cf(x) + f(y) = c\widehat{x}(f) + \widehat{y}(f)$$
$$= (c\widehat{x} + \widehat{y})(f).$$

Therefore

$$\psi(cx + y) = c\widehat{x} + \widehat{y} = c\psi(x) + \psi(y).$$

(b) ψ is one-to-one: Suppose that $\psi(x)$ is the zero functional on V^* for some $x \in V$. Then $\widehat{x}(f) = 0$ for every $f \in V^*$. By the previous lemma, we conclude that $x = 0$.

(c) ψ is an isomorphism: This follows from (b) and the fact that $\dim(V) = \dim(V^{**})$. ∎

Corollary. *Let* V *be a finite-dimensional vector space with dual space* V^*. *Then every ordered basis for* V^* *is the dual basis for some basis for* V.

Proof. Let $\{f_1, f_2, \ldots, f_n\}$ be an ordered basis for V^*. We may combine Theorems 2.24 and 2.26 to conclude that for this basis for V^* there exists a dual basis $\{\widehat{x}_1, \widehat{x}_2, \ldots, \widehat{x}_n\}$ in V^{**}, that is, $\delta_{ij} = \widehat{x}_i(f_j) = f_j(x_i)$ for all i and j. Thus $\{f_1, f_2, \ldots, f_n\}$ is the dual basis of $\{x_1, x_2, \ldots, x_n\}$. ∎

Although many of the ideas of this section, (e.g., the existence of a dual space), can be extended to the case where V is not finite-dimensional, only a finite-dimensional vector space is isomorphic to its double dual via the map $x \rightarrow \widehat{x}$. In fact, for infinite-dimensional vector spaces, no two of V, V^*, and V^{**} are isomorphic.

EXERCISES

1. Label the following statements as true or false. Assume that all vector spaces are finite-dimensional.

    (a) Every linear transformation is a linear functional.
    (b) A linear functional defined on a field may be represented as a 1×1 matrix.
    (c) Every vector space is isomorphic to its dual space.
    (d) Every vector space is the dual of some other vector space.
    (e) If T is an isomorphism from V onto V^* and β is a finite ordered basis for V, then $T(\beta) = \beta^*$.
    (f) If T is a linear transformation from V to W, then the domain of $(T^t)^t$ is V^{**}.
    (g) If V is isomorphic to W, then V^* is isomorphic to W^*.

(h) The derivative of a function may be considered as a linear functional on the vector space of differentiable functions.

2. For the following functions f on a vector space V, determine which are linear functionals.

(a) $V = P(R)$; $f(p(x)) = 2p'(0) + p''(1)$, where $'$ denotes differentiation
(b) $V = R^2$; $f(x, y) = (2x, 4y)$
(c) $V = M_{2 \times 2}(F)$; $f(A) = \text{tr}(A)$
(d) $V = R^3$; $f(x, y, z) = x^2 + y^2 + z^2$
(e) $V = P(R)$; $f(p(x)) = \int_0^1 p(t)\, dt$
(f) $V = M_{2 \times 2}(F)$; $f(A) = A_{11}$

3. For each of the following vector spaces V and bases β, find explicit formulas for vectors of the dual basis β^* for V^*, as in Example 4.

(a) $V = R^3$; $\beta = \{(1, 0, 1), (1, 2, 1), (0, 0, 1)\}$
(b) $V = P_2(R)$; $\beta = \{1, x, x^2\}$

4. Let $V = R^3$, and define $f_1, f_2, f_3 \in V^*$ as follows:

$$f_1(x, y, z) = x - 2y, \quad f_2(x, y, z) = x + y + z, \quad f_3(x, y, z) = y - 3z.$$

Prove that $\{f_1, f_2, f_3\}$ is a basis for V^*, and then find a basis for V for which it is the dual basis.

5. Let $V = P_1(R)$, and, for $p(x) \in V$, define $f_1, f_2 \in V^*$ by

$$f_1(p(x)) = \int_0^1 p(t)\, dt \quad \text{and} \quad f_2(p(x)) = \int_0^2 p(t)\, dt.$$

Prove that $\{f_1, f_2\}$ is a basis for V^*, and find a basis for V for which it is the dual basis.

6. Define $f \in (R^2)^*$ by $f(x, y) = 2x + y$ and $T: R^2 \to R^2$ by $T(x, y) = (3x + 2y, x)$.

(a) Compute $T^t(f)$.
(b) Compute $[T^t]_{\beta^*}$, where β is the standard ordered basis for R^2 and $\beta^* = \{f_1, f_2\}$ is the dual basis, by finding scalars a, b, c, and d such that $T^t(f_1) = af_1 + cf_2$ and $T^t(f_2) = bf_1 + df_2$.
(c) Compute $[T]_\beta$ and $([T]_\beta)^t$, and compare your results with (b).

7. Let $V = P_1(R)$ and $W = R^2$ with respective standard ordered bases β and γ. Define $T: V \to W$ by

$$T(p(x)) = (p(0) - 2p(1), p(0) + p'(0)),$$

where $p'(x)$ is the derivative of $p(x)$.

(a) For $f \in W^*$ defined by $f(a, b) = a - 2b$, compute $T^t(f)$.

(b) Compute $[T^t]_{\gamma^*}^{\beta^*}$ without appealing to Theorem 2.25.

(c) Compute $[T]_{\beta}^{\gamma}$ and its transpose, and compare your results with (b).

8. Show that every plane through the origin in R^3 may be identified with the null space of a vector in $(R^3)^*$. State an analogous result for R^2.

9. Prove that a function $T: F^n \to F^m$ is linear if and only if there exist $f_1, f_2, \ldots, f_m \in (F^n)^*$ such that $T(x) = (f_1(x), f_2(x), \ldots, f_m(x))$ for all $x \in F^n$. *Hint:* If T is linear, define $f_i(x) = (g_i T)(x)$ for $x \in F^n$; that is, $f_i = T^t(g_i)$ for $1 \le i \le m$, where $\{g_1, g_2, \ldots, g_m\}$ is the dual basis of the standard ordered basis for F^m.

10. Let $V = P_n(F)$, and let $c_0, c_1, \ldots, c_n$ be distinct scalars in F.

(a) For $0 \le i \le n$, define $f_i \in V^*$ by $f_i(p(x)) = p(c_i)$. Prove that $\{f_0, f_1, \ldots, f_n\}$ is a basis for V^*. *Hint:* Apply any linear combination of this set that equals the zero transformation to $p(x) = (x - c_1)(x - c_2) \cdots (x - c_n)$, and deduce that the first coefficient is zero.

(b) Use the corollary to Theorem 2.26 and (a) to show that there exist unique polynomials $p_0(x), p_1(x), \ldots, p_n(x)$ such that $p_i(c_j) = \delta_{ij}$ for $0 \le i \le n$. These polynomials are the Lagrange polynomials defined in Section 1.6.

(c) For any scalars $a_0, a_1, \ldots, a_n$ (not necessarily distinct), deduce that there exists a unique polynomial $q(x)$ of degree at most n such that $q(c_i) = a_i$ for $0 \le i \le n$. In fact,

$$q(x) = \sum_{i=0}^{n} a_i p_i(x).$$

(d) Deduce the Lagrange interpolation formula:

$$p(x) = \sum_{i=0}^{n} p(c_i) p_i(x)$$

for any $p(x) \in V$.

(e) Prove that

$$\int_a^b p(t)\, dt = \sum_{i=0}^{n} p(c_i) d_i,$$

where

$$d_i = \int_a^b p_i(t)\, dt.$$

Suppose now that

$$c_i = a + \frac{i(b-a)}{n} \quad \text{for } i = 0, 1, \dots, n.$$

For $n = 1$, the preceding result yields the trapezoidal rule for evaluating the definite integral of a polynomial. For $n = 2$, this result yields Simpson's rule for evaluating the definite integral of a polynomial.

11. Let V and W be finite-dimensional vector spaces over F, and let ψ_1 and ψ_2 be the isomorphisms between V and V^{**} and W and W^{**}, respectively, as defined in Theorem 2.26. Let $\mathsf{T}\colon \mathsf{V} \to \mathsf{W}$ be linear, and define $\mathsf{T}^{tt} = (\mathsf{T}^t)^t$. Prove that the diagram depicted in Figure 2.6 commutes (i.e., prove that $\psi_2\mathsf{T} = \mathsf{T}^{tt}\psi_1$).

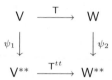

Figure 2.6

12. Let V be a finite-dimensional vector space with the ordered basis β. Prove that $\psi(\beta) = \beta^{**}$, where ψ is defined in Theorem 2.26.

In Exercises 13 through 17, V denotes a finite-dimensional vector space over F. For every subset S of V, define the **annihilator** S^0 of S as

$$S^0 = \{\mathsf{f} \in \mathsf{V}^* : \mathsf{f}(x) = 0 \text{ for all } x \in S\}.$$

13. **(a)** Prove that S^0 is a subspace of V^*.
  (b) If W is a subspace of V and $x \notin \mathsf{W}$, prove that there exists $\mathsf{f} \in \mathsf{W}^0$ such that $\mathsf{f}(x) \neq 0$.
  (c) Prove that $(S^0)^0 = \text{span}(\psi(S))$, where ψ is defined as in Theorem 2.26.
  (d) For subspaces W_1 and W_2, prove that $\mathsf{W}_1 = \mathsf{W}_2$ if and only if $\mathsf{W}_1^0 = \mathsf{W}_2^0$.
  (e) For subspaces W_1 and W_2, show that $(\mathsf{W}_1 + \mathsf{W}_2)^0 = \mathsf{W}_1^0 \cap \mathsf{W}_2^0$.

14. Prove that if W is a subspace of V, then $\dim(\mathsf{W}) + \dim(\mathsf{W}^0) = \dim(\mathsf{V})$. *Hint:* Extend an ordered basis $\{x_1, x_2, \dots, x_k\}$ of W to an ordered basis $\beta = \{x_1, x_2, \dots, x_n\}$ of V. Let $\beta^* = \{\mathsf{f}_1, \mathsf{f}_2, \dots, \mathsf{f}_n\}$. Prove that $\{\mathsf{f}_{k+1}, \mathsf{f}_{k+2}, \dots, \mathsf{f}_n\}$ is a basis for W^0.

15. Suppose that W is a finite-dimensional vector space and that $T: V \to W$ is linear. Prove that $N(T^t) = (R(T))^0$.

16. Use Exercises 14 and 15 to deduce that $\text{rank}(L_{A^t}) = \text{rank}(L_A)$ for any $A \in M_{m \times n}(F)$.

17. Let T be a linear operator on V, and let W be a subspace of V. Prove that W is T-invariant (as defined in the exercises of Section 2.1) if and only if W^0 is T^t-invariant.

18. Let V be a nonzero vector space over a field F, and let S be a basis for V. (By the corollary to Theorem 1.13 (p. 60) in Section 1.7, every vector space has a basis.) Let $\Phi: V^* \to \mathcal{L}(S, F)$ be the mapping defined by $\Phi(f) = f_S$, the restriction of f to S. Prove that Φ is an isomorphism. *Hint:* Apply Exercise 34 of Section 2.1.

19. Let V be a nonzero vector space, and let W be a proper subspace of V (i.e., $W \neq V$). Prove that there exists a nonzero linear functional $f \in V^*$ such that $f(x) = 0$ for all $x \in W$. *Hint:* For the infinite-dimensional case, use Exercise 34 of Section 2.1 as well as results about extending linearly independent sets to bases in Section 1.7.

20. Let V and W be nonzero vector spaces over the same field, and let $T: V \to W$ be a linear transformation.

   (a) Prove that T is onto if and only if T^t is one-to-one.
   (b) Prove that T^t is onto if and only if T is one-to-one.

   Hint: Parts of the proof require the result of Exercise 19 for the infinite-dimensional case.

2.7* HOMOGENEOUS LINEAR DIFFERENTIAL EQUATIONS WITH CONSTANT COEFFICIENTS

As an introduction to this section, consider the following physical problem. A weight of mass m is attached to a vertically suspended spring that is allowed to stretch until the forces acting on the weight are in equilibrium. Suppose that the weight is now motionless and impose an xy-coordinate system with the weight at the origin and the spring lying on the positive y-axis (see Figure 2.7).

Suppose that at a certain time, say $t = 0$, the weight is lowered a distance s along the y-axis and released. The spring then begins to oscillate.

We describe the motion of the spring. At any time $t \geq 0$, let $F(t)$ denote the force acting on the weight and $y(t)$ denote the position of the weight along the y-axis. For example, $y(0) = -s$. The second derivative of y with respect

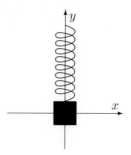

Figure 2.7

to time, $y''(t)$, is the acceleration of the weight at time t; hence, by Newton's second law of motion,

$$F(t) = my''(t). \tag{1}$$

It is reasonable to assume that the force acting on the weight is due totally to the tension of the spring, and that this force satisfies Hooke's law: *The force acting on the weight is proportional to its displacement from the equilibrium position, but acts in the opposite direction.* If $k > 0$ is the proportionality constant, then Hooke's law states that

$$F(t) = -ky(t). \tag{2}$$

Combining (1) and (2), we obtain $my'' = -ky$ or

$$y'' + \frac{k}{m}y = 0. \tag{3}$$

The expression (3) is an example of a *differential equation*. A **differential equation** in an unknown function $y = y(t)$ is an equation involving y, t, and derivatives of y. If the differential equation is of the form

$$a_n y^{(n)} + a_{n-1} y^{(n-1)} + \cdots + a_1 y^{(1)} + a_0 y = f, \tag{4}$$

where $a_0, a_1, \ldots, a_n$ and f are functions of t and $y^{(k)}$ denotes the kth derivative of y, then the equation is said to be **linear**. The functions a_i are called the **coefficients** of the differential equation (4). Thus (3) is an example of a linear differential equation in which the coefficients are constants and the function f is identically zero. When f is identically zero, (4) is called **homogeneous**.

In this section, we apply the linear algebra we have studied to solve homogeneous linear differential equations with constant coefficients. If $a_n \neq 0$, we say that differential equation (4) is of **order** n. In this case, we divide both sides by a_n to obtain a new, but equivalent, equation

$$y^{(n)} + b_{n-1}y^{(n-1)} + \cdots + b_1 y^{(1)} + b_0 y = 0,$$

where $b_i = a_i/a_n$ for $i = 0, 1, \ldots, n - 1$. Because of this observation, we always assume that the coefficient a_n in (4) is 1.

A **solution** to (4) is a function that when substituted for y reduces (4) to an identity.

Example 1

The function $y(t) = \sin\sqrt{k/m}\,t$ is a solution to (3) since

$$y''(t) + \frac{k}{m}y(t) = -\frac{k}{m}\sin\sqrt{\frac{k}{m}}\,t + \frac{k}{m}\sin\sqrt{\frac{k}{m}}\,t = 0$$

for all t. Notice, however, that substituting $y(t) = t$ into (3) yields

$$y''(t) + \frac{k}{m}y(t) = \frac{k}{m}t,$$

which is not identically zero. Thus $y(t) = t$ is not a solution to (3).   ◆

In our study of differential equations, it is useful to regard solutions as complex-valued functions of a real variable even though the solutions that are meaningful to us in a physical sense are real-valued. The convenience of this viewpoint will become clear later. Thus we are concerned with the vector space $\mathcal{F}(R, C)$ (as defined in Example 3 of Section 1.2). In order to consider complex-valued functions of a real variable as solutions to differential equations, we must define what it means to differentiate such functions. Given a complex-valued function $x \in \mathcal{F}(R, C)$ of a real variable t, there exist unique real-valued functions x_1 and x_2 of t, such that

$$x(t) = x_1(t) + ix_2(t) \quad \text{for} \quad t \in R,$$

where i is the imaginary number such that $i^2 = -1$. We call x_1 the **real part** and x_2 the **imaginary part** of x.

Definitions. Given a function $x \in \mathcal{F}(R, C)$ with real part x_1 and imaginary part x_2, we say that x is **differentiable** if x_1 and x_2 are differentiable. If x is differentiable, we define the **derivative** x' of x by

$$x' = x_1' + ix_2'.$$

We illustrate some computations with complex-valued functions in the following example.

Example 2

Suppose that $x(t) = \cos 2t + i \sin 2t$. Then

$$x'(t) = -2 \sin 2t + 2i \cos 2t.$$

We next find the real and imaginary parts of x^2. Since

$$x^2(t) = (\cos 2t + i \sin 2t)^2 = (\cos^2 2t - \sin^2 2t) + i(2 \sin 2t \cos 2t)$$
$$= \cos 4t + i \sin 4t,$$

the real part of $x^2(t)$ is $\cos 4t$, and the imaginary part is $\sin 4t$.   ◆

The next theorem indicates that we may limit our investigations to a vector space considerably smaller than $\mathcal{F}(R, C)$. Its proof, which is illustrated in Example 3, involves a simple induction argument, which we omit.

Theorem 2.27. *Any solution to a homogeneous linear differential equation with constant coefficients has derivatives of all orders; that is, if x is a solution to such an equation, then $x^{(k)}$ exists for every positive integer k.*

Example 3

To illustrate Theorem 2.27, consider the equation

$$y^{(2)} + 4y = 0.$$

Clearly, to qualify as a solution, a function y must have two derivatives. If y is a solution, however, then

$$y^{(2)} = -4y.$$

Thus since $y^{(2)}$ is a constant multiple of a function y that has two derivatives, $y^{(2)}$ must have two derivatives. Hence $y^{(4)}$ exists; in fact,

$$y^{(4)} = -4y^{(2)}.$$

Since $y^{(4)}$ is a constant multiple of a function that we have shown has at least two derivatives, it also has at least two derivatives; hence $y^{(6)}$ exists. Continuing in this manner, we can show that any solution has derivatives of all orders.   ◆

Definition. *We use C^∞ to denote the set of all functions in $\mathcal{F}(R, C)$ that have derivatives of all orders.*

It is a simple exercise to show that C^∞ is a subspace of $\mathcal{F}(R, C)$ and hence a vector space over C. In view of Theorem 2.27, it is this vector space that is of interest to us. For $x \in C^\infty$, the derivative x' of x also lies in C^∞. We can use the derivative operation to define a mapping $\mathsf{D} \colon C^\infty \to C^\infty$ by

$$\mathsf{D}(x) = x' \quad \text{for } x \in C^\infty.$$

It is easy to show that D is a linear operator. More generally, consider any polynomial over C of the form

$$p(t) = a_n t^n + a_{n-1} t^{n-1} + \cdots + a_1 t + a_0.$$

If we define

$$p(\mathsf{D}) = a_n \mathsf{D}^n + a_{n-1} \mathsf{D}^{n-1} + \cdots + a_1 \mathsf{D} + a_0 \mathsf{I},$$

then $p(\mathsf{D})$ is a linear operator on C^∞. (See Appendix E.)

Definitions. *For any polynomial $p(t)$ over C of positive degree, $p(\mathsf{D})$ is called a **differential operator**. The **order** of the differential operator $p(\mathsf{D})$ is the degree of the polynomial $p(t)$.*

Differential operators are useful since they provide us with a means of reformulating a differential equation in the context of linear algebra. Any homogeneous linear differential equation with constant coefficients,

$$y^{(n)} + a_{n-1} y^{(n-1)} + \cdots + a_1 y^{(1)} + a_0 y = 0,$$

can be rewritten using differential operators as

$$(\mathsf{D}^n + a_{n-1} \mathsf{D}^{n-1} + \cdots + a_1 \mathsf{D} + a_0 \mathsf{I})(y) = 0.$$

Definition. *Given the differential equation above, the complex polynomial*

$$p(t) = t^n + a_{n-1} t^{n-1} + \cdots + a_1 t + a_0$$

*is called the **auxiliary polynomial** associated with the equation.*

For example, (3) has the auxiliary polynomial

$$p(t) = t^2 + \frac{k}{m}.$$

Any homogeneous linear differential equation with constant coefficients can be rewritten as

$$p(\mathsf{D})(y) = 0,$$

where $p(t)$ is the auxiliary polynomial associated with the equation. Clearly, this equation implies the following theorem.

Theorem 2.28. *The set of all solutions to a homogeneous linear differential equation with constant coefficients coincides with the null space of $p(\mathsf{D})$, where $p(t)$ is the auxiliary polynomial associated with the equation.*

Proof. Exercise.   ∎

Corollary. *The set of all solutions to a homogeneous linear differential equation with constant coefficients is a subspace of C^∞.*

In view of the preceding corollary, we call the set of solutions to a homogeneous linear differential equation with constant coefficients the **solution space** of the equation. A practical way of describing such a space is in terms of a basis. We now examine a certain class of functions that is of use in finding bases for these solution spaces.

For a real number s, we are familiar with the real number e^s, where e is the unique number whose natural logarithm is 1 (i.e., $\ln e = 1$). We know, for instance, certain properties of exponentiation, namely,

$$e^{s+t} = e^s e^t \quad \text{and} \quad e^{-t} = \frac{1}{e^t}$$

for any real numbers s and t. We now extend the definition of powers of e to include complex numbers in such a way that these properties are preserved.

Definition. *Let $c = a + ib$ be a complex number with real part a and imaginary part b. Define*

$$e^c = e^a(\cos b + i \sin b).$$

The special case

$$e^{ib} = \cos b + i \sin b$$

*is called **Euler's formula**.*

For example, for $c = 2 + i(\pi/3)$,

$$e^c = e^2 \left(\cos \frac{\pi}{3} + i \sin \frac{\pi}{3} \right) = e^2 \left(\frac{1}{2} + i \frac{\sqrt{3}}{2} \right).$$

Clearly, if c is real ($b = 0$), then we obtain the usual result: $e^c = e^a$. Using the approach of Example 2, we can show by the use of trigonometric identities that

$$e^{c+d} = e^c e^d \quad \text{and} \quad e^{-c} = \frac{1}{e^c}$$

for any complex numbers c and d.

Definition.   A function $f: R \rightarrow C$ defined by $f(t) = e^{ct}$ for a fixed complex number c is called an **exponential function**.

The derivative of an exponential function, as described in the next theorem, is consistent with the real version. The proof involves a straightforward computation, which we leave as an exercise.

Theorem 2.29. For any exponential function $f(t) = e^{ct}$, $f'(t) = ce^{ct}$.

Proof. Exercise. ∎

We can use exponential functions to describe all solutions to a homogeneous linear differential equation of order 1. Recall that the **order** of such an equation is the degree of its auxiliary polynomial. Thus an equation of order 1 is of the form

$$y' + a_0 y = 0. \tag{5}$$

Theorem 2.30. The solution space for (5) is of dimension 1 and has $\{e^{-a_0 t}\}$ as a basis.

Proof. Clearly (5) has $e^{-a_0 t}$ as a solution. Suppose that $x(t)$ is any solution to (5). Then

$$x'(t) = -a_0 x(t) \quad \text{for all } t \in R.$$

Define

$$z(t) = e^{a_0 t} x(t).$$

Differentiating z yields

$$z'(t) = (e^{a_0 t})' x(t) + e^{a_0 t} x'(t) = a_0 e^{a_0 t} x(t) - a_0 e^{a_0 t} x(t) = 0.$$

(Notice that the familiar product rule for differentiation holds for complex-valued functions of a real variable. A justification of this involves a lengthy, although direct, computation.)

Since z' is identically zero, z is a constant function. (Again, this fact, well known for real-valued functions, is also true for complex-valued functions. The proof, which relies on the real case, involves looking separately at the real and imaginary parts of z.) Thus there exists a complex number k such that

$$z(t) = e^{a_0 t} x(t) = k \quad \text{for all } t \in R.$$

So

$$x(t) = ke^{-a_0 t}.$$

We conclude that any solution to (5) is a linear combination of $e^{-a_0 t}$. ∎

Another way of stating Theorem 2.30 is as follows.

Corollary. *For any complex number c, the null space of the differential operator $D - cI$ has $\{e^{ct}\}$ as a basis.*

We next concern ourselves with differential equations of order greater than one. Given an nth order homogeneous linear differential equation with constant coefficients,

$$y^{(n)} + a_{n-1}y^{(n-1)} + \cdots + a_1 y^{(1)} + a_0 y = 0,$$

its auxiliary polynomial

$$p(t) = t^n + a_{n-1}t^{n-1} + \cdots + a_1 t + a_0$$

factors into a product of polynomials of degree 1, that is,

$$p(t) = (t - c_1)(t - c_2) \cdots (t - c_n),$$

where $c_1, c_2, \ldots, c_n$ are (not necessarily distinct) complex numbers. (This follows from the fundamental theorem of algebra in Appendix D.) Thus

$$p(D) = (D - c_1 I)(D - c_2 I) \cdots (D - c_n I).$$

The operators $D - c_i I$ commute, and so, by Exercise 9, we have that

$$N(D - c_i I) \subseteq N(p(D)) \quad \text{for all } i.$$

Since $N(p(D))$ coincides with the solution space of the given differential equation, we can deduce the following result from the preceding corollary.

Theorem 2.31. *Let $p(t)$ be the auxiliary polynomial for a homogeneous linear differential equation with constant coefficients. For any complex number c, if c is a zero of $p(t)$, then e^{ct} is a solution to the differential equation.*

Example 4

Given the differential equation

$$y'' - 3y' + 2y = 0,$$

its auxiliary polynomial is

$$p(t) = t^2 - 3t + 2 = (t - 1)(t - 2).$$

Hence, by Theorem 2.31, e^t and e^{2t} are solutions to the differential equation because $c = 1$ and $c = 2$ are zeros of $p(t)$. Since the solution space of the differential equation is a subspace of C^∞, $\text{span}(\{e^t, e^{2t}\})$ lies in the solution space. It is a simple matter to show that $\{e^t, e^{2t}\}$ is linearly independent. Thus if we can show that the solution space is two-dimensional, we can conclude that $\{e^t, e^{2t}\}$ is a basis for the solution space. This result is a consequence of the next theorem.   ♦

Theorem 2.32. *For any differential operator* $p(\mathsf{D})$ *of order* n, *the null space of* $p(\mathsf{D})$ *is an* n-*dimensional subspace of* C^∞.

As a preliminary to the proof of Theorem 2.32, we establish two lemmas.

Lemma 1. *The differential operator* $\mathsf{D} - c\mathsf{I} \colon \mathsf{C}^\infty \to \mathsf{C}^\infty$ *is onto for any complex number* c.

Proof. Let $v \in \mathsf{C}^\infty$. We wish to find a $u \in \mathsf{C}^\infty$ such that $(\mathsf{D} - c\mathsf{I})u = v$. Let $w(t) = v(t)e^{-ct}$ for $t \in R$. Clearly, $w \in \mathsf{C}^\infty$ because both v and e^{-ct} lie in C^∞. Let w_1 and w_2 be the real and imaginary parts of w. Then w_1 and w_2 are continuous because they are differentiable. Hence they have antiderivatives, say, W_1 and W_2, respectively. Let $W \colon R \to C$ be defined by

$$W(t) = W_1(t) + iW_2(t) \quad \text{for } t \in R.$$

Then $W \in \mathsf{C}^\infty$, and the real and imaginary parts of W are W_1 and W_2, respectively. Furthermore, $W' = w$. Finally, let $u \colon R \to C$ be defined by $u(t) = W(t)e^{ct}$ for $t \in R$. Clearly $u \in \mathsf{C}^\infty$, and since

$$
\begin{aligned}
(\mathsf{D} - c\mathsf{I})u(t) &= u'(t) - cu(t) \\
&= W'(t)e^{ct} + W(t)ce^{ct} - cW(t)e^{ct} \\
&= w(t)e^{ct} \\
&= v(t)e^{-ct}e^{ct} \\
&= v(t),
\end{aligned}
$$

we have $(\mathsf{D} - c\mathsf{I})u = v$. ∎

Lemma 2. *Let* V *be a vector space, and suppose that* T *and* U *are linear operators on* V *such that* U *is onto and the null spaces of* T *and* U *are finite-dimensional. Then the null space of* TU *is finite-dimensional, and*

$$\dim(\mathsf{N}(\mathsf{TU})) = \dim(\mathsf{N}(\mathsf{T})) + \dim(\mathsf{N}(\mathsf{U})).$$

Proof. Let $p = \dim(\mathsf{N}(\mathsf{T}))$, $q = \dim(\mathsf{N}(\mathsf{U}))$, and $\{u_1, u_2, \ldots, u_p\}$ and $\{v_1, v_2, \ldots, v_q\}$ be bases for $\mathsf{N}(\mathsf{T})$ and $\mathsf{N}(\mathsf{U})$, respectively. Since U is onto, we can choose for each i $(1 \le i \le p)$ a vector $w_i \in \mathsf{V}$ such that $\mathsf{U}(w_i) = u_i$. Note that the w_i's are distinct. Furthermore, for any i and j, $w_i \ne v_j$, for otherwise $u_i = \mathsf{U}(w_i) = \mathsf{U}(v_j) = 0$—a contradiction. Hence the set

$$\beta = \{w_1, w_2, \ldots, w_p, v_1, v_2, \ldots, v_q\}$$

contains $p+q$ distinct vectors. To complete the proof of the lemma, it suffices to show that β is a basis for $\mathsf{N}(\mathsf{TU})$.

We first show that β generates $\mathsf{N(TU)}$. Since for any w_i and v_j in β, $\mathsf{TU}(w_i) = \mathsf{T}(u_i) = 0$ and $\mathsf{TU}(v_j) = \mathsf{T}(0) = 0$, it follows that $\beta \subseteq \mathsf{N(TU)}$. Now suppose that $v \in \mathsf{N(TU)}$. Then $0 = \mathsf{TU}(v) = \mathsf{T}(\mathsf{U}(v))$. Thus $\mathsf{U}(v) \in \mathsf{N(T)}$. So there exist scalars $a_1, a_2, \dots, a_p$ such that

$$\begin{aligned} \mathsf{U}(v) &= a_1 u_1 + a_2 u_2 + \cdots + a_p u_p \\ &= a_1 U(w_1) + a_2 U(w_2) + \cdots + a_p U(w_p) \\ &= \mathsf{U}(a_1 w_1 + a_2 w_2 + \cdots + a_p w_p). \end{aligned}$$

Hence

$$\mathsf{U}(v - (a_1 w_1 + a_2 w_2 + \cdots + a_p w_p)) = 0.$$

Consequently, $v - (a_1 w_1 + a_2 w_2 + \cdots + a_p w_p)$ lies in $\mathsf{N(U)}$. It follows that there exist scalars $b_1, b_2, \dots, b_q$ such that

$$v - (a_1 w_1 + a_2 w_2 + \cdots + a_p w_p) = b_1 v_1 + b_2 v_2 + \cdots + b_q v_q$$

or

$$v = a_1 w_1 + a_2 w_2 + \cdots + a_p w_p + b_1 v_1 + b_2 v_2 + \cdots + b_q v_q.$$

Therefore β spans $\mathsf{N(TU)}$.

To prove that β is linearly independent, let $a_1, a_2, \dots, a_p, b_1, b_2, \dots, b_q$ be any scalars such that

$$a_1 w_1 + a_2 w_2 + \cdots + a_p w_p + b_1 v_1 + b_2 v_2 + \cdots + b_q v_q = 0. \qquad (6)$$

Applying U to both sides of (6), we obtain

$$a_1 u_1 + a_2 u_2 + \cdots + a_p u_p = 0.$$

Since $\{u_1, u_2, \dots, u_p\}$ is linearly independent, the a_i's are all zero. Thus (6) reduces to

$$b_1 v_1 + b_2 v_2 + \cdots + b_q v_q = 0.$$

Again, the linear independence of $\{v_1, v_2, \dots, v_q\}$ implies that the b_i's are all zero. We conclude that β is a basis for $\mathsf{N(TU)}$. Hence $\mathsf{N(TU)}$ is finite-dimensional, and $\dim(\mathsf{N(TU)}) = p + q = \dim(\mathsf{N(T)}) + \dim(\mathsf{N(U)})$. ∎

Proof of Theorem 2.32. The proof is by mathematical induction on the order of the differential operator $p(\mathsf{D})$. The first-order case coincides with Theorem 2.30. For some integer $n > 1$, suppose that Theorem 2.32 holds for any differential operator of order less than n, and consider a differential operator $p(\mathsf{D})$ of order n. The polynomial $p(t)$ can be factored into a product of two polynomials as follows:

$$p(t) = q(t)(t - c),$$

where $q(t)$ is a polynomial of degree $n - 1$ and c is a complex number. Thus the given differential operator may be rewritten as

$$p(\mathsf{D}) = q(\mathsf{D})(\mathsf{D} - c\mathsf{I}).$$

Now, by Lemma 1, $\mathsf{D} - c\mathsf{I}$ is onto, and by the corollary to Theorem 2.30, $\dim(\mathsf{N}(\mathsf{D} - c\mathsf{I})) = 1$. Also, by the induction hypothesis, $\dim(\mathsf{N}(q(\mathsf{D})) = n - 1$. Thus, by Lemma 2, we conclude that

$$\dim(\mathsf{N}(p(\mathsf{D}))) = \dim(\mathsf{N}(q(\mathsf{D}))) + \dim(\mathsf{N}(\mathsf{D} - c\mathsf{I}))$$
$$= (n - 1) + 1 = n. \qquad \blacksquare$$

Corollary. *The solution space of any nth-order homogeneous linear differential equation with constant coefficients is an n-dimensional subspace of* C^∞.

The corollary to Theorem 2.32 reduces the problem of finding all solutions to an nth-order homogeneous linear differential equation with constant coefficients to finding a set of n linearly independent solutions to the equation. By the results of Chapter 1, any such set must be a basis for the solution space. The next theorem enables us to find a basis quickly for many such equations. Hints for its proof are provided in the exercises.

Theorem 2.33. *Given n distinct complex numbers* $c_1, c_2, \ldots, c_n$, *the set of exponential functions* $\{e^{c_1 t}, e^{c_2 t}, \ldots, e^{c_n t}\}$ *is linearly independent.*

Proof. Exercise. (See Exercise 10.) $\qquad \blacksquare$

Corollary. *For any nth-order homogeneous linear differential equation with constant coefficients, if the auxiliary polynomial has n distinct zeros* $c_1, c_2, \ldots, c_n$, *then* $\{e^{c_1 t}, e^{c_2 t}, \ldots, e^{c_n t}\}$ *is a basis for the solution space of the differential equation.*

Proof. Exercise. (See Exercise 10.) $\qquad \blacksquare$

Example 5

We find all solutions to the differential equation

$$y'' + 5y' + 4y = 0.$$

Since the auxiliary polynomial factors as $(t + 4)(t + 1)$, it has two distinct zeros, -1 and -4. Thus $\{e^{-t}, e^{-4t}\}$ is a basis for the solution space. So any solution to the given equation is of the form

$$y(t) = b_1 e^{-t} + b_2 e^{-4t}$$

for unique scalars b_1 and b_2.   ◆

Example 6

We find all solutions to the differential equation

$$y'' + 9y = 0.$$

The auxiliary polynomial $t^2 + 9$ factors as $(t - 3i)(t + 3i)$ and hence has distinct zeros $c_1 = 3i$ and $c_2 = -3i$. Thus $\{e^{3it}, e^{-3it}\}$ is a basis for the solution space. Since

$$\cos 3t = \frac{1}{2}(e^{3it} + e^{-3it}) \quad \text{and} \quad \sin 3t = \frac{1}{2i}(e^{3it} - e^{-3it}),$$

it follows from Exercise 7 that $\{\cos 3t, \sin 3t\}$ is also a basis for this solution space. This basis has an advantage over the original one because it consists of the familiar sine and cosine functions and makes no reference to the imaginary number i. Using this latter basis, we see that any solution to the given equation is of the form

$$y(t) = b_1 \cos 3t + b_2 \sin 3t$$

for unique scalars b_1 and b_2.   ◆

Next consider the differential equation

$$y'' + 2y' + y = 0,$$

for which the auxiliary polynomial is $(t + 1)^2$. By Theorem 2.31, e^{-t} is a solution to this equation. By the corollary to Theorem 2.32, its solution space is two-dimensional. In order to obtain a basis for the solution space, we need a solution that is linearly independent of e^{-t}. The reader can verify that te^{-t} is a such a solution. The following lemma extends this result.

Lemma. *For a given complex number c and positive integer n, suppose that $(t - c)^n$ is the auxiliary polynomial of a homogeneous linear differential equation with constant coefficients. Then the set*

$$\beta = \{e^{ct}, te^{ct}, \ldots, t^{n-1}e^{ct}\}$$

is a basis for the solution space of the equation.

Proof. Since the solution space is n-dimensional, we need only show that β is linearly independent and lies in the solution space. First, observe that for any positive integer k,

$$(\mathsf{D} - c\mathsf{I})(t^k e^{ct}) = kt^{k-1}e^{ct} + ct^k e^{ct} - ct^k e^{ct}$$
$$= kt^{k-1}e^{ct}.$$

Hence for $k < n$,

$$(\mathsf{D} - c\mathsf{I})^n(t^k e^{ct}) = 0.$$

It follows that β is a subset of the solution space.

We next show that β is linearly independent. Consider any linear combination of vectors in β such that

$$b_0 e^{ct} + b_1 t e^{ct} + \cdots + b_{n-1}t^{n-1}e^{ct} = 0 \tag{7}$$

for some scalars $b_0, b_1, \ldots, b_{n-1}$. Dividing by e^{ct} in (7), we obtain

$$b_0 + b_1 t + \cdots + b_{n-1}t^{n-1} = 0. \tag{8}$$

Thus the left side of (8) must be the zero polynomial function. We conclude that the coefficients $b_0, b_1, \ldots, b_{n-1}$ are all zero. So β is linearly independent and hence is a basis for the solution space. ∎

Example 7

We find all solutions to the differential equation

$$y^{(4)} - 4y^{(3)} + 6y^{(2)} - 4y^{(1)} + y = 0.$$

Since the auxiliary polynomial is

$$t^4 - 4t^3 + 6t^2 - 4t + 1 = (t - 1)^4,$$

we can immediately conclude by the preceding lemma that $\{e^t, te^t, t^2 e^t, t^3 e^t\}$ is a basis for the solution space. So any solution y to the given differential equation is of the form

$$y(t) = b_1 e^t + b_2 t e^t + b_3 t^2 e^t + b_4 t^3 e^t$$

for unique scalars b_1, b_2, b_3, and b_4. ◆

The most general situation is stated in the following theorem.

Theorem 2.34. *Given a homogeneous linear differential equation with constant coefficients and auxiliary polynomial*

$$(t - c_1)^{n_1}(t - c_2)^{n_2} \cdots (t - c_k)^{n_k},$$

where $n_1, n_2, \ldots, n_k$ are positive integers and $c_1, c_2, \ldots, c_k$ are distinct complex numbers, the following set is a basis for the solution space of the equation:

$$\{e^{c_1 t}, te^{c_1 t}, \ldots, t^{n_1 - 1}e^{c_1 t}, \ldots, e^{c_k t}, te^{c_k t}, \ldots, t^{n_k - 1}e^{c_k t}\}.$$

Proof. Exercise.   ∎

Example 8

The differential equation

$$y^{(3)} - 4y^{(2)} + 5y^{(1)} - 2y = 0$$

has the auxiliary polynomial

$$t^3 - 4t^2 + 5t - 2 = (t-1)^2(t-2).$$

By Theorem 2.34, $\{e^t, te^t, e^{2t}\}$ is a basis for the solution space of the differential equation. Thus any solution y has the form

$$y(t) = b_1 e^t + b_2 t e^t + b_3 e^{2t}$$

for unique scalars b_1, b_2, and b_3.   ◆

EXERCISES

1. Label the following statements as true or false.

    (a) The set of solutions to an nth-order homogeneous linear differential equation with constant coefficients is an n-dimensional subspace of C^∞.

    (b) The solution space of a homogeneous linear differential equation with constant coefficients is the null space of a differential operator.

    (c) The auxiliary polynomial of a homogeneous linear differential equation with constant coefficients is a solution to the differential equation.

    (d) Any solution to a homogeneous linear differential equation with constant coefficients is of the form ae^{ct} or $at^k e^{ct}$, where a and c are complex numbers and k is a positive integer.

    (e) Any linear combination of solutions to a given homogeneous linear differential equation with constant coefficients is also a solution to the given equation.

    (f) For any homogeneous linear differential equation with constant coefficients having auxiliary polynomial $p(t)$, if $c_1, c_2, \ldots, c_k$ are the distinct zeros of $p(t)$, then $\{e^{c_1 t}, e^{c_2 t}, \ldots, e^{c_k t}\}$ is a basis for the solution space of the given differential equation.

    (g) Given any polynomial $p(t) \in P(C)$, there exists a homogeneous linear differential equation with constant coefficients whose auxiliary polynomial is $p(t)$.

2. For each of the following parts, determine whether the statement is true or false. Justify your claim with either a proof or a counterexample, whichever is appropriate.

(a) Any finite-dimensional subspace of C^∞ is the solution space of a homogeneous linear differential equation with constant coefficients.

(b) There exists a homogeneous linear differential equation with constant coefficients whose solution space has the basis $\{t, t^2\}$.

(c) For any homogeneous linear differential equation with constant coefficients, if x is a solution to the equation, so is its derivative x'.

Given two polynomials $p(t)$ and $q(t)$ in $P(C)$, if $x \in N(p(D))$ and $y \in N(q(D))$, then

(d) $x + y \in N(p(D)q(D))$.

(e) $xy \in N(p(D)q(D))$.

3. Find a basis for the solution space of each of the following differential equations.

(a) $y'' + 2y' + y = 0$

(b) $y''' = y'$

(c) $y^{(4)} - 2y^{(2)} + y = 0$

(d) $y'' + 2y' + y = 0$

(e) $y^{(3)} - y^{(2)} + 3y^{(1)} + 5y = 0$

4. Find a basis for each of the following subspaces of C^∞.

(a) $N(D^2 - D - I)$

(b) $N(D^3 - 3D^2 + 3D - I)$

(c) $N(D^3 + 6D^2 + 8D)$

5. Show that C^∞ is a subspace of $\mathcal{F}(R, C)$.

6. (a) Show that $D \colon C^\infty \to C^\infty$ is a linear operator.

(b) Show that any differential operator is a linear operator on C^∞.

7. Prove that if $\{x, y\}$ is a basis for a vector space over C, then so is

$$\left\{ \frac{1}{2}(x + y), \frac{1}{2i}(x - y) \right\}.$$

8. Consider a second-order homogeneous linear differential equation with constant coefficients in which the auxiliary polynomial has distinct conjugate complex roots $a + ib$ and $a - ib$, where $a, b \in R$. Show that $\{e^{at} \cos bt, e^{at} \sin bt\}$ is a basis for the solution space.

9. Suppose that $\{U_1, U_2, \ldots, U_n\}$ is a collection of pairwise commutative linear operators on a vector space V (i.e., operators such that $U_i U_j = U_j U_i$ for all i, j). Prove that, for any i $(1 \le i \le n)$,

$$N(U_i) \subseteq N(U_1 U_2 \cdots U_n).$$

10. Prove Theorem 2.33 and its corollary. *Hint:* Suppose that

$$b_1 e^{c_1 t} + b_2 e^{c_2 t} + \cdots + b_n e^{c_n t} = 0 \quad \text{(where the } c_i\text{'s are distinct)}.$$

To show the b_i's are zero, apply mathematical induction on n as follows. Verify the theorem for $n = 1$. Assuming that the theorem is true for $n - 1$ functions, apply the operator $D - c_n I$ to both sides of the given equation to establish the theorem for n distinct exponential functions.

11. Prove Theorem 2.34. *Hint:* First verify that the alleged basis lies in the solution space. Then verify that this set is linearly independent by mathematical induction on k as follows. The case $k = 1$ is the lemma to Theorem 2.34. Assuming that the theorem holds for $k - 1$ distinct c_i's, apply the operator $(D - c_k I)^{n_k}$ to any linear combination of the alleged basis that equals 0.

12. Let V be the solution space of an nth-order homogeneous linear differential equation with constant coefficients having auxiliary polynomial $p(t)$. Prove that if $p(t) = g(t) h(t)$, where $g(t)$ and $h(t)$ are polynomials of positive degree, then

$$N(h(D)) = R(g(D_V)) = g(D)(V),$$

where $D_V \colon V \to V$ is defined by $D_V(x) = x'$ for $x \in V$. *Hint:* First prove $g(D)(V) \subseteq N(h(D))$. Then prove that the two spaces have the same finite dimension.

13. A differential equation

$$y^{(n)} + a_{n-1} y^{(n-1)} + \cdots + a_1 y^{(1)} + a_0 y = x$$

is called a **nonhomogeneous** linear differential equation with constant coefficients if the a_i's are constant and x is a function that is not identically zero.

(a) Prove that for any $x \in C^\infty$ there exists $y \in C^\infty$ such that y is a solution to the differential equation. *Hint:* Use Lemma 1 to Theorem 2.32 to show that for any polynomial $p(t)$, the linear operator $p(D) \colon C^\infty \to C^\infty$ is onto.

(b) Let V be the solution space for the homogeneous linear equation

$$y^{(n)} + a_{n-1}y^{(n-1)} + \cdots + a_1 y^{(1)} + a_0 y = 0.$$

Prove that if z is any solution to the associated nonhomogeneous linear differential equation, then the set of all solutions to the nonhomogeneous linear differential equation is

$$\{z + y \colon y \in V\}.$$

14. Given any nth-order homogeneous linear differential equation with constant coefficients, prove that, for any solution x and any $t_0 \in R$, if $x(t_0) = x'(t_0) = \cdots = x^{(n-1)}(t_0) = 0$, then $x = 0$ (the zero function). *Hint:* Use mathematical induction on n as follows. First prove the conclusion for the case $n = 1$. Next suppose that it is true for equations of order $n - 1$, and consider an nth-order differential equation with auxiliary polynomial $p(t)$. Factor $p(t) = q(t)(t - c)$, and let $z = q((D))x$. Show that $z(t_0) = 0$ and $z' - cz = 0$ to conclude that $z = 0$. Now apply the induction hypothesis.

15. Let V be the solution space of an nth-order homogeneous linear differential equation with constant coefficients. Fix $t_0 \in R$, and define a mapping $\Phi \colon V \to C^n$ by

$$\Phi(x) = \begin{pmatrix} x(t_0) \\ x'(t_0) \\ \vdots \\ x^{(n-1)}(t_0) \end{pmatrix} \qquad \text{for each } x \text{ in V.}$$

(a) Prove that Φ is linear and its null space is the zero subspace of V. Deduce that Φ is an isomorphism. *Hint:* Use Exercise 14.
(b) Prove the following: For any nth-order homogeneous linear differential equation with constant coefficients, any $t_0 \in R$, and any complex numbers $c_0, c_1, \ldots, c_{n-1}$ (not necessarily distinct), there exists exactly one solution, x, to the given differential equation such that $x(t_0) = c_0$ and $x^{(k)}(t_0) = c_k$ for $k = 1, 2, \ldots n - 1$.

16. *Pendular Motion.* It is well known that the motion of a pendulum is approximated by the differential equation

$$\theta'' + \frac{g}{l}\theta = 0,$$

where $\theta(t)$ is the angle in radians that the pendulum makes with a vertical line at time t (see Figure 2.8), interpreted so that θ is positive if the pendulum is to the right and negative if the pendulum is to the

Figure 2.8

left of the vertical line as viewed by the reader. Here l is the length of the pendulum and g is the magnitude of acceleration due to gravity. The variable t and constants l and g must be in compatible units (e.g., t in seconds, l in meters, and g in meters per second per second).

(a) Express an arbitrary solution to this equation as a linear combination of two real-valued solutions.

(b) Find the unique solution to the equation that satisfies the conditions

$$\theta(0) = \theta_0 > 0 \quad \text{and} \quad \theta'(0) = 0.$$

(The significance of these conditions is that at time $t = 0$ the pendulum is released from a position displaced from the vertical by θ_0.)

(c) Prove that it takes $2\pi \sqrt{l/g}$ units of time for the pendulum to make one circuit back and forth. (This time is called the **period** of the pendulum.)

17. *Periodic Motion of a Spring without Damping.* Find the general solution to (3), which describes the periodic motion of a spring, ignoring frictional forces.

18. *Periodic Motion of a Spring with Damping.* The ideal periodic motion described by solutions to (3) is due to the ignoring of frictional forces. In reality, however, there is a frictional force acting on the motion that is proportional to the speed of motion, but that acts in the opposite direction. The modification of (3) to account for the frictional force, called the *damping force*, is given by

$$my'' + ry' + ky = 0,$$

where $r > 0$ is the proportionality constant.

(a) Find the general solution to this equation.

(b) Find the unique solution in (a) that satisfies the initial conditions $y(0) = 0$ and $y'(0) = v_0$, the initial velocity.

(c) For $y(t)$ as in (b), show that the amplitude of the oscillation decreases to zero; that is, prove that $\lim_{t \to \infty} y(t) = 0$.

19. In our study of differential equations, we have regarded solutions as complex-valued functions even though functions that are useful in describing physical motion are real-valued. Justify this approach.

20. The following parts, which do not involve linear algebra, are included for the sake of completeness.

(a) Prove Theorem 2.27. *Hint:* Use mathematical induction on the number of derivatives possessed by a solution.

(b) For any $c, d \in C$, prove that

$$e^{c+d} = c^c e^d \quad \text{and} \quad e^{-c} = \frac{1}{e^c}.$$

(c) Prove Theorem 2.28.

(d) Prove Theorem 2.29.

(e) Prove the product rule for differentiating complex-valued functions of a real variable: For any differentiable functions x and y in $\mathcal{F}(R, C)$, the product xy is differentiable and

$$(xy)' = x'y + xy'.$$

Hint: Apply the rules of differentiation to the real and imaginary parts of xy.

(f) Prove that if $x \in \mathcal{F}(R, C)$ and $x' = 0$, then x is a constant function.

INDEX OF DEFINITIONS FOR CHAPTER 2

3

Elementary Matrix Operations and Systems of Linear Equations

*T*his chapter is devoted to two related objectives:

1. the study of certain "rank-preserving" operations on matrices;
2. the application of these operations and the theory of linear transformations to the solution of systems of linear equations.

As a consequence of objective 1, we obtain a simple method for computing the rank of a linear transformation between finite-dimensional vector spaces by applying these rank-preserving matrix operations to a matrix that represents that transformation.

Solving a system of linear equations is probably the most important application of linear algebra. The familiar method of elimination for solving systems of linear equations, which was discussed in Section 1.4, involves the elimination of variables so that a simpler system can be obtained. The technique by which the variables are eliminated utilizes three types of operations:

1. interchanging any two equations in the system;
2. multiplying any equation in the system by a nonzero constant;
3. adding a multiple of one equation to another.

In Section 3.3, we express a system of linear equations as a single matrix equation. In this representation of the system, the three operations above are the "elementary row operations" for matrices. These operations provide a convenient computational method for determining all solutions to a system of linear equations.

3.1 ELEMENTARY MATRIX OPERATIONS AND ELEMENTARY MATRICES

In this section, we define the elementary operations that are used throughout the chapter. In subsequent sections, we use these operations to obtain simple computational methods for determining the rank of a linear transformation and the solution of a system of linear equations. There are two types of elementary matrix operations—row operations and column operations. As we will see, the row operations are more useful. They arise from the three operations that can be used to eliminate variables in a system of linear equations.

Definitions. *Let A be an $m \times n$ matrix. Any one of the following three operations on the rows [columns] of A is called an **elementary row [column] operation**:*

(1) *interchanging any two rows [columns] of A;*
(2) *multiplying any row [column] of A by a nonzero scalar;*
(3) *adding any scalar multiple of a row [column] of A to another row [column].*

*Any of these three operations is called an **elementary operation**. Elementary operations are of **type 1**, **type 2**, or **type 3** depending on whether they are obtained by (1), (2), or (3).*

Example 1

Let

$$A = \begin{pmatrix} 1 & 2 & 3 & 4 \\ 2 & 1 & -1 & 3 \\ 4 & 0 & 1 & 2 \end{pmatrix}.$$

Interchanging the second row of A with the first row is an example of an elementary row operation of type 1. The resulting matrix is

$$B = \begin{pmatrix} 2 & 1 & -1 & 3 \\ 1 & 2 & 3 & 4 \\ 4 & 0 & 1 & 2 \end{pmatrix}.$$

Multiplying the second column of A by 3 is an example of an elementary column operation of type 2. The resulting matrix is

$$C = \begin{pmatrix} 1 & 6 & 3 & 4 \\ 2 & 3 & -1 & 3 \\ 4 & 0 & 1 & 2 \end{pmatrix}.$$

Adding 4 times the third row of A to the first row is an example of an elementary row operation of type 3. In this case, the resulting matrix is

$$M = \begin{pmatrix} 17 & 2 & 7 & 12 \\ 2 & 1 & -1 & 3 \\ 4 & 0 & 1 & 2 \end{pmatrix}. \quad \blacklozenge$$

Notice that if a matrix Q can be obtained from a matrix P by means of an elementary row operation, then P can be obtained from Q by an elementary row operation of the same type. (See Exercise 8.) So, in Example 1, A can be obtained from M by adding -4 times the third row of M to the first row of M.

Definition.   An $n \times n$ **elementary matrix** is a matrix obtained by performing an elementary operation on I_n. The elementary matrix is said to be of **type 1, 2,** or **3** according to whether the elementary operation performed on I_n is a type 1, 2, or 3 operation, respectively.

For example, interchanging the first two rows of I_3 produces the elementary matrix

$$E = \begin{pmatrix} 0 & 1 & 0 \\ 1 & 0 & 0 \\ 0 & 0 & 1 \end{pmatrix}.$$

Note that E can also be obtained by interchanging the first two columns of I_3. In fact, any elementary matrix can be obtained in at least two ways— either by performing an elementary row operation on I_n or by performing an elementary column operation on I_n. (See Exercise 4.) Similarly,

$$\begin{pmatrix} 1 & 0 & -2 \\ 0 & 1 & 0 \\ 0 & 0 & 1 \end{pmatrix}$$

is an elementary matrix since it can be obtained from I_3 by an elementary column operation of type 3 (adding -2 times the first column of I_3 to the third column) or by an elementary row operation of type 3 (adding -2 times the third row to the first row).

Our first theorem shows that performing an elementary row operation on a matrix is equivalent to multiplying the matrix by an elementary matrix.

Theorem 3.1. Let $A \in \mathsf{M}_{m \times n}(F)$, and suppose that B is obtained from A by performing an elementary row [column] operation. Then there exists an $m \times m$ [$n \times n$] elementary matrix E such that $B = EA$ [$B = AE$]. In fact, E is obtained from I_m [I_n] by performing the same elementary row [column] operation as that which was performed on A to obtain B. Conversely, if E is an elementary $m \times m$ $[n \times n]$ matrix, then EA $[AE]$ is the matrix obtained from A by performing the same elementary row [column] operation as that which produces E from I_m $[I_n]$.

The proof, which we omit, requires verifying Theorem 3.1 for each type of elementary row operation. The proof for column operations can then be obtained by using the matrix transpose to transform a column operation into a row operation. The details are left as an exercise. (See Exercise 7.)

The next example illustrates the use of the theorem.

Example 2

Consider the matrices A and B in Example 1. In this case, B is obtained from A by interchanging the first two rows of A. Performing this same operation on I_3, we obtain the elementary matrix

$$E = \begin{pmatrix} 0 & 1 & 0 \\ 1 & 0 & 0 \\ 0 & 0 & 1 \end{pmatrix}.$$

Note that $EA = B$.

In the second part of Example 1, C is obtained from A by multiplying the second column of A by 3. Performing this same operation on I_4, we obtain the elementary matrix

$$E = \begin{pmatrix} 1 & 0 & 0 & 0 \\ 0 & 3 & 0 & 0 \\ 0 & 0 & 1 & 0 \\ 0 & 0 & 0 & 1 \end{pmatrix}.$$

Observe that $AE = C$. ◆

It is a useful fact that the inverse of an elementary matrix is also an elementary matrix.

Theorem 3.2. *Elementary matrices are invertible, and the inverse of an elementary matrix is an elementary matrix of the same type.*

Proof. Let E be an elementary $n \times n$ matrix. Then E can be obtained by an elementary row operation on I_n. By reversing the steps used to transform I_n into E, we can transform E back into I_n. The result is that I_n can be obtained from E by an elementary row operation of the same type. By Theorem 3.1, there is an elementary matrix $\overline{E}$ such that $\overline{E}E = I_n$. Therefore, by Exercise 10 of Section 2.4, E is invertible and $E^{-1} = \overline{E}$. ■

EXERCISES

1. Label the following statements as true or false.

(a) An elementary matrix is always square.

(b) The only entries of an elementary matrix are zeros and ones.

(c) The $n \times n$ identity matrix is an elementary matrix.

(d) The product of two $n \times n$ elementary matrices is an elementary matrix.

(e) The inverse of an elementary matrix is an elementary matrix.

(f) The sum of two $n \times n$ elementary matrices is an elementary matrix.

(g) The transpose of an elementary matrix is an elementary matrix.

(h) If B is a matrix that can be obtained by performing an elementary row operation on a matrix A, then B can also be obtained by performing an elementary column operation on A.

(i) If B is a matrix that can be obtained by performing an elementary row operation on a matrix A, then A can be obtained by performing an elementary row operation on B.

2. Let

$$A = \begin{pmatrix} 1 & 2 & 3 \\ 1 & 0 & 1 \\ 1 & -1 & 1 \end{pmatrix}, \ B = \begin{pmatrix} 1 & 0 & 3 \\ 1 & -2 & 1 \\ 1 & -3 & 1 \end{pmatrix}, \text{ and } C = \begin{pmatrix} 1 & 0 & 3 \\ 0 & -2 & -2 \\ 1 & -3 & 1 \end{pmatrix}.$$

Find an elementary operation that transforms A into B and an elementary operation that transforms B into C. By means of several additional operations, transform C into I_3.

3. Use the proof of Theorem 3.2 to obtain the inverse of each of the following elementary matrices.

(a) $\begin{pmatrix} 0 & 0 & 1 \\ 0 & 1 & 0 \\ 1 & 0 & 0 \end{pmatrix}$ **(b)** $\begin{pmatrix} 1 & 0 & 0 \\ 0 & 3 & 0 \\ 0 & 0 & 1 \end{pmatrix}$ **(c)** $\begin{pmatrix} 1 & 0 & 0 \\ 0 & 1 & 0 \\ -2 & 0 & 1 \end{pmatrix}$

4. Prove the assertion made on page 149: Any elementary $n \times n$ matrix can be obtained in at least two ways—either by performing an elementary row operation on I_n or by performing an elementary column operation on I_n.

5. Prove that E is an elementary matrix if and only if E^t is.

6. Let A be an $m \times n$ matrix. Prove that if B can be obtained from A by an elementary row [column] operation, then B^t can be obtained from A^t by the corresponding elementary column [row] operation.

7. Prove Theorem 3.1.

8. Prove that if a matrix Q can be obtained from a matrix P by an elementary row operation, then P can be obtained from Q by an elementary matrix of the same type. *Hint:* Treat each type of elementary row operation separately.

9. Prove that any elementary row [column] operation of type 1 can be obtained by a succession of three elementary row [column] operations of type 3 followed by one elementary row [column] operation of type 2.

10. Prove that any elementary row [column] operation of type 2 can be obtained by *dividing* some row [column] by a nonzero scalar.

11. Prove that any elementary row [column] operation of type 3 can be obtained by *subtracting* a multiple of some row [column] from another row [column].

12. Let A be an $m \times n$ matrix. Prove that there exists a sequence of elementary row operations of types 1 and 3 that transforms A into an upper triangular matrix.

3.2   THE RANK OF A MATRIX AND MATRIX INVERSES

In this section, we define the *rank* of a matrix. We then use elementary operations to compute the rank of a matrix and a linear transformation. The section concludes with a procedure for computing the inverse of an invertible matrix.

Definition. *If $A \in \mathsf{M}_{m \times n}(F)$, we define the **rank** of A, denoted $\mathrm{rank}(A)$, to be the rank of the linear transformation $\mathsf{L}_A \colon F^n \to F^m$.*

Many results about the rank of a matrix follow immediately from the corresponding facts about a linear transformation. An important result of this type, which follows from Fact 3 (p. 100) and Corollary 2 to Theorem 2.18 (p. 102), is that *an $n \times n$ matrix is invertible if and only if its rank is n.*

Every matrix A is the matrix representation of the linear transformation L_A with respect to the appropriate standard ordered bases. Thus the rank of the linear transformation L_A is the same as the rank of one of its matrix representations, namely, A. The next theorem extends this fact to any matrix representation of any linear transformation defined on finite-dimensional vector spaces.

Theorem 3.3. *Let $\mathsf{T} \colon \mathsf{V} \to \mathsf{W}$ be a linear transformation between finite-dimensional vector spaces, and let β and γ be ordered bases for V and W, respectively. Then $\mathrm{rank}(\mathsf{T}) = \mathrm{rank}([\mathsf{T}]_\beta^\gamma)$.*

Proof. This is a restatement of Exercise 20 of Section 2.4.   ∎

Now that the problem of finding the rank of a linear transformation has been reduced to the problem of finding the rank of a matrix, we need a result that allows us to perform rank-preserving operations on matrices. The next theorem and its corollary tell us how to do this.

Theorem 3.4. *Let A be an $m \times n$ matrix. If P and Q are invertible $m \times m$ and $n \times n$ matrices, respectively, then*
  (a) $\operatorname{rank}(AQ) = \operatorname{rank}(A)$,
  (b) $\operatorname{rank}(PA) = \operatorname{rank}(A)$,
and therefore,
  (c) $\operatorname{rank}(PAQ) = \operatorname{rank}(A)$.

Proof. First observe that

$$\mathsf{R}(\mathsf{L}_{AQ}) = \mathsf{R}(\mathsf{L}_A\mathsf{L}_Q) = \mathsf{L}_A\mathsf{L}_Q(\mathsf{F}^n) = \mathsf{L}_A(\mathsf{L}_Q(\mathsf{F}^n)) = \mathsf{L}_A(\mathsf{F}^n) = \mathsf{R}(\mathsf{L}_A)$$

since L_Q is onto. Therefore

$$\operatorname{rank}(AQ) = \dim(\mathsf{R}(\mathsf{L}_{AQ})) = \dim(\mathsf{R}(\mathsf{L}_A)) = \operatorname{rank}(A).$$

This establishes (a). To establish (b), apply Exercise 17 of Section 2.4 to $\mathsf{T} = \mathsf{L}_P$. We omit the details. Finally, applying (a) and (b), we have

$$\operatorname{rank}(PAQ) = \operatorname{rank}(PA) = \operatorname{rank}(A). \qquad \blacksquare$$

Corollary. *Elementary row and column operations on a matrix are rank-preserving.*

Proof. If B is obtained from a matrix A by an elementary row operation, then there exists an elementary matrix E such that $B = EA$. By Theorem 3.2 (p. 150), E is invertible, and hence $\operatorname{rank}(B) = \operatorname{rank}(A)$ by Theorem 3.4. The proof that elementary column operations are rank-preserving is left as an exercise. $\qquad \blacksquare$

Now that we have a class of matrix operations that preserve rank, we need a way of examining a transformed matrix to ascertain its rank. The next theorem is the first of several in this direction.

Theorem 3.5. *The rank of any matrix equals the maximum number of its linearly independent columns; that is, the rank of a matrix is the dimension of the subspace generated by its columns.*

Proof. For any $A \in \mathsf{M}_{m \times n}(F)$,

$$\operatorname{rank}(A) = \operatorname{rank}(\mathsf{L}_A) = \dim(\mathsf{R}(\mathsf{L}_A)).$$

Let β be the standard ordered basis for F^n. Then β spans F^n and hence, by Theorem 2.2 (p. 68),

$$R(L_A) = \text{span}(L_A(\beta)) = \text{span}\left(\{L_A(e_1), L_A(e_2), \ldots, L_A(e_n)\}\right).$$

But, for any j, we have seen in Theorem 2.13(b) (p. 90) that $L_A(e_j) = Ae_j = a_j$, where a_j the jth column of A. Hence

$$R(L_A) = \text{span}\left(\{a_1, a_2, \ldots, a_n\}\right).$$

Thus

$$\text{rank}(A) = \dim(R(L_A)) = \dim(\text{span}\left(\{a_1, a_2, \ldots, a_n\}\right)). \qquad \blacksquare$$

Example 1

Let

$$A = \begin{pmatrix} 1 & 0 & 1 \\ 0 & 1 & 1 \\ 1 & 0 & 1 \end{pmatrix}.$$

Observe that the first and second columns of A are linearly independent and that the third column is a linear combination of the first two. Thus

$$\text{rank}(A) = \dim\left(\text{span}\left(\left\{\begin{pmatrix} 1 \\ 0 \\ 1 \end{pmatrix}, \begin{pmatrix} 0 \\ 1 \\ 0 \end{pmatrix}, \begin{pmatrix} 1 \\ 1 \\ 1 \end{pmatrix}\right\}\right)\right) = 2. \qquad \blacklozenge$$

To compute the rank of a matrix A, it is frequently useful to postpone the use of Theorem 3.5 until A has been suitably modified by means of appropriate elementary row and column operations so that the number of linearly independent columns is obvious. The corollary to Theorem 3.4 guarantees that the rank of the modified matrix is the same as the rank of A. One such modification of A can be obtained by using elementary row and column operations to introduce zero entries. The next example illustrates this procedure.

Example 2

Let

$$A = \begin{pmatrix} 1 & 2 & 1 \\ 1 & 0 & 3 \\ 1 & 1 & 2 \end{pmatrix}.$$

If we subtract the first row of A from rows 2 and 3 (type 3 elementary row operations), the result is

$$\begin{pmatrix} 1 & 2 & 1 \\ 0 & -2 & 2 \\ 0 & -1 & 1 \end{pmatrix}.$$

If we now subtract twice the first column from the second and subtract the first column from the third (type 3 elementary column operations), we obtain

$$\begin{pmatrix} 1 & 0 & 0 \\ 0 & -2 & 2 \\ 0 & -1 & 1 \end{pmatrix}.$$

It is now obvious that the maximum number of linearly independent columns of this matrix is 2. Hence the rank of A is 2.   ◆

   The next theorem uses this process to transform a matrix into a particularly simple form. The power of this theorem can be seen in its corollaries.

   Theorem 3.6. *Let A be an $m \times n$ matrix of rank r. Then $r \leq m$, $r \leq n$, and, by means of a finite number of elementary row and column operations, A can be transformed into the matrix*

$$D = \begin{pmatrix} I_r & O_1 \\ O_2 & O_3 \end{pmatrix},$$

where O_1, O_2, and O_3 are zero matrices. Thus $D_{ii} = 1$ for $i \leq r$ and $D_{ij} = 0$ otherwise.

   Theorem 3.6 and its corollaries are quite important. Its proof, though easy to understand, is tedious to read. As an aid in following the proof, we first consider an example.

Example 3

Consider the matrix

$$A = \begin{pmatrix} 0 & 2 & 4 & 2 & 2 \\ 4 & 4 & 4 & 8 & 0 \\ 8 & 2 & 0 & 10 & 2 \\ 6 & 3 & 2 & 9 & 1 \end{pmatrix}.$$

By means of a succession of elementary row and column operations, we can transform A into a matrix D as in Theorem 3.6. We list many of the intermediate matrices, but on several occasions a matrix is transformed from the preceding one by means of several elementary operations. The number above each arrow indicates how many elementary operations are involved. Try to identify the nature of each elementary operation (row or column and type) in the following matrix transformations.

$$\begin{pmatrix} 0 & 2 & 4 & 2 & 2 \\ 4 & 4 & 4 & 8 & 0 \\ 8 & 2 & 0 & 10 & 2 \\ 6 & 3 & 2 & 9 & 1 \end{pmatrix} \xrightarrow{1} \begin{pmatrix} 4 & 4 & 4 & 8 & 0 \\ 0 & 2 & 4 & 2 & 2 \\ 8 & 2 & 0 & 10 & 2 \\ 6 & 3 & 2 & 9 & 1 \end{pmatrix} \xrightarrow{1} \begin{pmatrix} 1 & 1 & 1 & 2 & 0 \\ 0 & 2 & 4 & 2 & 2 \\ 8 & 2 & 0 & 10 & 2 \\ 6 & 3 & 2 & 9 & 1 \end{pmatrix} \xrightarrow{2}$$

$$\begin{pmatrix} 1 & 1 & 1 & 2 & 0 \\ 0 & 2 & 4 & 2 & 2 \\ 0 & -6 & -8 & -6 & 2 \\ 0 & -3 & -4 & -3 & 1 \end{pmatrix} \xrightarrow{3} \begin{pmatrix} 1 & 0 & 0 & 0 & 0 \\ 0 & 2 & 4 & 2 & 2 \\ 0 & -6 & -8 & -6 & 2 \\ 0 & -3 & -4 & -3 & 1 \end{pmatrix} \xrightarrow{1}$$

$$\begin{pmatrix} 1 & 0 & 0 & 0 & 0 \\ 0 & 1 & 2 & 1 & 1 \\ 0 & -6 & -8 & -6 & 2 \\ 0 & -3 & -4 & -3 & 1 \end{pmatrix} \xrightarrow{2} \begin{pmatrix} 1 & 0 & 0 & 0 & 0 \\ 0 & 1 & 2 & 1 & 1 \\ 0 & 0 & 4 & 0 & 8 \\ 0 & 0 & 2 & 0 & 4 \end{pmatrix} \xrightarrow{3} \begin{pmatrix} 1 & 0 & 0 & 0 & 0 \\ 0 & 1 & 0 & 0 & 0 \\ 0 & 0 & 4 & 0 & 8 \\ 0 & 0 & 2 & 0 & 4 \end{pmatrix} \xrightarrow{1}$$

$$\begin{pmatrix} 1 & 0 & 0 & 0 & 0 \\ 0 & 1 & 0 & 0 & 0 \\ 0 & 0 & 1 & 0 & 2 \\ 0 & 0 & 2 & 0 & 4 \end{pmatrix} \xrightarrow{1} \begin{pmatrix} 1 & 0 & 0 & 0 & 0 \\ 0 & 1 & 0 & 0 & 0 \\ 0 & 0 & 1 & 0 & 2 \\ 0 & 0 & 0 & 0 & 0 \end{pmatrix} \xrightarrow{1} \begin{pmatrix} 1 & 0 & 0 & 0 & 0 \\ 0 & 1 & 0 & 0 & 0 \\ 0 & 0 & 1 & 0 & 0 \\ 0 & 0 & 0 & 0 & 0 \end{pmatrix} = D$$

By the corollary to Theorem 3.4, $\mathrm{rank}(A) = \mathrm{rank}(D)$. Clearly, however, $\mathrm{rank}(D) = 3$; so $\mathrm{rank}(A) = 3$. ◆

Note that the first two elementary operations in Example 3 result in a 1 in the 1,1 position, and the next several operations (type 3) result in 0's everywhere in the first row and first column except for the 1,1 position. Subsequent elementary operations do not change the first row and first column. With this example in mind, we proceed with the proof of Theorem 3.6.

Proof of Theorem 3.6. If A is the zero matrix, $r = 0$ by Exercise 3. In this case, the conclusion follows with $D = A$.

Now suppose that $A \ne O$ and $r = \mathrm{rank}(A)$; then $r > 0$. The proof is by mathematical induction on m, the number of rows of A.

Suppose that $m = 1$. By means of at most one type 1 column operation and at most one type 2 column operation, A can be transformed into a matrix with a 1 in the 1,1 position. By means of at most $n - 1$ type 3 column operations, this matrix can in turn be transformed into the matrix

$$\begin{pmatrix} 1 & 0 & \cdots & 0 \end{pmatrix}.$$

Note that there is one linearly independent column in D. So $\mathrm{rank}(D) = \mathrm{rank}(A) = 1$ by the corollary to Theorem 3.4 and by Theorem 3.5. Thus the theorem is established for $m = 1$.

Next assume that the theorem holds for any matrix with at most $m - 1$ rows (for some $m > 1$). We must prove that the theorem holds for any matrix with m rows.

Suppose that A is any $m \times n$ matrix. If $n = 1$, Theorem 3.6 can be established in a manner analogous to that for $m = 1$ (see Exercise 10).

We now suppose that $n > 1$. Since $A \ne O$, $A_{ij} \ne 0$ for some i, j. By means of at most one elementary row and at most one elementary column operation (each of type 1), we can move the nonzero entry to the 1,1 position (just as was done in Example 3). By means of at most one additional type 2 operation, we can assure a 1 in the 1,1 position. (Look at the second operation in Example 3.) By means of at most $m-1$ type 3 row operations and at most $n-1$ type 3 column operations, we can eliminate all nonzero entries in the first row and the first column with the exception of the 1 in the 1,1 position. (In Example 3, we used two row and three column operations to do this.)

Thus, with a finite number of elementary operations, A can be transformed into a matrix

$$B = \left(\begin{array}{c|ccc} 1 & 0 & \cdots & 0 \\ \hline 0 & & & \\ \vdots & & B' & \\ 0 & & & \end{array} \right),$$

where B' is an $(m-1) \times (n-1)$ matrix. In Example 3, for instance,

$$B' = \begin{pmatrix} 2 & 4 & 2 & 2 \\ -6 & -8 & -6 & 2 \\ -3 & -4 & -3 & 1 \end{pmatrix}.$$

By Exercise 11, B' has rank one less than B. Since $\operatorname{rank}(A) = \operatorname{rank}(B) = r$, $\operatorname{rank}(B') = r - 1$. Therefore $r - 1 \leq m - 1$ and $r - 1 \leq n - 1$ by the induction hypothesis. Hence $r \leq m$ and $r \leq n$.

Also by the induction hypothesis, B' can be transformed by a finite number of elementary row and column operations into the $(m-1) \times (n-1)$ matrix D' such that

$$D' = \begin{pmatrix} I_{r-1} & O_4 \\ O_5 & O_6 \end{pmatrix},$$

where O_4, O_5, and O_6 are zero matrices. That is, D' consists of all zeros except for its first $r - 1$ diagonal entries, which are ones. Let

$$D = \left(\begin{array}{c|ccc} 1 & 0 & \cdots & 0 \\ \hline 0 & & & \\ \vdots & & D' & \\ 0 & & & \end{array} \right).$$

We see that the theorem now follows once we show that D can be obtained from B by means of a finite number of elementary row and column operations. However this follows by repeated applications of Exercise 12.

Thus, since A can be transformed into B and B can be transformed into D, each by a finite number of elementary operations, A can be transformed into D by a finite number of elementary operations.

Finally, since D' contains ones as its first $r-1$ diagonal entries, D contains ones as its first r diagonal entries and zeros elsewhere. This establishes the theorem. ∎

Corollary 1. *Let A be an $m \times n$ matrix of rank r. Then there exist invertible matrices B and C of sizes $m \times m$ and $n \times n$, respectively, such that $D = BAC$, where*

$$D = \begin{pmatrix} I_r & O_1 \\ O_2 & O_3 \end{pmatrix}$$

is the $m \times n$ matrix in which O_1, O_2, and O_3 are zero matrices.

Proof. By Theorem 3.6, A can be transformed by means of a finite number of elementary row and column operations into the matrix D. We can appeal to Theorem 3.1 (p. 149) each time we perform an elementary operation. Thus there exist elementary $m \times m$ matrices $E_1, E_2, \ldots, E_p$ and elementary $n \times n$ matrices $G_1, G_2, \ldots, G_q$ such that

$$D = E_p E_{p-1} \cdots E_2 E_1 A G_1 G_2 \cdots G_q.$$

By Theorem 3.2 (p. 150), each E_j and G_j is invertible. Let $B = E_p E_{p-1} \cdots E_1$ and $C = G_1 G_2 \cdots G_q$. Then B and C are invertible by Exercise 4 of Section 2.4, and $D = BAC$. ∎

Corollary 2. *Let A be an $m \times n$ matrix. Then*
(a) $\operatorname{rank}(A^t) = \operatorname{rank}(A)$.
(b) *The rank of any matrix equals the maximum number of its linearly independent rows; that is, the rank of a matrix is the dimension of the subspace generated by its rows.*
(c) *The rows and columns of any matrix generate subspaces of the same dimension, numerically equal to the rank of the matrix.*

Proof. (a) By Corollary 1, there exist invertible matrices B and C such that $D = BAC$, where D satisfies the stated conditions of the corollary. Taking transposes, we have

$$D^t = (BAC)^t = C^t A^t B^t.$$

Since B and C are invertible, so are B^t and C^t by Exercise 5 of Section 2.4. Hence by Theorem 3.4,

$$\operatorname{rank}(A^t) = \operatorname{rank}(C^t A^t B^t) = \operatorname{rank}(D^t).$$

Suppose that $r = \operatorname{rank}(A)$. Then D^t is an $n \times m$ matrix with the form of the matrix D in Corollary 1, and hence $\operatorname{rank}(D^t) = r$ by Theorem 3.5. Thus

$$\operatorname{rank}(A^t) = \operatorname{rank}(D^t) = r = \operatorname{rank}(A).$$

This establishes (a).

The proofs of (b) and (c) are left as exercises. (See Exercise 13.) ∎

Corollary 3. *Every invertible matrix is a product of elementary matrices.*

Proof. If A is an invertible $n \times n$ matrix, then $\text{rank}(A) = n$. Hence the matrix D in Corollary 1 equals I_n, and there exist invertible matrices B and C such that $I_n = BAC$.

As in the proof of Corollary 1, note that $B = E_p E_{p-1} \cdots E_1$ and $C = G_1 G_2 \cdots G_q$, where the E_i's and G_i's are elementary matrices. Thus $A = B^{-1} I_n C^{-1} = B^{-1} C^{-1}$, so that

$$A = E_1^{-1} E_2^{-1} \cdots E_p^{-1} G_q^{-1} G_{q-1}^{-1} \cdots G_1^{-1}.$$

The inverses of elementary matrices are elementary matrices, however, and hence A is the product of elementary matrices. ∎

We now use Corollary 2 to relate the rank of a matrix product to the rank of each factor. Notice how the proof exploits the relationship between the rank of a matrix and the rank of a linear transformation.

Theorem 3.7. *Let $\mathsf{T}\colon \mathsf{V} \to \mathsf{W}$ and $\mathsf{U}\colon \mathsf{W} \to \mathsf{Z}$ be linear transformations on finite-dimensional vector spaces V, W, and Z, and let A and B be matrices such that the product AB is defined. Then*
  (a) $\text{rank}(\mathsf{UT}) \leq \text{rank}(\mathsf{U})$.
  (b) $\text{rank}(\mathsf{UT}) \leq \text{rank}(\mathsf{T})$.
  (c) $\text{rank}(AB) \leq \text{rank}(A)$.
  (d) $\text{rank}(AB) \leq \text{rank}(B)$.

Proof. We prove these items in the order: (a), (c), (d), and (b).
(a) Clearly, $\mathsf{R}(\mathsf{T}) \subseteq \mathsf{W}$. Hence

$$\mathsf{R}(\mathsf{UT}) = \mathsf{UT}(\mathsf{V}) = \mathsf{U}(\mathsf{T}(\mathsf{V})) = \mathsf{U}(\mathsf{R}(\mathsf{T})) \subseteq \mathsf{U}(\mathsf{W}) = \mathsf{R}(\mathsf{U}).$$

Thus

$$\text{rank}(\mathsf{UT}) = \dim(\mathsf{R}(\mathsf{UT})) \leq \dim(\mathsf{R}(\mathsf{U})) = \text{rank}(\mathsf{U}).$$

(c) By (a),

$$\text{rank}(AB) = \text{rank}(\mathsf{L}_{AB}) = \text{rank}(\mathsf{L}_A \mathsf{L}_B) \leq \text{rank}(\mathsf{L}_A) = \text{rank}(A).$$

(d) By (c) and Corollary 2 to Theorem 3.6,

$$\text{rank}(AB) = \text{rank}((AB)^t) = \text{rank}(B^t A^t) \leq \text{rank}(B^t) = \text{rank}(B).$$

(b) Let α, β, and γ be ordered bases for V, W, and Z, respectively, and let $A' = [\mathsf{U}]_\beta^\gamma$ and $B' = [\mathsf{T}]_\alpha^\beta$. Then $A'B' = [\mathsf{UT}]_\alpha^\gamma$ by Theorem 2.11 (p. 88). Hence, by Theorem 3.3 and (d),

$$\text{rank}(\mathsf{UT}) = \text{rank}(A'B') \leq \text{rank}(B') = \text{rank}(\mathsf{T}).$$ ∎

It is important to be able to compute the rank of any matrix. We can use the corollary to Theorem 3.4, Theorems 3.5 and 3.6, and Corollary 2 to Theorem 3.6 to accomplish this goal.

The object is to perform elementary row and column operations on a matrix to "simplify" it (so that the transformed matrix has many zero entries) to the point where a simple observation enables us to determine how many linearly independent rows or columns the matrix has, and thus to determine its rank.

Example 4

(a) Let

$$A = \begin{pmatrix} 1 & 2 & 1 & 1 \\ 1 & 1 & -1 & 1 \end{pmatrix}.$$

Note that the first and second rows of A are linearly independent since one is not a multiple of the other. Thus $\text{rank}(A) = 2$.

(b) Let

$$A = \begin{pmatrix} 1 & 3 & 1 & 1 \\ 1 & 0 & 1 & 1 \\ 0 & 3 & 0 & 0 \end{pmatrix}.$$

In this case, there are several ways to proceed. Suppose that we begin with an elementary row operation to obtain a zero in the 2,1 position. Subtracting the first row from the second row, we obtain

$$\begin{pmatrix} 1 & 3 & 1 & 1 \\ 0 & -3 & 0 & 0 \\ 0 & 3 & 0 & 0 \end{pmatrix}.$$

Now note that the third row is a multiple of the second row, and the first and second rows are linearly independent. Thus $\text{rank}(A) = 2$.

As an alternative method, note that the first, third, and fourth columns of A are identical and that the first and second columns of A are linearly independent. Hence $\text{rank}(A) = 2$.

(c) Let

$$A = \begin{pmatrix} 1 & 2 & 3 & 1 \\ 2 & 1 & 1 & 1 \\ 1 & -1 & 1 & 0 \end{pmatrix}.$$

Using elementary row operations, we can transform A as follows:

$$A \longrightarrow \begin{pmatrix} 1 & 2 & 3 & 1 \\ 0 & -3 & -5 & -1 \\ 0 & -3 & -2 & -1 \end{pmatrix} \longrightarrow \begin{pmatrix} 1 & 2 & 3 & 1 \\ 0 & -3 & -5 & -1 \\ 0 & 0 & 3 & 0 \end{pmatrix}.$$

It is clear that the last matrix has three linearly independent rows and hence has rank 3.   ◆

In summary, perform row and column operations until the matrix is simplified enough so that the maximum number of linearly independent rows or columns is obvious.

The Inverse of a Matrix

We have remarked that an $n \times n$ matrix is invertible if and only if its rank is n. Since we know how to compute the rank of any matrix, we can always test a matrix to determine whether it is invertible. We now provide a simple technique for computing the inverse of a matrix that utilizes elementary row operations.

Definition.   Let A and B be $m \times n$ and $m \times p$ matrices, respectively. By the **augmented matrix** $(A|B)$, we mean the $m \times (n+p)$ matrix $(A \ B)$, that is, the matrix whose first n columns are the columns of A, and whose last p columns are the columns of B.

Let A be an invertible $n \times n$ matrix, and consider the $n \times 2n$ augmented matrix $C = (A|I_n)$. By Exercise 15, we have

$$A^{-1}C = (A^{-1}A|A^{-1}I_n) = (I_n|A^{-1}). \tag{1}$$

By Corollary 3 to Theorem 3.6, A^{-1} is the product of elementary matrices, say $A^{-1} = E_p E_{p-1} \cdots E_1$. Thus (1) becomes

$$E_p E_{p-1} \cdots E_1(A|I_n) = A^{-1}C = (I_n|A^{-1}).$$

Because multiplying a matrix on the left by an elementary matrix transforms the matrix by an elementary row operation (Theorem 3.1 p. 149), we have the following result: *If A is an invertible $n \times n$ matrix, then it is possible to transform the matrix $(A|I_n)$ into the matrix $(I_n|A^{-1})$ by means of a finite number of elementary row operations.*

Conversely, suppose that A is invertible and that, for some $n \times n$ matrix B, the matrix $(A|I_n)$ can be transformed into the matrix $(I_n|B)$ by a finite number of elementary row operations. Let $E_1, E_2, \ldots, E_p$ be the elementary matrices associated with these elementary row operations as in Theorem 3.1; then

$$E_p E_{p-1} \cdots E_1(A|I_n) = (I_n|B). \tag{2}$$

Letting $M = E_p E_{p-1} \cdots E_1$, we have from (2) that

$$(MA|M) = M(A|I_n) = (I_n|B).$$

Hence $MA = I_n$ and $M = B$. It follows that $M = A^{-1}$. So $B = A^{-1}$. Thus we have the following result: *If A is an invertible $n \times n$ matrix, and the matrix $(A|I_n)$ is transformed into a matrix of the form $(I_n|B)$ by means of a finite number of elementary row operations, then $B = A^{-1}$.*

If, on the other hand, A is an $n \times n$ matrix that is not invertible, then $\text{rank}(A) < n$. Hence any attempt to transform $(A|I_n)$ into a matrix of the form $(I_n|B)$ by means of elementary row operations must fail because otherwise A can be transformed into I_n using the same row operations. This is impossible, however, because elementary row operations preserve rank. In fact, A can be transformed into a matrix with a row containing only zero entries, yielding the following result: *If A is an $n \times n$ matrix that is not invertible, then any attempt to transform $(A|I_n)$ into a matrix of the form $(I_n|B)$ produces a row whose first n entries are zeros.*

The next two examples demonstrate these comments.

Example 5

We determine whether the matrix

$$A = \begin{pmatrix} 0 & 2 & 4 \\ 2 & 4 & 2 \\ 3 & 3 & 1 \end{pmatrix}$$

is invertible, and if it is, we compute its inverse.

We attempt to use elementary row operations to transform

$$(A|I) = \begin{pmatrix} 0 & 2 & 4 & | & 1 & 0 & 0 \\ 2 & 4 & 2 & | & 0 & 1 & 0 \\ 3 & 3 & 1 & | & 0 & 0 & 1 \end{pmatrix}$$

into a matrix of the form $(I|B)$. One method for accomplishing this transformation is to change each column of A successively, beginning with the first column, into the corresponding column of I. Since we need a nonzero entry in the 1,1 position, we begin by interchanging rows 1 and 2. The result is

$$\begin{pmatrix} 2 & 4 & 2 & | & 0 & 1 & 0 \\ 0 & 2 & 4 & | & 1 & 0 & 0 \\ 3 & 3 & 1 & | & 0 & 0 & 1 \end{pmatrix}.$$

In order to place a 1 in the 1,1 position, we must multiply the first row by $\frac{1}{2}$; this operation yields

$$\begin{pmatrix} 1 & 2 & 1 & | & 0 & \frac{1}{2} & 0 \\ 0 & 2 & 4 & | & 1 & 0 & 0 \\ 3 & 3 & 1 & | & 0 & 0 & 1 \end{pmatrix}.$$

We now complete work in the first column by adding -3 times row 1 to row 3 to obtain

$$\left(\begin{array}{ccc|ccc} 1 & 2 & 1 & 0 & \frac{1}{2} & 0 \\ 0 & 2 & 4 & 1 & 0 & 0 \\ 0 & -3 & -2 & 0 & -\frac{3}{2} & 1 \end{array}\right).$$

In order to change the second column of the preceding matrix into the second column of I, we multiply row 2 by $\frac{1}{2}$ to obtain a 1 in the 2,2 position. This operation produces

$$\left(\begin{array}{ccc|ccc} 1 & 2 & 1 & 0 & \frac{1}{2} & 0 \\ 0 & 1 & 2 & \frac{1}{2} & 0 & 0 \\ 0 & -3 & -2 & 0 & -\frac{3}{2} & 1 \end{array}\right).$$

We now complete our work on the second column by adding -2 times row 2 to row 1 and 3 times row 2 to row 3. The result is

$$\left(\begin{array}{ccc|ccc} 1 & 0 & -3 & -1 & \frac{1}{2} & 0 \\ 0 & 1 & 2 & \frac{1}{2} & 0 & 0 \\ 0 & 0 & 4 & \frac{3}{2} & -\frac{3}{2} & 1 \end{array}\right).$$

Only the third column remains to be changed. In order to place a 1 in the 3,3 position, we multiply row 3 by $\frac{1}{4}$; this operation yields

$$\left(\begin{array}{ccc|ccc} 1 & 0 & -3 & -1 & \frac{1}{2} & 0 \\ 0 & 1 & 2 & \frac{1}{2} & 0 & 0 \\ 0 & 0 & 1 & \frac{3}{8} & -\frac{3}{8} & \frac{1}{4} \end{array}\right).$$

Adding appropriate multiples of row 3 to rows 1 and 2 completes the process and gives

$$\left(\begin{array}{ccc|ccc} 1 & 0 & 0 & \frac{1}{8} & -\frac{5}{8} & \frac{3}{4} \\ 0 & 1 & 0 & -\frac{1}{4} & \frac{3}{4} & -\frac{1}{2} \\ 0 & 0 & 1 & \frac{3}{8} & -\frac{3}{8} & \frac{1}{4} \end{array}\right).$$

Thus A is invertible, and

$$A^{-1} = \left(\begin{array}{ccc} \frac{1}{8} & -\frac{5}{8} & \frac{3}{4} \\ -\frac{1}{4} & \frac{3}{4} & -\frac{1}{2} \\ \frac{3}{8} & -\frac{3}{8} & \frac{1}{4} \end{array}\right). \quad \blacklozenge$$

Example 6

We determine whether the matrix

$$A = \begin{pmatrix} 1 & 2 & 1 \\ 2 & 1 & -1 \\ 1 & 5 & 4 \end{pmatrix}$$

is invertible, and if it is, we compute its inverse. Using a strategy similar to the one used in Example 5, we attempt to use elementary row operations to transform

$$(A|I) = \begin{pmatrix} 1 & 2 & 1 & | & 1 & 0 & 0 \\ 2 & 1 & -1 & | & 0 & 1 & 0 \\ 1 & 5 & 4 & | & 0 & 0 & 1 \end{pmatrix}$$

into a matrix of the form $(I|B)$. We first add -2 times row 1 to row 2 and -1 times row 1 to row 3. We then add row 2 to row 3. The result,

$$\begin{pmatrix} 1 & 2 & 1 & | & 1 & 0 & 0 \\ 2 & 1 & -1 & | & 0 & 1 & 0 \\ 1 & 5 & 4 & | & 0 & 0 & 1 \end{pmatrix} \longrightarrow \begin{pmatrix} 1 & 2 & 1 & | & 1 & 0 & 0 \\ 0 & -3 & -3 & | & -2 & 1 & 0 \\ 0 & 3 & 3 & | & -1 & 0 & 1 \end{pmatrix}$$

$$\longrightarrow \begin{pmatrix} 1 & 2 & 1 & | & 1 & 0 & 0 \\ 0 & -3 & -3 & | & -2 & 1 & 0 \\ 0 & 0 & 0 & | & -3 & 1 & 1 \end{pmatrix},$$

is a matrix with a row whose first 3 entries are zeros. Therefore A is not invertible.   ◆

Being able to test for invertibility and compute the inverse of a matrix allows us, with the help of Theorem 2.18 (p. 101) and its corollaries, to test for invertibility and compute the inverse of a linear transformation. The next example demonstrates this technique.

Example 7

Let $T: P_2(R) \to P_2(R)$ be defined by $T(f(x)) = f(x) + f'(x) + f''(x)$, where $f'(x)$ and $f''(x)$ denote the first and second derivatives of $f(x)$. We use Corollary 1 of Theorem 2.18 (p. 102) to test T for invertibility and compute the inverse if T is invertible. Taking β to be the standard ordered basis of $P_2(R)$, we have

$$[T]_\beta = \begin{pmatrix} 1 & 1 & 2 \\ 0 & 1 & 2 \\ 0 & 0 & 1 \end{pmatrix}.$$

Using the method of Examples 5 and 6, we can show that $[T]_\beta$ is invertible with inverse

$$([T]_\beta)^{-1} = \begin{pmatrix} 1 & -1 & 0 \\ 0 & 1 & -2 \\ 0 & 0 & 1 \end{pmatrix}.$$

Thus T is invertible, and $([T]_\beta)^{-1} = [T^{-1}]_\beta$. Hence by Theorem 2.14 (p. 91), we have

$$[T^{-1}(a_0 + a_1 x + a_2 x^2)]_\beta = \begin{pmatrix} 1 & -1 & 0 \\ 0 & 1 & -2 \\ 0 & 0 & 1 \end{pmatrix} \begin{pmatrix} a_0 \\ a_1 \\ a_2 \end{pmatrix}$$

$$= \begin{pmatrix} a_0 - a_1 \\ a_1 - 2a_2 \\ a_2 \end{pmatrix}.$$

Therefore

$$T^{-1}(a_0 + a_1 x + a_2 x^2) = (a_0 - a_1) + (a_1 - 2a_2)x + a_2 x^2. \quad \blacklozenge$$

EXERCISES

1. Label the following statements as true or false.

   (a) The rank of a matrix is equal to the number of its nonzero columns.
   (b) The product of two matrices always has rank equal to the lesser of the ranks of the two matrices.
   (c) The $m \times n$ zero matrix is the only $m \times n$ matrix having rank 0.
   (d) Elementary row operations preserve rank.
   (e) Elementary column operations do not necessarily preserve rank.
   (f) The rank of a matrix is equal to the maximum number of linearly independent rows in the matrix.
   (g) The inverse of a matrix can be computed exclusively by means of elementary row operations.
   (h) The rank of an $n \times n$ matrix is at most n.
   (i) An $n \times n$ matrix having rank n is invertible.

2. Find the rank of the following matrices.

   (a) $\begin{pmatrix} 1 & 1 & 0 \\ 0 & 1 & 1 \\ 1 & 1 & 0 \end{pmatrix}$ (b) $\begin{pmatrix} 1 & 1 & 0 \\ 2 & 1 & 1 \\ 1 & 1 & 1 \end{pmatrix}$ (c) $\begin{pmatrix} 1 & 0 & 2 \\ 1 & 1 & 4 \end{pmatrix}$

(d) $\begin{pmatrix} 1 & 2 & 1 \\ 2 & 4 & 2 \end{pmatrix}$

(e) $\begin{pmatrix} 1 & 2 & 3 & 1 & 1 \\ 1 & 4 & 0 & 1 & 2 \\ 0 & 2 & -3 & 0 & 1 \\ 1 & 0 & 0 & 0 & 0 \end{pmatrix}$

(f) $\begin{pmatrix} 1 & 2 & 0 & 1 & 1 \\ 2 & 4 & 1 & 3 & 0 \\ 3 & 6 & 2 & 5 & 1 \\ -4 & -8 & 1 & -3 & 1 \end{pmatrix}$

(g) $\begin{pmatrix} 1 & 1 & 0 & 1 \\ 2 & 2 & 0 & 2 \\ 1 & 1 & 0 & 1 \\ 1 & 1 & 0 & 1 \end{pmatrix}$

3. Prove that for any $m \times n$ matrix A, $\text{rank}(A) = 0$ if and only if A is the zero matrix.

4. Use elementary row and column operations to transform each of the following matrices into a matrix D satisfying the conditions of Theorem 3.6, and then determine the rank of each matrix.

(a) $\begin{pmatrix} 1 & 1 & 1 & 2 \\ 2 & 0 & -1 & 2 \\ 1 & 1 & 1 & 2 \end{pmatrix}$ (b) $\begin{pmatrix} 2 & 1 \\ -1 & 2 \\ 2 & 1 \end{pmatrix}$

5. For each of the following matrices, compute the rank and the inverse if it exists.

(a) $\begin{pmatrix} 1 & 2 \\ 1 & 1 \end{pmatrix}$ (b) $\begin{pmatrix} 1 & 2 \\ 2 & 4 \end{pmatrix}$ (c) $\begin{pmatrix} 1 & 2 & 1 \\ 1 & 3 & 4 \\ 2 & 3 & -1 \end{pmatrix}$

(d) $\begin{pmatrix} 0 & -2 & 4 \\ 1 & 1 & -1 \\ 2 & 4 & -5 \end{pmatrix}$ (e) $\begin{pmatrix} 1 & 2 & 1 \\ -1 & 1 & 2 \\ 1 & 0 & 1 \end{pmatrix}$ (f) $\begin{pmatrix} 1 & 2 & 1 \\ 1 & 0 & 1 \\ 1 & 1 & 1 \end{pmatrix}$

(g) $\begin{pmatrix} 1 & 2 & 1 & 0 \\ 2 & 5 & 5 & 1 \\ -2 & -3 & 0 & 3 \\ 3 & 4 & -2 & -3 \end{pmatrix}$ (h) $\begin{pmatrix} 1 & 0 & 1 & 1 \\ 1 & 1 & -1 & 2 \\ 2 & 0 & 1 & 0 \\ 0 & -1 & 1 & -3 \end{pmatrix}$

6. For each of the following linear transformations T, determine whether T is invertible, and compute T^{-1} if it exists.

(a) T: $P_2(R) \rightarrow P_2(R)$ defined by $T(f(x)) = f''(x) + 2f'(x) - f(x)$.
(b) T: $P_2(R) \rightarrow P_2(R)$ defined by $T(f(x)) = (x + 1)f'(x)$.
(c) T: $R^3 \rightarrow R^3$ defined by

$$T(a_1, a_2, a_3) = (a_1 + 2a_2 + a_3, -a_1 + a_2 + 2a_3, a_1 + a_3).$$

(d) $T: R^3 \to P_2(R)$ defined by

$$T(a_1, a_2, a_3) = (a_1 + a_2 + a_3) + (a_1 - a_2 + a_3)x + a_1 x^2.$$

(e) $T: P_2(R) \to R^3$ defined by $T(f(x)) = (f(-1), f(0), f(1))$.

(f) $T: M_{2 \times 2}(R) \to R^4$ defined by

$$T(A) = (\operatorname{tr}(A), \operatorname{tr}(A^t), \operatorname{tr}(EA), \operatorname{tr}(AE)),$$

where

$$E = \begin{pmatrix} 0 & 1 \\ 1 & 0 \end{pmatrix}.$$

7. Express the invertible matrix

$$\begin{pmatrix} 1 & 2 & 1 \\ 1 & 0 & 1 \\ 1 & 1 & 2 \end{pmatrix}$$

   as a product of elementary matrices.

8. Let A be an $m \times n$ matrix. Prove that if c is any nonzero scalar, then $\operatorname{rank}(cA) = \operatorname{rank}(A)$.

9. Complete the proof of the corollary to Theorem 3.4 by showing that elementary column operations preserve rank.

10. Prove Theorem 3.6 for the case that A is an $m \times 1$ matrix.

11. Let

$$B = \begin{pmatrix} 1 & 0 & \cdots & 0 \\ 0 & & & \\ \vdots & & B' & \\ 0 & & & \end{pmatrix},$$

   where B' is an $m \times n$ submatrix of B. Prove that if $\operatorname{rank}(B) = r$, then $\operatorname{rank}(B') = r - 1$.

12. Let B' and D' be $m \times n$ matrices, and let B and D be $(m+1) \times (n+1)$ matrices respectively defined by

$$B = \begin{pmatrix} 1 & 0 & \cdots & 0 \\ 0 & & & \\ \vdots & & B' & \\ 0 & & & \end{pmatrix} \quad \text{and} \quad D = \begin{pmatrix} 1 & 0 & \cdots & 0 \\ 0 & & & \\ \vdots & & D' & \\ 0 & & & \end{pmatrix}.$$

   Prove that if B' can be transformed into D' by an elementary row [column] operation, then B can be transformed into D by an elementary row [column] operation.

13. Prove (b) and (c) of Corollary 2 to Theorem 3.6.

14. Let $T, U: V \to W$ be linear transformations.
  (a) Prove that $R(T+U) \subseteq R(T)+R(U)$. (See the definition of the sum of subsets of a vector space on page 22.)
  (b) Prove that if W is finite-dimensional, then $\text{rank}(T+U) \leq \text{rank}(T)+\text{rank}(U)$.
  (c) Deduce from (b) that $\text{rank}(A+B) \leq \text{rank}(A)+\text{rank}(B)$ for any $m \times n$ matrices A and B.

15. Suppose that A and B are matrices having n rows. Prove that $M(A|B) = (MA|MB)$ for any $m \times n$ matrix M.

16. Supply the details to the proof of (b) of Theorem 3.4.

17. Prove that if B is a 3×1 matrix and C is a 1×3 matrix, then the 3×3 matrix BC has rank at most 1. Conversely, show that if A is any 3×3 matrix having rank 1, then there exist a 3×1 matrix B and a 1×3 matrix C such that $A = BC$.

18. Let A be an $m \times n$ matrix and B be an $n \times p$ matrix. Prove that AB can be written as a sum of n matrices of rank at most one.

19. Let A be an $m \times n$ matrix with rank m and B be an $n \times p$ matrix with rank n. Determine the rank of AB. Justify your answer.

20. Let

$$A = \begin{pmatrix} 1 & 0 & -1 & 2 & 1 \\ -1 & 1 & 3 & -1 & 0 \\ -2 & 1 & 4 & -1 & 3 \\ 3 & -1 & -5 & 1 & -6 \end{pmatrix}.$$

  (a) Find a 5×5 matrix M with rank 2 such that $AM = O$, where O is the 4×5 zero matrix.
  (b) Suppose that B is a 5×5 matrix such that $AB = O$. Prove that $\text{rank}(B) \leq 2$.

21. Let A be an $m \times n$ matrix with rank m. Prove that there exists an $n \times m$ matrix B such that $AB = I_m$.

22. Let B be an $n \times m$ matrix with rank m. Prove that there exists an $m \times n$ matrix A such that $AB = I_m$.

3.3   SYSTEMS OF LINEAR EQUATIONS—THEORETICAL ASPECTS

This section and the next are devoted to the study of systems of linear equations, which arise naturally in both the physical and social sciences. In this section, we apply results from Chapter 2 to describe the solution sets of systems of linear equations as subsets of a vector space. In Section 3.4, elementary row operations are used to provide a computational method for finding all solutions to such systems.

The system of equations

$$(S) \qquad \begin{aligned} a_{11}x_1 + a_{12}x_2 + \cdots + a_{1n}x_n &= b_1 \\ a_{21}x_1 + a_{22}x_2 + \cdots + a_{2n}x_n &= b_2 \\ &\vdots \\ a_{m1}x_1 + a_{m2}x_2 + \cdots + a_{mn}x_n &= b_m, \end{aligned}$$

where a_{ij} and b_i $(1 \le i \le m$ and $1 \le j \le n)$ are scalars in a field F and $x_1, x_2, \ldots, x_n$ are n variables taking values in F, is called a **system of** m **linear equations in** n **unknowns over the field** F.

The $m \times n$ matrix

$$A = \begin{pmatrix} a_{11} & a_{12} & \cdots & a_{1n} \\ a_{21} & a_{22} & \cdots & a_{2n} \\ \vdots & \vdots & & \vdots \\ a_{m1} & a_{m2} & \cdots & a_{mn} \end{pmatrix}$$

is called the **coefficient matrix** of the system (S).

If we let

$$x = \begin{pmatrix} x_1 \\ x_2 \\ \vdots \\ x_n \end{pmatrix} \quad \text{and} \quad b = \begin{pmatrix} b_1 \\ b_2 \\ \vdots \\ b_m \end{pmatrix},$$

then the system (S) may be rewritten as a single matrix equation

$$Ax = b.$$

To exploit the results that we have developed, we often consider a system of linear equations as a single matrix equation.

A **solution** to the system (S) is an n-tuple

$$s = \begin{pmatrix} s_1 \\ s_2 \\ \vdots \\ s_n \end{pmatrix} \in \mathsf{F}^n$$

such that $As = b$. The set of all solutions to the system (S) is called the **solution set** of the system. System (S) is called **consistent** if its solution set is nonempty; otherwise it is called **inconsistent**.

Example 1

(a) Consider the system

$$x_1 + x_2 = 3$$
$$x_1 - x_2 = 1.$$

By use of familiar techniques, we can solve the preceding system and conclude that there is only one solution: $x_1 = 2$, $x_2 = 1$; that is,

$$s = \begin{pmatrix} 2 \\ 1 \end{pmatrix}.$$

In matrix form, the system can be written

$$\begin{pmatrix} 1 & 1 \\ 1 & -1 \end{pmatrix} \begin{pmatrix} x_1 \\ x_2 \end{pmatrix} = \begin{pmatrix} 3 \\ 1 \end{pmatrix};$$

so

$$A = \begin{pmatrix} 1 & 1 \\ 1 & -1 \end{pmatrix} \quad \text{and} \quad B = \begin{pmatrix} 3 \\ 1 \end{pmatrix}.$$

(b) Consider

$$2x_1 + 3x_2 + x_3 = 1$$
$$x_1 - x_2 + 2x_3 = 6;$$

that is,

$$\begin{pmatrix} 2 & 3 & 1 \\ 1 & -1 & 2 \end{pmatrix} \begin{pmatrix} x_1 \\ x_2 \\ x_3 \end{pmatrix} = \begin{pmatrix} 1 \\ 6 \end{pmatrix}.$$

This system has many solutions, such as

$$s = \begin{pmatrix} -6 \\ 2 \\ 7 \end{pmatrix} \quad \text{and} \quad s = \begin{pmatrix} 8 \\ -4 \\ -3 \end{pmatrix}.$$

(c) Consider

$$x_1 + x_2 = 0$$
$$x_1 + x_2 = 1;$$

that is,

$$\begin{pmatrix} 1 & 1 \\ 1 & 1 \end{pmatrix} \begin{pmatrix} x_1 \\ x_2 \end{pmatrix} = \begin{pmatrix} 0 \\ 1 \end{pmatrix}.$$

It is evident that this system has no solutions. Thus we see that a system of linear equations can have one, many, or no solutions.   ◆

We must be able to recognize when a system has a solution and then be able to describe all its solutions. This section and the next are devoted to this end.

We begin our study of systems of linear equations by examining the class of *homogeneous* systems of linear equations. Our first result (Theorem 3.8) shows that the set of solutions to a homogeneous system of m linear equations in n unknowns forms a subspace of F^n. We can then apply the theory of vector spaces to this set of solutions. For example, a basis for the solution space can be found, and any solution can be expressed as a linear combination of the vectors in the basis.

Definitions.   *A system $Ax = b$ of m linear equations in n unknowns is said to be **homogeneous** if $b = 0$. Otherwise the system is said to be **nonhomogeneous**.*

Any homogeneous system has at least one solution, namely, the zero vector. The next result gives further information about the set of solutions to a homogeneous system.

Theorem 3.8. *Let $Ax = 0$ be a homogeneous system of m linear equations in n unknowns over a field F. Let K denote the set of all solutions to $Ax = 0$. Then $\mathsf{K} = \mathsf{N}(\mathsf{L}_A)$; hence K is a subspace of F^n of dimension $n - \mathrm{rank}(\mathsf{L}_A) = n - \mathrm{rank}(A)$.*

Proof. Clearly, $\mathsf{K} = \{s \in F^n : As = 0\} = \mathsf{N}(\mathsf{L}_A)$. The second part now follows from the dimension theorem (p. 70). ∎

Corollary. *If $m < n$, the system $Ax = 0$ has a nonzero solution.*

Proof. Suppose that $m < n$. Then $\mathrm{rank}(A) = \mathrm{rank}(\mathsf{L}_A) \leq m$. Hence

$$\dim(\mathsf{K}) = n - \mathrm{rank}(\mathsf{L}_A) \geq n - m > 0,$$

where $\mathsf{K} = \mathsf{N}(\mathsf{L}_A)$. Since $\dim(\mathsf{K}) > 0$, $\mathsf{K} \neq \{0\}$. Thus there exists a nonzero vector $s \in \mathsf{K}$; so s is a nonzero solution to $Ax = 0$. ∎

Example 2

(a) Consider the system

$$\begin{aligned} x_1 + 2x_2 + x_3 &= 0 \\ x_1 - x_2 - x_3 &= 0. \end{aligned}$$

Let

$$A = \begin{pmatrix} 1 & 2 & 1 \\ 1 & -1 & -1 \end{pmatrix}$$

be the coefficient matrix of this system. It is clear that $\text{rank}(A) = 2$. If K is the solution set of this system, then $\dim(K) = 3 - 2 = 1$. Thus any nonzero solution constitutes a basis for K. For example, since

$$\begin{pmatrix} 1 \\ -2 \\ 3 \end{pmatrix}$$

is a solution to the given system,

$$\left\{ \begin{pmatrix} 1 \\ -2 \\ 3 \end{pmatrix} \right\}$$

is a basis for K. Thus any vector in K is of the form

$$t \begin{pmatrix} 1 \\ -2 \\ 3 \end{pmatrix} = \begin{pmatrix} t \\ -2t \\ 3t \end{pmatrix},$$

where $t \in R$.

(b) Consider the system $x_1 - 2x_2 + x_3 = 0$ of one equation in three unknowns. If $A = \begin{pmatrix} 1 & -2 & 1 \end{pmatrix}$ is the coefficient matrix, then $\text{rank}(A) = 1$. Hence if K is the solution set, then $\dim(K) = 3 - 1 = 2$. Note that

$$\begin{pmatrix} 2 \\ 1 \\ 0 \end{pmatrix} \quad \text{and} \quad \begin{pmatrix} -1 \\ 0 \\ 1 \end{pmatrix}$$

are linearly independent vectors in K. Thus they constitute a basis for K, so that

$$K = \left\{ t_1 \begin{pmatrix} 2 \\ 1 \\ 0 \end{pmatrix} + t_2 \begin{pmatrix} -1 \\ 0 \\ 1 \end{pmatrix} : t_1, t_2 \in R \right\}. \quad \blacklozenge$$

In Section 3.4, explicit computational methods for finding a basis for the solution set of a homogeneous system are discussed.

We now turn to the study of nonhomogeneous systems. Our next result shows that the solution set of a nonhomogeneous system $Ax = b$ can be described in terms of the solution set of the homogeneous system $Ax = 0$. We refer to the equation $Ax = 0$ as the **homogeneous system corresponding to** $Ax = b$.

Theorem 3.9. Let K be the solution set of a system of linear equations $Ax = b$, and let K_H be the solution set of the corresponding homogeneous system $Ax = 0$. Then for any solution s to $Ax = b$

$$K = \{s\} + K_H = \{s + k : k \in K_H\}.$$

Proof. Let s be any solution to $Ax = b$. We must show that $K = \{s\} + \mathsf{K_H}$. If $w \in K$, then $Aw = b$. Hence

$$A(w - s) = Aw - As = b - b = 0.$$

So $w - s \in \mathsf{K_H}$. Thus there exists $k \in \mathsf{K_H}$ such that $w - s = k$. It follows that $w = s + k \in \{s\} + \mathsf{K_H}$, and therefore

$$K \subseteq \{s\} + \mathsf{K_H}.$$

Conversely, suppose that $w \in \{s\} + \mathsf{K_H}$; then $w = s + k$ for some $k \in \mathsf{K_H}$. But then $Aw = A(s + k) = As + Ak = b + 0 = b$; so $w \in K$. Therefore $\{s\} + \mathsf{K_H} \subseteq K$, and thus $K = \{s\} + \mathsf{K_H}$. ∎

Example 3

(a) Consider the system

$$\begin{aligned} x_1 + 2x_2 + x_3 &= 7 \\ x_1 - x_2 - x_3 &= -4. \end{aligned}$$

The corresponding homogeneous system is the system in Example 2(a). It is easily verified that

$$s = \begin{pmatrix} 1 \\ 1 \\ 4 \end{pmatrix}$$

is a solution to the preceding nonhomogeneous system. So the solution set of the system is

$$\mathsf{K} = \left\{ \begin{pmatrix} 1 \\ 1 \\ 4 \end{pmatrix} + t \begin{pmatrix} 1 \\ -2 \\ 3 \end{pmatrix} : t \in R \right\}$$

by Theorem 3.9.

(b) Consider the system $x_1 - 2x_2 + x_3 = 4$. The corresponding homogeneous system is the system in Example 2(b). Since

$$s = \begin{pmatrix} 4 \\ 0 \\ 0 \end{pmatrix}$$

is a solution to the given system, the solution set K can be written as

$$\mathsf{K} = \left\{ \begin{pmatrix} 4 \\ 0 \\ 0 \end{pmatrix} + t_1 \begin{pmatrix} 2 \\ 1 \\ 0 \end{pmatrix} + t_2 \begin{pmatrix} -1 \\ 0 \\ 1 \end{pmatrix} : t_1, t_2 \in R \right\}. \quad \blacklozenge$$

The following theorem provides us with a means of computing solutions to certain systems of linear equations.

Theorem 3.10. *Let* $Ax = b$ *be a system of* n *linear equations in* n *unknowns. If* A *is invertible, then the system has exactly one solution, namely,* $A^{-1}b$. *Conversely, if the system has exactly one solution, then* A *is invertible.*

Proof. Suppose that A is invertible. Substituting $A^{-1}b$ into the system, we have $A(A^{-1}b) = (AA^{-1})b = b$. Thus $A^{-1}b$ is a solution. If s is an arbitrary solution, then $As = b$. Multiplying both sides by A^{-1} gives $s = A^{-1}b$. Thus the system has one and only one solution, namely, $A^{-1}b$.

Conversely, suppose that the system has exactly one solution s. Let $\mathsf{K_H}$ denote the solution set for the corresponding homogeneous system $Ax = 0$. By Theorem 3.9, $\{s\} = \{s\} + \mathsf{K_H}$. But this is so only if $\mathsf{K_H} = \{0\}$. Thus $\mathsf{N}(\mathsf{L}_A) = \{0\}$, and hence A is invertible. ∎

Example 4

Consider the following system of three linear equations in three unknowns:

$$
\begin{aligned}
2x_2 + 4x_3 &= 2 \\
2x_1 + 4x_2 + 2x_3 &= 3 \\
3x_1 + 3x_2 + x_3 &= 1.
\end{aligned}
$$

In Example 5 of Section 3.2, we computed the inverse of the coefficient matrix A of this system. Thus the system has exactly one solution, namely,

$$
\begin{pmatrix} x_1 \\ x_2 \\ x_3 \end{pmatrix} = A^{-1}b = \begin{pmatrix} \frac{1}{8} & -\frac{5}{8} & \frac{3}{4} \\ -\frac{1}{4} & \frac{3}{4} & -\frac{1}{2} \\ \frac{3}{8} & -\frac{3}{8} & \frac{1}{4} \end{pmatrix} \begin{pmatrix} 2 \\ 3 \\ 1 \end{pmatrix} = \begin{pmatrix} -\frac{7}{8} \\ \frac{5}{4} \\ -\frac{1}{8} \end{pmatrix}. \quad \blacklozenge
$$

We use this technique for solving systems of linear equations having invertible coefficient matrices in the application that concludes this section.

In Example 1(c), we saw a system of linear equations that has no solutions. We now establish a criterion for determining when a system has solutions. This criterion involves the rank of the coefficient matrix of the system $Ax = b$ and the rank of the matrix $(A|b)$. The matrix $(A|b)$ is called the **augmented matrix of the system** $Ax = b$.

Theorem 3.11. *Let* $Ax = b$ *be a system of linear equations. Then the system is consistent if and only if* $\text{rank}(A) = \text{rank}(A|b)$.

Proof. To say that $Ax = b$ has a solution is equivalent to saying that $b \in \mathsf{R}(\mathsf{L}_A)$. (See Exercise 9.) In the proof of Theorem 3.5 (p. 153), we saw that

$$
\mathsf{R}(\mathsf{L}_A) = \text{span}(\{a_1, a_2, \ldots, a_n\}),
$$

the span of the columns of A. Thus $Ax = b$ has a solution if and only if $b \in \text{span}(\{a_1, a_2, \ldots, a_n\})$. But $b \in \text{span}(\{a_1, a_2, \ldots, a_n\})$ if and only if $\text{span}(\{a_1, a_2, \ldots, a_n\}) = \text{span}(\{a_1, a_2, \ldots, a_n, b\})$. This last statement is equivalent to

$$\dim(\text{span}(\{a_1, a_2, \ldots, a_n\})) = \dim(\text{span}(\{a_1, a_2, \ldots, a_n, b\})).$$

So by Theorem 3.5, the preceding equation reduces to

$$\text{rank}(A) = \text{rank}(A|b). \qquad \blacksquare$$

Example 5

Recall the system of equations

$$x_1 + x_2 = 0$$
$$x_1 + x_2 = 1$$

in Example 1(c).

Since

$$A = \begin{pmatrix} 1 & 1 \\ 1 & 1 \end{pmatrix} \quad \text{and} \quad (A|b) = \begin{pmatrix} 1 & 1 & 0 \\ 1 & 1 & 1 \end{pmatrix},$$

$\text{rank}(A) = 1$ and $\text{rank}(A|b) = 2$. Because the two ranks are unequal, the system has no solutions. ◆

Example 6

We can use Theorem 3.11 to determine whether $(3, 3, 2)$ is in the range of the linear transformation $\mathsf{T} \colon \mathsf{R}^3 \to \mathsf{R}^3$ defined by

$$\mathsf{T}(a_1, a_2, a_3) = (a_1 + a_2 + a_3, a_1 - a_2 + a_3, a_1 + a_3).$$

Now $(3, 3, 2) \in \mathsf{R}(\mathsf{T})$ if and only if there exists a vector $s = (x_1, x_2, x_3)$ in R^3 such that $\mathsf{T}(s) = (3, 3, 2)$. Such a vector s must be a solution to the system

$$x_1 + x_2 + x_3 = 3$$
$$x_1 - x_2 + x_3 = 3$$
$$x_1 \qquad + x_3 = 2.$$

Since the ranks of the coefficient matrix and the augmented matrix of this system are 2 and 3, respectively, it follows that this system has no solutions. Hence $(3, 3, 2) \notin \mathsf{R}(\mathsf{T})$. ◆

An Application

In 1973, Wassily Leontief won the Nobel prize in economics for his work in developing a mathematical model that can be used to describe various economic phenomena. We close this section by applying some of the ideas we have studied to illustrate two special cases of his work.

We begin by considering a simple society composed of three people (industries)—a farmer who grows all the food, a tailor who makes all the clothing, and a carpenter who builds all the housing. We assume that each person sells to and buys from a central pool and that everything produced is consumed. Since no commodities either enter or leave the system, this case is referred to as the **closed model**.

Each of these three individuals consumes all three of the commodities produced in the society. Suppose that the proportion of each of the commodities consumed by each person is given in the following table. Notice that each of the columns of the table must sum to 1.

	Food	Clothing	Housing
Farmer	0.40	0.20	0.20
Tailor	0.10	0.70	0.20
Carpenter	0.50	0.10	0.60

Let p_1, p_2, and p_3 denote the incomes of the farmer, tailor, and carpenter, respectively. To ensure that this society survives, we require that the consumption of each individual equals his or her income. Note that the farmer consumes 20% of the clothing. Because the total cost of all clothing is p_2, the tailor's income, the amount spent by the farmer on clothing is $0.20p_2$. Moreover, the amount spent by the farmer on food, clothing, and housing must equal the farmer's income, and so we obtain the equation

$$0.40p_1 + 0.20p_2 + 0.20p_3 = p_1.$$

Similar equations describing the expenditures of the tailor and carpenter produce the following system of linear equations:

$$0.40p_1 + 0.20p_2 + 0.20p_3 = p_1$$
$$0.10p_1 + 0.70p_2 + 0.20p_3 = p_2$$
$$0.50p_1 + 0.10p_2 + 0.60p_3 = p_3.$$

This system can be written as $Ap = p$, where

$$p = \begin{pmatrix} p_1 \\ p_2 \\ p_3 \end{pmatrix}$$

and A is the coefficient matrix of the system. In this context, A is called the **input–output (or consumption) matrix**, and $Ap = p$ is called the **equilibrium condition**.

For vectors $b = (b_1, b_2, \ldots, b_n)$ and $c = (c_1, c_2, \ldots, c_n)$ in $\mathbb{R}^n$, we use the notation $b \geq c$ $[b > c]$ to mean $b_i \geq c_i$ $[b_i > c_i]$ for all i. The vector b is called **nonnegative [positive]** if $b \geq 0$ $[b > 0]$.

At first, it may seem reasonable to replace the equilibrium condition by the inequality $Ap \leq p$, that is, the requirement that consumption not exceed production. But, in fact, $Ap \leq p$ implies that $Ap = p$ in the closed model. For otherwise, there exists a k for which

$$p_k > \sum_j A_{kj} p_j.$$

Hence, since the columns of A sum to 1,

$$\sum_i p_i > \sum_i \sum_j A_{ij} p_j = \sum_j \left(\sum_i A_{ij} \right) p_j = \sum_j p_j,$$

which is a contradiction.

One solution to the homogeneous system $(I - A)x = 0$, which is equivalent to the equilibrium condition, is

$$p = \begin{pmatrix} 0.25 \\ 0.35 \\ 0.40 \end{pmatrix}.$$

We may interpret this to mean that the society survives if the farmer, tailor, and carpenter have incomes in the proportions $25 : 35 : 40$ (or $5 : 7 : 8$).

Notice that we are not simply interested in any nonzero solution to the system, but in one that is nonnegative. Thus we must consider the question of whether the system $(I - A)x = 0$ has a nonnegative solution, where A is a matrix with nonnegative entries whose columns sum to 1. A useful theorem in this direction (whose proof may be found in "Applications of Matrices to Economic Models and Social Science Relationships," by Ben Noble, *Proceedings of the Summer Conference for College Teachers on Applied Mathematics*, 1971, CUPM, Berkeley, California) is stated below.

Theorem 3.12. Let A be an $n \times n$ input–output matrix having the form

$$A = \begin{pmatrix} B & C \\ D & E \end{pmatrix},$$

where D is a $1 \times (n-1)$ positive vector and C is an $(n-1) \times 1$ positive vector. Then $(I - A)x = 0$ has a one-dimensional solution set that is generated by a nonnegative vector.

Observe that any input–output matrix with all positive entries satisfies the hypothesis of this theorem. The following matrix does also:

$$\begin{pmatrix} 0.75 & 0.50 & 0.65 \\ 0 & 0.25 & 0.35 \\ 0.25 & 0.25 & 0 \end{pmatrix}.$$

In the **open model**, we assume that there is an outside demand for each of the commodities produced. Returning to our simple society, let x_1, x_2, and x_3 be the monetary values of food, clothing, and housing produced with respective outside demands d_1, d_2, and d_3. Let A be the 3×3 matrix such that A_{ij} represents the amount (in a fixed monetary unit such as the dollar) of commodity i required to produce one monetary unit of commodity j. Then the value of the surplus of food in the society is

$$x_1 - (A_{11}x_1 + A_{12}x_2 + A_{13}x_3),$$

that is, the value of food produced minus the value of food consumed while producing the three commodities. The assumption that everything produced is consumed gives us a similar equilibrium condition for the open model, namely, that the surplus of each of the three commodities must equal the corresponding outside demands. Hence

$$x_i - \sum_{j=1}^{3} A_{ij}x_j = d_i \quad \text{for } i = 1, 2, 3.$$

In general, we must find a nonnegative solution to $(I - A)x = d$, where A is a matrix with nonnegative entries such that the sum of the entries of each column of A does not exceed one, and $d \geq 0$. It is easy to see that if $(I - A)^{-1}$ exists and is nonnegative, then the desired solution is $(I - A)^{-1}d$.

Recall that for a real number a, the series $1 + a + a^2 + \cdots$ converges to $(1 - a)^{-1}$ if $|a| < 1$. Similarly, it can be shown (using the concept of convergence of matrices developed in Section 5.3) that the series $I + A + A^2 + \cdots$ converges to $(I - A)^{-1}$ if $\{A^n\}$ converges to the zero matrix. In this case, $(I - A)^{-1}$ is nonnegative since the matrices $I, A, A^2, \ldots$ are nonnegative.

To illustrate the open model, suppose that 30 cents worth of food, 10 cents worth of clothing, and 30 cents worth of housing are required for the production of \$1 worth of food. Similarly, suppose that 20 cents worth of food, 40 cents worth of clothing, and 20 cents worth of housing are required for the production of \$1 of clothing. Finally, suppose that 30 cents worth of food, 10 cents worth of clothing, and 30 cents worth of housing are required for the production of \$1 worth of housing. Then the input–output matrix is

$$A = \begin{pmatrix} 0.30 & 0.20 & 0.30 \\ 0.10 & 0.40 & 0.10 \\ 0.30 & 0.20 & 0.30 \end{pmatrix};$$

so

$$I - A = \begin{pmatrix} 0.70 & -0.20 & -0.30 \\ -0.10 & 0.60 & -0.10 \\ -0.30 & -0.20 & 0.70 \end{pmatrix} \quad \text{and} \quad (I - A)^{-1} = \begin{pmatrix} 2.0 & 1.0 & 1.0 \\ 0.5 & 2.0 & 0.5 \\ 1.0 & 1.0 & 2.0 \end{pmatrix}.$$

Since $(I-A)^{-1}$ is nonnegative, we can find a (unique) nonnegative solution to $(I - A)x = d$ for any demand d. For example, suppose that there are outside demands for \$30 billion in food, \$20 billion in clothing, and \$10 billion in housing. If we set

$$d = \begin{pmatrix} 30 \\ 20 \\ 10 \end{pmatrix},$$

then

$$x = (I - A)^{-1}d = \begin{pmatrix} 90 \\ 60 \\ 70 \end{pmatrix}.$$

So a gross production of \$90 billion of food, \$60 billion of clothing, and \$70 billion of housing is necessary to meet the required demands.

EXERCISES

1. Label the following statements as true or false.
    (a) Any system of linear equations has at least one solution.
    (b) Any system of linear equations has at most one solution.
    (c) Any homogeneous system of linear equations has at least one solution.
    (d) Any system of n linear equations in n unknowns has at most one solution.
    (e) Any system of n linear equations in n unknowns has at least one solution.
    (f) If the homogeneous system corresponding to a given system of linear equations has a solution, then the given system has a solution.
    (g) If the coefficient matrix of a homogeneous system of n linear equations in n unknowns is invertible, then the system has no nonzero solutions.
    (h) The solution set of any system of m linear equations in n unknowns is a subspace of F^n.

2. For each of the following homogeneous systems of linear equations, find the dimension of and a basis for the solution set.

(a) $\begin{aligned} x_1 + 3x_2 &= 0 \\ 2x_1 + 6x_2 &= 0 \end{aligned}$ 　　　(b) $\begin{aligned} x_1 + x_2 - x_3 &= 0 \\ 4x_1 + x_2 - 2x_3 &= 0 \end{aligned}$

(c) $\begin{aligned} x_1 + 2x_2 - x_3 &= 0 \\ 2x_1 + x_2 + x_3 &= 0 \end{aligned}$ 　　　(d) $\begin{aligned} 2x_1 + x_2 - x_3 &= 0 \\ x_1 - x_2 + x_3 &= 0 \\ x_1 + 2x_2 - 2x_3 &= 0 \end{aligned}$

(e) $x_1 + 2x_2 - 3x_3 + x_4 = 0$ 　　　(f) $\begin{aligned} x_1 + 2x_2 &= 0 \\ x_1 - x_2 &= 0 \end{aligned}$

(g) $\begin{aligned} x_1 + 2x_2 + x_3 + x_4 &= 0 \\ x_2 - x_3 + x_4 &= 0 \end{aligned}$

3. Using the results of Exercise 2, find all solutions to the following systems.

(a) $\begin{aligned} x_1 + 3x_2 &= 5 \\ 2x_1 + 6x_2 &= 10 \end{aligned}$ 　　　(b) $\begin{aligned} x_1 + x_2 - x_3 &= 1 \\ 4x_1 + x_2 - 2x_3 &= 3 \end{aligned}$

(c) $\begin{aligned} x_1 + 2x_2 - x_3 &= 3 \\ 2x_1 + x_2 + x_3 &= 6 \end{aligned}$ 　　　(d) $\begin{aligned} 2x_1 + x_2 - x_3 &= 5 \\ x_1 - x_2 + x_3 &= 1 \\ x_1 + 2x_2 - 2x_3 &= 4 \end{aligned}$

(e) $x_1 + 2x_2 - 3x_3 + x_4 = 1$ 　　　(f) $\begin{aligned} x_1 + 2x_2 &= 5 \\ x_1 - x_2 &= -1 \end{aligned}$

(g) $\begin{aligned} x_1 + 2x_2 + x_3 + x_4 &= 1 \\ x_2 - x_3 + x_4 &= 1 \end{aligned}$

4. For each system of linear equations with the invertible coefficient matrix A,

   (1) Compute A^{-1}.

   (2) Use A^{-1} to solve the system.

   (a) $\begin{aligned} x_1 + 3x_2 &= 4 \\ 2x_1 + 5x_2 &= 3 \end{aligned}$ 　　　(b) $\begin{aligned} x_1 + 2x_2 - x_3 &= 5 \\ x_1 + x_2 + x_3 &= 1 \\ 2x_1 - 2x_2 + x_3 &= 4 \end{aligned}$

5. Give an example of a system of n linear equations in n unknowns with infinitely many solutions.

6. Let $T: R^3 \to R^2$ be defined by $T(a, b, c) = (a + b, 2a - c)$. Determine $T^{-1}(1, 11)$.

7. Determine which of the following systems of linear equations has a solution.

(a)
$$x_1 + x_2 - x_3 + 2x_4 = 2$$
$$x_1 + x_2 + 2x_3 \qquad = 1$$
$$2x_1 + 2x_2 + x_3 + 2x_4 = 4$$

(b)
$$x_1 + x_2 - x_3 = 1$$
$$2x_1 + x_2 + 3x_3 = 2$$

(c)
$$x_1 + 2x_2 + 3x_3 = 1$$
$$x_1 + x_2 - x_3 = 0$$
$$x_1 + 2x_2 + x_3 = 3$$

(d)
$$x_1 + x_2 + 3x_3 - x_4 = 0$$
$$x_1 + x_2 + x_3 + x_4 = 1$$
$$x_1 - 2x_2 + x_3 - x_4 = 1$$
$$4x_1 + x_2 + 8x_3 - x_4 = 0$$

(e)
$$x_1 + 2x_2 - x_3 = 1$$
$$2x_1 + x_2 + 2x_3 = 3$$
$$x_1 - 4x_2 + 7x_3 = 4$$

8. Let $\mathsf{T}: \mathsf{R}^3 \to \mathsf{R}^3$ be defined by $\mathsf{T}(a, b, c) = (a + b, b - 2c, a + 2c)$. For each vector v in R^3, determine whether $v \in \mathsf{R}(\mathsf{T})$.

(a) $v = (1, 3, -2)$ (b) $v = (2, 1, 1)$

9. Prove that the system of linear equations $Ax = b$ has a solution if and only if $b \in \mathsf{R}(\mathsf{L}_A)$.

10. Prove or give a counterexample to the following statement: If the coefficient matrix of a system of m linear equations in n unknowns has rank m, then the system has a solution.

11. In the closed model of Leontief with food, clothing, and housing as the basic industries, suppose that the input–output matrix is

$$A = \begin{pmatrix} \frac{7}{16} & \frac{1}{2} & \frac{3}{16} \\ \frac{5}{16} & \frac{1}{6} & \frac{5}{16} \\ \frac{1}{4} & \frac{1}{3} & \frac{1}{2} \end{pmatrix}.$$

At what ratio must the farmer, tailor, and carpenter produce in order for equilibrium to be attained?

12. A certain economy consists of two sectors: goods and services. Suppose that 60% of all goods and 30% of all services are used in the production of goods. What proportion of the total economic output is used in the production of goods?

13. In the notation of the open model of Leontief, suppose that

$$A = \begin{pmatrix} \frac{1}{2} & \frac{1}{5} \\ \frac{1}{3} & \frac{1}{5} \end{pmatrix} \qquad \text{and} \qquad d = \begin{pmatrix} 2 \\ 5 \end{pmatrix}$$

are the input–output matrix and the demand vector, respectively. How much of each commodity must be produced to satisfy this demand?

14. A certain economy consisting of the two sectors of goods and services supports a defense system that consumes $90 billion worth of goods and $20 billion worth of services from the economy but does not contribute to economic production. Suppose that 50 cents worth of goods and 20 cents worth of services are required to produce $1 worth of goods and that 30 cents worth of of goods and 60 cents worth of services are required to produce $1 worth of services. What must the total output of the economic system be to support this defense system?

3.4   SYSTEMS OF LINEAR EQUATIONS— COMPUTATIONAL ASPECTS

In Section 3.3, we obtained a necessary and sufficient condition for a system of linear equations to have solutions (Theorem 3.11 p. 174) and learned how to express the solutions to a nonhomogeneous system in terms of solutions to the corresponding homogeneous system (Theorem 3.9 p. 172). The latter result enables us to determine all the solutions to a given system if we can find one solution to the given system and a basis for the solution set of the corresponding homogeneous system. In this section, we use elementary row operations to accomplish these two objectives simultaneously. The essence of this technique is to transform a given system of linear equations into a system having the same solutions, but which is easier to solve (as in Section 1.4).

Definition.   *Two systems of linear equations are called* **equivalent** *if they have the same solution set.*

The following theorem and corollary give a useful method for obtaining equivalent systems.

Theorem 3.13. *Let* $Ax = b$ *be a system of* m *linear equations in* n *unknowns, and let* C *be an invertible* $m \times m$ *matrix. Then the system* $(CA)x = Cb$ *is equivalent to* $Ax = b$.

Proof. Let K be the solution set for $Ax = b$ and K' the solution set for $(CA)x = Cb$. If $w \in K$, then $Aw = b$. So $(CA)w = Cb$, and hence $w \in K'$. Thus $K \subseteq K'$.

Conversely, if $w \in K'$, then $(CA)w = Cb$. Hence

$$Aw = C^{-1}(CAw) = C^{-1}(Cb) = b;$$

so $w \in K$. Thus $K' \subseteq K$, and therefore, $K = K'$.   ∎

Corollary. *Let* $Ax = b$ *be a system of* m *linear equations in* n *unknowns. If* $(A'|b')$ *is obtained from* $(A|b)$ *by a finite number of elementary row operations, then the system* $A'x = b'$ *is equivalent to the original system.*

Proof. Suppose that $(A'|b')$ is obtained from $(A|b)$ by elementary row operations. These may be executed by multiplying $(A|b)$ by elementary $m \times m$ matrices $E_1, E_2, \ldots, E_p$. Let $C = E_p \cdots E_2 E_1$; then

$$(A'|b') = C(A|b) = (CA|Cb).$$

Since each E_i is invertible, so is C. Now $A' = CA$ and $b' = Cb$. Thus by Theorem 3.13, the system $A'x = b'$ is equivalent to the system $Ax = b$. ∎

We now describe a method for solving any system of linear equations. Consider, for example, the system of linear equations

$$\begin{aligned}
3x_1 + 2x_2 + 3x_3 - 2x_4 &= 1 \\
x_1 + x_2 + x_3 \phantom{{}- 2x_4} &= 3 \\
x_1 + 2x_2 + x_3 - x_4 &= 2.
\end{aligned}$$

First, we form the augmented matrix

$$\begin{pmatrix} 3 & 2 & 3 & -2 & 1 \\ 1 & 1 & 1 & 0 & 3 \\ 1 & 2 & 1 & -1 & 2 \end{pmatrix}.$$

By using elementary row operations, we transform the augmented matrix into an upper triangular matrix in which the first nonzero entry of each row is 1, and it occurs in a column to the right of the first nonzero entry of each preceding row. (Recall that matrix A is upper triangular if $A_{ij} = 0$ whenever $i > j$.)

1. *In the leftmost nonzero column, create a 1 in the first row.* In our example, we can accomplish this step by interchanging the first and third rows. The resulting matrix is

$$\begin{pmatrix} 1 & 2 & 1 & -1 & 2 \\ 1 & 1 & 1 & 0 & 3 \\ 3 & 2 & 3 & -2 & 1 \end{pmatrix}.$$

2. *By means of type 3 row operations, use the first row to obtain zeros in the remaining positions of the leftmost nonzero column.* In our example, we must add -1 times the first row to the second row and then add -3 times the first row to the third row to obtain

$$\begin{pmatrix} 1 & 2 & 1 & -1 & 2 \\ 0 & -1 & 0 & 1 & 1 \\ 0 & -4 & 0 & 1 & -5 \end{pmatrix}.$$

3. *Create a 1 in the next row in the leftmost possible column, without using previous row(s).* In our example, the second column is the leftmost

possible column, and we can obtain a 1 in the second row, second column by multiplying the second row by -1. This operation produces

$$\begin{pmatrix} 1 & 2 & 1 & -1 & \bigm| & 2 \\ 0 & 1 & 0 & -1 & \bigm| & -1 \\ 0 & -4 & 0 & 1 & \bigm| & -5 \end{pmatrix}.$$

4. *Now use type 3 elementary row operations to obtain zeros below the 1 created in the preceding step.* In our example, we must add four times the second row to the third row. The resulting matrix is

$$\begin{pmatrix} 1 & 2 & 1 & -1 & \bigm| & 2 \\ 0 & 1 & 0 & -1 & \bigm| & -1 \\ 0 & 0 & 0 & -3 & \bigm| & -9 \end{pmatrix}.$$

5. *Repeat steps 3 and 4 on each succeeding row until no nonzero rows remain.* (This creates zeros above the first nonzero entry in each row.) In our example, this can be accomplished by multiplying the third row by $-\frac{1}{3}$. This operation produces

$$\begin{pmatrix} 1 & 2 & 1 & -1 & \bigm| & 2 \\ 0 & 1 & 0 & -1 & \bigm| & -1 \\ 0 & 0 & 0 & 1 & \bigm| & 3 \end{pmatrix}.$$

We have now obtained the desired matrix. To complete the simplification of the augmented matrix, we must make the first nonzero entry in each row the only nonzero entry in its column. (This corresponds to eliminating certain unknowns from all but one of the equations.)

6. *Work upward, beginning with the last nonzero row, and add multiples of each row to the rows above.* (This creates zeros above the first nonzero entry in each row.) In our example, the third row is the last nonzero row, and the first nonzero entry of this row lies in column 4. Hence we add the third row to the first and second rows to obtain zeros in row 1, column 4 and row 2, column 4. The resulting matrix is

$$\begin{pmatrix} 1 & 2 & 1 & 0 & \bigm| & 5 \\ 0 & 1 & 0 & 0 & \bigm| & 2 \\ 0 & 0 & 0 & 1 & \bigm| & 3 \end{pmatrix}.$$

7. *Repeat the process described in step 6 for each preceding row until it is performed with the second row, at which time the reduction process is complete.* In our example, we must add -2 times the second row to the first row in order to make the first row, second column entry become zero. This operation produces

$$\begin{pmatrix} 1 & 0 & 1 & 0 & \bigm| & 1 \\ 0 & 1 & 0 & 0 & \bigm| & 2 \\ 0 & 0 & 0 & 1 & \bigm| & 3 \end{pmatrix}.$$

We have now obtained the desired reduction of the augmented matrix. This matrix corresponds to the system of linear equations

$$\begin{aligned} x_1 + \quad x_3 \quad &= 1 \\ x_2 \qquad &= 2 \\ x_4 &= 3. \end{aligned}$$

Recall that, by the corollary to Theorem 3.13, this system is equivalent to the original system. But this system is easily solved. Obviously $x_2 = 2$ and $x_4 = 3$. Moreover, x_1 and x_3 can have any values provided their sum is 1. Letting $x_3 = t$, we then have $x_1 = 1 - t$. Thus an arbitrary solution to the original system has the form

$$\begin{pmatrix} 1 - t \\ 2 \\ t \\ 3 \end{pmatrix} = \begin{pmatrix} 1 \\ 2 \\ 0 \\ 3 \end{pmatrix} + t \begin{pmatrix} -1 \\ 0 \\ 1 \\ 0 \end{pmatrix}.$$

Observe that

$$\left\{ \begin{pmatrix} -1 \\ 0 \\ 1 \\ 0 \end{pmatrix} \right\}$$

is a basis for the homogeneous system of equations corresponding to the given system.

In the preceding example we performed elementary row operations on the augmented matrix of the system until we obtained the augmented matrix of a system having properties 1, 2, and 3 on page 27. Such a matrix has a special name.

Definition.   *A matrix is said to be in* **reduced row echelon form** *if the following three conditions are satisfied.*

(a) *Any row containing a nonzero entry precedes any row in which all the entries are zero (if any).*

(b) *The first nonzero entry in each row is the only nonzero entry in its column.*

(c) *The first nonzero entry in each row is 1 and it occurs in a column to the right of the first nonzero entry in the preceding row.*

Example 1

(a) The matrix on page 184 is in reduced row echelon form. Note that the first nonzero entry of each row is 1 and that the column containing each such entry has all zeros otherwise. Also note that each time we move downward to a new row, we must move to the right one or more columns to find the first nonzero entry of the new row.

   (b) The matrix

$$\begin{pmatrix} 1 & 1 & 0 \\ 0 & 1 & 0 \\ 1 & 0 & 1 \end{pmatrix},$$

is *not* in reduced row echelon form, because the first column, which contains the first nonzero entry in row 1, contains another nonzero entry. Similarly, the matrix

$$\begin{pmatrix} 0 & 1 & 0 & 2 \\ 1 & 0 & 0 & 1 \\ 0 & 0 & 1 & 1 \end{pmatrix},$$

is not in reduced row echelon form, because the first nonzero entry of the second row is not to the right of the first nonzero entry of the first row. Finally, the matrix

$$\begin{pmatrix} 2 & 0 & 0 \\ 0 & 1 & 0 \end{pmatrix},$$

is not in reduced row echelon form, because the first nonzero entry of the first row is not 1.   ◆

   It can be shown (see the corollary to Theorem 3.16) that the reduced row echelon form of a matrix is unique; that is, if different sequences of elementary row operations are used to transform a matrix into matrices Q and Q' in reduced row echelon form, then $Q = Q'$. Thus, although there are many different sequences of elementary row operations that can be used to transform a given matrix into reduced row echelon form, they all produce the same result.

   The procedure described on pages 183–185 for reducing an augmented matrix to reduced row echelon form is called **Gaussian elimination**. It consists of two separate parts.

   1. In the *forward pass* (steps 1-5), the augmented matrix is transformed into an upper triangular matrix in which the first nonzero entry of each row is 1, and it occurs in a column to the right of the first nonzero entry of each preceding row.
   2. In the *backward pass* or *back-substitution* (steps 6-7), the upper triangular matrix is transformed into reduced row echelon form by making the first nonzero entry of each row the only nonzero entry of its column.

Of all the methods for transforming a matrix into its reduced row echelon form, Gaussian elimination requires the fewest arithmetic operations. (For large matrices, it requires approximately 50% fewer operations than the Gauss-Jordan method, in which the matrix is transformed into reduced row echelon form by using the first nonzero entry in each row to make zero all other entries in its column.) Because of this efficiency, Gaussian elimination is the preferred method when solving systems of linear equations on a computer. In this context, the Gaussian elimination procedure is usually modified in order to minimize roundoff errors. Since discussion of these techniques is inappropriate here, readers who are interested in such matters are referred to books on numerical analysis.

When a matrix is in reduced row echelon form, the corresponding system of linear equations is easy to solve. We present below a procedure for solving any system of linear equations for which the augmented matrix is in reduced row echelon form. First, however, we note that every matrix can be transformed into reduced row echelon form by Gaussian elimination. In the forward pass, we satisfy conditions (a) and (c) in the definition of reduced row echelon form and thereby make zero all entries below the first nonzero entry in each row. Then in the backward pass, we make zero all entries above the first nonzero entry in each row, thereby satisfying condition (b) in the definition of reduced row echelon form.

Theorem 3.14. *Gaussian elimination transforms any matrix into its reduced row echelon form.*

We now describe a method for solving a system in which the augmented matrix is in reduced row echelon form. To illustrate this procedure, we consider the system

$$\begin{aligned} 2x_1 + 3x_2 + x_3 + 4x_4 - 9x_5 &= 17 \\ x_1 + x_2 + x_3 + x_4 - 3x_5 &= 6 \\ x_1 + x_2 + x_3 + 2x_4 - 5x_5 &= 8 \\ 2x_1 + 2x_2 + 2x_3 + 3x_4 - 8x_5 &= 14, \end{aligned}$$

for which the augmented matrix is

$$\left(\begin{array}{ccccc|c} 2 & 3 & 1 & 4 & -9 & 17 \\ 1 & 1 & 1 & 1 & -3 & 6 \\ 1 & 1 & 1 & 2 & -5 & 8 \\ 2 & 2 & 2 & 3 & -8 & 14 \end{array} \right).$$

Applying Gaussian elimination to the augmented matrix of the system produces the following sequence of matrices.

$$\left(\begin{array}{ccccc|c} 2 & 3 & 1 & 4 & -9 & 17 \\ 1 & 1 & 1 & 1 & -3 & 6 \\ 1 & 1 & 1 & 2 & -5 & 8 \\ 2 & 2 & 2 & 3 & -8 & 14 \end{array} \right) \longrightarrow \left(\begin{array}{ccccc|c} 1 & 1 & 1 & 1 & -3 & 6 \\ 2 & 3 & 1 & 4 & -9 & 17 \\ 1 & 1 & 1 & 2 & -5 & 8 \\ 2 & 2 & 2 & 3 & -8 & 14 \end{array} \right) \longrightarrow$$

$$\begin{pmatrix} 1 & 1 & 1 & 1 & -3 & | & 6 \\ 0 & 1 & -1 & 2 & -3 & | & 5 \\ 0 & 0 & 0 & 1 & -2 & | & 2 \\ 0 & 0 & 0 & 1 & -2 & | & 2 \end{pmatrix} \longrightarrow \begin{pmatrix} 1 & 1 & 1 & 1 & -3 & | & 6 \\ 0 & 1 & -1 & 2 & -3 & | & 5 \\ 0 & 0 & 0 & 1 & -2 & | & 2 \\ 0 & 0 & 0 & 0 & 0 & | & 0 \end{pmatrix} \longrightarrow$$

$$\begin{pmatrix} 1 & 1 & 1 & 0 & -1 & | & 4 \\ 0 & 1 & -1 & 0 & 1 & | & 1 \\ 0 & 0 & 0 & 1 & -2 & | & 2 \\ 0 & 0 & 0 & 0 & 0 & | & 0 \end{pmatrix} \longrightarrow \begin{pmatrix} 1 & 0 & 2 & 0 & -2 & | & 3 \\ 0 & 1 & -1 & 0 & 1 & | & 1 \\ 0 & 0 & 0 & 1 & -2 & | & 2 \\ 0 & 0 & 0 & 0 & 0 & | & 0 \end{pmatrix}.$$

The system of linear equations corresponding to this last matrix is

$$\begin{aligned} x_1 \quad + 2x_3 \quad - 2x_5 &= 3 \\ x_2 - \quad x_3 \quad + \quad x_5 &= 1 \\ x_4 - 2x_5 &= 2. \end{aligned}$$

Notice that we have ignored the last row since it consists entirely of zeros.

To solve a system for which the augmented matrix is in reduced row echelon form, divide the variables into two sets. The first set consists of those variables that appear as leftmost variables in one of the equations of the system (in this case the set is $\{x_1, x_2, x_4\}$). The second set consists of all the remaining variables (in this case, $\{x_3, x_5\}$). To each variable in the second set, assign a parametric value $t_1, t_2, \ldots$ ($x_3 = t_1$, $x_5 = t_2$), and then solve for the variables of the first set in terms of those in the second set:

$$\begin{aligned} x_1 &= -2x_3 + 2x_5 + 3 = -2t_1 + 2t_2 + 3 \\ x_2 &= \quad x_3 - \quad x_5 + 1 = \quad t_1 - \quad t_2 + 1 \\ x_4 &= \quad\quad 2x_5 + 2 = \quad\quad 2t_2 + 2. \end{aligned}$$

Thus an arbitrary solution is of the form

$$\begin{pmatrix} x_1 \\ x_2 \\ x_3 \\ x_4 \\ x_5 \end{pmatrix} = \begin{pmatrix} -2t_1 + 2t_2 + 3 \\ t_1 - t_2 + 1 \\ t_1 \\ 2t_2 + 2 \\ t_2 \end{pmatrix} = \begin{pmatrix} 3 \\ 1 \\ 0 \\ 2 \\ 0 \end{pmatrix} + t_1 \begin{pmatrix} -2 \\ 1 \\ 1 \\ 0 \\ 0 \end{pmatrix} + t_2 \begin{pmatrix} 2 \\ -1 \\ 0 \\ 2 \\ 1 \end{pmatrix},$$

where $t_1, t_2 \in R$. Notice that

$$\left\{ \begin{pmatrix} -2 \\ 1 \\ 1 \\ 0 \\ 0 \end{pmatrix}, \begin{pmatrix} 2 \\ -1 \\ 0 \\ 2 \\ 1 \end{pmatrix} \right\}$$

is a basis for the solution set of the corresponding homogeneous system of equations and

$$\begin{pmatrix} 3 \\ 1 \\ 0 \\ 2 \\ 0 \end{pmatrix}$$

is a particular solution to the original system.

Therefore, in simplifying the augmented matrix of the system to reduced row echelon form, we are in effect simultaneously finding a particular solution to the original system and a basis for the solution set of the associated homogeneous system. Moreover, this procedure detects when a system is inconsistent, for by Exercise 3, solutions exist if and only if, in the reduction of the augmented matrix to reduced row echelon form, we do not obtain a row in which the only nonzero entry lies in the last column.

Thus to use this procedure for solving a system $Ax = b$ of m linear equations in n unknowns, we need only begin to transform the augmented matrix $(A|b)$ into its reduced row echelon form $(A'|b')$ by means of Gaussian elimination. If a row is obtained in which the only nonzero entry lies in the last column, then the original system is inconsistent. Otherwise, discard any zero rows from $(A'|b')$, and write the corresponding system of equations. Solve this system as described above to obtain an arbitrary solution of the form

$$s = s_0 + t_1 u_1 + t_2 u_2 + \cdots + t_{n-r} u_{n-r},$$

where r is the number of nonzero rows in A' ($r \leq m$). The preceding equation is called a **general solution** of the system $Ax = b$. It expresses an arbitrary solution s of $Ax = b$ in terms of $n - r$ parameters. The following theorem states that s cannot be expressed in fewer than $n - r$ parameters.

Theorem 3.15. *Let $Ax = b$ be a system of r nonzero equations in n unknowns. Suppose that $\operatorname{rank}(A) = \operatorname{rank}(A|b)$ and that $(A|b)$ is in reduced row echelon form. Then*

(a) $\operatorname{rank}(A) = r$.

(b) *If the general solution obtained by the procedure above is of the form*

$$s = s_0 + t_1 u_1 + t_2 u_2 + \cdots + t_{n-r} u_{n-r},$$

then $\{u_1, u_2, \ldots, u_{n-r}\}$ is a basis for the solution set of the corresponding homogeneous system, and s_0 is a solution to the original system.

Proof. Since $(A|b)$ is in reduced row echelon form, $(A|b)$ must have r nonzero rows. Clearly these rows are linearly independent by the definition of the reduced row echelon form, and so $\operatorname{rank}(A|b) = r$. Thus $\operatorname{rank}(A) = r$.

Let K be the solution set for $Ax = b$, and let $\mathsf{K_H}$ be the solution set for $Ax = 0$. Setting $t_1 = t_2 = \cdots = t_{n-r} = 0$, we see that $s = s_0 \in K$. But by Theorem 3.9 (p. 172), $K = \{s_0\} + \mathsf{K_H}$. Hence

$$\mathsf{K_H} = \{-s_0\} + K = \text{span}(\{u_1, u_2, \ldots, u_{n-r}\}).$$

Because $\text{rank}(A) = r$, we have $\dim(\mathsf{K_H}) = n - r$. Thus since $\dim(\mathsf{K_H}) = n - r$ and $\mathsf{K_H}$ is generated by a set $\{u_1, u_2, \ldots, u_{n-r}\}$ containing at most $n - r$ vectors, we conclude that this set is a basis for $\mathsf{K_H}$. ∎

An Interpretation of the Reduced Row Echelon Form

Let A be an $m \times n$ matrix with columns $a_1, a_2, \ldots, a_n$, and let B be the reduced row echelon form of A. Denote the columns of B by $b_1, b_2, \ldots, b_n$. If the rank of A is r, then the rank of B is also r by the corollary to Theorem 3.4 (p. 153). Because B is in reduced row echelon form, no nonzero row of B can be a linear combination of the other rows of B. Hence B must have exactly r nonzero rows, and if $r \geq 1$, the vectors $e_1, e_2, \ldots, e_r$ must occur among the columns of B. For $i = 1, 2, \ldots, r$, let j_i denote a column number of B such that $b_{j_i} = e_i$. We claim that $a_{j_1}, a_{j_2}, \ldots, a_{j_r}$, the columns of A corresponding to these columns of B, are linearly independent. For suppose that there are scalars $c_1, c_2, \ldots, c_r$ such that

$$c_1 a_{j_1} + c_2 a_{j_2} + \cdots + c_r a_{j_r} = 0.$$

Because B can be obtained from A by a sequence of elementary row operations, there exists (as in the proof of the corollary to Theorem 3.13) an invertible $m \times m$ matrix M such that $MA = B$. Multiplying the preceding equation by M yields

$$c_1 M a_{j_1} + c_2 M a_{j_2} + \cdots + c_r M a_{j_r} = 0.$$

Since $M a_{j_i} = b_{j_i} = e_i$, it follows that

$$c_1 e_1 + c_2 e_2 + \cdots + c_r e_r = 0.$$

Hence $c_1 = c_2 = \cdots = c_r = 0$, proving that the vectors $a_{j_1}, a_{j_2}, \ldots, a_{j_r}$ are linearly independent.

Because B has only r nonzero rows, every column of B has the form

$$\begin{pmatrix} d_1 \\ d_2 \\ \vdots \\ d_r \\ 0 \\ \vdots \\ 0 \end{pmatrix}$$

for scalars $d_1, d_2, \ldots, d_r$. The corresponding column of A must be

$$M^{-1}(d_1 e_1 + d_2 e_2 + \cdots + d_r e_r) = d_1 M^{-1} e_1 + d_2 M^{-1} e_2 + \cdots + d_r M^{-1} e_r$$
$$= d_1 M^{-1} b_{j_1} + d_2 M^{-1} b_{j_2} + \cdots + d_r M^{-1} b_{j_r}$$
$$= d_1 a_{j_1} + d_2 a_{j_2} + \cdots + d_r a_{j_r}.$$

The next theorem summarizes these results.

Theorem 3.16. *Let A be an $m \times n$ matrix of rank r, where $r > 0$, and let B be the reduced row echelon form of A. Then*
  (a) *The number of nonzero rows in B is r.*
  (b) *For each $i = 1, 2, \ldots, r$, there is a column b_{j_i} of B such that $b_{j_i} = e_i$.*
  (c) *The columns of A numbered $j_1, j_2, \ldots, j_r$ are linearly independent.*
  (d) *For each $k = 1, 2, \ldots n$, if column k of B is $d_1 e_1 + d_2 e_2 + \cdots + d_r e_r$, then column k of A is $d_1 a_{j_1} + d_2 a_{j_2} + \cdots + d_r a_{j_r}$.*

Corollary. *The reduced row echelon form of a matrix is unique.*

Proof. Exercise. (See Exercise 15.) ∎

Example 2

Let

$$A = \begin{pmatrix} 2 & 4 & 6 & 2 & 4 \\ 1 & 2 & 3 & 1 & 1 \\ 2 & 4 & 8 & 0 & 0 \\ 3 & 6 & 7 & 5 & 9 \end{pmatrix}.$$

The reduced row echelon form of A is

$$B = \begin{pmatrix} 1 & 2 & 0 & 4 & 0 \\ 0 & 0 & 1 & -1 & 0 \\ 0 & 0 & 0 & 0 & 1 \\ 0 & 0 & 0 & 0 & 0 \end{pmatrix}.$$

Since B has three nonzero rows, the rank of A is 3. The first, third, and fifth columns of B are e_1, e_2, and e_3; so Theorem 3.16(c) asserts that the first, third, and fifth columns of A are linearly independent.

Let the columns of A be denoted a_1, a_2, a_3, a_4, and a_5. Because the second column of B is $2e_1$, it follows from Theorem 3.16(d) that $a_2 = 2a_1$, as is easily checked. Moreover, since the fourth column of B is $4e_1 + (-1)e_2$, the same result shows that

$$a_4 = 4a_1 + (-1)a_3. \quad \blacklozenge$$

In Example 6 of Section 1.6, we extracted a basis for R^3 from the generating set

$$S = \{(2, -3, 5), (8, -12, 20), (1, 0, -2), (0, 2, -1), (7, 2, 0)\}.$$

The procedure described there can be streamlined by using Theorem 3.16. We begin by noting that if S were linearly independent, then S would be a basis for R^3. In this case, it is clear that S is linearly dependent because S contains more than $\dim(\mathsf{R}^3) = 3$ vectors. Nevertheless, it is instructive to consider the calculation that is needed to determine whether S is linearly dependent or linearly independent. Recall that S is linearly dependent if there are scalars c_1, c_2, c_3, c_4, and c_5, not all zero, such that

$$c_1(2, -3, 5) + c_2(8, -12, 20) + c_3(1, 0, -2) + c_4(0, 2, -1) + c_5(7, 2, 0) = (0, 0, 0).$$

Thus S is linearly dependent if and only if the system of linear equations

$$\begin{array}{rrrrrr} 2c_1 + & 8c_2 + & c_3 & & + 7c_5 = 0 \\ -3c_1 - & 12c_2 & & + 2c_4 + & 2c_5 = 0 \\ 5c_1 + & 20c_2 - & 2c_3 - & c_4 & = 0 \end{array}$$

has a nonzero solution. The augmented matrix of this system of equations is

$$A = \begin{pmatrix} 2 & 8 & 1 & 0 & 7 & 0 \\ -3 & -12 & 0 & 2 & 2 & 0 \\ 5 & 20 & -2 & -1 & 0 & 0 \end{pmatrix},$$

and its reduced row echelon form is

$$B = \begin{pmatrix} 1 & 4 & 0 & 0 & 2 & 0 \\ 0 & 0 & 1 & 0 & 3 & 0 \\ 0 & 0 & 0 & 1 & 4 & 0 \end{pmatrix}.$$

Using the technique described earlier in this section, we can find nonzero solutions of the preceding system, confirming that S is linearly dependent. However, Theorem 3.16(c) gives us additional information. Since the first, third, and fourth columns of B are e_1, e_2, and e_3, we conclude that the first, third, and fourth columns of A are linearly independent. But the columns of A other than the last column (which is the zero vector) are vectors in S. Hence

$$\beta = \{(2, -3, 5), (1, 0, -2), (0, 2, -1)\}$$

is a linearly independent subset of S. If follows from (b) of Corollary 2 to the replacement theorem (p. 47) that β is a basis for R^3.

Because every finite-dimensional vector space over F is isomorphic to F^n for some n, a similar approach can be used to reduce any finite generating set to a basis. This technique is illustrated in the next example.

Example 3

The set

$$S = \{2 + x + 2x^2 + 3x^3, 4 + 2x + 4x^2 + 6x^3, 6 + 3x + 8x^2 + 7x^3, 2 + x + 5x^3, 4 + x + 9x^3\}$$

generates a subspace V of $P_3(R)$. To find a subset of S that is a basis for V, we consider the subset

$$S' = \{(2, 1, 2, 3), (4, 2, 4, 6), (6, 3, 8, 7), (2, 1, 0, 5), (4, 1, 0, 9)\}$$

consisting of the images of the polynomials in S under the standard representation of $P_3(R)$ with respect to the standard ordered basis. Note that the 4×5 matrix in which the columns are the vectors in S' is the matrix A in Example 2. From the reduced row echelon form of A, which is the matrix B in Example 2, we see that the first, third, and fifth columns of A are linearly independent and the second and fourth columns of A are linear combinations of the first, third, and fifth columns. Hence

$$\{(2, 1, 2, 3), (6, 3, 8, 7), (4, 1, 0, 9)\}$$

is a basis for the subspace of R^4 that is generated by S'. It follows that

$$\{2 + x + 2x^2 + 3x^3, 6 + 3x + 8x^2 + 7x^3, 4 + x + 9x^3\}$$

is a basis for the subspace V of $P_3(R)$.   ♦

   We conclude this section by describing a method for extending a linearly independent subset S of a finite-dimensional vector space V to a basis for V. Recall that this is always possible by (c) of Corollary 2 to the replacement theorem (p. 47). Our approach is based on the replacement theorem and assumes that we can find an explicit basis β for V. Let S' be the ordered set consisting of the vectors in S followed by those in β. Since $\beta \subseteq S'$, the set S' generates V. We can then apply the technique described above to reduce this generating set to a basis for V containing S.

Example 4

Let

$$V = \{(x_1, x_2, x_3, x_4, x_5) \in R^5 : x_1 + 7x_2 + 5x_3 - 4x_4 + 2x_5 = 0\}.$$

It is easily verified that V is a subspace of R^5 and that

$$S = \{(-2, 0, 0, -1, -1), (1, 1, -2, -1, -1), (-5, 1, 0, 1, 1)\}$$

is a linearly independent subset of V.

To extend S to a basis for V, we first obtain a basis β for V. To do so, we solve the system of linear equations that defines V. Since in this case V is defined by a single equation, we need only write the equation as

$$x_1 = -7x_2 - 5x_3 + 4x_4 - 2x_5$$

and assign parametric values to x_2, x_3, x_4, and x_5. If $x_2 = t_1$, $x_3 = t_2$, $x_4 = t_3$, and $x_5 = t_4$, then the vectors in V have the form

$$\begin{aligned}(x_1,x_2,x_3,x_4,x_5) &= (-7t_1 - 5t_2 + 4t_3 - 2t_4, t_1, t_2, t_3, t_4) \\ &= t_1(-7,1,0,0,0) + t_2(-5,0,1,0,0) + t_3(4,0,0,1,0) + t_4(-2,0,0,0,1).\end{aligned}$$

Hence

$$\beta = \{(-7,1,0,0,0), (-5,0,1,0,0), (4,0,0,1,0), (-2,0,0,0,1)\}$$

is a basis for V by Theorem 3.15.

The matrix whose columns consist of the vectors in S followed by those in β is

$$\begin{pmatrix} -2 & 1 & -5 & -7 & -5 & 4 & -2 \\ 0 & 1 & 1 & 1 & 0 & 0 & 0 \\ 0 & -2 & 0 & 0 & 1 & 0 & 0 \\ -1 & -1 & 1 & 0 & 0 & 1 & 0 \\ -1 & -1 & 1 & 0 & 0 & 0 & 1 \end{pmatrix},$$

and its reduced row echelon form is

$$\begin{pmatrix} 1 & 0 & 0 & 1 & 1 & 0 & -1 \\ 0 & 1 & 0 & 0 & -.5 & 0 & 0 \\ 0 & 0 & 1 & 1 & .5 & 0 & 0 \\ 0 & 0 & 0 & 0 & 0 & 1 & -1 \\ 0 & 0 & 0 & 0 & 0 & 0 & 0 \end{pmatrix}.$$

Thus

$$\{(-2,0,0,-1,-1), (1,1,-2,-1,-1), (-5,1,0,1,1), (4,0,0,1,0)\}$$

is a basis for V containing S.   ◆

EXERCISES

1. Label the following statements as true or false.

(a) If $(A'|b')$ is obtained from $(A|b)$ by a finite sequence of elementary column operations, then the systems $Ax = b$ and $A'x = b'$ are equivalent.

(b) If $(A'|b')$ is obtained from $(A|b)$ by a finite sequence of elementary row operations, then the systems $Ax = b$ and $A'x = b'$ are equivalent.

(c) If A is an $n \times n$ matrix with rank n, then the reduced row echelon form of A is I_n.

(d) Any matrix can be put in reduced row echelon form by means of a finite sequence of elementary row operations.

(e) If $(A|b)$ is in reduced row echelon form, then the system $Ax = b$ is consistent.

(f) Let $Ax = b$ be a system of m linear equations in n unknowns for which the augmented matrix is in reduced row echelon form. If this system is consistent, then the dimension of the solution set of $Ax = 0$ is $n - r$, where r equals the number of nonzero rows in A.

(g) If a matrix A is transformed by elementary row operations into a matrix A' in reduced row echelon form, then the number of nonzero rows in A' equals the rank of A.

2. Use Gaussian elimination to solve the following systems of linear equations.

(a)
$$\begin{aligned}
x_1 + 2x_2 - \quad x_3 &= -1 \\
2x_1 + 2x_2 + \quad x_3 &= 1 \\
3x_1 + 5x_2 - 2x_3 &= -1
\end{aligned}$$

(b)
$$\begin{aligned}
x_1 - 2x_2 - \quad x_3 &= 1 \\
2x_1 - 3x_2 + \quad x_3 &= 6 \\
3x_1 - 5x_2 \phantom{{}+ x_3} &= 7 \\
x_1 \phantom{{}- 5x_2} + 5x_3 &= 9
\end{aligned}$$

(c)
$$\begin{aligned}
x_1 + 2x_2 \phantom{{}+ x_3} + \quad 2x_4 &= 6 \\
3x_1 + 5x_2 - \quad x_3 + \quad 6x_4 &= 17 \\
2x_1 + 4x_2 + \quad x_3 + \quad 2x_4 &= 12 \\
2x_1 \phantom{{}+ 4x_2} - 7x_3 + 11x_4 &= 7
\end{aligned}$$

(d)
$$\begin{aligned}
x_1 - \quad x_2 - 2x_3 + \quad 3x_4 &= -7 \\
2x_1 - \quad x_2 + 6x_3 + \quad 6x_4 &= -2 \\
-2x_1 + \quad x_2 - 4x_3 - \quad 3x_4 &= 0 \\
3x_1 - 2x_2 + 9x_3 + 10x_4 &= -5
\end{aligned}$$

(e)
$$\begin{aligned}
x_1 - 4x_2 - x_3 + \quad x_4 &= 3 \\
2x_1 - 8x_2 + x_3 - 4x_4 &= 9 \\
-x_1 + 4x_2 - 2x_3 + 5x_4 &= -6
\end{aligned}$$

(f)
$$\begin{aligned}
x_1 + 2x_2 - x_3 + 3x_4 &= 2 \\
2x_1 + 4x_2 - x_3 + 6x_4 &= 5 \\
x_2 \phantom{{}- x_3} + 2x_4 &= 3
\end{aligned}$$

(g)
$$\begin{aligned}
2x_1 - 2x_2 - \quad x_3 + 6x_4 - 2x_5 &= 1 \\
x_1 - \quad x_2 + \quad x_3 + 2x_4 - \quad x_5 &= 2 \\
4x_1 - 4x_2 + 5x_3 + 7x_4 - \quad x_5 &= 6
\end{aligned}$$

(h)
$$\begin{aligned}
3x_1 - \quad x_2 + x_3 - \quad x_4 + 2x_5 &= 5 \\
x_1 - \quad x_2 - x_3 - 2x_4 - \quad x_5 &= 2 \\
5x_1 - 2x_2 + x_3 - 3x_4 + 3x_5 &= 10 \\
2x_1 - \quad x_2 \phantom{{}+ x_3} - 2x_4 + \quad x_5 &= 5
\end{aligned}$$

(i)
$$\begin{aligned} 3x_1 - x_2 + 2x_3 + 4x_4 + \ x_5 &= \ 2 \\ x_1 - x_2 + 2x_3 + 3x_4 + \ x_5 &= -1 \\ 2x_1 - 3x_2 + 6x_3 + 9x_4 + 4x_5 &= -5 \\ 7x_1 - 2x_2 + 4x_3 + 8x_4 + \ x_5 &= \ 6 \end{aligned}$$

(j)
$$\begin{aligned} 2x_1 \quad\quad + 3x_3 \quad\quad - 4x_5 &= \ 5 \\ 3x_1 - 4x_2 + 8x_3 + 3x_4 \quad\quad &= \ 8 \\ x_1 - x_2 + 2x_3 + \ x_4 - \ x_5 &= \ 2 \\ -2x_1 + 5x_2 - 9x_3 - 3x_4 - 5x_5 &= -8 \end{aligned}$$

3. Suppose that the augmented matrix of a system $Ax = b$ is transformed into a matrix $(A'|b')$ in reduced row echelon form by a finite sequence of elementary row operations.

(a) Prove that $\text{rank}(A') \neq \text{rank}(A'|b')$ if and only if $(A'|b')$ contains a row in which the only nonzero entry lies in the last column.

(b) Deduce that $Ax = b$ is consistent if and only if $(A'|b')$ contains no row in which the only nonzero entry lies in the last column.

4. For each of the systems that follow, apply Exercise 3 to determine whether the system is consistent. If the system is consistent, find all solutions. Finally, find a basis for the solution set of the corresponding homogeneous system.

(a)
$$\begin{aligned} x_1 + 2x_2 - \ x_3 + \ x_4 &= 2 \\ 2x_1 + \ x_2 + \ x_3 - \ x_4 &= 3 \\ x_1 + 2x_2 - 3x_3 + 2x_4 &= 2 \end{aligned}$$

(b)
$$\begin{aligned} x_1 + x_2 - 3x_3 + x_4 &= -2 \\ x_1 + x_2 + \ x_3 - x_4 &= \ 2 \\ x_1 + x_2 - \ x_3 \quad\quad &= \ 0 \end{aligned}$$

(c)
$$\begin{aligned} x_1 + x_2 - 3x_3 + x_4 &= 1 \\ x_1 + x_2 + \ x_3 - x_4 &= 2 \\ x_1 + x_2 - \ x_3 \quad\quad &= 0 \end{aligned}$$

5. Let the reduced row echelon form of A be

$$\begin{pmatrix} 1 & 0 & 2 & 0 & -2 \\ 0 & 1 & -5 & 0 & -3 \\ 0 & 0 & 0 & 1 & 6 \end{pmatrix}.$$

Determine A if the first, second, and fourth columns of A are

$$\begin{pmatrix} 1 \\ -1 \\ 3 \end{pmatrix}, \quad \begin{pmatrix} 0 \\ -1 \\ 1 \end{pmatrix}, \quad \text{and} \quad \begin{pmatrix} 1 \\ -2 \\ 0 \end{pmatrix},$$

respectively.

6. Let the reduced row echelon form of A be

$$\begin{pmatrix} 1 & -3 & 0 & 4 & 0 & 5 \\ 0 & 0 & 1 & 3 & 0 & 2 \\ 0 & 0 & 0 & 0 & 1 & -1 \\ 0 & 0 & 0 & 0 & 0 & 0 \end{pmatrix}.$$

Determine A if the first, third, and sixth columns of A are

$$\begin{pmatrix} 1 \\ -2 \\ -1 \\ 3 \end{pmatrix}, \quad \begin{pmatrix} -1 \\ 1 \\ 2 \\ -4 \end{pmatrix}, \quad \text{and} \quad \begin{pmatrix} 3 \\ -9 \\ 2 \\ 5 \end{pmatrix},$$

respectively.

7. It can be shown that the vectors $u_1 = (2, -3, 1)$, $u_2 = (1, 4, -2)$, $u_3 = (-8, 12, -4)$, $u_4 = (1, 37, -17)$, and $u_5 = (-3, -5, 8)$ generate R^3. Find a subset of $\{u_1, u_2, u_3, u_4, u_5\}$ that is a basis for R^3.

8. Let W denote the subspace of R^5 consisting of all vectors having coordinates that sum to zero. The vectors

$$u_1 = (2, -3, 4, -5, 2), \qquad u_2 = (-6, 9, -12, 15, -6),$$
$$u_3 = (3, -2, 7, -9, 1), \qquad u_4 = (2, -8, 2, -2, 6),$$
$$u_5 = (-1, 1, 2, 1, -3), \qquad u_6 = (0, -3, -18, 9, 12),$$
$$u_7 = (1, 0, -2, 3, -2), \qquad \text{and} \qquad u_8 = (2, -1, 1, -9, 7)$$

generate W. Find a subset of $\{u_1, u_2, \ldots, u_8\}$ that is a basis for W.

9. Let W be the subspace of $\mathsf{M}_{2\times 2}(R)$ consisting of the symmetric 2×2 matrices. The set

$$S = \left\{ \begin{pmatrix} 0 & -1 \\ -1 & 1 \end{pmatrix}, \begin{pmatrix} 1 & 2 \\ 2 & 3 \end{pmatrix}, \begin{pmatrix} 2 & 1 \\ 1 & 9 \end{pmatrix}, \begin{pmatrix} 1 & -2 \\ -2 & 4 \end{pmatrix}, \begin{pmatrix} -1 & 2 \\ 2 & -1 \end{pmatrix} \right\}$$

generates W. Find a subset of S that is a basis for W.

10. Let

$$\mathsf{V} = \{(x_1, x_2, x_3, x_4, x_5) \in \mathsf{R}^5 : x_1 - 2x_2 + 3x_3 - x_4 + 2x_5 = 0\}.$$

    (a) Show that $S = \{(0, 1, 1, 1, 0)\}$ is a linearly independent subset of V.

    (b) Extend S to a basis for V.

11. Let V be as in Exercise 10.

    (a) Show that $S = \{(1, 2, 1, 0, 0)\}$ is a linearly independent subset of V.

    (b) Extend S to a basis for V.

12. Let V denote the set of all solutions to the system of linear equations

$$x_1 - x_2 \qquad + 2x_4 - 3x_5 + \ x_6 = 0$$
$$2x_1 - x_2 - x_3 + 3x_4 - 4x_5 + 4x_6 = 0.$$

(a) Show that $S = \{(0, -1, 0, 1, 1, 0), (1, 0, 1, 1, 1, 0)\}$ is a linearly independent subset of V.

(b) Extend S to a basis for V.

13. Let V be as in Exercise 12.

(a) Show that $S = \{(1, 0, 1, 1, 1, 0), (0, 2, 1, 1, 0, 0)\}$ is a linearly independent subset of V.

(b) Extend S to a basis for V.

14. If $(A|b)$ is in reduced row echelon form, prove that A is also in reduced row echelon form.

15. Prove the corollary to Theorem 3.16: The reduced row echelon form of a matrix is unique.

INDEX OF DEFINITIONS FOR CHAPTER 3

4

Determinants

4.1 Determinants of Order 2
4.2 Determinants of Order n
4.3 Properties of Determinants
4.4 Summary — Important Facts about Determinants
4.5* A Characterization of the Determinant

The determinant, which has played a prominent role in the theory of linear algebra, is a special scalar-valued function defined on the set of square matrices. Although it still has a place in the study of linear algebra and its applications, its role is less central than in former times. Yet no linear algebra book would be complete without a systematic treatment of the determinant, and we present one here. However, the main use of determinants in this book is to compute and establish the properties of eigenvalues, which we discuss in Chapter 5.

Although the determinant is not a linear transformation on $\mathsf{M}_{n \times n}(F)$ for $n > 1$, it does possess a kind of linearity (called *n-linearity*) as well as other properties that are examined in this chapter. In Section 4.1, we consider the determinant on the set of 2×2 matrices and derive its important properties and develop an efficient computational procedure. To illustrate the important role that determinants play in geometry, we also include optional material that explores the applications of the determinant to the study of area and orientation. In Sections 4.2 and 4.3, we extend the definition of the determinant to all square matrices and derive its important properties and develop an efficient computational procedure. For the reader who prefers to treat determinants lightly, Section 4.4 contains the essential properties that are needed in later chapters. Finally, Section 4.5, which is optional, offers an axiomatic approach to determinants by showing how to characterize the determinant in terms of three key properties.

4.1 DETERMINANTS OF ORDER 2

In this section, we define the determinant of a 2×2 matrix and investigate its geometric significance in terms of area and orientation.

Definition. *If*

$$A = \begin{pmatrix} a & b \\ c & d \end{pmatrix}$$

is a 2×2 *matrix with entries from a field* F, *then we define the* **determinant** *of* A, *denoted* $\det(A)$ *or* $|A|$, *to be the scalar* $ad - bc$.

Example 1

For the matrices

$$A = \begin{pmatrix} 1 & 2 \\ 3 & 4 \end{pmatrix} \quad \text{and} \quad B = \begin{pmatrix} 3 & 2 \\ 6 & 4 \end{pmatrix}$$

in $\mathsf{M}_{2\times 2}(R)$, we have

$$\det(A) = 1 \cdot 4 - 2 \cdot 3 = -2 \text{ and } \det(B) = 3 \cdot 4 - 2 \cdot 6 = 0. \quad \blacklozenge$$

For the matrices A and B in Example 1, we have

$$A + B = \begin{pmatrix} 4 & 4 \\ 9 & 8 \end{pmatrix},$$

and so

$$\det(A + B) = 4 \cdot 8 - 4 \cdot 9 = -4.$$

Since $\det(A + B) \neq \det(A) + \det(B)$, the function $\det \colon \mathsf{M}_{2\times 2}(R) \to R$ is *not* a linear transformation. Nevertheless, the determinant does possess an important linearity property, which is explained in the following theorem.

Theorem 4.1. *The function* $\det \colon \mathsf{M}_{2\times 2}(F) \to F$ *is a linear function of each row of a* 2×2 *matrix when the other row is held fixed. That is, if* u, v, *and* w *are in* F^2 *and* k *is a scalar, then*

$$\det \begin{pmatrix} u + kv \\ w \end{pmatrix} = \det \begin{pmatrix} u \\ w \end{pmatrix} + k \det \begin{pmatrix} v \\ w \end{pmatrix}$$

and

$$\det \begin{pmatrix} w \\ u + kv \end{pmatrix} = \det \begin{pmatrix} w \\ u \end{pmatrix} + k \det \begin{pmatrix} w \\ v \end{pmatrix}.$$

Proof. Let $u = (a_1, a_2)$, $v = (b_1, b_2)$, and $w = (c_1, c_2)$ be in F^2 and k be a scalar. Then

$$\det \begin{pmatrix} u \\ w \end{pmatrix} + k \det \begin{pmatrix} v \\ w \end{pmatrix} = \det \begin{pmatrix} a_1 & a_2 \\ c_1 & c_2 \end{pmatrix} + k \det \begin{pmatrix} b_1 & b_2 \\ c_1 & c_2 \end{pmatrix}$$

$$= (a_1 c_2 - a_2 c_1) + k(b_1 c_2 - b_2 c_1)$$
$$= (a_1 + k b_1) c_2 - (a_2 + k b_2) c_1$$
$$= \det \begin{pmatrix} a_1 + k b_1 & a_2 + k b_2 \\ c_1 & c_2 \end{pmatrix}$$
$$= \det \begin{pmatrix} u + k v \\ w \end{pmatrix}.$$

A similar calculation shows that

$$\det \begin{pmatrix} w \\ u \end{pmatrix} + k \det \begin{pmatrix} w \\ v \end{pmatrix} = \det \begin{pmatrix} w \\ u + k v \end{pmatrix}. \qquad \blacksquare$$

For the 2×2 matrices A and B in Example 1, it is easily checked that A is invertible but B is not. Note that $\det(A) \neq 0$ but $\det(B) = 0$. We now show that this property is true in general.

Theorem 4.2. *Let $A \in \mathsf{M}_{2\times 2}(F)$. Then the determinant of A is nonzero if and only if A is invertible. Moreover, if A is invertible, then*

$$A^{-1} = \frac{1}{\det(A)} \begin{pmatrix} A_{22} & -A_{12} \\ -A_{21} & A_{11} \end{pmatrix}.$$

Proof. If $\det(A) \neq 0$, then we can define a matrix

$$M = \frac{1}{\det(A)} \begin{pmatrix} A_{22} & -A_{12} \\ -A_{21} & A_{11} \end{pmatrix}.$$

A straightforward calculation shows that $AM = MA = I$, and so A is invertible and $M = A^{-1}$.

Conversely, suppose that A is invertible. A remark on page 152 shows that the rank of

$$A = \begin{pmatrix} A_{11} & A_{12} \\ A_{21} & A_{22} \end{pmatrix}$$

must be 2. Hence $A_{11} \neq 0$ or $A_{21} \neq 0$. If $A_{11} \neq 0$, add $-A_{21}/A_{11}$ times row 1 of A to row 2 to obtain the matrix

$$\begin{pmatrix} A_{11} & A_{12} \\ 0 & A_{22} - \dfrac{A_{12} A_{21}}{A_{11}} \end{pmatrix}.$$

Because elementary row operations are rank-preserving by the corollary to Theorem 3.4 (p. 153), it follows that

$$A_{22} - \frac{A_{12} A_{21}}{A_{11}} \neq 0.$$

Therefore $\det(A) = A_{11}A_{22} - A_{12}A_{21} \neq 0$. On the other hand, if $A_{21} \neq 0$, we see that $\det(A) \neq 0$ by adding $-A_{11}/A_{21}$ times row 2 of A to row 1 and applying a similar argument. Thus, in either case, $\det(A) \neq 0$. ∎

In Sections 4.2 and 4.3, we extend the definition of the determinant to $n \times n$ matrices and show that Theorem 4.2 remains true in this more general context. In the remainder of this section, which can be omitted if desired, we explore the geometric significance of the determinant of a 2×2 matrix. In particular, we show the importance of the sign of the determinant in the study of orientation.

The Area of a Parallelogram

By the **angle** between two vectors in R^2, we mean the angle with measure θ $(0 \leq \theta < \pi)$ that is formed by the vectors having the same magnitude and direction as the given vectors but emanating from the origin. (See Figure 4.1.)

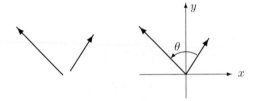

Figure 4.1: Angle between two vectors in R^2

If $\beta = \{u, v\}$ is an ordered basis for R^2, we define the **orientation** of β to be the real number

$$
O\begin{pmatrix} u \\ v \end{pmatrix} = \frac{\det\begin{pmatrix} u \\ v \end{pmatrix}}{\left| \det\begin{pmatrix} u \\ v \end{pmatrix} \right|}.
$$

(The denominator of this fraction is nonzero by Theorem 4.2.) Clearly

$$
O\begin{pmatrix} u \\ v \end{pmatrix} = \pm 1.
$$

Notice that

$$
O\begin{pmatrix} e_1 \\ e_2 \end{pmatrix} = 1 \quad \text{and} \quad O\begin{pmatrix} e_1 \\ -e_2 \end{pmatrix} = -1.
$$

Recall that a coordinate system $\{u, v\}$ is called **right-handed** if u can be rotated in a counterclockwise direction through an angle θ $(0 < \theta < \pi)$

to coincide with v. Otherwise $\{u, v\}$ is called a **left-handed** system. (See Figure 4.2.) In general (see Exercise 12),

A right-handed coordinate system   A left-handed coordinate system

Figure 4.2

$$O\begin{pmatrix} u \\ v \end{pmatrix} = 1$$

if and only if the ordered basis $\{u, v\}$ forms a right-handed coordinate system. For convenience, we also define

$$O\begin{pmatrix} u \\ v \end{pmatrix} = 1$$

if $\{u, v\}$ is linearly dependent.

Any ordered set $\{u, v\}$ in R^2 determines a parallelogram in the following manner. Regarding u and v as arrows emanating from the origin of R^2, we call the parallelogram having u and v as adjacent sides the **parallelogram determined by u and v.** (See Figure 4.3.) Observe that if the set $\{u, v\}$

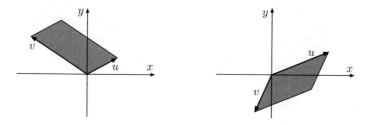

Figure 4.3: **Parallelograms determined by u and v**

is linearly dependent (i.e., if u and v are parallel), then the "parallelogram" determined by u and v is actually a line segment, which we consider to be a degenerate parallelogram having area zero.

There is an interesting relationship between

$$A \begin{pmatrix} u \\ v \end{pmatrix},$$

the area of the parallelogram determined by u and v, and

$$\det \begin{pmatrix} u \\ v \end{pmatrix},$$

which we now investigate. Observe first, however, that since

$$\det \begin{pmatrix} u \\ v \end{pmatrix}$$

may be negative, we cannot expect that

$$A \begin{pmatrix} u \\ v \end{pmatrix} = \det \begin{pmatrix} u \\ v \end{pmatrix}.$$

But we can prove that

$$A \begin{pmatrix} u \\ v \end{pmatrix} = O \begin{pmatrix} u \\ v \end{pmatrix} \cdot \det \begin{pmatrix} u \\ v \end{pmatrix},$$

from which it follows that

$$A \begin{pmatrix} u \\ v \end{pmatrix} = \left| \det \begin{pmatrix} u \\ v \end{pmatrix} \right|.$$

Our argument that

$$A \begin{pmatrix} u \\ v \end{pmatrix} = O \begin{pmatrix} u \\ v \end{pmatrix} \cdot \det \begin{pmatrix} u \\ v \end{pmatrix}$$

employs a technique that, although somewhat indirect, can be generalized to R^n. First, since

$$O \begin{pmatrix} u \\ v \end{pmatrix} = \pm 1,$$

we may multiply both sides of the desired equation by

$$O \begin{pmatrix} u \\ v \end{pmatrix}$$

to obtain the equivalent form

$$O \begin{pmatrix} u \\ v \end{pmatrix} \cdot A \begin{pmatrix} u \\ v \end{pmatrix} = \det \begin{pmatrix} u \\ v \end{pmatrix}.$$

We establish this equation by verifying that the three conditions of Exercise 11 are satisfied by the function

$$\delta\begin{pmatrix} u \\ v \end{pmatrix} = O\begin{pmatrix} u \\ v \end{pmatrix} \cdot A\begin{pmatrix} u \\ v \end{pmatrix}.$$

(a) We begin by showing that for any real number c

$$\delta\begin{pmatrix} u \\ cv \end{pmatrix} = c \cdot \delta\begin{pmatrix} u \\ v \end{pmatrix}.$$

Observe that this equation is valid if $c = 0$ because

$$\delta\begin{pmatrix} u \\ cv \end{pmatrix} = O\begin{pmatrix} u \\ 0 \end{pmatrix} \cdot A\begin{pmatrix} u \\ 0 \end{pmatrix} = 1 \cdot 0 = 0.$$

So assume that $c \neq 0$. Regarding cv as the base of the parallelogram determined by u and cv, we see that

$$A\begin{pmatrix} u \\ cv \end{pmatrix} = \text{base} \times \text{altitude} = |c|(\text{length of } v)(\text{altitude}) = |c| \cdot A\begin{pmatrix} u \\ v \end{pmatrix},$$

since the altitude h of the parallelogram determined by u and cv is the same as that in the parallelogram determined by u and v. (See Figure 4.4.) Hence

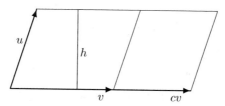

$$u \qquad h \qquad v \qquad cv$$

Figure 4.4

$$\delta\begin{pmatrix} u \\ cv \end{pmatrix} = O\begin{pmatrix} u \\ cv \end{pmatrix} \cdot A\begin{pmatrix} u \\ cv \end{pmatrix} = \left[\frac{c}{|c|} \cdot O\begin{pmatrix} u \\ v \end{pmatrix}\right]\left[|c| \cdot A\begin{pmatrix} u \\ v \end{pmatrix}\right]$$

$$= c \cdot O\begin{pmatrix} u \\ v \end{pmatrix} \cdot A\begin{pmatrix} u \\ v \end{pmatrix} = c \cdot \delta\begin{pmatrix} u \\ v \end{pmatrix}.$$

A similar argument shows that

$$\delta\begin{pmatrix} cu \\ v \end{pmatrix} = c \cdot \delta\begin{pmatrix} u \\ v \end{pmatrix}.$$

We next prove that

$$\delta \begin{pmatrix} u \\ au + bw \end{pmatrix} = b \cdot \delta \begin{pmatrix} u \\ w \end{pmatrix}$$

for any $u, w \in \mathsf{R}^2$ and any real numbers a and b. Because the parallelograms determined by u and w and by u and $u + w$ have a common base u and the same altitude (see Figure 4.5), it follows that

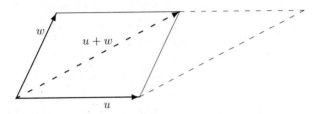

Figure 4.5

$$A \begin{pmatrix} u \\ w \end{pmatrix} = A \begin{pmatrix} u \\ u + w \end{pmatrix}.$$

If $a = 0$, then

$$\delta \begin{pmatrix} u \\ au + bw \end{pmatrix} = \delta \begin{pmatrix} u \\ bw \end{pmatrix} = b \cdot \delta \begin{pmatrix} u \\ w \end{pmatrix}$$

by the first paragraph of (a). Otherwise, if $a \neq 0$, then

$$\delta \begin{pmatrix} u \\ au + bw \end{pmatrix} = a \cdot \delta \begin{pmatrix} u \\ u + \dfrac{b}{a}w \end{pmatrix} = a \cdot \delta \begin{pmatrix} u \\ \dfrac{b}{a}w \end{pmatrix} = b \cdot \delta \begin{pmatrix} u \\ w \end{pmatrix}.$$

So the desired conclusion is obtained in either case.

We are now able to show that

$$\delta \begin{pmatrix} u \\ v_1 + v_2 \end{pmatrix} = \delta \begin{pmatrix} u \\ v_1 \end{pmatrix} + \delta \begin{pmatrix} u \\ v_2 \end{pmatrix}$$

for all $u, v_1, v_2 \in \mathsf{R}^2$. Since the result is immediate if $u = 0$, we assume that $u \neq 0$. Choose any vector $w \in \mathsf{R}^2$ such that $\{u, w\}$ is linearly independent. Then for any vectors $v_1, v_2 \in \mathsf{R}^2$ there exist scalars a_i and b_i such that $v_i = a_i u + b_i w$ $(i = 1, 2)$. Thus

$$\delta \begin{pmatrix} u \\ v_1 + v_2 \end{pmatrix} = \delta \begin{pmatrix} u \\ (a_1 + a_2)u + (b_1 + b_2)w \end{pmatrix} = (b_1 + b_2)\delta \begin{pmatrix} u \\ w \end{pmatrix}$$

$$= \delta \begin{pmatrix} u \\ a_1 u + b_1 w \end{pmatrix} + \delta \begin{pmatrix} u \\ a_2 u + b_2 w \end{pmatrix} = \delta \begin{pmatrix} u \\ v_1 \end{pmatrix} + \delta \begin{pmatrix} u \\ v_2 \end{pmatrix}.$$

A similar argument shows that

$$\delta \begin{pmatrix} u_1 + u_2 \\ v \end{pmatrix} = \delta \begin{pmatrix} u_1 \\ v \end{pmatrix} + \delta \begin{pmatrix} u_2 \\ v \end{pmatrix}$$

for all $u_1, u_2, v \in \mathsf{R}^2$.

(b) Since

$$A \begin{pmatrix} u \\ u \end{pmatrix} = 0, \quad \text{it follows that} \quad \delta \begin{pmatrix} u \\ u \end{pmatrix} = O \begin{pmatrix} u \\ u \end{pmatrix} \cdot A \begin{pmatrix} u \\ u \end{pmatrix} = 0$$

for any $u \in \mathsf{R}^2$.

(c) Because the parallelogram determined by e_1 and e_2 is the unit square,

$$\delta \begin{pmatrix} e_1 \\ e_2 \end{pmatrix} = O \begin{pmatrix} e_1 \\ e_2 \end{pmatrix} \cdot A \begin{pmatrix} e_1 \\ e_2 \end{pmatrix} = 1 \cdot 1 = 1.$$

Therefore δ satisfies the three conditions of Exercise 11, and hence $\delta = \det$. So the area of the parallelogram determined by u and v equals

$$O \begin{pmatrix} u \\ v \end{pmatrix} \cdot \det \begin{pmatrix} u \\ v \end{pmatrix}.$$

Thus we see, for example, that the area of the parallelogram determined by $u = (-1, 5)$ and $v = (4, -2)$ is

$$\left| \det \begin{pmatrix} u \\ v \end{pmatrix} \right| = \left| \det \begin{pmatrix} -1 & 5 \\ 4 & -2 \end{pmatrix} \right| = 18.$$

EXERCISES

1. Label the following statements as true or false.

   (a) The function $\det \colon \mathsf{M}_{2\times 2}(F) \to F$ is a linear transformation.

   (b) The determinant of a 2×2 matrix is a linear function of each row of the matrix when the other row is held fixed.

   (c) If $A \in \mathsf{M}_{2\times 2}(F)$ and $\det(A) = 0$, then A is invertible.

   (d) If u and v are vectors in R^2 emanating from the origin, then the area of the parallelogram having u and v as adjacent sides is

$$\det \begin{pmatrix} u \\ v \end{pmatrix}.$$

(e) A coordinate system is right-handed if and only if its orientation equals 1.

2. Compute the determinants of the following matrices in $M_{2\times2}(R)$.

   (a) $\begin{pmatrix} 6 & -3 \\ 2 & 4 \end{pmatrix}$ **(b)** $\begin{pmatrix} -5 & 2 \\ 6 & 1 \end{pmatrix}$ **(c)** $\begin{pmatrix} 8 & 0 \\ 3 & -1 \end{pmatrix}$

3. Compute the determinants of the following matrices in $M_{2\times2}(C)$.

   (a) $\begin{pmatrix} -1+i & 1-4i \\ 3+2i & 2-3i \end{pmatrix}$ **(b)** $\begin{pmatrix} 5-2i & 6+4i \\ -3+i & 7i \end{pmatrix}$ **(c)** $\begin{pmatrix} 2i & 3 \\ 4 & 6i \end{pmatrix}$

4. For each of the following pairs of vectors u and v in R^2, compute the area of the parallelogram determined by u and v.

   (a) $u = (3,-2)$ and $v = (2,5)$
   (b) $u = (1,3)$ and $v = (-3,1)$
   (c) $u = (4,-1)$ and $v = (-6,-2)$
   (d) $u = (3,4)$ and $v = (2,-6)$

5. Prove that if B is the matrix obtained by interchanging the rows of a 2×2 matrix A, then $\det(B) = -\det(A)$.

6. Prove that if the two columns of $A \in M_{2\times2}(F)$ are identical, then $\det(A) = 0$.

7. Prove that $\det(A^t) = \det(A)$ for any $A \in M_{2\times2}(F)$.

8. Prove that if $A \in M_{2\times2}(F)$ is upper triangular, then $\det(A)$ equals the product of the diagonal entries of A.

9. Prove that $\det(AB) = \det(A) \cdot \det(B)$ for any $A, B \in M_{2\times2}(F)$.

10. The **classical adjoint** of a 2×2 matrix $A \in M_{2\times2}(F)$ is the matrix

$$C = \begin{pmatrix} A_{22} & -A_{12} \\ -A_{21} & A_{11} \end{pmatrix}.$$

   Prove that
   (a) $CA = AC = [\det(A)]I$.
   (b) $\det(C) = \det(A)$.
   (c) The classical adjoint of A^t is C^t.
   (d) If A is invertible, then $A^{-1} = [\det(A)]^{-1}C$.

11. Let $\delta: M_{2\times2}(F) \to F$ be a function with the following three properties.

   (i) δ is a linear function of each row of the matrix when the other row is held fixed.
   (ii) If the two rows of $A \in M_{2\times2}(F)$ are identical, then $\delta(A) = 0$.

  (iii) If I is the 2×2 identity matrix, then $\delta(I) = 1$.

Prove that $\delta(A) = \det(A)$ for all $A \in \mathsf{M}_{2\times2}(F)$. (This result is generalized in Section 4.5.)

12. Let $\{u, v\}$ be an ordered basis for $\mathbb{R}^2$. Prove that

$$O\begin{pmatrix} u \\ v \end{pmatrix} = 1$$

if and only if $\{u, v\}$ forms a right-handed coordinate system. *Hint:* Recall the definition of a rotation given in Example 2 of Section 2.1.

4.2 DETERMINANTS OF ORDER n

In this section, we extend the definition of the determinant to $n \times n$ matrices for $n \geq 3$. For this definition, it is convenient to introduce the following notation: Given $A \in \mathsf{M}_{n\times n}(F)$, for $n \geq 2$, denote the $(n-1) \times (n-1)$ matrix obtained from A by deleting row i and column j by $\tilde{A}_{ij}$. Thus for

$$A = \begin{pmatrix} 1 & 2 & 3 \\ 4 & 5 & 6 \\ 7 & 8 & 9 \end{pmatrix} \in \mathsf{M}_{3\times3}(R),$$

we have

$$\tilde{A}_{11} = \begin{pmatrix} 5 & 6 \\ 8 & 9 \end{pmatrix}, \qquad \tilde{A}_{13} = \begin{pmatrix} 4 & 5 \\ 7 & 8 \end{pmatrix}, \quad \text{and} \quad \tilde{A}_{32} = \begin{pmatrix} 1 & 3 \\ 4 & 6 \end{pmatrix},$$

and for

$$B = \begin{pmatrix} 1 & -1 & 2 & -1 \\ -3 & 4 & 1 & -1 \\ 2 & -5 & -3 & 8 \\ -2 & 6 & -4 & 1 \end{pmatrix} \in \mathsf{M}_{4\times4}(R),$$

we have

$$\tilde{B}_{23} = \begin{pmatrix} 1 & -1 & -1 \\ 2 & -5 & 8 \\ -2 & 6 & 1 \end{pmatrix} \quad \text{and} \quad \tilde{B}_{42} = \begin{pmatrix} 1 & 2 & -1 \\ -3 & 1 & -1 \\ 2 & -3 & 8 \end{pmatrix}.$$

Definitions. *Let $A \in \mathsf{M}_{n\times n}(F)$. If $n = 1$, so that $A = (A_{11})$, we define* $\det(A) = A_{11}$. *For $n \geq 2$, we define $\det(A)$ recursively as*

$$\det(A) = \sum_{j=1}^{n} (-1)^{1+j} A_{1j} \cdot \det(\tilde{A}_{1j}).$$

The scalar $\det(A)$ is called the **determinant** of A and is also denoted by $|A|$. The scalar

$$(-1)^{i+j} \det(\tilde{A}_{ij})$$

is called the **cofactor** of the entry of A in row i, column j.

Letting

$$c_{ij} = (-1)^{i+j} \det(\tilde{A}_{ij})$$

denote the cofactor of the row i, column j entry of A, we can express the formula for the determinant of A as

$$\det(A) = A_{11}c_{11} + A_{12}c_{12} + \cdots + A_{1n}c_{1n}.$$

Thus the determinant of A equals the sum of the products of each entry in row 1 of A multiplied by its cofactor. This formula is called **cofactor expansion along the first row** of A. Note that, for 2×2 matrices, this definition of the determinant of A agrees with the one given in Section 4.1 because

$$\det(A) = A_{11}(-1)^{1+1} \det(\tilde{A}_{11}) + A_{12}(-1)^{1+2} \det(\tilde{A}_{12}) = A_{11}A_{22} - A_{12}A_{21}.$$

Example 1

Let

$$A = \begin{pmatrix} 1 & 3 & -3 \\ -3 & -5 & 2 \\ -4 & 4 & -6 \end{pmatrix} \in \mathsf{M}_{3\times3}(R).$$

Using cofactor expansion along the first row of A, we obtain

$$\begin{aligned}
\det(A) &= (-1)^{1+1} A_{11} \cdot \det(\tilde{A}_{11}) + (-1)^{1+2} A_{12} \cdot \det(\tilde{A}_{12}) \\
&\quad + (-1)^{1+3} A_{13} \cdot \det(\tilde{A}_{13}) \\
&= (-1)^2 (1) \cdot \det\begin{pmatrix} -5 & 2 \\ 4 & -6 \end{pmatrix} + (-1)^3 (3) \cdot \begin{pmatrix} -3 & 2 \\ -4 & -6 \end{pmatrix} \\
&\quad + (-1)^4 (-3) \cdot \det\begin{pmatrix} -3 & -5 \\ -4 & 4 \end{pmatrix} \\
&= 1\,[-5(-6) - 2(4)] - 3\,[-3(-6) - 2(-4)] - 3\,[-3(4) - (-5)(-4)] \\
&= 1(22) - 3(26) - 3(-32) \\
&= 40. \quad \blacklozenge
\end{aligned}$$

Example 2

Let

$$B = \begin{pmatrix} 0 & 1 & 3 \\ -2 & -3 & -5 \\ 4 & -4 & 4 \end{pmatrix} \in \mathsf{M}_{3\times3}(R).$$

Using cofactor expansion along the first row of B, we obtain

$$\det(B) = (-1)^{1+1} B_{11} \cdot \det(\tilde{B}_{11}) + (-1)^{1+2} B_{12} \cdot \det(\tilde{B}_{12})$$
$$+ (-1)^{1+3} B_{13} \cdot \det(\tilde{B}_{13})$$
$$= (-1)^2 (0) \cdot \det \begin{pmatrix} -3 & -5 \\ -4 & 4 \end{pmatrix} + (-1)^3 (1) \cdot \det \begin{pmatrix} -2 & -5 \\ 4 & 4 \end{pmatrix}$$
$$+ (-1)^4 (3) \cdot \det \begin{pmatrix} -2 & -3 \\ 4 & -4 \end{pmatrix}.$$
$$= 0 - 1\left[-2(4) - (-5)(4)\right] + 3\left[-2(-4) - (-3)(4)\right]$$
$$= 0 - 1(12) + 3(20)$$
$$= 48. \quad \blacklozenge$$

Example 3

Let

$$C = \begin{pmatrix} 2 & 0 & 0 & 1 \\ 0 & 1 & 3 & -3 \\ -2 & -3 & -5 & 2 \\ 4 & -4 & 4 & -6 \end{pmatrix} \in \mathsf{M}_{4\times4}(R).$$

Using cofactor expansion along the first row of C and the results of Examples 1 and 2, we obtain

$$\det(C) = (-1)^2 (2) \cdot \det(\tilde{C}_{11}) + (-1)^3 (0) \cdot \det(\tilde{C}_{12})$$
$$+ (-1)^4 (0) \cdot \det(\tilde{C}_{13}) + (-1)^5 (1) \cdot \det(\tilde{C}_{14})$$

$$= (-1)^2 (2) \cdot \det \begin{pmatrix} 1 & 3 & -3 \\ -3 & -5 & 2 \\ -4 & 4 & -6 \end{pmatrix} + 0 + 0$$

$$+ (-1)^5 (1) \cdot \det \begin{pmatrix} 0 & 1 & 3 \\ -2 & -3 & -5 \\ 4 & -4 & 4 \end{pmatrix}$$

$$= 2(40) + 0 + 0 - 1(48)$$
$$= 32. \quad \blacklozenge$$

Example 4

The determinant of the $n \times n$ identity matrix is 1. We prove this assertion by mathematical induction on n. The result is clearly true for the 1×1 identity matrix. Assume that the determinant of the $(n-1) \times (n-1)$ identity matrix is 1 for some $n \geq 2$, and let I denote the $n \times n$ identity matrix. Using cofactor expansion along the first row of I, we obtain

$$\det(I) = (-1)^2(1)\cdot \det(\tilde{I}_{11}) + (-1)^3(0)\cdot \det(\tilde{I}_{12}) + \cdots$$
$$+ (-1)^{1+n}(0)\cdot \det(\tilde{I}_{1n})$$
$$= 1(1) + 0 + \cdots + 0$$
$$= 1$$

because $\tilde{I}_{11}$ is the $(n-1) \times (n-1)$ identity matrix. This shows that the determinant of the $n \times n$ identity matrix is 1, and so the determinant of any identity matrix is 1 by the principle of mathematical induction.   ◆

As is illustrated in Example 3, the calculation of a determinant using the recursive definition is extremely tedious, even for matrices as small as 4×4. Later in this section, we present a more efficient method for evaluating determinants, but we must first learn more about them.

Recall from Theorem 4.1 (p. 200) that, although the determinant of a 2×2 matrix is *not* a linear transformation, it is a linear function of each row when the other row is held fixed. We now show that a similar property is true for determinants of any size.

Theorem 4.3. *The determinant of an $n \times n$ matrix is a linear function of each row when the remaining rows are held fixed. That is, for $1 \leq r \leq n$, we have*

$$\det \begin{pmatrix} a_1 \\ \vdots \\ a_{r-1} \\ u + kv \\ a_{r+1} \\ \vdots \\ a_n \end{pmatrix} = \det \begin{pmatrix} a_1 \\ \vdots \\ a_{r-1} \\ u \\ a_{r+1} \\ \vdots \\ a_n \end{pmatrix} + k \det \begin{pmatrix} a_1 \\ \vdots \\ a_{r-1} \\ v \\ a_{r+1} \\ \vdots \\ a_n \end{pmatrix}$$

whenever k is a scalar and u, v, and each a_i are row vectors in F^n.

Proof. The proof is by mathematical induction on n. The result is immediate if $n = 1$. Assume that for some integer $n \geq 2$ the determinant of any $(n-1) \times (n-1)$ matrix is a linear function of each row when the remaining rows are held fixed. Let A be an $n \times n$ matrix with rows $a_1, a_2, \ldots, a_n$, respectively, and suppose that for some r $(1 \le r \le n)$, we have $a_r = u + kv$ for some $u, v \in \mathsf{F}^n$ and some scalar k. Let $u = (b_1, b_2, \ldots, b_n)$ and $v = (c_1, c_2, \ldots, c_n)$, and let B and C be the matrices obtained from A by replacing row r of A by u and v, respectively. We must prove that $\det(A) = \det(B) + k \det(C)$. We leave the proof of this fact to the reader for the case $r = 1$. For $r > 1$ and $1 \le j \le n$, the rows of $\tilde{A}_{1j}$, $\tilde{B}_{1j}$, and $\tilde{C}_{1j}$ are the same except for row $r - 1$. Moreover, row $r - 1$ of $\tilde{A}_{1j}$ is

$$(b_1 + kc_1, \ldots, b_{j-1} + kc_{j-1}, b_{j+1} + kc_{j+1}, \ldots, b_n + kc_n),$$

which is the sum of row $r - 1$ of $\tilde{B}_{1j}$ and k times row $r - 1$ of $\tilde{C}_{1j}$. Since $\tilde{B}_{1j}$ and $\tilde{C}_{1j}$ are $(n - 1) \times (n - 1)$ matrices, we have

$$\det(\tilde{A}_{1j}) = \det(\tilde{B}_{1j}) + k \det(\tilde{C}_{1j})$$

by the induction hypothesis. Thus since $A_{1j} = B_{1j} = C_{1j}$, we have

$$\det(A) = \sum_{j=1}^{n} (-1)^{1+j} A_{1j} \cdot \det(\tilde{A}_{1j})$$

$$= \sum_{j=1}^{n} (-1)^{1+j} A_{1j} \cdot \left[\det(\tilde{B}_{1j}) + k \det(\tilde{C}_{1j}) \right]$$

$$= \sum_{j=1}^{n} (-1)^{1+j} A_{1j} \cdot \det(\tilde{B}_{1j}) + k \sum_{j=1}^{n} (-1)^{1+j} A_{1j} \cdot \det(\tilde{C}_{1j})$$

$$= \det(B) + k \det(C).$$

This shows that the theorem is true for $n \times n$ matrices, and so the theorem is true for all square matrices by mathematical induction. ∎

Corollary. If $A \in \mathsf{M}_{n \times n}(F)$ has a row consisting entirely of zeros, then $\det(A) = 0$.

Proof. See Exercise 24. ∎

The definition of a determinant requires that the determinant of a matrix be evaluated by cofactor expansion along the first row. Our next theorem shows that the determinant of a square matrix can be evaluated by cofactor expansion along *any* row. Its proof requires the following technical result.

Lemma. Let $B \in \mathsf{M}_{n \times n}(F)$, where $n \ge 2$. If row i of B equals e_k for some k $(1 \le k \le n)$, then $\det(B) = (-1)^{i+k} \det(\tilde{B}_{ik})$.

Proof. The proof is by mathematical induction on n. The lemma is easily proved for $n = 2$. Assume that for some integer $n \geq 3$, the lemma is true for $(n-1) \times (n-1)$ matrices, and let B be an $n \times n$ matrix in which row i of B equals e_k for some k $(1 \leq k \leq n)$. The result follows immediately from the definition of the determinant if $i = 1$. Suppose therefore that $1 < i \leq n$. For each $j \neq k$ $(1 \leq j \leq n)$, let C_{ij} denote the $(n-2) \times (n-2)$ matrix obtained from B by deleting rows 1 and i and columns j and k. For each j, row $i-1$ of $\tilde{B}_{1j}$ is the following vector in F^{n-1}:

$$\begin{cases} e_{k-1} & \text{if } j < k \\ 0 & \text{if } j = k \\ e_k & \text{if } j > k. \end{cases}$$

Hence by the induction hypothesis and the corollary to Theorem 4.3, we have

$$\det(\tilde{B}_{1j}) = \begin{cases} (-1)^{(i-1)+(k-1)} \det(C_{ij}) & \text{if } j < k \\ 0 & \text{if } j = k \\ (-1)^{(i-1)+k} \det(C_{ij}) & \text{if } j > k. \end{cases}$$

Therefore

$$\det(B) = \sum_{j=1}^{n} (-1)^{1+j} B_{1j} \cdot \det(\tilde{B}_{1j})$$

$$= \sum_{j<k} (-1)^{1+j} B_{1j} \cdot \det(\tilde{B}_{1j}) + \sum_{j>k} (-1)^{1+j} B_{1j} \cdot \det(\tilde{B}_{1j})$$

$$= \sum_{j<k} (-1)^{1+j} B_{1j} \cdot \left[(-1)^{(i-1)+(k-1)} \det(C_{ij}) \right]$$

$$+ \sum_{j>k} (-1)^{1+j} B_{1j} \cdot \left[(-1)^{(i-1)+k} \det(C_{ij}) \right]$$

$$= (-1)^{i+k} \left[\sum_{j<k} (-1)^{1+j} B_{1j} \cdot \det(C_{ij}) \right.$$

$$\left. + \sum_{j>k} (-1)^{1+(j-1)} B_{1j} \cdot \det(C_{ij}) \right].$$

Because the expression inside the preceding bracket is the cofactor expansion of $\tilde{B}_{ik}$ along the first row, it follows that

$$\det(B) = (-1)^{i+k} \det(\tilde{B}_{ik}).$$

This shows that the lemma is true for $n \times n$ matrices, and so the lemma is true for all square matrices by mathematical induction. ∎

We are now able to prove that cofactor expansion along any row can be used to evaluate the determinant of a square matrix.

Theorem 4.4. *The determinant of a square matrix can be evaluated by cofactor expansion along any row. That is, if $A \in \mathsf{M}_{n \times n}(F)$, then for any integer i $(1 \le i \le n)$,*

$$\det(A) = \sum_{j=1}^{n} (-1)^{i+j} A_{ij} \cdot \det(\tilde{A}_{ij}).$$

Proof. Cofactor expansion along the first row of A gives the determinant of A by definition. So the result is true if $i = 1$. Fix $i > 1$. Row i of A can be written as $\sum_{j=1}^{n} A_{ij} e_j$. For $1 \le j \le n$, let B_j denote the matrix obtained from A by replacing row i of A by e_j. Then by Theorem 4.3 and the lemma, we have

$$\det(A) = \sum_{j=1}^{n} A_{ij} \det(B_j) = \sum_{j=1}^{n} (-1)^{i+j} A_{ij} \cdot \det(\tilde{A}_{ij}). \qquad \blacksquare$$

Corollary. *If $A \in \mathsf{M}_{n \times n}(F)$ has two identical rows, then $\det(A) = 0$.*

Proof. The proof is by mathematical induction on n. We leave the proof of the result to the reader in the case that $n = 2$. Assume that for some integer $n \ge 3$, it is true for $(n-1) \times (n-1)$ matrices, and let rows r and s of $A \in \mathsf{M}_{n \times n}(F)$ be identical for $r \neq s$. Because $n \ge 3$, we can choose an integer i $(1 \le i \le n)$ other than r and s. Now

$$\det(A) = \sum_{j=1}^{n} (-1)^{i+j} A_{ij} \cdot \det(\tilde{A}_{ij})$$

by Theorem 4.4. Since each $\tilde{A}_{ij}$ is an $(n-1) \times (n-1)$ matrix with two identical rows, the induction hypothesis implies that each $\det(\tilde{A}_{ij}) = 0$, and hence $\det(A) = 0$. This completes the proof for $n \times n$ matrices, and so the lemma is true for all square matrices by mathematical induction. $\blacksquare$

It is possible to evaluate determinants more efficiently by combining cofactor expansion with the use of elementary row operations. Before such a process can be developed, we need to learn what happens to the determinant of a matrix if we perform an elementary row operation on that matrix. Theorem 4.3 provides this information for elementary row operations of type 2 (those in which a row is multiplied by a nonzero scalar). Next we turn our attention to elementary row operations of type 1 (those in which two rows are interchanged).

Theorem 4.5. *If $A \in \mathsf{M}_{n \times n}(F)$ and B is a matrix obtained from A by interchanging any two rows of A, then $\det(B) = -\det(A)$.*

Proof. Let the rows of $A \in \mathsf{M}_{n \times n}(F)$ be $a_1, a_2, \dots, a_n$, and let B be the matrix obtained from A by interchanging rows r and s, where $r < s$. Thus

$$
A = \begin{pmatrix} a_1 \\ \vdots \\ a_r \\ \vdots \\ a_s \\ \vdots \\ a_n \end{pmatrix} \quad \text{and} \quad B = \begin{pmatrix} a_1 \\ \vdots \\ a_s \\ \vdots \\ a_r \\ \vdots \\ a_n \end{pmatrix}.
$$

Consider the matrix obtained from A by replacing rows r and s by $a_r + a_s$. By the corollary to Theorem 4.4 and Theorem 4.3, we have

$$
0 = \det \begin{pmatrix} a_1 \\ \vdots \\ a_r + a_s \\ \vdots \\ a_r + a_s \\ \vdots \\ a_n \end{pmatrix} = \det \begin{pmatrix} a_1 \\ \vdots \\ a_r \\ \vdots \\ a_r + a_s \\ \vdots \\ a_n \end{pmatrix} + \det \begin{pmatrix} a_1 \\ \vdots \\ a_s \\ \vdots \\ a_r + a_s \\ \vdots \\ a_n \end{pmatrix}
$$

$$
= \det \begin{pmatrix} a_1 \\ \vdots \\ a_r \\ \vdots \\ a_r \\ \vdots \\ a_n \end{pmatrix} + \det \begin{pmatrix} a_1 \\ \vdots \\ a_r \\ \vdots \\ a_s \\ \vdots \\ a_n \end{pmatrix} + \det \begin{pmatrix} a_1 \\ \vdots \\ a_s \\ \vdots \\ a_r \\ \vdots \\ a_n \end{pmatrix} + \det \begin{pmatrix} a_1 \\ \vdots \\ a_s \\ \vdots \\ a_s \\ \vdots \\ a_n \end{pmatrix}
$$

$$
= 0 + \det(A) + \det(B) + 0.
$$

Therefore $\det(B) = -\det(A)$. ∎

We now complete our investigation of how an elementary row operation affects the determinant of a matrix by showing that elementary row operations of type 3 do not change the determinant of a matrix.

Theorem 4.6. *Let $A \in \mathsf{M}_{n \times n}(F)$, and let B be a matrix obtained by adding a multiple of one row of A to another row of A. Then $\det(B) = \det(A)$.*

Proof. Suppose that B is the $n \times n$ matrix obtained from A by adding k times row r to row s, where $r \neq s$. Let the rows of A be $a_1, a_2, \ldots, a_n$, and the rows of B be $b_1, b_2, \ldots, b_n$. Then $b_i = a_i$ for $i \neq s$ and $b_s = a_s + ka_r$. Let C be the matrix obtained from A by replacing row s with a_r. Applying Theorem 4.3 to row s of B, we obtain

$$\det(B) = \det(A) + k \det(C) = \det(A)$$

because $\det(C) = 0$ by the corollary to Theorem 4.4.   ∎

In Theorem 4.2 (p. 201), we proved that a 2×2 matrix is invertible if and only if its determinant is nonzero. As a consequence of Theorem 4.6, we can prove half of the promised generalization of this result in the following corollary. The converse is proved in the corollary to Theorem 4.7.

Corollary. *If $A \in \mathsf{M}_{n \times n}(F)$ has rank less than n, then $\det(A) = 0$.*

Proof. If the rank of A is less than n, then the rows $a_1, a_2, \ldots, a_n$ of A are linearly dependent. By Exercise 14 of Section 1.5, some row of A, say, row r, is a linear combination of the other rows. So there exist scalars c_i such that

$$a_r = c_1 a_1 + \cdots + c_{r-1} a_{r-1} + c_{r+1} a_{r+1} + \cdots + c_n a_n.$$

Let B be the matrix obtained from A by adding $-c_i$ times row i to row r for each $i \neq r$. Then row r of B consists entirely of zeros, and so $\det(B) = 0$. But by Theorem 4.6, $\det(B) = \det(A)$. Hence $\det(A) = 0$.   ∎

The following rules summarize the effect of an elementary row operation on the determinant of a matrix $A \in \mathsf{M}_{n \times n}(F)$.

(a) If B is a matrix obtained by interchanging any two rows of A, then $\det(B) = -\det(A)$.
(b) If B is a matrix obtained by multiplying a row of A by a nonzero scalar k, then $\det(B) = k \det(A)$.
(c) If B is a matrix obtained by adding a multiple of one row of A to another row of A, then $\det(B) = \det(A)$.

These facts can be used to simplify the evaluation of a determinant. Consider, for instance, the matrix in Example 1:

$$A = \begin{pmatrix} 1 & 3 & -3 \\ -3 & -5 & 2 \\ -4 & 4 & -6 \end{pmatrix}.$$

Adding 3 times row 1 of A to row 2 and 4 times row 1 to row 3, we obtain

$$M = \begin{pmatrix} 1 & 4 & -3 \\ 0 & 4 & -7 \\ 0 & 16 & -18 \end{pmatrix}.$$

Since M was obtained by performing two type 3 elementary row operations on A, we have $\det(A) = \det(M)$. The cofactor expansion of M along the first row gives

$$\det(M) = (-1)^{1+1}(1)\cdot \det(\tilde{M}_{11}) + (-1)^{1+2}(4)\cdot \det(\tilde{M}_{12})$$
$$+ (-1)^{1+3}(-3)\cdot \det(\tilde{M}_{13}).$$

Both $\tilde{M}_{12}$ and $\tilde{M}_{13}$ have a column consisting entirely of zeros, and so $\det(\tilde{M}_{12}) = \det(\tilde{M}_{13}) = 0$ by the corollary to Theorem 4.6. Hence

$$\det(M) = (-1)^{1+1}(1)\cdot \det(\tilde{M}_{11})$$
$$= (-1)^{1+1}(1)\cdot \det \begin{pmatrix} 4 & -7 \\ 16 & -18 \end{pmatrix}$$
$$= 1[4(-18) - (-7)(16)] = 40.$$

Thus with the use of two elementary row operations of type 3, we have reduced the computation of $\det(A)$ to the evaluation of one determinant of a 2×2 matrix.

But we can do even better. If we add -4 times row 2 of M to row 3 (another elementary row operation of type 3), we obtain

$$P = \begin{pmatrix} 1 & 4 & -3 \\ 0 & 4 & -7 \\ 0 & 0 & 10 \end{pmatrix}.$$

Evaluating $\det(P)$ by cofactor expansion along the first row, we have

$$\det(P) = (-1)^{1+1}(1)\cdot \det(\tilde{P}_{11})$$
$$= (-1)^{1+1}(1)\cdot \det \begin{pmatrix} 4 & -7 \\ 0 & 10 \end{pmatrix} = 1\cdot 4\cdot 10 = 40,$$

as described earlier. Since $\det(A) = \det(M) = \det(P)$, it follows that $\det(A) = 40$.

The preceding calculation of $\det(P)$ illustrates an important general fact. *The determinant of an upper triangular matrix is the product of its diagonal entries.* (See Exercise 23.) By using elementary row operations of types 1 and 3 only, we can transform any square matrix into an upper triangular matrix, and so we can easily evaluate the determinant of any square matrix. The next two examples illustrate this technique.

Example 5

To evaluate the determinant of the matrix

$$B = \begin{pmatrix} 0 & 1 & 3 \\ -2 & -3 & -5 \\ 4 & -4 & 4 \end{pmatrix}$$

in Example 2, we must begin with a row interchange. Interchanging rows 1 and 2 of B produces

$$C = \begin{pmatrix} -2 & -3 & -5 \\ 0 & 1 & 3 \\ 4 & -4 & 4 \end{pmatrix}.$$

By means of a sequence of elementary row operations of type 3, we can transform C into an upper triangular matrix:

$$\begin{pmatrix} -2 & -3 & -5 \\ 0 & 1 & 3 \\ 4 & -4 & 4 \end{pmatrix} \longrightarrow \begin{pmatrix} -2 & -3 & -5 \\ 0 & 1 & 3 \\ 0 & -10 & -6 \end{pmatrix} \longrightarrow \begin{pmatrix} -2 & -3 & -5 \\ 0 & 1 & 3 \\ 0 & 0 & 24 \end{pmatrix}.$$

Thus $\det(C) = -2 \cdot 1 \cdot 24 = -48$. Since C was obtained from B by an interchange of rows, it follows that

$$\det(B) = -\det(C) = 48. \quad \blacklozenge$$

Example 6

The technique in Example 5 can be used to evaluate the determinant of the matrix

$$C = \begin{pmatrix} 2 & 0 & 0 & 1 \\ 0 & 1 & 3 & -3 \\ -2 & -3 & -5 & 2 \\ 4 & -4 & 4 & -6 \end{pmatrix}$$

in Example 3. This matrix can be transformed into an upper triangular matrix by means of the following sequence of elementary row operations of type 3:

$$\begin{pmatrix} 2 & 0 & 0 & 1 \\ 0 & 1 & 3 & -3 \\ -2 & -3 & -5 & 2 \\ 4 & -4 & 4 & -6 \end{pmatrix} \longrightarrow \begin{pmatrix} 2 & 0 & 0 & 1 \\ 0 & 1 & 3 & -3 \\ 0 & -3 & -5 & 3 \\ 0 & -4 & 4 & -8 \end{pmatrix} \longrightarrow \begin{pmatrix} 2 & 0 & 0 & 1 \\ 0 & 1 & 3 & -3 \\ 0 & 0 & 4 & -6 \\ 0 & 0 & 16 & -20 \end{pmatrix}$$

$$\longrightarrow \begin{pmatrix} 2 & 0 & 0 & 1 \\ 0 & 1 & 3 & -3 \\ 0 & 0 & 4 & -6 \\ 0 & 0 & 0 & 4 \end{pmatrix}.$$

Thus $\det(C) = 2 \cdot 1 \cdot 4 \cdot 4 = 32$. $\quad \blacklozenge$

Using elementary row operations to evaluate the determinant of a matrix, as illustrated in Example 6, is far more efficient than using cofactor expansion. Consider first the evaluation of a 2×2 matrix. Since

$$\det \begin{pmatrix} a & b \\ c & d \end{pmatrix} = ad - bc,$$

the evaluation of the determinant of a 2×2 matrix requires 2 multiplications (and 1 subtraction). For $n \geq 3$, evaluating the determinant of an $n \times n$ matrix by cofactor expansion along any row expresses the determinant as a sum of n products involving determinants of $(n-1) \times (n-1)$ matrices. Thus in all, the evaluation of the determinant of an $n \times n$ matrix by cofactor expansion along any row requires over $n!$ multiplications, whereas evaluating the determinant of an $n \times n$ matrix by elementary row operations as in Examples 5 and 6 can be shown to require only $(n^3 + 2n - 3)/3$ multiplications. To evaluate the determinant of a 20×20 matrix, which is not large by present standards, cofactor expansion along a row requires over $20! \approx 2.4 \times 10^{18}$ multiplications. Thus it would take a computer performing one billion multiplications per second over 77 years to evaluate the determinant of a 20×20 matrix by this method. By contrast, the method using elementary row operations requires only 2679 multiplications for this calculation and would take the same computer less than three-millionths of a second! It is easy to see why most computer programs for evaluating the determinant of an arbitrary matrix do not use cofactor expansion.

In this section, we have defined the determinant of a square matrix in terms of cofactor expansion along the first row. We then showed that the determinant of a square matrix can be evaluated using cofactor expansion along any row. In addition, we showed that the determinant possesses a number of special properties, including properties that enable us to calculate $\det(B)$ from $\det(A)$ whenever B is a matrix obtained from A by means of an elementary row operation. These properties enable us to evaluate determinants much more efficiently. In the next section, we continue this approach to discover additional properties of determinants.

EXERCISES

1. Label the following statements as true or false.

   (a) The function $\det \colon \mathsf{M}_{n \times n}(F) \to F$ is a linear transformation.
   (b) The determinant of a square matrix can be evaluated by cofactor expansion along any row.
   (c) If two rows of a square matrix A are identical, then $\det(A) = 0$.
   (d) If B is a matrix obtained from a square matrix A by interchanging any two rows, then $\det(B) = -\det(A)$.
   (e) If B is a matrix obtained from a square matrix A by multiplying a row of A by a scalar, then $\det(B) = \det(A)$.
   (f) If B is a matrix obtained from a square matrix A by adding k times row i to row j, then $\det(B) = k \det(A)$.
   (g) If $A \in \mathsf{M}_{n \times n}(F)$ has rank n, then $\det(A) = 0$.
   (h) The determinant of an upper triangular matrix equals the product of its diagonal entries.

2. Find the value of k that satisfies the following equation:

$$\det \begin{pmatrix} 3a_1 & 3a_2 & 3a_3 \\ 3b_1 & 3b_2 & 3b_3 \\ 3c_1 & 3c_2 & 3c_3 \end{pmatrix} = k \det \begin{pmatrix} a_1 & a_2 & a_3 \\ b_1 & b_2 & b_3 \\ c_1 & c_2 & c_3 \end{pmatrix}.$$

3. Find the value of k that satisfies the following equation:

$$\det \begin{pmatrix} 2a_1 & 2a_2 & 2a_3 \\ 3b_1 + 5c_1 & 3b_2 + 5c_2 & 3b_3 + 5c_3 \\ 7c_1 & 7c_2 & 7c_3 \end{pmatrix} = k \det \begin{pmatrix} a_1 & a_2 & a_3 \\ b_1 & b_2 & b_3 \\ c_1 & c_2 & c_3 \end{pmatrix}.$$

4. Find the value of k that satisfies the following equation:

$$\det \begin{pmatrix} b_1 + c_1 & b_2 + c_2 & b_3 + c_3 \\ a_1 + c_1 & a_2 + c_2 & a_3 + c_3 \\ a_1 + b_1 & a_2 + b_2 & a_3 + b_3 \end{pmatrix} = k \det \begin{pmatrix} a_1 & a_2 & a_3 \\ b_1 & b_2 & b_3 \\ c_1 & c_2 & c_3 \end{pmatrix}.$$

In Exercises 5–12, evaluate the determinant of the given matrix by cofactor expansion along the indicated row.

5.
$$\begin{pmatrix} 0 & 1 & 2 \\ -1 & 0 & -3 \\ 2 & 3 & 0 \end{pmatrix}$$
along the first row

6.
$$\begin{pmatrix} 1 & 0 & 2 \\ 0 & 1 & 5 \\ -1 & 3 & 0 \end{pmatrix}$$
along the first row

7.
$$\begin{pmatrix} 0 & 1 & 2 \\ -1 & 0 & -3 \\ 2 & 3 & 0 \end{pmatrix}$$
along the second row

8.
$$\begin{pmatrix} 1 & 0 & 2 \\ 0 & 1 & 5 \\ -1 & 3 & 0 \end{pmatrix}$$
along the third row

9.
$$\begin{pmatrix} 0 & 1+i & 2 \\ -2i & 0 & 1-i \\ 3 & 4i & 0 \end{pmatrix}$$
along the third row

10.
$$\begin{pmatrix} i & 2+i & 0 \\ -1 & 3 & 2i \\ 0 & -1 & 1-i \end{pmatrix}$$
along the second row

11.
$$\begin{pmatrix} 0 & 2 & 1 & 3 \\ 1 & 0 & -2 & 2 \\ 3 & -1 & 0 & 1 \\ -1 & 1 & 2 & 0 \end{pmatrix}$$
along the fourth row

12.
$$\begin{pmatrix} 1 & -1 & 2 & -1 \\ -3 & 4 & 1 & -1 \\ 2 & -5 & -3 & 8 \\ -2 & 6 & -4 & 1 \end{pmatrix}$$
along the fourth row

In Exercises 13–22, evaluate the determinant of the given matrix by any legitimate method.

13. $\begin{pmatrix} 0 & 0 & 1 \\ 0 & 2 & 3 \\ 4 & 5 & 6 \end{pmatrix}$ **14.** $\begin{pmatrix} 2 & 3 & 4 \\ 5 & 6 & 0 \\ 7 & 0 & 0 \end{pmatrix}$

15. $\begin{pmatrix} 1 & 2 & 3 \\ 4 & 5 & 6 \\ 7 & 8 & 9 \end{pmatrix}$ **16.** $\begin{pmatrix} -1 & 3 & 2 \\ 4 & -8 & 1 \\ 2 & 2 & 5 \end{pmatrix}$

17. $\begin{pmatrix} 0 & 1 & 1 \\ 1 & 2 & -5 \\ 6 & -4 & 3 \end{pmatrix}$ **18.** $\begin{pmatrix} 1 & -2 & 3 \\ -1 & 2 & -5 \\ 3 & -1 & 2 \end{pmatrix}$

19. $\begin{pmatrix} i & 2 & -1 \\ 3 & 1+i & 2 \\ -2i & 1 & 4-i \end{pmatrix}$ **20.** $\begin{pmatrix} -1 & 2+i & 3 \\ 1-i & i & 1 \\ 3i & 2 & -1+i \end{pmatrix}$

21. $\begin{pmatrix} 1 & 0 & -2 & 3 \\ -3 & 1 & 1 & 2 \\ 0 & 4 & -1 & 1 \\ 2 & 3 & 0 & 1 \end{pmatrix}$ **22.** $\begin{pmatrix} 1 & -2 & 3 & -12 \\ -5 & 12 & -14 & 19 \\ -9 & 22 & -20 & 31 \\ -4 & 9 & -14 & 15 \end{pmatrix}$

23. Prove that the determinant of an upper triangular matrix is the product of its diagonal entries.

24. Prove the corollary to Theorem 4.3.

25. Prove that $\det(kA) = k^n \det(A)$ for any $A \in \mathsf{M}_{n \times n}(F)$.

26. Let $A \in \mathsf{M}_{n \times n}(F)$. Under what conditions is $\det(-A) = \det(A)$?

27. Prove that if $A \in \mathsf{M}_{n \times n}(F)$ has two identical columns, then $\det(A) = 0$.

28. Compute $\det(E_i)$ if E_i is an elementary matrix of type i.

29.† Prove that if E is an elementary matrix, then $\det(E^t) = \det(E)$.

30. Let the rows of $A \in \mathsf{M}_{n \times n}(F)$ be $a_1, a_2, \ldots, a_n$, and let B be the matrix in which the rows are $a_n, a_{n-1}, \ldots, a_1$. Calculate $\det(B)$ in terms of $\det(A)$.

4.3   PROPERTIES OF DETERMINANTS

In Theorem 3.1, we saw that performing an elementary row operation on a matrix can be accomplished by multiplying the matrix by an elementary matrix. This result is very useful in studying the effects on the determinant of applying a sequence of elementary row operations. Because the determinant of the $n \times n$ identity matrix is 1 (see Example 4 in Section 4.2), we can interpret the statements on page 217 as the following facts about the determinants of elementary matrices.

(a) If E is an elementary matrix obtained by interchanging any two rows of I, then $\det(E) = -1$.
(b) If E is an elementary matrix obtained by multiplying some row of I by the nonzero scalar k, then $\det(E) = k$.
(c) If E is an elementary matrix obtained by adding a multiple of some row of I to another row, then $\det(E) = 1$.

We now apply these facts about determinants of elementary matrices to prove that the determinant is a *multiplicative* function.

Theorem 4.7. *For any* $A, B \in \mathsf{M}_{n \times n}(F)$, $\det(AB) = \det(A) \cdot \det(B)$.

Proof. We begin by establishing the result when A is an elementary matrix. If A is an elementary matrix obtained by interchanging two rows of I, then $\det(A) = -1$. But by Theorem 3.1 (p. 149), AB is a matrix obtained by interchanging two rows of B. Hence by Theorem 4.5 (p. 216), $\det(AB) = -\det(B) = \det(A) \cdot \det(B)$. Similar arguments establish the result when A is an elementary matrix of type 2 or type 3. (See Exercise 18.)

If A is an $n \times n$ matrix with rank less than n, then $\det(A) = 0$ by the corollary to Theorem 4.6 (p. 216). Since $\operatorname{rank}(AB) \leq \operatorname{rank}(A) < n$ by Theorem 3.7 (p. 159), we have $\det(AB) = 0$. Thus $\det(AB) = \det(A) \cdot \det(B)$ in this case.

On the other hand, if A has rank n, then A is invertible and hence is the product of elementary matrices (Corollary 3 to Theorem 3.6 p. 159), say, $A = E_m \cdots E_2 E_1$. The first paragraph of this proof shows that

$$
\begin{aligned}
\det(AB) &= \det(E_m \cdots E_2 E_1 B) \\
&= \det(E_m) \cdot \det(E_{m-1} \cdots E_2 E_1 B) \\
&\quad \vdots \\
&= \det(E_m) \cdot \cdots \cdot \det(E_2) \cdot \det(E_1) \cdot \det(B) \\
&= \det(E_m \cdots E_2 E_1) \cdot \det(B) \\
&= \det(A) \cdot \det(B).
\end{aligned}
$$
∎

Corollary. *A matrix* $A \in \mathsf{M}_{n \times n}(F)$ *is invertible if and only if* $\det(A) \neq 0$. *Furthermore, if* A *is invertible, then* $\det(A^{-1}) = \dfrac{1}{\det(A)}$.

Proof. If $A \in \mathsf{M}_{n \times n}(F)$ is not invertible, then the rank of A is less than n. So $\det(A) = 0$ by the corollary to Theorem 4.6 (p, 217). On the other hand, if $A \in \mathsf{M}_{n \times n}(F)$ is invertible, then

$$
\det(A) \cdot \det(A^{-1}) = \det(AA^{-1}) = \det(I) = 1
$$

by Theorem 4.7. Hence $\det(A) \neq 0$ and $\det(A^{-1}) = \dfrac{1}{\det(A)}$. ∎

In our discussion of determinants until now, we have used only the rows of a matrix. For example, the recursive definition of a determinant involved cofactor expansion along a *row*, and the more efficient method developed in Section 4.2 used elementary *row* operations. Our next result shows that the determinants of A and A^t are always equal. Since the rows of A are the columns of A^t, this fact enables us to translate any statement about determinants that involves the rows of a matrix into a corresponding statement that involves its columns.

Theorem 4.8. *For any $A \in \mathsf{M}_{n \times n}(F)$, $\det(A^t) = \det(A)$.*

Proof. If A is not invertible, then $\operatorname{rank}(A) < n$. But $\operatorname{rank}(A^t) = \operatorname{rank}(A)$ by Corollary 2 to Theorem 3.6 (p. 158), and so A^t is not invertible. Thus $\det(A^t) = 0 = \det(A)$ in this case.

On the other hand, if A is invertible, then A is a product of elementary matrices, say $A = E_m \cdots E_2 E_1$. Since $\det(E_i) = \det(E_i^t)$ for every i by Exercise 29 of Section 4.2, by Theorem 4.7 we have

$$
\begin{aligned}
\det(A^t) &= \det(E_1^t E_2^t \cdots E_m^t) \\
&= \det(E_1^t) \cdot \det(E_2^t) \cdot \cdots \cdot \det(E_m^t) \\
&= \det(E_1) \cdot \det(E_2) \cdot \cdots \cdot \det(E_m) \\
&= \det(E_m) \cdot \cdots \cdot \det(E_2) \cdot \det(E_1) \\
&= \det(E_m \cdots E_2 E_1) \\
&= \det(A).
\end{aligned}
$$

Thus, in either case, $\det(A^t) = \det(A)$. ∎

Among the many consequences of Theorem 4.8 are that determinants can be evaluated by cofactor expansion along a column, and that elementary column operations can be used as well as elementary row operations in evaluating a determinant. (The effect on the determinant of performing an elementary column operation is the same as the effect of performing the corresponding elementary row operation.) We conclude our discussion of determinant properties with a well-known result that relates determinants to the solutions of certain types of systems of linear equations.

Theorem 4.9 (Cramer's Rule).   *Let $Ax = b$ be the matrix form of a system of n linear equations in n unknowns, where $x = (x_1, x_2, \ldots, x_n)^t$. If $\det(A) \neq 0$, then this system has a unique solution, and for each k ($k = 1, 2, \ldots, n$),*

$$
x_k = \frac{\det(M_k)}{\det(A)},
$$

where M_k is the $n \times n$ matrix obtained from A by replacing column k of A by b.

Proof. If $\det(A) \neq 0$, then the system $Ax = b$ has a unique solution by the corollary to Theorem 4.7 and Theorem 3.10 (p. 174). For each integer k $(1 \leq k \leq n)$, let a_k denote the kth column of A and X_k denote the matrix obtained from the $n \times n$ identity matrix by replacing column k by x. Then by Theorem 2.13 (p. 90), AX_k is the $n \times n$ matrix whose ith column is

$$Ae_i = a_i \text{ if } i \neq k \qquad \text{and} \qquad Ax = b \text{ if } i = k.$$

Thus $AX_k = M_k$. Evaluating X_k by cofactor expansion along row k produces

$$\det(X_k) = x_k \cdot \det(I_{n-1}) = x_k.$$

Hence by Theorem 4.7,

$$\det(M_k) = \det(AX_k) = \det(A) \cdot \det(X_k) = \det(A) \cdot x_k.$$

Therefore

$$x_k = [\det(A)]^{-1} \cdot \det(M_k). \qquad \blacksquare$$

Example 1

We illustrate Theorem 4.9 by using Cramer's rule to solve the following system of linear equations:

$$\begin{aligned}
x_1 + 2x_2 + 3x_3 &= 2 \\
x_1 \qquad\quad + x_3 &= 3 \\
x_2 + x_2 - x_3 &= 1.
\end{aligned}$$

The matrix form of this system of linear equations is $Ax = b$, where

$$A = \begin{pmatrix} 1 & 2 & 3 \\ 1 & 0 & 1 \\ 1 & 1 & -1 \end{pmatrix} \qquad \text{and} \qquad b = \begin{pmatrix} 2 \\ 3 \\ 1 \end{pmatrix}.$$

Because $\det(A) = 6 \neq 0$, Cramer's rule applies. Using the notation of Theorem 4.9, we have

$$x_1 = \frac{\det(M_1)}{\det(A)} = \frac{\det \begin{pmatrix} 2 & 2 & 3 \\ 3 & 0 & 1 \\ 1 & 1 & -1 \end{pmatrix}}{\det(A)} = \frac{15}{6} = \frac{5}{2},$$

$$x_2 = \frac{\det(M_2)}{\det(A)} = \frac{\det \begin{pmatrix} 1 & 2 & 3 \\ 1 & 3 & 1 \\ 1 & 1 & -1 \end{pmatrix}}{\det(A)} = \frac{-6}{6} = -1,$$

and

$$x_3 = \frac{\det(M_3)}{\det(A)} = \frac{\det \begin{pmatrix} 1 & 2 & 2 \\ 1 & 0 & 3 \\ 1 & 1 & 1 \end{pmatrix}}{\det(A)} = \frac{3}{6} = \frac{1}{2}.$$

Thus the unique solution to the given system of linear equations is

$$(x_1, x_2, x_3) = \left(\frac{5}{2}, -1, \frac{1}{2} \right). \quad \blacklozenge$$

In applications involving systems of linear equations, we sometimes need to know that there is a solution in which the unknowns are integers. In this situation, Cramer's rule can be useful because it implies that a system of linear equations with integral coefficients has an integral solution if the determinant of its coefficient matrix is ± 1. On the other hand, Cramer's rule is not useful for computation because it requires evaluating $n + 1$ determinants of $n \times n$ matrices to solve a system of n linear equations in n unknowns. The amount of computation to do this is far greater than that required to solve the system by the method of Gaussian elimination, which was discussed in Section 3.4. Thus Cramer's rule is primarily of theoretical and aesthetic interest, rather than of computational value.

As in Section 4.1, it is possible to interpret the determinant of a matrix $A \in \mathsf{M}_{n \times n}(R)$ geometrically. If the rows of A are $a_1, a_2, \ldots, a_n$, respectively, then $|\det(A)|$ is the **n-dimensional volume** (the generalization of area in R^2 and volume in R^3) of the parallelepiped having the vectors $a_1, a_2, \ldots, a_n$ as adjacent sides. (For a proof of a more generalized result, see Jerrold E. Marsden and Michael J. Hoffman, *Elementary Classical Analysis*, W.H. Freeman and Company, New York, 1993, p. 524.)

Example 2

The volume of the parallelepiped having the vectors $a_1 = (1, -2, 1)$, $a_2 = (1, 0, -1)$, and $a_3 = (1, 1, 1)$ as adjacent sides is

$$\left| \det \begin{pmatrix} 1 & -2 & 1 \\ 1 & 0 & -1 \\ 1 & 1 & 1 \end{pmatrix} \right| = 6.$$

Note that the object in question is a rectangular parallelepiped (see Figure 4.6) with sides of lengths $\sqrt{6}$, $\sqrt{2}$, and $\sqrt{3}$. Hence by the familiar formula for volume, its volume should be $\sqrt{6} \cdot \sqrt{2} \cdot \sqrt{3} = 6$, as the determinant calculation shows. $\blacklozenge$

In our earlier discussion of the geometric significance of the determinant formed from the vectors in an ordered basis for R^2, we also saw that this

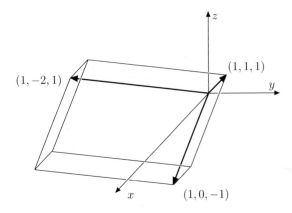

Figure 4.6: Parallelepiped determined by three vectors in $\mathbb{R}^3$.

determinant is positive if and only if the basis induces a right-handed coordinate system. A similar statement is true in $\mathbb{R}^n$. Specifically, if γ is any ordered basis for $\mathbb{R}^n$ and β is the standard ordered basis for $\mathbb{R}^n$, then γ induces a *right-handed coordinate system* if and only if $\det(Q) > 0$, where Q is the change of coordinate matrix changing γ-coordinates into β-coordinates. Thus, for instance,

$$\gamma = \left\{ \begin{pmatrix} 1 \\ 1 \\ 0 \end{pmatrix}, \begin{pmatrix} 1 \\ -1 \\ 0 \end{pmatrix}, \begin{pmatrix} 0 \\ 0 \\ 1 \end{pmatrix} \right\}$$

induces a left-handed coordinate system in $\mathbb{R}^3$ because

$$\det \begin{pmatrix} 1 & 1 & 0 \\ 1 & -1 & 0 \\ 0 & 0 & 1 \end{pmatrix} = -2 < 0,$$

whereas

$$\gamma' = \left\{ \begin{pmatrix} 1 \\ 2 \\ 0 \end{pmatrix}, \begin{pmatrix} -2 \\ 1 \\ 0 \end{pmatrix}, \begin{pmatrix} 0 \\ 0 \\ 1 \end{pmatrix} \right\}$$

induces a right-handed coordinate system in $\mathbb{R}^3$ because

$$\det \begin{pmatrix} 1 & -2 & 0 \\ 2 & 1 & 0 \\ 0 & 0 & 1 \end{pmatrix} = 5 > 0.$$

More generally, if β and γ are two ordered bases for R^n, then the coordinate systems induced by β and γ have the same **orientation** (either both are right-handed or both are left-handed) if and only if $\det(Q) > 0$, where Q is the change of coordinate matrix changing γ-coordinates into β-coordinates.

EXERCISES

1. Label the following statements as true or false.

   (a) If E is an elementary matrix, then $\det(E) = \pm 1$.
   (b) For any $A, B \in \mathsf{M}_{n \times n}(F)$, $\det(AB) = \det(A) \cdot \det(B)$.
   (c) A matrix $M \in \mathsf{M}_{n \times n}(F)$ is invertible if and only if $\det(M) = 0$.
   (d) A matrix $M \in \mathsf{M}_{n \times n}(F)$ has rank n if and only if $\det(M) \neq 0$.
   (e) For any $A \in \mathsf{M}_{n \times n}(F)$, $\det(A^t) = -\det(A)$.
   (f) The determinant of a square matrix can be evaluated by cofactor expansion along any column.
   (g) Every system of n linear equations in n unknowns can be solved by Cramer's rule.
   (h) Let $Ax = b$ be the matrix form of a system of n linear equations in n unknowns, where $x = (x_1, x_2, \ldots, x_n)^t$. If $\det(A) \neq 0$ and if M_k is the $n \times n$ matrix obtained from A by replacing row k of A by b^t, then the unique solution of $Ax = b$ is

   $$x_k = \frac{\det(M_k)}{\det(A)} \quad \text{for } k = 1, 2, \ldots, n.$$

In Exercises 2–7, use Cramer's rule to solve the given system of linear equations.

2. $\begin{aligned} a_{11}x_1 + a_{12}x_2 &= b_1 \\ a_{21}x_1 + a_{22}x_2 &= b_2 \\ \text{where } a_{11}a_{22} &- a_{12}a_{21} \neq 0 \end{aligned}$

3. $\begin{aligned} 2x_1 + x_2 - 3x_3 &= 5 \\ x_1 - 2x_2 + x_3 &= 10 \\ 3x_1 + 4x_2 - 2x_3 &= 0 \end{aligned}$

4. $\begin{aligned} 2x_1 + x_2 - 3x_3 &= 1 \\ x_1 - 2x_2 + x_3 &= 0 \\ 3x_1 + 4x_2 - 2x_3 &= -5 \end{aligned}$

5. $\begin{aligned} x_1 - x_2 + 4x_3 &= -4 \\ -8x_1 + 3x_2 + x_3 &= 8 \\ 2x_1 - x_2 + x_3 &= 0 \end{aligned}$

6. $\begin{aligned} x_1 - x_2 + 4x_3 &= -2 \\ -8x_1 + 3x_2 + x_3 &= 0 \\ 2x_1 - x_2 + x_3 &= 6 \end{aligned}$

7. $\begin{aligned} 3x_1 + x_2 + x_3 &= 4 \\ -2x_1 - x_2 &= 12 \\ x_1 + 2x_2 + x_3 &= -8 \end{aligned}$

8. Use Theorem 4.8 to prove a result analogous to Theorem 4.3 (p. 212), but for columns.

9. Prove that an upper triangular $n \times n$ matrix is invertible if and only if all its diagonal entries are nonzero.

10. A matrix $M \in M_{n \times n}(C)$ is called **nilpotent** if, for some positive integer k, $M^k = O$, where O is the $n \times n$ zero matrix. Prove that if M is nilpotent, then $\det(M) = 0$.

11. A matrix $M \in M_{n \times n}(C)$ is called **skew-symmetric** if $M^t = -M$. Prove that if M is skew-symmetric and n is odd, then M is not invertible. What happens if n is even?

12. A matrix $Q \in M_{n \times n}(R)$ is called **orthogonal** if $QQ^t = I$. Prove that if Q is orthogonal, then $\det(Q) = \pm 1$.

13. For $M \in M_{n \times n}(C)$, let $\overline{M}$ be the matrix such that $(\overline{M})_{ij} = \overline{M_{ij}}$ for all i, j, where $\overline{M_{ij}}$ is the complex conjugate of M_{ij}.

   (a) Prove that $\det(\overline{M}) = \overline{\det(M)}$.
   (b) A matrix $Q \in M_{n \times n}(C)$ is called **unitary** if $QQ^* = I$, where $Q^* = \overline{Q^t}$. Prove that if Q is a unitary matrix, then $|\det(Q)| = 1$.

14. Let $\beta = \{u_1, u_2, \ldots, u_n\}$ be a subset of F^n containing n distinct vectors, and let B be the matrix in $M_{n \times n}(F)$ having u_j as column j. Prove that β is a basis for F^n if and only if $\det(B) \neq 0$.

15.[†] Prove that if $A, B \in M_{n \times n}(F)$ are similar, then $\det(A) = \det(B)$.

16. Use determinants to prove that if $A, B \in M_{n \times n}(F)$ are such that $AB = I$, then A is invertible (and hence $B = A^{-1}$).

17. Let $A, B \in M_{n \times n}(F)$ be such that $AB = -BA$. Prove that if n is odd and F is not a field of characteristic two, then A or B is not invertible.

18. Complete the proof of Theorem 4.7 by showing that if A is an elementary matrix of type 2 or type 3, then $\det(AB) = \det(A) \cdot \det(B)$.

19. A matrix $A \in M_{n \times n}(F)$ is called **lower triangular** if $A_{ij} = 0$ for $1 \leq i < j \leq n$. Suppose that A is a lower triangular matrix. Describe $\det(A)$ in terms of the entries of A.

20. Suppose that $M \in M_{n \times n}(F)$ can be written in the form

$$M = \begin{pmatrix} A & B \\ O & I \end{pmatrix},$$

   where A is a square matrix. Prove that $\det(M) = \det(A)$.

21.[†] Prove that if $M \in M_{n \times n}(F)$ can be written in the form

$$M = \begin{pmatrix} A & B \\ O & C \end{pmatrix},$$

   where A and C are square matrices, then $\det(M) = \det(A) \cdot \det(C)$.

22. Let $T: P_n(F) \to F^{n+1}$ be the linear transformation defined in Exercise 22 of Section 2.4 by $T(f) = (f(c_0), f(c_1), \ldots, f(c_n))$, where $c_0, c_1, \ldots, c_n$ are distinct scalars in an infinite field F. Let β be the standard ordered basis for $P_n(F)$ and γ be the standard ordered basis for F^{n+1}.

(a) Show that $M = [T]_\beta^\gamma$ has the form

$$\begin{pmatrix} 1 & c_0 & c_0^2 & \cdots & c_0^n \\ 1 & c_1 & c_1^2 & \cdots & c_1^n \\ \vdots & \vdots & \vdots & & \vdots \\ 1 & c_n & c_n^2 & \cdots & c_n^n \end{pmatrix}.$$

A matrix with this form is called a **Vandermonde matrix**.

(b) Use Exercise 22 of Section 2.4 to prove that $\det(M) \neq 0$.

(c) Prove that

$$\det(M) = \prod_{0 \leq i < j \leq n} (c_j - c_i),$$

the product of all terms of the form $c_j - c_i$ for $0 \leq i < j \leq n$.

23. Let $A \in M_{n\times n}(F)$ be nonzero. For any m $(1 \leq m \leq n)$, an $m \times m$ **submatrix** is obtained by deleting any $n - m$ rows and any $n - m$ columns of A.

(a) Let k $(1 \leq k \leq n)$ denote the largest integer such that some $k \times k$ submatrix has a nonzero determinant. Prove that $\text{rank}(A) = k$.

(b) Conversely, suppose that $\text{rank}(A) = k$. Prove that there exists a $k \times k$ submatrix with a nonzero determinant.

24. Let $A \in M_{n\times n}(F)$ have the form

$$A = \begin{pmatrix} 0 & 0 & 0 & \cdots & 0 & a_0 \\ -1 & 0 & 0 & \cdots & 0 & a_1 \\ 0 & -1 & 0 & \cdots & 0 & a_2 \\ \vdots & \vdots & \vdots & & \vdots & \vdots \\ 0 & 0 & 0 & \cdots & -1 & a_{n-1} \end{pmatrix}.$$

Compute $\det(A + tI)$, where I is the $n \times n$ identity matrix.

25. Let c_{jk} denote the cofactor of the row j, column k entry of the matrix $A \in M_{n\times n}(F)$.

(a) Prove that if B is the matrix obtained from A by replacing column k by e_j, then $\det(B) = c_{jk}$.

(b) Show that for $1 \leq j \leq n$, we have

$$A \begin{pmatrix} c_{j1} \\ c_{j2} \\ \vdots \\ c_{jn} \end{pmatrix} = \det(A) \cdot e_j.$$

Hint: Apply Cramer's rule to $Ax = e_j$.

(c) Deduce that if C is the $n \times n$ matrix such that $C_{ij} = c_{ji}$, then $AC = [\det(A)]I$.

(d) Show that if $\det(A) \neq 0$, then $A^{-1} = [\det(A)]^{-1}C$.

The following definition is used in Exercises 26–27.

Definition. *The **classical adjoint** of a square matrix A is the transpose of the matrix whose ij-entry is the ij-cofactor of A.*

26. Find the classical adjoint of each of the following matrices.

(a) $\begin{pmatrix} A_{11} & A_{12} \\ A_{21} & A_{22} \end{pmatrix}$ **(b)** $\begin{pmatrix} 4 & 0 & 0 \\ 0 & 4 & 0 \\ 0 & 0 & 4 \end{pmatrix}$

(c) $\begin{pmatrix} -4 & 0 & 0 \\ 0 & 2 & 0 \\ 0 & 0 & 5 \end{pmatrix}$ **(d)** $\begin{pmatrix} 3 & 6 & 7 \\ 0 & 4 & 8 \\ 0 & 0 & 5 \end{pmatrix}$

(e) $\begin{pmatrix} 1-i & 0 & 0 \\ 4 & 3i & 0 \\ 2i & 1+4i & -1 \end{pmatrix}$ **(f)** $\begin{pmatrix} 7 & 1 & 4 \\ 6 & -3 & 0 \\ -3 & 5 & -2 \end{pmatrix}$

(g) $\begin{pmatrix} -1 & 2 & 5 \\ 8 & 0 & -3 \\ 4 & 6 & 1 \end{pmatrix}$ **(h)** $\begin{pmatrix} 3 & 2+i & 0 \\ -1+i & 0 & i \\ 0 & 1 & 3-2i \end{pmatrix}$

27. Let C be the classical adjoint of $A \in M_{n \times n}(F)$. Prove the following statements.

(a) $\det(C) = [\det(A)]^{n-1}$.

(b) C^t is the classical adjoint of A^t.

(c) If A is an invertible upper triangular matrix, then C and A^{-1} are both upper triangular matrices.

28. Let $y_1, y_2, \ldots, y_n$ be linearly independent functions in C^∞. For each $y \in C^\infty$, define $T(y) \in C^\infty$ by

$$[T(y)](t) = \det \begin{pmatrix} y(t) & y_1(t) & y_2(t) & \cdots & y_n(t) \\ y'(t) & y_1'(t) & y_2'(t) & \cdots & y_n'(t) \\ \vdots & \vdots & \vdots & & \vdots \\ y^{(n)}(t) & y_1^{(n)}(t) & y_2^{(n)}(t) & \cdots & y_n^{(n)}(t) \end{pmatrix}.$$

The preceding determinant is called the **Wronskian** of $y, y_1, \ldots, y_n$.

(a) Prove that $\mathsf{T} \colon \mathsf{C}^\infty \to \mathsf{C}^\infty$ is a linear transformation.

(b) Prove that $\mathsf{N}(\mathsf{T})$ contains $\mathrm{span}(\{y_1, y_2, \ldots, y_n\})$.

4.4   SUMMARY—IMPORTANT FACTS ABOUT DETERMINANTS

In this section, we summarize the important properties of the determinant needed for the remainder of the text. The results contained in this section have been derived in Sections 4.2 and 4.3; consequently, the facts presented here are stated without proofs.

The **determinant** of an $n \times n$ matrix A having entries from a field F is a scalar in F, denoted by $\det(A)$ or $|A|$, and can be computed in the following manner:

1. If A is 1×1, then $\det(A) = A_{11}$, the single entry of A.

2. If A is 2×2, then $\det(A) = A_{11}A_{22} - A_{12}A_{21}$. For example,

$$\det \begin{pmatrix} -1 & 2 \\ 5 & 3 \end{pmatrix} = (-1)(3) - (2)(5) = -13.$$

3. If A is $n \times n$ for $n > 2$, then

$$\det(A) = \sum_{j=1}^{n} (-1)^{i+j} A_{ij} \cdot \det(\tilde{A}_{ij})$$

(if the determinant is evaluated by the entries of row i of A) or

$$\det(A) = \sum_{i=1}^{n} (-1)^{i+j} A_{ij} \cdot \det(\tilde{A}_{ij})$$

(if the determinant is evaluated by the entries of column j of A), where $\tilde{A}_{ij}$ is the $(n-1) \times (n-1)$ matrix obtained by deleting row i and column j from A.

In the formulas above, the scalar $(-1)^{i+j} \det(\tilde{A}_{ij})$ is called the **cofactor** of the row i column j entry of A. In this language, the determinant of A is evaluated as the sum of terms obtained by multiplying each entry of some row or column of A by the cofactor of that entry. Thus $\det(A)$ is expressed in terms of n determinants of $(n-1) \times (n-1)$ matrices. These determinants are then evaluated in terms of determinants of $(n-2) \times (n-2)$ matrices, and so forth, until 2×2 matrices are obtained. The determinants of the 2×2 matrices are then evaluated as in item 2.

Let us consider two examples of this technique in evaluating the determinant of the 4×4 matrix

$$A = \begin{pmatrix} 2 & 1 & 1 & 5 \\ 1 & 1 & -4 & -1 \\ 2 & 0 & -3 & 1 \\ 3 & 6 & 1 & 2 \end{pmatrix}.$$

To evaluate the determinant of A by expanding along the fourth row, we must know the cofactors of each entry of that row. The cofactor of $A_{41} = 3$ is $(-1)^{4+1} \det(B)$, where

$$B = \begin{pmatrix} 1 & 1 & 5 \\ 1 & -4 & -1 \\ 0 & -3 & 1 \end{pmatrix}.$$

Let us evaluate this determinant by expanding along the first column. We have

$$\det(B) = (-1)^{1+1}(1) \det \begin{pmatrix} -4 & -1 \\ -3 & 1 \end{pmatrix} + (-1)^{2+1}(1) \det \begin{pmatrix} 1 & 5 \\ -3 & 1 \end{pmatrix}$$

$$+ (-1)^{3+1}(0) \det \begin{pmatrix} 1 & 5 \\ -4 & -1 \end{pmatrix}$$

$$= 1(1)[(-4)(1) - (-1)(-3)] + (-1)(1)[(1)(1) - (5)(-3)] + 0$$

$$= -7 - 16 + 0 = -23.$$

Thus the cofactor of A_{41} is $(-1)^5(-23) = 23$. Similarly, the cofactors of A_{42}, A_{43}, and A_{44} are 8, 11, and -13, respectively. We can now evaluate the determinant of A by multiplying each entry of the fourth row by its cofactor; this gives

$$\det(A) = 3(23) + 6(8) + 1(11) + 2(-13) = 102.$$

For the sake of comparison, let us also compute the determinant of A by expansion along the second column. The reader should verify that the cofactors of A_{12}, A_{22}, and A_{42} are -14, 40, and 8, respectively. Thus

$$\det(A) = (-1)^{1+2}(1) \det \begin{pmatrix} 1 & -4 & -1 \\ 2 & -3 & 1 \\ 3 & 1 & 2 \end{pmatrix} + (-1)^{2+2}(1) \det \begin{pmatrix} 2 & 1 & 5 \\ 2 & -3 & 1 \\ 3 & 1 & 2 \end{pmatrix}$$

$$+ (-1)^{3+2}(0) \det \begin{pmatrix} 2 & 1 & 5 \\ 1 & -4 & -1 \\ 3 & 1 & 2 \end{pmatrix} + (-1)^{4+2}(6) \det \begin{pmatrix} 2 & 1 & 5 \\ 1 & -4 & -1 \\ 2 & -3 & 1 \end{pmatrix}$$

$$= 14 + 40 + 0 + 48 = 102.$$

Of course, the fact that the value 102 is obtained again is no surprise since the value of the determinant of A is independent of the choice of row or column used in the expansion.

Observe that the computation of $\det(A)$ is easier when expanded along the second column than when expanded along the fourth row. The difference is the presence of a zero in the second column, which makes it unnecessary to evaluate one of the cofactors (the cofactor of A_{32}). For this reason, it is beneficial to evaluate the determinant of a matrix by expanding along a row or column of the matrix that contains the largest number of zero entries. In fact, it is often helpful to introduce zeros into the matrix by means of elementary row operations before computing the determinant. This technique utilizes the first three properties of the determinant.

Properties of the Determinant

1. If B is a matrix obtained by interchanging any two rows or interchanging any two columns of an $n \times n$ matrix A, then $\det(B) = -\det(A)$.

2. If B is a matrix obtained by multiplying each entry of some row or column of an $n \times n$ matrix A by a scalar k, then $\det(B) = k \cdot \det(A)$.

3. If B is a matrix obtained from an $n \times n$ matrix A by adding a multiple of row i to row j or a multiple of column i to column j for $i \neq j$, then $\det(B) = \det(A)$.

As an example of the use of these three properties in evaluating determinants, let us compute the determinant of the 4×4 matrix A considered previously. Our procedure is to introduce zeros into the second column of A by employing property 3, and then to expand along that column. (The elementary row operations used here consist of adding multiples of row 1 to rows 2 and 4.) This procedure yields

$$\det(A) = \det \begin{pmatrix} 2 & 1 & 1 & 5 \\ 1 & 1 & -4 & -1 \\ 2 & 0 & -3 & 1 \\ 3 & 6 & 1 & 2 \end{pmatrix} = \det \begin{pmatrix} 2 & 1 & 1 & 5 \\ -1 & 0 & -5 & -6 \\ 2 & 0 & -3 & 1 \\ -9 & 0 & -5 & -28 \end{pmatrix}$$

$$= 1(-1)^{1+2} \det \begin{pmatrix} -1 & -5 & -6 \\ 2 & -3 & 1 \\ -9 & -5 & -28 \end{pmatrix}.$$

The resulting determinant of a 3×3 matrix can be evaluated in the same manner: Use type 3 elementary row operations to introduce two zeros into the first column, and then expand along that column. This results in the value -102. Therefore

$$\det(A) = 1(-1)^{1+2}(-102) = 102.$$

The reader should compare this calculation of $\det(A)$ with the preceding ones to see how much less work is required when properties 1, 2, and 3 are employed.

In the chapters that follow, we often have to evaluate the determinant of matrices having special forms. The next two properties of the determinant are useful in this regard:

4. The determinant of an upper triangular matrix is the product of its diagonal entries. In particular, $\det(I) = 1$.
5. If two rows (or columns) of a matrix are identical, then the determinant of the matrix is zero.

As an illustration of property 4, notice that

$$\det \begin{pmatrix} -3 & 1 & 2 \\ 0 & 4 & 5 \\ 0 & 0 & -6 \end{pmatrix} = (-3)(4)(-6) = 72.$$

Property 4 provides an efficient method for evaluating the determinant of a matrix:

(a) Use Gaussian elimination and properties 1, 2, and 3 above to reduce the matrix to an upper triangular matrix.

(b) Compute the product of the diagonal entries.

For instance,

$$\det \begin{pmatrix} 1 & -1 & 2 & 1 \\ 2 & -1 & -1 & 4 \\ -4 & 5 & -10 & -6 \\ 3 & -2 & 10 & -1 \end{pmatrix} = \det \begin{pmatrix} 1 & -1 & 2 & 1 \\ 0 & 1 & -5 & 2 \\ 0 & 1 & -2 & -2 \\ 0 & 1 & 4 & -4 \end{pmatrix}$$

$$= \det \begin{pmatrix} 1 & -1 & 2 & 1 \\ 0 & 1 & -5 & 2 \\ 0 & 0 & 3 & -4 \\ 0 & 0 & 9 & -6 \end{pmatrix} = \det \begin{pmatrix} 1 & -1 & 2 & 1 \\ 0 & 1 & -5 & 2 \\ 0 & 0 & 3 & -4 \\ 0 & 0 & 0 & 6 \end{pmatrix}$$

$$= 1 \cdot 1 \cdot 3 \cdot 6 = 18.$$

The next three properties of the determinant are used frequently in later chapters. Indeed, perhaps the most significant property of the determinant is that it provides a simple characterization of invertible matrices. (See property 7.)

6. For any $n \times n$ matrices A and B, $\det(AB) = \det(A) \cdot \det(B)$.

7. An $n \times n$ matrix A is invertible if and only if $\det(A) \neq 0$. Furthermore, if A is invertible, then $\det(A^{-1}) = \dfrac{1}{\det(A)}$.

8. For any $n \times n$ matrix A, the determinants of A and A^t are equal.

For example, property 7 guarantees that the matrix A on page 233 is invertible because $\det(A) = 102$.

The final property, stated as Exercise 15 of Section 4.3, is used in Chapter 5. It is a simple consequence of properties 6 and 7.

9. If A and B are similar matrices, then $\det(A) = \det(B)$.

EXERCISES

1. Label the following statements as true or false.

   (a) The determinant of a square matrix may be computed by expanding the matrix along any row or column.

   (b) In evaluating the determinant of a matrix, it is wise to expand along a row or column containing the largest number of zero entries.

   (c) If two rows or columns of A are identical, then $\det(A) = 0$.

   (d) If B is a matrix obtained by interchanging two rows or two columns of A, then $\det(B) = \det(A)$.

   (e) If B is a matrix obtained by multiplying each entry of some row or column of A by a scalar, then $\det(B) = \det(A)$.

   (f) If B is a matrix obtained from A by adding a multiple of some row to a different row, then $\det(B) = \det(A)$.

   (g) The determinant of an upper triangular $n \times n$ matrix is the product of its diagonal entries.

   (h) For every $A \in \mathsf{M}_{n \times n}(F)$, $\det(A^t) = -\det(A)$.

   (i) If $A, B \in \mathsf{M}_{n \times n}(F)$, then $\det(AB) = \det(A) \cdot \det(B)$.

   (j) If Q is an invertible matrix, then $\det(Q^{-1}) = [\det(Q)]^{-1}$.

   (k) A matrix Q is invertible if and only if $\det(Q) \neq 0$.

2. Evaluate the determinant of the following 2×2 matrices.

   (a) $\begin{pmatrix} 4 & -5 \\ 2 & 3 \end{pmatrix}$ $\qquad$ (b) $\begin{pmatrix} -1 & 7 \\ 3 & 8 \end{pmatrix}$

   (c) $\begin{pmatrix} 2+i & -1+3i \\ 1-2i & 3-i \end{pmatrix}$ $\qquad$ (d) $\begin{pmatrix} 3 & 4i \\ -6i & 2i \end{pmatrix}$

3. Evaluate the determinant of the following matrices in the manner indicated.

(a) $\begin{pmatrix} 0 & 1 & 2 \\ -1 & 0 & -3 \\ 2 & 3 & 0 \end{pmatrix}$

along the first row

(b) $\begin{pmatrix} 1 & 0 & 2 \\ 0 & 1 & 5 \\ -1 & 3 & 0 \end{pmatrix}$

along the first column

(c) $\begin{pmatrix} 0 & 1 & 2 \\ -1 & 0 & -3 \\ 2 & 3 & 0 \end{pmatrix}$

along the second column

(d) $\begin{pmatrix} 1 & 0 & 2 \\ 0 & 1 & 5 \\ -1 & 3 & 0 \end{pmatrix}$

along the third row

(e) $\begin{pmatrix} 0 & 1+i & 2 \\ -2i & 0 & 1-i \\ 3 & 4i & 0 \end{pmatrix}$

along the third row

(f) $\begin{pmatrix} i & 2+i & 0 \\ -1 & 3 & 2i \\ 0 & -1 & 1-i \end{pmatrix}$

along the third column

(g) $\begin{pmatrix} 0 & 2 & 1 & 3 \\ 1 & 0 & -2 & 2 \\ 3 & -1 & 0 & 1 \\ -1 & 1 & 2 & 0 \end{pmatrix}$

along the fourth column

(h) $\begin{pmatrix} 1 & -1 & 2 & -1 \\ -3 & 4 & 1 & -1 \\ 2 & -5 & -3 & 8 \\ -2 & 6 & -4 & 1 \end{pmatrix}$

along the fourth row

4. Evaluate the determinant of the following matrices by any legitimate method.

(a) $\begin{pmatrix} 1 & 2 & 3 \\ 4 & 5 & 6 \\ 7 & 8 & 9 \end{pmatrix}$

(b) $\begin{pmatrix} -1 & 3 & 2 \\ 4 & -8 & 1 \\ 2 & 2 & 5 \end{pmatrix}$

(c) $\begin{pmatrix} 0 & 1 & 1 \\ 1 & 2 & -5 \\ 6 & -4 & 3 \end{pmatrix}$

(d) $\begin{pmatrix} 1 & -2 & 3 \\ -1 & 2 & -5 \\ 3 & -1 & 2 \end{pmatrix}$

(e) $\begin{pmatrix} i & 2 & -1 \\ 3 & 1+i & 2 \\ -2i & 1 & 4-i \end{pmatrix}$

(f) $\begin{pmatrix} -1 & 2+i & 3 \\ 1-i & i & 1 \\ 3i & 2 & -1+i \end{pmatrix}$

(g) $\begin{pmatrix} 1 & 0 & -2 & 3 \\ -3 & 1 & 1 & 2 \\ 0 & 4 & -1 & 1 \\ 2 & 3 & 0 & 1 \end{pmatrix}$

(h) $\begin{pmatrix} 1 & -2 & 3 & -12 \\ -5 & 12 & -14 & 19 \\ -9 & 22 & -20 & 31 \\ -4 & 9 & -14 & 15 \end{pmatrix}$

5. Suppose that $M \in \mathsf{M}_{n \times n}(F)$ can be written in the form

$$M = \begin{pmatrix} A & B \\ O & I \end{pmatrix},$$

where A is a square matrix. Prove that $\det(M) = \det(A)$.

6.[†] Prove that if $M \in \mathsf{M}_{n \times n}(F)$ can be written in the form

$$M = \begin{pmatrix} A & B \\ O & C \end{pmatrix},$$

where A and C are square matrices, then $\det(M) = \det(A) \cdot \det(C)$.

4.5* A CHARACTERIZATION OF THE DETERMINANT

In Sections 4.2 and 4.3, we showed that the determinant possesses a number of properties. In this section, we show that three of these properties completely characterize the determinant; that is, the only function $\delta \colon \mathsf{M}_{n \times n}(F) \to F$ having these three properties is the determinant. This characterization of the determinant is the one used in Section 4.1 to establish the relationship between $\det \begin{pmatrix} u \\ v \end{pmatrix}$ and the area of the parallelogram determined by u and v. The first of these properties that characterize the determinant is the one described in Theorem 4.3 (p. 212).

Definition. *A function $\delta \colon \mathsf{M}_{n \times n}(F) \to F$ is called an **n-linear function** if it is a linear function of each row of an $n \times n$ matrix when the remaining $n - 1$ rows are held fixed, that is, δ is n-linear if, for every $r = 1, 2, \ldots, n$, we have*

$$\delta \begin{pmatrix} a_1 \\ \vdots \\ a_{r-1} \\ u + kv \\ a_{r+1} \\ \vdots \\ a_n \end{pmatrix} = \delta \begin{pmatrix} a_1 \\ \vdots \\ a_{r-1} \\ u \\ a_{r+1} \\ \vdots \\ a_n \end{pmatrix} + k\delta \begin{pmatrix} a_1 \\ \vdots \\ a_{r-1} \\ v \\ a_{r+1} \\ \vdots \\ a_n \end{pmatrix}$$

whenever k is a scalar and u, v, and each a_i are vectors in F^n.

Example 1

The function $\delta \colon \mathsf{M}_{n \times n}(F) \to F$ defined by $\delta(A) = 0$ for each $A \in \mathsf{M}_{n \times n}(F)$ is an n-linear function.   ◆

Example 2

For $1 \le j \le n$, define $\delta_j \colon \mathsf{M}_{n \times n}(F) \to F$ by $\delta_j(A) = A_{1j} A_{2j} \cdots A_{nj}$ for each $A \in \mathsf{M}_{n \times n}(F)$; that is, $\delta_j(A)$ equals the product of the entries of column j of

A. Let $A \in \mathsf{M}_{n \times n}(F)$, $a_i = (A_{i1}, A_{i2}, \ldots, A_{in})$, and $v = (b_1, b_2, \ldots, b_n) \in F^n$. Then each δ_j is an n-linear function because, for any scalar k, we have

$$
\delta \begin{pmatrix} a_1 \\ \vdots \\ a_{r-1} \\ a_r + kv \\ a_{r+1} \\ \vdots \\ a_n \end{pmatrix} = A_{1j} \cdots A_{(r-1)j}(A_{rj} + kb_j)A_{(r+1)j} \cdots A_{nj}
$$

$$
\begin{aligned}
&= A_{1j} \cdots A_{(r-1)j} A_{rj} A_{(r+1)j} \cdots A_{nj} \\
&\quad + A_{1j} \cdots A_{(r-1)j}(kb_j)A_{(r+1)j} \cdots A_{nj} \\
&= A_{1j} \cdots A_{(r-1)j} A_{rj} A_{(r+1)j} \cdots A_{nj} \\
&\quad + k(A_{1j} \cdots A_{(r-1)j} b_j A_{(r+1)j} \cdots A_{nj})
\end{aligned}
$$

$$
= \delta \begin{pmatrix} a_1 \\ \vdots \\ a_{r-1} \\ a_r \\ a_{r+1} \\ \vdots \\ a_n \end{pmatrix} + k\delta \begin{pmatrix} a_1 \\ \vdots \\ a_{r-1} \\ v \\ a_{r+1} \\ \vdots \\ a_n \end{pmatrix}. \qquad \blacklozenge
$$

Example 3

The function $\delta \colon \mathsf{M}_{n \times n}(F) \to F$ defined for each $A \in \mathsf{M}_{n \times n}(F)$ by $\delta(A) = A_{11} A_{22} \cdots A_{nn}$ (i.e., $\delta(A)$ equals the product of the diagonal entries of A) is an n-linear function. $\blacklozenge$

Example 4

The function $\delta \colon \mathsf{M}_{n \times n}(R) \to R$ defined for each $A \in \mathsf{M}_{n \times n}(R)$ by $\delta(A) = \mathrm{tr}(A)$ is not an n-linear function for $n \geq 2$. For if I is the $n \times n$ identity matrix and A is the matrix obtained by multiplying the first row of I by 2, then $\delta(A) = n + 1 \neq 2n = 2 \cdot \delta(I)$. $\blacklozenge$

Theorem 4.3 (p. 212) asserts that the determinant is an n-linear function. For our purposes this is the most important example of an n-linear function. Now we introduce the second of the properties used in the characterization of the determinant.

Definition. *An n-linear function $\delta \colon \mathsf{M}_{n \times n}(F) \to F$ is called **alternating** if, for each $A \in \mathsf{M}_{n \times n}(F)$, we have $\delta(A) = 0$ whenever two adjacent rows of A are identical.*

Theorem 4.10. *Let $\delta \colon \mathsf{M}_{n \times n}(F) \to F$ be an alternating n-linear function.*
(a) *If $A \in \mathsf{M}_{n \times n}(F)$ and B is a matrix obtained from A by interchanging any two rows of A, then $\delta(B) = -\delta(A)$.*
(b) *If $A \in \mathsf{M}_{n \times n}(F)$ has two identical rows, then $\delta(A) = 0$.*

Proof. (a) Let $A \in \mathsf{M}_{n \times n}(F)$, and let B be the matrix obtained from A by interchanging rows r and s, where $r < s$. We first establish the result in the case that $s = r + 1$. Because $\delta \colon \mathsf{M}_{n \times n}(F) \to F$ is an n-linear function that is alternating, we have

$$0 = \delta \begin{pmatrix} a_1 \\ \vdots \\ a_r + a_{r+1} \\ a_r + a_{r+1} \\ \vdots \\ a_n \end{pmatrix} = \delta \begin{pmatrix} a_1 \\ \vdots \\ a_r \\ a_r + a_{r+1} \\ \vdots \\ a_n \end{pmatrix} + \delta \begin{pmatrix} a_1 \\ \vdots \\ a_{r+1} \\ a_r + a_{r+1} \\ \vdots \\ a_n \end{pmatrix}$$

$$= \delta \begin{pmatrix} a_1 \\ \vdots \\ a_r \\ a_r \\ \vdots \\ a_n \end{pmatrix} + \delta \begin{pmatrix} a_1 \\ \vdots \\ a_r \\ a_{r+1} \\ \vdots \\ a_n \end{pmatrix} + \delta \begin{pmatrix} a_1 \\ \vdots \\ a_{r+1} \\ a_r \\ \vdots \\ a_n \end{pmatrix} + \delta \begin{pmatrix} a_1 \\ \vdots \\ a_{r+1} \\ a_{r+1} \\ \vdots \\ a_n \end{pmatrix}$$

$$= 0 + \delta(A) + \delta(B) + 0.$$

Thus $\delta(B) = -\delta(A)$.

Next suppose that $s > r + 1$, and let the rows of A be $a_1, a_2, \dots, a_n$. Beginning with a_r and a_{r+1}, successively interchange a_r with the row that follows it until the rows are in the sequence

$$a_1, a_2, \dots, a_{r-1}, a_{r+1}, \dots, a_s, a_r, a_{s+1}, \dots, a_n.$$

In all, $s - r$ interchanges of adjacent rows are needed to produce this sequence. Then successively interchange a_s with the row that precedes it until the rows are in the order

$$a_1, a_2, \dots, a_{r-1}, a_s, a_{r+1}, \dots, a_{s-1}, a_r, a_{s+1}, \dots, a_n.$$

This process requires an additional $s - r - 1$ interchanges of adjacent rows and produces the matrix B. It follows from the preceding paragraph that

$$\delta(B) = (-1)^{(s-r)+(s-r-1)}\delta(A) = -\delta(A).$$

(b) Suppose that rows r and s of $A \in \mathsf{M}_{n \times n}(F)$ are identical, where $r < s$. If $s = r + 1$, then $\delta(A) = 0$ because δ is alternating and two adjacent rows of A are identical. If $s > r + 1$, let B be the matrix obtained from A by interchanging rows $r + 1$ and s. Then $\delta(B) = 0$ because two adjacent rows of B are identical. But $\delta(B) = -\delta(A)$ by (a). Hence $\delta(A) = 0$. ∎

Corollary 1. Let $\delta \colon \mathsf{M}_{n \times n}(F) \to F$ be an alternating n-linear function. If B is a matrix obtained from $A \in \mathsf{M}_{n \times n}(F)$ by adding a multiple of some row of A to another row, then $\delta(B) = \delta(A)$.

Proof. Let B be obtained from $A \in \mathsf{M}_{n \times n}(F)$ by adding k times row i of A to row j, where $j \neq i$, and let C be obtained from A by replacing row j of A by row i of A. Then the rows of A, B, and C are identical except for row j. Moreover, row j of B is the sum of row j of A and k times row j of C. Since δ is an n-linear function and C has two identical rows, it follows that

$$\delta(B) = \delta(A) + k\delta(C) = \delta(A) + k \cdot 0 = \delta(A).$$ ∎

The next result now follows as in the proof of the corollary to Theorem 4.6 (p. 216). (See Exercise 11.)

Corollary 2. Let $\delta \colon \mathsf{M}_{n \times n}(F) \to F$ be an alternating n-linear function. If $M \in \mathsf{M}_{n \times n}(F)$ has rank less than n, then $\delta(M) = 0$.

Proof. Exercise. ∎

Corollary 3. Let $\delta \colon \mathsf{M}_{n \times n}(F) \to F$ be an alternating n-linear function, and let E_1, E_2, and E_3 in $\mathsf{M}_{n \times n}(F)$ be elementary matrices of types 1, 2, and 3, respectively. Suppose that E_2 is obtained by multiplying some row of I by the nonzero scalar k. Then $\delta(E_1) = -\delta(I)$, $\delta(E_2) = k \cdot \delta(I)$, and $\delta(E_3) = \delta(I)$.

Proof. Exercise. ∎

We wish to show that under certain circumstances, the only alternating n-linear function $\delta \colon \mathsf{M}_{n \times n}(F) \to F$ is the determinant, that is, $\delta(A) = \det(A)$ for all $A \in \mathsf{M}_{n \times n}(F)$. In view of Corollary 3 to Theorem 4.10 and the facts on page 223 about the determinant of an elementary matrix, this can happen only if $\delta(I) = 1$. Hence the third condition that is used in the characterization of the determinant is that the determinant of the $n \times n$ identity matrix is 1. Before we can establish the desired characterization of the determinant, we must first show that an alternating n-linear function δ such that $\delta(I) = 1$ is a multiplicative function. The proof of this result is identical to the proof of Theorem 4.7 (p. 223), and so it is omitted. (See Exercise 12.)

Theorem 4.11. Let $\delta \colon \mathsf{M}_{n \times n}(F) \to F$ be an alternating n-linear function such that $\delta(I) = 1$. For any $A, B \in \mathsf{M}_{n \times n}(F)$, we have $\delta(AB) = \delta(A) \cdot \delta(B)$.

Proof. Exercise.   ∎

Theorem 4.12. *If* $\delta\colon \mathsf{M}_{n\times n}(F) \to F$ *is an alternating n-linear function such that* $\delta(I) = 1$, *then* $\delta(A) = \det(A)$ *for every* $A \in \mathsf{M}_{n\times n}(F)$.

Proof. Let $\delta\colon \mathsf{M}_{n\times n}(F) \to F$ be an alternating n-linear function such that $\delta(I) = 1$, and let $A \in \mathsf{M}_{n\times n}(F)$. If A has rank less than n, then by Corollary 2 to Theorem 4.10, $\delta(A) = 0$. Since the corollary to Theorem 4.6 (p. 217) gives $\det(A) = 0$, we have $\delta(A) = \det(A)$ in this case. If, on the other hand, A has rank n, then A is invertible and hence is the product of elementary matrices (Corollary 3 to Theorem 3.6 p. 159), say $A = E_m \cdots E_2 E_1$. Since $\delta(I) = 1$, it follows from Corollary 3 to Theorem 4.10 and the facts on page 223 that $\delta(E) = \det(E)$ for every elementary matrix E. Hence by Theorems 4.11 and 4.7 (p. 223), we have

$$\begin{aligned}
\delta(A) &= \delta(E_m \cdots E_2 E_1) \\
&= \delta(E_m) \cdot \cdots \cdot \delta(E_2) \cdot \delta(E_1) \\
&= \det(E_m) \cdot \cdots \cdot \det(E_2) \cdot \det(E_1) \\
&= \det(E_m \cdots E_2 E_1) \\
&= \det(A).
\end{aligned}$$

∎

Theorem 4.12 provides the desired characterization of the determinant: It is the unique function $\delta\colon \mathsf{M}_{n\times n}(F) \to F$ that is n-linear, is alternating, and has the property that $\delta(I) = 1$.

EXERCISES

1. Label the following statements as true or false.
   (a) Any n-linear function $\delta\colon \mathsf{M}_{n\times n}(F) \to F$ is a linear transformation.
   (b) Any n-linear function $\delta\colon \mathsf{M}_{n\times n}(F) \to F$ is a linear function of each row of an $n \times n$ matrix when the other $n-1$ rows are held fixed.
   (c) If $\delta\colon \mathsf{M}_{n\times n}(F) \to F$ is an alternating n-linear function and the matrix $A \in \mathsf{M}_{n\times n}(F)$ has two identical rows, then $\delta(A) = 0$.
   (d) If $\delta\colon \mathsf{M}_{n\times n}(F) \to F$ is an alternating n-linear function and B is obtained from $A \in \mathsf{M}_{n\times n}(F)$ by interchanging two rows of A, then $\delta(B) = \delta(A)$.
   (e) There is a unique alternating n-linear function $\delta\colon \mathsf{M}_{n\times n}(F) \to F$.
   (f) The function $\delta\colon \mathsf{M}_{n\times n}(F) \to F$ defined by $\delta(A) = 0$ for every $A \in \mathsf{M}_{n\times n}(F)$ is an alternating n-linear function.

2. Determine all the 1-linear functions $\delta\colon \mathsf{M}_{1\times 1}(F) \to F$.

Determine which of the functions $\delta\colon \mathsf{M}_{3\times 3}(F) \to F$ in Exercises 3–10 are 3-linear functions. Justify each answer.

3. $\delta(A) = k$, where k is any nonzero scalar

4. $\delta(A) = A_{22}$

5. $\delta(A) = A_{11} A_{23} A_{32}$

6. $\delta(A) = A_{11} + A_{23} + A_{32}$

7. $\delta(A) = A_{11} A_{21} A_{32}$

8. $\delta(A) = A_{11} A_{31} A_{32}$

9. $\delta(A) = A_{11}^2 A_{22}^2 A_{33}^2$

10. $\delta(A) = A_{11} A_{22} A_{33} - A_{11} A_{21} A_{32}$

11. Prove Corollaries 2 and 3 of Theorem 4.10.

12. Prove Theorem 4.11.

13. Prove that det: $\mathsf{M}_{2 \times 2}(F) \to F$ is a 2-linear function of the *columns* of a matrix.

14. Let $a, b, c, d \in F$. Prove that the function $\delta: \mathsf{M}_{2 \times 2}(F) \to F$ defined by $\delta(A) = A_{11} A_{22} a + A_{11} A_{21} b + A_{12} A_{22} c + A_{12} A_{21} d$ is a 2-linear function.

15. Prove that $\delta: \mathsf{M}_{2 \times 2}(F) \to F$ is a 2-linear function if and only if it has the form

$$\delta(A) = A_{11} A_{22} a + A_{11} A_{21} b + A_{12} A_{22} c + A_{12} A_{21} d$$

for some scalars $a, b, c, d \in F$.

16. Prove that if $\delta: \mathsf{M}_{n \times n}(F) \to F$ is an alternating n-linear function, then there exists a scalar k such that $\delta(A) = k \det(A)$ for all $A \in \mathsf{M}_{n \times n}(F)$.

17. Prove that a linear combination of two n-linear functions is an n-linear function, where the sum and scalar product of n-linear functions are as defined in Example 3 of Section 1.2 (p. 9).

18. Prove that the set of all n-linear functions over a field F is a vector space over F under the operations of function addition and scalar multiplication as defined in Example 3 of Section 1.2 (p. 9).

19. Let $\delta: \mathsf{M}_{n \times n}(F) \to F$ be an n-linear function and F a field that does not have characteristic two. Prove that if $\delta(B) = -\delta(A)$ whenever B is obtained from $A \in \mathsf{M}_{n \times n}(F)$ by interchanging any two rows of A, then $\delta(M) = 0$ whenever $M \in \mathsf{M}_{n \times n}(F)$ has two identical rows.

20. Give an example to show that the implication in Exercise 19 need not hold if F has characteristic two.

INDEX OF DEFINITIONS FOR CHAPTER 4

5

Diagonalization

This chapter is concerned with the so-called *diagonalization problem*. For a given linear operator T on a finite-dimensional vector space V, we seek answers to the following questions.

1. Does there exist an ordered basis β for V such that $[T]_\beta$ is a diagonal matrix?
2. If such a basis exists, how can it be found?

Since computations involving diagonal matrices are simple, an affirmative answer to question 1 leads us to a clearer understanding of how the operator T acts on V, and an answer to question 2 enables us to obtain easy solutions to many practical problems that can be formulated in a linear algebra context. We consider some of these problems and their solutions in this chapter; see, for example, Section 5.3.

A solution to the diagonalization problem leads naturally to the concepts of *eigenvalue* and *eigenvector*. Aside from the important role that these concepts play in the diagonalization problem, they also prove to be useful tools in the study of many nondiagonalizable operators, as we will see in Chapter 7.

5.1 EIGENVALUES AND EIGENVECTORS

In Example 3 of Section 2.5, we were able to obtain a formula for the reflection of R^2 about the line $y = 2x$. The key to our success was to find a basis β' for which $[T]_{\beta'}$ is a diagonal matrix. We now introduce the name for an operator or matrix that has such a basis.

Definitions. *A linear operator T on a finite-dimensional vector space V is called* **diagonalizable** *if there is an ordered basis β for V such that $[T]_\beta$*

is a *diagonal matrix. A square matrix A is called* **diagonalizable** *if L_A is diagonalizable.*

We want to determine when a linear operator T on a finite-dimensional vector space V is diagonalizable and, if so, how to obtain an ordered basis $\beta = \{v_1, v_2, \dots, v_n\}$ for V such that $[T]_\beta$ is a diagonal matrix. Note that, if $D = [T]_\beta$ is a diagonal matrix, then for each vector $v_j \in \beta$, we have

$$T(v_j) = \sum_{i=1}^{n} D_{ij} v_i = D_{jj} v_j = \lambda_j v_j,$$

where $\lambda_j = D_{jj}$.

Conversely, if $\beta = \{v_1, v_2, \dots, v_n\}$ is an ordered basis for V such that $T(v_j) = \lambda_j v_j$ for some scalars $\lambda_1, \lambda_2, \dots, \lambda_n$, then clearly

$$[T]_\beta = \begin{pmatrix} \lambda_1 & 0 & \cdots & 0 \\ 0 & \lambda_2 & \cdots & 0 \\ \vdots & \vdots & & \vdots \\ 0 & 0 & \cdots & \lambda_n \end{pmatrix}.$$

In the preceding paragraph, each vector v in the basis β satisfies the condition that $T(v) = \lambda v$ for some scalar λ. Moreover, because v lies in a basis, v is nonzero. These computations motivate the following definitions.

Definitions. *Let T be a linear operator on a vector space V. A nonzero vector $v \in V$ is called an* **eigenvector** *of T if there exists a scalar λ such that $T(v) = \lambda v$. The scalar λ is called the* **eigenvalue** *corresponding to the eigenvector v.*

Let A be in $M_{n \times n}(F)$. A nonzero vector $v \in F^n$ is called an **eigenvector** *of A if v is an eigenvector of L_A; that is, if $Av = \lambda v$ for some scalar λ. The scalar λ is called the* **eigenvalue** *of A corresponding to the eigenvector v.*

The words *characteristic vector* and *proper vector* are also used in place of *eigenvector*. The corresponding terms for *eigenvalue* are *characteristic value* and *proper value*.

Note that a vector is an eigenvector of a matrix A if and only if it is an eigenvector of L_A. Likewise, a scalar λ is an eigenvalue of A if and only if it is an eigenvalue of L_A. Using the terminology of eigenvectors and eigenvalues, we can summarize the preceding discussion as follows.

Theorem 5.1. *A linear operator T on a finite-dimensional vector space V is diagonalizable if and only if there exists an ordered basis β for V consisting of eigenvectors of T. Furthermore, if T is diagonalizable, $\beta = \{v_1, v_2, \dots, v_n\}$ is an ordered basis of eigenvectors of T, and $D = [T]_\beta$, then D is a diagonal matrix and D_{jj} is the eigenvalue corresponding to v_j for $1 \le j \le n$.*

To *diagonalize* a matrix or a linear operator is to find a basis of eigenvectors and the corresponding eigenvalues.

Before continuing our study of the diagonalization problem, we consider three examples of eigenvalues and eigenvectors.

Example 1

Let

$$A = \begin{pmatrix} 1 & 3 \\ 4 & 2 \end{pmatrix}, \quad v_1 = \begin{pmatrix} 1 \\ -1 \end{pmatrix}, \quad \text{and} \quad v_2 = \begin{pmatrix} 3 \\ 4 \end{pmatrix}.$$

Since

$$\mathsf{L}_A(v_1) = \begin{pmatrix} 1 & 3 \\ 4 & 2 \end{pmatrix} \begin{pmatrix} 1 \\ -1 \end{pmatrix} = \begin{pmatrix} -2 \\ 2 \end{pmatrix} = -2 \begin{pmatrix} 1 \\ -1 \end{pmatrix} = -2v_1,$$

v_1 is an eigenvector of L_A, and hence of A. Here $\lambda_1 = -2$ is the eigenvalue corresponding to v_1. Furthermore,

$$\mathsf{L}_A(v_2) = \begin{pmatrix} 1 & 3 \\ 4 & 2 \end{pmatrix} \begin{pmatrix} 3 \\ 4 \end{pmatrix} = \begin{pmatrix} 15 \\ 20 \end{pmatrix} = 5 \begin{pmatrix} 3 \\ 4 \end{pmatrix} = 5v_2,$$

and so v_2 is an eigenvector of L_A, and hence of A, with the corresponding eigenvalue $\lambda_2 = 5$. Note that $\beta = \{v_1, v_2\}$ is an ordered basis for R^2 consisting of eigenvectors of both A and L_A, and therefore A and L_A are diagonalizable. Moreover, by Theorem 5.1,

$$[\mathsf{L}_A]_\beta = \begin{pmatrix} -2 & 0 \\ 0 & 5 \end{pmatrix}. \quad \blacklozenge$$

Example 2

Let T be the linear operator on R^2 that rotates each vector in the plane through an angle of $\pi/2$. It is clear geometrically that for any nonzero vector v, the vectors v and $\mathsf{T}(v)$ are not collinear; hence $\mathsf{T}(v)$ is not a multiple of v. Therefore T has no eigenvectors and, consequently, no eigenvalues. Thus there exist operators (and matrices) with no eigenvalues or eigenvectors. Of course, such operators and matrices are not diagonalizable. $\blacklozenge$

Example 3

Let $\mathsf{C}^\infty(R)$ denote the set of all functions $f \colon R \to R$ having derivatives of all orders. (Thus $\mathsf{C}^\infty(R)$ includes the polynomial functions, the sine and cosine functions, the exponential functions, etc.) Clearly, $\mathsf{C}^\infty(R)$ is a subspace of the vector space $\mathcal{F}(R, R)$ of all functions from R to R as defined in Section 1.2. Let $\mathsf{T} \colon \mathsf{C}^\infty(R) \to \mathsf{C}^\infty(R)$ be the function defined by $\mathsf{T}(f) = f'$, the derivative of f. It is easily verified that T is a linear operator on $\mathsf{C}^\infty(R)$. We determine the eigenvalues and eigenvectors of T.

Suppose that f is an eigenvector of T with corresponding eigenvalue λ. Then $f' = \mathsf{T}(f) = \lambda f$. This is a first-order differential equation whose solutions are of the form $f(t) = ce^{\lambda t}$ for some constant c. Consequently, every real number λ is an eigenvalue of T, and λ corresponds to eigenvectors of the form $ce^{\lambda t}$ for $c \neq 0$. Note that for $\lambda = 0$, the eigenvectors are the nonzero constant functions.   ◆

In order to obtain a basis of eigenvectors for a matrix (or a linear operator), we need to be able to determine its eigenvalues and eigenvectors. The following theorem gives us a method for computing eigenvalues.

Theorem 5.2. *Let $A \in \mathsf{M}_{n \times n}(F)$. Then a scalar λ is an eigenvalue of A if and only if $\det(A - \lambda I_n) = 0$.*

Proof. A scalar λ is an eigenvalue of A if and only if there exists a nonzero vector $v \in F^n$ such that $Av = \lambda v$, that is, $(A - \lambda I_n)(v) = 0$. By Theorem 2.5 (p. 71), this is true if and only if $A - \lambda I_n$ is not invertible. However, this result is equivalent to the statement that $\det(A - \lambda I_n) = 0$.   ∎

Definition. *Let $A \in \mathsf{M}_{n \times n}(F)$. The polynomial $f(t) = \det(A - tI_n)$ is called the* **characteristic polynomial** [1] *of A.*

Theorem 5.2 states that the eigenvalues of a matrix are the zeros of its characteristic polynomial. When determining the eigenvalues of a matrix or a linear operator, we normally compute its characteristic polynomial, as in the next example.

Example 4

To find the eigenvalues of

$$A = \begin{pmatrix} 1 & 1 \\ 4 & 1 \end{pmatrix} \in \mathsf{M}_{2 \times 2}(R),$$

we compute its characteristic polynomial:

$$\det(A - tI_2) = \det \begin{pmatrix} 1 - t & 1 \\ 4 & 1 - t \end{pmatrix} = t^2 - 2t - 3 = (t - 3)(t + 1).$$

It follows from Theorem 5.2 that the only eigenvalues of A are 3 and -1.   ◆

---

[1]The observant reader may have noticed that the entries of the matrix $A - tI_n$ are not scalars in the field F. They are, however, scalars in another field $F(t)$, the field of quotients of polynomials in t with coefficients from F. Consequently, any results proved about determinants in Chapter 4 remain valid in this context.

It is easily shown that similar matrices have the same characteristic polynomial (see Exercise 12). This fact enables us to define the characteristic polynomial of a linear operator as follows.

Definition. *Let* T *be a linear operator on an n-dimensional vector space* V *with ordered basis* β. *We define the* **characteristic polynomial** $f(t)$ *of* T *to be the characteristic polynomial of* $A = [T]_\beta$. *That is,*

$$f(t) = \det(A - tI_n).$$

The remark preceding this definition shows that the definition is independent of the choice of ordered basis β. Thus if T is a linear operator on a finite-dimensional vector space V and β is an ordered basis for V, then λ is an eigenvalue of T if and only if λ is an eigenvalue of $[T]_\beta$. We often denote the characteristic polynomial of an operator T by $\det(T - tI)$.

Example 5

Let T be the linear operator on $P_2(R)$ defined by $T(f(x)) = f(x) + (x+1)f'(x)$, let β be the standard ordered basis for $P_2(R)$, and let $A = [T]_\beta$. Then

$$A = \begin{pmatrix} 1 & 1 & 0 \\ 0 & 2 & 2 \\ 0 & 0 & 3 \end{pmatrix}.$$

The characteristic polynomial of T is

$$\det(A - tI_3) = \det \begin{pmatrix} 1-t & 1 & 0 \\ 0 & 2-t & 2 \\ 0 & 0 & 3-t \end{pmatrix}$$

$$= (1-t)(2-t)(3-t)$$
$$= -(t-1)(t-2)(t-3).$$

Hence λ is an eigenvalue of T (or A) if and only if $\lambda = 1, 2$, or 3.   ◆

Examples 4 and 5 suggest that the characteristic polynomial of an $n \times n$ matrix A is a polynomial of degree n. The next theorem tells us even more. It can be proved by a straightforward induction argument.

Theorem 5.3. *Let* $A \in M_{n \times n}(F)$.
(a) *The characteristic polynomial of* A *is a polynomial of degree* n *with leading coefficient* $(-1)^n$.
(b) A *has at most* n *distinct eigenvalues.*

Proof. Exercise.   ∎

Theorem 5.2 enables us to determine all the eigenvalues of a matrix or a linear operator on a finite-dimensional vector space provided that we can compute the zeros of the characteristic polynomial. Our next result gives us a procedure for determining the eigenvectors corresponding to a given eigenvalue.

Theorem 5.4. *Let* T *be a linear operator on a vector space* V, *and let* λ *be an eigenvalue of* T. *A vector* $v \in V$ *is an eigenvector of* T *corresponding to* λ *if and only if* $v \neq 0$ *and* $v \in N(T - \lambda I)$.

Proof. Exercise.   ∎

Example 6

To find all the eigenvectors of the matrix

$$A = \begin{pmatrix} 1 & 1 \\ 4 & 1 \end{pmatrix}$$

in Example 4, recall that A has two eigenvalues, $\lambda_1 = 3$ and $\lambda_2 = -1$. We begin by finding all the eigenvectors corresponding to $\lambda_1 = 3$. Let

$$B_1 = A - \lambda_1 I = \begin{pmatrix} 1 & 1 \\ 4 & 1 \end{pmatrix} - \begin{pmatrix} 3 & 0 \\ 0 & 3 \end{pmatrix} = \begin{pmatrix} -2 & 1 \\ 4 & -2 \end{pmatrix}.$$

Then

$$x = \begin{pmatrix} x_1 \\ x_2 \end{pmatrix} \in R^2$$

is an eigenvector corresponding to $\lambda_1 = 3$ if and only if $x \neq 0$ and $x \in N(L_{B_1})$; that is, $x \neq 0$ and

$$\begin{pmatrix} -2 & 1 \\ 4 & -2 \end{pmatrix} \begin{pmatrix} x_1 \\ x_2 \end{pmatrix} = \begin{pmatrix} -2x_1 + x_2 \\ 4x_1 - 2x_2 \end{pmatrix} = \begin{pmatrix} 0 \\ 0 \end{pmatrix}.$$

Clearly the set of all solutions to this equation is

$$\left\{ t \begin{pmatrix} 1 \\ 2 \end{pmatrix} : t \in R \right\}.$$

Hence x is an eigenvector corresponding to $\lambda_1 = 3$ if and only if

$$x = t \begin{pmatrix} 1 \\ 2 \end{pmatrix} \qquad \text{for some } t \neq 0.$$

Now suppose that x is an eigenvector of A corresponding to $\lambda_2 = -1$. Let

$$B_2 = A - \lambda_2 I = \begin{pmatrix} 1 & 1 \\ 4 & 1 \end{pmatrix} - \begin{pmatrix} -1 & 0 \\ 0 & -1 \end{pmatrix} = \begin{pmatrix} 2 & 1 \\ 4 & 2 \end{pmatrix}.$$

Then

$$x = \begin{pmatrix} x_1 \\ x_2 \end{pmatrix} \in \mathsf{N}(\mathsf{L}_{B_2})$$

if and only if x is a solution to the system

$$\begin{aligned} 2x_1 + \ x_2 &= 0 \\ 4x_1 + 2x_2 &= 0. \end{aligned}$$

Hence

$$\mathsf{N}(\mathsf{L}_{B_2}) = \left\{ t \begin{pmatrix} 1 \\ -2 \end{pmatrix} : t \in R \right\}.$$

Thus x is an eigenvector corresponding to $\lambda_2 = -1$ if and only if

$$x = t \begin{pmatrix} 1 \\ -2 \end{pmatrix} \qquad \text{for some } t \neq 0.$$

Observe that

$$\left\{ \begin{pmatrix} 1 \\ 2 \end{pmatrix}, \begin{pmatrix} 1 \\ -2 \end{pmatrix} \right\}$$

is a basis for R^2 consisting of eigenvectors of A. Thus L_A, and hence A, is diagonalizable.   ◆

Suppose that β is a basis for F^n consisting of eigenvectors of A. The corollary to Theorem 2.23 assures us that if Q is the $n \times n$ matrix whose columns are the vectors in β, then $Q^{-1}AQ$ is a diagonal matrix. In Example 6, for instance, if

$$Q = \begin{pmatrix} 1 & 1 \\ 2 & -2 \end{pmatrix},$$

then

$$Q^{-1}AQ = \begin{pmatrix} 3 & 0 \\ 0 & -1 \end{pmatrix}.$$

Of course, the diagonal entries of this matrix are the eigenvalues of A that correspond to the respective columns of Q.

To find the eigenvectors of a linear operator T on an n-dimensional vector space, select an ordered basis β for V and let $A = [\mathsf{T}]_\beta$. Figure 5.1 is the special case of Figure 2.2 in Section 2.4 in which $\mathsf{V} = \mathsf{W}$ and $\beta = \gamma$. Recall that for $v \in \mathsf{V}$, $\phi_\beta(v) = [v]_\beta$, the coordinate vector of v relative to β. We show that $v \in \mathsf{V}$ is an eigenvector of T corresponding to λ if and only if $\phi_\beta(v)$

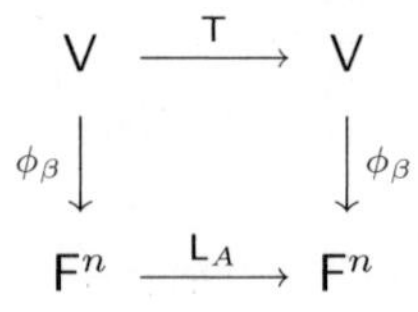

$$V \xrightarrow{\mathsf{T}} V$$

$$\phi_\beta \downarrow \qquad\qquad \downarrow \phi_\beta$$

$$\mathsf{F}^n \xrightarrow{\mathsf{L}_A} \mathsf{F}^n$$

Figure 5.1

is an eigenvector of A corresponding to λ. Suppose that v is an eigenvector of T corresponding to λ. Then $\mathsf{T}(v) = \lambda v$. Hence

$$A\phi_\beta(v) = \mathsf{L}_A \phi_\beta(v) = \phi_\beta \mathsf{T}(v) = \phi_\beta(\lambda v) = \lambda \phi_\beta(v).$$

Now $\phi_\beta(v) \neq 0$, since ϕ_β is an isomorphism; hence $\phi_\beta(v)$ is an eigenvector of A. This argument is reversible, and so we can establish that if $\phi_\beta(v)$ is an eigenvector of A corresponding to λ, then v is an eigenvector of T corresponding to λ. (See Exercise 13.)

An equivalent formulation of the result discussed in the preceding paragraph is that for an eigenvalue λ of A (and hence of T), a vector $y \in \mathsf{F}^n$ is an eigenvector of A corresponding to λ if and only if $\phi_\beta^{-1}(y)$ is an eigenvector of T corresponding to λ.

Thus we have reduced the problem of finding the eigenvectors of a linear operator on a finite-dimensional vector space to the problem of finding the eigenvectors of a matrix. The next example illustrates this procedure.

Example 7

Let T be the linear operator on $\mathsf{P}_2(R)$ defined in Example 5, and let β be the standard ordered basis for $\mathsf{P}_2(R)$. Recall that T has eigenvalues 1, 2, and 3 and that

$$A = [\mathsf{T}]_\beta = \begin{pmatrix} 1 & 1 & 0 \\ 0 & 2 & 2 \\ 0 & 0 & 3 \end{pmatrix}.$$

We consider each eigenvalue separately.

Let $\lambda_1 = 1$, and define

$$B_1 = A - \lambda_1 I = \begin{pmatrix} 0 & 1 & 0 \\ 0 & 1 & 2 \\ 0 & 0 & 2 \end{pmatrix}.$$

Then

$$x = \begin{pmatrix} x_1 \\ x_2 \\ x_3 \end{pmatrix} \in R^3$$

is an eigenvector corresponding to $\lambda_1 = 1$ if and only if $x \neq 0$ and $x \in \mathsf{N}(\mathsf{L}_{B_1})$; that is, x is a nonzero solution to the system

$$
\begin{aligned}
x_2 \qquad\quad &= 0 \\
x_2 + 2x_3 &= 0 \\
2x_3 &= 0.
\end{aligned}
$$

Notice that this system has three unknowns, x_1, x_2, and x_3, but one of these, x_1, does not actually appear in the system. Since the values of x_1 do not affect the system, we assign x_1 a parametric value, say $x_1 = a$, and solve the system for x_2 and x_3. Clearly, $x_2 = x_3 = 0$, and so the eigenvectors of A corresponding to $\lambda_1 = 1$ are of the form

$$
a \begin{pmatrix} 1 \\ 0 \\ 0 \end{pmatrix} = ae_1
$$

for $a \neq 0$. Consequently, the eigenvectors of T corresponding to $\lambda_1 = 1$ are of the form

$$
\phi_\beta^{-1}(ae_1) = a\phi_\beta^{-1}(e_1) = a \cdot 1 = a
$$

for any $a \neq 0$. Hence the nonzero constant polynomials are the eigenvectors of T corresponding to $\lambda_1 = 1$.

Next let $\lambda_2 = 2$, and define

$$
B_2 = A - \lambda_2 I = \begin{pmatrix} -1 & 1 & 0 \\ 0 & 0 & 2 \\ 0 & 0 & 1 \end{pmatrix}.
$$

It is easily verified that

$$
\mathsf{N}(\mathsf{L}_{B_2}) = \left\{ a \begin{pmatrix} 1 \\ 1 \\ 0 \end{pmatrix} : a \in R \right\},
$$

and hence the eigenvectors of T corresponding to $\lambda_2 = 2$ are of the form

$$
\phi_\beta^{-1} \left(a \begin{pmatrix} 1 \\ 1 \\ 0 \end{pmatrix} \right) = a\phi_\beta^{-1}(e_1 + e_2) = a(1 + x)
$$

for $a \neq 0$.

Finally, consider $\lambda_3 = 3$ and

$$
B_3 = A - \lambda_3 I = \begin{pmatrix} -2 & 1 & 0 \\ 0 & -1 & 2 \\ 0 & 0 & 0 \end{pmatrix}.
$$

Since

$$
N(L_{B_3}) = \left\{ a \begin{pmatrix} 1 \\ 2 \\ 1 \end{pmatrix} : a \in R \right\},
$$

the eigenvectors of T corresponding to $\lambda_3 = 3$ are of the form

$$
\phi_\beta^{-1} \left(a \begin{pmatrix} 1 \\ 2 \\ 1 \end{pmatrix} \right) = a\phi_\beta^{-1}(e_1 + 2e_2 + e_3) = a(1 + 2x + x^2)
$$

for $a \neq 0$.

For each eigenvalue, select the corresponding eigenvector with $a = 1$ in the preceding descriptions to obtain $\gamma = \{1, 1+x, 1+2x+x^2\}$, which is an ordered basis for $P_2(R)$ consisting of eigenvectors of T. Thus T is diagonalizable, and

$$
[T]_\gamma = \begin{pmatrix} 1 & 0 & 0 \\ 0 & 2 & 0 \\ 0 & 0 & 3 \end{pmatrix}. \quad \blacklozenge
$$

We close this section with a geometric description of how a linear operator T acts on an eigenvector in the context of a vector space V over R. Let v be an eigenvector of T and λ be the corresponding eigenvalue. We can think of $W = \text{span}(\{v\})$, the one-dimensional subspace of V spanned by v, as a line in V that passes through 0 and v. For any $w \in W$, $w = cv$ for some scalar c, and hence

$$
T(w) = T(cv) = cT(v) = c\lambda v = \lambda w;
$$

so T acts on the vectors in W by multiplying each such vector by λ. There are several possible ways for T to act on the vectors in W, depending on the value of λ. We consider several cases. (See Figure 5.2.)

CASE 1. If $\lambda > 1$, then T moves vectors in W farther from 0 by a factor of λ.

CASE 2. If $\lambda = 1$, then T acts as the identity operator on W.

CASE 3. If $0 < \lambda < 1$, then T moves vectors in W closer to 0 by a factor of λ.

CASE 4. If $\lambda = 0$, then T acts as the zero transformation on W.

CASE 5. If $\lambda < 0$, then T reverses the orientation of W; that is, T moves vectors in W from one side of 0 to the other.

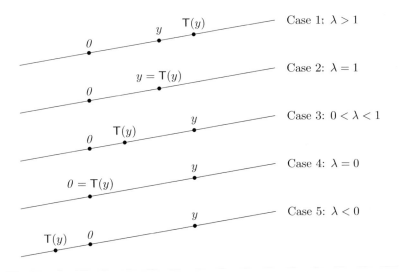

Figure 5.2: The action of T on W = span({x}) when x is an eigenvector of T.

To illustrate these ideas, we consider the linear operators in Examples 3, 4, and 2 of Section 2.1.

For the operator T on R^2 defined by $T(a_1, a_2) = (a_1, -a_2)$, the reflection about the x-axis, e_1 and e_2 are eigenvectors of T with corresponding eigenvalues 1 and -1, respectively. Since e_1 and e_2 span the x-axis and the y-axis, respectively, T acts as the identity on the x-axis and reverses the orientation of the y-axis.

For the operator T on R^2 defined by $T(a_1, a_2) = (a_1, 0)$, the projection on the x-axis, e_1 and e_2 are eigenvectors of T with corresponding eigenvalues 1 and 0, respectively. Thus, T acts as the identity on the x-axis and as the zero operator on the y-axis.

Finally, we generalize Example 2 of this section by considering the operator that rotates the plane through the angle θ, which is defined by

$$T_\theta(a_1, a_2) = (a_1 \cos\theta - a_2 \sin\theta, a_1 \sin\theta + a_2 \cos\theta).$$

Suppose that $0 < \theta < \pi$. Then for any nonzero vector v, the vectors v and $T_\theta(v)$ are not collinear, and hence T_θ maps no one-dimensional subspace of R^2 into itself. But this implies that T_θ has no eigenvectors and therefore no eigenvalues. To confirm this conclusion, we note that the characteristic polynomial of T_θ is

$$\det(T_\theta - tI) = \det \begin{pmatrix} \cos\theta - t & -\sin\theta \\ \sin\theta & \cos\theta - t \end{pmatrix} = t^2 - (2\cos\theta)t + 1,$$

which has no real zeros because, for $0 < \theta < \pi$, the discriminant $4\cos^2\theta - 4$ is negative.

EXERCISES

1. Label the following statements as true or false.

   (a) Every linear operator on an n-dimensional vector space has n distinct eigenvalues.

   (b) If a real matrix has one eigenvector, then it has an infinite number of eigenvectors.

   (c) There exists a square matrix with no eigenvectors.

   (d) Eigenvalues must be nonzero scalars.

   (e) Any two eigenvectors are linearly independent.

   (f) The sum of two eigenvalues of a linear operator T is also an eigenvalue of T.

   (g) Linear operators on infinite-dimensional vector spaces never have eigenvalues.

   (h) An $n \times n$ matrix A with entries from a field F is similar to a diagonal matrix if and only if there is a basis for F^n consisting of eigenvectors of A.

   (i) Similar matrices always have the same eigenvalues.

   (j) Similar matrices always have the same eigenvectors.

   (k) The sum of two eigenvectors of an operator T is always an eigenvector of T.

2. For each of the following linear operators T on a vector space V and ordered bases β, compute $[\mathsf{T}]_\beta$, and determine whether β is a basis consisting of eigenvectors of T.

   (a) $V = \mathsf{R}^2$, $\mathsf{T}\begin{pmatrix} a \\ b \end{pmatrix} = \begin{pmatrix} 10a - 6b \\ 17a - 10b \end{pmatrix}$, and $\beta = \left\{ \begin{pmatrix} 1 \\ 2 \end{pmatrix}, \begin{pmatrix} 2 \\ 3 \end{pmatrix} \right\}$

   (b) $V = \mathsf{P}_1(R)$, $\mathsf{T}(a + bx) = (6a - 6b) + (12a - 11b)x$, and $\beta = \{3 + 4x, 2 + 3x\}$

   (c) $V = \mathsf{R}^3$, $\mathsf{T}\begin{pmatrix} a \\ b \\ c \end{pmatrix} = \begin{pmatrix} 3a + 2b - 2c \\ -4a - 3b + 2c \\ -c \end{pmatrix}$, and

   $$\beta = \left\{ \begin{pmatrix} 0 \\ 1 \\ 1 \end{pmatrix}, \begin{pmatrix} 1 \\ -1 \\ 0 \end{pmatrix}, \begin{pmatrix} 1 \\ 0 \\ 2 \end{pmatrix} \right\}$$

   (d) $V = \mathsf{P}_2(R)$, $\mathsf{T}(a + bx + cx^2) =$

   $$(-4a + 2b - 2c) - (7a + 3b + 7c)x + (7a + b + 5c)x^2,$$

   and $\beta = \{x - x^2, -1 + x^2, -1 - x + x^2\}$

(e) $V = P_3(R)$, $T(a + bx + cx^2 + dx^3) =$

$$-d + (-c + d)x + (a + b - 2c)x^2 + (-b + c - 2d)x^3,$$

and $\beta = \{1 - x + x^3, 1 + x^2, 1, x + x^2\}$

(f) $V = M_{2\times 2}(R)$, $T\begin{pmatrix} a & b \\ c & d \end{pmatrix} = \begin{pmatrix} -7a - 4b + 4c - 4d & b \\ -8a - 4b + 5c - 4d & d \end{pmatrix}$, and

$$\beta = \left\{ \begin{pmatrix} 1 & 0 \\ 1 & 0 \end{pmatrix}, \begin{pmatrix} -1 & 2 \\ 0 & 0 \end{pmatrix}, \begin{pmatrix} 1 & 0 \\ 2 & 0 \end{pmatrix}, \begin{pmatrix} -1 & 0 \\ 0 & 2 \end{pmatrix} \right\}$$

3. For each of the following matrices $A \in M_{n\times n}(F)$,

   (i) Determine all the eigenvalues of A.

   (ii) For each eigenvalue λ of A, find the set of eigenvectors corresponding to λ.

   (iii) If possible, find a basis for F^n consisting of eigenvectors of A.

   (iv) If successful in finding such a basis, determine an invertible matrix Q and a diagonal matrix D such that $Q^{-1}AQ = D$.

   (a) $A = \begin{pmatrix} 1 & 2 \\ 3 & 2 \end{pmatrix}$ for $F = R$

   (b) $A = \begin{pmatrix} 0 & -2 & -3 \\ -1 & 1 & -1 \\ 2 & 2 & 5 \end{pmatrix}$ for $F = R$

   (c) $A = \begin{pmatrix} i & 1 \\ 2 & -i \end{pmatrix}$ for $F = C$

   (d) $A = \begin{pmatrix} 2 & 0 & -1 \\ 4 & 1 & -4 \\ 2 & 0 & -1 \end{pmatrix}$ for $F = R$

4. For each linear operator T on V, find the eigenvalues of T and an ordered basis β for V such that $[T]_\beta$ is a diagonal matrix.

   (a) $V = R^2$ and $T(a, b) = (-2a + 3b, -10a + 9b)$
   (b) $V = R^3$ and $T(a, b, c) = (7a - 4b + 10c, 4a - 3b + 8c, -2a + b - 2c)$
   (c) $V = R^3$ and $T(a, b, c) = (-4a + 3b - 6c, 6a - 7b + 12c, 6a - 6b + 11c)$
   (d) $V = P_1(R)$ and $T(ax + b) = (-6a + 2b)x + (-6a + b)$
   (e) $V = P_2(R)$ and $T(f(x)) = xf'(x) + f(2)x + f(3)$
   (f) $V = P_3(R)$ and $T(f(x)) = f(x) + f(2)x$
   (g) $V = P_3(R)$ and $T(f(x)) = xf'(x) + f''(x) - f(2)$
   (h) $V = M_{2\times 2}(R)$ and $T\begin{pmatrix} a & b \\ c & d \end{pmatrix} = \begin{pmatrix} d & b \\ c & a \end{pmatrix}$

(i) $V = M_{2\times2}(R)$ and $T\begin{pmatrix} a & b \\ c & d \end{pmatrix} = \begin{pmatrix} c & d \\ a & b \end{pmatrix}$

(j) $V = M_{2\times2}(R)$ and $T(A) = A^t + 2 \cdot \text{tr}(A) \cdot I_2$

5. Prove Theorem 5.4.

6. Let T be a linear operator on a finite-dimensional vector space V, and let β be an ordered basis for V. Prove that λ is an eigenvalue of T if and only if λ is an eigenvalue of $[T]_\beta$.

7. Let T be a linear operator on a finite-dimensional vector space V. We define the **determinant** of T, denoted $\det(T)$, as follows: Choose any ordered basis β for V, and define $\det(T) = \det([T]_\beta)$.

    (a) Prove that the preceding definition is independent of the choice of an ordered basis for V. That is, prove that if β and γ are two ordered bases for V, then $\det([T]_\beta) = \det([T]_\gamma)$.
    (b) Prove that T is invertible if and only if $\det(T) \neq 0$.
    (c) Prove that if T is invertible, then $\det(T^{-1}) = [\det(T)]^{-1}$.
    (d) Prove that if U is also a linear operator on V, then $\det(TU) = \det(T) \cdot \det(U)$.
    (e) Prove that $\det(T - \lambda I_V) = \det([T]_\beta - \lambda I)$ for any scalar λ and any ordered basis β for V.

8. (a) Prove that a linear operator T on a finite-dimensional vector space is invertible if and only if zero is not an eigenvalue of T.
    (b) Let T be an invertible linear operator. Prove that a scalar λ is an eigenvalue of T if and only if λ^{-1} is an eigenvalue of T^{-1}.
    (c) State and prove results analogous to (a) and (b) for matrices.

9. Prove that the eigenvalues of an upper triangular matrix M are the diagonal entries of M.

10. Let V be a finite-dimensional vector space, and let λ be any scalar.

    (a) For any ordered basis β for V, prove that $[\lambda I_V]_\beta = \lambda I$.
    (b) Compute the characteristic polynomial of λI_V.
    (c) Show that λI_V is diagonalizable and has only one eigenvalue.

11. A **scalar matrix** is a square matrix of the form λI for some scalar λ; that is, a scalar matrix is a diagonal matrix in which all the diagonal entries are equal.

    (a) Prove that if a square matrix A is similar to a scalar matrix λI, then $A = \lambda I$.
    (b) Show that a diagonalizable matrix having only one eigenvalue is a scalar matrix.

(c) Prove that $\begin{pmatrix} 1 & 1 \\ 0 & 1 \end{pmatrix}$ is not diagonalizable.

12. (a) Prove that similar matrices have the same characteristic polynomial.

(b) Show that the definition of the characteristic polynomial of a linear operator on a finite-dimensional vector space V is independent of the choice of basis for V.

13. Let T be a linear operator on a finite-dimensional vector space V over a field F, let β be an ordered basis for V, and let $A = [\mathsf{T}]_\beta$. In reference to Figure 5.1, prove the following.

(a) If $v \in \mathsf{V}$ and $\phi_\beta(v)$ is an eigenvector of A corresponding to the eigenvalue λ, then v is an eigenvector of T corresponding to λ.

(b) If λ is an eigenvalue of A (and hence of T), then a vector $y \in F^n$ is an eigenvector of A corresponding to λ if and only if $\phi_\beta^{-1}(y)$ is an eigenvector of T corresponding to λ.

14.[†] For any square matrix A, prove that A and A^t have the same characteristic polynomial (and hence the same eigenvalues).

15.[†] (a) Let T be a linear operator on a vector space V, and let x be an eigenvector of T corresponding to the eigenvalue λ. For any positive integer m, prove that x is an eigenvector of T^m corresponding to the eigenvalue λ^m.

(b) State and prove the analogous result for matrices.

16. (a) Prove that similar matrices have the same trace. *Hint:* Use Exercise 13 of Section 2.3.

(b) How would you define the trace of a linear operator on a finite-dimensional vector space? Justify that your definition is well-defined.

17. Let T be the linear operator on $\mathsf{M}_{n \times n}(R)$ defined by $\mathsf{T}(A) = A^t$.

(a) Show that ± 1 are the only eigenvalues of T.

(b) Describe the eigenvectors corresponding to each eigenvalue of T.

(c) Find an ordered basis β for $\mathsf{M}_{2 \times 2}(R)$ such that $[\mathsf{T}]_\beta$ is a diagonal matrix.

(d) Find an ordered basis β for $\mathsf{M}_{n \times n}(R)$ such that $[\mathsf{T}]_\beta$ is a diagonal matrix for $n > 2$.

18. Let $A, B \in \mathsf{M}_{n \times n}(C)$.

(a) Prove that if B is invertible, then there exists a scalar $c \in C$ such that $A + cB$ is not invertible. *Hint:* Examine $\det(A + cB)$.

(b) Find nonzero 2×2 matrices A and B such that both A and $A+cB$ are invertible for all $c \in C$.

19.[†] Let A and B be similar $n \times n$ matrices. Prove that there exists an n-dimensional vector space V, a linear operator T on V, and ordered bases β and γ for V such that $A = [T]_\beta$ and $B = [T]_\gamma$. *Hint:* Use Exercise 14 of Section 2.5.

20. Let A be an $n \times n$ matrix with characteristic polynomial

$$f(t) = (-1)^n t^n + a_{n-1}t^{n-1} + \cdots + a_1 t + a_0.$$

Prove that $f(0) = a_0 = \det(A)$. Deduce that A is invertible if and only if $a_0 \neq 0$.

21. Let A and $f(t)$ be as in Exercise 20.

(a) Prove that $f(t) = (A_{11} - t)(A_{22} - t) \cdots (A_{nn} - t) + q(t)$, where $q(t)$ is a polynomial of degree at most $n-2$. *Hint:* Apply mathematical induction to n.

(b) Show that $\text{tr}(A) = (-1)^{n-1}a_{n-1}$.

22.[†] (a) Let T be a linear operator on a vector space V over the field F, and let $g(t)$ be a polynomial with coefficients from F. Prove that if x is an eigenvector of T with corresponding eigenvalue λ, then $g(T)(x) = g(\lambda)x$. That is, x is an eigenvector of $g(T)$ with corresponding eigenvalue $g(\lambda)$.

(b) State and prove a comparable result for matrices.

(c) Verify (b) for the matrix A in Exercise 3(a) with polynomial $g(t) = 2t^2 - t + 1$, eigenvector $x = \begin{pmatrix} 2 \\ 3 \end{pmatrix}$, and corresponding eigenvalue $\lambda = 4$.

23. Use Exercise 22 to prove that if $f(t)$ is the characteristic polynomial of a diagonalizable linear operator T, then $f(T) = T_0$, the zero operator. (In Section 5.4 we prove that this result does not depend on the diagonalizability of T.)

24. Use Exercise 21(a) to prove Theorem 5.3.

25. Prove Corollaries 1 and 2 of Theorem 5.3.

26. Determine the number of distinct characteristic polynomials of matrices in $M_{2\times2}(Z_2)$.

5.2   DIAGONALIZABILITY

In Section 5.1, we presented the diagonalization problem and observed that not all linear operators or matrices are diagonalizable. Although we are able to diagonalize operators and matrices and even obtain a necessary and sufficient condition for diagonalizability (Theorem 5.1 p. 246), we have not yet solved the diagonalization problem. What is still needed is a simple test to determine whether an operator or a matrix can be diagonalized, as well as a method for actually finding a basis of eigenvectors. In this section, we develop such a test and method.

In Example 6 of Section 5.1, we obtained a basis of eigenvectors by choosing one eigenvector corresponding to each eigenvalue. In general, such a procedure does not yield a basis, but the following theorem shows that any set constructed in this manner is linearly independent.

Theorem 5.5. *Let* T *be a linear operator on a vector space* V, *and let* $\lambda_1, \lambda_2, \ldots, \lambda_k$ *be distinct eigenvalues of* T. *If* $v_1, v_2, \ldots, v_k$ *are eigenvectors of* T *such that* λ_i *corresponds to* v_i $(1 \le i \le k)$, *then* $\{v_1, v_2, \ldots, v_k\}$ *is linearly independent.*

Proof. The proof is by mathematical induction on k. Suppose that $k = 1$. Then $v_1 \ne 0$ since v_1 is an eigenvector, and hence $\{v_1\}$ is linearly independent. Now assume that the theorem holds for $k - 1$ distinct eigenvalues, where $k - 1 \ge 1$, and that we have k eigenvectors $v_1, v_2, \ldots, v_k$ corresponding to the distinct eigenvalues $\lambda_1, \lambda_2, \ldots, \lambda_k$. We wish to show that $\{v_1, v_2, \ldots, v_k\}$ is linearly independent. Suppose that $a_1, a_2, \ldots, a_k$ are scalars such that

$$a_1 v_1 + a_2 v_2 + \cdots + a_k v_k = 0. \tag{1}$$

Applying $\mathsf{T} - \lambda_k \mathsf{I}$ to both sides of (1), we obtain

$$a_1(\lambda_1 - \lambda_k)v_1 + a_2(\lambda_2 - \lambda_k)v_2 + \cdots + a_{k-1}(\lambda_{k-1} - \lambda_k)v_{k-1} = 0.$$

By the induction hypothesis $\{v_1, v_2, \ldots, v_{k-1}\}$ is linearly independent, and hence

$$a_1(\lambda_1 - \lambda_k) = a_2(\lambda_2 - \lambda_k) = \cdots = a_{k-1}(\lambda_{k-1} - \lambda_k) = 0.$$

Since $\lambda_1, \lambda_2, \ldots, \lambda_k$ are distinct, it follows that $\lambda_i - \lambda_k \ne 0$ for $1 \le i \le k - 1$. So $a_1 = a_2 = \cdots = a_{k-1} = 0$, and (1) therefore reduces to $a_k v_k = 0$. But $v_k \ne 0$ and therefore $a_k = 0$. Consequently $a_1 = a_2 = \cdots = a_k = 0$, and it follows that $\{v_1, v_2, \ldots, v_k\}$ is linearly independent. ∎

Corollary. *Let* T *be a linear operator on an* n-*dimensional vector space* V. *If* T *has* n *distinct eigenvalues, then* T *is diagonalizable.*

Proof. Suppose that T has n distinct eigenvalues $\lambda_1, \ldots, \lambda_n$. For each i choose an eigenvector v_i corresponding to λ_i. By Theorem 5.5, $\{v_1, \ldots, v_n\}$ is linearly independent, and since $\dim(V) = n$, this set is a basis for V. Thus, by Theorem 5.1 (p. 246), T is diagonalizable. ∎

Example 1

Let

$$A = \begin{pmatrix} 1 & 1 \\ 1 & 1 \end{pmatrix} \in M_{2 \times 2}(R).$$

The characteristic polynomial of A (and hence of L_A) is

$$\det(A - tI) = \det \begin{pmatrix} 1 - t & 1 \\ 1 & 1 - t \end{pmatrix} = t(t - 2),$$

and thus the eigenvalues of L_A are 0 and 2. Since L_A is a linear operator on the two-dimensional vector space R^2, we conclude from the preceding corollary that L_A (and hence A) is diagonalizable. ◆

The converse of Theorem 5.5 is false. That is, it is not true that if T is diagonalizable, then it has n distinct eigenvalues. For example, the identity operator is diagonalizable even though it has only one eigenvalue, namely, $\lambda = 1$.

We have seen that diagonalizability requires the existence of eigenvalues. Actually, diagonalizability imposes a stronger condition on the characteristic polynomial.

Definition. *A polynomial $f(t)$ in $P(F)$ **splits over** F if there are scalars $c, a_1, \ldots, a_n$ (not necessarily distinct) in F such that*

$$f(t) = c(t - a_1)(t - a_2) \cdots (t - a_n).$$

For example, $t^2 - 1 = (t + 1)(t - 1)$ splits over R, but $(t^2 + 1)(t - 2)$ does not split over R because $t^2 + 1$ cannot be factored into a product of linear factors. However, $(t^2 + 1)(t - 2)$ does split over C because it factors into the product $(t + i)(t - i)(t - 2)$. If $f(t)$ is the characteristic polynomial of a linear operator or a matrix over a field F, then the statement that $f(t)$ splits is understood to mean that it splits over F.

Theorem 5.6. *The characteristic polynomial of any diagonalizable linear operator splits.*

Proof. Let T be a diagonalizable linear operator on the n-dimensional vector space V, and let β be an ordered basis for V such that $[T]_\beta = D$ is a diagonal matrix. Suppose that

$$
D = \begin{pmatrix}
\lambda_1 & 0 & \cdots & 0 \\
0 & \lambda_2 & \cdots & 0 \\
\vdots & \vdots & & \vdots \\
0 & 0 & \cdots & \lambda_n
\end{pmatrix},
$$

and let $f(t)$ be the characteristic polynomial of T. Then

$$
f(t) = \det(D - tI) = \det \begin{pmatrix}
\lambda_1 - t & 0 & \cdots & 0 \\
0 & \lambda_2 - t & \cdots & 0 \\
\vdots & \vdots & & \vdots \\
0 & 0 & \cdots & \lambda_n - t
\end{pmatrix}
$$

$$
= (\lambda_1 - t)(\lambda_2 - t)\cdots(\lambda_n - t) = (-1)^n (t - \lambda_1)(t - \lambda_2)\cdots(t - \lambda_n). \quad\blacksquare
$$

From this theorem, it is clear that if T is a diagonalizable linear operator on an n-dimensional vector space that fails to have distinct eigenvalues, then the characteristic polynomial of T must have repeated zeros.

The converse of Theorem 5.6 is false; that is, the characteristic polynomial of T may split, but T need not be diagonalizable. (See Example 3, which follows.) The following concept helps us determine when an operator whose characteristic polynomial splits is diagonalizable.

Definition. *Let λ be an eigenvalue of a linear operator or matrix with characteristic polynomial $f(t)$. The **(algebraic) multiplicity** of λ is the largest positive integer k for which $(t - \lambda)^k$ is a factor of $f(t)$.*

Example 2

Let

$$
A = \begin{pmatrix}
3 & 1 & 0 \\
0 & 3 & 4 \\
0 & 0 & 4
\end{pmatrix},
$$

which has characteristic polynomial $f(t) = -(t - 3)^2(t - 4)$. Hence $\lambda = 3$ is an eigenvalue of A with multiplicity 2, and $\lambda = 4$ is an eigenvalue of A with multiplicity 1. $\blacklozenge$

If T is a diagonalizable linear operator on a finite-dimensional vector space V, then there is an ordered basis β for V consisting of eigenvectors of T. We know from Theorem 5.1 (p. 246) that $[T]_\beta$ is a diagonal matrix in which the diagonal entries are the eigenvalues of T. Since the characteristic polynomial of T is $\det([T]_\beta - tI)$, it is easily seen that each eigenvalue of T must occur as a diagonal entry of $[T]_\beta$ exactly as many times as its multiplicity. Hence

β contains as many (linearly independent) eigenvectors corresponding to an eigenvalue as the multiplicity of that eigenvalue. So the number of linearly independent eigenvectors corresponding to a given eigenvalue is of interest in determining whether an operator can be diagonalized. Recalling from Theorem 5.4 (p. 250) that the eigenvectors of T corresponding to the eigenvalue λ are the nonzero vectors in the null space of $T - \lambda I$, we are led naturally to the study of this set.

Definition. *Let* T *be a linear operator on a vector space* V, *and let* λ *be an eigenvalue of* T. *Define* $E_\lambda = \{x \in V : T(x) = \lambda x\} = N(T - \lambda I_V)$. *The set* E_λ *is called the* **eigenspace** *of* T *corresponding to the eigenvalue* λ. *Analogously, we define the* **eigenspace** *of a square matrix* A *to be the eigenspace of* L_A.

Clearly, E_λ is a subspace of V consisting of the zero vector and the eigenvectors of T corresponding to the eigenvalue λ. The maximum number of linearly independent eigenvectors of T corresponding to the eigenvalue λ is therefore the dimension of E_λ. Our next result relates this dimension to the multiplicity of λ.

Theorem 5.7. *Let* T *be a linear operator on a finite-dimensional vector space* V, *and let* λ *be an eigenvalue of* T *having multiplicity* m. *Then* $1 \le \dim(E_\lambda) \le m$.

Proof. Choose an ordered basis $\{v_1, v_2, \ldots, v_p\}$ for E_λ, extend it to an ordered basis $\beta = \{v_1, v_2, \ldots, v_p, v_{p+1}, \ldots, v_n\}$ for V, and let $A = [T]_\beta$. Observe that v_i $(1 \le i \le p)$ is an eigenvector of T corresponding to λ, and therefore

$$A = \begin{pmatrix} \lambda I_p & B \\ O & C \end{pmatrix}.$$

By Exercise 21 of Section 4.3, the characteristic polynomial of T is

$$f(t) = \det(A - tI_n) = \det \begin{pmatrix} (\lambda - t)I_p & B \\ O & C - tI_{n-p} \end{pmatrix}$$

$$= \det((\lambda - t)I_p) \det(C - tI_{n-p})$$

$$= (\lambda - t)^p g(t),$$

where $g(t)$ is a polynomial. Thus $(\lambda - t)^p$ is a factor of $f(t)$, and hence the multiplicity of λ is at least p. But $\dim(E_\lambda) = p$, and so $\dim(E_\lambda) \le m$. ∎

Example 3

Let T be the linear operator on $P_2(R)$ defined by $T(f(x)) = f'(x)$. The matrix representation of T with respect to the standard ordered basis β for

$P_2(R)$ is

$$[T]_\beta = \begin{pmatrix} 0 & 1 & 0 \\ 0 & 0 & 2 \\ 0 & 0 & 0 \end{pmatrix}.$$

Consequently, the characteristic polynomial of T is

$$\det([T]_\beta - tI) = \det \begin{pmatrix} -t & 1 & 0 \\ 0 & -t & 2 \\ 0 & 0 & -t \end{pmatrix} = -t^3.$$

Thus T has only one eigenvalue ($\lambda = 0$) with multiplicity 3. Solving $T(f(x)) = f'(x) = 0$ shows that $E_\lambda = N(T - \lambda I) = N(T)$ is the subspace of $P_2(R)$ consisting of the constant polynomials. So $\{1\}$ is a basis for E_λ, and therefore $\dim(E_\lambda) = 1$. Consequently, there is no basis for $P_2(R)$ consisting of eigenvectors of T, and therefore T is not diagonalizable.   ◆

Example 4

Let T be the linear operator on R^3 defined by

$$T \begin{pmatrix} a_1 \\ a_2 \\ a_3 \end{pmatrix} = \begin{pmatrix} 4a_1 & + a_3 \\ 2a_1 + 3a_2 + 2a_3 \\ a_1 & + 4a_3 \end{pmatrix}.$$

We determine the eigenspace of T corresponding to each eigenvalue. Let β be the standard ordered basis for R^3. Then

$$[T]_\beta = \begin{pmatrix} 4 & 0 & 1 \\ 2 & 3 & 2 \\ 1 & 0 & 4 \end{pmatrix},$$

and hence the characteristic polynomial of T is

$$\det([T]_\beta - tI) = \det \begin{pmatrix} 4-t & 0 & 1 \\ 2 & 3-t & 2 \\ 1 & 0 & 4-t \end{pmatrix} = -(t-5)(t-3)^2.$$

So the eigenvalues of T are $\lambda_1 = 5$ and $\lambda_2 = 3$ with multiplicities 1 and 2, respectively.

Since

$$E_{\lambda_1} = N(T - \lambda_1 I) = \left\{ \begin{pmatrix} x_1 \\ x_2 \\ x_3 \end{pmatrix} \in R^3 : \begin{pmatrix} -1 & 0 & 1 \\ 2 & -2 & 2 \\ 1 & 0 & -1 \end{pmatrix} \begin{pmatrix} x_1 \\ x_2 \\ x_3 \end{pmatrix} = \begin{pmatrix} 0 \\ 0 \\ 0 \end{pmatrix} \right\},$$

E_{λ_1} is the solution space of the system of linear equations

$$\begin{aligned} -x_1 \quad\quad\ + \ x_3 &= 0 \\ 2x_1 - 2x_2 + 2x_3 &= 0 \\ x_1 \quad\quad\ - \ x_3 &= 0. \end{aligned}$$

It is easily seen (using the techniques of Chapter 3) that

$$\left\{ \begin{pmatrix} 1 \\ 2 \\ 1 \end{pmatrix} \right\}$$

is a basis for E_{λ_1}. Hence $\dim(E_{\lambda_1}) = 1$.

Similarly, $E_{\lambda_2} = N(T - \lambda_2 I)$ is the solution space of the system

$$\begin{aligned} x_1 + \ x_3 &= 0 \\ 2x_1 + 2x_3 &= 0 \\ x_1 + \ x_3 &= 0. \end{aligned}$$

Since the unknown x_2 does not appear in this system, we assign it a parametric value, say, $x_2 = s$, and solve the system for x_1 and x_3, introducing another parameter t. The result is the general solution to the system

$$\begin{pmatrix} x_1 \\ x_2 \\ x_3 \end{pmatrix} = s \begin{pmatrix} 0 \\ 1 \\ 0 \end{pmatrix} + t \begin{pmatrix} -1 \\ 0 \\ 1 \end{pmatrix}, \quad \text{for } s, t \in R.$$

It follows that

$$\left\{ \begin{pmatrix} 0 \\ 1 \\ 0 \end{pmatrix}, \begin{pmatrix} -1 \\ 0 \\ 1 \end{pmatrix} \right\}$$

is a basis for E_{λ_2}, and $\dim(E_{\lambda_2}) = 2$.

In this case, the multiplicity of each eigenvalue λ_i is equal to the dimension of the corresponding eigenspace E_{λ_i}. Observe that the union of the two bases just derived, namely,

$$\left\{ \begin{pmatrix} 1 \\ 2 \\ 1 \end{pmatrix}, \begin{pmatrix} 0 \\ 1 \\ 0 \end{pmatrix}, \begin{pmatrix} -1 \\ 0 \\ 1 \end{pmatrix} \right\},$$

is linearly independent and hence is a basis for R^3 consisting of eigenvectors of T. Consequently, T is diagonalizable. ◆

Examples 3 and 4 suggest that an operator whose characteristic polynomial splits is diagonalizable if and only if the dimension of each eigenspace is equal to the multiplicity of the corresponding eigenvalue. This is indeed true, as we now show. We begin with the following lemma, which is a slight variation of Theorem 5.5.

Lemma. *Let* T *be a linear operator, and let* $\lambda_1, \lambda_2, \ldots, \lambda_k$ *be distinct eigenvalues of* T. *For each* $i = 1, 2, \ldots, k$, *let* $v_i \in \mathsf{E}_{\lambda_i}$, *the eigenspace corresponding to* λ_i. *If*

$$v_1 + v_2 + \cdots + v_k = 0,$$

then $v_i = 0$ *for all* i.

Proof. Suppose otherwise. By renumbering if necessary, suppose that, for $1 \leq m \leq k$, we have $v_i \neq 0$ for $1 \leq i \leq m$, and $v_i = 0$ for $i > m$. Then, for each $i \leq m$, v_i is an eigenvector of T corresponding to λ_i and

$$v_1 + v_2 + \cdots + v_m = 0.$$

But this contradicts Theorem 5.5, which states that these v_i's are linearly independent. We conclude, therefore, that $v_i = 0$ for all i. $\blacksquare$

Theorem 5.8. *Let* T *be a linear operator on a vector space* V, *and let* $\lambda_1, \lambda_2, \ldots, \lambda_k$ *be distinct eigenvalues of* T. *For each* $i = 1, 2, \ldots, k$, *let* S_i *be a finite linearly independent subset of the eigenspace* E_{λ_i}. *Then* $S = S_1 \cup S_2 \cup \cdots \cup S_k$ *is a linearly independent subset of* V.

Proof. Suppose that for each i

$$S_i = \{v_{i1}, v_{i2}, \ldots, v_{in_i}\}.$$

Then $S = \{v_{ij} : 1 \leq j \leq n_i, \text{ and } 1 \leq i \leq k\}$. Consider any scalars $\{a_{ij}\}$ such that

$$\sum_{i=1}^{k} \sum_{j=1}^{n_i} a_{ij} v_{ij} = 0.$$

For each i, let

$$w_i = \sum_{j=1}^{n_i} a_{ij} v_{ij}.$$

Then $w_i \in \mathsf{E}_{\lambda_i}$ for each i, and $w_1 + \cdots + w_k = 0$. Therefore, by the lemma, $w_i = 0$ for all i. But each S_i is linearly independent, and hence $a_{ij} = 0$ for all j. We conclude that S is linearly independent. $\blacksquare$

Theorem 5.8 tells us how to construct a linearly independent subset of eigenvectors, namely, by collecting bases for the individual eigenspaces. The next theorem tells us when the resulting set is a basis for the entire space.

Theorem 5.9. *Let* T *be a linear operator on a finite-dimensional vector space* V *such that the characteristic polynomial of* T *splits. Let* $\lambda_1, \lambda_2, \ldots, \lambda_k$ *be the distinct eigenvalues of* T. *Then*
  (a) T *is diagonalizable if and only if the multiplicity of* λ_i *is equal to* $\dim(\mathsf{E}_{\lambda_i})$ *for all* i.
  (b) *If* T *is diagonalizable and* β_i *is an ordered basis for* E_{λ_i} *for each* i, *then* $\beta = \beta_1 \cup \beta_2 \cup \cdots \cup \beta_k$ *is an ordered basis*[2] *for* V *consisting of eigenvectors of* T.

Proof. For each i, let m_i denote the multiplicity of λ_i, $d_i = \dim(\mathsf{E}_{\lambda_i})$, and $n = \dim(\mathsf{V})$.

First, suppose that T is diagonalizable. Let β be a basis for V consisting of eigenvectors of T. For each i, let $\beta_i = \beta \cap \mathsf{E}_{\lambda_i}$, the set of vectors in β that are eigenvectors corresponding to λ_i, and let n_i denote the number of vectors in β_i. Then $n_i \leq d_i$ for each i because β_i is a linearly independent subset of a subspace of dimension d_i, and $d_i \leq m_i$ by Theorem 5.7. The n_i's sum to n because β contains n vectors. The m_i's also sum to n because the degree of the characteristic polynomial of T is equal to the sum of the multiplicities of the eigenvalues. Thus

$$n = \sum_{i=1}^{k} n_i \leq \sum_{i=1}^{k} d_i \leq \sum_{i=1}^{k} m_i = n.$$

It follows that

$$\sum_{i=1}^{k} (m_i - d_i) = 0.$$

Since $(m_i - d_i) \geq 0$ for all i, we conclude that $m_i = d_i$ for all i.

Conversely, suppose that $m_i = d_i$ for all i. We simultaneously show that T is diagonalizable and prove (b). For each i, let β_i be an ordered basis for E_{λ_i}, and let $\beta = \beta_1 \cup \beta_2 \cup \cdots \cup \beta_k$. By Theorem 5.8, β is linearly independent. Furthermore, since $d_i = m_i$ for all i, β contains

$$\sum_{i=1}^{k} d_i = \sum_{i=1}^{k} m_i = n$$

---

[2]We regard $\beta_1 \cup \beta_2 \cup \cdots \cup \beta_k$ as an ordered basis in the natural way—the vectors in β_1 are listed first (in the same order as in β_1), then the vectors in β_2 (in the same order as in β_2), etc.

vectors. Therefore β is an ordered basis for V consisting of eigenvectors of V, and we conclude that T is diagonalizable. ∎

This theorem completes our study of the diagonalization problem. We summarize our results.

Test for Diagonalization

Let T be a linear operator on an n-dimensional vector space V. Then T is diagonalizable if and only if both of the following conditions hold.

1. The characteristic polynomial of T splits.
2. For each eigenvalue λ of T, the multiplicity of λ equals $n - \text{rank}(T - \lambda I)$.

These same conditions can be used to test if a square matrix A is diagonalizable because diagonalizability of A is equivalent to diagonalizability of the operator L_A.

If T is a diagonalizable operator and $\beta_1, \beta_2, \dots, \beta_k$ are ordered bases for the eigenspaces of T, then the union $\beta = \beta_1 \cup \beta_2 \cup \dots \cup \beta_k$ is an ordered basis for V consisting of eigenvectors of T, and hence $[T]_\beta$ is a diagonal matrix.

When testing T for diagonalizability, it is usually easiest to choose a convenient basis α for V and work with $B = [T]_\alpha$. If the characteristic polynomial of B splits, then use condition 2 above to check if the multiplicity of each *repeated* eigenvalue of B equals $n - \text{rank}(B - \lambda I)$. (By Theorem 5.7, condition 2 is automatically satisfied for eigenvalues of multiplicity 1.) If so, then B, and hence T, is diagonalizable.

If T is diagonalizable and a basis β for V consisting of eigenvectors of T is desired, then we first find a basis for each eigenspace of B. The union of these bases is a basis γ for F^n consisting of eigenvectors of B. Each vector in γ is the coordinate vector relative to α of an eigenvector of T. The set consisting of these n eigenvectors of T is the desired basis β.

Furthermore, if A is an $n \times n$ diagonalizable matrix, we can use the corollary to Theorem 2.23 (p. 115) to find an invertible $n \times n$ matrix Q and a diagonal $n \times n$ matrix D such that $Q^{-1}AQ = D$. The matrix Q has as its columns the vectors in a basis of eigenvectors of A, and D has as its jth diagonal entry the eigenvalue of A corresponding to the jth column of Q.

We now consider some examples illustrating the preceding ideas.

Example 5

We test the matrix

$$A = \begin{pmatrix} 3 & 1 & 0 \\ 0 & 3 & 0 \\ 0 & 0 & 4 \end{pmatrix} \in \mathsf{M}_{3\times 3}(R)$$

for diagonalizability.

The characteristic polynomial of A is $\det(A-tI) = -(t-4)(t-3)^2$, which splits, and so condition 1 of the test for diagonalization is satisfied. Also A has eigenvalues $\lambda_1 = 4$ and $\lambda_2 = 3$ with multiplicities 1 and 2, respectively. Since λ_1 has multiplicity 1, condition 2 is satisfied for λ_1. Thus we need only test condition 2 for λ_2. Because

$$A - \lambda_2 I = \begin{pmatrix} 0 & 1 & 0 \\ 0 & 0 & 0 \\ 0 & 0 & 1 \end{pmatrix}$$

has rank 2, we see that $3 - \text{rank}(A - \lambda_2 I) = 1$, which is not the multiplicity of λ_2. Thus condition 2 fails for λ_2, and A is therefore not diagonalizable. ◆

Example 6

Let T be the linear operator on $\mathsf{P}_2(R)$ defined by

$$\mathsf{T}(f(x)) = f(1) + f'(0)x + (f'(0) + f''(0))x^2.$$

We first test T for diagonalizability. Let α denote the standard ordered basis for $\mathsf{P}_2(R)$ and $B = [\mathsf{T}]_\alpha$. Then

$$B = \begin{pmatrix} 1 & 1 & 1 \\ 0 & 1 & 0 \\ 0 & 1 & 2 \end{pmatrix}.$$

The characteristic polynomial of B, and hence of T, is $-(t-1)^2(t-2)$, which splits. Hence condition 1 of the test for diagonalization is satisfied. Also B has the eigenvalues $\lambda_1 = 1$ and $\lambda_2 = 2$ with multiplicities 2 and 1, respectively. Condition 2 is satisfied for λ_2 because it has multiplicity 1. So we need only verify condition 2 for $\lambda_1 = 1$. For this case,

$$3 - \text{rank}(B - \lambda_1 I) = 3 - \text{rank} \begin{pmatrix} 0 & 1 & 1 \\ 0 & 0 & 0 \\ 0 & 1 & 1 \end{pmatrix} = 3 - 1 = 2,$$

which is equal to the multiplicity of λ_1. Therefore T is diagonalizable.

We now find an ordered basis γ for R^3 of eigenvectors of B. We consider each eigenvalue separately.

The eigenspace corresponding to $\lambda_1 = 1$ is

$$\mathsf{E}_{\lambda_1} = \left\{ \begin{pmatrix} x_1 \\ x_2 \\ x_3 \end{pmatrix} \in R^3 : \begin{pmatrix} 0 & 1 & 1 \\ 0 & 0 & 0 \\ 0 & 1 & 1 \end{pmatrix} \begin{pmatrix} x_1 \\ x_2 \\ x_3 \end{pmatrix} = 0 \right\},$$

which is the solution space for the system

$$x_2 + x_3 = 0,$$

and has

$$\gamma_1 = \left\{ \begin{pmatrix} 1 \\ 0 \\ 0 \end{pmatrix}, \begin{pmatrix} 0 \\ -1 \\ 1 \end{pmatrix} \right\}$$

as a basis.

The eigenspace corresponding to $\lambda_2 = 2$ is

$$E_{\lambda_2} = \left\{ \begin{pmatrix} x_1 \\ x_2 \\ x_3 \end{pmatrix} \in \mathsf{R}^3 : \begin{pmatrix} -1 & 1 & 1 \\ 0 & -1 & 0 \\ 0 & 1 & 0 \end{pmatrix} \begin{pmatrix} x_1 \\ x_2 \\ x_3 \end{pmatrix} = 0 \right\},$$

which is the solution space for the system

$$\begin{aligned} -x_1 + x_2 + x_3 &= 0 \\ x_2 \qquad\quad &= 0, \end{aligned}$$

and has

$$\gamma_2 = \left\{ \begin{pmatrix} 1 \\ 0 \\ 1 \end{pmatrix} \right\}$$

as a basis.

Let

$$\gamma = \gamma_1 \cup \gamma_2 = \left\{ \begin{pmatrix} 1 \\ 0 \\ 0 \end{pmatrix}, \begin{pmatrix} 0 \\ -1 \\ 1 \end{pmatrix}, \begin{pmatrix} 1 \\ 0 \\ 1 \end{pmatrix} \right\}.$$

Then γ is an ordered basis for R^3 consisting of eigenvectors of B.

Finally, observe that the vectors in γ are the coordinate vectors relative to α of the vectors in the set

$$\beta = \{ 1, -x + x^2, 1 + x^2 \},$$

which is an ordered basis for $\mathsf{P}_2(R)$ consisting of eigenvectors of T. Thus

$$[\mathsf{T}]_\beta = \begin{pmatrix} 1 & 0 & 0 \\ 0 & 1 & 0 \\ 0 & 0 & 2 \end{pmatrix}. \quad \blacklozenge$$

Our next example is an application of diagonalization that is of interest in Section 5.3.

Example 7

Let

$$A = \begin{pmatrix} 0 & -2 \\ 1 & 3 \end{pmatrix}.$$

We show that A is diagonalizable and find a 2×2 matrix Q such that $Q^{-1}AQ$ is a diagonal matix. We then show how to use this result to compute A^n for any positive integer n.

First observe that the characteristic polynomial of A is $(t-1)(t-2)$, and hence A has two distinct eigenvalues, $\lambda_1 = 1$ and $\lambda_2 = 2$. By applying the corollary to Theorem 5.5 to the operator L_A, we see that A is diagonalizable. Moreover,

$$\gamma_1 = \left\{ \begin{pmatrix} -2 \\ 1 \end{pmatrix} \right\} \quad \text{and} \quad \gamma_2 = \left\{ \begin{pmatrix} -1 \\ 1 \end{pmatrix} \right\}$$

are bases for the eigenspaces E_{λ_1} and E_{λ_2}, respectively. Therefore

$$\gamma = \gamma_1 \cup \gamma_2 = \left\{ \begin{pmatrix} -2 \\ 1 \end{pmatrix}, \begin{pmatrix} -1 \\ 1 \end{pmatrix} \right\}$$

is an ordered basis for R^2 consisting of eigenvectors of A. Let

$$Q = \begin{pmatrix} -2 & -1 \\ 1 & 1 \end{pmatrix},$$

the matrix whose columns are the vectors in γ. Then, by the corollary to Theorem 2.23 (p. 115),

$$D = Q^{-1}AQ = [\mathsf{L}_A]_\beta = \begin{pmatrix} 1 & 0 \\ 0 & 2 \end{pmatrix}.$$

To find A^n for any positive integer n, observe that $A = QDQ^{-1}$. Therefore

$$\begin{aligned}
A^n &= (QDQ^{-1})^n \\
&= (QDQ^{-1})(QDQ^{-1})\cdots(QDQ^{-1}) \\
&= QD^nQ^{-1} \\
&= Q \begin{pmatrix} 1^n & 0 \\ 0 & 2^n \end{pmatrix} Q^{-1} \\
&= \begin{pmatrix} -2 & -1 \\ 1 & 1 \end{pmatrix} \begin{pmatrix} 1 & 0 \\ 0 & 2^n \end{pmatrix} \begin{pmatrix} -1 & -1 \\ 1 & 2 \end{pmatrix} = \begin{pmatrix} 2-2^n & 2-2^{n+1} \\ -1+2^n & -1+2^{n+1} \end{pmatrix}. \quad \blacklozenge
\end{aligned}$$

We now consider an application that uses diagonalization to solve a system of differential equations.

Systems of Differential Equations

Consider the system of differential equations

$$
\begin{aligned}
x_1' &= 3x_1 + x_2 + x_3 \\
x_2' &= 2x_1 + 4x_2 + 2x_3 \\
x_3' &= -x_1 - x_2 + x_3,
\end{aligned}
$$

where, for each i, $x_i = x_i(t)$ is a differentiable real-valued function of the real variable t. Clearly, this system has a solution, namely, the solution in which each $x_i(t)$ is the zero function. We determine all of the solutions to this system.

Let $x\colon R \to R^3$ be the function defined by

$$
x(t) = \begin{pmatrix} x_1(t) \\ x_2(t) \\ x_3(t) \end{pmatrix}.
$$

The **derivative** of x, denoted x', is defined by

$$
x'(t) = \begin{pmatrix} x_1'(t) \\ x_2'(t) \\ x_3'(t) \end{pmatrix}.
$$

Let

$$
A = \begin{pmatrix} 3 & 1 & 1 \\ 2 & 4 & 2 \\ -1 & -1 & 1 \end{pmatrix}
$$

be the coefficient matrix of the given system, so that we can rewrite the system as the matrix equation $x' = Ax$.

It can be verified that for

$$
Q = \begin{pmatrix} -1 & 0 & -1 \\ 0 & -1 & -2 \\ 1 & 1 & 1 \end{pmatrix} \quad \text{and} \quad D = \begin{pmatrix} 2 & 0 & 0 \\ 0 & 2 & 0 \\ 0 & 0 & 4 \end{pmatrix},
$$

we have $Q^{-1}AQ = D$. Substitute $A = QDQ^{-1}$ into $x' = Ax$ to obtain $x' = QDQ^{-1}x$ or, equivalently, $Q^{-1}x' = DQ^{-1}x$. The function $y\colon R \to R^3$ defined by $y(t) = Q^{-1}x(t)$ can be shown to be differentiable, and $y' = Q^{-1}x'$ (see Exercise 16). Hence the original system can be written as $y' = Dy$.

Since D is a diagonal matrix, the system $y' = Dy$ is easy to solve. Setting

$$y(t) = \begin{pmatrix} y_1(t) \\ y_2(t) \\ y_3(t) \end{pmatrix},$$

we can rewrite $y' = Dy$ as

$$\begin{pmatrix} y_1'(t) \\ y_2'(t) \\ y_3'(t) \end{pmatrix} = \begin{pmatrix} 2 & 0 & 0 \\ 0 & 2 & 0 \\ 0 & 0 & 4 \end{pmatrix} \begin{pmatrix} y_1(t) \\ y_2(t) \\ y_3(t) \end{pmatrix} = \begin{pmatrix} 2y_1(t) \\ 2y_2(t) \\ 4y_3(t) \end{pmatrix}.$$

The three equations

$$y_1' = 2y_1$$
$$y_2' = 2y_2$$
$$y_3' = 4y_3$$

are independent of each other, and thus can be solved individually. It is easily seen (as in Example 3 of Section 5.1) that the general solution to these equations is $y_1(t) = c_1 e^{2t}$, $y_2(t) = c_2 e^{2t}$, and $y_3(t) = c_3 e^{4t}$, where c_1, c_2, and c_3 are arbitrary constants. Finally,

$$\begin{pmatrix} x_1(t) \\ x_2(t) \\ x_3(t) \end{pmatrix} = x(t) = Qy(t) = \begin{pmatrix} -1 & 0 & -1 \\ 0 & -1 & -2 \\ 1 & 1 & 1 \end{pmatrix} \begin{pmatrix} c_1 e^{2t} \\ c_2 e^{2t} \\ c_3 e^{4t} \end{pmatrix}$$

$$= \begin{pmatrix} -c_1 e^{2t} & & - & c_3 e^{4t} \\ & - c_2 e^{2t} & - & 2c_3 e^{4t} \\ c_1 e^{2t} & + c_2 e^{2t} & + & c_3 e^{4t} \end{pmatrix}$$

yields the general solution of the original system. Note that this solution can be written as

$$x(t) = e^{2t} \left[c_1 \begin{pmatrix} -1 \\ 0 \\ 1 \end{pmatrix} + c_2 \begin{pmatrix} 0 \\ -1 \\ 1 \end{pmatrix} \right] + e^{4t} \left[c_3 \begin{pmatrix} -1 \\ -2 \\ 1 \end{pmatrix} \right].$$

The expressions in brackets are arbitrary vectors in E_{λ_1} and E_{λ_2}, respectively, where $\lambda_1 = 2$ and $\lambda_2 = 4$. Thus the general solution of the original system is $x(t) = e^{2t} z_1 + e^{4t} z_2$, where $z_1 \in E_{\lambda_1}$ and $z_2 \in E_{\lambda_2}$. This result is generalized in Exercise 15.

Direct Sums*

Let T be a linear operator on a finite-dimensional vector space V. There is a way of decomposing V into simpler subspaces that offers insight into the behavior of T. This approach is especially useful in Chapter 7, where we study nondiagonalizable linear operators. In the case of diagonalizable operators, the simpler subspaces are the eigenspaces of the operator.

Definition. *Let* $W_1, W_2, \ldots, W_k$ *be subspaces of a vector space* V. *We define the* **sum** *of these subspaces to be the set*

$$\{v_1 + v_2 + \cdots + v_k \colon v_i \in W_i \text{ for } 1 \le i \le k\},$$

which we denote by $W_1 + W_2 + \cdots + W_k$ *or* $\displaystyle\sum_{i=1}^{k} W_i$.

It is a simple exercise to show that the sum of subspaces of a vector space is also a subspace.

Example 8

Let $V = R^3$, let W_1 denote the xy-plane, and let W_2 denote the yz-plane. Then $R^3 = W_1 + W_2$ because, for any vector $(a, b, c) \in R^3$, we have

$$(a, b, c) = (a, 0, 0) + (0, b, c),$$

where $(a, 0, 0) \in W_1$ and $(0, b, c) \in W_2$.   ◆

Notice that in Example 8 the representation of (a, b, c) as a sum of vectors in W_1 and W_2 is not unique. For example, $(a, b, c) = (a, b, 0) + (0, 0, c)$ is another representation. Because we are often interested in sums for which representations are unique, we introduce a condition that assures this outcome. The definition of *direct sum* that follows is a generalization of the definition given in the exercises of Section 1.3.

Definition. *Let* $W_1, W_2, \ldots, W_k$ *be subspaces of a vector space* V. *We call* V *the* **direct sum** *of the subspaces* $W_1, W_2, \ldots, W_k$ *and write* $V = W_1 \oplus W_2 \oplus \cdots \oplus W_k$, *if*

$$V = \sum_{i=1}^{k} W_i$$

and

$$W_j \cap \sum_{i \ne j} W_i = \{0\} \quad \text{for each } j \ (1 \le j \le k).$$

Example 9

Let $V = R^4$, $W_1 = \{(a, b, 0, 0) \colon a, b, \in R\}$, $W_2 = \{(0, 0, c, 0) \colon c \in R\}$, and $W_3 = \{(0, 0, 0, d) \colon d \in R\}$. For any $(a, b, c, d) \in V$,

$$(a, b, c, d) = (a, b, 0, 0) + (0, 0, c, 0) + (0, 0, 0, d) \in W_1 + W_2 + W_3.$$

Thus

$$V = \sum_{i=1}^{3} W_i.$$

To show that V is the direct sum of W_1, W_2, and W_3, we must prove that $W_1 \cap (W_2 + W_3) = W_2 \cap (W_1 + W_3) = W_3 \cap (W_1 + W_2) = \{0\}$. But these equalities are obvious, and so $V = W_1 \oplus W_2 \oplus W_3$.   ◆

Our next result contains several conditions that are equivalent to the definition of a direct sum.

Theorem 5.10. *Let* $W_1, W_2, \ldots, W_k$ *be subspaces of a finite-dimensional vector space* V. *The following conditions are equivalent.*

(a) $V = W_1 \oplus W_2 \oplus \cdots \oplus W_k$.

(b) $V = \displaystyle\sum_{i=1}^{k} W_i$ *and, for any vectors* $v_1, v_2, \ldots, v_k$ *such that* $v_i \in W_i$
$(1 \le i \le k)$, *if* $v_1 + v_2 + \cdots + v_k = 0$, *then* $v_i = 0$ *for all* i.

(c) *Each vector* $v \in V$ *can be uniquely written as* $v = v_1 + v_2 + \cdots + v_k$, *where* $v_i \in W_i$.

(d) *If* γ_i *is an ordered basis for* W_i $(1 \le i \le k)$, *then* $\gamma_1 \cup \gamma_2 \cup \cdots \cup \gamma_k$ *is an ordered basis for* V.

(e) *For each* $i = 1, 2, \ldots, k$, *there exists an ordered basis* γ_i *for* W_i *such that* $\gamma_1 \cup \gamma_2 \cup \cdots \cup \gamma_k$ *is an ordered basis for* V.

Proof. Assume (a). We prove (b). Clearly

$$V = \sum_{i=1}^{k} W_i.$$

Now suppose that $v_1, v_2, \ldots, v_k$ are vectors such that $v_i \in W_i$ for all i and $v_1 + v_2 + \cdots + v_k = 0$. Then for any j

$$-v_j = \sum_{i \ne j} v_i \in \sum_{i \ne j} W_i.$$

But $-v_j \in W_j$ and hence

$$-v_j \in W_j \cap \sum_{i \ne j} W_i = \{0\}.$$

So $v_j = 0$, proving (b).

Now assume (b). We prove (c). Let $v \in V$. By (b), there exist vectors $v_1, v_2, \ldots, v_k$ such that $v_i \in W_i$ and $v = v_1 + v_2 + \cdots + v_k$. We must show that this representation is unique. Suppose also that $v = w_1 + w_2 + \cdots + w_k$, where $w_i \in W_i$ for all i. Then

$$(v_1 - w_1) + (v_2 - w_2) + \cdots + (v_k - w_k) = 0.$$

But $v_i - w_i \in W_i$ for all i, and therefore $v_i - w_i = 0$ for all i by (b). Thus $v_i = w_i$ for all i, proving the uniqueness of the representation.

Now assume (c). We prove (d). For each i, let γ_i be an ordered basis for W_i. Since

$$V = \sum_{i=1}^{k} W_i$$

by (c), it follows that $\gamma_1 \cup \gamma_2 \cup \cdots \cup \gamma_k$ generates V. To show that this set is linearly independent, consider vectors $v_{ij} \in \gamma_i$ $(j = 1, 2, \ldots, m_i$ and $i = 1, 2, \ldots, k)$ and scalars a_{ij} such that

$$\sum_{i,j} a_{ij} v_{ij} = 0.$$

For each i, set

$$w_i = \sum_{j=1}^{m_i} a_{ij} v_{ij}.$$

Then for each i, $w_i \in \operatorname{span}(\gamma_i) = W_i$ and

$$w_1 + w_2 + \cdots + w_k = \sum_{i,j} a_{ij} v_{ij} = 0.$$

Since $0 \in W_i$ for each i and $0 + 0 + \cdots + 0 = w_1 + w_2 + \cdots + w_k$, (c) implies that $w_i = 0$ for all i. Thus

$$0 = w_i = \sum_{j=1}^{m_i} a_{ij} v_{ij}$$

for each i. But each γ_i is linearly independent, and hence $a_{ij} = 0$ for all i and j. Consequently $\gamma_1 \cup \gamma_2 \cup \cdots \cup \gamma_k$ is linearly independent and therefore is a basis for V.

Clearly (e) follows immediately from (d).

Finally, we assume (e) and prove (a). For each i, let γ_i be an ordered basis for W_i such that $\gamma_1 \cup \gamma_2 \cup \cdots \cup \gamma_k$ is an ordered basis for V. Then

$$V = \operatorname{span}(\gamma_1 \cup \gamma_2 \cup \cdots \cup \gamma_k)$$

$$= \operatorname{span}(\gamma_1) + \operatorname{span}(\gamma_2) + \cdots + \operatorname{span}(\gamma_k) = \sum_{i=1}^{k} \mathsf{W}_i$$

by repeated applications of Exercise 14 of Section 1.4. Fix j $(1 \leq j \leq k)$, and suppose that, for some nonzero vector $v \in \mathsf{V}$,

$$v \in \mathsf{W}_j \cap \sum_{i \neq j} \mathsf{W}_i.$$

Then

$$v \in \mathsf{W}_j = \operatorname{span}(\gamma_j) \quad \text{and} \quad v \in \sum_{i \neq j} \mathsf{W}_i = \operatorname{span}\left(\bigcup_{i \neq j} \gamma_i\right).$$

Hence v is a nontrivial linear combination of both γ_j and $\left(\bigcup_{i \neq j} \gamma_i\right)$, so that v can be expressed as a linear combination of $\gamma_1 \cup \gamma_2 \cup \cdots \cup \gamma_k$ in more than one way. But these representations contradict Theorem 1.8 (p. 43), and so we conclude that

$$\mathsf{W}_j \cap \sum_{i \neq j} \mathsf{W}_i = \{0\},$$

proving (a). ∎

With the aid of Theorem 5.10, we are able to characterize diagonalizability in terms of direct sums.

Theorem 5.11. *A linear operator* T *on a finite-dimensional vector space* V *is diagonalizable if and only if* V *is the direct sum of the eigenspaces of* T.

Proof. Let $\lambda_1, \lambda_2, \ldots, \lambda_k$ be the distinct eigenvalues of T.

First suppose that T is diagonalizable, and for each i choose an ordered basis γ_i for the eigenspace E_{λ_i}. By Theorem 5.9, $\gamma_1 \cup \gamma_2 \cup \cdots \cup \gamma_k$ is a basis for V, and hence V is a direct sum of the E_{λ_i}'s by Theorem 5.10.

Conversely, suppose that V is a direct sum of the eigenspaces of T. For each i, choose an ordered basis γ_i of E_{λ_i}. By Theorem 5.10, the union $\gamma_1 \cup \gamma_2 \cup \cdots \cup \gamma_k$ is a basis for V. Since this basis consists of eigenvectors of T, we conclude that T is diagonalizable. ∎

Example 10

Let T be the linear operator on R^4 defined by

$$\mathsf{T}(a, b, c, d) = (a, b, 2c, 3d).$$

It is easily seen that T is diagonalizable with eigenvalues $\lambda_1 = 1$, $\lambda_2 = 2$, and $\lambda_3 = 3$. Furthermore, the corresponding eigenspaces coincide with the subspaces W_1, W_2, and W_3 of Example 9. Thus Theorem 5.11 provides us with another proof that $\mathsf{R}^4 = \mathsf{W}_1 \oplus \mathsf{W}_2 \oplus \mathsf{W}_3$.   ◆

EXERCISES

1. Label the following statements as true or false.

   (a) Any linear operator on an n-dimensional vector space that has fewer than n distinct eigenvalues is not diagonalizable.

   (b) Two distinct eigenvectors corresponding to the same eigenvalue are always linearly dependent.

   (c) If λ is an eigenvalue of a linear operator T, then each vector in E_λ is an eigenvector of T.

   (d) If λ_1 and λ_2 are distinct eigenvalues of a linear operator T, then $\mathsf{E}_{\lambda_1} \cap \mathsf{E}_{\lambda_2} = \{0\}$.

   (e) Let $A \in \mathsf{M}_{n \times n}(F)$ and $\beta = \{v_1, v_2, \ldots, v_n\}$ be an ordered basis for F^n consisting of eigenvectors of A. If Q is the $n \times n$ matrix whose jth column is v_j $(1 \le j \le n)$, then $Q^{-1}AQ$ is a diagonal matrix.

   (f) A linear operator T on a finite-dimensional vector space is diagonalizable if and only if the multiplicity of each eigenvalue λ equals the dimension of E_λ.

   (g) Every diagonalizable linear operator on a nonzero vector space has at least one eigenvalue.

   The following two items relate to the optional subsection on direct sums.

   (h) If a vector space is the direct sum of subspaces $\mathsf{W}_1, \mathsf{W}_2, \ldots, \mathsf{W}_k$, then $\mathsf{W}_i \cap \mathsf{W}_j = \{0\}$ for $i \ne j$.

   (i) If

$$\mathsf{V} = \sum_{i=1}^{k} \mathsf{W}_i \quad \text{and} \quad \mathsf{W}_i \cap \mathsf{W}_j = \{0\} \quad \text{for } i \ne j,$$

   then $\mathsf{V} = \mathsf{W}_1 \oplus \mathsf{W}_2 \oplus \cdots \oplus \mathsf{W}_k$.

2. For each of the following matrices $A \in \mathsf{M}_{n \times n}(R)$, test A for diagonalizability, and if A is diagonalizable, find an invertible matrix Q and a diagonal matrix D such that $Q^{-1}AQ = D$.

   (a) $\begin{pmatrix} 1 & 2 \\ 0 & 1 \end{pmatrix}$
   (b) $\begin{pmatrix} 1 & 3 \\ 3 & 1 \end{pmatrix}$
   (c) $\begin{pmatrix} 1 & 4 \\ 3 & 2 \end{pmatrix}$

   (d) $\begin{pmatrix} 7 & -4 & 0 \\ 8 & -5 & 0 \\ 6 & -6 & 3 \end{pmatrix}$
   (e) $\begin{pmatrix} 0 & 0 & 1 \\ 1 & 0 & -1 \\ 0 & 1 & 1 \end{pmatrix}$
   (f) $\begin{pmatrix} 1 & 1 & 0 \\ 0 & 1 & 2 \\ 0 & 0 & 3 \end{pmatrix}$

(g) $\begin{pmatrix} 3 & 1 & 1 \\ 2 & 4 & 2 \\ -1 & -1 & 1 \end{pmatrix}$

3. For each of the following linear operators T on a vector space V, test T for diagonalizability, and if T is diagonalizable, find a basis β for V such that $[\mathsf{T}]_\beta$ is a diagonal matrix.

   (a) $\mathsf{V} = \mathsf{P}_3(R)$ and T is defined by $\mathsf{T}(f(x)) = f'(x) + f''(x)$, respectively.

   (b) $\mathsf{V} = \mathsf{P}_2(R)$ and T is defined by $\mathsf{T}(ax^2 + bx + c) = cx^2 + bx + a$.

   (c) $\mathsf{V} = R^3$ and T is defined by

   $$\mathsf{T}\begin{pmatrix} a_1 \\ a_2 \\ a_3 \end{pmatrix} = \begin{pmatrix} a_2 \\ -a_1 \\ 2a_3 \end{pmatrix}.$$

   (d) $\mathsf{V} = \mathsf{P}_2(R)$ and T is defined by $\mathsf{T}(f(x)) = f(0) + f(1)(x + x^2)$.

   (e) $\mathsf{V} = C^2$ and T is defined by $\mathsf{T}(z, w) = (z + iw, iz + w)$.

   (f) $\mathsf{V} = \mathsf{M}_{2\times2}(R)$ and T is defined by $\mathsf{T}(A) = A^t$.

4. Prove the matrix version of the corollary to Theorem 5.5: If $A \in \mathsf{M}_{n\times n}(F)$ has n distinct eigenvalues, then A is diagonalizable.

5. State and prove the matrix version of Theorem 5.6.

6. (a) Justify the test for diagonalizability and the method for diagonalization stated in this section.

   (b) Formulate the results in (a) for matrices.

7. For

   $$A = \begin{pmatrix} 1 & 4 \\ 2 & 3 \end{pmatrix} \in \mathsf{M}_{2\times2}(R),$$

   find an expression for A^n, where n is an arbitrary positive integer.

8. Suppose that $A \in \mathsf{M}_{n\times n}(F)$ has two distinct eigenvalues, λ_1 and λ_2, and that $\dim(\mathsf{E}_{\lambda_1}) = n - 1$. Prove that A is diagonalizable.

9. Let T be a linear operator on a finite-dimensional vector space V, and suppose there exists an ordered basis β for V such that $[\mathsf{T}]_\beta$ is an upper triangular matrix.

   (a) Prove that the characteristic polynomial for T splits.

   (b) State and prove an analogous result for matrices.

   The converse of (a) is treated in Exercise 32 of Section 5.4.

10. Let T be a linear operator on a finite-dimensional vector space V with the distinct eigenvalues $\lambda_1, \lambda_2, \ldots, \lambda_k$ and corresponding multiplicities $m_1, m_2, \ldots, m_k$. Suppose that β is a basis for V such that $[T]_\beta$ is an upper triangular matrix. Prove that the diagonal entries of $[T]_\beta$ are $\lambda_1, \lambda_2, \ldots, \lambda_k$ and that each λ_i occurs m_i times $(1 \le i \le k)$.

11. Let A be an $n \times n$ matrix that is similar to an upper triangular matrix and has the distinct eigenvalues $\lambda_1, \lambda_2, \ldots, \lambda_k$ with corresponding multiplicities $m_1, m_2, \ldots, m_k$. Prove the following statements.

   (a) $\operatorname{tr}(A) = \sum_{i=1}^{k} m_i \lambda_i$

   (b) $\det(A) = (\lambda_1)^{m_1}(\lambda_2)^{m_2} \cdots (\lambda_k)^{m_k}$.

12. Let T be an invertible linear operator on a finite-dimensional vector space V.

   (a) Recall that for any eigenvalue λ of T, λ^{-1} is an eigenvalue of T^{-1} (Exercise 8 of Section 5.1). Prove that the eigenspace of T corresponding to λ is the same as the eigenspace of T^{-1} corresponding to λ^{-1}.

   (b) Prove that if T is diagonalizable, then T^{-1} is diagonalizable.

13. Let $A \in M_{n \times n}(F)$. Recall from Exercise 14 of Section 5.1 that A and A^t have the same characteristic polynomial and hence share the same eigenvalues with the same multiplicities. For any eigenvalue λ of A and A^t, let E_λ and E_λ' denote the corresponding eigenspaces for A and A^t, respectively.

   (a) Show by way of example that for a given common eigenvalue, these two eigenspaces need not be the same.

   (b) Prove that for any eigenvalue λ, $\dim(E_\lambda) = \dim(E_\lambda')$.

   (c) Prove that if A is diagonalizable, then A^t is also diagonalizable.

14. Find the general solution to each system of differential equations.

   (a) $\begin{aligned} x' &= x + y \\ y' &= 3x - y \end{aligned}$ (b) $\begin{aligned} x_1' &= 8x_1 + 10x_2 \\ x_2' &= -5x_1 - 7x_2 \end{aligned}$

   (c) $\begin{aligned} x_1' &= x_1 \quad\;\; + x_3 \\ x_2' &= \quad x_2 + x_3 \\ x_3' &= \qquad\quad 2x_3 \end{aligned}$

15. Let

$$A = \begin{pmatrix} a_{11} & a_{12} & \cdots & a_{1n} \\ a_{21} & a_{22} & \cdots & a_{2n} \\ \vdots & \vdots & & \vdots \\ a_{n1} & a_{n2} & \cdots & a_{nn} \end{pmatrix}$$

be the coefficient matrix of the system of differential equations

$$x_1' = a_{11}x_1 + a_{12}x_2 + \cdots + a_{1n}x_n$$
$$x_2' = a_{21}x_1 + a_{22}x_2 + \cdots + a_{2n}x_n$$
$$\vdots$$
$$x_n' = a_{n1}x_1 + a_{n2}x_2 + \cdots + a_{nn}x_n.$$

Suppose that A is diagonalizable and that the distinct eigenvalues of A are $\lambda_1, \lambda_2, \ldots, \lambda_k$. Prove that a differentiable function $x \colon R \to \mathsf{R}^n$ is a solution to the system if and only if x is of the form

$$x(t) = e^{\lambda_1 t}z_1 + e^{\lambda_2 t}z_2 + \cdots + e^{\lambda_k t}z_k,$$

where $z_i \in \mathsf{E}_{\lambda_i}$ for $i = 1, 2, \ldots, k$. Use this result to prove that the set of solutions to the system is an n-dimensional real vector space.

16. Let $C \in \mathsf{M}_{m \times n}(R)$, and let Y be an $n \times p$ matrix of differentiable functions. Prove $(CY)' = CY'$, where $(Y')_{ij} = Y_{ij}'$ for all i, j.

Exercises 17 through 19 are concerned with *simultaneous diagonalization*.

Definitions. *Two linear operators* T *and* U *on a finite-dimensional vector space* V *are called* **simultaneously diagonalizable** *if there exists an ordered basis* β *for* V *such that both* $[\mathsf{T}]_\beta$ *and* $[\mathsf{U}]_\beta$ *are diagonal matrices. Similarly,* $A, B \in \mathsf{M}_{n \times n}(F)$ *are called* **simultaneously diagonalizable** *if there exists an invertible matrix* $Q \in \mathsf{M}_{n \times n}(F)$ *such that both* $Q^{-1}AQ$ *and* $Q^{-1}BQ$ *are diagonal matrices.*

17. **(a)** Prove that if T and U are simultaneously diagonalizable linear operators on a finite-dimensional vector space V, then the matrices $[\mathsf{T}]_\beta$ and $[\mathsf{U}]_\beta$ are simultaneously diagonalizable for any ordered basis β.

    (b) Prove that if A and B are simultaneously diagonalizable matrices, then L_A and L_B are simultaneously diagonalizable linear operators.

18. **(a)** Prove that if T and U are simultaneously diagonalizable operators, then T and U commute (i.e., $\mathsf{TU} = \mathsf{UT}$).

    (b) Show that if A and B are simultaneously diagonalizable matrices, then A and B commute.

    The converses of (a) and (b) are established in Exercise 25 of Section 5.4.

19. Let T be a diagonalizable linear operator on a finite-dimensional vector space, and let m be any positive integer. Prove that T and T^m are simultaneously diagonalizable.

Exercises 20 through 23 are concerned with direct sums.

20. Let $W_1, W_2, \ldots, W_k$ be subspaces of a finite-dimensional vector space V such that

$$\sum_{i=1}^{k} W_i = V.$$

Prove that V is the direct sum of $W_1, W_2, \ldots, W_k$ if and only if

$$\dim(V) = \sum_{i=1}^{k} \dim(W_i).$$

21. Let V be a finite-dimensional vector space with a basis β, and let $\beta_1, \beta_2, \ldots, \beta_k$ be a partition of β (i.e., $\beta_1, \beta_2, \ldots, \beta_k$ are subsets of β such that $\beta = \beta_1 \cup \beta_2 \cup \cdots \cup \beta_k$ and $\beta_i \cap \beta_j = \varnothing$ if $i \neq j$). Prove that $V = \text{span}(\beta_1) \oplus \text{span}(\beta_2) \oplus \cdots \oplus \text{span}(\beta_k)$.

22. Let T be a linear operator on a finite-dimensional vector space V, and suppose that the distinct eigenvalues of T are $\lambda_1, \lambda_2, \ldots, \lambda_k$. Prove that

$$\text{span}(\{x \in V \colon x \text{ is an eigenvector of } T\}) = E_{\lambda_1} \oplus E_{\lambda_2} \oplus \cdots \oplus E_{\lambda_k}.$$

23. Let $W_1, W_2, K_1, K_2, \ldots, K_p, M_1, M_2, \ldots, M_q$ be subspaces of a vector space V such that $W_1 = K_1 \oplus K_2 \oplus \cdots \oplus K_p$ and $W_2 = M_1 \oplus M_2 \oplus \cdots \oplus M_q$. Prove that if $W_1 \cap W_2 = \{0\}$, then

$$W_1 + W_2 = W_1 \oplus W_2 = K_1 \oplus K_2 \oplus \cdots \oplus K_p \oplus M_1 \oplus M_2 \oplus \cdots \oplus M_q.$$

5.3* MATRIX LIMITS AND MARKOV CHAINS

In this section, we apply what we have learned thus far in Chapter 5 to study the *limit* of a sequence of powers $A, A^2, \ldots, A^n, \ldots$, where A is a square matrix with complex entries. Such sequences and their limits have practical applications in the natural and social sciences.

We assume familiarity with limits of sequences of real numbers. The limit of a sequence of complex numbers $\{z_m \colon m = 1, 2, \ldots\}$ can be defined in terms of the limits of the sequences of the real and imaginary parts: If $z_m = r_m + i s_m$, where r_m and s_m are real numbers, and i is the imaginary number such that $i^2 = -1$, then

$$\lim_{m \to \infty} z_m = \lim_{m \to \infty} r_m + i \lim_{m \to \infty} s_m,$$

provided that $\lim_{m \to \infty} r_m$ and $\lim_{m \to \infty} s_m$ exist.

Definition. *Let $L, A_1, A_2, \ldots$ be $n \times p$ matrices having complex entries. The sequence $A_1, A_2, \ldots$ is said to **converge** to the $n \times p$ matrix L, called the **limit** of the sequence, if*

$$\lim_{m \to \infty} (A_m)_{ij} = L_{ij}$$

for all $1 \le i \le n$ and $1 \le j \le p$. To designate that L is the limit of the sequence, we write

$$\lim_{m \to \infty} A_m = L.$$

Example 1

If

$$A_m = \begin{pmatrix} 1 - \frac{1}{m} & \left(-\frac{3}{4}\right)^m & \frac{3m^2}{m^2+1} + i \left(\frac{2m+1}{m-1}\right) \\ \left(\frac{i}{2}\right)^m & 2 & \left(1 + \frac{1}{m}\right)^m \end{pmatrix},$$

then

$$\lim_{m \to \infty} A_m = \begin{pmatrix} 1 & 0 & 3 + 2i \\ 0 & 2 & e \end{pmatrix},$$

where e is the base of the natural logarithm.   ♦

A simple, but important, property of matrix limits is contained in the next theorem. Note the analogy with the familiar property of limits of sequences of real numbers that asserts that if $\lim_{m \to \infty} a_m$ exists, then

$$\lim_{m \to \infty} ca_m = c \left(\lim_{m \to \infty} a_m \right).$$

Theorem 5.12. *Let $A_1, A_2, \ldots$ be a sequence of $n \times p$ matrices with complex entries that converges to the matrix L. Then for any $P \in \mathsf{M}_{r \times n}(C)$ and $Q \in \mathsf{M}_{p \times s}(C)$,*

$$\lim_{m \to \infty} PA_m = PL \quad \text{and} \quad \lim_{m \to \infty} A_m Q = LQ.$$

Proof. For any i $(1 \le i \le r)$ and j $(1 \le j \le p)$,

$$\lim_{m \to \infty} (PA_m)_{ij} = \lim_{m \to \infty} \sum_{k=1}^{n} P_{ik}(A_m)_{kj}$$

$$= \sum_{k=1}^{n} P_{ik} \cdot \lim_{m\to\infty} (A_m)_{kj} = \sum_{k=1}^{n} P_{ik} L_{kj} = (PL)_{ij}.$$

Hence $\lim_{m\to\infty} PA_m = PL$. The proof that $\lim_{m\to\infty} A_m Q = LQ$ is similar. ∎

Corollary. Let $A \in M_{n\times n}(C)$ be such that $\lim_{m\to\infty} A^m = L$. Then for any invertible matrix $Q \in M_{n\times n}(C)$,

$$\lim_{m\to\infty} (QAQ^{-1})^m = QLQ^{-1}.$$

Proof. Since

$$(QAQ^{-1})^m = (QAQ^{-1})(QAQ^{-1})\cdots(QAQ^{-1}) = QA^m Q^{-1},$$

we have

$$\lim_{m\to\infty} (QAQ^{-1})^m = \lim_{m\to\infty} QA^m Q^{-1} = Q\left(\lim_{m\to\infty} A^m\right) Q^{-1} = QLQ^{-1}$$

by applying Theorem 5.12 twice. ∎

In the discussion that follows, we frequently encounter the set

$$S = \{\lambda \in C : |\lambda| < 1 \text{ or } \lambda = 1\}.$$

Geometrically, this set consists of the complex number 1 and the interior of the unit disk (the disk of radius 1 centered at the origin). This set is of interest because if λ is a complex number, then $\lim_{m\to\infty} \lambda^n$ exists if and only if $\lambda \in S$. This fact, which is obviously true if λ is real, can be shown to be true for complex numbers also.

The following important result gives necessary and sufficient conditions for the existence of the type of limit under consideration.

Theorem 5.13. *Let A be a square matrix with complex entries. Then $\lim_{m\to\infty} A^m$ exists if and only if both of the following conditions hold.*

(a) *Every eigenvalue of A is contained in S.*
(b) *If 1 is an eigenvalue of A, then the dimension of the eigenspace corresponding to 1 equals the multiplicity of 1 as an eigenvalue of A.*

One proof of this theorem, which relies on the theory of Jordan canonical forms (Section 7.2), can be found in Exercise 19 of Section 7.2. A second proof, which makes use of Schur's theorem (Theorem 6.14 of Section 6.4), can be found in the article by S. H. Friedberg and A. J. Insel, "Convergence of matrix powers," *Int. J. Math. Educ. Sci. Technol.*, 1992, Vol. 23, no. 5, pp. 765-769.

The necessity of condition (a) is easily justified. For suppose that λ is an eigenvalue of A such that $\lambda \notin S$. Let v be an eigenvector of A corresponding to λ. Regarding v as an $n \times 1$ matrix, we see that

$$\lim_{m\to\infty} (A^m v) = \left(\lim_{m\to\infty} A^m\right) v = Lv$$

by Theorem 5.12, where $L = \lim\limits_{m\to\infty} A^m$. But $\lim\limits_{m\to\infty} (A^m v) = \lim\limits_{m\to\infty} (\lambda^m v)$ diverges because $\lim\limits_{m\to\infty} \lambda^m$ does not exist. Hence if $\lim\limits_{m\to\infty} A^m$ exists, then condition (a) of Theorem 5.13 must hold.

Although we are unable to prove the necessity of condition (b) here, we consider an example for which this condition fails. Observe that the characteristic polynomial for the matrix

$$B = \begin{pmatrix} 1 & 1 \\ 0 & 1 \end{pmatrix}$$

is $(t - 1)^2$, and hence B has eigenvalue $\lambda = 1$ with multiplicity 2. It can easily be verified that $\dim(\mathsf{E}_\lambda) = 1$, so that condition (b) of Theorem 5.13 is violated. A simple mathematical induction argument can be used to show that

$$B^m = \begin{pmatrix} 1 & m \\ 0 & 1 \end{pmatrix},$$

and therefore that $\lim\limits_{m\to\infty} B^m$ does not exist. We see in Chapter 7 that if A is a matrix for which condition (b) fails, then A is similar to a matrix whose upper left 2×2 submatrix is precisely this matrix B.

In most of the applications involving matrix limits, the matrix is diagonalizable, and so condition (b) of Theorem 5.13 is automatically satisfied. In this case, Theorem 5.13 reduces to the following theorem, which can be proved using our previous results.

Theorem 5.14. Let $A \in \mathsf{M}_{n\times n}(C)$ satisfy the following two conditions.

(i) Every eigenvalue of A is contained in S.

(ii) A is diagonalizable.

Then $\lim\limits_{m\to\infty} A^m$ exists.

Proof. Since A is diagonalizable, there exists an invertible matrix Q such that $Q^{-1}AQ = D$ is a diagonal matrix. Suppose that

$$D = \begin{pmatrix} \lambda_1 & 0 & \cdots & 0 \\ 0 & \lambda_2 & \cdots & 0 \\ \vdots & \vdots & & \vdots \\ 0 & 0 & \cdots & \lambda_n \end{pmatrix}.$$

Because $\lambda_1, \lambda_2, \ldots, \lambda_n$ are the eigenvalues of A, condition (i) requires that for each i, either $\lambda_i = 1$ or $|\lambda_i| < 1$. Thus

$$\lim_{m \to \infty} \lambda_i{}^m = \begin{cases} 1 & \text{if } \lambda_i = 1 \\ 0 & \text{otherwise.} \end{cases}$$

But since

$$D^m = \begin{pmatrix} \lambda_1{}^m & 0 & \cdots & 0 \\ 0 & \lambda_2{}^m & \cdots & 0 \\ \vdots & \vdots & & \vdots \\ 0 & 0 & \cdots & \lambda_n{}^m \end{pmatrix},$$

the sequence $D, D^2, \ldots$ converges to a limit L. Hence

$$\lim_{m \to \infty} A^m = \lim_{m \to \infty} (QDQ^{-1})^m = QLQ^{-1}$$

by the corollary to Theorem 5.12.   ∎

The technique for computing $\lim\limits_{m \to \infty} A^m$ used in the proof of Theorem 5.14 can be employed in actual computations, as we now illustrate. Let

$$A = \begin{pmatrix} \frac{7}{4} & -\frac{9}{4} & -\frac{15}{4} \\ \frac{3}{4} & \frac{7}{4} & \frac{3}{4} \\ \frac{3}{4} & -\frac{9}{4} & -\frac{11}{4} \end{pmatrix}.$$

Using the methods in Sections 5.1 and 5.2, we obtain

$$Q = \begin{pmatrix} 1 & 3 & -1 \\ -3 & -2 & 1 \\ 2 & 3 & -1 \end{pmatrix} \quad \text{and} \quad D = \begin{pmatrix} 1 & 0 & 0 \\ 0 & -\frac{1}{2} & 0 \\ 0 & 0 & \frac{1}{4} \end{pmatrix}$$

such that $Q^{-1}AQ = D$. Hence

$$\lim_{m \to \infty} A^m = \lim_{m \to \infty} (QDQ^{-1})^m = \lim_{m \to \infty} QD^mQ^{-1} = Q \left(\lim_{m \to \infty} D^m \right) Q^{-1}$$

$$= \begin{pmatrix} 1 & 3 & -1 \\ -3 & -2 & 1 \\ 2 & 3 & -1 \end{pmatrix} \left[\lim_{m \to \infty} \begin{pmatrix} 1 & 0 & 0 \\ 0 & (-\frac{1}{2})^m & 0 \\ 0 & 0 & (\frac{1}{4})^m \end{pmatrix} \right] \begin{pmatrix} -1 & 0 & 1 \\ -1 & 1 & 2 \\ -5 & 3 & 7 \end{pmatrix}$$

$$= \begin{pmatrix} 1 & 3 & -1 \\ -3 & -2 & 1 \\ 2 & 3 & -1 \end{pmatrix} \begin{pmatrix} 1 & 0 & 0 \\ 0 & 0 & 0 \\ 0 & 0 & 0 \end{pmatrix} \begin{pmatrix} -1 & 0 & 1 \\ -1 & 1 & 2 \\ -5 & 3 & 7 \end{pmatrix} = \begin{pmatrix} -1 & 0 & 1 \\ 3 & 0 & -3 \\ -2 & 0 & 2 \end{pmatrix}.$$

Next, we consider an application that uses the limit of powers of a matrix. Suppose that the population of a certain metropolitan area remains constant but there is a continual movement of people between the city and the suburbs. Specifically, let the entries of the following matrix A represent the probabilities that someone living in the city or in the suburbs on January 1 will be living in each region on January 1 of the next year.

$$
\begin{array}{cc}
& \begin{array}{cc} \text{Currently} & \text{Currently} \\ \text{living in} & \text{living in} \\ \text{the city} & \text{the suburbs} \end{array}
\end{array}
$$

$$
\begin{array}{l}
\text{Living next year in the city} \\
\text{Living next year in the suburbs}
\end{array}
\left(\begin{array}{cc} 0.90 & 0.02 \\ 0.10 & 0.98 \end{array} \right) = A
$$

For instance, the probability that someone living in the city (on January 1) will be living in the suburbs next year (on January 1) is 0.10. Notice that since the entries of A are probabilities, they are nonnegative. Moreover, the assumption of a constant population in the metropolitan area requires that the sum of the entries of each column of A be 1.

Any square matrix having these two properties (nonnegative entries and columns that sum to 1) is called a **transition matrix** or a **stochastic matrix**. For an arbitrary $n \times n$ transition matrix M, the rows and columns correspond to n **states**, and the entry M_{ij} represents the probability of moving from state j to state i in one **stage**.

In our example, there are two states (residing in the city and residing in the suburbs). So, for example, A_{21} is the probability of moving from the city to the suburbs in one stage, that is, in one year. We now determine the

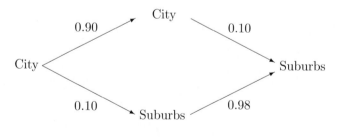

Figure 5.3

probability that a city resident will be living in the suburbs after 2 years. There are two different ways in which such a move can be made: remaining in the city for 1 year and then moving to the suburbs, or moving to the suburbs during the first year and remaining there the second year. (See

Figure 5.3.) The probability that a city dweller remains in the city for the first year is 0.90, whereas the probability that the city dweller moves to the suburbs during the first year is 0.10. Hence the probability that a city dweller stays in the city for the first year and then moves to the suburbs during the second year is the product $(0.90)(0.10)$. Likewise, the probability that a city dweller moves to the suburbs in the first year and remains in the suburbs during the second year is the product $(0.10)(0.98)$. Thus the probability that a city dweller will be living in the suburbs after 2 years is the sum of these products, $(0.90)(0.10) + (0.10)(0.98) = 0.188$. Observe that this number is obtained by the same calculation as that which produces $(A^2)_{21}$, and hence $(A^2)_{21}$ represents the probability that a city dweller will be living in the suburbs after 2 years. In general, for any transition matrix M, the entry $(M^m)_{ij}$ represents the probability of moving from state j to state i in m stages.

Suppose additionally that 70% of the 2000 population of the metropolitan area lived in the city and 30% lived in the suburbs. We record these data as a column vector:

$$\text{Proportion of city dwellers} \atop \text{Proportion of suburb residents} \qquad \begin{pmatrix} 0.70 \\ 0.30 \end{pmatrix} = P.$$

Notice that the rows of P correspond to the states of residing in the city and residing in the suburbs, respectively, and that these states are listed in the same order as the listing in the transition matrix A. Observe also that the column vector P contains nonnegative entries that sum to 1; such a vector is called a **probability vector**. In this terminology, each column of a transition matrix is a probability vector. It is often convenient to regard the entries of a transition matrix or a probability vector as proportions or percentages instead of probabilities, as we have already done with the probability vector P.

In the vector AP, the first coordinate is the sum $(0.90)(0.70)+(0.02)(0.30)$. The first term of this sum, $(0.90)(0.70)$, represents the proportion of the 2000 metropolitan population that remained in the city during the next year, and the second term, $(0.02)(0.30)$, represents the proportion of the 2000 metropolitan population that moved into the city during the next year. Hence the first coordinate of AP represents the proportion of the metropolitan population that was living in the city in 2001. Similarly, the second coordinate of

$$AP = \begin{pmatrix} 0.636 \\ 0.364 \end{pmatrix}$$

represents the proportion of the metropolitan population that was living in the suburbs in 2001. This argument can be easily extended to show that the coordinates of

$$A^2 P = A(AP) = \begin{pmatrix} 0.57968 \\ 0.42032 \end{pmatrix}$$

represent the proportions of the metropolitan population that were living in each location in 2002. In general, the coordinates of $A^m P$ represent the proportion of the metropolitan population that will be living in the city and suburbs, respectively, after m stages (m years after 2000).

Will the city eventually be depleted if this trend continues? In view of the preceding discussion, it is natural to define the eventual proportion of the city dwellers and suburbanites to be the first and second coordinates, respectively, of $\lim_{m \to \infty} A^m P$. We now compute this limit. It is easily shown that A is diagonalizable, and so there is an invertible matrix Q and a diagonal matrix D such that $Q^{-1}AQ = D$. In fact,

$$Q = \begin{pmatrix} \frac{1}{6} & -\frac{1}{6} \\ \frac{5}{6} & \frac{1}{6} \end{pmatrix} \quad \text{and} \quad D = \begin{pmatrix} 1 & 0 \\ 0 & 0.88 \end{pmatrix}.$$

Therefore

$$L = \lim_{m \to \infty} A^m = \lim_{m \to \infty} QD^mQ^{-1} = Q \begin{pmatrix} 1 & 0 \\ 0 & 0 \end{pmatrix} Q^{-1} = \begin{pmatrix} \frac{1}{6} & \frac{1}{6} \\ \frac{5}{6} & \frac{5}{6} \end{pmatrix}.$$

Consequently

$$\lim_{m \to \infty} A^m P = LP = \begin{pmatrix} \frac{1}{6} \\ \frac{5}{6} \end{pmatrix}.$$

Thus, eventually, $\frac{1}{6}$ of the population will live in the city and $\frac{5}{6}$ will live in the suburbs each year. Note that the vector LP satisfies $A(LP) = LP$. Hence LP is both a probability vector and an eigenvector of A corresponding to the eigenvalue 1. Since the eigenspace of A corresponding to the eigenvalue 1 is one-dimensional, there is only one such vector, and LP is independent of the initial choice of probability vector P. (See Exercise 15.) For example, had the 2000 metropolitan population consisted entirely of city dwellers, the limiting outcome would be the same.

In analyzing the city–suburb problem, we gave probabilistic interpretations of A^2 and AP, showing that A^2 is a transition matrix and AP is a probability vector. In fact, the product of any two transition matrices is a transition matrix, and the product of any transition matrix and probability vector is a probability vector. A proof of these facts is a simple corollary of the next theorem, which characterizes transition matrices and probability vectors.

Theorem 5.15. *Let M be an $n \times n$ matrix having real nonnegative entries, let v be a column vector in $\mathbb{R}^n$ having nonnegative coordinates, and let $u \in \mathbb{R}^n$ be the column vector in which each coordinate equals 1. Then*

(a) M is a transition matrix if and only if $M^t u = u$;

(b) v is a probability vector if and only if $u^t v = (1)$.

Proof. Exercise.   ■

Corollary.

(a) *The product of two $n \times n$ transition matrices is an $n \times n$ transition matrix. In particular, any power of a transition matrix is a transition matrix.*

(b) *The product of a transition matrix and a probability vector is a probability vector.*

Proof. Exercise.   ■

The city–suburb problem is an example of a process in which elements of a set are each classified as being in one of several fixed states that can switch over time. In general, such a process is called a **stochastic process**. The switching to a particular state is described by a probability, and in general this probability depends on such factors as the state in question, the time in question, some or all of the previous states in which the object has been (including the current state), and the states that other objects are in or have been in.

For instance, the object could be an American voter, and the state of the object could be his or her preference of political party; or the object could be a molecule of H_2O, and the states could be the three physical states in which H_2O can exist (solid, liquid, and gas). In these examples, all four of the factors mentioned above influence the probability that an object is in a particular state at a particular time.

If, however, the probability that an object in one state changes to a different state in a fixed interval of time depends only on the two states (and not on the time, earlier states, or other factors), then the stochastic process is called a **Markov process**. If, in addition, the number of possible states is finite, then the Markov process is called a **Markov chain**. We treated the city–suburb example as a two-state Markov chain. Of course, a Markov process is usually only an idealization of reality because the probabilities involved are almost never constant over time.

With this in mind, we consider another Markov chain. A certain community college would like to obtain information about the likelihood that students in various categories will graduate. The school classifies a student as a sophomore or a freshman depending on the number of credits that the student has earned. Data from the school indicate that, from one fall semester to the next, 40% of the sophomores will graduate, 30% will remain sophomores, and 30% will quit permanently. For freshmen, the data show that 10% will graduate by next fall, 50% will become sophomores, 20% will remain freshmen, and 20% will quit permanently. During the present year,

50% of the students at the school are sophomores and 50% are freshmen. Assuming that the trend indicated by the data continues indefinitely, the school
would like to know

1. the percentage of the present students who will graduate, the percentage
   who will be sophomores, the percentage who will be freshmen, and the
   percentage who will quit school permanently by next fall;
2. the same percentages as in item 1 for the fall semester two years hence;
   and
3. the probability that one of its present students will eventually graduate.

The preceding paragraph describes a four-state Markov chain with the
following states:

1. having graduated
2. being a sophomore
3. being a freshman
4. having quit permanently.

The given data provide us with the transition matrix

$$A = \begin{pmatrix} 1 & 0.4 & 0.1 & 0 \\ 0 & 0.3 & 0.5 & 0 \\ 0 & 0 & 0.2 & 0 \\ 0 & 0.3 & 0.2 & 1 \end{pmatrix}$$

of the Markov chain. (Notice that students who have graduated or have quit
permanently are assumed to remain indefinitely in those respective states.
Thus a freshman who quits the school and returns during a later semester
is not regarded as having changed states—the student is assumed to have
remained in the state of being a freshman during the time he or she was not
enrolled.) Moreover, we are told that the present distribution of students is
half in each of states 2 and 3 and none in states 1 and 4. The vector

$$P = \begin{pmatrix} 0 \\ 0.5 \\ 0.5 \\ 0 \end{pmatrix}$$

that describes the initial probability of being in each state is called the **initial
probability vector** for the Markov chain.

To answer question 1, we must determine the probabilities that a present
student will be in each state by next fall. As we have seen, these probabilities
are the coordinates of the vector

$$AP = \begin{pmatrix} 1 & 0.4 & 0.1 & 0 \\ 0 & 0.3 & 0.5 & 0 \\ 0 & 0 & 0.2 & 0 \\ 0 & 0.3 & 0.2 & 1 \end{pmatrix} \begin{pmatrix} 0 \\ 0.5 \\ 0.5 \\ 0 \end{pmatrix} = \begin{pmatrix} 0.25 \\ 0.40 \\ 0.10 \\ 0.25 \end{pmatrix}.$$

Hence by next fall, 25% of the present students will graduate, 40% will be sophomores, 10% will be freshmen, and 25% will quit the school permanently. Similarly,

$$A^2 P = A(AP) = \begin{pmatrix} 1 & 0.4 & 0.1 & 0 \\ 0 & 0.3 & 0.5 & 0 \\ 0 & 0 & 0.2 & 0 \\ 0 & 0.3 & 0.2 & 1 \end{pmatrix} \begin{pmatrix} 0.25 \\ 0.40 \\ 0.10 \\ 0.25 \end{pmatrix} = \begin{pmatrix} 0.42 \\ 0.17 \\ 0.02 \\ 0.39 \end{pmatrix}$$

provides the information needed to answer question 2: within two years 42% of the present students will graduate, 17% will be sophomores, 2% will be freshmen, and 39% will quit school.

Finally, the answer to question 3 is provided by the vector LP, where $L = \lim_{m \to \infty} A^m$. For the matrices

$$Q = \begin{pmatrix} 1 & 4 & 19 & 0 \\ 0 & -7 & -40 & 0 \\ 0 & 0 & 8 & 0 \\ 0 & 3 & 13 & 1 \end{pmatrix} \quad \text{and} \quad D = \begin{pmatrix} 1 & 0 & 0 & 0 \\ 0 & 0.3 & 0 & 0 \\ 0 & 0 & 0.2 & 0 \\ 0 & 0 & 0 & 1 \end{pmatrix},$$

we have $Q^{-1}AQ = D$. Thus

$$L = \lim_{m \to \infty} A^m = Q \left(\lim_{m \to \infty} D^m \right) Q^{-1}$$

$$= \begin{pmatrix} 1 & 4 & 19 & 0 \\ 0 & -7 & -40 & 0 \\ 0 & 0 & 8 & 0 \\ 0 & 3 & 13 & 1 \end{pmatrix} \begin{pmatrix} 1 & 0 & 0 & 0 \\ 0 & 0 & 0 & 0 \\ 0 & 0 & 0 & 0 \\ 0 & 0 & 0 & 1 \end{pmatrix} \begin{pmatrix} 1 & \frac{4}{7} & \frac{27}{56} & 0 \\ 0 & -\frac{1}{7} & -\frac{5}{7} & 0 \\ 0 & 0 & \frac{1}{8} & 0 \\ 0 & \frac{3}{7} & \frac{29}{56} & 1 \end{pmatrix}$$

$$= \begin{pmatrix} 1 & \frac{4}{7} & \frac{27}{56} & 0 \\ 0 & 0 & 0 & 0 \\ 0 & 0 & 0 & 0 \\ 0 & \frac{3}{7} & \frac{29}{56} & 1 \end{pmatrix}.$$

So

$$LP = \begin{pmatrix} 1 & \frac{4}{7} & \frac{27}{56} & 0 \\ 0 & 0 & 0 & 0 \\ 0 & 0 & 0 & 0 \\ 0 & \frac{3}{7} & \frac{29}{56} & 1 \end{pmatrix} \begin{pmatrix} 0 \\ 0.5 \\ 0.5 \\ 0 \end{pmatrix} = \begin{pmatrix} \frac{59}{112} \\ 0 \\ 0 \\ \frac{53}{112} \end{pmatrix},$$

and hence the probability that one of the present students will graduate is $\frac{59}{112}$.

In the preceding two examples, we saw that $\lim_{m \to \infty} A^m P$, where A is the transition matrix and P is the initial probability vector of the Markov chain, gives the eventual proportions in each state. In general, however, the limit of powers of a transition matrix need not exist. For example, if

$$M = \begin{pmatrix} 0 & 1 \\ 1 & 0 \end{pmatrix},$$

then $\lim_{m \to \infty} M^m$ does not exist because odd powers of M equal M and even powers of M equal I. The reason that the limit fails to exist is that condition (a) of Theorem 5.13 does not hold for M (-1 is an eigenvalue). In fact, it can be shown (see Exercise 20 of Section 7.2) that the only transition matrices A such that $\lim_{m \to \infty} A^m$ does not exist are precisely those matrices for which condition (a) of Theorem 5.13 fails to hold.

But even if the limit of powers of the transition matrix exists, the computation of the limit may be quite difficult. (The reader is encouraged to work Exercise 6 to appreciate the truth of the last sentence.) Fortunately, there is a large and important class of transition matrices for which this limit exists and is easily computed—this is the class of *regular* transition matrices.

Definition. *A transition matrix is called **regular** if some power of the matrix contains only positive entries.*

Example 2

The transition matrix

$$\begin{pmatrix} 0.90 & 0.02 \\ 0.10 & 0.98 \end{pmatrix}$$

of the Markov chain used in the city–suburb problem is clearly regular because each entry is positive. On the other hand, the transition matrix

$$A = \begin{pmatrix} 1 & 0.4 & 0.1 & 0 \\ 0 & 0.3 & 0.5 & 0 \\ 0 & 0 & 0.2 & 0 \\ 0 & 0.3 & 0.2 & 1 \end{pmatrix}$$

of the Markov chain describing community college enrollments is not regular because the first column of A^m is

$$\begin{pmatrix} 1 \\ 0 \\ 0 \\ 0 \end{pmatrix}$$

for any power m.

Observe that a regular transition matrix may contain zero entries. For example,

$$M = \begin{pmatrix} 0.9 & 0.5 & 0 \\ 0 & 0.5 & 0.4 \\ 0.1 & 0 & 0.6 \end{pmatrix}$$

is regular because every entry of M^2 is positive.   ◆

The remainder of this section is devoted to proving that, for a regular transition matrix A, the limit of the sequence of powers of A exists and has identical columns. From this fact, it is easy to compute this limit. In the course of proving this result, we obtain some interesting bounds for the magnitudes of eigenvalues of any square matrix. These bounds are given in terms of the sum of the absolute values of the rows and columns of the matrix. The necessary terminology is introduced in the definitions that follow.

Definitions.   Let $A \in \mathsf{M}_{n \times n}(C)$. For $1 \leq i, j \leq n$, define $\rho_i(A)$ to be the sum of the absolute values of the entries of row i of A, and define $\nu_j(A)$ to be equal to the sum of the absolute values of the entries of column j of A. Thus

$$\rho_i(A) = \sum_{j=1}^{n} |A_{ij}| \quad \text{for } i = 1, 2, \dots n$$

and

$$\nu_j(A) = \sum_{i=1}^{n} |A_{ij}| \quad \text{for } j = 1, 2, \dots n.$$

The **row sum** of A, denoted $\rho(A)$, and the **column sum** of A, denoted $\nu(A)$, are defined as

$$\rho(A) = \max\{\rho_i(A) : 1 \leq i \leq n\} \quad \text{and} \quad \nu(A) = \max\{\nu_j(A) : 1 \leq j \leq n\}.$$

Example 3

For the matrix

$$A = \begin{pmatrix} 1 & -i & 3 - 4i \\ -2 + i & 0 & 6 \\ 3 & 2 & i \end{pmatrix},$$

$\rho_1(A) = 7$, $\rho_2(A) = 6 + \sqrt{5}$, $\rho_3(A) = 6$, $\nu_1(A) = 4 + \sqrt{5}$, $\nu_2(A) = 3$, and $\nu_3(A) = 12$. Hence $\rho(A) = 6 + \sqrt{5}$ and $\nu(A) = 12$.   ◆

Our next results show that the smaller of $\rho(A)$ and $\nu(A)$ is an upper bound for the absolute values of eigenvalues of A. In the preceding example, for instance, A has no eigenvalue with absolute value greater than $6 + \sqrt{5}$.

To obtain a geometric view of the following theorem, we introduce some terminology. For an $n \times n$ matrix A, we define the ith **Gerschgorin disk** C_i to be the disk in the complex plane with center A_{ii} and radius $r_i = \rho_i(A) - |A_{ii}|$; that is,

$$C_i = \{z \in C \colon |z - A_{ii}| < r_i\}.$$

For example, consider the matrix

$$A = \begin{pmatrix} 1 + 2i & 1 \\ 2i & -3 \end{pmatrix}.$$

For this matrix, C_1 is the disk with center $1 + 2i$ and radius 1, and C_2 is the disk with center -3 and radius 2. (See Figure 5.4.)

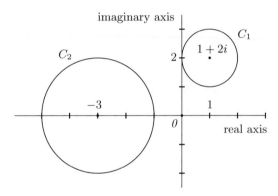

Figure 5.4

Gershgorin's disk theorem, stated below, tells us that all the eigenvalues of A are located within these two disks. In particular, we see that 0 is *not* an eigenvalue, and hence by Exercise 8(c) of section 5.1, A is invertible.

Theorem 5.16 (Gerschgorin's Disk Theorem). *Let $A \in \mathsf{M}_{n \times n}(C)$. Then every eigenvalue of A is contained in a Gerschgorin disk.*

Proof. Let λ be an eigenvalue of A with the corresponding eigenvector

$$v = \begin{pmatrix} v_1 \\ v_2 \\ \vdots \\ v_n \end{pmatrix}.$$

Then v satisfies the matrix equation $Av = \lambda v$, which can be written

$$\sum_{j=1}^{n} A_{ij} v_j = \lambda v_i \quad (i = 1, 2, \ldots, n). \tag{2}$$

Suppose that v_k is the coordinate of v having the largest absolute value; note that $v_k \neq 0$ because v is an eigenvector of A.

We show that λ lies in C_k, that is, $|\lambda - A_{kk}| \leq r_k$. For $i = k$, it follows from (2) that

$$\left| \lambda v_k - A_{kk} v_k \right| = \left| \sum_{j=1}^{n} A_{kj} v_j - A_{kk} v_k \right| = \left| \sum_{j \neq k} A_{kj} v_j \right|$$

$$\leq \sum_{j \neq k} |A_{kj}| |v_j| \leq \sum_{j \neq k} |A_{kj}| |v_k|$$

$$= |v_k| \sum_{j \neq k} |A_{kj}| = |v_k| r_k.$$

Thus

$$|v_k| |\lambda - A_{kk}| \leq |v_k| r_k;$$

so

$$|\lambda - A_{kk}| \leq r_k$$

because $|v_k| > 0$.   ∎

Corollary 1. *Let λ be any eigenvalue of $A \in \mathsf{M}_{n \times n}(C)$. Then $|\lambda| \leq \rho(A)$.*

Proof. By Gerschgorin's disk theorem, $|\lambda - A_{kk}| \leq r_k$ for some k. Hence

$$|\lambda| = |(\lambda - A_{kk}) + A_{kk}| \leq |\lambda - A_{kk}| + |A_{kk}|$$

$$\leq r_k + |A_{kk}| = \rho_k(A) \leq \rho(A).$$ ∎

Corollary 2. *Let λ be any eigenvalue of $A \in \mathsf{M}_{n \times n}(C)$. Then*

$$|\lambda| \leq \min\{\rho(A), \nu(A)\}.$$

Proof. Since $|\lambda| \leq \rho(A)$ by Corollary 1, it suffices to show that $|\lambda| \leq \nu(A)$. By Exercise 14 of Section 5.1, λ is an eigenvalue of A^t, and so $|\lambda| \leq \rho(A^t)$ by Corollary 2. But the rows of A^t are the columns of A; consequently $\rho(A^t) = \nu(A)$. Therefore $|\lambda| \leq \nu(A)$.   ∎

The next corollary is immediate from Corollary 2.

Corollary 3. *If λ is an eigenvalue of a transition matrix, then $|\lambda| \leq 1$.*

The next result asserts that the upper bound in Corollary 3 is attained.

Theorem 5.17. *Every transition matrix has 1 as an eigenvalue.*

Proof. Let A be an $n \times n$ transition matrix, and let $u \in \mathsf{R}^n$ be the column vector in which each coordinate is 1. Then $A^t u = u$ by Theorem 5.15, and hence u is an eigenvector of A^t corresponding to the eigenvalue 1. But since A and A^t have the same eigenvalues, it follows that 1 is also an eigenvalue of A. ∎

Suppose that A is a transition matrix for which some eigenvector corresponding to the eigenvalue 1 has only nonnegative coordinates. Then some multiple of this vector is a probability vector P as well as an eigenvector of A corresponding to eigenvalue 1. It is interesting to observe that if P is the initial probability vector of a Markov chain having A as its transition matrix, then the Markov chain is completely static. For in this situation, $A^m P = P$ for every positive integer m; hence the probability of being in each state never changes. Consider, for instance, the city–suburb problem with

$$P = \begin{pmatrix} \frac{1}{6} \\ \frac{5}{6} \end{pmatrix}.$$

Theorem 5.18. *Let $A \in \mathsf{M}_{n \times n}(C)$ be a matrix in which each entry is positive, and let λ be an eigenvalue of A such that $|\lambda| = \rho(A)$. Then $\lambda = \rho(A)$ and $\{u\}$ is a basis for E_λ, where $u \in C^n$ is the column vector in which each coordinate equals 1.*

Proof. Let v be an eigenvector of A corresponding to λ, with coordinates $v_1, v_2, \ldots, v_n$. Suppose that v_k is the coordinate of v having the largest absolute value, and let $b = |v_k|$. Then

$$|\lambda|b = |\lambda||v_k| = |\lambda v_k| = \left| \sum_{j=1}^{n} A_{kj} v_j \right| \leq \sum_{j=1}^{n} |A_{kj} v_j|$$

$$= \sum_{j=1}^{n} |A_{kj}||v_j| \leq \sum_{j=1}^{n} |A_{kj}|b = \rho_k(A)b \leq \rho(A)b. \tag{3}$$

Since $|\lambda| = \rho(A)$, the three inequalities in (3) are actually equalities; that is,

(a) $\left| \sum_{j=1}^{n} A_{kj} v_j \right| = \sum_{j=1}^{n} |A_{kj} v_j|,$

(b) $\displaystyle\sum_{j=1}^{n} |A_{kj}||v_j| = \sum_{j=1}^{n} |A_{kj}|b$, and

(c) $\rho_k(A) = \rho(A)$.

We see in Exercise 15(b) of Section 6.1 that (a) holds if and only if all the terms $A_{kj}v_j$ $(j = 1, 2, \ldots, n)$ are nonnegative multiples of some nonzero complex number z. Without loss of generality, we assume that $|z| = 1$. Thus there exist nonnegative real numbers $c_1, c_2, \ldots, c_n$ such that

$$A_{kj}v_j = c_j z. \qquad (4)$$

By (b) and the assumption that $A_{kj} \neq 0$ for all k and j, we have

$$|v_j| = b \quad \text{for } j = 1, 2, \ldots, n. \qquad (5)$$

Combining (4) and (5), we obtain

$$b = |v_j| = \left| \frac{c_j}{A_{kj}} z \right| = \frac{c_j}{A_{kj}} \quad \text{for } j = 1, 2, \ldots, n,$$

and therefore by (4), we have $v_j = bz$ for all j. So

$$v = \begin{pmatrix} v_1 \\ v_2 \\ \vdots \\ v_n \end{pmatrix} = \begin{pmatrix} bz \\ bz \\ \vdots \\ bz \end{pmatrix} = bzu,$$

and hence $\{u\}$ is a basis for E_λ.

Finally, observe that all of the entries of Au are positive because the same is true for the entries of both A and u. But $Au = \lambda u$, and hence $\lambda > 0$. Therefore, $\lambda = |\lambda| = \rho(A)$. ∎

Corollary 1. Let $A \in \mathsf{M}_{n \times n}(C)$ be a matrix in which each entry is positive, and let λ be an eigenvalue of A such that $|\lambda| = \nu(A)$. Then $\lambda = \nu(A)$, and the dimension of $\mathsf{E}_\lambda = 1$.

Proof. Exercise. ∎

Corollary 2. Let $A \in \mathsf{M}_{n \times n}(C)$ be a transition matrix in which each entry is positive, and let λ be any eigenvalue of A other than 1. Then $|\lambda| < 1$. Moreover, the eigenspace corresponding to the eigenvalue 1 has dimension 1.

Proof. Exercise. ∎

Our next result extends Corollary 2 to regular transition matrices and thus shows that regular transition matrices satisfy condition (a) of Theorems 5.13 and 5.14.

Theorem 5.19. *Let A be a regular transition matrix, and let λ be an eigenvalue of A. Then*
(a) $|\lambda| \leq 1$.
(b) *If $|\lambda| = 1$, then $\lambda = 1$, and $\dim(\mathsf{E}_\lambda) = 1$.*

Proof. Statement (a) was proved as Corollary 3 to Theorem 5.16.

(b) Since A is regular, there exists a positive integer s such that A^s has only positive entries. Because A is a transition matrix and the entries of A^s are positive, the entries of $A^{s+1} = A^s(A)$ are positive. Suppose that $|\lambda| = 1$. Then λ^s and λ^{s+1} are eigenvalues of A^s and A^{s+1}, respectively, having absolute value 1. So by Corollary 2 to Theorem 5.18, $\lambda^s = \lambda^{s+1} = 1$. Thus $\lambda = 1$. Let E_λ and E'_λ denote the eigenspaces of A and A^s, respectively, corresponding to $\lambda = 1$. Then $\mathsf{E}_\lambda \subseteq \mathsf{E}'_\lambda$ and, by Corollary 2 to Theorem 5.18, $\dim(\mathsf{E}'_\lambda) = 1$. Hence $\mathsf{E}_\lambda = \mathsf{E}'_\lambda$, and $\dim(\mathsf{E}_\lambda) = 1$. ∎

Corollary. *Let A be a regular transition matrix that is diagonalizable. Then $\lim_{m \to \infty} A^m$ exists.*

The preceding corollary, which follows immediately from Theorems 5.19 and 5.14, is not the best possible result. In fact, it can be shown that if A is a regular transition matrix, then the multiplicity of 1 as an eigenvalue of A is 1. Thus, by Theorem 5.7 (p. 264), condition (b) of Theorem 5.13 is satisfied. So if A is a regular transition matrix, $\lim_{m \to \infty} A^m$ exists regardless of whether A is or is not diagonalizable. As with Theorem 5.13, however, the fact that the multiplicity of 1 as an eigenvalue of A is 1 cannot be proved at this time. Nevertheless, we state this result here (leaving the proof until Exercise 20 of Section 7.2) and deduce further facts about $\lim_{m \to \infty} A^m$ when A is a regular transition matrix.

Theorem 5.20. *Let A be an $n \times n$ regular transition matrix. Then*
(a) *The multiplicity of 1 as an eigenvalue of A is 1.*
(b) $\lim_{m \to \infty} A^m$ *exists.*
(c) $L = \lim_{m \to \infty} A^m$ *is a transition matrix.*
(d) $AL = LA = L$.
(e) *The columns of L are identical. In fact, each column of L is equal to the unique probability vector v that is also an eigenvector of A corresponding to the eigenvalue 1.*
(f) *For any probability vector w, $\lim_{m \to \infty} (A^m w) = v$.*

Proof. (a) See Exercise 21 of Section 7.2.

(b) This follows from (a) and Theorems 5.19 and 5.13.

(c) By Theorem 5.15, we must show that $u^t L = u^t$. Now A^m is a transition matrix by the corollary to Theorem 5.15, so

$$u^t L = u^t \lim_{m \to \infty} A^m = \lim_{m \to \infty} u^t A^m = \lim_{m \to \infty} u^t = u^t,$$

and it follows that L is a transition matrix.

(d) By Theorem 5.12,

$$AL = A \lim_{m \to \infty} A^m = \lim_{m \to \infty} AA^m = \lim_{m \to \infty} A^{m+1} = L.$$

Similarly, $LA = L$.

(e) Since $AL = L$ by (d), each column of L is an eigenvector of A corresponding to the eigenvalue 1. Moreover, by (c), each column of L is a probability vector. Thus, by (a), each column of L is equal to the unique probability vector v corresponding to the eigenvalue 1 of A.

(f) Let w be any probability vector, and set $y = \lim_{m \to \infty} A^m w = Lw$. Then y is a probability vector by the corollary to Theorem 5.15, and also $Ay = ALw = Lw = y$ by (d). Hence y is also an eigenvector corresponding to the eigenvalue 1 of A. So $y = v$ by (e).   ∎

Definition.   *The vector v in Theorem 5.20(e) is called the **fixed probability vector** or **stationary vector** of the regular transition matrix A.*

Theorem 5.20 can be used to deduce information about the eventual distribution in each state of a Markov chain having a regular transition matrix.

Example 4

A survey in Persia showed that on a particular day 50% of the Persians preferred a loaf of bread, 30% preferred a jug of wine, and 20% preferred "thou beside me in the wilderness." A subsequent survey 1 month later yielded the following data: Of those who preferred a loaf of bread on the first survey, 40% continued to prefer a loaf of bread, 10% now preferred a jug of wine, and 50% preferred "thou"; of those who preferred a jug of wine on the first survey, 20% now preferred a loaf of bread, 70% continued to prefer a jug of wine, and 10% now preferred "thou"; of those who preferred "thou" on the first survey, 20% now preferred a loaf of bread, 20% now preferred a jug of wine, and 60% continued to prefer "thou."

Assuming that this trend continues, the situation described in the preceding paragraph is a three-state Markov chain in which the states are the three possible preferences. We can predict the percentage of Persians in each state for each month following the original survey. Letting the first, second, and third states be preferences for bread, wine, and "thou", respectively, we see that the probability vector that gives the initial probability of being in each state is

$$P = \begin{pmatrix} 0.50 \\ 0.30 \\ 0.20 \end{pmatrix},$$

and the transition matrix is

$$A = \begin{pmatrix} 0.40 & 0.20 & 0.20 \\ 0.10 & 0.70 & 0.20 \\ 0.50 & 0.10 & 0.60 \end{pmatrix}.$$

The probabilities of being in each state m months after the original survey are the coordinates of the vector $A^m P$. The reader may check that

$$AP = \begin{pmatrix} 0.30 \\ 0.30 \\ 0.40 \end{pmatrix}, \quad A^2 P = \begin{pmatrix} 0.26 \\ 0.32 \\ 0.42 \end{pmatrix}, \quad A^3 P = \begin{pmatrix} 0.252 \\ 0.334 \\ 0.414 \end{pmatrix}, \quad \text{and } A^4 P = \begin{pmatrix} 0.2504 \\ 0.3418 \\ 0.4078 \end{pmatrix}.$$

Note the apparent convergence of $A^m P$.

Since A is regular, the long-range prediction concerning the Persians' preferences can be found by computing the fixed probability vector for A. This vector is the unique probability vector v such that $(A - I)v = 0$. Letting

$$v = \begin{pmatrix} v_1 \\ v_2 \\ v_3 \end{pmatrix},$$

we see that the matrix equation $(A - I)v = 0$ yields the following system of linear equations:

$$-0.60v_1 + 0.20v_2 + 0.20v_3 = 0$$
$$0.10v_1 - 0.30v_2 + 0.20v_3 = 0$$
$$0.50v_1 + 0.10v_2 - 0.40v_3 = 0.$$

It is easily shown that

$$\begin{pmatrix} 5 \\ 7 \\ 8 \end{pmatrix}$$

is a basis for the solution space of this system. Hence the unique fixed probability vector for A is

$$\begin{pmatrix} \frac{5}{5+7+8} \\ \frac{7}{5+7+8} \\ \frac{8}{5+7+8} \end{pmatrix} = \begin{pmatrix} 0.25 \\ 0.35 \\ 0.40 \end{pmatrix}.$$

Thus, in the long run, 25% of the Persians prefer a loaf of bread, 35% prefer a jug of wine, and 40% prefer "thou beside me in the wilderness."

Note that if

$$Q = \begin{pmatrix} 5 & 0 & -3 \\ 7 & -1 & -1 \\ 8 & 1 & 4 \end{pmatrix},$$

then

$$Q^{-1}AQ = \begin{pmatrix} 1 & 0 & 0 \\ 0 & 0.5 & 0 \\ 0 & 0 & 0.2 \end{pmatrix}.$$

So

$$\lim_{m \to \infty} A^m = Q \left[\lim_{m \to \infty} \begin{pmatrix} 1 & 0 & 0 \\ 0 & 0.5 & 0 \\ 0 & 0 & 0.2 \end{pmatrix}^m \right] Q^{-1} = Q \begin{pmatrix} 1 & 0 & 0 \\ 0 & 0 & 0 \\ 0 & 0 & 0 \end{pmatrix} Q^{-1}$$

$$= \begin{pmatrix} 0.25 & 0.25 & 0.25 \\ 0.35 & 0.35 & 0.35 \\ 0.40 & 0.40 & 0.40 \end{pmatrix}. \quad \blacklozenge$$

Example 5

Farmers in Lamron plant one crop per year—either corn, soybeans, or wheat. Because they believe in the necessity of rotating their crops, these farmers do not plant the same crop in successive years. In fact, of the total acreage on which a particular crop is planted, exactly half is planted with each of the other two crops during the succeeding year. This year, 300 acres of corn, 200 acres of soybeans, and 100 acres of wheat were planted.

The situation just described is another three-state Markov chain in which the three states correspond to the planting of corn, soybeans, and wheat, respectively. In this problem, however, the amount of land devoted to each crop, rather than the percentage of the total acreage (600 acres), is given. By converting these amounts into fractions of the total acreage, we see that the transition matrix A and the initial probability vector P of the Markov chain are

$$A = \begin{pmatrix} 0 & \frac{1}{2} & \frac{1}{2} \\ \frac{1}{2} & 0 & \frac{1}{2} \\ \frac{1}{2} & \frac{1}{2} & 0 \end{pmatrix} \quad \text{and} \quad P = \begin{pmatrix} \frac{300}{600} \\ \frac{200}{600} \\ \frac{100}{600} \end{pmatrix} = \begin{pmatrix} \frac{1}{2} \\ \frac{1}{3} \\ \frac{1}{6} \end{pmatrix}.$$

The fraction of the total acreage devoted to each crop in m years is given by the coordinates of $A^m P$, and the eventual proportions of the total acreage used for each crop are the coordinates of $\lim_{m \to \infty} A^m P$. Thus the eventual amounts of land devoted to each crop are found by multiplying this limit by the total acreage; that is, the eventual amounts of land used for each crop are the coordinates of $600 \cdot \lim\limits_{m \to \infty} A^m P$.

Since A is a regular transition matrix, Theorem 5.20 shows that $\lim\limits_{m \to \infty} A^m$ is a matrix L in which each column equals the unique fixed probability vector for A. It is easily seen that the fixed probability vector for A is

$$\begin{pmatrix} \frac{1}{3} \\ \frac{1}{3} \\ \frac{1}{3} \end{pmatrix}.$$

Hence

$$L = \begin{pmatrix} \frac{1}{3} & \frac{1}{3} & \frac{1}{3} \\ \frac{1}{3} & \frac{1}{3} & \frac{1}{3} \\ \frac{1}{3} & \frac{1}{3} & \frac{1}{3} \end{pmatrix};$$

so

$$600 \cdot \lim_{m \to \infty} A^m P = 600 L P = \begin{pmatrix} 200 \\ 200 \\ 200 \end{pmatrix}.$$

Thus, in the long run, we expect 200 acres of each crop to be planted each year. (For a direct computation of $600 \cdot \lim\limits_{m \to \infty} A^m P$, see Exercise 14.)   ◆

In this section, we have concentrated primarily on the theory of regular transition matrices. There is another interesting class of transition matrices that can be represented in the form

$$\begin{pmatrix} I & B \\ O & C \end{pmatrix},$$

where I is an identity matrix and O is a zero matrix. (Such transition matrices are not regular since the lower left block remains O in any power of the matrix.) The states corresponding to the identity submatrix are called **absorbing states** because such a state is never left once it is entered. A Markov chain is called an **absorbing Markov chain** if it is possible to go from an arbitrary state into an absorbing state in a finite number of stages. Observe that the Markov chain that describes the enrollment pattern in a community college is an absorbing Markov chain with states 1 and 4 as its absorbing states. Readers interested in learning more about absorbing Markov chains are referred to *Introduction to Finite Mathematics* (third edition) by

J. Kemeny, J. Snell, and G. Thompson (Prentice-Hall, Inc., Englewood Cliffs, N. J., 1974) or *Discrete Mathematical Models* by Fred S. Roberts (Prentice-Hall, Inc., Englewood Cliffs, N. J., 1976).

An Application

In species that reproduce sexually, the characteristics of an offspring with respect to a particular genetic trait are determined by a pair of genes, one inherited from each parent. The genes for a particular trait are of two types, which are denoted by G and g. The gene G represents the dominant characteristic, and g represents the recessive characteristic. Offspring with genotypes GG or Gg exhibit the dominant characteristic, whereas offspring with genotype gg exhibit the recessive characteristic. For example, in humans, brown eyes are a dominant characteristic and blue eyes are the corresponding recessive characteristic; thus the offspring with genotypes GG or Gg are brown-eyed, whereas those of type gg are blue-eyed.

Let us consider the probability of offspring of each genotype for a male parent of genotype Gg. (We assume that the population under consideration is large, that mating is random with respect to genotype, and that the distribution of each genotype within the population is independent of sex and life expectancy.) Let

$$P = \begin{pmatrix} p \\ q \\ r \end{pmatrix}$$

denote the proportion of the adult population with genotypes GG, Gg, and gg, respectively, at the start of the experiment. This experiment describes a three-state Markov chain with the following transition matrix:

Genotype of female parent

		GG	Gg	gg	
Genotype	GG	$\frac{1}{2}$	$\frac{1}{4}$	0	
of	Gg	$\frac{1}{2}$	$\frac{1}{2}$	$\frac{1}{2}$	$= B.$
offspring	gg	0	$\frac{1}{4}$	$\frac{1}{2}$	

It is easily checked that B^2 contains only positive entries; so B is regular. Thus, by permitting only males of genotype Gg to reproduce, the proportion of offspring in the population having a certain genotype will stabilize at the fixed probability vector for B, which is

$$\begin{pmatrix} \frac{1}{4} \\ \frac{1}{2} \\ \frac{1}{4} \end{pmatrix}.$$

Now suppose that similar experiments are to be performed with males of genotypes GG and gg. As already mentioned, these experiments are three-state Markov chains with transition matrices

$$A = \begin{pmatrix} 1 & \frac{1}{2} & 0 \\ 0 & \frac{1}{2} & 1 \\ 0 & 0 & 0 \end{pmatrix} \quad \text{and} \quad C = \begin{pmatrix} 0 & 0 & 0 \\ 1 & \frac{1}{2} & 0 \\ 0 & \frac{1}{2} & 1 \end{pmatrix},$$

respectively. In order to consider the case where all male genotypes are permitted to reproduce, we must form the transition matrix $M = pA + qB + rC$, which is the linear combination of A, B, and C weighted by the proportion of males of each genotype. Thus

$$M = \begin{pmatrix} p + \frac{1}{2}q & \frac{1}{2}p + \frac{1}{4}q & 0 \\ \frac{1}{2}q + r & \frac{1}{2}p + \frac{1}{2}q + \frac{1}{2}r & p + \frac{1}{2}q \\ 0 & \frac{1}{4}q + \frac{1}{2}r & \frac{1}{2}q + r \end{pmatrix}.$$

To simplify the notation, let $a = p + \frac{1}{2}q$ and $b = \frac{1}{2}q + r$. (The numbers a and b represent the proportions of G and g genes, respectively, in the population.) Then

$$M = \begin{pmatrix} a & \frac{1}{2}a & 0 \\ b & \frac{1}{2} & a \\ 0 & \frac{1}{2}b & b \end{pmatrix},$$

where $a + b = p + q + r = 1$.

Let p', q', and r' denote the proportions of the first-generation offspring having genotypes GG, Gg, and gg, respectively. Then

$$\begin{pmatrix} p' \\ q' \\ r' \end{pmatrix} = MP = \begin{pmatrix} ap + \frac{1}{2}aq \\ bp + \frac{1}{2}q + ar \\ \frac{1}{2}bq + br \end{pmatrix} = \begin{pmatrix} a^2 \\ 2ab \\ b^2 \end{pmatrix}.$$

In order to consider the effects of unrestricted matings among the first-generation offspring, a new transition matrix $\widetilde{M}$ must be determined based upon the distribution of first-generation genotypes. As before, we find that

$$\widetilde{M} = \begin{pmatrix} p' + \frac{1}{2}q' & \frac{1}{2}p' + \frac{1}{4}q' & 0 \\ \frac{1}{2}q' + r' & \frac{1}{2}p' + \frac{1}{2}q' + \frac{1}{2}r' & p' + \frac{1}{2}q' \\ 0 & \frac{1}{4}q' + \frac{1}{2}r' & \frac{1}{2}q' + r' \end{pmatrix} = \begin{pmatrix} a' & \frac{1}{2}a' & 0 \\ b' & \frac{1}{2} & a' \\ 0 & \frac{1}{2}b' & b' \end{pmatrix},$$

where $a' = p' + \frac{1}{2}q'$ and $b' = \frac{1}{2}q' + r'$. However

$$a' = a^2 + \frac{1}{2}(2ab) = a(a+b) = a \quad \text{and} \quad b' = \frac{1}{2}(2ab) + b^2 = b(a+b) = b.$$

Thus $\widetilde{M} = M$; so the distribution of second-generation offspring among the three genotypes is

$$\widetilde{M}(MP) = M^2 P = \begin{pmatrix} a^3 + a^2 b \\ a^2 b + ab + ab^2 \\ ab^2 + b^3 \end{pmatrix} = \begin{pmatrix} a^2(a+b) \\ ab(a+1+b) \\ b^2(a+b) \end{pmatrix} = \begin{pmatrix} a^2 \\ 2ab \\ b^2 \end{pmatrix}$$

$$= MP,$$

the same as the first-generation offspring. In other words, MP is the fixed probability vector for M, and genetic equilibrium is achieved in the population after only one generation. (This result is called the *Hardy–Weinberg law*.) Notice that in the important special case that $a = b$ (or equivalently, that $p = r$), the distribution at equilibrium is

$$MP = \begin{pmatrix} a^2 \\ 2ab \\ b^2 \end{pmatrix} = \begin{pmatrix} \frac{1}{4} \\ \frac{1}{2} \\ \frac{1}{4} \end{pmatrix}.$$

EXERCISES

1. Label the following statements as true or false.

   (a) If $A \in \mathsf{M}_{n \times n}(C)$ and $\lim\limits_{m \to \infty} A^m = L$, then, for any invertible matrix $Q \in \mathsf{M}_{n \times n}(C)$, we have $\lim\limits_{m \to \infty} QA^m Q^{-1} = QLQ^{-1}$.

   (b) If 2 is an eigenvalue of $A \in \mathsf{M}_{n \times n}(C)$, then $\lim\limits_{m \to \infty} A^m$ does not exist.

   (c) Any vector

   $$\begin{pmatrix} x_1 \\ x_2 \\ \vdots \\ x_n \end{pmatrix} \in \mathsf{R}^n$$

   such that $x_1 + x_2 + \cdots + x_n = 1$ is a probability vector.

   (d) The sum of the entries of each row of a transition matrix equals 1.

   (e) The product of a transition matrix and a probability vector is a probability vector.

(f) Let z be any complex number such that $|z| < 1$. Then the matrix

$$\begin{pmatrix} 1 & z & -1 \\ z & 1 & 1 \\ -1 & 1 & z \end{pmatrix}$$

does not have 3 as an eigenvalue.

(g) Every transition matrix has 1 as an eigenvalue.

(h) No transition matrix can have -1 as an eigenvalue.

(i) If A is a transition matrix, then $\lim_{m \to \infty} A^m$ exists.

(j) If A is a regular transition matrix, then $\lim_{m \to \infty} A^m$ exists and has rank 1.

2. Determine whether $\lim_{m \to \infty} A^m$ exists for each of the following matrices A, and compute the limit if it exists.

(a) $\begin{pmatrix} 0.1 & 0.7 \\ 0.7 & 0.1 \end{pmatrix}$ **(b)** $\begin{pmatrix} -1.4 & 0.8 \\ -2.4 & 1.8 \end{pmatrix}$ **(c)** $\begin{pmatrix} 0.4 & 0.7 \\ 0.6 & 0.3 \end{pmatrix}$

(d) $\begin{pmatrix} -1.8 & 4.8 \\ -0.8 & 2.2 \end{pmatrix}$ **(e)** $\begin{pmatrix} -2 & -1 \\ 4 & 3 \end{pmatrix}$ **(f)** $\begin{pmatrix} 2.0 & -0.5 \\ 3.0 & -0.5 \end{pmatrix}$

(g) $\begin{pmatrix} -1.8 & 0 & -1.4 \\ -5.6 & 1 & -2.8 \\ 2.8 & 0 & 2.4 \end{pmatrix}$ **(h)** $\begin{pmatrix} 3.4 & -0.2 & 0.8 \\ 3.9 & 1.8 & 1.3 \\ -16.5 & -2.0 & -4.5 \end{pmatrix}$

(i) $\begin{pmatrix} -\frac{1}{2} - 2i & 4i & \frac{1}{2} + 5i \\ 1 + 2i & -3i & -1 - 4i \\ -1 - 2i & 4i & 1 + 5i \end{pmatrix}$

(j) $\begin{pmatrix} \dfrac{-26 + i}{3} & \dfrac{-28 - 4i}{3} & 28 \\[2ex] \dfrac{-7 + 2i}{3} & \dfrac{-5 + i}{3} & 7 - 2i \\[2ex] \dfrac{-13 + 6i}{6} & \dfrac{-5 + 6i}{6} & \dfrac{35 - 20i}{6} \end{pmatrix}$

3. Prove that if $A_1, A_2, \ldots$ is a sequence of $n \times p$ matrices with complex entries such that $\lim_{m \to \infty} A_m = L$, then $\lim_{m \to \infty} (A_m)^t = L^t$.

4. Prove that if $A \in M_{n \times n}(C)$ is diagonalizable and $L = \lim_{m \to \infty} A^m$ exists, then either $L = I_n$ or $\text{rank}(L) < n$.

5. Find 2×2 matrices A and B having real entries such that $\lim_{m \to \infty} A^m$, $\lim_{m \to \infty} B^m$, and $\lim_{m \to \infty} (AB)^m$ all exist, but

$$\lim_{m \to \infty} (AB)^m \neq (\lim_{m \to \infty} A^m)(\lim_{m \to \infty} B^m).$$

6. A hospital trauma unit has determined that 30% of its patients are ambulatory and 70% are bedridden at the time of arrival at the hospital. A month after arrival, 60% of the ambulatory patients have recovered, 20% remain ambulatory, and 20% have become bedridden. After the same amount of time, 10% of the bedridden patients have recovered, 20% have become ambulatory, 50% remain bedridden, and 20% have died. Determine the percentages of patients who have recovered, are ambulatory, are bedridden, and have died 1 month after arrival. Also determine the eventual percentages of patients of each type.

7. A player begins a game of chance by placing a marker in box 2, marked *Start*. (See Figure 5.5.) A die is rolled, and the marker is moved one square to the left if a 1 or a 2 is rolled and one square to the right if a 3, 4, 5, or 6 is rolled. This process continues until the marker lands in square 1, in which case the player wins the game, or in square 4, in which case the player loses the game. What is the probability of winning this game? *Hint:* Instead of diagonalizing the appropriate transition matrix

Win	Start		Lose
1	2	3	4

Figure 5.5

A, it is easier to represent e_2 as a linear combination of eigenvectors of A and then apply A^n to the result.

8. Which of the following transition matrices are regular?

(a) $\begin{pmatrix} 0.2 & 0.3 & 0.5 \\ 0.3 & 0.2 & 0.5 \\ 0.5 & 0.5 & 0 \end{pmatrix}$ (b) $\begin{pmatrix} 0.5 & 0 & 1 \\ 0.5 & 0 & 0 \\ 0 & 1 & 0 \end{pmatrix}$ (c) $\begin{pmatrix} 0.5 & 0 & 0 \\ 0.5 & 0 & 1 \\ 0 & 1 & 0 \end{pmatrix}$

(d) $\begin{pmatrix} 0.5 & 0 & 1 \\ 0.5 & 1 & 0 \\ 0 & 0 & 0 \end{pmatrix}$ (e) $\begin{pmatrix} \frac{1}{3} & 0 & 0 \\ \frac{1}{3} & 1 & 0 \\ \frac{1}{3} & 0 & 1 \end{pmatrix}$ (f) $\begin{pmatrix} 1 & 0 & 0 \\ 0 & 0.7 & 0.2 \\ 0 & 0.3 & 0.8 \end{pmatrix}$

$$\text{(g)} \quad \begin{pmatrix} 0 & \frac{1}{2} & 0 & 0 \\ \frac{1}{2} & 0 & 0 & 0 \\ \frac{1}{4} & \frac{1}{4} & 1 & 0 \\ \frac{1}{4} & \frac{1}{4} & 0 & 1 \end{pmatrix} \qquad \text{(h)} \quad \begin{pmatrix} \frac{1}{4} & \frac{1}{4} & 0 & 0 \\ \frac{1}{4} & \frac{1}{4} & 0 & 0 \\ \frac{1}{4} & \frac{1}{4} & 1 & 0 \\ \frac{1}{4} & \frac{1}{4} & 0 & 1 \end{pmatrix}$$

9. Compute $\lim_{m\to\infty} A^m$ if it exists, for each matrix A in Exercise 8.

10. Each of the matrices that follow is a regular transition matrix for a three-state Markov chain. In all cases, the initial probability vector is

$$P = \begin{pmatrix} 0.3 \\ 0.3 \\ 0.4 \end{pmatrix}.$$

For each transition matrix, compute the proportions of objects in each state after two stages and the eventual proportions of objects in each state by determining the fixed probability vector.

(a) $\begin{pmatrix} 0.6 & 0.1 & 0.1 \\ 0.1 & 0.9 & 0.2 \\ 0.3 & 0 & 0.7 \end{pmatrix}$ (b) $\begin{pmatrix} 0.8 & 0.1 & 0.2 \\ 0.1 & 0.8 & 0.2 \\ 0.1 & 0.1 & 0.6 \end{pmatrix}$ (c) $\begin{pmatrix} 0.9 & 0.1 & 0.1 \\ 0.1 & 0.6 & 0.1 \\ 0 & 0.3 & 0.8 \end{pmatrix}$

(d) $\begin{pmatrix} 0.4 & 0.2 & 0.2 \\ 0.1 & 0.7 & 0.2 \\ 0.5 & 0.1 & 0.6 \end{pmatrix}$ (e) $\begin{pmatrix} 0.5 & 0.3 & 0.2 \\ 0.2 & 0.5 & 0.3 \\ 0.3 & 0.2 & 0.5 \end{pmatrix}$ (f) $\begin{pmatrix} 0.6 & 0 & 0.4 \\ 0.2 & 0.8 & 0.2 \\ 0.2 & 0.2 & 0.4 \end{pmatrix}$

11. In 1940, a county land-use survey showed that 10% of the county land was urban, 50% was unused, and 40% was agricultural. Five years later, a follow-up survey revealed that 70% of the urban land had remained urban, 10% had become unused, and 20% had become agricultural. Likewise, 20% of the unused land had become urban, 60% had remained unused, and 20% had become agricultural. Finally, the 1945 survey showed that 20% of the agricultural land had become unused while 80% remained agricultural. Assuming that the trends indicated by the 1945 survey continue, compute the percentages of urban, unused, and agricultural land in the county in 1950 and the corresponding eventual percentages.

12. A diaper liner is placed in each diaper worn by a baby. If, after a diaper change, the liner is soiled, then it is discarded and replaced by a new liner. Otherwise, the liner is washed with the diapers and reused, except that each liner is discarded and replaced after its third use (even if it has never been soiled). The probability that the baby will soil any diaper liner is one-third. If there are only new diaper liners at first, eventually what proportions of the diaper liners being used will be new,

once used, and twice used? *Hint:* Assume that a diaper liner ready for use is in one of three states: new, once used, and twice used. After its use, it then transforms into one of the three states described.

13. In 1975, the automobile industry determined that 40% of American car owners drove large cars, 20% drove intermediate-sized cars, and 40% drove small cars. A second survey in 1985 showed that 70% of the large-car owners in 1975 still owned large cars in 1985, but 30% had changed to an intermediate-sized car. Of those who owned intermediate-sized cars in 1975, 10% had switched to large cars, 70% continued to drive intermediate-sized cars, and 20% had changed to small cars in 1985. Finally, of the small-car owners in 1975, 10% owned intermediate-sized cars and 90% owned small cars in 1985. Assuming that these trends continue, determine the percentages of Americans who own cars of each size in 1995 and the corresponding eventual percentages.

14. Show that if A and P are as in Example 5, then

$$A^m = \begin{pmatrix} r_m & r_{m+1} & r_{m+1} \\ r_{m+1} & r_m & r_{m+1} \\ r_{m+1} & r_{m+1} & r_m \end{pmatrix},$$

where

$$r_m = \frac{1}{3}\left[1 + \frac{(-1)^m}{2^{m-1}}\right].$$

Deduce that

$$600(A^m P) = A^m \begin{pmatrix} 300 \\ 200 \\ 100 \end{pmatrix} = \begin{pmatrix} 200 + \dfrac{(-1)^m}{2^m}(100) \\ 200 \\ 200 + \dfrac{(-1)^{m+1}}{2^m}(100) \end{pmatrix}.$$

15. Prove that if a 1-dimensional subspace W of R^n contains a nonzero vector with all nonnegative entries, then W contains a unique probability vector.

16. Prove Theorem 5.15 and its corollary.

17. Prove the two corollaries of Theorem 5.18.

18. Prove the corollary of Theorem 5.19.

19. Suppose that M and M' are $n \times n$ transition matrices.

(a) Prove that if M is regular, N is any $n \times n$ transition matrix, and c is a real number such that $0 < c \leq 1$, then $cM + (1 - c)N$ is a regular transition matrix.

(b) Suppose that for all i, j, we have that $M'_{ij} > 0$ whenever $M_{ij} > 0$. Prove that there exists a transition matrix N and a real number c with $0 < c \leq 1$ such that $M' = cM + (1 - c)N$.

(c) Deduce that if the nonzero entries of M and M' occur in the same positions, then M is regular if and only if M' is regular.

The following definition is used in Exercises 20–24.

Definition. For $A \in \mathsf{M}_{n \times n}(C)$, define $e^A = \lim\limits_{m \to \infty} B_m$, where

$$B_m = I + A + \frac{A^2}{2!} + \cdots + \frac{A^m}{m!}$$

(see Exercise 22). Thus e^A is the sum of the infinite series

$$I + A + \frac{A^2}{2!} + \frac{A^3}{3!} + \cdots,$$

and B_m is the mth partial sum of this series. (Note the analogy with the power series

$$e^a = 1 + a + \frac{a^2}{2!} + \frac{a^3}{3!} + \cdots,$$

which is valid for all complex numbers a.)

20. Compute e^O and e^I, where O and I denote the $n \times n$ zero and identity matrices, respectively.

21. Let $P^{-1}AP = D$ be a diagonal matrix. Prove that $e^A = Pe^D P^{-1}$.

22. Let $A \in \mathsf{M}_{n \times n}(C)$ be diagonalizable. Use the result of Exercise 21 to show that e^A exists. (Exercise 21 of Section 7.2 shows that e^A exists for every $A \in \mathsf{M}_{n \times n}(C)$.)

23. Find $A, B \in \mathsf{M}_{2 \times 2}(R)$ such that $e^A e^B \neq e^{A+B}$.

24. Prove that a differentiable function $x \colon R \to R^n$ is a solution to the system of differential equations defined in Exercise 15 of Section 5.2 if and only if $x(t) = e^{tA}v$ for some $v \in R^n$, where A is defined in that exercise.

5.4   INVARIANT SUBSPACES AND THE CAYLEY–HAMILTON THEOREM

In Section 5.1, we observed that if v is an eigenvector of a linear operator T, then T maps the span of $\{v\}$ into itself. Subspaces that are mapped into themselves are of great importance in the study of linear operators (see, e.g., Exercises 28–32 of Section 2.1).

Definition. *Let* T *be a linear operator on a vector space* V*. A subspace* W *of* V *is called a* T**-invariant subspace** *of* V *if* $\mathsf{T}(\mathsf{W}) \subseteq \mathsf{W}$*, that is, if* $\mathsf{T}(v) \in \mathsf{W}$ *for all* $v \in \mathsf{W}$*.*

Example 1

Suppose that T is a linear operator on a vector space V. Then the following subspaces of V are T-invariant:

1. $\{0\}$
2. V
3. $\mathsf{R}(\mathsf{T})$
4. $\mathsf{N}(\mathsf{T})$
5. E_λ, for any eigenvalue λ of T.

The proofs that these subspaces are T-invariant are left as exercises. (See Exercise 3.)   ◆

Example 2

Let T be the linear operator on R^3 defined by

$$\mathsf{T}(a, b, c) = (a + b, b + c, 0).$$

Then the xy-plane $= \{(x, y, 0) \colon x, y \in R\}$ and the x-axis $= \{(x, 0, 0) \colon x \in R\}$ are T-invariant subspaces of R^3.   ◆

Let T be a linear operator on a vector space V, and let x be a nonzero vector in V. The subspace

$$\mathsf{W} = \operatorname{span}(\{x, \mathsf{T}(x), \mathsf{T}^2(x), \ldots\})$$

is called the T**-cyclic subspace** of V **generated by** x. It is a simple matter to show that W is T-invariant. In fact, W is the "smallest" T-invariant subspace of V containing x. That is, any T-invariant subspace of V containing x must also contain W (see Exercise 11). Cyclic subspaces have various uses. We apply them in this section to establish the Cayley–Hamilton theorem. In Exercise 31, we outline a method for using cyclic subspaces to compute the characteristic polynomial of a linear operator without resorting to determinants. Cyclic subspaces also play an important role in Chapter 7, where we study matrix representations of nondiagonalizable linear operators.

Example 3

Let T be the linear operator on R^3 defined by

$$\mathsf{T}(a,b,c) = (-b + c, a + c, 3c).$$

We determine the T-cyclic subspace generated by $e_1 = (1,0,0)$. Since

$$\mathsf{T}(e_1) = \mathsf{T}(1,0,0) = (0,1,0) = e_2$$

and

$$\mathsf{T}^2(e_1) = \mathsf{T}(\mathsf{T}(e_1)) = \mathsf{T}(e_2) = (-1,0,0) = -e_1,$$

it follows that

$$\mathrm{span}(\{e_1, \mathsf{T}(e_1), \mathsf{T}^2(e_1), \ldots\}) = \mathrm{span}(\{e_1, e_2\}) = \{(s,t,0)\colon s,t \in R\}. \quad \blacklozenge$$

Example 4

Let T be the linear operator on $\mathsf{P}(R)$ defined by $\mathsf{T}(f(x)) = f'(x)$. Then the T-cyclic subspace generated by x^2 is $\mathrm{span}(\{x^2, 2x, 2\}) = \mathsf{P}_2(R)$. $\quad \blacklozenge$

The existence of a T-invariant subspace provides the opportunity to define a new linear operator whose domain is this subspace. If T is a linear operator on V and W is a T-invariant subspace of V, then the restriction T_W of T to W (see Appendix B) is a mapping from W to W, and it follows that T_W is a linear operator on W (see Exercise 7). As a linear operator, T_W inherits certain properties from its parent operator T. The following result illustrates one way in which the two operators are linked.

Theorem 5.21. *Let* T *be a linear operator on a finite-dimensional vector space* V, *and let* W *be a* T-*invariant subspace of* V. *Then the characteristic polynomial of* T_W *divides the characteristic polynomial of* T.

Proof. Choose an ordered basis $\gamma = \{v_1, v_2, \ldots, v_k\}$ for W, and extend it to an ordered basis $\beta = \{v_1, v_2, \ldots, v_k, v_{k+1}, \ldots, v_n\}$ for V. Let $A = [\mathsf{T}]_\beta$ and $B_1 = [\mathsf{T}_\mathsf{W}]_\gamma$. Then, by Exercise 12, A can be written in the form

$$A = \begin{pmatrix} B_1 & B_2 \\ O & B_3 \end{pmatrix}.$$

Let $f(t)$ be the characteristic polynomial of T and $g(t)$ the characteristic polynomial of T_W. Then

$$f(t) = \det(A - tI_n) = \det \begin{pmatrix} B_1 - tI_k & B_2 \\ O & B_3 - tI_{n-k} \end{pmatrix} = g(t) \cdot \det(B_3 - tI_{n-k})$$

by Exercise 21 of Section 4.3. Thus $g(t)$ divides $f(t)$. $\quad \blacksquare$

Example 5

Let T be the linear operator on R^4 defined by

$$T(a, b, c, d) = (a + b + 2c - d, b + d, 2c - d, c + d),$$

and let $W = \{(t, s, 0, 0): t, s \in R\}$. Observe that W is a T-invariant subspace of R^4 because, for any vector $(a, b, 0, 0) \in R^4$,

$$T(a, b, 0, 0) = (a + b, b, 0, 0) \in W.$$

Let $\gamma = \{e_1, e_2\}$, which is an ordered basis for W. Extend γ to the standard ordered basis β for R^4. Then

$$B_1 = [T_W]_\gamma = \begin{pmatrix} 1 & 1 \\ 0 & 1 \end{pmatrix} \quad \text{and} \quad A = [T]_\beta = \begin{pmatrix} 1 & 1 & 2 & -1 \\ 0 & 1 & 0 & 1 \\ 0 & 0 & 2 & -1 \\ 0 & 0 & 1 & 1 \end{pmatrix}$$

in the notation of Theorem 5.21. Let $f(t)$ be the characteristic polynomial of T and $g(t)$ be the characteristic polynomial of T_W. Then

$$f(t) = \det(A - tI_4) = \det \begin{pmatrix} 1-t & 1 & 2 & -1 \\ 0 & 1-t & 0 & 1 \\ 0 & 0 & 2-t & -1 \\ 0 & 0 & 1 & 1-t \end{pmatrix}$$

$$= \det \begin{pmatrix} 1-t & 1 \\ 0 & 1-t \end{pmatrix} \cdot \det \begin{pmatrix} 2-t & -1 \\ 1 & 1-t \end{pmatrix}$$

$$= g(t) \cdot \det \begin{pmatrix} 2-t & -1 \\ 1 & 1-t \end{pmatrix}. \quad \blacklozenge$$

In view of Theorem 5.21, we may use the characteristic polynomial of T_W to gain information about the characteristic polynomial of T itself. In this regard, cyclic subspaces are useful because the characteristic polynomial of the restriction of a linear operator T to a cyclic subspace is readily computable.

Theorem 5.22. Let T be a linear operator on a finite-dimensional vector space V, and let W denote the T-cyclic subspace of V generated by a nonzero vector $v \in V$. Let $k = \dim(W)$. Then

(a) $\{v, T(v), T^2(v), \ldots, T^{k-1}(v)\}$ is a basis for W.

(b) If $a_0 v + a_1 T(v) + \cdots + a_{k-1} T^{k-1}(v) + T^k(v) = 0$, then the characteristic polynomial of T_W is $f(t) = (-1)^k (a_0 + a_1 t + \cdots + a_{k-1} t^{k-1} + t^k)$.

Proof. (a) Since $v \neq 0$, the set $\{v\}$ is linearly independent. Let j be the largest positive integer for which

$$\beta = \{v, \mathsf{T}(v), \ldots, \mathsf{T}^{j-1}(v)\}$$

is linearly independent. Such a j must exist because V is finite-dimensional. Let $\mathsf{Z} = \mathrm{span}(\beta)$. Then β is a basis for Z. Furthermore, $\mathsf{T}^{j}(v) \in \mathsf{Z}$ by Theorem 1.7 (p. 39). We use this information to show that Z is a T-invariant subspace of V. Let $w \in \mathsf{Z}$. Since w is a linear combination of the vectors of β, there exist scalars $b_0, b_1, \ldots, b_{j-1}$ such that

$$w = b_0 v + b_1 \mathsf{T}(v) + \cdots + b_{j-1}\mathsf{T}^{j-1}(v),$$

and hence

$$\mathsf{T}(w) = b_0 \mathsf{T}(v) + b_1 \mathsf{T}^2(v) + \cdots + b_{j-1}\mathsf{T}^{j}(v).$$

Thus $\mathsf{T}(w)$ is a linear combination of vectors in Z, and hence belongs to Z. So Z is T-invariant. Furthermore, $v \in \mathsf{Z}$. By Exercise 11, W is the smallest T-invariant subspace of V that contains v, so that $\mathsf{W} \subseteq \mathsf{Z}$. Clearly, $\mathsf{Z} \subseteq \mathsf{W}$, and so we conclude that $\mathsf{Z} = \mathsf{W}$. It follows that β is a basis for W, and therefore $\dim(\mathsf{W}) = j$. Thus $j = k$. This proves (a).

(b) Now view β (from (a)) as an ordered basis for W. Let $a_0, a_1, \ldots, a_{k-1}$ be the scalars such that

$$a_0 v + a_1 \mathsf{T}(v) + \cdots + a_{k-1}\mathsf{T}^{k-1}(v) + \mathsf{T}^{k}(v) = 0.$$

Observe that

$$[\mathsf{T_W}]_\beta = \begin{pmatrix} 0 & 0 & \cdots & 0 & -a_0 \\ 1 & 0 & \cdots & 0 & -a_1 \\ \vdots & \vdots & & \vdots & \vdots \\ 0 & 0 & \cdots & 1 & -a_{k-1} \end{pmatrix},$$

which has the characteristic polynomial

$$f(t) = (-1)^k(a_0 + a_1 t + \cdots + a_{k-1}t^{k-1} + t^k)$$

by Exercise 19. Thus $f(t)$ is the characteristic polynomial of $\mathsf{T_W}$, proving (b). ∎

Example 6

Let T be the linear operator of Example 3, and let $\mathsf{W} = \mathrm{span}(\{e_1, e_2\})$, the T-cyclic subspace generated by e_1. We compute the characteristic polynomial $f(t)$ of $\mathsf{T_W}$ in two ways: by means of Theorem 5.22 and by means of determinants.

(a) *By means of Theorem 5.22.* From Example 3, we have that $\{e_1, e_2\}$ is a cycle that generates W, and that $T^2(e_1) = -e_1$. Hence

$$1e_1 + 0T(e_1) + T^2(e_1) = 0.$$

Therefore, by Theorem 5.22(b),

$$f(t) = (-1)^2(1 + 0t + t^2) = t^2 + 1.$$

(b) *By means of determinants.* Let $\beta = \{e_1, e_2\}$, which is an ordered basis for W. Since $T(e_1) = e_2$ and $T(e_2) = -e_1$, we have

$$[T_W]_\beta = \begin{pmatrix} 0 & -1 \\ 1 & 0 \end{pmatrix}$$

and therefore,

$$f(t) = \det \begin{pmatrix} -t & -1 \\ 1 & -t \end{pmatrix} = t^2 + 1. \quad \blacklozenge$$

The Cayley–Hamilton Theorem

As an illustration of the importance of Theorem 5.22, we prove a well-known result that is used in Chapter 7. The reader should refer to Appendix E for the definition of $f(T)$, where T is a linear operator and $f(x)$ is a polynomial.

Theorem 5.23 (Cayley–Hamilton). *Let* T *be a linear operator on a finite-dimensional vector space* V, *and let* $f(t)$ *be the characteristic polynomial of* T. *Then* $f(T) = T_0$, *the zero transformation. That is,* T *"satisfies" its characteristic equation.*

Proof. We show that $f(T)(v) = 0$ for all $v \in V$. This is obvious if $v = 0$ because $f(T)$ is linear; so suppose that $v \neq 0$. Let W be the T-cyclic subspace generated by v, and suppose that $\dim(W) = k$. By Theorem 5.22(a), there exist scalars $a_0, a_1, \ldots, a_{k-1}$ such that

$$a_0 v + a_1 T(v) + \cdots + a_{k-1} T^{k-1}(v) + T^k(v) = 0.$$

Hence Theorem 5.22(b) implies that

$$g(t) = (-1)^k(a_0 + a_1 t + \cdots + a_{k-1}t^{k-1} + t^k)$$

is the characteristic polynomial of T_W. Combining these two equations yields

$$g(T)(v) = (-1)^k(a_0 I + a_1 T + \cdots + a_{k-1}T^{k-1} + T^k)(v) = 0.$$

By Theorem 5.21, $g(t)$ divides $f(t)$; hence there exists a polynomial $q(t)$ such that $f(t) = q(t)g(t)$. So

$$f(T)(v) = q(T)g(T)(v) = q(T)(g(T)(v)) = q(T)(0) = 0. \quad \blacksquare$$

Example 7

Let T be the linear operator on R^2 defined by $\mathsf{T}(a, b) = (a + 2b, -2a + b)$, and let $\beta = \{e_1, e_2\}$. Then

$$A = \begin{pmatrix} 1 & 2 \\ -2 & 1 \end{pmatrix},$$

where $A = [\mathsf{T}]_\beta$. The characteristic polynomial of T is, therefore,

$$f(t) = \det(A - tI) = \det \begin{pmatrix} 1 - t & 2 \\ -2 & 1 - t \end{pmatrix} = t^2 - 2t + 5.$$

It is easily verified that $\mathsf{T}_0 = f(\mathsf{T}) = \mathsf{T}^2 - 2\mathsf{T} + 5\mathsf{I}$. Similarly,

$$f(A) = A^2 - 2A + 5I = \begin{pmatrix} -3 & 4 \\ -4 & -3 \end{pmatrix} + \begin{pmatrix} -2 & -4 \\ 4 & -2 \end{pmatrix} + \begin{pmatrix} 5 & 0 \\ 0 & 5 \end{pmatrix}$$

$$= \begin{pmatrix} 0 & 0 \\ 0 & 0 \end{pmatrix}. \quad \blacklozenge$$

Example 7 suggests the following result.

Corollary (Cayley–Hamilton Theorem for Matrices). *Let A be an $n \times n$ matrix, and let $f(t)$ be the characteristic polynomial of A. Then $f(A) = O$, the $n \times n$ zero matrix.*

Proof. See Exercise 15.   ∎

Invariant Subspaces and Direct Sums*[3]

It is useful to decompose a finite-dimensional vector space V into a direct sum of as many T-invariant subspaces as possible because the behavior of T on V can be inferred from its behavior on the direct summands. For example, T is diagonalizable if and only if V can be decomposed into a direct sum of one-dimensional T-invariant subspaces (see Exercise 36). In Chapter 7, we consider alternate ways of decomposing V into direct sums of T-invariant subspaces if T is not diagonalizable. We proceed to gather a few facts about direct sums of T-invariant subspaces that are used in Section 7.4. The first of these facts is about characteristic polynomials.

Theorem 5.24. *Let T be a linear operator on a finite-dimensional vector space V, and suppose that $\mathsf{V} = \mathsf{W}_1 \oplus \mathsf{W}_2 \oplus \cdots \oplus \mathsf{W}_k$, where W_i is a T-invariant subspace of V for each i $(1 \le i \le k)$. Suppose that $f_i(t)$ is the characteristic polynomial of $\mathsf{T}_{\mathsf{W}_i}$ $(1 \le i \le k)$. Then $f_1(t) \cdot f_2(t) \cdot \cdots \cdot f_k(t)$ is the characteristic polynomial of T.*

---

[3]This subsection uses optional material on direct sums from Section 5.2.

Proof. The proof is by mathematical induction on k. In what follows, $f(t)$ denotes the characteristic polynomial of T. Suppose first that $k = 2$. Let β_1 be an ordered basis for W_1, β_2 an ordered basis for W_2, and $\beta = \beta_1 \cup \beta_2$. Then β is an ordered basis for V by Theorem 5.10(d) (p. 276). Let $A = [\mathsf{T}]_\beta$, $B_1 = [\mathsf{T}_{W_1}]_{\beta_1}$, and $B_2 = [\mathsf{T}_{W_2}]_{\beta_2}$. By Exercise 34, it follows that

$$A = \begin{pmatrix} B_1 & O \\ O' & B_2 \end{pmatrix},$$

where O and O' are zero matrices of the appropriate sizes. Then

$$f(t) = \det(A - tI) = \det(B_1 - tI) \cdot \det(B_2 - tI) = f_1(t) \cdot f_2(t)$$

as in the proof of Theorem 5.21, proving the result for $k = 2$.

Now assume that the theorem is valid for $k-1$ summands, where $k-1 \geq 2$, and suppose that V is a direct sum of k subspaces, say,

$$V = W_1 \oplus W_2 \oplus \cdots \oplus W_k.$$

Let $W = W_1 + W_2 + \cdots + W_{k-1}$. It is easily verified that W is T-invariant and that $V = W \oplus W_k$. So by the case for $k = 2$, $f(t) = g(t) \cdot f_k(t)$, where $g(t)$ is the characteristic polynomial of T_W. Clearly $W = W_1 \oplus W_2 \oplus \cdots \oplus W_{k-1}$, and therefore $g(t) = f_1(t) \cdot f_2(t) \cdot \cdots \cdot f_{k-1}(t)$ by the induction hypothesis. We conclude that $f(t) = g(t) \cdot f_k(t) = f_1(t) \cdot f_2(t) \cdot \cdots \cdot f_k(t)$. ∎

As an illustration of this result, suppose that T is a diagonalizable linear operator on a finite-dimensional vector space V with distinct eigenvalues $\lambda_1, \lambda_2, \ldots, \lambda_k$. By Theorem 5.11 (p. 278), V is a direct sum of the eigenspaces of T. Since each eigenspace is T-invariant, we may view this situation in the context of Theorem 5.24. For each eigenvalue λ_i, the restriction of T to E_{λ_i} has characteristic polynomial $(\lambda_i - t)^{m_i}$, where m_i is the dimension of E_{λ_i}. By Theorem 5.24, the characteristic polynomial $f(t)$ of T is the product

$$f(t) = (\lambda_1 - t)^{m_1} (\lambda_2 - t)^{m_2} \cdots (\lambda_k - t)^{m_k}.$$

It follows that the multiplicity of each eigenvalue is equal to the dimension of the corresponding eigenspace, as expected.

Example 8

Let T be the linear operator on R^4 defined by

$$\mathsf{T}(a, b, c, d) = (2a - b, a + b, c - d, c + d),$$

and let $W_1 = \{(s, t, 0, 0) \colon s, t \in R\}$ and $W_2 = \{(0, 0, s, t) \colon s, t \in R\}$. Notice that W_1 and W_2 are each T-invariant and that $R^4 = W_1 \oplus W_2$. Let $\beta_1 = \{e_1, e_2\}$, $\beta_2 = \{e_3, e_4\}$, and $\beta = \beta_1 \cup \beta_2 = \{e_1, e_2, e_3, e_4\}$. Then β_1 is an ordered basis for W_1, β_2 is an ordered basis for W_2, and β is an ordered basis for R^4. Let $A = [T]_\beta$, $B_1 = [T_{W_1}]_{\beta_1}$, and $B_2 = [T_{W_2}]_{\beta_2}$. Then

$$B_1 = \begin{pmatrix} 2 & -1 \\ 1 & 1 \end{pmatrix}, \qquad B_2 = \begin{pmatrix} 1 & -1 \\ 1 & 1 \end{pmatrix},$$

and

$$A = \begin{pmatrix} B_1 & O \\ O & B_2 \end{pmatrix} = \begin{pmatrix} 2 & -1 & 0 & 0 \\ 1 & 1 & 0 & 0 \\ 0 & 0 & 1 & -1 \\ 0 & 0 & 1 & 1 \end{pmatrix}.$$

Let $f(t)$, $f_1(t)$, and $f_2(t)$ denote the characteristic polynomials of T, T_{W_1}, and T_{W_2}, respectively. Then

$$f(t) = \det(A - tI) = \det(B_1 - tI) \cdot \det(B_2 - tI) = f_1(t) \cdot f_2(t). \qquad \blacklozenge$$

The matrix A in Example 8 can be obtained by joining the matrices B_1 and B_2 in the manner explained in the next definition.

Definition. Let $B_1 \in \mathsf{M}_{m \times m}(F)$, and let $B_2 \in \mathsf{M}_{n \times n}(F)$. We define the **direct sum** of B_1 and B_2, denoted $B_1 \oplus B_2$, as the $(m+n) \times (m+n)$ matrix A such that

$$A_{ij} = \begin{cases} (B_1)_{ij} & \text{for } 1 \leq i, j \leq m \\ (B_2)_{(i-m),(j-m)} & \text{for } m+1 \leq i, j \leq n+m \\ 0 & \text{otherwise.} \end{cases}$$

If $B_1, B_2, \ldots, B_k$ are square matrices with entries from F, then we define the **direct sum** of $B_1, B_2, \ldots, B_k$ recursively by

$$B_1 \oplus B_2 \oplus \cdots \oplus B_k = (B_1 \oplus B_2 \oplus \cdots \oplus B_{k-1}) \oplus B_k.$$

If $A = B_1 \oplus B_2 \oplus \cdots \oplus B_k$, then we often write

$$A = \begin{pmatrix} B_1 & O & \cdots & O \\ O & B_2 & \cdots & O \\ \vdots & \vdots & & \vdots \\ O & O & \cdots & B_k \end{pmatrix}.$$

Example 9

Let

$$B_1 = \begin{pmatrix} 1 & 2 \\ 1 & 1 \end{pmatrix}, \quad B_2 = (3), \quad \text{and} \quad B_3 = \begin{pmatrix} 1 & 2 & 1 \\ 1 & 2 & 3 \\ 1 & 1 & 1 \end{pmatrix}.$$

Then

$$B_1 \oplus B_2 \oplus B_3 = \begin{pmatrix} 1 & 2 & 0 & 0 & 0 & 0 \\ 1 & 1 & 0 & 0 & 0 & 0 \\ 0 & 0 & 3 & 0 & 0 & 0 \\ 0 & 0 & 0 & 1 & 2 & 1 \\ 0 & 0 & 0 & 1 & 2 & 3 \\ 0 & 0 & 0 & 1 & 1 & 1 \end{pmatrix}. \quad \blacklozenge$$

The final result of this section relates direct sums of matrices to direct sums of invariant subspaces. It is an extension of Exercise 34 to the case $k \geq 2$.

Theorem 5.25. Let T be a linear operator on a finite-dimensional vector space V, and let $\mathsf{W}_1, \mathsf{W}_2, \ldots, \mathsf{W}_k$ be T-invariant subspaces of V such that $\mathsf{V} = \mathsf{W}_1 \oplus \mathsf{W}_2 \oplus \cdots \oplus \mathsf{W}_k$. For each i, let β_i be an ordered basis for W_i, and let $\beta = \beta_1 \cup \beta_2 \cup \cdots \cup \beta_k$. Let $A = [\mathsf{T}]_\beta$ and $B_i = [\mathsf{T}_{\mathsf{W}_i}]_{\beta_i}$ for $i = 1, 2, \ldots, k$. Then $A = B_1 \oplus B_2 \oplus \cdots \oplus B_k$.

Proof. See Exercise 35.   ∎

EXERCISES

1. Label the following statements as true or false.

   (a) There exists a linear operator T with no T-invariant subspace.

   (b) If T is a linear operator on a finite-dimensional vector space V and W is a T-invariant subspace of V, then the characteristic polynomial of T_W divides the characteristic polynomial of T.

   (c) Let T be a linear operator on a finite-dimensional vector space V, and let v and w be in V. If W is the T-cyclic subspace generated by v, W' is the T-cyclic subspace generated by w, and $\mathsf{W} = \mathsf{W}'$, then $v = w$.

   (d) If T is a linear operator on a finite-dimensional vector space V, then for any $v \in \mathsf{V}$ the T-cyclic subspace generated by v is the same as the T-cyclic subspace generated by $\mathsf{T}(v)$.

   (e) Let T be a linear operator on an n-dimensional vector space. Then there exists a polynomial $g(t)$ of degree n such that $g(\mathsf{T}) = \mathsf{T}_0$.

   (f) Any polynomial of degree n with leading coefficient $(-1)^n$ is the characteristic polynomial of some linear operator.

   (g) If T is a linear operator on a finite-dimensional vector space V, and if V is the direct sum of k T-invariant subspaces, then there is an ordered basis β for V such that $[\mathsf{T}]_\beta$ is a direct sum of k matrices.

2. For each of the following linear operators T on the vector space V, determine whether the given subspace W is a T-invariant subspace of V.

   (a) $V = P_3(R)$, $T(f(x)) = f'(x)$, and $W = P_2(R)$
   (b) $V = P(R)$, $T(f(x)) = xf(x)$, and $W = P_2(R)$
   (c) $V = R^3$, $T(a, b, c) = (a + b + c, a + b + c, a + b + c)$, and $W = \{(t, t, t) : t \in R\}$
   (d) $V = C([0, 1])$, $T(f(t)) = \left[\int_0^1 f(x)\, dx\right] t$, and $W = \{f \in V : f(t) = at + b \text{ for some } a \text{ and } b\}$
   (e) $V = M_{2 \times 2}(R)$, $T(A) = \begin{pmatrix} 0 & 1 \\ 1 & 0 \end{pmatrix} A$, and $W = \{A \in V : A^t = A\}$

3. Let T be a linear operator on a finite-dimensional vector space V. Prove that the following subspaces are T-invariant.

   (a) $\{0\}$ and V
   (b) $N(T)$ and $R(T)$
   (c) E_λ, for any eigenvalue λ of T

4. Let T be a linear operator on a vector space V, and let W be a T-invariant subspace of V. Prove that W is $g(T)$-invariant for any polynomial $g(t)$.

5. Let T be a linear operator on a vector space V. Prove that the intersection of any collection of T-invariant subspaces of V is a T-invariant subspace of V.

6. For each linear operator T on the vector space V, find an ordered basis for the T-cyclic subspace generated by the vector z.

   (a) $V = R^4$, $T(a, b, c, d) = (a + b, b - c, a + c, a + d)$, and $z = e_1$.
   (b) $V = P_3(R)$, $T(f(x)) = f''(x)$, and $z = x^3$.
   (c) $V = M_{2 \times 2}(R)$, $T(A) = A^t$, and $z = \begin{pmatrix} 0 & 1 \\ 1 & 0 \end{pmatrix}$.
   (d) $V = M_{2 \times 2}(R)$, $T(A) = \begin{pmatrix} 1 & 1 \\ 2 & 2 \end{pmatrix} A$, and $z = \begin{pmatrix} 0 & 1 \\ 1 & 0 \end{pmatrix}$.

7. Prove that the restriction of a linear operator T to a T-invariant subspace is a linear operator on that subspace.

8. Let T be a linear operator on a vector space with a T-invariant subspace W. Prove that if v is an eigenvector of T_W with corresponding eigenvalue λ, then the same is true for T.

9. For each linear operator T and cyclic subspace W in Exercise 6, compute the characteristic polynomial of T_W in two ways, as in Example 6.

10. For each linear operator in Exercise 6, find the characteristic polynomial $f(t)$ of T, and verify that the characteristic polynomial of T_W (computed in Exercise 9) divides $f(t)$.

11. Let T be a linear operator on a vector space V, let v be a nonzero vector in V, and let W be the T-cyclic subspace of V generated by v. Prove that

   (a) W is T-invariant.
   (b) Any T-invariant subspace of V containing v also contains W.

12. Prove that $A = \begin{pmatrix} B_1 & B_2 \\ O & B_3 \end{pmatrix}$ in the proof of Theorem 5.21.

13. Let T be a linear operator on a vector space V, let v be a nonzero vector in V, and let W be the T-cyclic subspace of V generated by v. For any $w \in V$, prove that $w \in W$ if and only if there exists a polynomial $g(t)$ such that $w = g(T)(v)$.

14. Prove that the polynomial $g(t)$ of Exercise 13 can always be chosen so that its degree is less than or equal to $\dim(W)$.

15. Use the Cayley–Hamilton theorem (Theorem 5.23) to prove its corollary for matrices. *Warning:* If $f(t) = \det(A - tI)$ is the characteristic polynomial of A, it is tempting to "prove" that $f(A) = O$ by saying "$f(A) = \det(A - AI) = \det(O) = 0$." But this argument is nonsense. Why?

16. Let T be a linear operator on a finite-dimensional vector space V.

   (a) Prove that if the characteristic polynomial of T splits, then so does the characteristic polynomial of the restriction of T to any T-invariant subspace of V.
   (b) Deduce that if the characteristic polynomial of T splits, then any nontrivial T-invariant subspace of V contains an eigenvector of T.

17. Let A be an $n \times n$ matrix. Prove that

$$\dim(\operatorname{span}(\{I_n, A, A^2, \ldots\})) \leq n.$$

18. Let A be an $n \times n$ matrix with characteristic polynomial

$$f(t) = (-1)^n t^n + a_{n-1} t^{n-1} + \cdots + a_1 t + a_0.$$

   (a) Prove that A is invertible if and only if $a_0 \neq 0$.
   (b) Prove that if A is invertible, then

$$A^{-1} = (-1/a_0)[(-1)^n A^{n-1} + a_{n-1} A^{n-2} + \cdots + a_1 I_n].$$

(c) Use (b) to compute A^{-1} for

$$A = \begin{pmatrix} 1 & 2 & 1 \\ 0 & 2 & 3 \\ 0 & 0 & -1 \end{pmatrix}.$$

19. Let A denote the $k \times k$ matrix

$$\begin{pmatrix} 0 & 0 & \cdots & 0 & -a_0 \\ 1 & 0 & \cdots & 0 & -a_1 \\ 0 & 1 & \cdots & 0 & -a_2 \\ \vdots & \vdots & & \vdots & \vdots \\ 0 & 0 & \cdots & 0 & -a_{k-2} \\ 0 & 0 & \cdots & 1 & -a_{k-1} \end{pmatrix},$$

where $a_0, a_1, \ldots, a_{k-1}$ are arbitrary scalars. Prove that the characteristic polynomial of A is

$$(-1)^k (a_0 + a_1 t + \cdots + a_{k-1} t^{k-1} + t^k).$$

Hint: Use mathematical induction on k, expanding the determinant along the first row.

20. Let T be a linear operator on a vector space V, and suppose that V is a T-cyclic subspace of itself. Prove that if U is a linear operator on V, then $UT = TU$ if and only if $U = g(T)$ for some polynomial $g(t)$. *Hint:* Suppose that V is generated by v. Choose $g(t)$ according to Exercise 13 so that $g(T)(v) = U(v)$.

21. Let T be a linear operator on a two-dimensional vector space V. Prove that either V is a T-cyclic subspace of itself or $T = cI$ for some scalar c.

22. Let T be a linear operator on a two-dimensional vector space V and suppose that $T \neq cI$ for any scalar c. Show that if U is any linear operator on V such that $UT = TU$, then $U = g(T)$ for some polynomial $g(t)$.

23. Let T be a linear operator on a finite-dimensional vector space V, and let W be a T-invariant subspace of V. Suppose that $v_1, v_2, \ldots, v_k$ are eigenvectors of T corresponding to distinct eigenvalues. Prove that if $v_1 + v_2 + \cdots + v_k$ is in W, then $v_i \in W$ for all i. *Hint:* Use mathematical induction on k.

24. Prove that the restriction of a diagonalizable linear operator T to any nontrivial T-invariant subspace is also diagonalizable. *Hint:* Use the result of Exercise 23.

25. (a) Prove the converse to Exercise 18(a) of Section 5.2: If T and U are diagonalizable linear operators on a finite-dimensional vector space V such that $UT = TU$, then T and U are simultaneously diagonalizable. (See the definitions in the exercises of Section 5.2.) *Hint:* For any eigenvalue λ of T, show that E_λ is U-invariant, and apply Exercise 24 to obtain a basis for E_λ of eigenvectors of U.

(b) State and prove a matrix version of (a).

26. Let T be a linear operator on an n-dimensional vector space V such that T has n distinct eigenvalues. Prove that V is a T-cyclic subspace of itself. *Hint:* Use Exercise 23 to find a vector v such that $\{v, \mathsf{T}(v), \ldots, \mathsf{T}^{n-1}(v)\}$ is linearly independent.

Exercises 27 through 32 require familiarity with quotient spaces as defined in Exercise 31 of Section 1.3. Before attempting these exercises, the reader should first review the other exercises treating quotient spaces: Exercise 35 of Section 1.6, Exercise 40 of Section 2.1, and Exercise 24 of Section 2.4.

For the purposes of Exercises 27 through 32, T is a fixed linear operator on a finite-dimensional vector space V, and W is a nonzero T-invariant subspace of V. We require the following definition.

Definition. *Let* T *be a linear operator on a vector space* V, *and let* W *be a T-invariant subspace of* V. *Define* $\overline{\mathsf{T}} \colon \mathsf{V}/\mathsf{W} \to \mathsf{V}/\mathsf{W}$ *by*

$$\overline{\mathsf{T}}(v + \mathsf{W}) = \mathsf{T}(v) + \mathsf{W} \quad \text{for any } v + \mathsf{W} \in \mathsf{V}/\mathsf{W}.$$

27. (a) Prove that $\overline{\mathsf{T}}$ is well defined. That is, show that $\overline{\mathsf{T}}(v + \mathsf{W}) = \overline{\mathsf{T}}(v' + \mathsf{W})$ whenever $v + \mathsf{W} = v' + \mathsf{W}$.

(b) Prove that $\overline{\mathsf{T}}$ is a linear operator on V/W.

(c) Let $\eta \colon \mathsf{V} \to \mathsf{V}/\mathsf{W}$ be the linear transformation defined in Exercise 40 of Section 2.1 by $\eta(v) = v + \mathsf{W}$. Show that the diagram of Figure 5.6 commutes; that is, prove that $\eta\mathsf{T} = \overline{\mathsf{T}}\eta$. (This exercise does not require the assumption that V is finite-dimensional.)

$$
\begin{array}{ccc}
\mathsf{V} & \xrightarrow{\ \mathsf{T}\ } & \mathsf{V} \\
{\scriptstyle \eta}\downarrow & & \downarrow{\scriptstyle \eta} \\
\mathsf{V}/\mathsf{W} & \xrightarrow{\ \overline{\mathsf{T}}\ } & \mathsf{V}/\mathsf{W}
\end{array}
$$

Figure 5.6

28. Let $f(t)$, $g(t)$, and $h(t)$ be the characteristic polynomials of T, $\mathsf{T_W}$, and $\overline{\mathsf{T}}$, respectively. Prove that $f(t) = g(t)h(t)$. *Hint:* Extend an ordered basis $\gamma = \{v_1, v_2, \ldots, v_k\}$ for W to an ordered basis $\beta = \{v_1, v_2, \ldots, v_k, v_{k+1}, \ldots, v_n\}$ for V. Then show that the collection of cosets $\alpha = \{v_{k+1} + W, v_{k+2} + W, \ldots, v_n + W\}$ is an ordered basis for V/W, and prove that

$$[\mathsf{T}]_\beta = \begin{pmatrix} B_1 & B_2 \\ O & B_3 \end{pmatrix},$$

where $B_1 = [\mathsf{T}]_\gamma$ and $B_3 = [\overline{\mathsf{T}}]_\alpha$.

29. Use the hint in Exercise 28 to prove that if T is diagonalizable, then so is $\overline{\mathsf{T}}$.

30. Prove that if both $\mathsf{T_W}$ and $\overline{\mathsf{T}}$ are diagonalizable and have no common eigenvalues, then T is diagonalizable.

The results of Theorem 5.22 and Exercise 28 are useful in devising methods for computing characteristic polynomials without the use of determinants. This is illustrated in the next exercise.

31. Let $A = \begin{pmatrix} 1 & 1 & -3 \\ 2 & 3 & 4 \\ 1 & 2 & 1 \end{pmatrix}$, let $\mathsf{T} = \mathsf{L}_A$, and let W be the cyclic subspace of R^3 generated by e_1.

   (a) Use Theorem 5.22 to compute the characteristic polynomial of $\mathsf{T_W}$.
   (b) Show that $\{e_2 + W\}$ is a basis for R^3/W, and use this fact to compute the characteristic polynomial of $\overline{\mathsf{T}}$.
   (c) Use the results of (a) and (b) to find the characteristic polynomial of A.

32. Prove the converse to Exercise 9(a) of Section 5.2: If the characteristic polynomial of T splits, then there is an ordered basis β for V such that $[\mathsf{T}]_\beta$ is an upper triangular matrix. *Hints:* Apply mathematical induction to $\dim(V)$. First prove that T has an eigenvector v, let $W = \mathrm{span}(\{v\})$, and apply the induction hypothesis to $\overline{\mathsf{T}} \colon V/W \to V/W$. Exercise 35(b) of Section 1.6 is helpful here.

Exercises 33 through 40 are concerned with direct sums.

33. Let T be a linear operator on a vector space V, and let $W_1, W_2, \ldots, W_k$ be T-invariant subspaces of V. Prove that $W_1 + W_2 + \cdots + W_k$ is also a T-invariant subspace of V.

34. Give a direct proof of Theorem 5.25 for the case $k = 2$. (This result is used in the proof of Theorem 5.24.)

35. Prove Theorem 5.25. *Hint:* Begin with Exercise 34 and extend it using mathematical induction on k, the number of subspaces.

36. Let T be a linear operator on a finite-dimensional vector space V. Prove that T is diagonalizable if and only if V is the direct sum of one-dimensional T-invariant subspaces.

37. Let T be a linear operator on a finite-dimensional vector space V, and let $\mathsf{W}_1, \mathsf{W}_2, \ldots, \mathsf{W}_k$ be T-invariant subspaces of V such that $\mathsf{V} = \mathsf{W}_1 \oplus \mathsf{W}_2 \oplus \cdots \oplus \mathsf{W}_k$. Prove that

$$\det(\mathsf{T}) = \det(\mathsf{T}_{\mathsf{W}_1}) \det(\mathsf{T}_{\mathsf{W}_2}) \cdots \det(\mathsf{T}_{\mathsf{W}_k}).$$

38. Let T be a linear operator on a finite-dimensional vector space V, and let $\mathsf{W}_1, \mathsf{W}_2, \ldots, \mathsf{W}_k$ be T-invariant subspaces of V such that $\mathsf{V} = \mathsf{W}_1 \oplus \mathsf{W}_2 \oplus \cdots \oplus \mathsf{W}_k$. Prove that T is diagonalizable if and only if $\mathsf{T}_{\mathsf{W}_i}$ is diagonalizable for all i.

39. Let $\mathcal{C}$ be a collection of diagonalizable linear operators on a finite-dimensional vector space V. Prove that there is an ordered basis β such that $[\mathsf{T}]_\beta$ is a diagonal matrix for all $\mathsf{T} \in \mathcal{C}$ if and only if the operators of $\mathcal{C}$ commute under composition. (This is an extension of Exercise 25.) *Hints for the case that the operators commute:* The result is trivial if each operator has only one eigenvalue. Otherwise, establish the general result by mathematical induction on $\dim(\mathsf{V})$, using the fact that V is the direct sum of the eigenspaces of some operator in $\mathcal{C}$ that has more than one eigenvalue.

40. Let $B_1, B_2, \ldots, B_k$ be square matrices with entries in the same field, and let $A = B_1 \oplus B_2 \oplus \cdots \oplus B_k$. Prove that the characteristic polynomial of A is the product of the characteristic polynomials of the B_i's.

41. Let

$$A = \begin{pmatrix} 1 & 2 & \cdots & n \\ n+1 & n+2 & \cdots & 2n \\ \vdots & \vdots & & \vdots \\ n^2 - n + 1 & n^2 - n + 2 & \cdots & n^2 \end{pmatrix}.$$

Find the characteristic polynomial of A. *Hint:* First prove that A has rank 2 and that $\operatorname{span}(\{(1, 1, \ldots, 1), (1, 2, \ldots, n)\})$ is L_A-invariant.

42. Let $A \in \mathsf{M}_{n \times n}(R)$ be the matrix defined by $A_{ij} = 1$ for all i and j. Find the characteristic polynomial of A.

INDEX OF DEFINITIONS FOR CHAPTER 5

6

Inner Product Spaces

*M*ost applications of mathematics are involved with the concept of measurement and hence of the magnitude or relative size of various quantities. So it is not surprising that the fields of real and complex numbers, which have a built-in notion of distance, should play a special role. Except for Section 6.8, we assume that all vector spaces are over the field F, where F denotes either R or C. (See Appendix D for properties of complex numbers.)

We introduce the idea of distance or length into vector spaces via a much richer structure, the so-called *inner product space* structure. This added structure provides applications to geometry (Sections 6.5 and 6.11), physics (Section 6.9), conditioning in systems of linear equations (Section 6.10), least squares (Section 6.3), and quadratic forms (Section 6.8).

6.1 INNER PRODUCTS AND NORMS

Many geometric notions such as angle, length, and perpendicularity in R^2 and R^3 may be extended to more general real and complex vector spaces. All of these ideas are related to the concept of *inner product*.

Definition. *Let* V *be a vector space over* F. *An* **inner product** *on* V *is a function that assigns, to every ordered pair of vectors* x *and* y *in* V, *a* scalar in F, denoted $\langle x, y \rangle$, such that for all x, y, and z in V and all c in F, the following hold:

(a) $\langle x + z, y \rangle = \langle x, y \rangle + \langle z, y \rangle$.

(b) $\langle cx, y \rangle = c \langle x, y \rangle$.

(c) $\overline{\langle x, y \rangle} = \langle y, x \rangle$, where the bar denotes complex conjugation.

(d) $\langle x, x \rangle > 0$ if $x \neq 0$.

Note that (c) reduces to $\langle x, y \rangle = \langle y, x \rangle$ if $F = R$. Conditions (a) and (b) simply require that the inner product be linear in the first component.

It is easily shown that if $a_1, a_2, \ldots, a_n \in F$ and $y, v_1, v_2, \ldots, v_n \in V$, then

$$\left\langle \sum_{i=1}^{n} a_i v_i, y \right\rangle = \sum_{i=1}^{n} a_i \langle v_i, y \rangle.$$

Example 1

For $x = (a_1, a_2, \ldots, a_n)$ and $y = (b_1, b_2, \ldots, b_n)$ in F^n, define

$$\langle x, y \rangle = \sum_{i=1}^{n} a_i \overline{b_i}.$$

The verification that $\langle \cdot, \cdot \rangle$ satisfies conditions (a) through (d) is easy. For example, if $z = (c_1, c_2, \ldots, c_n)$, we have for (a)

$$\langle x + z, y \rangle = \sum_{i=1}^{n} (a_i + c_i) \overline{b_i} = \sum_{i=1}^{n} a_i \overline{b_i} + \sum_{i=1}^{n} c_i \overline{b_i}$$

$$= \langle x, y \rangle + \langle z, y \rangle.$$

Thus, for $x = (1 + i, 4)$ and $y = (2 - 3i, 4 + 5i)$ in C^2,

$$\langle x, y \rangle = (1 + i)(2 + 3i) + 4(4 - 5i) = 15 - 15i. \quad \blacklozenge$$

The inner product in Example 1 is called the **standard inner product** on F^n. When $F = R$ the conjugations are not needed, and in early courses this standard inner product is usually called the *dot product* and is denoted by $x \cdot y$ instead of $\langle x, y \rangle$.

Example 2

If $\langle x, y \rangle$ is any inner product on a vector space V and $r > 0$, we may define another inner product by the rule $\langle x, y \rangle' = r \langle x, y \rangle$. If $r \leq 0$, then (d) would not hold. $\quad \blacklozenge$

Example 3

Let $V = C([0,1])$, the vector space of real-valued continuous functions on $[0,1]$. For $f, g \in V$, define $\langle f, g \rangle = \int_0^1 f(t)g(t)\, dt$. Since the preceding integral is linear in f, (a) and (b) are immediate, and (c) is trivial. If $f \neq 0$, then f^2 is bounded away from zero on some subinterval of $[0,1]$ (continuity is used here), and hence $\langle f, f \rangle = \int_0^1 [f(t)]^2\, dt > 0$.   ◆

Definition.   Let $A \in M_{m \times n}(F)$. We define the **conjugate transpose** or **adjoint** of A to be the $n \times m$ matrix A^* such that $(A^*)_{ij} = \overline{A_{ji}}$ for all i, j.

Example 4

Let

$$A = \begin{pmatrix} i & 1 + 2i \\ 2 & 3 + 4i \end{pmatrix}.$$

Then

$$A^* = \begin{pmatrix} -i & 2 \\ 1 - 2i & 3 - 4i \end{pmatrix}.   ◆$$

Notice that if x and y are viewed as column vectors in F^n, then $\langle x, y \rangle = y^* x$.

The conjugate transpose of a matrix plays a very important role in the remainder of this chapter. In the case that A has real entries, A^* is simply the transpose of A.

Example 5

Let $V = M_{n \times n}(F)$, and define $\langle A, B \rangle = \text{tr}(B^* A)$ for $A, B \in V$. (Recall that the trace of a matrix A is defined by $\text{tr}(A) = \sum_{i=1}^n A_{ii}$.) We verify that (a) and (d) of the definition of inner product hold and leave (b) and (c) to the reader. For this purpose, let $A, B, C \in V$. Then (using Exercise 6 of Section 1.3)

$$\langle A + B, C \rangle = \text{tr}(C^*(A + B)) = \text{tr}(C^*A + C^*B)$$
$$= \text{tr}(C^*A) + \text{tr}(C^*B) = \langle A, C \rangle + \langle B, C \rangle.$$

Also

$$\langle A, A \rangle = \text{tr}(A^*A) = \sum_{i=1}^n (A^*A)_{ii} = \sum_{i=1}^n \sum_{k=1}^n (A^*)_{ik} A_{ki}$$

$$= \sum_{i=1}^n \sum_{k=1}^n \overline{A_{ki}} A_{ki} = \sum_{i=1}^n \sum_{k=1}^n |A_{ki}|^2.$$

Now if $A \neq O$, then $A_{ki} \neq 0$ for some k and i. So $\langle A, A \rangle > 0$.   ◆

The inner product on $M_{n \times n}(F)$ in Example 5 is called the **Frobenius inner product**.

A vector space V over F endowed with a specific inner product is called an **inner product space**. If $F = C$, we call V a **complex inner product space**, whereas if $F = R$, we call V a **real inner product space**.

It is clear that if V has an inner product $\langle x, y \rangle$ and W is a subspace of V, then W is also an inner product space when the same function $\langle x, y \rangle$ is restricted to the vectors $x, y \in$ W.

Thus Examples 1, 3, and 5 also provide examples of inner product spaces. *For the remainder of this chapter, F^n denotes the inner product space with the standard inner product as defined in Example 1. Likewise, $M_{n \times n}(F)$ denotes the inner product space with the Frobenius inner product as defined in Example 5.* The reader is cautioned that two distinct inner products on a given vector space yield two distinct inner product spaces. For instance, it can be shown that both

$$\langle f(x), g(x) \rangle_1 = \int_0^1 f(t)g(t)\, dt \quad \text{and} \quad \langle f(x), g(x) \rangle_2 = \int_{-1}^1 f(t)g(t)\, dt$$

are inner products on the vector space $P(R)$. Even though the underlying vector space is the same, however, these two inner products yield two different inner product spaces. For example, the polynomials $f(x) = x$ and $g(x) = x^2$ are orthogonal in the second inner product space, but not in the first.

A very important inner product space that resembles $C([0, 1])$ is the space H of continuous complex-valued functions defined on the interval $[0, 2\pi]$ with the inner product

$$\langle f, g \rangle = \frac{1}{2\pi} \int_0^{2\pi} f(t)\overline{g(t)}\, dt.$$

The reason for the constant $1/2\pi$ will become evident later. This inner product space, which arises often in the context of physical situations, is examined more closely in later sections.

At this point, we mention a few facts about integration of complex-valued functions. First, the imaginary number i can be treated as a constant under the integration sign. Second, every complex-valued function f may be written as $f = f_1 + if_2$, where f_1 and f_2 are real-valued functions. Thus we have

$$\int f = \int f_1 + i \int f_2 \quad \text{and} \quad \overline{\int f} = \int \overline{f}.$$

From these properties, as well as the assumption of continuity, it follows that H is an inner product space (see Exercise 16(a)).

Some properties that follow easily from the definition of an inner product are contained in the next theorem.

Theorem 6.1. *Let* V *be an inner product space. Then for* $x, y, z \in V$ *and* $c \in F$, *the following statements are true.*

(a) $\langle x, y + z \rangle = \langle x, y \rangle + \langle x, z \rangle.$

(b) $\langle x, cy \rangle = \bar{c} \langle x, y \rangle.$

(c) $\langle x, 0 \rangle = \langle 0, x \rangle = 0.$

(d) $\langle x, x \rangle = 0$ *if and only if* $x = 0.$

(e) *If* $\langle x, y \rangle = \langle x, z \rangle$ *for all* $x \in V$, *then* $y = z.$

Proof. (a) We have

$$\langle x, y + z \rangle = \overline{\langle y + z, x \rangle} = \overline{\langle y, x \rangle + \langle z, x \rangle}$$
$$= \overline{\langle y, x \rangle} + \overline{\langle z, x \rangle} = \langle x, y \rangle + \langle x, z \rangle.$$

The proofs of (b), (c), (d), and (e) are left as exercises. ∎

The reader should observe that (a) and (b) of Theorem 6.1 show that the inner product is **conjugate linear** in the second component.

In order to generalize the notion of length in R^3 to arbitrary inner product spaces, we need only observe that the length of $x = (a, b, c) \in R^3$ is given by $\sqrt{a^2 + b^2 + c^2} = \sqrt{\langle x, x \rangle}$. This leads to the following definition.

Definition. *Let* V *be an inner product space. For* $x \in V$, *we define the* **norm** *or* **length** *of* x *by* $\|x\| = \sqrt{\langle x, x \rangle}.$

Example 6

Let $V = F^n$. If $x = (a_1, a_2 \ldots, a_n)$, then

$$\|x\| = \|(a_1, a_2 \ldots, a_n)\| = \left[\sum_{i=1}^{n} |a_i|^2 \right]^{1/2}$$

is the Euclidean definition of length. Note that if $n = 1$, we have $\|a\| = |a|$. ◆

As we might expect, the well-known properties of Euclidean length in R^3 hold in general, as shown next.

Theorem 6.2. *Let* V *be an inner product space over* F. *Then for all* $x, y \in V$ *and* $c \in F$, *the following statements are true.*

(a) $\|cx\| = |c| \cdot \|x\|.$

(b) $\|x\| = 0$ *if and only if* $x = 0$. *In any case,* $\|x\| \geq 0.$

(c) *(Cauchy–Schwarz Inequality)* $|\langle x, y \rangle| \leq \|x\| \cdot \|y\|.$

(d) *(Triangle Inequality)* $\|x + y\| \leq \|x\| + \|y\|.$

Proof. We leave the proofs of (a) and (b) as exercises.

(c) If $y = 0$, then the result is immediate. So assume that $y \neq 0$. For any $c \in F$, we have

$$0 \leq \|x - cy\|^2 = \langle x - cy, x - cy \rangle = \langle x, x - cy \rangle - c \langle y, x - cy \rangle$$
$$= \langle x, x \rangle - \bar{c} \langle x, y \rangle - c \langle y, x \rangle + c\bar{c} \langle y, y \rangle .$$

In particular, if we set

$$c = \frac{\langle x, y \rangle}{\langle y, y \rangle},$$

the inequality becomes

$$0 \leq \langle x, x \rangle - \frac{|\langle x, y \rangle|^2}{\langle y, y \rangle} = \|x\|^2 - \frac{|\langle x, y \rangle|^2}{\|y\|^2},$$

from which (c) follows.

(d) We have

$$\|x + y\|^2 = \langle x + y, x + y \rangle = \langle x, x \rangle + \langle y, x \rangle + \langle x, y \rangle + \langle y, y \rangle$$
$$= \|x\|^2 + 2\Re \langle x, y \rangle + \|y\|^2$$
$$\leq \|x\|^2 + 2|\langle x, y \rangle| + \|y\|^2$$
$$\leq \|x\|^2 + 2\|x\| \cdot \|y\| + \|y\|^2$$
$$= (\|x\| + \|y\|)^2,$$

where $\Re \langle x, y \rangle$ denotes the real part of the complex number $\langle x, y \rangle$. Note that we used (c) to prove (d). ∎

The case when equality results in (c) and (d) is considered in Exercise 15.

Example 7

For F^n, we may apply (c) and (d) of Theorem 6.2 to the standard inner product to obtain the following well-known inequalities:

$$\left| \sum_{i=1}^{n} a_i \bar{b}_i \right| \leq \left[\sum_{i=1}^{n} |a_i|^2 \right]^{1/2} \left[\sum_{i=1}^{n} |b_i|^2 \right]^{1/2}$$

and

$$\left[\sum_{i=1}^{n} |a_i + b_i|^2 \right]^{1/2} \leq \left[\sum_{i=1}^{n} |a_i|^2 \right]^{1/2} + \left[\sum_{i=1}^{n} |b_i|^2 \right]^{1/2} . \quad \blacklozenge$$

The reader may recall from earlier courses that, for x and y in R^3 or R^2, we have that $\langle x, y \rangle = \|x\| \cdot \|y\| \cos \theta$, where θ $(0 \le \theta \le \pi)$ denotes the angle between x and y. This equation implies (c) immediately since $|\cos \theta| \le 1$. Notice also that nonzero vectors x and y are perpendicular if and only if $\cos \theta = 0$, that is, if and only if $\langle x, y \rangle = 0$.

We are now at the point where we can generalize the notion of perpendicularity to arbitrary inner product spaces.

Definitions. *Let* V *be an inner product space. Vectors* x *and* y *in* V *are* **orthogonal (perpendicular)** *if* $\langle x, y \rangle = 0$*. A subset* S *of* V *is* **orthogonal** *if any two distinct vectors in* S *are orthogonal. A vector* x *in* V *is a* **unit vector** *if* $\|x\| = 1$*. Finally, a subset* S *of* V *is* **orthonormal** *if* S *is orthogonal and consists entirely of unit vectors.*

Note that if $S = \{v_1, v_2, \ldots\}$, then S is orthonormal if and only if $\langle v_i, v_j \rangle = \delta_{ij}$, where δ_{ij} denotes the Kronecker delta. Also, observe that multiplying vectors by nonzero scalars does not affect their orthogonality and that if x is any nonzero vector, then $(1/\|x\|)x$ is a unit vector. The process of multiplying a nonzero vector by the reciprocal of its length is called **normalizing**.

Example 8

In F^3, $\{(1, 1, 0), (1, -1, 1), (-1, 1, 2)\}$ is an orthogonal set of nonzero vectors, but it is not orthonormal; however, if we normalize the vectors in the set, we obtain the orthonormal set

$$\left\{ \frac{1}{\sqrt{2}}(1, 1, 0), \frac{1}{\sqrt{3}}(1, -1, 1), \frac{1}{\sqrt{6}}(-1, 1, 2) \right\}. \quad \blacklozenge$$

Our next example is of an infinite orthonormal set that is important in analysis. This set is used in later examples in this chapter.

Example 9

Recall the inner product space H (defined on page 332). We introduce an important orthonormal subset S of H. For what follows, i is the imaginary number such that $i^2 = -1$. For any integer n, let $f_n(t) = e^{int}$, where $0 \le t \le 2\pi$. (Recall that $e^{int} = \cos nt + i \sin nt$.) Now define $S = \{f_n : n \text{ is an integer}\}$.

Clearly S is a subset of H. Using the property that $\overline{e^{it}} = e^{-it}$ for every real number t, we have, for $m \ne n$,

$$\langle f_m, f_n \rangle = \frac{1}{2\pi} \int_0^{2\pi} e^{imt} \overline{e^{int}} \, dt = \frac{1}{2\pi} \int_0^{2\pi} e^{i(m-n)t} \, dt$$

$$= \frac{1}{2\pi(m-n)} e^{i(m-n)t} \Big|_0^{2\pi} = 0.$$

Also,

$$\langle f_n, f_n \rangle = \frac{1}{2\pi} \int_0^{2\pi} e^{i(n-n)t}\, dt = \frac{1}{2\pi} \int_0^{2\pi} 1\, dt = 1.$$

In other words, $\langle f_m, f_n \rangle = \delta_{mn}$.   ◆

EXERCISES

1. Label the following statements as true or false.

   (a) An inner product is a scalar-valued function on the set of ordered pairs of vectors.

   (b) An inner product space must be over the field of real or complex numbers.

   (c) An inner product is linear in both components.

   (d) There is exactly one inner product on the vector space R^n.

   (e) The triangle inequality only holds in finite-dimensional inner product spaces.

   (f) Only square matrices have a conjugate-transpose.

   (g) If x, y, and z are vectors in an inner product space such that $\langle x, y \rangle = \langle x, z \rangle$, then $y = z$.

   (h) If $\langle x, y \rangle = 0$ for all x in an inner product space, then $y = 0$.

2. Let $x = (2, 1+i, i)$ and $y = (2-i, 2, 1+2i)$ be vectors in C^3. Compute $\langle x, y \rangle$, $\|x\|$, $\|y\|$, and $\|x+y\|$. Then verify both the Cauchy–Schwarz inequality and the triangle inequality.

3. In $\mathsf{C}([0,1])$, let $f(t) = t$ and $g(t) = e^t$. Compute $\langle f, g \rangle$ (as defined in Example 3), $\|f\|$, $\|g\|$, and $\|f+g\|$. Then verify both the Cauchy–Schwarz inequality and the triangle inequality.

4. (a) Complete the proof in Example 5 that $\langle \cdot, \cdot \rangle$ is an inner product (the Frobenius inner product) on $\mathsf{M}_{n \times n}(F)$.

   (b) Use the Frobenius inner product to compute $\|A\|$, $\|B\|$, and $\langle A, B \rangle$ for

   $$A = \begin{pmatrix} 1 & 2+i \\ 3 & i \end{pmatrix} \quad \text{and} \quad B = \begin{pmatrix} 1+i & 0 \\ i & -i \end{pmatrix}.$$

5. In C^2, show that $\langle x, y \rangle = xAy^*$ is an inner product, where

   $$A = \begin{pmatrix} 1 & i \\ -i & 2 \end{pmatrix}.$$

   Compute $\langle x, y \rangle$ for $x = (1-i, 2+3i)$ and $y = (2+i, 3-2i)$.

6. Complete the proof of Theorem 6.1.

7. Complete the proof of Theorem 6.2.

8. Provide reasons why each of the following is not an inner product on the given vector spaces.

   (a) $\langle (a,b),(c,d) \rangle = ac - bd$ on R^2.
   (b) $\langle A,B \rangle = \text{tr}(A + B)$ on $M_{2\times 2}(R)$.
   (c) $\langle f(x), g(x) \rangle = \int_0^1 f'(t)g(t)\, dt$ on $P(R)$, where $'$ denotes differentiation.

9. Let β be a basis for a finite-dimensional inner product space.

   (a) Prove that if $\langle x, z \rangle = 0$ for all $z \in \beta$, then $x = 0$.
   (b) Prove that if $\langle x, z \rangle = \langle y, z \rangle$ for all $z \in \beta$, then $x = y$.

10.† Let V be an inner product space, and suppose that x and y are orthogonal vectors in V. Prove that $\|x + y\|^2 = \|x\|^2 + \|y\|^2$. Deduce the Pythagorean theorem in R^2.

11. Prove the *parallelogram law* on an inner product space V; that is, show that

$$\|x + y\|^2 + \|x - y\|^2 = 2\|x\|^2 + 2\|y\|^2 \quad \text{for all } x, y \in V.$$

What does this equation state about parallelograms in R^2?

12.† Let $\{v_1, v_2, \ldots, v_k\}$ be an orthogonal set in V, and let $a_1, a_2, \ldots, a_k$ be scalars. Prove that

$$\left\| \sum_{i=1}^{k} a_i v_i \right\|^2 = \sum_{i=1}^{k} |a_i|^2 \|v_i\|^2.$$

13. Suppose that $\langle \cdot, \cdot \rangle_1$ and $\langle \cdot, \cdot \rangle_2$ are two inner products on a vector space V. Prove that $\langle \cdot, \cdot \rangle = \langle \cdot, \cdot \rangle_1 + \langle \cdot, \cdot \rangle_2$ is another inner product on V.

14. Let A and B be $n \times n$ matrices, and let c be a scalar. Prove that $(A + cB)^* = A^* + \bar{c}B^*$.

15. (a) Prove that if V is an inner product space, then $|\langle x, y \rangle| = \|x\| \cdot \|y\|$ if and only if one of the vectors x or y is a multiple of the other. *Hint:* If the identity holds and $y \neq 0$, let

$$a = \frac{\langle x, y \rangle}{\|y\|^2},$$

and let $z = x - ay$. Prove that y and z are orthogonal and

$$|a| = \frac{\|x\|}{\|y\|}.$$

Then apply Exercise 10 to $\|x\|^2 = \|ay + z\|^2$ to obtain $\|z\| = 0$.

    (b) Derive a similar result for the equality $\|x + y\| = \|x\| + \|y\|$, and generalize it to the case of n vectors.

16. (a) Show that the vector space H with $\langle \cdot, \cdot \rangle$ defined on page 332 is an inner product space.

    (b) Let $V = C([0, 1])$, and define

$$\langle f, g \rangle = \int_0^{1/2} f(t)g(t)\, dt.$$

Is this an inner product on V?

17. Let T be a linear operator on an inner product space V, and suppose that $\|T(x)\| = \|x\|$ for all x. Prove that T is one-to-one.

18. Let V be a vector space over F, where $F = R$ or $F = C$, and let W be an inner product space over F with inner product $\langle \cdot, \cdot \rangle$. If $T: V \rightarrow W$ is linear, prove that $\langle x, y \rangle' = \langle T(x), T(y) \rangle$ defines an inner product on V if and only if T is one-to-one.

19. Let V be an inner product space. Prove that

    (a) $\|x \pm y\|^2 = \|x\|^2 \pm 2\Re \langle x, y \rangle + \|y\|^2$ for all $x, y \in V$, where $\Re \langle x, y \rangle$ denotes the real part of the complex number $\langle x, y \rangle$.

    (b) $|\, \|x\| - \|y\| \,| \le \|x - y\|$ for all $x, y \in V$.

20. Let V be an inner product space over F. Prove the *polar identities*: For all $x, y \in V$,

    (a) $\langle x, y \rangle = \frac{1}{4}\|x + y\|^2 - \frac{1}{4}\|x - y\|^2$ if $F = R$;

    (b) $\langle x, y \rangle = \frac{1}{4}\sum_{k=1}^{4} i^k \|x + i^k y\|^2$ if $F = C$, where $i^2 = -1$.

21. Let A be an $n \times n$ matrix. Define

$$A_1 = \frac{1}{2}(A + A^*) \quad \text{and} \quad A_2 = \frac{1}{2i}(A - A^*).$$

    (a) Prove that $A_1^* = A_1$, $A_2^* = A_2$, and $A = A_1 + iA_2$. Would it be reasonable to define A_1 and A_2 to be the real and imaginary parts, respectively, of the matrix A?

    (b) Let A be an $n \times n$ matrix. Prove that the representation in (a) is unique. That is, prove that if $A = B_1 + iB_2$, where $B_1^* = B_1$ and $B_2^* = B_2$, then $B_1 = A_1$ and $B_2 = A_2$.

22. Let V be a real or complex vector space (possibly infinite-dimensional), and let β be a basis for V. For $x, y \in$ V there exist $v_1, v_2, \ldots, v_n \in \beta$ such that

$$x = \sum_{i=1}^{n} a_i v_i \quad \text{and} \quad y = \sum_{i=1}^{n} b_i v_i.$$

Define

$$\langle x, y \rangle = \sum_{i=1}^{n} a_i \bar{b}_i.$$

(a) Prove that $\langle \cdot, \cdot \rangle$ is an inner product on V and that β is an orthonormal basis for V. Thus every real or complex vector space may be regarded as an inner product space.
(b) Prove that if V $= \mathsf{R}^n$ or V $= \mathsf{C}^n$ and β is the standard ordered basis, then the inner product defined above is the standard inner product.

23. Let V $= \mathsf{F}^n$, and let $A \in \mathsf{M}_{n \times n}(F)$.

(a) Prove that $\langle x, Ay \rangle = \langle A^* x, y \rangle$ for all $x, y \in$ V.
(b) Suppose that for some $B \in \mathsf{M}_{n \times n}(F)$, we have $\langle x, Ay \rangle = \langle Bx, y \rangle$ for all $x, y \in$ V. Prove that $B = A^*$.
(c) Let α be the standard ordered basis for V. For any orthonormal basis β for V, let Q be the $n \times n$ matrix whose columns are the vectors in β. Prove that $Q^* = Q^{-1}$.
(d) Define linear operators T and U on V by $\mathsf{T}(x) = Ax$ and $\mathsf{U}(x) = A^* x$. Show that $[\mathsf{U}]_\beta = [\mathsf{T}]_\beta^*$ for any orthonormal basis β for V.

The following definition is used in Exercises 24–27.

Definition. *Let* V *be a vector space over* F*, where* F *is either* R *or* C*. Regardless of whether* V *is or is not an inner product space, we may still define a norm* $\|\cdot\|$ *as a real-valued function on* V *satisfying the following three conditions for all* $x, y \in$ V *and* $a \in F$:

(1) $\|x\| \geq 0$, *and* $\|x\| = 0$ *if and only if* $x = 0$.
(2) $\|ax\| = |a| \cdot \|x\|$.
(3) $\|x + y\| \leq \|x\| + \|y\|$.

24. Prove that the following are norms on the given vector spaces V.

(a) V $= \mathsf{M}_{m \times n}(F)$; $\|A\| = \max_{i,j} |A_{ij}|$ for all $A \in$ V
(b) V $= \mathsf{C}([0, 1])$; $\|f\| = \max_{t \in [0,1]} |f(t)|$ for all $f \in$ V

(c) $V = C([0,1]);$ $\|f\| = \displaystyle\int_0^1 |f(t)|\,dt$ for all $f \in V$

(d) $V = R^2;$ $\|(a,b)\| = \max\{|a|,|b|\}$ for all $(a,b) \in V$

25. Use Exercise 20 to show that there is no inner product $\langle\cdot,\cdot\rangle$ on R^2 such that $\|x\|^2 = \langle x,x\rangle$ for all $x \in R^2$ if the norm is defined as in Exercise 24(d).

26. Let $\|\cdot\|$ be a norm on a vector space V, and define, for each ordered pair of vectors, the scalar $d(x,y) = \|x - y\|$, called the **distance** between x and y. Prove the following results for all $x,y,z \in V$.

    (a) $d(x,y) \geq 0.$
    (b) $d(x,y) = d(y,x).$
    (c) $d(x,y) \leq d(x,z) + d(z,y).$
    (d) $d(x,x) = 0.$
    (e) $d(x,y) \neq 0$ if $x \neq y.$

27. Let $\|\cdot\|$ be a norm on a real vector space V satisfying the parallelogram law given in Exercise 11. Define

$$\langle x,y\rangle = \frac{1}{4}\left[\|x + y\|^2 - \|x - y\|^2\right].$$

Prove that $\langle\cdot,\cdot\rangle$ defines an inner product on V such that $\|x\|^2 = \langle x,x\rangle$ for all $x \in V$.

Hints:

(a) Prove $\langle x,2y\rangle = 2\langle x,y\rangle$ for all $x,y \in V$.
(b) Prove $\langle x+u,y\rangle = \langle x,y\rangle + \langle u,y\rangle$ for all $x,u,y \in V$.
(c) Prove $\langle nx,y\rangle = n\langle x,y\rangle$ for every positive integer n and every $x,y \in V$.
(d) Prove $m\langle \frac{1}{m}x,y\rangle = \langle x,y\rangle$ for every positive integer m and every $x,y \in V$.
(e) Prove $\langle rx,y\rangle = r\langle x,y\rangle$ for every rational number r and every $x,y \in V$.
(f) Prove $|\langle x,y\rangle| \leq \|x\|\|y\|$ for every $x,y \in V$. *Hint:* Condition (3) in the definition of norm can be helpful.
(g) Prove that for every $c \in R$, every rational number r, and every $x,y \in V$,

$$|c\langle x,y\rangle - \langle cx,y\rangle| = |(c-r)\langle x,y\rangle - \langle(c-r)x,y\rangle| \leq 2|c-r|\|x\|\|y\|.$$

(h) Use the fact that for any $c \in R$, $|c - r|$ can be made arbitrarily small, where r varies over the set of rational numbers, to establish item (b) of the definition of *inner product*.

28. Let V be a complex inner product space with an inner product $\langle \cdot, \cdot \rangle$. Let $[\cdot, \cdot]$ be the real-valued function such that $[x, y]$ is the real part of the complex number $\langle x, y \rangle$ for all $x, y \in V$. Prove that $[\cdot, \cdot]$ is an inner product for V, where V is regarded as a vector space over R. Prove, furthermore, that $[x, ix] = 0$ for all $x \in V$.

29. Let V be a vector space over C, and suppose that $[\cdot, \cdot]$ is a real inner product on V, where V is regarded as a vector space over R, such that $[x, ix] = 0$ for all $x \in V$. Let $\langle \cdot, \cdot \rangle$ be the complex-valued function defined by

$$\langle x, y \rangle = [x, y] + i[x, iy] \qquad \text{for } x, y \in V.$$

Prove that $\langle \cdot, \cdot \rangle$ is a complex inner product on V.

30. Let $\|\cdot\|$ be a norm (as defined in Exercise 24) on a complex vector space V satisfying the parallelogram law given in Exercise 11. Prove that there is an inner product $\langle \cdot, \cdot \rangle$ on V such that $\|x\|^2 = \langle x, x \rangle$ for all $x \in V$.

Hint: Apply Exercise 27 to V regarded as a vector space over R. Then apply Exercise 29.

6.2   THE GRAM–SCHMIDT ORTHOGONALIZATION PROCESS AND ORTHOGONAL COMPLEMENTS

In previous chapters, we have seen the special role of the standard ordered bases for C^n and R^n. The special properties of these bases stem from the fact that the basis vectors form an orthonormal set. Just as bases are the building blocks of vector spaces, bases that are also orthonormal sets are the building blocks of inner product spaces. We now name such bases.

Definition.   *Let V be an inner product space. A subset of V is an* **orthonormal basis** *for V if it is an ordered basis that is orthonormal.*

Example 1

The standard ordered basis for F^n is an orthonormal basis for F^n.   ◆

Example 2

The set

$$\left\{ \left(\frac{1}{\sqrt{5}}, \frac{2}{\sqrt{5}} \right), \left(\frac{2}{\sqrt{5}}, \frac{-1}{\sqrt{5}} \right) \right\}$$

is an orthonormal basis for R^2.   ◆

The next theorem and its corollaries illustrate why orthonormal sets and, in particular, orthonormal bases are so important.

Theorem 6.3. *Let* V *be an inner product space and* $S = \{v_1, v_2, \ldots, v_k\}$ *be an orthogonal subset of* V *consisting of nonzero vectors. If* $y \in \text{span}(S)$, *then*

$$y = \sum_{i=1}^{k} \frac{\langle y, v_i \rangle}{\|v_i\|^2} v_i.$$

Proof. Write $y = \sum_{i=1}^{k} a_i v_i$, where $a_1, a_2, \ldots, a_k \in F$. Then, for $1 \leq j \leq k$, we have

$$\langle y, v_j \rangle = \left\langle \sum_{i=1}^{k} a_i v_i, v_j \right\rangle = \sum_{i=1}^{k} a_i \langle v_i, v_j \rangle = a_j \langle v_j, v_j \rangle = a_j \|v_j\|^2.$$

So $a_j = \dfrac{\langle y, v_j \rangle}{\|v_j\|^2}$, and the result follows. ∎

The next corollary follows immediately from Theorem 6.3.

Corollary 1. *If, in addition to the hypotheses of Theorem 6.3,* S *is orthonormal and* $y \in \text{span}(S)$, *then*

$$y = \sum_{i=1}^{k} \langle y, v_i \rangle v_i.$$

If V possesses a finite orthonormal basis, then Corollary 1 allows us to compute the coefficients in a linear combination very easily. (See Example 3.)

Corollary 2. *Let* V *be an inner product space, and let* S *be an orthogonal subset of* V *consisting of nonzero vectors. Then* S *is linearly independent.*

Proof. Suppose that $v_1, v_2, \ldots, v_k \in S$ and

$$\sum_{i=1}^{k} a_i v_i = 0.$$

As in the proof of Theorem 6.3 with $y = 0$, we have $a_j = \langle 0, v_j \rangle / \|v_j\|^2 = 0$ for all j. So S is linearly independent. ∎

Example 3

By Corollary 2, the orthonormal set

$$\left\{ \frac{1}{\sqrt{2}}(1,1,0), \frac{1}{\sqrt{3}}(1,-1,1), \frac{1}{\sqrt{6}}(-1,1,2) \right\}$$

obtained in Example 8 of Section 6.1 is an orthonormal basis for R^3. Let $x = (2,1,3)$. The coefficients given by Corollary 1 to Theorem 6.3 that express x as a linear combination of the basis vectors are

$$a_1 = \frac{1}{\sqrt{2}}(2+1) = \frac{3}{\sqrt{2}}, \quad a_2 = \frac{1}{\sqrt{3}}(2-1+3) = \frac{4}{\sqrt{3}},$$

and

$$a_3 = \frac{1}{\sqrt{6}}(-2+1+6) = \frac{5}{\sqrt{6}}.$$

As a check, we have

$$(2,1,3) = \frac{3}{2}(1,1,0) + \frac{4}{3}(1,-1,1) + \frac{5}{6}(-1,1,2). \quad \blacklozenge$$

Corollary 2 tells us that the vector space H in Section 6.1 contains an infinite linearly independent set, and hence H is not a finite-dimensional vector space.

Of course, we have not yet shown that every finite-dimensional inner product space possesses an orthonormal basis. The next theorem takes us most of the way in obtaining this result. It tells us how to construct an orthogonal set from a linearly independent set of vectors in such a way that both sets generate the same subspace.

Before stating this theorem, let us consider a simple case. Suppose that $\{w_1, w_2\}$ is a linearly independent subset of an inner product space (and hence a basis for some two-dimensional subspace). We want to construct an orthogonal set from $\{w_1, w_2\}$ that spans the same subspace. Figure 6.1 suggests that the set $\{v_1, v_2\}$, where $v_1 = w_1$ and $v_2 = w_2 - cw_1$, has this property if c is chosen so that v_2 is orthogonal to W_1.

To find c, we need only solve the following equation:

$$0 = \langle v_2, w_1 \rangle = \langle w_2 - cw_1, w_1 \rangle = \langle w_2, w_1 \rangle - c \langle w_1, w_1 \rangle.$$

So

$$c = \frac{\langle w_2, w_1 \rangle}{\|w_1\|^2}.$$

Thus

$$v_2 = w_2 - \frac{\langle w_2, w_1 \rangle}{\|w_1\|^2} w_1.$$

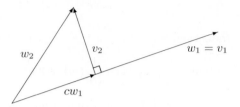

Figure 6.1

The next theorem shows us that this process can be extended to any finite linearly independent subset.

Theorem 6.4. *Let* V *be an inner product space and* $S = \{w_1, w_2, \ldots, w_n\}$ *be a linearly independent subset of* V. *Define* $S' = \{v_1, v_2, \ldots, v_n\}$, *where* $v_1 = w_1$ *and*

$$v_k = w_k - \sum_{j=1}^{k-1} \frac{\langle w_k, v_j \rangle}{\|v_j\|^2} v_j \quad \text{for } 2 \le k \le n. \tag{1}$$

Then S' *is an orthogonal set of nonzero vectors such that* $\operatorname{span}(S') = \operatorname{span}(S)$.

Proof. The proof is by mathematical induction on n, the number of vectors in S. For $k = 1, 2, \ldots, n$, let $S_k = \{w_1, w_2, \ldots, w_k\}$. If $n = 1$, then the theorem is proved by taking $S_1' = S_1$; i.e., $v_1 = w_1 \ne 0$. Assume then that the set $S_{k-1}' = \{v_1, v_2, \ldots, v_{k-1}\}$ with the desired properties has been constructed by the repeated use of (1). We show that the set $S_k' = \{v_1, v_2, \ldots, v_{k-1}, v_k\}$ also has the desired properties, where v_k is obtained from S_{k-1}' by (1). If $v_k = 0$, then (1) implies that $w_k \in \operatorname{span}(S_{k-1}') = \operatorname{span}(S_{k-1})$, which contradicts the assumption that S_k is linearly independent. For $1 \le i \le k - 1$, it follows from (1) that

$$\langle v_k, v_i \rangle = \langle w_k, v_i \rangle - \sum_{j=1}^{k-1} \frac{\langle w_k, v_j \rangle}{\|v_j\|^2} \langle v_j, v_i \rangle = \langle w_k, v_i \rangle - \frac{\langle w_k, v_i \rangle}{\|v_i\|^2} \|v_i\|^2 = 0,$$

since $\langle v_j, v_i \rangle = 0$ if $i \ne j$ by the induction assumption that S_{k-1}' is orthogonal. Hence S_k' is an orthogonal set of nonzero vectors. Now, by (1), we have that $\operatorname{span}(S_k') \subseteq \operatorname{span}(S_k)$. But by Corollary 2 to Theorem 6.3, S_k' is linearly independent; so $\dim(\operatorname{span}(S_k')) = \dim(\operatorname{span}(S_k)) = k$. Therefore $\operatorname{span}(S_k') = \operatorname{span}(S_k)$. ∎

The construction of $\{v_1, v_2, \ldots, v_n\}$ by the use of Theorem 6.4 is called the **Gram–Schmidt process**.

Example 4

In R^4, let $w_1 = (1, 0, 1, 0)$, $w_2 = (1, 1, 1, 1)$, and $w_3 = (0, 1, 2, 1)$. Then $\{w_1, w_2, w_3\}$ is linearly independent. We use the Gram–Schmidt process to compute the orthogonal vectors v_1, v_2, and v_3, and then we normalize these vectors to obtain an orthonormal set.

Take $v_1 = w_1 = (1, 0, 1, 0)$. Then

$$v_2 = w_2 - \frac{\langle w_2, v_1 \rangle}{\|v_1\|^2} v_1$$

$$= (1, 1, 1, 1) - \frac{2}{2}(1, 0, 1, 0)$$

$$= (0, 1, 0, 1).$$

Finally,

$$v_3 = w_3 - \frac{\langle w_3, v_1 \rangle}{\|v_1\|^2} v_1 - \frac{\langle w_3, v_2 \rangle}{\|v_2\|^2} v_2$$

$$= (0, 1, 2, 1) - \frac{2}{2}(1, 0, 1, 0) - \frac{2}{2}(0, 1, 0, 1)$$

$$= (-1, 0, 1, 0).$$

These vectors can be normalized to obtain the orthonormal basis $\{u_1, u_2, u_3\}$, where

$$u_1 = \frac{1}{\|v_1\|} v_1 = \frac{1}{\sqrt{2}}(1, 0, 1, 0),$$

$$u_2 = \frac{1}{\|v_2\|} v_2 = \frac{1}{\sqrt{2}}(0, 1, 0, 1),$$

and

$$u_3 = \frac{v_3}{\|v_3\|} = \frac{1}{\sqrt{2}}(-1, 0, 1, 0). \quad \blacklozenge$$

Example 5

Let $\mathsf{V} = \mathsf{P}(R)$ with the inner product $\langle f(x), g(x) \rangle = \int_{-1}^{1} f(t)g(t)\, dt$, and consider the subspace $\mathsf{P}_2(R)$ with the standard ordered basis β. We use the Gram–Schmidt process to replace β by an orthogonal basis $\{v_1, v_2, v_3\}$ for $\mathsf{P}_2(R)$, and then use this orthogonal basis to obtain an orthonormal basis for $\mathsf{P}_2(R)$.

Take $v_1 = 1$. Then $\|v_1\|^2 = \int_{-1}^{1} 1^2\, dt = 2$, and $\langle x, v_1 \rangle = \int_{-1}^{1} t \cdot 1\, dt = 0$. Thus

$$v_2 = x - \frac{\langle v_1, x \rangle}{\|v_1\|^2} = x - \frac{0}{2} = x.$$

Furthermore,

$$\langle x^2, v_1 \rangle = \int_{-1}^{1} t^2 \cdot 1 \, dt = \frac{2}{3} \quad \text{and} \quad \langle x^2, v_2 \rangle = \int_{-1}^{1} t^2 \cdot t \, dt = 0.$$

Therefore

$$v_3 = x^2 - \frac{\langle x^2, v_1 \rangle}{\|v_1\|^2} v_1 - \frac{\langle x^2, v_2 \rangle}{\|v_2\|^2} v_2$$

$$= x^2 - \frac{1}{3} \cdot 1 - 0 \cdot x$$

$$= x^2 - \frac{1}{3}.$$

We conclude that $\{1, x, x^2 - \frac{1}{3}\}$ is an orthogonal basis for $\mathsf{P}_2(R)$.

To obtain an orthonormal basis, we normalize v_1, v_2, and v_3 to obtain

$$u_1 = \frac{1}{\sqrt{\int_{-1}^{1} 1^2 \, dt}} = \frac{1}{\sqrt{2}},$$

$$u_2 = \frac{x}{\sqrt{\int_{-1}^{1} t^2 \, dt}} = \sqrt{\frac{3}{2}} x,$$

and similarly,

$$u_3 = \frac{v_3}{\|v_3\|} = \sqrt{\frac{5}{8}} (3x^2 - 1).$$

Thus $\{u_1, u_2, u_3\}$ is the desired orthonormal basis for $\mathsf{P}_2(R)$.   ◆

If we continue applying the Gram–Schmidt orthogonalization process to the basis $\{1, x, x^2, \ldots\}$ for $\mathsf{P}(R)$, we obtain an orthogonal basis whose elements are called the *Legendre polynomials*. The orthogonal polynomials v_1, v_2, and v_3 in Example 5 are the first three Legendre polynomials.

The following result gives us a simple method of representing a vector as a linear combination of the vectors in an orthonormal basis.

Theorem 6.5. *Let* V *be a nonzero finite-dimensional inner product space. Then* V *has an orthonormal basis* β. *Furthermore, if* $\beta = \{v_1, v_2, \ldots, v_n\}$ *and* $x \in \mathsf{V}$, *then*

$$x = \sum_{i=1}^{n} \langle x, v_i \rangle v_i.$$

Proof. Let β_0 be an ordered basis for V. Apply Theorem 6.4 to obtain an orthogonal set β' of nonzero vectors with $\operatorname{span}(\beta') = \operatorname{span}(\beta_0) = \mathsf{V}$. By normalizing each vector in β', we obtain an orthonormal set β that generates V. By Corollary 2 to Theorem 6.3, β is linearly independent; therefore β is an orthonormal basis for V. The remainder of the theorem follows from Corollary 1 to Theorem 6.3. ∎

Example 6

We use Theorem 6.5 to represent the polynomial $f(x) = 1 + 2x + 3x^2$ as a linear combination of the vectors in the orthonormal basis $\{u_1, u_2, u_3\}$ for $\mathsf{P}_2(R)$ obtained in Example 5. Observe that

$$\langle f(x), u_1 \rangle = \int_{-1}^{1} \frac{1}{\sqrt{2}} (1 + 2t + 3t^2)\, dt = 2\sqrt{2},$$

$$\langle f(x), u_2 \rangle = \int_{-1}^{1} \sqrt{\frac{3}{2}} t (1 + 2t + 3t^2)\, dt = \frac{2\sqrt{6}}{3},$$

and

$$\langle f(x), u_3 \rangle = \int_{-1}^{1} \sqrt{\frac{5}{8}} (3t^2 - 1)(1 + 2t + 3t^2)\, dt = \frac{2\sqrt{10}}{5}.$$

Therefore $f(x) = 2\sqrt{2}\, u_1 + \dfrac{2\sqrt{6}}{3}\, u_2 + \dfrac{2\sqrt{10}}{5}\, u_3.$ ◆

Theorem 6.5 gives us a simple method for computing the entries of the matrix representation of a linear operator with respect to an orthonormal basis.

Corollary. *Let V be a finite-dimensional inner product space with an orthonormal basis $\beta = \{v_1, v_2, \ldots, v_n\}$. Let T be a linear operator on V, and let $A = [\mathsf{T}]_\beta$. Then for any i and j, $A_{ij} = \langle \mathsf{T}(v_j), v_i \rangle$.*

Proof. From Theorem 6.5, we have

$$\mathsf{T}(v_j) = \sum_{i=1}^{n} \langle \mathsf{T}(v_j), v_i \rangle\, v_i.$$

Hence $A_{ij} = \langle \mathsf{T}(v_j), v_i \rangle$. ∎

The scalars $\langle x, v_i \rangle$ given in Theorem 6.5 have been studied extensively for special inner product spaces. Although the vectors $v_1, v_2, \ldots, v_n$ were chosen from an orthonormal basis, we introduce a terminology associated with orthonormal sets β in more general inner product spaces.

Definition.   Let β be an orthonormal subset (possibly infinite) of an inner product space V, and let $x \in \text{V}$. We define the **Fourier coefficients** of x relative to β to be the scalars $\langle x, y \rangle$, where $y \in \beta$.

In the first half of the 19th century, the French mathematician Jean Baptiste Fourier was associated with the study of the scalars

$$\int_0^{2\pi} f(t) \sin nt \, dt \quad \text{and} \quad \int_0^{2\pi} f(t) \cos nt \, dt,$$

or more generally,

$$c_n = \frac{1}{2\pi} \int_0^{2\pi} f(t) e^{-int} \, dt,$$

for a function f. In the context of Example 9 of Section 6.1, we see that $c_n = \langle f, f_n \rangle$, where $f_n(t) = e^{int}$; that is, c_n is the nth Fourier coefficient for a continuous function $f \in \text{V}$ relative to S. These coefficients are the "classical" Fourier coefficients of a function, and the literature concerning the behavior of these coefficients is extensive. We learn more about these Fourier coefficients in the remainder of this chapter.

Example 7

Let $S = \{e^{int} : n \text{ is an integer}\}$. In Example 9 of Section 6.1, S was shown to be an orthonormal set in H. We compute the Fourier coefficients of $f(t) = t$ relative to S. Using integration by parts, we have, for $n \neq 0$,

$$\langle f, f_n \rangle = \frac{1}{2\pi} \int_0^{2\pi} t \overline{e^{int}} \, dt = \frac{1}{2\pi} \int_0^{2\pi} t e^{-int} \, dt = \frac{-1}{in},$$

and, for $n = 0$,

$$\langle f, 1 \rangle = \frac{1}{2\pi} \int_0^{2\pi} t(1) \, dt = \pi.$$

As a result of these computations, and using Exercise 16 of this section, we obtain an upper bound for the sum of a special infinite series as follows:

$$\|f\|^2 \geq \sum_{n=-k}^{-1} |\langle f, f_n \rangle|^2 + |\langle f, 1 \rangle|^2 + \sum_{n=1}^{k} |\langle f, f_n \rangle|^2$$

$$= \sum_{n=-k}^{-1} \frac{1}{n^2} + \pi^2 + \sum_{n=1}^{k} \frac{1}{n^2}$$

$$= 2 \sum_{n=1}^{k} \frac{1}{n^2} + \pi^2$$

for every k. Now, using the fact that $\|f\|^2 = \dfrac{4}{3}\pi^2$, we obtain

$$\frac{4}{3}\pi^2 \geq 2\sum_{n=1}^{k} \frac{1}{n^2} + \pi^2,$$

or

$$\frac{\pi^2}{6} \geq \sum_{n=1}^{k} \frac{1}{n^2}.$$

Because this inequality holds for all k, we may let $k \to \infty$ to obtain

$$\frac{\pi^2}{6} \geq \sum_{n=1}^{\infty} \frac{1}{n^2}.$$

Additional results may be produced by replacing f by other functions.   ◆

We are now ready to proceed with the concept of an *orthogonal comple-ment*.

Definition. *Let S be a nonempty subset of an inner product space* V. *We define* $S^\perp$ *(read "S perp") to be the set of all vectors in* V *that are orthogonal to every vector in S; that is,* $S^\perp = \{x \in \mathsf{V}\colon \langle x, y \rangle = 0 \text{ for all } y \in S\}$. *The set* $S^\perp$ *is called the* **orthogonal complement** *of S.*

It is easily seen that $S^\perp$ is a subspace of V for any subset S of V.

Example 8

The reader should verify that $\{0\}^\perp = \mathsf{V}$ and $\mathsf{V}^\perp = \{0\}$ for any inner product space V.   ◆

Example 9

If $\mathsf{V} = \mathsf{R}^3$ and $S = \{e_3\}$, then $S^\perp$ equals the xy-plane (see Exercise 5).   ◆

Exercise 18 provides an interesting example of an orthogonal complement in an infinite-dimensional inner product space.

Consider the problem in R^3 of finding the distance from a point P to a plane W. (See Figure 6.2.) Problems of this type arise in many settings. If we let y be the vector determined by 0 and P, we may restate the problem as follows: Determine the vector u in W that is "closest" to y. The desired distance is clearly given by $\|y - u\|$. Notice from the figure that the vector $z = y - u$ is orthogonal to every vector in W, and so $z \in \mathsf{W}^\perp$.

The next result presents a practical method of finding u in the case that W is a finite-dimensional subspace of an inner product space.

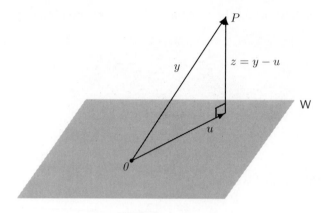

Figure 6.2

Theorem 6.6. *Let* W *be a finite-dimensional subspace of an inner product space* V, *and let* $y \in$ V. *Then there exist unique vectors* $u \in$ W *and* $z \in$ W$^\perp$ *such that* $y = u + z$. *Furthermore, if* $\{v_1, v_2, \ldots, v_k\}$ *is an orthonormal basis for* W, *then*

$$u = \sum_{i=1}^{k} \langle y, v_i \rangle \, v_i.$$

Proof. Let $\{v_1, v_2, \ldots, v_k\}$ be an orthonormal basis for W, let u be as defined in the preceding equation, and let $z = y - u$. Clearly $u \in$ W and $y = u + z$.

To show that $z \in$ W$^\perp$, it suffices to show, by Exercise 7, that z is orthogonal to each v_j. For any j, we have

$$\langle z, v_j \rangle = \left\langle \left(y - \sum_{i=1}^{k} \langle y, v_i \rangle v_i \right), v_j \right\rangle = \langle y, v_j \rangle - \sum_{i=1}^{k} \langle y, v_i \rangle \langle v_i, v_j \rangle$$

$$= \langle y, v_j \rangle - \langle y, v_j \rangle = 0.$$

To show uniqueness of u and z, suppose that $y = u + z = u' + z'$, where $u' \in$ W and $z' \in$ W$^\perp$. Then $u - u' = z' - z \in$ W $\cap$ W$^\perp$ = $\{0\}$. Therefore, $u = u'$ and $z = z'$.   ∎

Corollary. *In the notation of Theorem 6.6, the vector* u *is the unique vector in* W *that is "closest" to* y; *that is, for any* $x \in$ W, $\|y - x\| \geq \|y - u\|$, *and this inequality is an equality if and only if* $x = u$.

Proof. As in Theorem 6.6, we have that $y = u + z$, where $z \in$ W$^\perp$. Let $x \in$ W. Then $u - x$ is orthogonal to z, so, by Exercise 10 of Section 6.1, we have

$$\|y - x\|^2 = \|u + z - x\|^2 = \|(u - x) + z\|^2 = \|u - x\|^2 + \|z\|^2$$
$$\geq \|z\|^2 = \|y - u\|^2.$$

Now suppose that $\|y - x\| = \|y - u\|$. Then the inequality above becomes an equality, and therefore $\|u - x\|^2 + \|z\|^2 = \|z\|^2$. It follows that $\|u - x\| = 0$, and hence $x = u$. The proof of the converse is obvious. ∎

The vector u in the corollary is called the **orthogonal projection** of y on W. We will see the importance of orthogonal projections of vectors in the application to least squares in Section 6.3.

Example 10

Let $V = P_3(R)$ with the inner product

$$\langle f(x), g(x) \rangle = \int_{-1}^{1} f(t) g(t) \, dt \quad \text{for all } f(x), g(x) \in V.$$

We compute the orthogonal projection $f_1(x)$ of $f(x) = x^3$ on $P_2(R)$.

By Example 5,

$$\{u_1, u_2, u_3\} = \left\{ \frac{1}{\sqrt{2}}, \sqrt{\frac{3}{2}} x, \sqrt{\frac{5}{8}} (3x^2 - 1) \right\}$$

is an orthonormal basis for $P_2(R)$. For these vectors, we have

$$\langle f(x), u_1 \rangle = \int_{-1}^{1} t^3 \frac{1}{\sqrt{2}} \, dt = 0, \qquad \langle f(x), u_2 \rangle = \int_{-1}^{1} t^3 \sqrt{\frac{3}{2}} t \, dt = \frac{\sqrt{6}}{5},$$

and

$$\langle f(x), u_3 \rangle = \int_{-1}^{1} t^3 \sqrt{\frac{5}{8}} (3t^2 - 1) \, dt = 0.$$

Hence

$$f_1(x) = \langle f(x), u_1 \rangle u_1 + \langle f(x), u_2 \rangle u_2 + \langle f(x), u_3 \rangle u_3 = \frac{3}{5} x. \quad \blacklozenge$$

It was shown (Corollary 2 to the replacement theorem, p. 47) that any linearly independent set in a finite-dimensional vector space can be extended to a basis. The next theorem provides an interesting analog for an orthonormal subset of a finite-dimensional inner product space.

Theorem 6.7. *Suppose that* $S = \{v_1, v_2, \ldots, v_k\}$ *is an orthonormal set in an n-dimensional inner product space* V. *Then*

(a) *S can be extended to an orthonormal basis $\{v_1, v_2, \ldots, v_k, v_{k+1}, \ldots, v_n\}$ for* V.

(b) *If* $W = \text{span}(S)$, *then* $S_1 = \{v_{k+1}, v_{k+2}, \ldots, v_n\}$ *is an orthonormal basis for* $W^\perp$ *(using the preceding notation).*

(c) *If* W *is any subspace of* V, *then* $\dim(V) = \dim(W) + \dim(W^\perp)$.

Proof. (a) By Corollary 2 to the replacement theorem (p. 47), S can be extended to an ordered basis $S' = \{v_1, v_2, \ldots, v_k, w_{k+1}, \ldots, w_n\}$ for V. Now apply the Gram–Schmidt process to S'. The first k vectors resulting from this process are the vectors in S by Exercise 8, and this new set spans V. Normalizing the last $n - k$ vectors of this set produces an orthonormal set that spans V. The result now follows.

(b) Because S_1 is a subset of a basis, it is linearly independent. Since S_1 is clearly a subset of $W^\perp$, we need only show that it spans $W^\perp$. Note that, for any $x \in V$, we have

$$ x = \sum_{i=1}^{n} \langle x, v_i \rangle\, v_i. $$

If $x \in W^\perp$, then $\langle x, v_i \rangle = 0$ for $1 \le i \le k$. Therefore,

$$ x = \sum_{i=k+1}^{n} \langle x, v_i \rangle\, v_i \in \text{span}(S_1). $$

(c) Let W be a subspace of V. It is a finite-dimensional inner product space because V is, and so it has an orthonormal basis $\{v_1, v_2, \ldots, v_k\}$. By (a) and (b), we have

$$ \dim(V) = n = k + (n - k) = \dim(W) + \dim(W^\perp). \qquad \blacksquare $$

Example 11

Let $W = \text{span}(\{e_1, e_2\})$ in F^3. Then $x = (a, b, c) \in W^\perp$ if and only if $0 = \langle x, e_1 \rangle = a$ and $0 = \langle x, e_2 \rangle = b$. So $x = (0, 0, c)$, and therefore $W^\perp = \text{span}(\{e_3\})$. One can deduce the same result by noting that $e_3 \in W^\perp$ and, from (c), that $\dim(W^\perp) = 3 - 2 = 1$. ◆

EXERCISES

1. Label the following statements as true or false.

   (a) The Gram–Schmidt orthogonalization process allows us to construct an orthonormal set from an arbitrary set of vectors.

(b) Every nonzero finite-dimensional inner product space has an orthonormal basis.

(c) The orthogonal complement of any set is a subspace.

(d) If $\{v_1, v_2, \ldots, v_n\}$ is a basis for an inner product space V, then for any $x \in V$ the scalars $\langle x, v_i \rangle$ are the Fourier coefficients of x.

(e) An orthonormal basis must be an ordered basis.

(f) Every orthogonal set is linearly independent.

(g) Every orthonormal set is linearly independent.

2. In each part, apply the Gram–Schmidt process to the given subset S of the inner product space V to obtain an orthogonal basis for span(S). Then normalize the vectors in this basis to obtain an orthonormal basis β for span(S), and compute the Fourier coefficients of the given vector relative to β. Finally, use Theorem 6.5 to verify your result.

(a) $V = R^3$, $S = \{(1,0,1),(0,1,1),(1,3,3)\}$, and $x = (1,1,2)$

(b) $V = R^3$, $S = \{(1,1,1),(0,1,1),(0,0,1)\}$, and $x = (1,0,1)$

(c) $V = P_2(R)$ with the inner product $\langle f(x), g(x) \rangle = \int_0^1 f(t)g(t)\,dt$, $S = \{1, x, x^2\}$, and $h(x) = 1 + x$

(d) $V = \text{span}(S)$, where $S = \{(1,i,0),(1-i,2,4i)\}$, and $x = (3+i, 4i, -4)$

(e) $V = R^4$, $S = \{(2,-1,-2,4),(-2,1,-5,5),(-1,3,7,11)\}$, and $x = (-11,8,-4,18)$

(f) $V = R^4$, $S = \{(1,-2,-1,3),(3,6,3,-1),(1,4,2,8)\}$, and $x = (-1,2,1,1)$

(g) $V = M_{2\times2}(R)$, $S = \left\{\begin{pmatrix} 3 & 5 \\ -1 & 1 \end{pmatrix}, \begin{pmatrix} -1 & 9 \\ 5 & -1 \end{pmatrix}, \begin{pmatrix} 7 & -17 \\ 2 & -6 \end{pmatrix}\right\}$, and $A = \begin{pmatrix} -1 & 27 \\ -4 & 8 \end{pmatrix}$

(h) $V = M_{2\times2}(R)$, $S = \left\{\begin{pmatrix} 2 & 2 \\ 2 & 1 \end{pmatrix}, \begin{pmatrix} 11 & 4 \\ 2 & 5 \end{pmatrix}, \begin{pmatrix} 4 & -12 \\ 3 & -16 \end{pmatrix}\right\}$, and $A = \begin{pmatrix} 8 & 6 \\ 25 & -13 \end{pmatrix}$

(i) $V = \text{span}(S)$ with the inner product $\langle f, g \rangle = \int_0^\pi f(t)g(t)\,dt$, $S = \{\sin t, \cos t, 1, t\}$, and $h(t) = 2t + 1$

(j) $V = C^4$, $S = \{(1,i,2-i,-1),(2+3i,3i,1-i,2i),(-1+7i,6+10i,11-4i,3+4i)\}$, and $x = (-2+7i,6+9i,9-3i,4+4i)$

(k) $V = C^4$, $S = \{(-4,3-2i,i,1-4i),(-1-5i,5-4i,-3+5i,7-2i),(-27-i,-7-6i,-15+25i,-7-6i)\}$, and $x = (-13-7i,-12+3i,-39-11i,-26+5i)$

(1) $V = M_{2\times 2}(C)$, $S = \left\{ \begin{pmatrix} 1-i & -2-3i \\ 2+2i & 4+i \end{pmatrix}, \begin{pmatrix} 8i & 4 \\ -3-3i & -4+4i \end{pmatrix}, \right.$

$\left. \begin{pmatrix} -25-38i & -2-13i \\ 12-78i & -7+24i \end{pmatrix} \right\}$, and $A = \begin{pmatrix} -2+8i & -13+i \\ 10-10i & 9-9i \end{pmatrix}$

(m) $V = M_{2\times 2}(C)$, $S = \left\{ \begin{pmatrix} -1+i & -i \\ 2-i & 1+3i \end{pmatrix}, \begin{pmatrix} -1-7i & -9-8i \\ 1+10i & -6-2i \end{pmatrix}, \right.$

$\left. \begin{pmatrix} -11-132i & -34-31i \\ 7-126i & -71-5i \end{pmatrix} \right\}$, and $A = \begin{pmatrix} -7+5i & 3+18i \\ 9-6i & -3+7i \end{pmatrix}$

3. In R^2, let

$$\beta = \left\{ \left(\frac{1}{\sqrt{2}}, \frac{1}{\sqrt{2}} \right), \left(\frac{1}{\sqrt{2}}, \frac{-1}{\sqrt{2}} \right) \right\}.$$

Find the Fourier coefficients of $(3,4)$ relative to β.

4. Let $S = \{(1,0,i), (1,2,1)\}$ in C^3. Compute $S^{\perp}$.

5. Let $S_0 = \{x_0\}$, where x_0 is a nonzero vector in R^3. Describe $S_0^{\perp}$ geometrically. Now suppose that $S = \{x_1, x_2\}$ is a linearly independent subset of R^3. Describe $S^{\perp}$ geometrically.

6. Let V be an inner product space, and let W be a finite-dimensional subspace of V. If $x \notin W$, prove that there exists $y \in V$ such that $y \in W^{\perp}$, but $\langle x, y \rangle \neq 0$. *Hint:* Use Theorem 6.6.

7. Let β be a basis for a subspace W of an inner product space V, and let $z \in V$. Prove that $z \in W^{\perp}$ if and only if $\langle z, v \rangle = 0$ for every $v \in \beta$.

8. Prove that if $\{w_1, w_2, \ldots, w_n\}$ is an orthogonal set of nonzero vectors, then the vectors $v_1, v_2, \ldots, v_n$ derived from the Gram–Schmidt process satisfy $v_i = w_i$ for $i = 1, 2, \ldots, n$. *Hint:* Use mathematical induction.

9. Let $W = \text{span}(\{(i,0,1)\})$ in C^3. Find orthonormal bases for W and $W^{\perp}$.

10. Let W be a finite-dimensional subspace of an inner product space V. Prove that there exists a projection T on W along $W^{\perp}$ that satisfies $N(T) = W^{\perp}$. In addition, prove that $\|T(x)\| \leq \|x\|$ for all $x \in V$. *Hint:* Use Theorem 6.6 and Exercise 10 of Section 6.1. (Projections are defined in the exercises of Section 2.1.)

11. Let A be an $n \times n$ matrix with complex entries. Prove that $AA^* = I$ if and only if the rows of A form an orthonormal basis for C^n.

12. Prove that for any matrix $A \in M_{m\times n}(F)$, $(R(L_{A^*}))^{\perp} = N(L_A)$.

13. Let V be an inner product space, S and S_0 be subsets of V, and W be a finite-dimensional subspace of V. Prove the following results.

   (a) $S_0 \subseteq S$ implies that $S^\perp \subseteq S_0^\perp$.
   (b) $S \subseteq (S^\perp)^\perp$; so span$(S) \subseteq (S^\perp)^\perp$.
   (c) $\mathsf{W} = (\mathsf{W}^\perp)^\perp$. *Hint:* Use Exercise 6.
   (d) $\mathsf{V} = \mathsf{W} \oplus \mathsf{W}^\perp$. (See the exercises of Section 1.3.)

14. Let W_1 and W_2 be subspaces of a finite-dimensional inner product space. Prove that $(\mathsf{W}_1 + \mathsf{W}_2)^\perp = \mathsf{W}_1^\perp \cap \mathsf{W}_2^\perp$ and $(\mathsf{W}_1 \cap \mathsf{W}_2)^\perp = \mathsf{W}_1^\perp + \mathsf{W}_2^\perp$. (See the definition of the sum of subsets of a vector space on page 22.) *Hint for the second equation:* Apply Exercise 13(c) to the first equation.

15. Let V be a finite-dimensional inner product space over F.

   (a) *Parseval's Identity.* Let $\{v_1, v_2, \ldots, v_n\}$ be an orthonormal basis for V. For any $x, y \in \mathsf{V}$ prove that

$$\langle x, y \rangle = \sum_{i=1}^{n} \langle x, v_i \rangle \overline{\langle y, v_i \rangle}.$$

   (b) Use (a) to prove that if β is an orthonormal basis for V with inner product $\langle \cdot, \cdot \rangle$, then for any $x, y \in \mathsf{V}$

$$\langle \phi_\beta(x), \phi_\beta(y) \rangle' = \langle [x]_\beta, [y]_\beta \rangle' = \langle x, y \rangle,$$

   where $\langle \cdot, \cdot \rangle'$ is the standard inner product on F^n.

16. (a) *Bessel's Inequality.* Let V be an inner product space, and let $S = \{v_1, v_2, \ldots, v_n\}$ be an orthonormal subset of V. Prove that for any $x \in \mathsf{V}$ we have

$$\|x\|^2 \geq \sum_{i=1}^{n} |\langle x, v_i \rangle|^2.$$

   Hint: Apply Theorem 6.6 to $x \in \mathsf{V}$ and $\mathsf{W} = \text{span}(S)$. Then use Exercise 10 of Section 6.1.
   (b) In the context of (a), prove that Bessel's inequality is an equality if and only if $x \in \text{span}(S)$.

17. Let T be a linear operator on an inner product space V. If $\langle \mathsf{T}(x), y \rangle = 0$ for all $x, y \in \mathsf{V}$, prove that $\mathsf{T} = \mathsf{T}_0$. In fact, prove this result if the equality holds for all x and y in some basis for V.

18. Let $\mathsf{V} = C([-1, 1])$. Suppose that W_e and W_o denote the subspaces of V consisting of the even and odd functions, respectively. (See Exercise 22 of Section 1.3.) Prove that $W_e^\perp = W_o$, where the inner product on V is defined by

$$\langle f, g \rangle = \int_{-1}^{1} f(t)g(t)\, dt.$$

19. In each of the following parts, find the orthogonal projection of the given vector on the given subspace W of the inner product space V.

(a) $V = R^2$, $u = (2,6)$, and $W = \{(x, y): y = 4x\}$.

(b) $V = R^3$, $u = (2,1,3)$, and $W = \{(x, y, z): x + 3y - 2z = 0\}$.

(c) $V = P(R)$ with the inner product $\langle f(x), g(x) \rangle = \int_0^1 f(t)g(t)\, dt$, $h(x) = 4 + 3x - 2x^2$, and $W = P_1(R)$.

20. In each part of Exercise 19, find the distance from the given vector to the subspace W.

21. Let $V = C([-1,1])$ with the inner product $\langle f, g \rangle = \int_{-1}^{1} f(t)g(t)\, dt$, and let W be the subspace $P_2(R)$, viewed as a space of functions. Use the orthonormal basis obtained in Example 5 to compute the "best" (closest) second-degree polynomial approximation of the function $h(t) = e^t$ on the interval $[-1,1]$.

22. Let $V = C([0,1])$ with the inner product $\langle f, g \rangle = \int_0^1 f(t)g(t)\, dt$. Let W be the subspace spanned by the linearly independent set $\{t, \sqrt{t}\}$.

(a) Find an orthonormal basis for W.

(b) Let $h(t) = t^2$. Use the orthonormal basis obtained in (a) to obtain the "best" (closest) approximation of h in W.

23. Let V be the vector space defined in Example 5 of Section 1.2, the space of all sequences σ in F (where $F = R$ or $F = C$) such that $\sigma(n) \neq 0$ for only finitely many positive integers n. For $\sigma, \mu \in V$, we define $\langle \sigma, \mu \rangle = \sum_{n=1}^{\infty} \sigma(n)\overline{\mu(n)}$. Since all but a finite number of terms of the series are zero, the series converges.

(a) Prove that $\langle \cdot, \cdot \rangle$ is an inner product on V, and hence V is an inner product space.

(b) For each positive integer n, let e_n be the sequence defined by $e_n(k) = \delta_{n,k}$, where $\delta_{n,k}$ is the Kronecker delta. Prove that $\{e_1, e_2, \ldots\}$ is an orthonormal basis for V.

(c) Let $\sigma_n = e_1 + e_n$ and $W = \text{span}(\{\sigma_n : n \geq 2\})$.

   (i) Prove that $e_1 \notin W$, so $W \neq V$.

   (ii) Prove that $W^\perp = \{0\}$, and conclude that $W \neq (W^\perp)^\perp$.

Thus the assumption in Exercise 13(c) that W is finite-dimensional is essential.

6.3   THE ADJOINT OF A LINEAR OPERATOR

In Section 6.1, we defined the conjugate transpose A^* of a matrix A. For a linear operator T on an inner product space V, we now define a related linear operator on V called the *adjoint* of T, whose matrix representation with respect to any orthonormal basis β for V is $[T]_\beta^*$. The analogy between conjugation of complex numbers and adjoints of linear operators will become apparent. We first need a preliminary result.

Let V be an inner product space, and let $y \in V$. The function g: $V \to F$ defined by $g(x) = \langle x, y \rangle$ is clearly linear. More interesting is the fact that if V is finite-dimensional, every linear transformation from V into F is of this form.

Theorem 6.8. *Let V be a finite-dimensional inner product space over F, and let g: $V \to F$ be a linear transformation. Then there exists a unique vector $y \in V$ such that $g(x) = \langle x, y \rangle$ for all $x \in V$.*

Proof. Let $\beta = \{v_1, v_2, \dots, v_n\}$ be an orthonormal basis for V, and let

$$y = \sum_{i=1}^{n} \overline{g(v_i)} v_i.$$

Define h: $V \to F$ by $h(x) = \langle x, y \rangle$, which is clearly linear. Furthermore, for $1 \le j \le n$ we have

$$h(v_j) = \langle v_j, y \rangle = \left\langle v_j, \sum_{i=1}^{n} \overline{g(v_i)} v_i \right\rangle = \sum_{i=1}^{n} g(v_i) \langle v_j, v_i \rangle$$

$$= \sum_{i=1}^{n} g(v_i) \delta_{ji} = g(v_j).$$

Since g and h both agree on β, we have that $g = h$ by the corollary to Theorem 2.6 (p. 73).

To show that y is unique, suppose that $g(x) = \langle x, y' \rangle$ for all x. Then $\langle x, y \rangle = \langle x, y' \rangle$ for all x; so by Theorem 6.1(e) (p. 333), we have $y = y'$. ∎

Example 1

Define g: $R^2 \to R$ by $g(a_1, a_2) = 2a_1 + a_2$; clearly g is a linear transformation. Let $\beta = \{e_1, e_2\}$, and let $y = g(e_1)e_1 + g(e_2)e_2 = 2e_1 + e_2 = (2, 1)$, as in the proof of Theorem 6.8. Then $g(a_1, a_2) = \langle (a_1, a_2), (2, 1) \rangle = 2a_1 + a_2$. ♦

Theorem 6.9. *Let* V *be a finite-dimensional inner product space, and let* T *be a linear operator on* V. *Then there exists a unique function* $T^*\colon V \to V$ *such that* $\langle T(x), y \rangle = \langle x, T^*(y) \rangle$ *for all* $x, y \in V$. *Furthermore,* T^* *is linear.*

Proof. Let $y \in V$. Define $g\colon V \to F$ by $g(x) = \langle T(x), y \rangle$ for all $x \in V$. We first show that g is linear. Let $x_1, x_2 \in V$ and $c \in F$. Then

$$g(cx_1 + x_2) = \langle T(cx_1 + x_2), y \rangle = \langle cT(x_1) + T(x_2), y \rangle$$
$$= c\langle T(x_1), y \rangle + \langle T(x_2), y \rangle = cg(x_1) + g(x_2).$$

Hence g is linear.

We now apply Theorem 6.8 to obtain a unique vector $y' \in V$ such that $g(x) = \langle x, y' \rangle$; that is, $\langle T(x), y \rangle = \langle x, y' \rangle$ for all $x \in V$. Defining $T^*\colon V \to V$ by $T^*(y) = y'$, we have $\langle T(x), y \rangle = \langle x, T^*(y) \rangle$.

To show that T^* is linear, let $y_1, y_2 \in V$ and $c \in F$. Then for any $x \in V$, we have

$$\langle x, T^*(cy_1 + y_2) \rangle = \langle T(x), cy_1 + y_2 \rangle$$
$$= \bar{c}\langle T(x), y_1 \rangle + \langle T(x), y_2 \rangle$$
$$= \bar{c}\langle x, T^*(y_1) \rangle + \langle x, T^*(y_2) \rangle$$
$$= \langle x, cT^*(y_1) + T^*(y_2) \rangle .$$

Since x is arbitrary, $T^*(cy_1 + y_2) = cT^*(y_1) + T^*(y_2)$ by Theorem 6.1(e) (p. 333).

Finally, we need to show that T^* is unique. Suppose that $U\colon V \to V$ is linear and that it satisfies $\langle T(x), y \rangle = \langle x, U(y) \rangle$ for all $x, y \in V$. Then $\langle x, T^*(y) \rangle = \langle x, U(y) \rangle$ for all $x, y \in V$, so $T^* = U$. ∎

The linear operator T^* described in Theorem 6.9 is called the **adjoint** of the operator T. The symbol T^* is read "T star."

Thus T^* is the unique operator on V satisfying $\langle T(x), y \rangle = \langle x, T^*(y) \rangle$ for all $x, y \in V$. Note that we also have

$$\langle x, T(y) \rangle = \overline{\langle T(y), x \rangle} = \overline{\langle y, T^*(x) \rangle} = \langle T^*(x), y \rangle \, ;$$

so $\langle x, T(y) \rangle = \langle T^*(x), y \rangle$ for all $x, y \in V$. We may view these equations symbolically as adding a * to T when shifting its position inside the inner product symbol.

For an infinite-dimensional inner product space, the adjoint of a linear operator T may be defined to be the function T^* such that $\langle T(x), y \rangle = \langle x, T^*(y) \rangle$ for all $x, y \in V$, provided it exists. Although the uniqueness and linearity of T^* follow as before, the existence of the adjoint is not guaranteed (see Exercise 24). The reader should observe the necessity of the hypothesis of finite-dimensionality in the proof of Theorem 6.8. Many of the theorems we prove about adjoints, nevertheless, do not depend on V being finite-dimensional. *Thus, unless stated otherwise, for the remainder of this chapter we adopt the convention that a reference to the adjoint of a linear operator on an infinite-dimensional inner product space assumes its existence.*

Theorem 6.10 is a useful result for computing adjoints.

Theorem 6.10. *Let* V *be a finite-dimensional inner product space, and let β be an orthonormal basis for* V. *If* T *is a linear operator on* V, *then*

$$[\mathsf{T}^*]_\beta = [\mathsf{T}]_\beta^*.$$

Proof. Let $A = [\mathsf{T}]_\beta$, $B = [\mathsf{T}^*]_\beta$, and $\beta = \{v_1, v_2, \ldots, v_n\}$. Then from the corollary to Theorem 6.5 (p. 346), we have

$$B_{ij} = \langle \mathsf{T}^*(v_j), v_i \rangle = \overline{\langle v_i, \mathsf{T}^*(v_j) \rangle} = \overline{\langle \mathsf{T}(v_i), v_j \rangle} = \overline{A_{ji}} = (A^*)_{ij}.$$

Hence $B = A^*$.   ∎

Corollary. *Let* A *be an* $n \times n$ *matrix. Then* $\mathsf{L}_{A^*} = (\mathsf{L}_A)^*$.

Proof. If β is the standard ordered basis for F^n, then, by Theorem 2.16 (p. 93), we have $[\mathsf{L}_A]_\beta = A$. Hence $[(\mathsf{L}_A)^*]_\beta = [\mathsf{L}_A]_\beta^* = A^* = [\mathsf{L}_{A^*}]_\beta$, and so $(\mathsf{L}_A)^* = \mathsf{L}_{A^*}$.   ∎

As an illustration of Theorem 6.10, we compute the adjoint of a specific linear operator.

Example 2

Let T be the linear operator on C^2 defined by $\mathsf{T}(a_1, a_2) = (2ia_1 + 3a_2, a_1 - a_2)$. If β is the standard ordered basis for C^2, then

$$[\mathsf{T}]_\beta = \begin{pmatrix} 2i & 3 \\ 1 & -1 \end{pmatrix}.$$

So

$$[\mathsf{T}^*]_\beta = [\mathsf{T}]_\beta^* = \begin{pmatrix} -2i & 1 \\ 3 & -1 \end{pmatrix}.$$

Hence

$$\mathsf{T}^*(a_1, a_2) = (-2ia_1 + a_2, 3a_1 - a_2). \quad \blacklozenge$$

The following theorem suggests an analogy between the conjugates of complex numbers and the adjoints of linear operators.

Theorem 6.11. *Let* V *be an inner product space, and let* T *and* U *be linear operators on* V. *Then*

(a) $(\mathsf{T} + \mathsf{U})^* = \mathsf{T}^* + \mathsf{U}^*$;
(b) $(c\mathsf{T})^* = \bar{c}\,\mathsf{T}^*$ for any $c \in F$;
(c) $(\mathsf{TU})^* = \mathsf{U}^*\mathsf{T}^*$;
(d) $\mathsf{T}^{**} = \mathsf{T}$;
(e) $\mathsf{I}^* = \mathsf{I}$.

Proof. We prove (a) and (d); the rest are proved similarly. Let $x, y \in \mathsf{V}$.
(a) Because

$$\langle x, (\mathsf{T} + \mathsf{U})^*(y) \rangle = \langle (\mathsf{T} + \mathsf{U})(x), y \rangle = \langle \mathsf{T}(x) + \mathsf{U}(x), y \rangle$$
$$= \langle \mathsf{T}(x), y \rangle + \langle \mathsf{U}(x), y \rangle = \langle x, \mathsf{T}^*(y) \rangle + \langle x, \mathsf{U}^*(y) \rangle$$
$$= \langle x, \mathsf{T}^*(y) + \mathsf{U}^*(y) \rangle = \langle x, (\mathsf{T}^* + \mathsf{U}^*)(y) \rangle,$$

$\mathsf{T}^* + \mathsf{U}^*$ has the property unique to $(\mathsf{T} + \mathsf{U})^*$. Hence $\mathsf{T}^* + \mathsf{U}^* = (\mathsf{T} + \mathsf{U})^*$.
(d) Similarly, since

$$\langle x, \mathsf{T}(y) \rangle = \langle \mathsf{T}^*(x), y \rangle = \langle x, \mathsf{T}^{**}(y) \rangle,$$

(d) follows.   ∎

The same proof works in the infinite-dimensional case, provided that the existence of T^* and U^* is assumed.

Corollary. *Let A and B be $n \times n$ matrices. Then*
(a) $(A + B)^* = A^* + B^*$;
(b) $(cA)^* = \bar{c}A^*$ for all $c \in F$;
(c) $(AB)^* = B^*A^*$;
(d) $A^{**} = A$;
(e) $I^* = I$.

Proof. We prove only (c); the remaining parts can be proved similarly.
Since $\mathsf{L}_{(AB)^*} = (\mathsf{L}_{AB})^* = (\mathsf{L}_A\mathsf{L}_B)^* = (\mathsf{L}_B)^*(\mathsf{L}_A)^* = \mathsf{L}_{B^*}\mathsf{L}_{A^*} = \mathsf{L}_{B^*A^*}$, we
have $(AB)^* = B^*A^*$.   ∎

In the preceding proof, we relied on the corollary to Theorem 6.10. An alternative proof, which holds even for nonsquare matrices, can be given by appealing directly to the definition of the conjugate transpose of a matrix (see Exercise 5).

Least Squares Approximation

Consider the following problem: An experimenter collects data by taking measurements $y_1, y_2, \ldots, y_m$ at times $t_1, t_2, \ldots, t_m$, respectively. For example, he or she may be measuring unemployment at various times during some period. Suppose that the data $(t_1, y_1), (t_2, y_2), \ldots, (t_m, y_m)$ are plotted as points in the plane. (See Figure 6.3.) From this plot, the experimenter feels that there exists an essentially linear relationship between y and t, say $y = ct + d$, and would like to find the constants c and d so that the line $y = ct + d$ represents the best possible *fit* to the data collected. One such estimate of fit is to calculate the error E that represents the sum of the squares of the vertical distances from the points to the line; that is,

$$E = \sum_{i=1}^{m} (y_i - ct_i - d)^2.$$

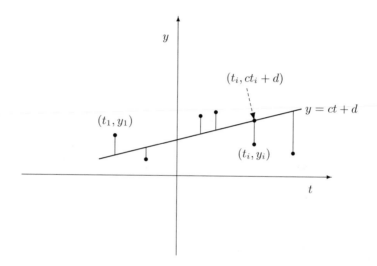

Figure 6.3

Thus the problem is reduced to finding the constants c and d that minimize E. (For this reason the line $y = ct + d$ is called the **least squares line**.) If we let

$$A = \begin{pmatrix} t_1 & 1 \\ t_2 & 1 \\ \vdots & \vdots \\ t_m & 1 \end{pmatrix}, \quad x = \begin{pmatrix} c \\ d \end{pmatrix}, \quad \text{and} \quad y = \begin{pmatrix} y_1 \\ y_2 \\ \vdots \\ y_m \end{pmatrix},$$

then it follows that $E = \|y - Ax\|^2$.

We develop a general method for finding an explicit vector $x_0 \in \mathsf{F}^n$ that minimizes E; that is, given an $m \times n$ matrix A, we find $x_0 \in \mathsf{F}^n$ such that $\|y - Ax_0\| \le \|y - Ax\|$ for all vectors $x \in \mathsf{F}^n$. This method not only allows us to find the linear function that best fits the data, but also, for any positive integer n, the best fit using a polynomial of degree at most n.

First, we need some notation and two simple lemmas. For $x, y \in \mathsf{F}^n$, let $\langle x, y \rangle_n$ denote the standard inner product of x and y in F^n. Recall that if x and y are regarded as column vectors, then $\langle x, y \rangle_n = y^* x$.

Lemma 1. *Let* $A \in \mathsf{M}_{m \times n}(F)$, $x \in \mathsf{F}^n$, *and* $y \in \mathsf{F}^m$. *Then*

$$\langle Ax, y \rangle_m = \langle x, A^* y \rangle_n.$$

Proof. By a generalization of the corollary to Theorem 6.11 (see Exercise 5(b)), we have

$$\langle Ax, y \rangle_m = y^*(Ax) = (y^* A)x = (A^* y)^* x = \langle x, A^* y \rangle_n. \qquad \blacksquare$$

Lemma 2. *Let* $A \in \mathsf{M}_{m \times n}(F)$. *Then* $\operatorname{rank}(A^* A) = \operatorname{rank}(A)$.

Proof. By the dimension theorem, we need only show that, for $x \in \mathsf{F}^n$, we have $A^* Ax = 0$ if and only if $Ax = 0$. Clearly, $Ax = 0$ implies that $A^* Ax = 0$. So assume that $A^* Ax = 0$. Then

$$0 = \langle A^* Ax, x \rangle_n = \langle Ax, A^{**} x \rangle_m = \langle Ax, Ax \rangle_m,$$

so that $Ax = 0$. $\qquad \blacksquare$

Corollary. *If* A *is an* $m \times n$ *matrix such that* $\operatorname{rank}(A) = n$, *then* $A^* A$ *is invertible.*

Now let A be an $m \times n$ matrix and $y \in \mathsf{F}^m$. Define $\mathsf{W} = \{Ax \colon x \in \mathsf{F}^n\}$; that is, $\mathsf{W} = \mathsf{R}(\mathsf{L}_A)$. By the corollary to Theorem 6.6 (p. 350), there exists a unique vector in W that is closest to y. Call this vector Ax_0, where $x_0 \in \mathsf{F}^n$. Then $\|Ax_0 - y\| \le \|Ax - y\|$ for all $x \in \mathsf{F}^n$; so x_0 has the property that $E = \|Ax_0 - y\|$ is minimal, as desired.

To develop a practical method for finding such an x_0, we note from Theorem 6.6 and its corollary that $Ax_0 - y \in \mathsf{W}^\perp$; so $\langle Ax, Ax_0 - y \rangle_m = 0$ for all $x \in \mathsf{F}^n$. Thus, by Lemma 1, we have that $\langle x, A^*(Ax_0 - y) \rangle_n = 0$ for all $x \in \mathsf{F}^n$; that is, $A^*(Ax_0 - y) = 0$. So we need only find a solution x_0 to $A^* Ax = A^* y$. If, in addition, we assume that $\operatorname{rank}(A) = n$, then by Lemma 2 we have $x_0 = (A^* A)^{-1} A^* y$. We summarize this discussion in the following theorem.

Theorem 6.12. *Let* $A \in \mathsf{M}_{m \times n}(F)$ *and* $y \in \mathsf{F}^m$. *Then there exists* $x_0 \in \mathsf{F}^n$ *such that* $(A^* A)x_0 = A^* y$ *and* $\|Ax_0 - y\| \le \|Ax - y\|$ *for all* $x \in \mathsf{F}^n$. *Furthermore, if* $\operatorname{rank}(A) = n$, *then* $x_0 = (A^* A)^{-1} A^* y$.

To return to our experimenter, let us suppose that the data collected are $(1, 2), (2, 3), (3, 5),$ and $(4, 7)$. Then

$$A = \begin{pmatrix} 1 & 1 \\ 2 & 1 \\ 3 & 1 \\ 4 & 1 \end{pmatrix} \quad \text{and} \quad y = \begin{pmatrix} 2 \\ 3 \\ 5 \\ 7 \end{pmatrix};$$

hence

$$A^*A = \begin{pmatrix} 1 & 2 & 3 & 4 \\ 1 & 1 & 1 & 1 \end{pmatrix} \begin{pmatrix} 1 & 1 \\ 2 & 1 \\ 3 & 1 \\ 4 & 1 \end{pmatrix} = \begin{pmatrix} 30 & 10 \\ 10 & 4 \end{pmatrix}.$$

Thus

$$(A^*A)^{-1} = \frac{1}{20} \begin{pmatrix} 4 & -10 \\ -10 & 30 \end{pmatrix}.$$

Therefore

$$\begin{pmatrix} c \\ d \end{pmatrix} = x_0 = \frac{1}{20} \begin{pmatrix} 4 & -10 \\ -10 & 30 \end{pmatrix} \begin{pmatrix} 1 & 2 & 3 & 4 \\ 1 & 1 & 1 & 1 \end{pmatrix} \begin{pmatrix} 2 \\ 3 \\ 5 \\ 7 \end{pmatrix} = \begin{pmatrix} 1.7 \\ 0 \end{pmatrix}.$$

It follows that the line $y = 1.7t$ is the least squares line. The error E may be computed directly as $\|Ax_0 - y\|^2 = 0.3$.

Suppose that the experimenter chose the times t_i $(1 \le i \le m)$ to satisfy

$$\sum_{i=1}^{m} t_i = 0.$$

Then the two columns of A would be orthogonal, so A^*A would be a diagonal matrix (see Exercise 19). In this case, the computations are greatly simplified.

In practice, the $m \times 2$ matrix A in our least squares application has rank equal to two, and hence A^*A is invertible by the corollary to Lemma 2. For, otherwise, the first column of A is a multiple of the second column, which consists only of ones. But this would occur only if the experimenter collects all the data at exactly one time.

Finally, the method above may also be applied if, for some k, the experimenter wants to fit a polynomial of degree at most k to the data. For instance, if a polynomial $y = ct^2 + dt + e$ of degree at most 2 is desired, the appropriate model is

$$x = \begin{pmatrix} c \\ d \\ e \end{pmatrix}, \quad y = \begin{pmatrix} y_1 \\ y_2 \\ \vdots \\ y_m \end{pmatrix}, \quad \text{and} \quad A = \begin{pmatrix} t_1^2 & t_1 & 1 \\ \vdots & \vdots & \vdots \\ t_m^2 & t_m & 1 \end{pmatrix}.$$

Minimal Solutions to Systems of Linear Equations

Even when a system of linear equations $Ax = b$ is consistent, there may be no unique solution. In such cases, it may be desirable to find a solution of minimal norm. A solution s to $Ax = b$ is called a **minimal solution** if $\|s\| \le \|u\|$ for all other solutions u. The next theorem assures that every consistent system of linear equations has a unique minimal solution and provides a method for computing it.

Theorem 6.13. *Let $A \in \mathsf{M}_{m \times n}(F)$ and $b \in F^m$. Suppose that $Ax = b$ is consistent. Then the following statements are true.*

(a) *There exists exactly one minimal solution s of $Ax = b$, and $s \in \mathsf{R}(\mathsf{L}_{A^*})$.*

(b) *The vector s is the only solution to $Ax = b$ that lies in $\mathsf{R}(\mathsf{L}_{A^*})$; that is, if u satisfies $(AA^*)u = b$, then $s = A^*u$.*

Proof. (a) For simplicity of notation, we let $\mathsf{W} = \mathsf{R}(\mathsf{L}_{A^*})$ and $\mathsf{W}' = \mathsf{N}(\mathsf{L}_A)$. Let x be any solution to $Ax = b$. By Theorem 6.6 (p. 350), $x = s + y$ for some $s \in \mathsf{W}$ and $y \in \mathsf{W}^\perp$. But $\mathsf{W}^\perp = \mathsf{W}'$ by Exercise 12, and therefore $b = Ax = As + Ay = As$. So s is a solution to $Ax = b$ that lies in W. To prove (a), we need only show that s is the unique minimal solution. Let v be any solution to $Ax = b$. By Theorem 3.9 (p. 172), we have that $v = s + u$, where $u \in \mathsf{W}'$. Since $s \in \mathsf{W}$, which equals $\mathsf{W}'^\perp$ by Exercise 12, we have

$$\|v\|^2 = \|s + u\|^2 = \|s\|^2 + \|u\|^2 \ge \|s\|^2$$

by Exercise 10 of Section 6.1. Thus s is a minimal solution. We can also see from the preceding calculation that if $\|v\| = \|s\|$, then $u = 0$; hence $v = s$. Therefore s is the unique minimal solution to $Ax = b$, proving (a).

(b) Assume that v is also a solution to $Ax = b$ that lies in W. Then

$$v - s \in \mathsf{W} \cap \mathsf{W}' = \mathsf{W} \cap \mathsf{W}^\perp = \{0\};$$

so $v = s$.

Finally, suppose that $(AA^*)u = b$, and let $v = A^*u$. Then $v \in \mathsf{W}$ and $Av = b$. Therefore $s = v = A^*u$ by the discussion above. ∎

Example 3

Consider the system

$$\begin{aligned} x + 2y + z &= 4 \\ x - y + 2z &= -11 \\ x + 5y &= 19. \end{aligned}$$

Let

$$A = \begin{pmatrix} 1 & 2 & 1 \\ 1 & -1 & 2 \\ 1 & 5 & 0 \end{pmatrix} \quad \text{and} \quad b = \begin{pmatrix} 4 \\ -11 \\ 19 \end{pmatrix}.$$

To find the minimal solution to this system, we must first find some solution u to $AA^*x = b$. Now

$$AA^* = \begin{pmatrix} 6 & 1 & 11 \\ 1 & 6 & -4 \\ 11 & -4 & 26 \end{pmatrix};$$

so we consider the system

$$\begin{aligned} 6x + \ y + 11z &= \ \ 4 \\ x + 6y - \ \ 4z &= -11 \\ 11x - 4y + 26z &= \ \ 19, \end{aligned}$$

for which one solution is

$$u = \begin{pmatrix} 1 \\ -2 \\ 0 \end{pmatrix}.$$

(Any solution will suffice.) Hence

$$s = A^*u = \begin{pmatrix} -1 \\ 4 \\ -3 \end{pmatrix}$$

is the minimal solution to the given system.   ◆

EXERCISES

1. Label the following statements as true or false. Assume that the underlying inner product spaces are finite-dimensional.

   (a) Every linear operator has an adjoint.
   (b) Every linear operator on V has the form $x \to \langle x, y \rangle$ for some $y \in V$.
   (c) For every linear operator T on V and every ordered basis β for V, we have $[T^*]_\beta = ([T]_\beta)^*$.
   (d) The adjoint of a linear operator is unique.
   (e) For any linear operators T and U and scalars a and b,

   $$(aT + bU)^* = aT^* + bU^*.$$

   (f) For any $n \times n$ matrix A, we have $(L_A)^* = L_{A^*}$.
   (g) For any linear operator T, we have $(T^*)^* = T$.

2. For each of the following inner product spaces V (over F) and linear transformations $g \colon V \to F$, find a vector y such that $g(x) = \langle x, y \rangle$ for all $x \in V$.

(a) $V = R^3$, $g(a_1, a_2, a_3) = a_1 - 2a_2 + 4a_3$

(b) $V = C^2$, $g(z_1, z_2) = z_1 - 2z_2$

(c) $V = P_2(R)$ with $\langle f, h \rangle = \int_0^1 f(t)h(t)\,dt$, $g(f) = f(0) + f'(1)$

3. For each of the following inner product spaces V and linear operators T on V, evaluate T^* at the given vector in V.

(a) $V = R^2$, $T(a, b) = (2a + b, a - 3b)$, $x = (3, 5)$.

(b) $V = C^2$, $T(z_1, z_2) = (2z_1 + iz_2, (1 - i)z_1)$, $x = (3 - i, 1 + 2i)$.

(c) $V = P_1(R)$ with $\langle f, g \rangle = \int_{-1}^1 f(t)g(t)\,dt$, $T(f) = f' + 3f$,

$f(t) = 4 - 2t$

4. Complete the proof of Theorem 6.11.

5. (a)   Complete the proof of the corollary to Theorem 6.11 by using Theorem 6.11, as in the proof of (c).

(b)   State a result for nonsquare matrices that is analogous to the corollary to Theorem 6.11, and prove it using a matrix argument.

6. Let T be a linear operator on an inner product space V. Let $U_1 = T + T^*$ and $U_2 = TT^*$. Prove that $U_1 = U_1^*$ and $U_2 = U_2^*$.

7. Give an example of a linear operator T on an inner product space V such that $N(T) \neq N(T^*)$.

8. Let V be a finite-dimensional inner product space, and let T be a linear operator on V. Prove that if T is invertible, then T^* is invertible and $(T^*)^{-1} = (T^{-1})^*$.

9. Prove that if $V = W \oplus W^\perp$ and T is the projection on W along $W^\perp$, then $T = T^*$. *Hint:* Recall that $N(T) = W^\perp$. (For definitions, see the exercises of Sections 1.3 and 2.1.)

10. Let T be a linear operator on an inner product space V. Prove that $\|T(x)\| = \|x\|$ for all $x \in V$ if and only if $\langle T(x), T(y) \rangle = \langle x, y \rangle$ for all $x, y \in V$. *Hint:* Use Exercise 20 of Section 6.1.

11. For a linear operator T on an inner product space V, prove that $T^*T = T_0$ implies $T = T_0$. Is the same result true if we assume that $TT^* = T_0$?

12. Let V be an inner product space, and let T be a linear operator on V. Prove the following results.

(a) $R(T^*)^\perp = N(T)$.

(b)   If V is finite-dimensional, then $R(T^*) = N(T)^\perp$. *Hint:* Use Exercise 13(c) of Section 6.2.

13. Let T be a linear operator on a finite-dimensional vector space V. Prove the following results.

    (a) $N(T^*T) = N(T)$. Deduce that $\text{rank}(T^*T) = \text{rank}(T)$.
    (b) $\text{rank}(T) = \text{rank}(T^*)$. Deduce from (a) that $\text{rank}(TT^*) = \text{rank}(T)$.
    (c) For any $n \times n$ matrix A, $\text{rank}(A^*A) = \text{rank}(AA^*) = \text{rank}(A)$.

14. Let V be an inner product space, and let $y, z \in$ V. Define T: V $\to$ V by $T(x) = \langle x, y \rangle z$ for all $x \in$ V. First prove that T is linear. Then show that T^* exists, and find an explicit expression for it.

The following definition is used in Exercises 15–17 and is an extension of the definition of the *adjoint* of a linear operator.

Definition. *Let* T: V $\to$ W *be a linear transformation, where* V *and* W *are finite-dimensional inner product spaces with inner products* $\langle \cdot, \cdot \rangle_1$ *and* $\langle \cdot, \cdot \rangle_2$, *respectively. A function* T^*: W $\to$ V *is called an* **adjoint** *of* T *if* $\langle T(x), y \rangle_2 = \langle x, T^*(y) \rangle_1$ *for all* $x \in$ V *and* $y \in$ W.

15. Let T: V $\to$ W be a linear transformation, where V and W are finite-dimensional inner product spaces with inner products $\langle \cdot, \cdot \rangle_1$ and $\langle \cdot, \cdot \rangle_2$, respectively. Prove the following results.

    (a) There is a unique adjoint T^* of T, and T^* is linear.
    (b) If β and γ are orthonormal bases for V and W, respectively, then $[T^*]_\gamma^\beta = ([T]_\beta^\gamma)^*$.
    (c) $\text{rank}(T^*) = \text{rank}(T)$.
    (d) $\langle T^*(x), y \rangle_1 = \langle x, T(y) \rangle_2$ for all $x \in$ W and $y \in$ V.
    (e) For all $x \in$ V, $T^*T(x) = 0$ if and only if $T(x) = 0$.

16. State and prove a result that extends the first four parts of Theorem 6.11 using the preceding definition.

17. Let T: V $\to$ W be a linear transformation, where V and W are finite-dimensional inner product spaces. Prove that $(R(T^*))^\perp = N(T)$, using the preceding definition.

18.† Let A be an $n \times n$ matrix. Prove that $\det(A^*) = \overline{\det(A)}$.

19. Suppose that A is an $m \times n$ matrix in which no two columns are identical. Prove that A^*A is a diagonal matrix if and only if every pair of columns of A is orthogonal.

20. For each of the sets of data that follows, use the least squares approximation to find the best fits with both (i) a linear function and (ii) a quadratic function. Compute the error E in both cases.

    (a) $\{(-3, 9), (-2, 6), (0, 2), (1, 1)\}$

(b) $\{(1,2),(3,4),(5,7),(7,9),(9,12)\}$
(c) $\{(-2,4),(-1,3),(0,1),(1,-1),(2,-3)\}$

21. In physics, *Hooke's law* states that (within certain limits) there is a
    linear relationship between the length x of a spring and the force y
    applied to (or exerted by) the spring. That is, $y = cx + d$, where c is
    called the **spring constant**. Use the following data to estimate the
    spring constant (the length is given in inches and the force is given in
    pounds).

Length	Force
x	y
3.5	1.0
4.0	2.2
4.5	2.8
5.0	4.3

22. Find the minimal solution to each of the following systems of linear
    equations.

    (a) $x + 2y - z = 12$

    (b) $\begin{aligned} x + 2y - z &= 1 \\ 2x + 3y + z &= 2 \\ 4x + 7y - z &= 4 \end{aligned}$

    (c) $\begin{aligned} x + y - z &= 0 \\ 2x - y + z &= 3 \\ x - y + z &= 2 \end{aligned}$

    (d) $\begin{aligned} x + y + z - w &= 1 \\ 2x - y \quad\;\; + w &= 1 \end{aligned}$

23. Consider the problem of finding the least squares line $y = ct + d$ corre-
    sponding to the m observations $(t_1, y_1), (t_2, y_2), \ldots, (t_m, y_m)$.

    (a) Show that the equation $(A^*A)x_0 = A^*y$ of Theorem 6.12 takes the
    form of the *normal equations*:

    $$\left(\sum_{i=1}^{m} t_i^2\right) c + \left(\sum_{i=1}^{m} t_i\right) d = \sum_{i=1}^{m} t_i y_i$$

    and

    $$\left(\sum_{i=1}^{m} t_i\right) c + md = \sum_{i=1}^{m} y_i.$$

    These equations may also be obtained from the error E by setting
    the partial derivatives of E with respect to both c and d equal to
    zero.

(b)   Use the second normal equation of (a) to show that the least squares line must pass through the *center of mass*, $(\bar{t}, \bar{y})$, where

$$\bar{t} = \frac{1}{m} \sum_{i=1}^{m} t_i \quad \text{and} \quad \bar{y} = \frac{1}{m} \sum_{i=1}^{m} y_i.$$

24.   Let V and $\{e_1, e_2, \ldots\}$ be defined as in Exercise 23 of Section 6.2. Define T: V → V by

$$\mathsf{T}(\sigma)(k) = \sum_{i=k}^{\infty} \sigma(i) \quad \text{for every positive integer } k.$$

Notice that the infinite series in the definition of T converges because $\sigma(i) \neq 0$ for only finitely many i.

(a)   Prove that T is a linear operator on V.
(b)   Prove that for any positive integer n, $\mathsf{T}(e_n) = \sum_{i=1}^{n} e_i$.
(c)   Prove that T has no adjoint. *Hint:* By way of contradiction, suppose that T^* exists. Prove that for any positive integer n, $\mathsf{T}^*(e_n)(k) \neq 0$ for infinitely many k.

6.4   NORMAL AND SELF-ADJOINT OPERATORS

We have seen the importance of diagonalizable operators in Chapter 5. For these operators, it is necessary and sufficient for the vector space V to possess a basis of eigenvectors. As V is an inner product space in this chapter, it is reasonable to seek conditions that guarantee that V has an orthonormal basis of eigenvectors. A very important result that helps achieve our goal is Schur's theorem (Theorem 6.14). The formulation that follows is in terms of linear operators. The next section contains the more familiar matrix form. We begin with a lemma.

Lemma. *Let* T *be a linear operator on a finite-dimensional inner product space* V. *If* T *has an eigenvector, then so does* T^*.

Proof. Suppose that v is an eigenvector of T with corresponding eigenvalue λ. Then for any $x \in \mathsf{V}$,

$$0 = \langle 0, x \rangle = \langle (\mathsf{T} - \lambda\mathsf{I})(v), x \rangle = \langle v, (\mathsf{T} - \lambda\mathsf{I})^*(x) \rangle = \langle v, (\mathsf{T}^* - \bar{\lambda}\mathsf{I})(x) \rangle,$$

and hence v is orthogonal to the range of $T^* - \bar{\lambda}\mathsf{I}$. So $T^* - \bar{\lambda}\mathsf{I}$ is not onto and hence is not one-to-one. Thus $T^* - \bar{\lambda}\mathsf{I}$ has a nonzero null space, and any nonzero vector in this null space is an eigenvector of T^* with corresponding eigenvalue $\bar{\lambda}$. ∎

Recall (see the exercises of Section 2.1 and see Section 5.4) that a subspace W of V is said to be **T-invariant** if $T(W)$ is contained in W. If W is T-invariant, we may define the restriction $T_W \colon W \to W$ by $T_W(x) = T(x)$ for all $x \in W$. It is clear that T_W is a linear operator on W. Recall from Section 5.2 that a polynomial is said to **split** if it factors into linear polynomials.

Theorem 6.14 (Schur).   *Let* T *be a linear operator on a finite-dimensional inner product space* V. *Suppose that the characteristic polynomial of* T *splits. Then there exists an orthonormal basis* β *for* V *such that the matrix* $[T]_\beta$ *is upper triangular.*

Proof. The proof is by mathematical induction on the dimension n of V. The result is immediate if $n = 1$. So suppose that the result is true for linear operators on $(n-1)$-dimensional inner product spaces whose characteristic polynomials split. By the lemma, we can assume that T^* has a unit eigenvector z. Suppose that $T^*(z) = \lambda z$ and that $W = \text{span}(\{z\})$. We show that $W^\perp$ is T-invariant. If $y \in W^\perp$ and $x = cz \in W$, then

$$\langle T(y), x \rangle = \langle T(y), cz \rangle = \langle y, T^*(cz) \rangle = \langle y, cT^*(z) \rangle = \langle y, c\lambda z \rangle$$
$$= \overline{c\lambda}\, \langle y, z \rangle = \overline{c\lambda}(0) = 0.$$

So $T(y) \in W^\perp$. It is easy to show (see Theorem 5.21 p. 314, or as a consequence of Exercise 6 of Section 4.4) that the characteristic polynomial of $T_{W^\perp}$ divides the characteristic polynomial of T and hence splits. By Theorem 6.7(c) (p. 352), $\dim(W^\perp) = n - 1$, so we may apply the induction hypothesis to $T_{W^\perp}$ and obtain an orthonormal basis γ of $W^\perp$ such that $[T_{W^\perp}]_\gamma$ is upper triangular. Clearly, $\beta = \gamma \cup \{z\}$ is an orthonormal basis for V such that $[T]_\beta$ is upper triangular. ∎

We now return to our original goal of finding an orthonormal basis of eigenvectors of a linear operator T on a finite-dimensional inner product space V. Note that if such an orthonormal basis β exists, then $[T]_\beta$ is a diagonal matrix, and hence $[T^*]_\beta = [T]_\beta^*$ is also a diagonal matrix. Because diagonal matrices commute, we conclude that T and T^* commute. Thus *if* V *possesses an orthonormal basis of eigenvectors of* T, *then* $TT^* = T^*T$.

Definitions.   *Let* V *be an inner product space, and let* T *be a linear operator on* V. *We say that* T *is* **normal** *if* $TT^* = T^*T$. *An* $n \times n$ *real or complex matrix* A *is* **normal** *if* $AA^* = A^*A$.

It follows immediately from Theorem 6.10 (p. 359) that T is normal if and only if $[T]_\beta$ is normal, where β is an orthonormal basis.

Example 1

Let $T: R^2 \to R^2$ be rotation by θ, where $0 < \theta < \pi$. The matrix representation of T in the standard ordered basis is given by

$$A = \begin{pmatrix} \cos\theta & -\sin\theta \\ \sin\theta & \cos\theta \end{pmatrix}.$$

Note that $AA^* = I = A^*A$; so A, and hence T, is normal.   ◆

Example 2

Suppose that A is a real *skew-symmetric* matrix; that is, $A^t = -A$. Then A is normal because both AA^t and A^tA are equal to $-A^2$.   ◆

Clearly, the operator T in Example 1 does not even possess one eigenvector. So in the case of a real inner product space, we see that normality is not sufficient to guarantee an orthonormal basis of eigenvectors. All is not lost, however. We show that normality suffices if V is a *complex* inner product space.

Before we prove the promised result for normal operators, we need some general properties of normal operators.

Theorem 6.15. *Let V be an inner product space, and let T be a normal operator on V. Then the following statements are true.*
  (a) $\|T(x)\| = \|T^*(x)\|$ *for all $x \in V$.*
  (b) $T - cI$ *is normal for every $c \in F$.*
  (c) *If x is an eigenvector of T, then x is also an eigenvector of T^*. In fact, if $T(x) = \lambda x$, then $T^*(x) = \overline{\lambda}x$.*
  (d) *If λ_1 and λ_2 are distinct eigenvalues of T with corresponding eigenvectors x_1 and x_2, then x_1 and x_2 are orthogonal.*

Proof. (a) For any $x \in V$, we have

$$\|T(x)\|^2 = \langle T(x), T(x)\rangle = \langle T^*T(x), x\rangle = \langle TT^*(x), x\rangle$$
$$= \langle T^*(x), T^*(x)\rangle = \|T^*(x)\|^2.$$

The proof of (b) is left as an exercise.

(c) Suppose that $T(x) = \lambda x$ for some $x \in V$. Let $U = T - \lambda I$. Then $U(x) = 0$, and U is normal by (b). Thus (a) implies that

$$0 = \|U(x)\| = \|U^*(x)\| = \|(T^* - \overline{\lambda}I)(x)\| = \|T^*(x) - \overline{\lambda}x\|.$$

Hence $T^*(x) = \overline{\lambda}x$. So x is an eigenvector of T^*.

(d) Let λ_1 and λ_2 be distinct eigenvalues of T with corresponding eigenvectors x_1 and x_2. Then, using (c), we have

$$\lambda_1 \langle x_1, x_2\rangle = \langle \lambda_1 x_1, x_2\rangle = \langle T(x_1), x_2\rangle = \langle x_1, T^*(x_2)\rangle$$

$$= \langle x_1, \overline{\lambda_2} x_2 \rangle = \lambda_2 \langle x_1, x_2 \rangle .$$

Since $\lambda_1 \neq \lambda_2$, we conclude that $\langle x_1, x_2 \rangle = 0$. ∎

Theorem 6.16. *Let* T *be a linear operator on a finite-dimensional complex inner product space* V. *Then* T *is normal if and only if there exists an orthonormal basis for* V *consisting of eigenvectors of* T.

Proof. Suppose that T is normal. By the fundamental theorem of algebra (Theorem D.4), the characteristic polynomial of T splits. So we may apply Schur's theorem to obtain an orthonormal basis $\beta = \{v_1, v_2, \ldots, v_n\}$ for V such that $[\mathsf{T}]_\beta = A$ is upper triangular. We know that v_1 is an eigenvector of T because A is upper triangular. Assume that $v_1, v_2, \ldots, v_{k-1}$ are eigenvectors of T. We claim that v_k is also an eigenvector of T. It then follows by mathematical induction on k that all of the v_i's are eigenvectors of T. Consider any $j < k$, and let λ_j denote the eigenvalue of T corresponding to v_j. By Theorem 6.15, $T^*(v_j) = \overline{\lambda}_j v_j$. Since A is upper triangular,

$$\mathsf{T}(v_k) = A_{1k} v_1 + A_{2k} v_2 + \cdots + A_{jk} v_j + \cdots + A_{kk} v_k.$$

Furthermore, by the corollary to Theorem 6.5 (p. 347),

$$A_{jk} = \langle \mathsf{T}(v_k), v_j \rangle = \langle v_k, \mathsf{T}^*(v_j) \rangle = \langle v_k, \overline{\lambda}_j v_j \rangle = \lambda_j \langle v_k, v_j \rangle = 0.$$

It follows that $\mathsf{T}(v_k) = A_{kk} v_k$, and hence v_k is an eigenvector of T. So by induction, all the vectors in β are eigenvectors of T.

The converse was already proved on page 370. ∎

Interestingly, as the next example shows, Theorem 6.16 does not extend to infinite-dimensional complex inner product spaces.

Example 3

Consider the inner product space H with the orthonormal set S from Example 9 in Section 6.1. Let $\mathsf{V} = \text{span}(S)$, and let T and U be the linear operators on V defined by $\mathsf{T}(f) = f_1 f$ and $\mathsf{U}(f) = f_{-1} f$. Then

$$\mathsf{T}(f_n) = f_{n+1} \quad \text{and} \quad \mathsf{U}(f_n) = f_{n-1}$$

for all integers n. Thus

$$\langle \mathsf{T}(f_m), f_n \rangle = \langle f_{m+1}, f_n \rangle = \delta_{(m+1),n} = \delta_{m,(n-1)} = \langle f_m, f_{n-1} \rangle = \langle f_m, \mathsf{U}(f_n) \rangle .$$

It follows that $\mathsf{U} = \mathsf{T}^*$. Furthermore, $\mathsf{TT}^* = \mathsf{I} = \mathsf{T}^*\mathsf{T}$; so T is normal.

We show that T has no eigenvectors. Suppose that f is an eigenvector of T, say, $\mathsf{T}(f) = \lambda f$ for some λ. Since V equals the span of S, we may write

$$f = \sum_{i=n}^{m} a_i f_i, \quad \text{where } a_m \neq 0.$$

Hence

$$\sum_{i=n}^{m} a_i f_{i+1} = \mathsf{T}(f) = \lambda f = \sum_{i=n}^{m} \lambda a_i f_i.$$

Since $a_m \neq 0$, we can write f_{m+1} as a linear combination of $f_n, f_{n+1}, \ldots, f_m$. But this is a contradiction because S is linearly independent.   ♦

Example 1 illustrates that normality is not sufficient to guarantee the existence of an orthonormal basis of eigenvectors for real inner product spaces. For real inner product spaces, we must replace normality by the stronger condition that $\mathsf{T} = \mathsf{T}^*$ in order to guarantee such a basis.

Definitions. *Let T be a linear operator on an inner product space V. We say that T is **self-adjoint (Hermitian)** if $\mathsf{T} = \mathsf{T}^*$. An $n \times n$ real or complex matrix A is **self-adjoint (Hermitian)** if $A = A^*$.*

It follows immediately that if β is an orthonormal basis, then T is self-adjoint if and only if $[\mathsf{T}]_\beta$ is self-adjoint. For real matrices, this condition reduces to the requirement that A be symmetric.

Before we state our main result for self-adjoint operators, we need some preliminary work.

By definition, a linear operator on a real inner product space has only real eigenvalues. The lemma that follows shows that the same can be said for self-adjoint operators on a complex inner product space. Similarly, the characteristic polynomial of every linear operator on a complex inner product space splits, and the same is true for self-adjoint operators on a real inner product space.

Lemma. *Let T be a self-adjoint operator on a finite-dimensional inner product space V. Then*
  (a) *Every eigenvalue of T is real.*
  (b) *Suppose that V is a real inner product space. Then the characteristic polynomial of T splits.*

Proof. (a) Suppose that $\mathsf{T}(x) = \lambda x$ for $x \neq 0$. Because a self-adjoint operator is also normal, we can apply Theorem 6.15(c) to obtain

$$\lambda x = \mathsf{T}(x) = \mathsf{T}^*(x) = \overline{\lambda} x.$$

So $\lambda = \overline{\lambda}$; that is, λ is real.

  (b) Let $n = \dim(\mathsf{V})$, β be an orthonormal basis for V, and $A = [\mathsf{T}]_\beta$. Then A is self-adjoint. Let T_A be the linear operator on C^n defined by $\mathsf{T}_A(x) = Ax$ for all $x \in \mathsf{C}^n$. Note that T_A is self-adjoint because $[\mathsf{T}_A]_\gamma = A$, where γ is the standard ordered (orthonormal) basis for C^n. So, by (a), the eigenvalues of T_A are real. By the fundamental theorem of algebra, the

Chap. 6   Inner Product Spaces

characteristic polynomial of T_A splits into factors of the form $t - \lambda$. Since each λ is real, the characteristic polynomial splits over R. But T_A has the same characteristic polynomial as A, which has the same characteristic polynomial as T. Therefore the characteristic polynomial of T splits. ∎

We are now able to establish one of the major results of this chapter.

Theorem 6.17. *Let* T *be a linear operator on a finite-dimensional real inner product space* V. *Then* T *is self-adjoint if and only if there exists an orthonormal basis* β *for* V *consisting of eigenvectors of* T.

Proof. Suppose that T is self-adjoint. By the lemma, we may apply Schur's theorem to obtain an orthonormal basis β for V such that the matrix $A = [\mathsf{T}]_\beta$ is upper triangular. But

$$A^* = [\mathsf{T}]_\beta^* = [\mathsf{T}^*]_\beta = [\mathsf{T}]_\beta = A.$$

So A and A^* are both upper triangular, and therefore A is a diagonal matrix. Thus β must consist of eigenvectors of T.

The converse is left as an exercise. ∎

Theorem 6.17 is used extensively in many areas of mathematics and statistics. We restate this theorem in matrix form in the next section.

Example 4

As we noted earlier, real symmetric matrices are self-adjoint, and self-adjoint matrices are normal. The following matrix A is complex and symmetric:

$$A = \begin{pmatrix} i & i \\ i & 1 \end{pmatrix} \quad \text{and} \quad A^* = \begin{pmatrix} -i & -i \\ -i & 1 \end{pmatrix}.$$

But A is not normal, because $(AA^*)_{12} = 1 + i$ and $(A^*A)_{12} = 1 - i$. Therefore complex symmetric matrices need not be normal. ◆

EXERCISES

1. Label the following statements as true or false. Assume that the underlying inner product spaces are finite-dimensional.

   (a) Every self-adjoint operator is normal.
   (b) Operators and their adjoints have the same eigenvectors.
   (c) If T is an operator on an inner product space V, then T is normal if and only if $[\mathsf{T}]_\beta$ is normal, where β is any ordered basis for V.
   (d) A real or complex matrix A is normal if and only if L_A is normal.
   (e) The eigenvalues of a self-adjoint operator must all be real.

(f) The identity and zero operators are self-adjoint.
(g) Every normal operator is diagonalizable.
(h) Every self-adjoint operator is diagonalizable.

2. For each linear operator T on an inner product space V, determine whether T is normal, self-adjoint, or neither. If possible, produce an orthonormal basis of eigenvectors of T for V and list the corresponding eigenvalues.

(a) $V = R^2$ and T is defined by $T(a,b) = (2a - 2b, -2a + 5b)$.
(b) $V = R^3$ and T is defined by $T(a,b,c) = (-a + b, 5b, 4a - 2b + 5c)$.
(c) $V = C^2$ and T is defined by $T(a,b) = (2a + ib, a + 2b)$.
(d) $V = P_2(R)$ and T is defined by $T(f) = f'$, where

$$\langle f, g \rangle = \int_0^1 f(t)g(t)\, dt.$$

(e) $V = M_{2\times 2}(R)$ and T is defined by $T(A) = A^t$.

(f) $V = M_{2\times 2}(R)$ and T is defined by $T\begin{pmatrix} a & b \\ c & d \end{pmatrix} = \begin{pmatrix} c & d \\ a & b \end{pmatrix}$.

3. Give an example of a linear operator T on R^2 and an ordered basis for R^2 that provides a counterexample to the statement in Exercise 1(c).

4. Let T and U be self-adjoint operators on an inner product space V. Prove that TU is self-adjoint if and only if $TU = UT$.

5. Prove (b) of Theorem 6.15.

6. Let V be a complex inner product space, and let T be a linear operator on V. Define

$$T_1 = \frac{1}{2}(T + T^*) \quad \text{and} \quad T_2 = \frac{1}{2i}(T - T^*).$$

(a) Prove that T_1 and T_2 are self-adjoint and that $T = T_1 + iT_2$.
(b) Suppose also that $T = U_1 + iU_2$, where U_1 and U_2 are self-adjoint. Prove that $U_1 = T_1$ and $U_2 = T_2$.
(c) Prove that T is normal if and only if $T_1 T_2 = T_2 T_1$.

7. Let T be a linear operator on an inner product space V, and let W be a T-invariant subspace of V. Prove the following results.

(a) If T is self-adjoint, then T_W is self-adjoint.
(b) $W^\perp$ is T^*-invariant.
(c) If W is both T- and T^*-invariant, then $(T_W)^* = (T^*)_W$.
(d) If W is both T- and T^*-invariant and T is normal, then T_W is normal.

8. Let T be a normal operator on a finite-dimensional complex inner product space V, and let W be a subspace of V. Prove that if W is T-invariant, then W is also T^*-invariant. *Hint:* Use Exercise 24 of Section 5.4.

9. Let T be a normal operator on a finite-dimensional inner product space V. Prove that $\mathsf{N}(\mathsf{T}) = \mathsf{N}(\mathsf{T}^*)$ and $\mathsf{R}(\mathsf{T}) = \mathsf{R}(\mathsf{T}^*)$. *Hint:* Use Theorem 6.15 and Exercise 12 of Section 6.3.

10. Let T be a self-adjoint operator on a finite-dimensional inner product space V. Prove that for all $x \in \mathsf{V}$

$$\|\mathsf{T}(x) \pm ix\|^2 = \|\mathsf{T}(x)\|^2 + \|x\|^2.$$

Deduce that $\mathsf{T} - i\mathsf{I}$ is invertible and that $[(\mathsf{T} - i\mathsf{I})^{-1}]^* = (\mathsf{T} + i\mathsf{I})^{-1}$.

11. Assume that T is a linear operator on a complex (not necessarily finite-dimensional) inner product space V with an adjoint T^*. Prove the following results.

   (a) If T is self-adjoint, then $\langle \mathsf{T}(x), x \rangle$ is real for all $x \in \mathsf{V}$.
   (b) If T satisfies $\langle \mathsf{T}(x), x \rangle = 0$ for all $x \in \mathsf{V}$, then $\mathsf{T} = \mathsf{T}_0$. *Hint:* Replace x by $x + y$ and then by $x + iy$, and expand the resulting inner products.
   (c) If $\langle \mathsf{T}(x), x \rangle$ is real for all $x \in \mathsf{V}$, then $\mathsf{T} = \mathsf{T}^*$.

12. Let T be a normal operator on a finite-dimensional real inner product space V whose characteristic polynomial splits. Prove that V has an orthonormal basis of eigenvectors of T. Hence prove that T is self-adjoint.

13. An $n \times n$ real matrix A is said to be a **Gramian** matrix if there exists a real (square) matrix B such that $A = B^t B$. Prove that A is a Gramian matrix if and only if A is symmetric and all of its eigenvalues are non-negative. *Hint:* Apply Theorem 6.17 to $\mathsf{T} = \mathsf{L}_A$ to obtain an orthonormal basis $\{v_1, v_2, \ldots, v_n\}$ of eigenvectors with the associated eigenvalues $\lambda_1, \lambda_2, \ldots, \lambda_n$. Define the linear operator U by $\mathsf{U}(v_i) = \sqrt{\lambda_i} v_i$.

14. *Simultaneous Diagonalization.* Let V be a finite-dimensional real inner product space, and let U and T be self-adjoint linear operators on V such that $\mathsf{UT} = \mathsf{TU}$. Prove that there exists an orthonormal basis for V consisting of vectors that are eigenvectors of both U and T. (The complex version of this result appears as Exercise 10 of Section 6.6.) *Hint:* For any eigenspace $\mathsf{W} = \mathsf{E}_\lambda$ of T, we have that W is both T- and U-invariant. By Exercise 7, we have that $\mathsf{W}^\perp$ is both T- and U-invariant. Apply Theorem 6.17 and Theorem 6.6 (p. 350).

15. Let A and B be symmetric $n \times n$ matrices such that $AB = BA$. Use Exercise 14 to prove that there exists an orthogonal matrix P such that $P^t A P$ and $P^t B P$ are both diagonal matrices.

16. Prove the *Cayley–Hamilton theorem* for a complex $n \times n$ matrix A. That is, if $f(t)$ is the characteristic polynomial of A, prove that $f(A) = O$. *Hint:* Use Schur's theorem to show that A may be assumed to be upper triangular, in which case

$$f(t) = \prod_{i=1}^{n}(A_{ii} - t).$$

Now if $\mathsf{T} = \mathsf{L}_A$, we have $(A_{jj}\mathsf{I} - \mathsf{T})(e_j) \in \operatorname{span}(\{e_1, e_2, \dots, e_{j-1}\})$ for $j \geq 2$, where $\{e_1, e_2, \dots, e_n\}$ is the standard ordered basis for C^n. (The general case is proved in Section 5.4.)

The following definitions are used in Exercises 17 through 23.

Definitions. *A linear operator* T *on a finite-dimensional inner product space is called* **positive definite** *[***positive semidefinite***] if* T *is self-adjoint and* $\langle \mathsf{T}(x), x \rangle > 0$ $[\langle \mathsf{T}(x), x \rangle \geq 0]$ *for all* $x \neq 0$.

An $n \times n$ *matrix* A *with entries from* R *or* C *is called* **positive definite** *[***positive semidefinite***] if* L_A *is positive definite [positive semidefinite].*

17. Let T and U be a self-adjoint linear operators on an n-dimensional inner product space V, and let $A = [\mathsf{T}]_\beta$, where β is an orthonormal basis for V. Prove the following results.

  (a) T is positive definite [semidefinite] if and only if all of its eigenvalues are positive [nonnegative].

  (b) T is positive definite if and only if

$$\sum_{i,j} A_{ij} a_j \bar{a}_i > 0 \text{ for all nonzero } n\text{-tuples } (a_1, a_2, \dots, a_n).$$

  (c) T is positive semidefinite if and only if $A = B^*B$ for some square matrix B.

  (d) If T and U are positive semidefinite operators such that $\mathsf{T}^2 = \mathsf{U}^2$, then $\mathsf{T} = \mathsf{U}$.

  (e) If T and U are positive definite operators such that $\mathsf{TU} = \mathsf{UT}$, then TU is positive definite.

  (f) T is positive definite [semidefinite] if and only if A is positive definite [semidefinite].

Because of (f), results analogous to items (a) through (d) hold for matrices as well as operators.

18. Let $T: V \rightarrow W$ be a linear transformation, where V and W are finite-dimensional inner product spaces. Prove the following results.

   (a) T^*T and TT^* are positive semidefinite. (See Exercise 15 of Section 6.3.)
   (b) $\text{rank}(T^*T) = \text{rank}(TT^*) = \text{rank}(T)$.

19. Let T and U be positive definite operators on an inner product space V. Prove the following results.

   (a) $T + U$ is positive definite.
   (b) If $c > 0$, then cT is positive definite.
   (c) T^{-1} is positive definite.

20. Let V be an inner product space with inner product $\langle \cdot, \cdot \rangle$, and let T be a positive definite linear operator on V. Prove that $\langle x, y \rangle' = \langle T(x), y \rangle$ defines another inner product on V.

21. Let V be a finite-dimensional inner product space, and let T and U be self-adjoint operators on V such that T is positive definite. Prove that both TU and UT are diagonalizable linear operators that have only real eigenvalues. *Hint:* Show that UT is self-adjoint with respect to the inner product $\langle x, y \rangle' = \langle T(x), y \rangle$. To show that TU is self-adjoint, repeat the argument with T^{-1} in place of T.

22. This exercise provides a converse to Exercise 20. Let V be a finite-dimensional inner product space with inner product $\langle \cdot, \cdot \rangle$, and let $\langle \cdot, \cdot \rangle'$ be any other inner product on V.

   (a) Prove that there exists a unique linear operator T on V such that $\langle x, y \rangle' = \langle T(x), y \rangle$ for all x and y in V. *Hint:* Let $\beta = \{v_1, v_2, \ldots, v_n\}$ be an orthonormal basis for V with respect to $\langle \cdot, \cdot \rangle$, and define a matrix A by $A_{ij} = \langle v_j, v_i \rangle'$ for all i and j. Let T be the unique linear operator on V such that $[T]_\beta = A$.
   (b) Prove that the operator T of (a) is positive definite with respect to both inner products.

23. Let U be a diagonalizable linear operator on a finite-dimensional inner product space V such that all of the eigenvalues of U are real. Prove that there exist positive definite linear operators T_1 and T_1' and self-adjoint linear operators T_2 and T_2' such that $U = T_2 T_1 = T_1' T_2'$. *Hint:* Let $\langle \cdot, \cdot \rangle$ be the inner product associated with V, β a basis of eigenvectors for U, $\langle \cdot, \cdot \rangle'$ the inner product on V with respect to which β is orthonormal (see Exercise 22(a) of Section 6.1), and T_1 the positive definite operator according to Exercise 22. Show that U is self-adjoint with respect to $\langle \cdot, \cdot \rangle'$ and $U = T_1^{-1} U^* T_1$ (the adjoint is with respect to $\langle \cdot, \cdot \rangle$). Let $T_2 = T_1^{-1} U^*$.

24. This argument gives another proof of Schur's theorem. Let T be a linear operator on a finite dimensional inner product space V.

  (a) Suppose that β is an ordered basis for V such that $[\mathsf{T}]_\beta$ is an upper triangular matrix. Let γ be the orthonormal basis for V obtained by applying the Gram–Schmidt orthogonalization process to β and then normalizing the resulting vectors. Prove that $[\mathsf{T}]_\gamma$ is an upper triangular matrix.
  (b) Use Exercise 32 of Section 5.4 and (a) to obtain an alternate proof of Schur's theorem.

6.5   UNITARY AND ORTHOGONAL OPERATORS AND THEIR MATRICES

In this section, we continue our analogy between complex numbers and linear operators. Recall that the adjoint of a linear operator acts similarly to the conjugate of a complex number (see, for example, Theorem 6.11 p. 359). A complex number z has length 1 if $z\bar{z} = 1$. In this section, we study those linear operators T on an inner product space V such that $\mathsf{TT}^* = \mathsf{T}^*\mathsf{T} = \mathsf{I}$. We will see that these are precisely the linear operators that "preserve length" in the sense that $\|\mathsf{T}(x)\| = \|x\|$ for all $x \in \mathsf{V}$. As another characterization, we prove that, on a finite-dimensional complex inner product space, these are the normal operators whose eigenvalues all have absolute value 1.

In past chapters, we were interested in studying those functions that preserve the structure of the underlying space. In particular, linear operators preserve the operations of vector addition and scalar multiplication, and isomorphisms preserve all the vector space structure. It is now natural to consider those linear operators T on an inner product space that preserve length. We will see that this condition guarantees, in fact, that T preserves the inner product.

Definitions.   *Let T be a linear operator on a finite-dimensional inner product space V (over F). If $\|\mathsf{T}(x)\| = \|x\|$ for all $x \in \mathsf{V}$, we call T a **unitary operator** if $F = C$ and an **orthogonal operator** if $F = R$.*

It should be noted that, in the infinite-dimensional case, an operator satisfying the preceding norm requirement is generally called an **isometry**. If, in addition, the operator is onto (the condition guarantees one-to-one), then the operator is called a **unitary** or **orthogonal operator**.

Clearly, any rotation or reflection in R^2 preserves length and hence is an orthogonal operator. We study these operators in much more detail in Section 6.11.

Example 1

Let $h \in H$ satisfy $|h(x)| = 1$ for all x. Define the linear operator T on H by $T(f) = hf$. Then

$$\|T(f)\|^2 = \|hf\|^2 = \frac{1}{2\pi} \int_0^{2\pi} h(t)f(t)\overline{h(t)f(t)}\, dt = \|f\|^2$$

since $|h(t)|^2 = 1$ for all t. So T is a unitary operator.   ◆

Theorem 6.18. *Let* T *be a linear operator on a finite-dimensional inner product space* V. *Then the following statements are equivalent.*

(a) $TT^* = T^*T = I$.

(b) $\langle T(x), T(y) \rangle = \langle x, y \rangle$ *for all* $x, y \in V$.

(c) *If* β *is an orthonormal basis for* V, *then* $T(\beta)$ *is an orthonormal basis for* V.

(d) *There exists an orthonormal basis* β *for* V *such that* $T(\beta)$ *is an orthonormal basis for* V.

(e) $\|T(x)\| = \|x\|$ *for all* $x \in V$.

Thus all the conditions above are equivalent to the definition of a unitary or orthogonal operator. From (a), it follows that unitary or orthogonal operators are normal.

Before proving the theorem, we first prove a lemma. Compare this lemma to Exercise 11(b) of Section 6.4.

Lemma. *Let* U *be a self-adjoint operator on a finite-dimensional inner product space* V. *If* $\langle x, U(x) \rangle = 0$ *for all* $x \in V$, *then* $U = T_0$.

Proof. By either Theorem 6.16 (p. 372) or 6.17 (p. 374), we may choose an orthonormal basis β for V consisting of eigenvectors of U. If $x \in \beta$, then $U(x) = \lambda x$ for some λ. Thus

$$0 = \langle x, U(x) \rangle = \langle x, \lambda x \rangle = \overline{\lambda} \langle x, x \rangle\,;$$

so $\overline{\lambda} = 0$. Hence $U(x) = 0$ for all $x \in \beta$, and thus $U = T_0$.   ∎

Proof of Theorem 6.18. We prove first that (a) implies (b). Let $x, y \in V$. Then $\langle x, y \rangle = \langle T^*T(x), y \rangle = \langle T(x), T(y) \rangle$.

Second, we prove that (b) implies (c). Let $\beta = \{v_1, v_2, \ldots, v_n\}$ be an orthonormal basis for V; so $T(\beta) = \{T(v_1), T(v_2), \ldots, T(v_n)\}$. It follows that $\langle T(v_i), T(v_j) \rangle = \langle v_i, v_j \rangle = \delta_{ij}$. Therefore $T(\beta)$ is an orthonormal basis for V.

That (c) implies (d) is obvious.

Next we prove that (d) implies (e). Let $x \in V$, and let $\beta = \{v_1, v_2, \ldots, v_n\}$. Now

$$x = \sum_{i=1}^n a_i v_i$$

for some scalars a_i, and so

$$\|x\|^2 = \left\langle \sum_{i=1}^{n} a_i v_i, \sum_{j=1}^{n} a_j v_j \right\rangle = \sum_{i=1}^{n} \sum_{j=1}^{n} a_i \overline{a_j} \langle v_i, v_j \rangle$$

$$= \sum_{i=1}^{n} \sum_{j=1}^{n} a_i \overline{a_j} \delta_{ij} = \sum_{i=1}^{n} |a_i|^2$$

since β is orthonormal.

Applying the same manipulations to

$$\mathsf{T}(x) = \sum_{i=1}^{n} a_i \mathsf{T}(v_i)$$

and using the fact that $\mathsf{T}(\beta)$ is also orthonormal, we obtain

$$\|\mathsf{T}(x)\|^2 = \sum_{i=1}^{n} |a_i|^2.$$

Hence $\|\mathsf{T}(x)\| = \|x\|$.

Finally, we prove that (e) implies (a). For any $x \in \mathsf{V}$, we have

$$\langle x, x \rangle = \|x\|^2 = \|\mathsf{T}(x)\|^2 = \langle \mathsf{T}(x), \mathsf{T}(x) \rangle = \langle x, \mathsf{T}^* \mathsf{T}(x) \rangle.$$

So $\langle x, (\mathsf{I} - \mathsf{T}^* \mathsf{T})(x) \rangle = 0$ for all $x \in \mathsf{V}$. Let $\mathsf{U} = \mathsf{I} - \mathsf{T}^* \mathsf{T}$; then U is self-adjoint, and $\langle x, \mathsf{U}(x) \rangle = 0$ for all $x \in \mathsf{V}$. Hence, by the lemma, we have $\mathsf{T}_0 = \mathsf{U} = \mathsf{I} - \mathsf{T}^* \mathsf{T}$, and therefore $\mathsf{T}^* \mathsf{T} = \mathsf{I}$. Since V is finite-dimensional, we may use Exercise 10 of Section 2.4 to conclude that $\mathsf{T}\mathsf{T}^* = \mathsf{I}$. ∎

It follows immediately from the definition that every eigenvalue of a unitary or orthogonal operator has absolute value 1. In fact, even more is true.

Corollary 1.   *Let T be a linear operator on a finite-dimensional real inner product space V. Then V has an orthonormal basis of eigenvectors of T with corresponding eigenvalues of absolute value 1 if and only if T is both self-adjoint and orthogonal.*

Proof. Suppose that V has an orthonormal basis $\{v_1, v_2, \ldots, v_n\}$ such that $\mathsf{T}(v_i) = \lambda_i v_i$ and $|\lambda_i| = 1$ for all i. By Theorem 6.17 (p. 374), T is self-adjoint. Thus $(\mathsf{T}\mathsf{T}^*)(v_i) = \mathsf{T}(\lambda_i v_i) = \lambda_i \lambda_i v_i = \lambda_i^2 v_i = v_i$ for each i. So $\mathsf{T}\mathsf{T}^* = \mathsf{I}$, and again by Exercise 10 of Section 2.4, T is orthogonal by Theorem 6.18(a).

If T is self-adjoint, then, by Theorem 6.17, we have that V possesses an orthonormal basis $\{v_1, v_2, \ldots, v_n\}$ such that $\mathsf{T}(v_i) = \lambda_i v_i$ for all i. If T is also orthogonal, we have

$$|\lambda_i| \cdot \|v_i\| = \|\lambda_i v_i\| = \|\mathsf{T}(v_i)\| = \|v_i\|;$$

so $|\lambda_i| = 1$ for every i. ∎

Corollary 2. *Let* T *be a linear operator on a finite-dimensional complex inner product space* V. *Then* V *has an orthonormal basis of eigenvectors of* T *with corresponding eigenvalues of absolute value 1 if and only if* T *is unitary.*

Proof. The proof is similar to the proof of Corollary 1. ∎

Example 2

Let $T\colon R^2 \to R^2$ be a rotation by θ, where $0 < \theta < \pi$. It is clear geometrically that T "preserves length", that is, that $\|T(x)\| = \|x\|$ for all $x \in R^2$. The fact that rotations by a fixed angle preserve perpendicularity not only can be seen geometrically but now follows from (b) of Theorem 6.18. Perhaps the fact that such a transformation preserves the inner product is not so obvious; however, we obtain this fact from (b) also. Finally, an inspection of the matrix representation of T with respect to the standard ordered basis, which is

$$\begin{pmatrix} \cos\theta & -\sin\theta \\ \sin\theta & \cos\theta \end{pmatrix},$$

reveals that T is not self-adjoint for the given restriction on θ. As we mentioned earlier, this fact also follows from the geometric observation that T has no eigenvectors and from Theorem 6.15 (p. 371). It is seen easily from the preceding matrix that T^* is the rotation by $-\theta$. ◆

Definition. *Let* L *be a one-dimensional subspace of* R^2. *We may view* L *as a line in the plane through the origin. A linear operator* T *on* R^2 *is called a* **reflection** *of* R^2 *about* L *if* $T(x) = x$ *for all* $x \in L$ *and* $T(x) = -x$ *for all* $x \in L^\perp$.

As an example of a reflection, consider the operator defined in Example 3 of Section 2.5.

Example 3

Let T be a reflection of R^2 about a line L through the origin. We show that T is an orthogonal operator. Select vectors $v_1 \in L$ and $v_2 \in L^\perp$ such that $\|v_1\| = \|v_2\| = 1$. Then $T(v_1) = v_1$ and $T(v_2) = -v_2$. Thus v_1 and v_2 are eigenvectors of T with corresponding eigenvalues 1 and -1, respectively. Furthermore, $\{v_1, v_2\}$ is an orthonormal basis for R^2. It follows that T is an orthogonal operator by Corollary 1 to Theorem 6.18. ◆

We now examine the matrices that represent unitary and orthogonal transformations.

Definitions. *A square matrix* A *is called an an* **orthogonal matrix** *if* $A^t A = A A^t = I$ *and* **unitary** *if* $A^* A = A A^* = I$.

Since for a real matrix A we have $A^* = A^t$, a real unitary matrix is also orthogonal. In this case, we call A **orthogonal** rather than unitary.

Note that the condition $AA^* = I$ is equivalent to the statement that the rows of A form an orthonormal basis for F^n because

$$\delta_{ij} = I_{ij} = (AA^*)_{ij} = \sum_{k=1}^{n} A_{ik}(A^*)_{kj} = \sum_{k=1}^{n} A_{ik}\overline{A_{jk}},$$

and the last term represents the inner product of the ith and jth rows of A.

A similar remark can be made about the columns of A and the condition $A^*A = I$.

It also follows from the definition above and from Theorem 6.10 (p. 359) that a linear operator T on an inner product space V is unitary [orthogonal] if and only if $[\mathsf{T}]_\beta$ is unitary [orthogonal] for some orthonormal basis β for V.

Example 4

From Example 2, the matrix

$$\begin{pmatrix} \cos\theta & -\sin\theta \\ \sin\theta & \cos\theta \end{pmatrix}$$

is clearly orthogonal. One can easily see that the rows of the matrix form an orthonormal basis for R^2. Similarly, the columns of the matrix form an orthonormal basis for R^2.   ◆

Example 5

Let T be a reflection of R^2 about a line L through the origin, let β be the standard ordered basis for R^2, and let $A = [\mathsf{T}]_\beta$. Then $\mathsf{T} = \mathsf{L}_A$. Since T is an orthogonal operator and β is an orthonormal basis, A is an orthogonal matrix. We describe A.

Suppose that α is the angle from the positive x-axis to L. Let $v_1 = (\cos\alpha, \sin\alpha)$ and $v_2 = (-\sin\alpha, \cos\alpha)$. Then $\|v_1\| = \|v_2\| = 1$, $v_1 \in \mathsf{L}$, and $v_2 \in \mathsf{L}^\perp$. Hence $\gamma = \{v_1, v_2\}$ is an orthonormal basis for R^2. Because $\mathsf{T}(v_1) = v_1$ and $\mathsf{T}(v_2) = -v_2$, we have

$$[T]_\gamma = [\mathsf{L}_A]_\gamma = \begin{pmatrix} 1 & 0 \\ 0 & -1 \end{pmatrix}.$$

Let

$$Q = \begin{pmatrix} \cos\alpha & -\sin\alpha \\ \sin\alpha & \cos\alpha \end{pmatrix}.$$

By the corollary to Theorem 2.23 (p. 115),

$$A = Q[\mathsf{L}_A]_\gamma Q^{-1}$$

384 Chap. 6 Inner Product Spaces

$$= \begin{pmatrix} \cos\alpha & -\sin\alpha \\ \sin\alpha & \cos\alpha \end{pmatrix} \begin{pmatrix} 1 & 0 \\ 0 & -1 \end{pmatrix} \begin{pmatrix} \cos\alpha & \sin\alpha \\ -\sin\alpha & \cos\alpha \end{pmatrix}$$

$$= \begin{pmatrix} \cos^2\alpha - \sin^2\alpha & 2\sin\alpha\cos\alpha \\ 2\sin\alpha\cos\alpha & -(\cos^2\alpha - \sin^2\alpha) \end{pmatrix}$$

$$= \begin{pmatrix} \cos 2\alpha & \sin 2\alpha \\ \sin 2\alpha & -\cos 2\alpha \end{pmatrix}. \quad \blacklozenge$$

We know that, for a complex normal [real symmetric] matrix A, there exists an orthonormal basis β for F^n consisting of eigenvectors of A. Hence A is similar to a diagonal matrix D. By the corollary to Theorem 2.23 (p. 115), the matrix Q whose columns are the vectors in β is such that $D = Q^{-1}AQ$. But since the columns of Q are an orthonormal basis for F^n, it follows that Q is unitary [orthogonal]. In this case, we say that A is **unitarily equivalent** [**orthogonally equivalent**] to D. It is easily seen (see Exercise 18) that this relation is an equivalence relation on $\mathsf{M}_{n\times n}(C)$ [$\mathsf{M}_{n\times n}(R)$]. More generally, *A and B are unitarily equivalent [orthogonally equivalent] if and only if there exists a unitary [orthogonal] matrix P such that $A = P^*BP$.*

The preceding paragraph has proved half of each of the next two theorems.

Theorem 6.19. *Let A be a complex $n \times n$ matrix. Then A is normal if and only if A is unitarily equivalent to a diagonal matrix.*

Proof. By the preceding remarks, we need only prove that if A is unitarily equivalent to a diagonal matrix, then A is normal.

Suppose that $A = P^*DP$, where P is a unitary matrix and D is a diagonal matrix. Then

$$AA^* = (P^*DP)(P^*DP)^* = (P^*DP)(P^*D^*P) = P^*DID^*P = P^*DD^*P.$$

Similarly, $A^*A = P^*D^*DP$. Since D is a diagonal matrix, however, we have $DD^* = D^*D$. Thus $AA^* = A^*A$. $\blacksquare$

Theorem 6.20. *Let A be a real $n \times n$ matrix. Then A is symmetric if and only if A is orthogonally equivalent to a real diagonal matrix.*

Proof. The proof is similar to the proof of Theorem 6.19 and is left as an exercise. $\blacksquare$

Example 6

Let

$$A = \begin{pmatrix} 4 & 2 & 2 \\ 2 & 4 & 2 \\ 2 & 2 & 4 \end{pmatrix}.$$

Since A is symmetric, Theorem 6.20 tells us that A is orthogonally equivalent to a diagonal matrix. We find an orthogonal matrix P and a diagonal matrix D such that $P^t AP = D$.

To find P, we obtain an orthonormal basis of eigenvectors. It is easy to show that the eigenvalues of A are 2 and 8. The set $\{(-1, 1, 0), (-1, 0, 1)\}$ is a basis for the eigenspace corresponding to 2. Because this set is not orthogonal, we apply the Gram–Schmidt process to obtain the orthogonal set $\{(-1, 1, 0), -\frac{1}{2}(1, 1, -2)\}$. The set $\{(1, 1, 1)\}$ is a basis for the eigenspace corresponding to 8. Notice that $(1, 1, 1)$ is orthogonal to the preceding two vectors, as predicted by Theorem 6.15(d) (p. 371). Taking the union of these two bases and normalizing the vectors, we obtain the following orthonormal basis for R^3 consisting of eigenvectors of A:

$$\left\{ \frac{1}{\sqrt{2}}(-1, 1, 0), \frac{1}{\sqrt{6}}(1, 1, -2), \frac{1}{\sqrt{3}}(1, 1, 1) \right\}.$$

Thus one possible choice for P is

$$P = \begin{pmatrix} \frac{-1}{\sqrt{2}} & \frac{1}{\sqrt{6}} & \frac{1}{\sqrt{3}} \\ \frac{1}{\sqrt{2}} & \frac{1}{\sqrt{6}} & \frac{1}{\sqrt{3}} \\ 0 & \frac{-2}{\sqrt{6}} & \frac{1}{\sqrt{3}} \end{pmatrix}, \quad \text{and} \quad D = \begin{pmatrix} 2 & 0 & 0 \\ 0 & 2 & 0 \\ 0 & 0 & 8 \end{pmatrix}. \quad \blacklozenge$$

Because of Schur's theorem (Theorem 6.14 p. 370), the next result is immediate. As it is the matrix form of Schur's theorem, we also refer to it as Schur's theorem.

Theorem 6.21 (Schur). *Let $A \in \mathsf{M}_{n \times n}(F)$ be a matrix whose characteristic polynomial splits over F.*

(a) *If $F = C$, then A is unitarily equivalent to a complex upper triangular matrix.*

(b) *If $F = R$, then A is orthogonally equivalent to a real upper triangular matrix.*

Rigid Motions*

The purpose of this application is to characterize the so-called *rigid motions* of a finite-dimensional real inner product space. One may think intuitively of such a motion as a transformation that does not affect the shape of a figure under its action, hence the term *rigid*. The key requirement for such a transformation is that it preserves distances.

Definition. *Let V be a real inner product space. A function $f \colon \mathsf{V} \to \mathsf{V}$ is called a **rigid motion** if*

$$\|f(x) - f(y)\| = \|x - y\|$$

for all $x, y \in \mathsf{V}$.

For example, any orthogonal operator on a finite-dimensional real inner product space is a rigid motion.

Another class of rigid motions are the *translations*. A function $g \colon \mathsf{V} \to \mathsf{V}$, where V is a real inner product space, is called a **translation** if there exists a vector $v_0 \in \mathsf{V}$ such that $g(x) = x + v_0$ for all $x \in \mathsf{V}$. We say that g is the *translation by* v_0. It is a simple exercise to show that translations, as well as composites of rigid motions on a real inner product space, are also rigid motions. (See Exercise 22.) Thus an orthogonal operator on a finite-dimensional real inner product space V followed by a translation on V is a rigid motion on V. Remarkably, every rigid motion on V may be characterized in this way.

Theorem 6.22. *Let* $f \colon \mathsf{V} \to \mathsf{V}$ *be a rigid motion on a finite-dimensional real inner product space* V. *Then there exists a unique orthogonal operator* T *on* V *and a unique translation* g *on* V *such that* $f = g \circ \mathsf{T}$.

Any orthogonal operator is a special case of this composite, in which the translation is by *0*. Any translation is also a special case, in which the orthogonal operator is the identity operator.

Proof. Let $\mathsf{T} \colon \mathsf{V} \to \mathsf{V}$ be defined by

$$\mathsf{T}(x) = f(x) - f(0)$$

for all $x \in \mathsf{V}$. We show that T is an orthogonal operator, from which it follows that $f = g \circ \mathsf{T}$, where g is the translation by $f(0)$. Observe that T is the composite of f and the translation by $-f(0)$; hence T is a rigid motion. Furthermore, for any $x \in \mathsf{V}$

$$\|\mathsf{T}(x)\|^2 = \|f(x) - f(0)\|^2 = \|x - 0\|^2 = \|x\|^2,$$

and consequently $\|\mathsf{T}(x)\| = \|x\|$ for any $x \in \mathsf{V}$. Thus for any $x, y \in \mathsf{V}$,

$$\|T(x) - T(y)\|^2 = \|\mathsf{T}(x)\|^2 - 2\langle \mathsf{T}(x), \mathsf{T}(y)\rangle + \|\mathsf{T}(y)\|^2$$
$$= \|x\|^2 - 2\langle \mathsf{T}(x), \mathsf{T}(y)\rangle + \|y\|^2$$

and

$$\|x - y\|^2 = \|x\|^2 - 2\langle x, y\rangle + \|y\|^2.$$

But $\|T(x) - T(y)\|^2 = \|x - y\|^2$; so $\langle \mathsf{T}(x), \mathsf{T}(y)\rangle = \langle x, y\rangle$ for all $x, y \in \mathsf{V}$.

We are now in a position to show that T is a linear transformation. Let $x, y \in \mathsf{V}$, and let $a \in R$. Then

$$\|\mathsf{T}(x + ay) - \mathsf{T}(x) - a\mathsf{T}(y)\|^2 = \|[\mathsf{T}(x + ay) - \mathsf{T}(x)] - a\mathsf{T}(y)\|^2$$

$$= \|T(x + ay) - T(x)\|^2 + a^2\|T(y)\|^2 - 2a\langle T(x + ay) - T(x), T(y)\rangle$$
$$= \|(x + ay) - x\|^2 + a^2\|y\|^2 - 2a[\langle T(x + ay), T(y)\rangle - \langle T(x), T(y)\rangle]$$
$$= a^2\|y\|^2 + a^2\|y\|^2 - 2a[\langle x + ay, y\rangle - \langle x, y\rangle]$$
$$= 2a^2\|y\|^2 - 2a[\langle x, y\rangle + a\|y\|^2 - \langle x, y\rangle]$$
$$= 0.$$

Thus $T(x + ay) = T(x) + aT(y)$, and hence T is linear. Since T also preserves inner products, T is an orthogonal operator.

To prove uniqueness, suppose that u_0 and v_0 are in V and T and U are orthogonal operators on V such that

$$f(x) = T(x) + u_0 = U(x) + v_0$$

for all $x \in V$. Substituting $x = 0$ in the preceding equation yields $u_0 = v_0$, and hence the translation is unique. This equation, therefore, reduces to $T(x) = U(x)$ for all $x \in V$, and hence $T = U$. ∎

Orthogonal Operators on R^2

Because of Theorem 6.22, an understanding of rigid motions requires a characterization of orthogonal operators. The next result characterizes orthogonal operators on R^2. We postpone the case of orthogonal operators on more general spaces to Section 6.11.

Theorem 6.23. Let T be an orthogonal operator on R^2, and let $A = [T]_\beta$, where β is the standard ordered basis for R^2. Then exactly one of the following conditions is satisfied:
  (a) T is a rotation, and $\det(A) = 1$.
  (b) T is a reflection about a line through the origin, and $\det(A) = -1$.

Proof. Because T is an orthogonal operator, $T(\beta) = \{T(e_1), T(e_2)\}$ is an orthonormal basis for R^2 by Theorem 6.18(c). Since $T(e_1)$ is a unit vector, there is a unique angle θ, $0 \le \theta < 2\pi$, such that $T(e_1) = (\cos\theta, \sin\theta)$. Since $T(e_2)$ is a unit vector and is orthogonal to $T(e_1)$, there are only two possible choices for $T(e_2)$. Either

$$T(e_2) = (-\sin\theta, \cos\theta) \quad \text{or} \quad T(e_2) = (\sin\theta, -\cos\theta).$$

First, suppose that $T(e_2) = (-\sin\theta, \cos\theta)$. Then $A = \begin{pmatrix} \cos\theta & -\sin\theta \\ \sin\theta & \cos\theta \end{pmatrix}$.

It follows from Example 1 of Section 6.4 that T is a rotation by the angle θ. Also

$$\det(A) = \cos^2\theta + \sin^2\theta = 1.$$

Now suppose that $T(e_2) = (\sin\theta, -\cos\theta)$. Then $A = \begin{pmatrix} \cos\theta & \sin\theta \\ \sin\theta & -\cos\theta \end{pmatrix}$.

Comparing this matrix to the matrix A of Example 5, we see that T is the reflection of R^2 about a line L, so that $\alpha = \theta/2$ is the angle from the positive x-axis to L. Furthermore,

$$\det(A) = -\cos^2\theta - \sin^2\theta = -1. \qquad \blacksquare$$

Combining Theorems 6.22 and 6.23, we obtain the following characterization of rigid motions on R^2.

Corollary. *Any rigid motion on R^2 is either a rotation followed by a translation or a reflection about a line through the origin followed by a translation.*

Example 7

Let

$$A = \begin{pmatrix} \dfrac{1}{\sqrt{5}} & \dfrac{2}{\sqrt{5}} \\ \dfrac{2}{\sqrt{5}} & \dfrac{-1}{\sqrt{5}} \end{pmatrix}.$$

We show that L_A is the reflection of R^2 about a line L through the origin, and then describe L.

Clearly $AA^* = A^*A = I$, and therefore A is an orthogonal matrix. Hence L_A is an orthogonal operator. Furthermore,

$$\det(A) = -\frac{1}{5} - \frac{4}{5} = -1,$$

and thus L_A is a reflection of R^2 about a line L through the origin by Theorem 6.23. Since L is the one-dimensional eigenspace corresponding to the eigenvalue 1 of L_A, it suffices to find an eigenvector of L_A corresponding to 1. One such vector is $v = (2, \sqrt{5}-1)$. Thus L is the span of $\{v\}$. Alternatively, L is the line through the origin with slope $(\sqrt{5}-1)/2$, and hence is the line with the equation

$$y = \frac{\sqrt{5}-1}{2}x. \qquad \blacklozenge$$

Conic Sections

As an application of Theorem 6.20, we consider the quadratic equation

$$ax^2 + 2bxy + cy^2 + dx + ey + f = 0. \qquad (2)$$

For special choices of the coefficients in (2), we obtain the various conic sections. For example, if $a = c = 1$, $b = d = e = 0$, and $f = -1$, we obtain the circle $x^2 + y^2 = 1$ with center at the origin. The remaining conic sections, namely, the ellipse, parabola, and hyperbola, are obtained by other choices of the coefficients. If $b = 0$, then it is easy to graph the equation by the method of completing the square because the xy-term is absent. For example, the equation $x^2 + 2x + y^2 + 4y + 2 = 0$ may be rewritten as $(x+1)^2 + (y+2)^2 = 3$, which describes a circle with radius $\sqrt{3}$ and center at $(-1, -2)$ in the xy-coordinate system. If we consider the transformation of coordinates $(x, y) \rightarrow (x', y')$, where $x' = x + 1$ and $y' = y + 2$, then our equation simplifies to $(x')^2 + (y')^2 = 3$. This change of variable allows us to eliminate the x- and y-terms.

We now concentrate solely on the elimination of the xy-term. To accomplish this, we consider the expression

$$ax^2 + 2bxy + cy^2, \tag{3}$$

which is called the **associated quadratic form** of (2). Quadratic forms are studied in more generality in Section 6.8.

If we let

$$A = \begin{pmatrix} a & b \\ b & c \end{pmatrix} \quad \text{and} \quad X = \begin{pmatrix} x \\ y \end{pmatrix},$$

then (3) may be written as $X^t A X = \langle AX, X \rangle$. For example, the quadratic form $3x^2 + 4xy + 6y^2$ may be written as

$$X^t \begin{pmatrix} 3 & 2 \\ 2 & 6 \end{pmatrix} X.$$

The fact that A is symmetric is crucial in our discussion. For, by Theorem 6.20, we may choose an orthogonal matrix P and a diagonal matrix D with real diagonal entries λ_1 and λ_2 such that $P^t A P = D$. Now define

$$X' = \begin{pmatrix} x' \\ y' \end{pmatrix}$$

by $X' = P^t X$ or, equivalently, by $PX' = PP^t X = X$. Then

$$X^t A X = (PX')^t A(PX') = X'^t (P^t A P) X' = X'^t D X' = \lambda_1 (x')^2 + \lambda_2 (y')^2.$$

Thus the transformation $(x, y) \rightarrow (x', y')$ allows us to eliminate the xy-term in (3), and hence in (2).

Furthermore, since P is orthogonal, we have by Theorem 6.23 (with $\mathsf{T} = \mathsf{L}_P$) that $\det(P) = \pm 1$. If $\det(P) = -1$, we may interchange the columns of P to obtain a matrix Q. Because the columns of P form an orthonormal basis of eigenvectors of A, the same is true of the columns of Q. Therefore,

$$Q^t A Q = \begin{pmatrix} \lambda_2 & 0 \\ 0 & \lambda_1 \end{pmatrix}.$$

Notice that $\det(Q) = -\det(P) = 1$. So, if $\det(P) = -1$, we can take Q for our new P; consequently, we may always choose P so that $\det(P) = 1$. By Lemma 4 to Theorem 6.22 (with $\mathsf{T} = \mathsf{L}_P$), it follows that matrix P represents a rotation.

In summary, the xy-term in (2) may be eliminated by a rotation of the x-axis and y-axis to new axes x' and y' given by $X = PX'$, where P is an orthogonal matrix and $\det(P) = 1$. Furthermore, the coefficients of $(x')^2$ and $(y')^2$ are the eigenvalues of

$$A = \begin{pmatrix} a & b \\ b & c \end{pmatrix}.$$

This result is a restatement of a result known as the *principal axis theorem* for R^2. The arguments above, of course, are easily extended to quadratic equations in n variables. For example, in the case $n = 3$, by special choices of the coefficients, we obtain the quadratic surfaces—the elliptic cone, the ellipsoid, the hyperbolic paraboloid, etc.

As an illustration of the preceding transformation, consider the quadratic equation

$$2x^2 - 4xy + 5y^2 - 36 = 0,$$

for which the associated quadratic form is $2x^2 - 4xy + 5y^2$. In the notation we have been using,

$$A = \begin{pmatrix} 2 & -2 \\ -2 & 5 \end{pmatrix},$$

so that the eigenvalues of A are 1 and 6 with associated eigenvectors

$$\begin{pmatrix} 2 \\ 1 \end{pmatrix} \quad \text{and} \quad \begin{pmatrix} -1 \\ 2 \end{pmatrix}.$$

As expected (from Theorem 6.15(d) p. 371), these vectors are orthogonal. The corresponding orthonormal basis of eigenvectors

$$\beta = \left\{ \begin{pmatrix} \dfrac{2}{\sqrt{5}} \\ \dfrac{1}{\sqrt{5}} \end{pmatrix}, \begin{pmatrix} \dfrac{-1}{\sqrt{5}} \\ \dfrac{2}{\sqrt{5}} \end{pmatrix} \right\}$$

determines new axes x' and y' as in Figure 6.4. Hence if

$$P = \begin{pmatrix} \dfrac{2}{\sqrt{5}} & \dfrac{-1}{\sqrt{5}} \\ \dfrac{1}{\sqrt{5}} & \dfrac{2}{\sqrt{5}} \end{pmatrix} = \dfrac{1}{\sqrt{5}} \begin{pmatrix} 2 & -1 \\ 1 & 2 \end{pmatrix},$$

then

$$P^t A P = \begin{pmatrix} 1 & 0 \\ 0 & 6 \end{pmatrix}.$$

Under the transformation $X = PX'$ or

$$x = \frac{2}{\sqrt{5}}x' - \frac{1}{\sqrt{5}}y'$$
$$y = \frac{1}{\sqrt{5}}x' + \frac{2}{\sqrt{5}}y',$$

we have the new quadratic form $(x')^2 + 6(y')^2$. Thus the original equation $2x^2 - 4xy + 5y^2 = 36$ may be written in the form $(x')^2 + 6(y')^2 = 36$ relative to a new coordinate system with the x'- and y'-axes in the directions of the first and second vectors of β, respectively. It is clear that this equation represents

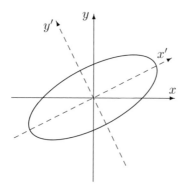

Figure 6.4

an ellipse. (See Figure 6.4.) Note that the preceding matrix P has the form

$$\begin{pmatrix} \cos\theta & -\sin\theta \\ \sin\theta & \cos\theta \end{pmatrix},$$

where $\theta = \cos^{-1}\dfrac{2}{\sqrt{5}} \approx 26.6°$. So P is the matrix representation of a rotation of R^2 through the angle θ. Thus the change of variable $X = PX'$ can be accomplished by this rotation of the x- and y-axes. There is another possibility for P, however. If the eigenvector of A corresponding to the eigenvalue 6 is taken to be $(1, -2)$ instead of $(-1, 2)$, and the eigenvalues are interchanged, then we obtain the matrix

$$\begin{pmatrix} \dfrac{1}{\sqrt{5}} & \dfrac{2}{\sqrt{5}} \\ \dfrac{-2}{\sqrt{5}} & \dfrac{1}{\sqrt{5}} \end{pmatrix},$$

which is the matrix representation of a rotation through the angle $\theta = \sin^{-1}\left(-\dfrac{2}{\sqrt{5}}\right) \approx -63.4°$. This possibility produces the same ellipse as the one in Figure 6.4, but interchanges the names of the x'- and y'-axes.

EXERCISES

1. Label the following statements as true or false. Assume that the underlying inner product spaces are finite-dimensional.

    (a) Every unitary operator is normal.
    (b) Every orthogonal operator is diagonalizable.
    (c) A matrix is unitary if and only if it is invertible.
    (d) If two matrices are unitarily equivalent, then they are also similar.
    (e) The sum of unitary matrices is unitary.
    (f) The adjoint of a unitary operator is unitary.
    (g) If T is an orthogonal operator on V, then $[\mathsf{T}]_\beta$ is an orthogonal matrix for any ordered basis β for V.
    (h) If all the eigenvalues of a linear operator are 1, then the operator must be unitary or orthogonal.
    (i) A linear operator may preserve the norm, but not the inner product.

2. For each of the following matrices A, find an orthogonal or unitary matrix P and a diagonal matrix D such that $P^*AP = D$.

    (a) $\begin{pmatrix} 1 & 2 \\ 2 & 1 \end{pmatrix}$ (b) $\begin{pmatrix} 0 & -1 \\ 1 & 0 \end{pmatrix}$ (c) $\begin{pmatrix} 2 & 3-3i \\ 3+3i & 5 \end{pmatrix}$

    (d) $\begin{pmatrix} 0 & 2 & 2 \\ 2 & 0 & 2 \\ 2 & 2 & 0 \end{pmatrix}$ (e) $\begin{pmatrix} 2 & 1 & 1 \\ 1 & 2 & 1 \\ 1 & 1 & 2 \end{pmatrix}$

3. Prove that the composite of unitary [orthogonal] operators is unitary [orthogonal].

4. For $z \in C$, define $T_z \colon C \to C$ by $T_z(u) = zu$. Characterize those z for which T_z is normal, self-adjoint, or unitary.

5. Which of the following pairs of matrices are unitarily equivalent?

(a) $\begin{pmatrix} 1 & 0 \\ 0 & 1 \end{pmatrix}$ and $\begin{pmatrix} 0 & 1 \\ 1 & 0 \end{pmatrix}$ (b) $\begin{pmatrix} 0 & 1 \\ 1 & 0 \end{pmatrix}$ and $\begin{pmatrix} 0 & \frac{1}{2} \\ \frac{1}{2} & 0 \end{pmatrix}$

(c) $\begin{pmatrix} 0 & 1 & 0 \\ -1 & 0 & 0 \\ 0 & 0 & 1 \end{pmatrix}$ and $\begin{pmatrix} 2 & 0 & 0 \\ 0 & -1 & 0 \\ 0 & 0 & 0 \end{pmatrix}$

(d) $\begin{pmatrix} 0 & 1 & 0 \\ -1 & 0 & 0 \\ 0 & 0 & 1 \end{pmatrix}$ and $\begin{pmatrix} 1 & 0 & 0 \\ 0 & i & 0 \\ 0 & 0 & -i \end{pmatrix}$

(e) $\begin{pmatrix} 1 & 1 & 0 \\ 0 & 2 & 2 \\ 0 & 0 & 3 \end{pmatrix}$ and $\begin{pmatrix} 1 & 0 & 0 \\ 0 & 2 & 0 \\ 0 & 0 & 3 \end{pmatrix}$

6. Let V be the inner product space of complex-valued continuous functions on $[0, 1]$ with the inner product

$$\langle f, g \rangle = \int_0^1 f(t)\overline{g(t)}\, dt.$$

Let $h \in V$, and define $T \colon V \to V$ by $T(f) = hf$. Prove that T is a unitary operator if and only if $|h(t)| = 1$ for $0 \le t \le 1$.

7. Prove that if T is a unitary operator on a finite-dimensional inner product space V, then T has a unitary *square root*; that is, there exists a unitary operator U such that $T = U^2$.

8. Let T be a self-adjoint linear operator on a finite-dimensional inner product space. Prove that $(T+iI)(T-iI)^{-1}$ is unitary using Exercise 10 of Section 6.4.

9. Let U be a linear operator on a finite-dimensional inner product space V. If $\|U(x)\| = \|x\|$ for all x in some orthonormal basis for V, must U be unitary? Justify your answer with a proof or a counterexample.

10. Let A be an $n \times n$ real symmetric or complex normal matrix. Prove that

$$\operatorname{tr}(A) = \sum_{i=1}^{n} \lambda_i \quad \text{and} \quad \operatorname{tr}(A^*A) = \sum_{i=1}^{n} |\lambda_i|^2,$$

where the λ_i's are the (not necessarily distinct) eigenvalues of A.

11. Find an orthogonal matrix whose first row is $(\frac{1}{3}, \frac{2}{3}, \frac{2}{3})$.

12. Let A be an $n \times n$ real symmetric or complex normal matrix. Prove that

$$\det(A) = \prod_{i=1}^{n} \lambda_i,$$

where the λ_i's are the (not necessarily distinct) eigenvalues of A.

13. Suppose that A and B are diagonalizable matrices. Prove or disprove that A is similar to B if and only if A and B are unitarily equivalent.

14. Prove that if A and B are unitarily equivalent matrices, then A is positive definite [semidefinite] if and only if B is positive definite [semidefinite]. (See the definitions in the exercises in Section 6.4.)

15. Let U be a unitary operator on an inner product space V, and let W be a finite-dimensional U-invariant subspace of V. Prove that

(a) $U(W) = W$;
(b) $W^{\perp}$ is U-invariant.

Contrast (b) with Exercise 16.

16. Find an example of a unitary operator U on an inner product space and a U-invariant subspace W such that $W^{\perp}$ is not U-invariant.

17. Prove that a matrix that is both unitary and upper triangular must be a diagonal matrix.

18. Show that "is unitarily equivalent to" is an equivalence relation on $M_{n \times n}(C)$.

19. Let W be a finite-dimensional subspace of an inner product space V. By Theorem 6.7 (p. 352) and the exercises of Section 1.3, $V = W \oplus W^{\perp}$. Define $U: V \to V$ by $U(v_1 + v_2) = v_1 - v_2$, where $v_1 \in W$ and $v_2 \in W^{\perp}$. Prove that U is a self-adjoint unitary operator.

20. Let V be a finite-dimensional inner product space. A linear operator U on V is called a **partial isometry** if there exists a subspace W of V such that $\|U(x)\| = \|x\|$ for all $x \in W$ and $U(x) = 0$ for all $x \in W^{\perp}$. Observe that W need *not* be U-invariant. Suppose that U is such an operator and $\{v_1, v_2, \ldots, v_k\}$ is an orthonormal basis for W. Prove the following results.

(a) $\langle U(x), U(y) \rangle = \langle x, y \rangle$ for all $x, y \in W$. *Hint:* Use Exercise 20 of Section 6.1.
(b) $\{U(v_1), U(v_2), \ldots, U(v_k)\}$ is an orthonormal basis for $R(U)$.

(c)   There exists an orthonormal basis γ for V such that the first k columns of $[U]_\gamma$ form an orthonormal set and the remaining columns are zero.

(d)   Let $\{w_1, w_2, \ldots, w_j\}$ be an orthonormal basis for $R(U)^\perp$ and $\beta = \{U(v_1), U(v_2), \ldots, U(v_k), w_1, \ldots, w_j\}$. Then β is an orthonormal basis for V.

(e)   Let T be the linear operator on V that satisfies $T(U(v_i)) = v_i$ $(1 \le i \le k)$ and $T(w_i) = 0$ $(1 \le i \le j)$. Then T is well defined, and $T = U^*$. *Hint:* Show that $\langle U(x), y \rangle = \langle x, T(y) \rangle$ for all $x, y \in \beta$. There are four cases.

(f) U^* is a partial isometry.

This exercise is continued in Exercise 9 of Section 6.6.

21. Let A and B be $n \times n$ matrices that are unitarily equivalent.

(a)   Prove that $\operatorname{tr}(A^*A) = \operatorname{tr}(B^*B)$.

(b)   Use (a) to prove that

$$\sum_{i,j=1}^{n} |A_{ij}|^2 = \sum_{i,j=1}^{n} |B_{ij}|^2.$$

(c)   Use (b) to show that the matrices

$$\begin{pmatrix} 1 & 2 \\ 2 & i \end{pmatrix} \qquad \text{and} \qquad \begin{pmatrix} i & 4 \\ 1 & 1 \end{pmatrix}$$

are *not* unitarily equivalent.

22. Let V be a real inner product space.

(a)   Prove that any translation on V is a rigid motion.

(b)   Prove that the composite of any two rigid motions on V is a rigid motion on V.

23. Prove the following variation of Theorem 6.22: If $f: V \rightarrow V$ is a rigid motion on a finite-dimensional real inner product space V, then there exists a unique orthogonal operator T on V and a unique translation g on V such that $f = T \circ g$.

24. Let T and U be orthogonal operators on R^2. Use Theorem 6.23 to prove the following results.

(a)   If T and U are both reflections about lines through the origin, then UT is a rotation.

(b)   If T is a rotation and U is a reflection about a line through the origin, then both UT and TU are reflections about lines through the origin.

25. Suppose that T and U are reflections of R^2 about the respective lines L and L' through the origin and that ϕ and ψ are the angles from the positive x-axis to L and L', respectively. By Exercise 24, UT is a rotation. Find its angle of rotation.

26. Suppose that T and U are orthogonal operators on R^2 such that T is the rotation by the angle ϕ and U is the reflection about the line L through the origin. Let ψ be the angle from the positive x-axis to L. By Exercise 24, both UT and TU are reflections about lines L_1 and L_2, respectively, through the origin.

    (a) Find the angle θ from the positive x-axis to L_1.
    (b) Find the angle θ from the positive x-axis to L_2.

27. Find new coordinates x', y' so that the following quadratic forms can be written as $\lambda_1(x')^2 + \lambda_2(y')^2$.

    (a) $x^2 + 4xy + y^2$
    (b) $2x^2 + 2xy + 2y^2$
    (c) $x^2 - 12xy - 4y^2$
    (d) $3x^2 + 2xy + 3y^2$
    (e) $x^2 - 2xy + y^2$

28. Consider the expression $X^t AX$, where $X^t = (x, y, z)$ and A is as defined in Exercise 2(e). Find a change of coordinates x', y', z' so that the preceding expression is of the form $\lambda_1(x')^2 + \lambda_2(y')^2 + \lambda_3(z')^2$.

29. *QR-Factorization.* Let $w_1, w_2, \ldots, w_n$ be linearly independent vectors in F^n, and let $v_1, v_2, \ldots, v_n$ be the orthogonal vectors obtained from $w_1, w_2, \ldots, w_n$ by the Gram–Schmidt process. Let $u_1, u_2, \ldots, u_n$ be the orthonormal basis obtained by normalizing the v_i's.

    (a) Solving (1) in Section 6.2 for w_k in terms of u_k, show that

    $$w_k = \|v_k\| u_k + \sum_{j=1}^{k-1} \langle w_k, u_j \rangle u_j \quad (1 \le k \le n).$$

    (b) Let A and Q denote the $n \times n$ matrices in which the kth columns are w_k and u_k, respectively. Define $R \in M_{n\times n}(F)$ by

    $$R_{jk} = \begin{cases} \|v_j\| & \text{if } j = k \\ \langle w_k, u_j \rangle & \text{if } j < k \\ 0 & \text{if } j > k. \end{cases}$$

    Prove $A = QR$.
    (c) Compute Q and R as in (b) for the 3×3 matrix whose columns are the vectors w_1, w_2, w_3, respectively, in Example 4 of Section 6.2.

(d) Since Q is unitary [orthogonal] and R is upper triangular in (b), we have shown that every invertible matrix is the product of a unitary [orthogonal] matrix and an upper triangular matrix. Suppose that $A \in \mathsf{M}_{n \times n}(F)$ is invertible and $A = Q_1 R_1 = Q_2 R_2$, where $Q_1, Q_2 \in \mathsf{M}_{n \times n}(F)$ are unitary and $R_1, R_2 \in \mathsf{M}_{n \times n}(F)$ are upper triangular. Prove that $D = R_2 R_1^{-1}$ is a unitary diagonal matrix. *Hint:* Use Exercise 17.

(e) The QR factorization described in (b) provides an orthogonalization method for solving a linear system $Ax = b$ when A is invertible. Decompose A to QR, by the Gram–Schmidt process or other means, where Q is unitary and R is upper triangular. Then $QRx = b$, and hence $Rx = Q^*b$. This last system can be easily solved since R is upper triangular. [1]

Use the orthogonalization method and (c) to solve the system

$$
\begin{aligned}
x_1 + 2x_2 + 2x_3 &= 1 \\
x_1 \qquad\;\; + 2x_3 &= 11 \\
x_2 + \;\; x_3 &= -1.
\end{aligned}
$$

30. Suppose that β and γ are ordered bases for an n-dimensional real [complex] inner product space V. Prove that if Q is an orthogonal [unitary] $n \times n$ matrix that changes γ-coordinates into β-coordinates, then β is orthonormal if and only if γ is orthonormal.

The following definition is used in Exercises 31 and 32.

Definition. Let V be a finite-dimensional complex [real] inner product space, and let u be a unit vector in V. Define the **Householder** operator $H_u \colon \mathsf{V} \to \mathsf{V}$ by $H_u(x) = x - 2 \langle x, u \rangle u$ for all $x \in \mathsf{V}$.

31. Let H_u be a Householder operator on a finite-dimensional inner product space V. Prove the following results.

   (a) H_u is linear.
   (b) $H_u(x) = x$ if and only if x is orthogonal to u.
   (c) $H_u(u) = -u$.
   (d) $H_u^* = H_u$ and $H_u^2 = I$, and hence H_u is a unitary [orthogonal] operator on V.

   (*Note:* If V is a real inner product space, then in the language of Section 6.11, H_u is a reflection.)

---

[1] At one time, because of its great stability, this method for solving large systems of linear equations with a computer was being advocated as a better method than Gaussian elimination even though it requires about three times as much work. (Later, however, J. H. Wilkinson showed that if Gaussian elimination is done "properly," then it is nearly as stable as the orthogonalization method.)

32. Let V be a finite-dimensional inner product space over F. Let x and y be linearly independent vectors in V such that $\|x\| = \|y\|$.

(a) If $F = C$, prove that there exists a unit vector u in V and a complex number θ with $|\theta| = 1$ such that $H_u(x) = \theta y$. *Hint:* Choose θ so that $\langle x, \theta y \rangle$ is real, and set $u = \dfrac{1}{\|x - \theta y\|}(x - \theta y)$.

(b) If $F = R$, prove that there exists a unit vector u in V such that $H_u(x) = y$.

6.6 ORTHOGONAL PROJECTIONS AND THE SPECTRAL THEOREM

In this section, we rely heavily on Theorems 6.16 (p. 372) and 6.17 (p. 374) to develop an elegant representation of a normal (if $F = C$) or a self-adjoint (if $F = R$) operator T on a finite-dimensional inner product space. We prove that T can be written in the form $\lambda_1 T_1 + \lambda_2 T_2 + \cdots + \lambda_k T_k$, where $\lambda_1, \lambda_2, \ldots, \lambda_k$ are the distinct eigenvalues of T and $T_1, T_2, \ldots, T_k$ are *orthogonal projections*. We must first develop some results about these special projections.

We assume that the reader is familiar with the results about direct sums developed at the end of Section 5.2. The special case where V is a direct sum of two subspaces is considered in the exercises of Section 1.3.

Recall from the exercises of Section 2.1 that if $V = W_1 \oplus W_2$, then a linear operator T on V is the **projection on W_1 along W_2** if, whenever $x = x_1 + x_2$, with $x_1 \in W_1$ and $x_2 \in W_2$, we have $T(x) = x_1$. By Exercise 26 of Section 2.1, we have

$$R(T) = W_1 = \{x \in V \colon T(x) = x\} \quad \text{and} \quad N(T) = W_2.$$

So $V = R(T) \oplus N(T)$. Thus there is no ambiguity if we refer to T as a "projection on W_1" or simply as a "projection." In fact, it can be shown (see Exercise 17 of Section 2.3) that T is a projection if and only if $T = T^2$. Because $V = W_1 \oplus W_2 = W_1 \oplus W_3$ does *not* imply that $W_2 = W_3$, we see that W_1 does not uniquely determine T. For an *orthogonal* projection T, however, T is uniquely determined by its range.

Definition. *Let V be an inner product space, and let $T \colon V \to V$ be a projection. We say that T is an **orthogonal projection** if $R(T)^{\perp} = N(T)$ and $N(T)^{\perp} = R(T)$.*

Note that by Exercise 13(c) of Section 6.2, if V is finite-dimensional, we need only assume that one of the preceding conditions holds. For example, if $R(T)^{\perp} = N(T)$, then $R(T) = R(T)^{\perp\perp} = N(T)^{\perp}$.

Now assume that W is a finite-dimensional subspace of an inner product space V. In the notation of Theorem 6.6 (p. 350), we can define a function

$T \colon V \to V$ by $T(y) = u$. It is easy to show that T is an orthogonal projection on W. We can say even more—there exists exactly one orthogonal projection on W. For if T and U are orthogonal projections on W, then $R(T) = W = R(U)$. Hence $N(T) = R(T)^{\perp} = R(U)^{\perp} = N(U)$, and since every projection is uniquely determined by its range and null space, we have $T = U$. We call T the **orthogonal projection** of V on W.

To understand the geometric difference between an arbitrary projection on W and the orthogonal projection on W, let $V = R^2$ and $W = \mathrm{span}\{(1,1)\}$. Define U and T as in Figure 6.5, where $T(v)$ is the foot of a perpendicular from v on the line $y = x$ and $U(a_1, a_2) = (a_1, a_1)$. Then T is the orthogonal projection of V on W, and U is a different projection on W. Note that $v - T(v) \in W^{\perp}$, whereas $v - U(v) \notin W^{\perp}$.

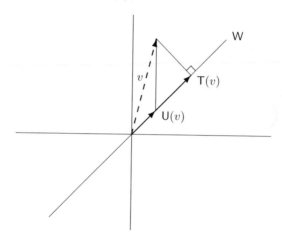

Figure 6.5

From Figure 6.5, we see that $T(v)$ is the "best approximation in W to v"; that is, if $w \in W$, then $\|w - v\| \geq \|T(v) - v\|$. In fact, this approximation property characterizes T. These results follow immediately from the corollary to Theorem 6.6 (p. 350).

As an application to Fourier analysis, recall the inner product space H and the orthonormal set S in Example 9 of Section 6.1. Define a **trigonometric polynomial of degree** n to be a function $g \in H$ of the form

$$g(t) = \sum_{j=-n}^{n} a_j f_j(t) = \sum_{j=-n}^{n} a_j e^{ijt},$$

where a_n or a_{-n} is nonzero.

Let $f \in H$. We show that the best approximation to f by a trigonometric polynomial of degree less than or equal to n is the trigonometric polynomial whose coefficients are the Fourier coefficients of f relative to the orthonormal set S. For this result, let $W = \text{span}(\{f_j : |j| \le n\})$, and let T be the orthogonal projection of H on W. The corollary to Theorem 6.6 (p. 350) tells us that the best approximation to f by a function in W is

$$T(f) = \sum_{j=-n}^{n} \langle f, f_j \rangle \, f_j.$$

An algebraic characterization of orthogonal projections follows in the next theorem.

Theorem 6.24. *Let V be an inner product space, and let T be a linear operator on V. Then T is an orthogonal projection if and only if T has an adjoint T^* and $T^2 = T = T^*$.*

Proof. Suppose that T is an orthogonal projection. Since $T^2 = T$ because T is a projection, we need only show that T^* exists and $T = T^*$. Now $V = R(T) \oplus N(T)$ and $R(T)^\perp = N(T)$. Let $x, y \in V$. Then we can write $x = x_1 + x_2$ and $y = y_1 + y_2$, where $x_1, y_1 \in R(T)$ and $x_2, y_2 \in N(T)$. Hence

$$\langle x, T(y) \rangle = \langle x_1 + x_2, y_1 \rangle = \langle x_1, y_1 \rangle + \langle x_2, y_1 \rangle = \langle x_1, y_1 \rangle$$

and

$$\langle T(x), y \rangle = \langle x_1, y_1 + y_2 \rangle = \langle x_1, y_1 \rangle + \langle x_1, y_2 \rangle = \langle x_1, y_1 \rangle .$$

So $\langle x, T(y) \rangle = \langle T(x), y \rangle$ for all $x, y \in V$; thus T^* exists and $T = T^*$.

Now suppose that $T^2 = T = T^*$. Then T is a projection by Exercise 17 of Section 2.3, and hence we must show that $R(T) = N(T)^\perp$ and $R(T)^\perp = N(T)$. Let $x \in R(T)$ and $y \in N(T)$. Then $x = T(x) = T^*(x)$, and so

$$\langle x, y \rangle = \langle T^*(x), y \rangle = \langle x, T(y) \rangle = \langle x, 0 \rangle = 0.$$

Therefore $x \in N(T)^\perp$, from which it follows that $R(T) \subseteq N(T)^\perp$.

Let $y \in N(T)^\perp$. We must show that $y \in R(T)$, that is, $T(y) = y$. Now

$$\|y - T(y)\|^2 = \langle y - T(y), y - T(y) \rangle$$
$$= \langle y, y - T(y) \rangle - \langle T(y), y - T(y) \rangle .$$

Since $y - T(y) \in N(T)$, the first term must equal zero. But also

$$\langle T(y), y - T(y) \rangle = \langle y, T^*(y - T(y)) \rangle = \langle y, T(y - T(y)) \rangle = \langle y, 0 \rangle = 0.$$

Thus $y - T(y) = 0$; that is, $y = T(y) \in R(T)$. Hence $R(T) = N(T)^\perp$.

Using the preceding results, we have $R(T)^\perp = N(T)^{\perp\perp} \supseteq N(T)$ by Exercise 13(b) of Section 6.2. Now suppose that $x \in R(T)^\perp$. For any $y \in V$, we have $\langle T(x), y \rangle = \langle x, T^*(y) \rangle = \langle x, T(y) \rangle = 0$. So $T(x) = 0$, and thus $x \in N(T)$. Hence $R(T)^\perp = N(T)$. ∎

Let V be a finite-dimensional inner product space, W be a subspace of V, and T be the orthogonal projection of V on W. We may choose an orthonormal basis $\beta = \{v_1, v_2, \dots, v_n\}$ for V such that $\{v_1, v_2, \dots, v_k\}$ is a basis for W. Then $[T]_\beta$ is a diagonal matrix with ones as the first k diagonal entries and zeros elsewhere. In fact, $[T]_\beta$ has the form

$$\begin{pmatrix} I_k & O_1 \\ O_2 & O_3 \end{pmatrix}.$$

If U is any projection on W, we may choose a basis γ for V such that $[U]_\gamma$ has the form above; however γ is not necessarily orthonormal.

We are now ready for the principal theorem of this section.

Theorem 6.25 (The Spectral Theorem). *Suppose that* T *is a linear operator on a finite-dimensional inner product space* V *over* F *with the distinct eigenvalues* $\lambda_1, \lambda_2, \dots, \lambda_k$. *Assume that* T *is normal if* $F = C$ *and that* T *is self-adjoint if* $F = R$. *For each* i *($1 \le i \le k$), let* W_i *be the eigenspace of* T *corresponding to the eigenvalue* λ_i, *and let* T_i *be the orthogonal projection of* V *on* W_i. *Then the following statements are true.*

(a) $V = W_1 \oplus W_2 \oplus \cdots \oplus W_k$.

(b) *If* W'_i *denotes the direct sum of the subspaces* W_j *for* $j \ne i$, *then* $W_i^\perp = W'_i$.

(c) $T_i T_j = \delta_{ij} T_i$ *for* $1 \le i, j \le k$.

(d) $I = T_1 + T_2 + \cdots + T_k$.

(e) $T = \lambda_1 T_1 + \lambda_2 T_2 + \cdots + \lambda_k T_k$.

Proof. (a) By Theorems 6.16 (p. 372) and 6.17 (p. 374), T is diagonalizable; so

$$V = W_1 \oplus W_2 \oplus \cdots \oplus W_k$$

by Theorem 5.11 (p. 278).

(b) If $x \in W_i$ and $y \in W_j$ for some $i \ne j$, then $\langle x, y \rangle = 0$ by Theorem 6.15(d) (p. 371). It follows easily from this result that $W'_i \subseteq W_i^\perp$. From (a), we have

$$\dim(W'_i) = \sum_{j \ne i} \dim(W_j) = \dim(V) - \dim(W_i).$$

On the other hand, we have $\dim(W_i^\perp) = \dim(V) - \dim(W_i)$ by Theorem 6.7(c) (p. 352). Hence $W'_i = W_i^\perp$, proving (b).

(c) The proof of (c) is left as an exercise.

(d) Since T_i is the orthogonal projection of V on W_i, it follows from (b) that $N(T_i) = R(T_i)^\perp = W_i^\perp = W'_i$. Hence, for $x \in V$, we have $x = x_1 + x_2 + \cdots + x_k$, where $T_i(x) = x_i \in W_i$, proving (d).

(e) For $x \in V$, write $x = x_1 + x_2 + \cdots + x_k$, where $x_i \in W_i$. Then

$$
\begin{aligned}
T(x) &= T(x_1) + T(x_2) + \cdots + T(x_k) \\
&= \lambda_1 x_1 + \lambda_2 x_2 + \cdots + \lambda_k x_k \\
&= \lambda_1 T_1(x) + \lambda_2 T_2(x) + \cdots + \lambda_k T_k(x) \\
&= (\lambda_1 T_1 + \lambda_2 T_2 + \cdots + \lambda_k T_k)(x). \quad\blacksquare
\end{aligned}
$$

The set $\{\lambda_1, \lambda_2, \ldots, \lambda_k\}$ of eigenvalues of T is called the **spectrum** of T, the sum $I = T_1 + T_2 + \cdots + T_k$ in (d) is called the **resolution of the identity operator** induced by T, and the sum $T = \lambda_1 T_1 + \lambda_2 T_2 + \cdots + \lambda_k T_k$ in (e) is called the **spectral decomposition** of T. The spectral decomposition of T is unique up to the order of its eigenvalues.

With the preceding notation, let β be the union of orthonormal bases of the W_i's and let $m_i = \dim(W_i)$. (Thus m_i is the multiplicity of λ_i.) Then $[T]_\beta$ has the form

$$
\begin{pmatrix}
\lambda_1 I_{m_1} & O & \cdots & O \\
O & \lambda_2 I_{m_2} & \cdots & O \\
\vdots & \vdots & & \vdots \\
O & O & \cdots & \lambda_k I_{m_k}
\end{pmatrix};
$$

that is, $[T]_\beta$ is a diagonal matrix in which the diagonal entries are the eigenvalues λ_i of T, and each λ_i is repeated m_i times. If $\lambda_1 T_1 + \lambda_2 T_2 + \cdots + \lambda_k T_k$ is the spectral decomposition of T, then it follows (from Exercise 7) that $g(T) = g(\lambda_1)T_1 + g(\lambda_2)T_2 + \cdots + g(\lambda_k)T_k$ for any polynomial g. This fact is used below.

We now list several interesting corollaries of the spectral theorem; many more results are found in the exercises. For what follows, we assume that T is a linear operator on a finite-dimensional inner product space V over F.

Corollary 1. *If $F = C$, then T is normal if and only if $T^* = g(T)$ for some polynomial g.*

Proof. Suppose first that T is normal. Let $T = \lambda_1 T_1 + \lambda_2 T_2 + \cdots + \lambda_k T_k$ be the spectral decomposition of T. Taking the adjoint of both sides of the preceding equation, we have $T^* = \overline{\lambda_1}T_1 + \overline{\lambda_2}T_2 + \cdots + \overline{\lambda_k}T_k$ since each T_i is self-adjoint. Using the Lagrange interpolation formula (see page 52), we may choose a polynomial g such that $g(\lambda_i) = \overline{\lambda_i}$ for $1 \le i \le k$. Then

$$
g(T) = g(\lambda_1)T_1 + g(\lambda_2)T_2 + \cdots + g(\lambda_k)T_k = \overline{\lambda_1}T_1 + \overline{\lambda_2}T_2 + \cdots + \overline{\lambda_k}T_k = T^*.
$$

Conversely, if $T^* = g(T)$ for some polynomial g, then T commutes with T^* since T commutes with every polynomial in T. So T is normal. $\blacksquare$

Corollary 2. *If $F = C$, then T is unitary if and only if T is normal and $|\lambda| = 1$ for every eigenvalue λ of T.*

Proof. If T is unitary, then T is normal and every eigenvalue of T has absolute value 1 by Corollary 2 to Theorem 6.18 (p. 382).

Let $\mathsf{T} = \lambda_1 \mathsf{T}_1 + \lambda_2 \mathsf{T}_2 + \cdots + \lambda_k \mathsf{T}_k$ be the spectral decomposition of T. If $|\lambda| = 1$ for every eigenvalue λ of T, then by (c) of the spectral theorem,

$$
\begin{aligned}
\mathsf{T}\mathsf{T}^* &= (\lambda_1 \mathsf{T}_1 + \lambda_2 \mathsf{T}_2 + \cdots + \lambda_k \mathsf{T}_k)(\overline{\lambda}_1 \mathsf{T}_1 + \overline{\lambda}_2 \mathsf{T}_2 + \cdots + \overline{\lambda}_k \mathsf{T}_k) \\
&= |\lambda_1|^2 \mathsf{T}_1 + |\lambda_2|^2 \mathsf{T}_2 + \cdots + |\lambda_k|^2 \mathsf{T}_k \\
&= \mathsf{T}_1 + \mathsf{T}_2 + \cdots + \mathsf{T}_k \\
&= \mathsf{I}.
\end{aligned}
$$

Hence T is unitary. ∎

Corollary 3. *If $F = C$ and T is normal, then T is self-adjoint if and only if every eigenvalue of T is real.*

Proof. Let $\mathsf{T} = \lambda_1 \mathsf{T}_1 + \lambda_2 \mathsf{T}_2 + \cdots + \lambda_k \mathsf{T}_k$ be the spectral decomposition of T. Suppose that every eigenvalue of T is real. Then

$$
\mathsf{T}^* = \overline{\lambda}_1 \mathsf{T}_1 + \overline{\lambda}_2 \mathsf{T}_2 + \cdots + \overline{\lambda}_k \mathsf{T}_k = \lambda_1 \mathsf{T}_1 + \lambda_2 \mathsf{T}_2 + \cdots + \lambda_k \mathsf{T}_k = \mathsf{T}.
$$

The converse has been proved in the lemma to Theorem 6.17 (p. 374). ∎

Corollary 4. *Let T be as in the spectral theorem with spectral decomposition $\mathsf{T} = \lambda_1 \mathsf{T}_1 + \lambda_2 \mathsf{T}_2 + \cdots + \lambda_k \mathsf{T}_k$. Then each T_j is a polynomial in T.*

Proof. Choose a polynomial g_j $(1 \le j \le k)$ such that $g_j(\lambda_i) = \delta_{ij}$. Then

$$
\begin{aligned}
g_j(\mathsf{T}) &= g_j(\lambda_1)\mathsf{T}_1 + g_j(\lambda_2)\mathsf{T}_2 + \cdots + g_j(\lambda_k)\mathsf{T}_k \\
&= \delta_{1j}\mathsf{T}_1 + \delta_{2j}\mathsf{T}_2 + \cdots + \delta_{kj}\mathsf{T}_k = \mathsf{T}_j. \qquad\blacksquare
\end{aligned}
$$

EXERCISES

1. Label the following statements as true or false. Assume that the underlying inner product spaces are finite-dimensional.

   (a) All projections are self-adjoint.

   (b) An orthogonal projection is uniquely determined by its range.

   (c) Every self-adjoint operator is a linear combination of orthogonal projections.

(d) If T is a projection on W, then $T(x)$ is the vector in W that is closest to x.

(e) Every orthogonal projection is a unitary operator.

2. Let $V = R^2$, $W = \text{span}(\{(1,2)\})$, and β be the standard ordered basis for V. Compute $[T]_\beta$, where T is the orthogonal projection of V on W. Do the same for $V = R^3$ and $W = \text{span}(\{(1,0,1)\})$.

3. For each of the matrices A in Exercise 2 of Section 6.5:

   (1) Verify that L_A possesses a spectral decomposition.
   (2) For each eigenvalue of L_A, explicitly define the orthogonal projection on the corresponding eigenspace.
   (3) Verify your results using the spectral theorem.

4. Let W be a finite-dimensional subspace of an inner product space V. Show that if T is the orthogonal projection of V on W, then $I - T$ is the orthogonal projection of V on $W^\perp$.

5. Let T be a linear operator on a finite-dimensional inner product space V.

   (a) If T is an orthogonal projection, prove that $\|T(x)\| \le \|x\|$ for all $x \in V$. Give an example of a projection for which this inequality does not hold. What can be concluded about a projection for which the inequality is actually an equality for all $x \in V$?
   (b) Suppose that T is a projection such that $\|T(x)\| \le \|x\|$ for $x \in V$. Prove that T is an orthogonal projection.

6. Let T be a normal operator on a finite-dimensional inner product space. Prove that if T is a projection, then T is also an orthogonal projection.

7. Let T be a normal operator on a finite-dimensional complex inner product space V. Use the spectral decomposition $\lambda_1 T_1 + \lambda_2 T_2 + \cdots + \lambda_k T_k$ of T to prove the following results.

   (a) If g is a polynomial, then

$$g(T) = \sum_{i=1}^{k} g(\lambda_i) T_i.$$

   (b) If $T^n = T_0$ for some n, then $T = T_0$.
   (c) Let U be a linear operator on V. Then U commutes with T if and only if U commutes with each T_i.
   (d) There exists a normal operator U on V such that $U^2 = T$.
   (e) T is invertible if and only if $\lambda_i \ne 0$ for $1 \le i \le k$.
   (f) T is a projection if and only if every eigenvalue of T is 1 or 0.

(g) $T = -T^*$ if and only if every λ_i is an imaginary number.

8. Use Corollary 1 of the spectral theorem to show that if T is a normal operator on a complex finite-dimensional inner product space and U is a linear operator that commutes with T, then U commutes with T^*.

9. Referring to Exercise 20 of Section 6.5, prove the following facts about a partial isometry U.

   (a) U^*U is an orthogonal projection on W.
   (b) $UU^*U = U$.

10. *Simultaneous diagonalization.* Let U and T be normal operators on a finite-dimensional complex inner product space V such that $TU = UT$. Prove that there exists an orthonormal basis for V consisting of vectors that are eigenvectors of both T and U. *Hint:* Use the hint of Exercise 14 of Section 6.4 along with Exercise 8.

11. Prove (c) of the spectral theorem.

6.7*   THE SINGULAR VALUE DECOMPOSITION AND THE PSEUDOINVERSE

In Section 6.4, we characterized normal operators on complex spaces and self-adjoint operators on real spaces in terms of orthonormal bases of eigenvectors and their corresponding eigenvalues (Theorems 6.16, p. 372, and 6.17, p. 374). In this section, we establish a comparable theorem whose scope is the entire class of linear transformations on both complex and real finite-dimensional inner product spaces—*the singular value theorem for linear transformations* (Theorem 6.26). There are similarities and differences among these theorems. All rely on the use of orthonormal bases and numerical invariants. However, because of its general scope, the singular value theorem is concerned with two (usually distinct) inner product spaces and with two (usually distinct) orthonormal bases. If the two spaces and the two bases are identical, then the transformation would, in fact, be a normal or self-adjoint operator. Another difference is that the numerical invariants in the singular value theorem, the *singular values*, are nonnegative, in contrast to their counterparts, the eigenvalues, for which there is no such restriction. This property is necessary to guarantee the uniqueness of singular values.

The singular value theorem encompasses both real and complex spaces. For brevity, in this section we use the terms *unitary operator* and *unitary matrix* to include orthogonal operators and orthogonal matrices in the context of real spaces. Thus any operator T for which $\langle T(x), T(y) \rangle = \langle x, y \rangle$, or any matrix A for which $\langle Ax, Ay \rangle = \langle x, y \rangle$, for all x and y is called *unitary* for the purposes of this section.

   In Exercise 15 of Section 6.3, the definition of the adjoint of an operator is extended to any linear transformation $\mathsf{T}\colon \mathsf{V} \to \mathsf{W}$, where V and W are finite-dimensional inner product spaces. By this exercise, the adjoint T^* of T is a linear transformation from W to V and $[\mathsf{T}^*]_\gamma^\beta = ([\mathsf{T}]_\beta^\gamma)^*$, where β and γ are orthonormal bases for V and W, respectively. Furthermore, the linear operator $\mathsf{T}^*\mathsf{T}$ on V is positive semidefinite and $\operatorname{rank}(\mathsf{T}^*\mathsf{T}) = \operatorname{rank}(\mathsf{T})$ by Exercise 18 of Section 6.4.

   With these facts in mind, we begin with the principal result.

Theorem 6.26 (Singular Value Theorem for Linear Transformations). *Let* V *and* W *be finite-dimensional inner product spaces, and let* $\mathsf{T}\colon \mathsf{V} \to \mathsf{W}$ *be a linear transformation of rank* r. *Then there exist orthonormal bases* $\{v_1, v_2, \dots, v_n\}$ *for* V *and* $\{u_1, u_2, \dots, u_m\}$ *for* W *and positive scalars* $\sigma_1 \geq \sigma_2 \geq \cdots \geq \sigma_r$ *such that*

$$
\mathsf{T}(v_i) = \begin{cases} \sigma_i u_i & \text{if } 1 \leq i \leq r \\ 0 & \text{if } i > r. \end{cases} \tag{4}
$$

Conversely, suppose that the preceding conditions are satisfied. Then for $1 \leq i \leq n$, v_i *is an eigenvector of* $\mathsf{T}^*\mathsf{T}$ *with corresponding eigenvalue* σ_i^2 *if* $1 \leq i \leq r$ *and* 0 *if* $i > r$. *Therefore the scalars* $\sigma_1, \sigma_2, \dots, \sigma_r$ *are uniquely determined by* T.

   Proof. We first establish the existence of the bases and scalars. By Exercises 18 of Section 6.4 and 15(d) of Section 6.3, $\mathsf{T}^*\mathsf{T}$ is a positive semidefinite linear operator of rank r on V; hence there is an orthonormal basis $\{v_1, v_2, \dots, v_n\}$ for V consisting of eigenvectors of $\mathsf{T}^*\mathsf{T}$ with corresponding eigenvalues λ_i, where $\lambda_1 \geq \lambda_2 \geq \cdots \geq \lambda_r > 0$, and $\lambda_i = 0$ for $i > r$. For $1 \leq i \leq r$, define $\sigma_i = \sqrt{\lambda_i}$ and $u_i = \dfrac{1}{\sigma_i}\mathsf{T}(v_i)$. We show that $\{u_1, u_2, \dots, u_r\}$ is an orthonormal subset of W. Suppose $1 \leq i, j \leq r$. Then

$$
\langle u_i, u_j \rangle = \left\langle \frac{1}{\sigma_i}\mathsf{T}(v_i), \frac{1}{\sigma_j}\mathsf{T}(v_j) \right\rangle
$$

$$
= \frac{1}{\sigma_i \sigma_j} \left\langle \mathsf{T}^*\mathsf{T}(v_i), v_j \right\rangle
$$

$$
= \frac{1}{\sigma_i \sigma_j} \left\langle \lambda_i v_i, v_j \right\rangle
$$

$$
= \frac{\sigma_i^2}{\sigma_i \sigma_j} \left\langle v_i, v_j \right\rangle
$$

$$
= \delta_{ij},
$$

and hence $\{u_1, u_2, \ldots, u_r\}$ is orthonormal. By Theorem 6.7(a) (p. 352), this set extends to an orthonormal basis $\{u_1, u_2, \ldots, u_r, \ldots, u_m\}$ for W. Clearly $\mathsf{T}(v_i) = \sigma_i u_i$ if $1 \leq i \leq r$. If $i > r$, then $\mathsf{T}^*\mathsf{T}(v_i) = 0$, and so $\mathsf{T}(v_i) = 0$ by Exercise 15(d) of Section 6.3.

To establish uniqueness, suppose that $\{v_1, v_2, \ldots, v_n\}$, $\{u_1, u_2, \ldots, u_m\}$, and $\sigma_1 \geq \sigma_2 \geq \cdots \geq \sigma_r > 0$ satisfy the properties stated in the first part of the theorem. Then for $1 \leq i \leq m$ and $1 \leq j \leq n$,

$$\langle \mathsf{T}^*(u_i), v_j \rangle = \langle u_i, \mathsf{T}(v_j) \rangle$$

$$= \begin{cases} \sigma_i & \text{if } i = j \leq r \\ 0 & \text{otherwise,} \end{cases}$$

and hence for any $1 \leq i \leq m$,

$$\mathsf{T}^*(u_i) = \sum_{j=1}^{n} \langle \mathsf{T}^*(u_i), v_j \rangle \, v_j = \begin{cases} \sigma_i v_i & \text{if } i = j \leq r \\ 0 & \text{otherwise.} \end{cases} \tag{5}$$

So for $i \leq r$,

$$\mathsf{T}^*\mathsf{T}(v_i) = \mathsf{T}^*(\sigma_i u_i) = \sigma_i \mathsf{T}^*(u_i) = \sigma_i^2 u_i$$

and $\mathsf{T}^*\mathsf{T}(v_i) = \mathsf{T}^*(0) = 0$ for $i > r$. Therefore each v_i is an eigenvector of $\mathsf{T}^*\mathsf{T}$ with corresponding eigenvalue σ_i^2 if $i \leq r$ and 0 if $i > r$. ∎

Definition. *The unique scalars* $\sigma_1, \sigma_2, \ldots, \sigma_r$ *in Theorem 6.26 are called the* **singular values** *of* T. *If* r *is less than both* m *and* n, *then the term singular value is extended to include* $\sigma_{r+1} = \cdots = \sigma_k = 0$, *where* k *is the minimum of* m *and* n.

Although the singular values of a linear transformation T are uniquely determined by T, the orthonormal bases given in the statement of Theorem 6.26 are not uniquely determined because there is more than one orthonormal basis of eigenvectors of $\mathsf{T}^*\mathsf{T}$.

In view of (5), the singular values of a linear transformation $\mathsf{T}: \mathsf{V} \to \mathsf{W}$ and its adjoint T^* are identical. Furthermore, the orthonormal bases for V and W given in Theorem 6.26 are simply reversed for T^*.

Example 1

Let $P_2(R)$ and $P_1(R)$ be the polynomial spaces with inner products defined by

$$\langle f(x), g(x) \rangle = \int_{-1}^{1} f(t)g(t) \, dt.$$

Let $T: P_2(R) \rightarrow P_1(R)$ be the linear transformation defined by $T(f(x)) = f'(x)$. Find orthonormal bases $\beta = \{v_1, v_2, v_3\}$ for $P_2(R)$ and $\gamma = \{u_1, u_2\}$ for $P_1(R)$ such that $T(v_i) = \sigma_i u_i$ for $i = 1, 2$ and $T(v_3) = 0$, where $\sigma_1 \geq \sigma_2 > 0$ are the nonzero singular values of T.

To facilitate the computations, we translate this problem into the corresponding problem for a matrix representation of T. Caution is advised here because not any matrix representation will do. Since the adjoint is defined in terms of inner products, we must use a matrix representation constructed from orthonormal bases for $P_2(R)$ and $P_1(R)$ to guarantee that the adjoint of the matrix representation of T is the same as the matrix representation of the adjoint of T. (See Exercise 15 of Section 6.3.) For this purpose, we use the results of Exercise 21(a) of Section 6.2 to obtain orthonormal bases

$$\alpha = \left\{ \frac{1}{\sqrt{2}}, \sqrt{\frac{3}{2}}\, x, \sqrt{\frac{5}{8}}\, (3x^2 - 1) \right\} \quad \text{and} \quad \alpha' = \left\{ \frac{1}{\sqrt{2}}, \sqrt{\frac{3}{2}}\, x \right\}$$

for $P_2(R)$ and $P_1(R)$, respectively.

Let

$$A = [T]_\alpha^{\alpha'} = \begin{pmatrix} 0 & \sqrt{3} & 0 \\ 0 & 0 & \sqrt{15} \end{pmatrix}.$$

Then

$$A^*A = \begin{pmatrix} 0 & 0 \\ \sqrt{3} & 0 \\ 0 & \sqrt{15} \end{pmatrix} \begin{pmatrix} 0 & \sqrt{3} & 0 \\ 0 & 0 & \sqrt{15} \end{pmatrix} = \begin{pmatrix} 0 & 0 & 0 \\ 0 & 3 & 0 \\ 0 & 0 & 15 \end{pmatrix},$$

which has eigenvalues (listed in descending order of size) $\lambda_1 = 15$, $\lambda_2 = 3$, and $\lambda_3 = 0$. These eigenvalues correspond, respectively, to the orthonormal eigenvectors $e_3 = (0, 0, 1)$, $e_2 = (0, 1, 0)$, and $e_1 = (1, 0, 0)$ in R^3. Translating everything into the context of T, $P_2(R)$, and $P_1(R)$, let

$$v_1 = \sqrt{\frac{5}{8}}\, (3x^2 - 1), \quad v_2 = \sqrt{\frac{3}{2}}\, x, \quad \text{and} \quad v_3 = \frac{1}{\sqrt{2}}.$$

Then $\beta = \{v_1, v_2, v_3\}$ is an orthonormal basis for $P_2(R)$ consisting of eigenvectors of T^*T with corresponding eigenvalues λ_1, λ_2, and λ_3. Now set $\sigma_1 = \sqrt{\lambda_1} = \sqrt{15}$ and $\sigma_2 = \sqrt{\lambda_2} = \sqrt{3}$, the nonzero singular values of T, and take

$$u_1 = \frac{1}{\sigma_1} T(v_1) = \sqrt{\frac{3}{2}}\, x \quad \text{and} \quad u_2 = \frac{1}{\sigma_2} T(v_2) = \frac{1}{\sqrt{2}},$$

to obtain the required basis $\gamma = \{u_1, u_2\}$ for $P_1(R)$.   ◆

We can use singular values to describe how a figure is distorted by a linear transformation. This is illustrated in the next example.

Example 2

Let T be an invertible linear operator on R^2 and $S = \{x \in \mathsf{R}^2 \colon \|x\| = 1\}$, the unit circle in R^2. We apply Theorem 6.26 to describe $S' = \mathsf{T}(S)$.

Since T is invertible, it has rank equal to 2 and hence has singular values $\sigma_1 \geq \sigma_2 > 0$. Let $\{v_1, v_2\}$ and $\beta = \{u_1, u_2\}$ be orthonormal bases for R^2 so that $\mathsf{T}(v_1) = \sigma_1 u_1$ and $\mathsf{T}(v_2) = \sigma_2 u_2$, as in Theorem 6.26. Then β determines a coordinate system, which we shall call the $x'y'$-coordinate system for R^2, where the x'-axis contains u_1 and the y'-axis contains u_2. For any vector $u \in \mathsf{R}^2$, if $u = x_1' u_1 + x_2' u_2$, then $[u]_\beta = \begin{pmatrix} x_1' \\ x_2' \end{pmatrix}$ is the coordinate vector of u relative to β. We characterize S' in terms of an equation relating x_1' and x_2'.

For any vector $v = x_1 v_1 + x_2 v_2 \in \mathsf{R}^2$, the equation $u = \mathsf{T}(v)$ means that

$$u = \mathsf{T}(x_1 v_1 + x_2 v_2) = x_1 \mathsf{T}(v_1) + x_2 \mathsf{T}(v_2) = x_1 \sigma_1 u_1 + x_2 \sigma_2 u_2.$$

Thus for $u = x_1' u_1 + x_2' u_2$, we have $x_1' = x_1 \sigma_1$ and $x_2' = x_2 \sigma_2$. Furthermore, $u \in S'$ if and only if $v \in S$ if and only if

$$\frac{(x_1')^2}{\sigma_1^2} + \frac{(x_2')^2}{\sigma_2^2} = x_1^2 + x_2^2 = 1.$$

If $\sigma_1 = \sigma_2$, this is the equation of a circle of radius σ_1, and if $\sigma_1 > \sigma_2$, this is the equation of an ellipse with major axis and minor axis oriented along the x'-axis and the y'-axis, respectively. (See Figure 6.6.)   ◆

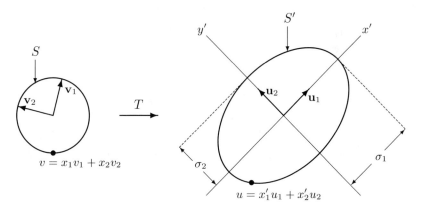

Figure 6.6

The singular value theorem for linear transformations is useful in its matrix form because we can perform numerical computations on matrices. We begin with the definition of the singular values of a matrix.

Definition. *Let A be an $m \times n$ matrix. We define the **singular values** of A to be the singular values of the linear transformation L_A.*

Theorem 6.27 (Singular Value Decomposition Theorem for Matrices).   *Let A be an $m \times n$ matrix of rank r with the positive singular values $\sigma_1 \geq \sigma_2 \geq \cdots \geq \sigma_r$, and let Σ be the $m \times n$ matrix defined by*

$$\Sigma_{ij} = \begin{cases} \sigma_i & \text{if } i = j \leq r \\ 0 & \text{otherwise.} \end{cases}$$

Then there exists an $m \times m$ unitary matrix U and an $n \times n$ unitary matrix V such that

$$A = U\Sigma V^*.$$

Proof. Let $T = \mathsf{L}_A \colon \mathsf{F}^n \to \mathsf{F}^m$. By Theorem 6.26, there exist orthonormal bases $\beta = \{v_1, v_2, \ldots, v_n\}$ for F^n and $\gamma = \{u_1, u_2, \ldots, u_m\}$ for F^m such that $\mathsf{T}(v_i) = \sigma_i u_i$ for $1 \leq i \leq r$ and $\mathsf{T}(v_i) = 0$ for $i > r$. Let U be the $m \times m$ matrix whose jth column is u_j for all j, and let V be the $n \times n$ matrix whose jth column is v_j for all j. Note that both U and V are unitary matrices.

By Theorem 2.13(a) (p. 90), the jth column of AV is $Av_j = \sigma_j u_j$. Observe that the jth column of Σ is $\sigma_j e_j$, where e_j is the jth standard vector of F^m. So by Theorem 2.13(a) and (b), the jth column of $U\Sigma$ is given by

$$U(\sigma_j e_j) = \sigma_j U(e_j) = \sigma_j u_j.$$

It follows that AV and $U\Sigma$ are $m \times n$ matrices whose corresponding columns are equal, and hence $AV = U\Sigma$. Therefore $A = AVV^* = U\Sigma V^*$. ∎

Definition. *Let A be an $m \times n$ matrix of rank r with positive singular values $\sigma_1 \geq \sigma_2 \geq \cdots \geq \sigma_r$. A factorization $A = U\Sigma V^*$ where U and V are unitary matrices and Σ is the $m \times n$ matrix defined as in Theorem 6.27 is called a **singular value decomposition** of A.*

In the proof of Theorem 6.27, the columns of V are the vectors in β, and the columns of U are the vectors in γ. Furthermore, the nonzero singular values of A are the same as those of L_A; hence they are the square roots of the nonzero eigenvalues of A^*A or of AA^*. (See Exercise 9.)

Example 3

We find a singular value decomposition for $A = \begin{pmatrix} 1 & 1 & -1 \\ 1 & 1 & -1 \end{pmatrix}$.

First observe that for

$$v_1 = \frac{1}{\sqrt{3}} \begin{pmatrix} 1 \\ 1 \\ -1 \end{pmatrix}, \quad v_2 = \frac{1}{\sqrt{2}} \begin{pmatrix} 1 \\ -1 \\ 0 \end{pmatrix}, \quad \text{and} \quad v_3 = \frac{1}{\sqrt{6}} \begin{pmatrix} 1 \\ 1 \\ 2 \end{pmatrix},$$

the set $\beta = \{v_1, v_2, v_3\}$ is an orthonormal basis for $\mathbf{R}^3$ consisting of eigenvectors of A^*A with corresponding eigenvalues $\lambda_1 = 6$, and $\lambda_2 = \lambda_3 = 0$. Consequently, $\sigma_1 = \sqrt{6}$ is the only nonzero singular value of A. Hence, as in the proof of Theorem 6.27, we let V be the matrix whose columns are the vectors in β. Then

$$\Sigma = \begin{pmatrix} \sqrt{6} & 0 & 0 \\ 0 & 0 & 0 \end{pmatrix} \quad \text{and} \quad V = \begin{pmatrix} \frac{1}{\sqrt{3}} & \frac{1}{\sqrt{2}} & \frac{1}{\sqrt{6}} \\ \frac{1}{\sqrt{3}} & \frac{-1}{\sqrt{2}} & \frac{1}{\sqrt{6}} \\ \frac{-1}{\sqrt{3}} & 0 & \frac{2}{\sqrt{6}} \end{pmatrix}.$$

Also, as in Theorem 6.27, we take

$$u_1 = \frac{1}{\sigma_i} \mathsf{L}_A(v_1) = \frac{1}{\sigma_i} A v_1 = \frac{1}{\sqrt{2}} \begin{pmatrix} 1 \\ 1 \end{pmatrix}.$$

Next choose $u_2 = \dfrac{1}{\sqrt{2}} \begin{pmatrix} 1 \\ -1 \end{pmatrix}$, a unit vector orthogonal to u_1, to obtain the orthonormal basis $\gamma = \{u_1, u_2\}$ for $\mathbf{R}^2$, and set

$$U = \begin{pmatrix} \frac{1}{\sqrt{2}} & \frac{1}{\sqrt{2}} \\ \frac{1}{\sqrt{2}} & \frac{-1}{\sqrt{2}} \end{pmatrix}.$$

Then $A = U\Sigma V^*$ is the desired singular value decomposition. ◆

The Polar Decomposition of a Square Matrix

A singular value decomposition of a matrix can be used to factor a square matrix in a manner analogous to the factoring of a complex number as the product of a complex number of length 1 and a nonnegative number. In the case of matrices, the complex number of length 1 is replaced by a unitary matrix, and the nonnegative number is replaced by a positive semidefinite matrix.

Theorem 6.28 (Polar Decomposition). *For any square matrix A, there exists a unitary matrix W and a positive semidefinite matrix P such that*

$$A = WP.$$

Furthermore, if A is invertible, then the representation is unique.

Proof. By Theorem 6.27, there exist unitary matrices U and V and a diagonal matrix Σ with nonnegative diagonal entries such that $A = U\Sigma V^*$. So

$$A = U\Sigma V^* = UV^*V\Sigma V^* = WP,$$

where $W = UV^*$ and $P = V\Sigma V^*$. Since W is the product of unitary matrices, W is unitary, and since Σ is positive semidefinite and P is unitarily equivalent to Σ, P is positive semidefinite by Exercise 14 of Section 6.5.

Now suppose that A is invertible and factors as the products

$$A = WP = ZQ,$$

where W and Z are unitary and P and Q are positive semidefinite. Since A is invertible, it follows that P and Q are positive definite and invertible, and therefore $Z^*W = QP^{-1}$. Thus QP^{-1} is unitary, and so

$$I = (QP^{-1})^*(QP^{-1}) = P^{-1}Q^2P^{-1}.$$

Hence $P^2 = Q^2$. Since both P and Q are positive definite, it follows that $P = Q$ by Exercise 17 of Section 6.4. Therefore $W = Z$, and consequently the factorization is unique. ∎

The factorization of a square matrix A as WP where W is unitary and P is positive semidefinite, is called a **polar decomposition** of A.

Example 4

To find the polar decomposition of $A = \begin{pmatrix} 11 & -5 \\ -2 & 10 \end{pmatrix}$, we begin by finding a singular value decomposition $U\Sigma V^*$ of A. The object is to find an orthonormal basis β for R^2 consisting of eigenvectors of A^*A. It can be shown that

$$v_1 = \frac{1}{\sqrt{2}} \begin{pmatrix} 1 \\ -1 \end{pmatrix} \qquad \text{and} \qquad v_2 = \frac{1}{\sqrt{2}} \begin{pmatrix} 1 \\ 1 \end{pmatrix}$$

are orthonormal eigenvectors of A^*A with corresponding eigenvalues $\lambda_1 = 200$ and $\lambda_2 = 50$. So $\beta = \{v_1, v_2\}$ is an appropriate basis. Thus $\sigma_1 = \sqrt{200} = 10\sqrt{2}$ and $\sigma_2 = \sqrt{50} = 5\sqrt{2}$ are the singular values of A. So we have

$$V = \begin{pmatrix} \frac{1}{\sqrt{2}} & \frac{1}{\sqrt{2}} \\ \frac{-1}{\sqrt{2}} & \frac{1}{\sqrt{2}} \end{pmatrix} \qquad \text{and} \qquad \Sigma = \begin{pmatrix} 10\sqrt{2} & 0 \\ 0 & 5\sqrt{2} \end{pmatrix}.$$

Next, we find the columns u_1 and u_2 of U:

$$u_1 = \frac{1}{\sigma_1} Av_1 = \frac{1}{5} \begin{pmatrix} 4 \\ -3 \end{pmatrix} \qquad \text{and} \qquad u_2 = \frac{1}{\sigma_2} Av_2 = \frac{1}{5} \begin{pmatrix} 3 \\ 4 \end{pmatrix}.$$

Thus

$$U = \begin{pmatrix} \frac{4}{5} & \frac{3}{5} \\ -\frac{3}{5} & \frac{4}{5} \end{pmatrix}.$$

Therefore, in the notation of Theorem 6.28, we have

$$W = UV^* = \begin{pmatrix} \frac{4}{5} & \frac{3}{5} \\ -\frac{3}{5} & \frac{4}{5} \end{pmatrix} \begin{pmatrix} \frac{1}{\sqrt{2}} & \frac{-1}{\sqrt{2}} \\ \frac{1}{\sqrt{2}} & \frac{1}{\sqrt{2}} \end{pmatrix} = \frac{1}{5\sqrt{2}} \begin{pmatrix} 7 & -1 \\ 1 & 7 \end{pmatrix},$$

and

$$P = V\Sigma V^* = \begin{pmatrix} \frac{1}{\sqrt{2}} & \frac{1}{\sqrt{2}} \\ \frac{-1}{\sqrt{2}} & \frac{1}{\sqrt{2}} \end{pmatrix} \begin{pmatrix} 10\sqrt{2} & 0 \\ 0 & 5\sqrt{2} \end{pmatrix} \begin{pmatrix} \frac{1}{\sqrt{2}} & \frac{-1}{\sqrt{2}} \\ \frac{1}{\sqrt{2}} & \frac{1}{\sqrt{2}} \end{pmatrix} = \frac{5}{\sqrt{2}} \begin{pmatrix} 3 & -1 \\ -1 & 3 \end{pmatrix}.$$

$\blacklozenge$

The Pseudoinverse

Let V and W be finite-dimensional inner product spaces over the same field, and let $T\colon V \to W$ be a linear transformation. It is desirable to have a linear transformation from W to V that captures some of the essence of an inverse of T even if T is not invertible. A simple approach to this problem is to focus on the "part" of T that is invertible, namely, the restriction of T to $N(T)^\perp$. Let $L\colon N(T)^\perp \to R(T)$ be the linear transformation defined by $L(x) = T(x)$ for all $x \in N(T)^\perp$. Then L is invertible, and we can use the inverse of L to construct a linear transformation from W to V that salvages some of the benefits of an inverse of T.

Definition. *Let* V *and* W *be finite-dimensional inner product spaces over the same field, and let* $T\colon V \to W$ *be a linear transformation. Let* $L\colon N(T)^\perp \to R(T)$ *be the linear transformation defined by* $L(x) = T(x)$ *for all* $x \in N(T)^\perp$. *The* **pseudoinverse** *(or Moore-Penrose generalized inverse) of* T, *denoted by* $T^\dagger$, *is defined as the unique linear transformation from* W *to* V *such that*

$$T^\dagger(y) = \begin{cases} L^{-1}(y) & \text{for } y \in R(T) \\ 0 & \text{for } y \in R(T)^\perp. \end{cases}$$

The pseudoinverse of a linear transformation T on a finite-dimensional inner product space exists even if T is not invertible. Furthermore, if T is invertible, then $T^\dagger = T^{-1}$ because $N(T)^\perp = V$, and L (as just defined) coincides with T.

As an extreme example, consider the zero transformation $T_0\colon V \to W$ between two finite-dimensional inner product spaces V and W. Then $R(T_0) = \{0\}$, and therefore $T^\dagger$ is the zero transformation from W to V.

We can use the singular value theorem to describe the pseudoinverse of a linear transformation. Suppose that V and W are finite-dimensional vector spaces and $T: V \to W$ is a linear transformation or rank r. Let $\{v_1, v_2, \ldots, v_n\}$ and $\{u_1, u_2, \ldots, u_m\}$ be orthonormal bases for V and W, respectively, and let $\sigma_1 \geq \sigma_2 \geq \cdots \geq \sigma_r$ be the nonzero singular values of T satisfying (4) in Theorem 6.26. Then $\{v_1, v_2, \ldots, v_r\}$ is a basis for $N(T)^\perp$, $\{v_{r+1}, v_{r+2}, \ldots, v_n\}$ is a basis for $N(T)$, $\{u_1, u_2, \ldots, u_r\}$ is a basis for $R(T)$, and $\{u_{r+1}, u_{r+2}, \ldots, u_m\}$ is a basis for $R(T)^\perp$. Let L be the restriction of T to $N(T)^\perp$, as in the definition of pseudoinverse. Then $L^{-1}(u_i) = \dfrac{1}{\sigma_i} v_i$ for $1 \leq i \leq r$. Therefore

$$T^\dagger(u_i) = \begin{cases} \dfrac{1}{\sigma_i} v_i & \text{if } 1 \leq i \leq r \\ 0 & \text{if } r < i \leq m. \end{cases} \tag{6}$$

Example 5

Let $T: P_2(R) \to P_1(R)$ be the linear transformation defined by $T(f(x)) = f'(x)$, as in Example 1. Let $\beta = \{v_1, v_2, v_3\}$ and $\gamma = \{u_1, u_2\}$ be the orthonormal bases for $P_2(R)$ and $P_1(R)$ in Example 1. Then $\sigma_1 = \sqrt{15}$ and $\sigma_2 = \sqrt{3}$ are the nonzero singular values of T. It follows that

$$T^\dagger\left(\sqrt{\frac{3}{2}} x\right) = T^\dagger(u_1) = \frac{1}{\sigma_1} v_1 = \frac{1}{\sqrt{15}} \sqrt{\frac{5}{8}} (3x^2 - 1),$$

and hence

$$T^\dagger(x) = \frac{1}{6}(3x^2 - 1).$$

Similarly, $T^\dagger(1) = x$. Thus, for any polynomial $a + bx \in P_1(R)$,

$$T^\dagger(a + bx) = aT^\dagger(1) + bT^\dagger(x) = ax + \frac{b}{6}(3x^2 - 1). \quad \blacklozenge$$

The Pseudoinverse of a Matrix

Let A be an $m \times n$ matrix. Then there exists a unique $n \times m$ matrix B such that $(L_A)^\dagger: F^m \to F^n$ is equal to the left-multiplication transformation L_B. We call B the **pseudoinverse** of A and denote it by $B = A^\dagger$. Thus

$$(L_A)^\dagger = L_{A^\dagger}.$$

Let A be an $m \times n$ matrix of rank r. The pseudoinverse of A can be computed with the aid of a singular value decomposition $A = U\Sigma V^*$. Let β and γ be the ordered bases whose vectors are the columns of V and U, respectively, and let $\sigma_1 \geq \sigma_2 \geq \cdots \geq \sigma_r$ be the nonzero singular values of A. Then β and γ are orthonormal bases for F^n and F^m, respectively, and (4) and (6) are satisfied for $\mathsf{T} = \mathsf{L}_A$. Reversing the roles of β and γ in the proof of Theorem 6.27, we obtain the following result.

Theorem 6.29. *Let A be an $m \times n$ matrix of rank r with a singular value decomposition $A = U\Sigma V^*$ and nonzero singular values $\sigma_1 \geq \sigma_2 \geq \cdots \geq \sigma_r$. Let $\Sigma^\dagger$ be the $n \times m$ matrix defined by*

$$\Sigma^\dagger_{ij} = \begin{cases} \dfrac{1}{\sigma_i} & \text{if } i = j \leq r \\ 0 & \text{otherwise.} \end{cases}$$

Then $A^\dagger = V\Sigma^\dagger U^*$, and this is a singular value decomposition of $A^\dagger$.

Notice that $\Sigma^\dagger$ as defined in Theorem 6.29 is actually the pseudoinverse of Σ.

Example 6

We find $A^\dagger$ for the matrix $A = \begin{pmatrix} 1 & 1 & -1 \\ 1 & 1 & -1 \end{pmatrix}$.

Since A is the matrix of Example 3, we can use the singular value decomposition obtained in that example:

$$A = U\Sigma V^* = \begin{pmatrix} \frac{1}{\sqrt{2}} & \frac{1}{\sqrt{2}} \\ \frac{1}{\sqrt{2}} & \frac{-1}{\sqrt{2}} \end{pmatrix} \begin{pmatrix} \sqrt{6} & 0 & 0 \\ 0 & 0 & 0 \end{pmatrix} \begin{pmatrix} \frac{1}{\sqrt{3}} & \frac{1}{\sqrt{2}} & \frac{1}{\sqrt{6}} \\ \frac{1}{\sqrt{3}} & \frac{-1}{\sqrt{2}} & \frac{1}{\sqrt{6}} \\ \frac{-1}{\sqrt{3}} & 0 & \frac{2}{\sqrt{6}} \end{pmatrix}^*.$$

By Theorem 6.29, we have

$$A^\dagger = V\Sigma^\dagger U^* = \begin{pmatrix} \frac{1}{\sqrt{3}} & \frac{1}{\sqrt{2}} & \frac{1}{\sqrt{6}} \\ \frac{1}{\sqrt{3}} & \frac{-1}{\sqrt{2}} & \frac{1}{\sqrt{6}} \\ \frac{-1}{\sqrt{3}} & 0 & \frac{2}{\sqrt{6}} \end{pmatrix} \begin{pmatrix} \frac{1}{\sqrt{6}} & 0 \\ 0 & 0 \\ 0 & 0 \end{pmatrix} \begin{pmatrix} \frac{1}{\sqrt{2}} & \frac{1}{\sqrt{2}} \\ \frac{1}{\sqrt{2}} & \frac{-1}{\sqrt{2}} \end{pmatrix} = \frac{1}{6} \begin{pmatrix} 1 & 1 \\ 1 & 1 \\ -1 & -1 \end{pmatrix}.$$

♦

Notice that the linear transformation T of Example 5 is L_A, where A is the matrix of Example 6, and that $\mathsf{T}^\dagger = \mathsf{L}_{A^\dagger}$.

The Pseudoinverse and Systems of Linear Equations

Let A be an $m \times n$ matrix with entries in F. Then for any $b \in \mathsf{F}^m$, the matrix equation $Ax = b$ is a system of linear equations, and so it either has no solutions, a unique solution, or infinitely many solutions. We know that the system has a unique solution for every $b \in F^m$ if and only if A is invertible, in which case the solution is given by $A^{-1}b$. Furthermore, if A is invertible, then $A^{-1} = A^\dagger$, and so the solution can be written as $x = A^\dagger b$. If, on the other hand, A is not invertible or the system $Ax = b$ is inconsistent, then $A^\dagger b$ still exists. We therefore pose the following question: In general, how is the vector $A^\dagger b$ related to the system of linear equations $Ax = b$?

In order to answer this question, we need the following lemma.

Lemma. *Let* V *and* W *be finite-dimensional inner product spaces, and let* $T : V \to W$ *be linear. Then*
(a) $T^\dagger T$ *is the orthogonal projection of* V *on* $N(T)^\perp$.
(b) $TT^\dagger$ *is the orthogonal projection of* W *on* $R(T)$.

Proof. As in the earlier discussion, we define $L : N(T)^\perp \to W$ by $L(x) = T(x)$ for all $x \in N(T)^\perp$. If $x \in N(T)^\perp$, then $T^\dagger T(x) = L^{-1}L(x) = x$, and if $x \in N(T)$, then $T^\dagger T(x) = T^\dagger(0) = 0$. Consequently $T^\dagger T$ is the orthogonal projection of V on $N(T)^\perp$. This proves (a).

The proof of (b) is similar and is left as an exercise. ∎

Theorem 6.30. *Consider the system of linear equations* $Ax = b$, *where* A *is an* $m \times n$ *matrix and* $b \in F^m$. *If* $z = A^\dagger b$, *then* z *has the following properties.*
(a) *If* $Ax = b$ *is consistent, then* z *is the unique solution to the system having minimum norm. That is,* z *is a solution to the system, and if* y *is any solution to the system, then* $\|z\| \leq \|y\|$ *with equality if and only if* $z = y$.
(b) *If* $Ax = b$ *is inconsistent, then* z *is the unique best approximation to a solution having minimum norm. That is,* $\|Az - b\| \leq \|Ay - b\|$ *for any* $y \in F^n$, *with equality if and only if* $Az = Ay$. *Furthermore, if* $Az = Ay$, *then* $\|z\| \leq \|y\|$ *with equality if and only if* $z = y$.

Proof. For convenience, let $T = L_A$.
(a) Suppose that $Ax = b$ is consistent, and let $z = A^\dagger b$. Observe that $b \in R(T)$, and therefore $Az = AA^\dagger b = TT^\dagger(b) = b$ by part (b) of the lemma. Thus z is a solution to the system. Now suppose that y is any solution to the system. Then

$$T^\dagger T(y) = A^\dagger Ay = A^\dagger b = z,$$

and hence z is the orthogonal projection of y on $N(T)^\perp$ by part (a) of the lemma. Therefore, by the corollary to Theorem 6.6 (p. 350), we have that $\|z\| \leq \|y\|$ with equality if and only if $z = y$.
(b) Suppose that $Ax = b$ is inconsistent. By the lemma, $Az = AA^\dagger b = TT^\dagger(b) = b$ is the orthogonal projection of b on $R(T)$; therefore, by the corollary to Theorem 6.6 (p. 350), Az is the vector in $R(T)$ nearest b. That is, if

Ay is any other vector in $\mathsf{R}(\mathsf{T})$, then $\|Az - b\| \leq \|Ay - b\|$ with equality if and only if $Az = Ay$.

Finally, suppose that y is any vector in F^n such that $Az = Ay = c$. Then

$$A^\dagger c = A^\dagger A z = A^\dagger A A^\dagger b = A^\dagger b = z$$

by Exercise 23; hence we may apply part (a) of this theorem to the system $Ax = c$ to conclude that $\|z\| \leq \|y\|$ with equality if and only if $z = y$. ∎

Note that the vector $z = A^\dagger b$ in Theorem 6.30 is the vector x_0 described in Theorem 6.12 that arises in the least squares application on pages 360–364.

Example 7

Consider the linear systems

$$
\begin{array}{ccc}
\begin{aligned}
x_1 + x_2 - x_3 &= 1 \\
x_1 + x_2 - x_3 &= 1
\end{aligned}
& \qquad \text{and} \qquad &
\begin{aligned}
x_1 + x_2 - x_3 &= 1 \\
x_1 + x_2 - x_3 &= 2.
\end{aligned}
\end{array}
$$

The first system has infinitely many solutions. Let $A = \begin{pmatrix} 1 & 1 & -1 \\ 1 & 1 & -1 \end{pmatrix}$, the coefficient matrix of the system, and let $b = \begin{pmatrix} 1 \\ 1 \end{pmatrix}$. By Example 6,

$$
A^\dagger = \frac{1}{6} \begin{pmatrix} 1 & 1 \\ 1 & 1 \\ -1 & -1 \end{pmatrix},
$$

and therefore

$$
z = A^\dagger b = \frac{1}{6} \begin{pmatrix} 1 & 1 \\ 1 & 1 \\ -1 & -1 \end{pmatrix} \begin{pmatrix} 1 \\ 1 \end{pmatrix} = \frac{1}{3} \begin{pmatrix} 1 \\ 1 \\ -1 \end{pmatrix}
$$

is the solution of minimal norm by Theorem 6.30(a).

The second system is obviously inconsistent. Let $b = \begin{pmatrix} 1 \\ 2 \end{pmatrix}$. Thus, although

$$
z = A^\dagger b = \frac{1}{6} \begin{pmatrix} 1 & 1 \\ 1 & 1 \\ -1 & -1 \end{pmatrix} \begin{pmatrix} 1 \\ 2 \end{pmatrix} = \frac{1}{2} \begin{pmatrix} 1 \\ 1 \\ -1 \end{pmatrix}
$$

is not a solution to the second system, it is the "best approximation" to a solution having minimum norm, as described in Theorem 6.30(b). ◆

EXERCISES

1. Label the following statements as true or false.

   (a) The singular values of any linear operator on a finite-dimensional vector space are also eigenvalues of the operator.

   (b) The singular values of any matrix A are the eigenvalues of A^*A.

   (c) For any matrix A and any scalar c, if σ is a singular value of A, then $|c|\sigma$ is a singular value of cA.

   (d) The singular values of any linear operator are nonnegative.

   (e) If λ is an eigenvalue of a self-adjoint matrix A, then λ is a singular value of A.

   (f) For any $m \times n$ matrix A and any $b \in \mathsf{F}^n$, the vector $A^\dagger b$ is a solution to $Ax = b$.

   (g) The pseudoinverse of any linear operator exists even if the operator is not invertible.

2. Let $\mathsf{T} \colon \mathsf{V} \to \mathsf{W}$ be a linear transformation of rank r, where V and W are finite-dimensional inner product spaces. In each of the following, find orthonormal bases $\{v_1, v_2, \ldots, v_n\}$ for V and $\{u_1, u_2, \ldots, u_m\}$ for W, and the nonzero singular values $\sigma_1 \geq \sigma_2 \geq \cdots \geq \sigma_r$ of T such that $\mathsf{T}(v_i) = \sigma_i u_i$ for $1 \leq i \leq r$.

   (a) $\mathsf{T} \colon \mathsf{R}^2 \to \mathsf{R}^3$ defined by $\mathsf{T}(x_1, x_2) = (x_1, x_1 + x_2, x_1 - x_2)$

   (b) $\mathsf{T} \colon \mathsf{P}_2(R) \to \mathsf{P}_1(R)$, where $\mathsf{T}(f(x)) = f''(x)$, and the inner products are defined as in Example 1

   (c) Let $\mathsf{V} = \mathsf{W} = \operatorname{span}(\{1, \sin x, \cos x\})$ with the inner product defined by $\langle f, g \rangle = \int_0^{2\pi} f(t)g(t)\, dt$, and T is defined by $\mathsf{T}(f) = f' + 2f$

   (d) $\mathsf{T} \colon \mathsf{C}^2 \to \mathsf{C}^2$ defined by $\mathsf{T}(z_1, z_2) = ((1-i)z_2, (1+i)z_1 + z_2)$

3. Find a singular value decomposition for each of the following matrices.

   (a) $\begin{pmatrix} 1 & 1 \\ 1 & 1 \\ -1 & -1 \end{pmatrix}$ (b) $\begin{pmatrix} 1 & 0 & 1 \\ 1 & 0 & -1 \end{pmatrix}$ (c) $\begin{pmatrix} 1 & 1 \\ 0 & 1 \\ 1 & 0 \\ 1 & 1 \end{pmatrix}$

   (d) $\begin{pmatrix} 1 & 1 & 1 \\ 1 & -1 & 0 \\ 1 & 0 & -1 \end{pmatrix}$ (e) $\begin{pmatrix} 1+i & 1 \\ 1-i & -i \end{pmatrix}$ (f) $\begin{pmatrix} 1 & 1 & 1 & 1 \\ 1 & 0 & -2 & 1 \\ 1 & -1 & 1 & 1 \end{pmatrix}$

4. Find a polar decomposition for each of the following matrices.

   (a) $\begin{pmatrix} 1 & 1 \\ 2 & -2 \end{pmatrix}$ (b) $\begin{pmatrix} 20 & 4 & 0 \\ 0 & 0 & 1 \\ 4 & 20 & 0 \end{pmatrix}$

5. Find an explicit formula for each of the following expressions.

(a) $T^\dagger(x_1, x_2, x_3)$, where T is the linear transformation of Exercise 2(a)
(b) $T^\dagger(a + bx + cx^2)$, where T is the linear transformation of Exercise 2(b)
(c) $T^\dagger(a + b \sin x + c \cos x)$, where T is the linear transformation of Exercise 2(c)
(d) $T^\dagger(z_1, z_2)$, where T is the linear transformation of Exercise 2(d)

6. Use the results of Exercise 3 to find the pseudoinverse of each of the following matrices.

(a) $\begin{pmatrix} 1 & 1 \\ 1 & 1 \\ -1 & -1 \end{pmatrix}$ (b) $\begin{pmatrix} 1 & 0 & 1 \\ 1 & 0 & -1 \end{pmatrix}$ (c) $\begin{pmatrix} 1 & 1 \\ 0 & 1 \\ 1 & 0 \\ 1 & 1 \end{pmatrix}$

(d) $\begin{pmatrix} 1 & 1 & 1 \\ 1 & -1 & 0 \\ 1 & 0 & -1 \end{pmatrix}$ (e) $\begin{pmatrix} 1+i & 1 \\ 1-i & -i \end{pmatrix}$ (f) $\begin{pmatrix} 1 & 1 & 1 & 1 \\ 1 & 0 & -2 & 1 \\ 1 & -1 & 1 & 1 \end{pmatrix}$

7. For each of the given linear transformations $T: V \to W$,
   (i) Describe the subspace Z_1 of V such that $T^\dagger T$ is the orthogonal projection of V on Z_1.
   (ii) Describe the subspace Z_2 of W such that $TT^\dagger$ is the orthogonal projection of W on Z_2.

   (a) T is the linear transformation of Exercise 2(a)
   (b) T is the linear transformation of Exercise 2(b)
   (c) T is the linear transformation of Exercise 2(c)
   (d) T is the linear transformation of Exercise 2(d)

8. For each of the given systems of linear equations,
   (i) If the system is consistent, find the unique solution having minimum norm.
   (ii) If the system is inconsistent, find the "best approximation to a solution" having minimum norm, as described in Theorem 6.30(b).
   (Use your answers to parts (a) and (f) of Exercise 6.)

   (a) $\begin{aligned} x_1 + \quad x_2 &= 1 \\ x_1 + \quad x_2 &= 2 \\ -x_1 + -x_2 &= 0 \end{aligned}$ (b) $\begin{aligned} x_1 + x_2 + \quad x_3 + x_4 &= 2 \\ x_1 \qquad\quad - 2x_3 + x_4 &= -1 \\ x_1 - x_2 + \quad x_3 + x_4 &= 2 \end{aligned}$

9. Let V and W be finite-dimensional inner product spaces over F, and suppose that $\{v_1, v_2, \ldots, v_n\}$ and $\{u_1, u_2, \ldots, u_m\}$ are orthonormal bases for V and W, respectively. Let $T: V \to W$ is a linear transformation of rank r, and suppose that $\sigma_1 \geq \sigma_2 \geq \cdots \geq \sigma_r > 0$ are such that

$$T(v_i) = \begin{cases} \sigma_i u_i & \text{if } 1 \leq i \leq r \\ 0 & \text{if } r < i. \end{cases}$$

(a) Prove that $\{u_1, u_2, \ldots, u_m\}$ is a set of eigenvectors of TT^* with corresponding eigenvalues $\lambda_1, \lambda_2, \ldots, \lambda_m$, where

$$\lambda_i = \begin{cases} \sigma_i^2 & \text{if } 1 \leq i \leq r \\ 0 & \text{if } r < i. \end{cases}$$

(b) Let A be an $m \times n$ matrix with real or complex entries. Prove that the nonzero singular values of A are the positive square roots of the nonzero eigenvalues of AA^*, including repetitions.

(c) Prove that TT^* and $\mathsf{T}^*\mathsf{T}$ have the same nonzero eigenvalues, including repetitions.

(d) State and prove a result for matrices analogous to (c).

10. Use Exercise 8 of Section 2.5 to obtain another proof of Theorem 6.27, the singular value decomposition theorem for matrices.

11. This exercise relates the singular values of a well-behaved linear operator or matrix to its eigenvalues.

(a) Let T be a normal linear operator on an n-dimensional inner product space with eigenvalues $\lambda_1, \lambda_2, \ldots, \lambda_n$. Prove that the singular values of T are $|\lambda_1|, |\lambda_2|, \ldots, |\lambda_n|$.

(b) State and prove a result for matrices analogous to (a).

12. Let A be a normal matrix with an orthonormal basis of eigenvectors $\beta = \{v_1, v_2, \ldots, v_n\}$ and corresponding eigenvalues $\lambda_1, \lambda_2, \ldots, \lambda_n$. Let V be the $n \times n$ matrix whose columns are the vectors in β. Prove that for each i there is a scalar θ_i of absolute value 1 such that if U is the $n \times n$ matrix with $\theta_i v_i$ as column i and Σ is the diagonal matrix such that $\Sigma_{ii} = |\lambda_i|$ for each i, then $U\Sigma V^*$ is a singular value decomposition of A.

13. Prove that if A is a positive semidefinite matrix, then the singular values of A are the same as the eigenvalues of A.

14. Prove that if A is a positive definite matrix and $A = U\Sigma V^*$ is a singular value decomposition of A, then $U = V$.

15. Let A be a square matrix with a polar decomposition $A = WP$.

(a) Prove that A is normal if and only if $WP^2 = P^2 W$.

(b) Use (a) to prove that A is normal if and only if $WP = PW$.

16. Let A be a square matrix. Prove an alternate form of the polar decomposition for A: There exists a unitary matrix W and a positive semidefinite matrix P such that $A = PW$.

17. Let T and U be linear operators on R^2 defined for all $(x_1, x_2) \in R^2$ by

$$T(x_1, x_2) = (x_1, 0) \quad \text{and} \quad U(x_1, x_2) = (x_1 + x_2, 0).$$

    (a) Prove that $(UT)^\dagger \neq T^\dagger U^\dagger$.
    (b) Exhibit matrices A and B such that AB is defined, but $(AB)^\dagger \neq B^\dagger A^\dagger$.

18. Let A be an $m \times n$ matrix. Prove the following results.

    (a) For any $m \times m$ unitary matrix G, $(GA)^\dagger = A^\dagger G^*$.
    (b) For any $n \times n$ unitary matrix H, $(AH)^\dagger = H^* A^\dagger$.

19. Let A be a matrix with real or complex entries. Prove the following results.

    (a) The nonzero singular values of A are the same as the nonzero singular values of A^*, which are the same as the nonzero singular values of A^t.
    (b) $(A^\dagger)^* = (A^*)^\dagger$.
    (c) $(A^\dagger)^t = (A^t)^\dagger$.

20. Let A be a square matrix such that $A^2 = O$. Prove that $(A^\dagger)^2 = O$.

21. Let V and W be finite-dimensional inner product spaces, and let T: V $\to$ W be linear. Prove the following results.

    (a) $TT^\dagger T = T$.
    (b) $T^\dagger TT^\dagger = T^\dagger$.
    (c) Both $T^\dagger T$ and $TT^\dagger$ are self-adjoint.

The preceding three statements are called the **Penrose conditions**, and they characterize the pseudoinverse of a linear transformation as shown in Exercise 22.

22. Let V and W be finite-dimensional inner product spaces. Let T: V $\to$ W and U: W $\to$ V be linear transformations such that $TUT = T$, $UTU = U$, and both UT and TU are self-adjoint. Prove that $U = T^\dagger$.

23. State and prove a result for matrices that is analogous to the result of Exercise 21.

24. State and prove a result for matrices that is analogous to the result of Exercise 22.

25. Let V and W be finite-dimensional inner product spaces, and let T: V $\to$ W be linear. Prove the following results.

    (a) If T is one-to-one, then T^*T is invertible and $T^\dagger = (T^*T)^{-1}T^*$.
    (b) If T is onto, then TT^* is invertible and $T^\dagger = T^*(TT^*)^{-1}$.

26. Let V and W be finite-dimensional inner product spaces with orthonor-
    mal bases β and γ, respectively, and let $T\colon V \to W$ be linear. Prove
    that $([T]_\beta^\gamma)^\dagger = [T^\dagger]_\gamma^\beta$.

27. Let V and W be finite-dimensional inner product spaces, and let
    $T\colon V \to W$ be a linear transformation. Prove part (b) of the lemma
    to Theorem 6.30: $TT^\dagger$ is the orthogonal projection of W on $R(T)$.

6.8* BILINEAR AND QUADRATIC FORMS

There is a certain class of scalar-valued functions of two variables defined on
a vector space that arises in the study of such diverse subjects as geometry
and multivariable calculus. This is the class of *bilinear forms*. We study the
basic properties of this class with a special emphasis on symmetric bilinear
forms, and we consider some of its applications to quadratic surfaces and
multivariable calculus.

Bilinear Forms

Definition. *Let V be a vector space over a field F. A function H from
the set $V \times V$ of ordered pairs of vectors to F is called a **bilinear form** on V
if H is linear in each variable when the other variable is held fixed; that is,
H is a bilinear form on V if*
   (a) $H(ax_1 + x_2, y) = aH(x_1, y) + H(x_2, y)$ for all $x_1, x_2, y \in V$ and $a \in F$
   (b) $H(x, ay_1 + y_2) = aH(x, y_1) + H(x, y_2)$ for all $x, y_1, y_2 \in V$ and $a \in F$.

We denote the set of all bilinear forms on V by $\mathcal{B}(V)$. Observe that an
inner product on a vector space is a bilinear form if the underlying field is
real, but not if the underlying field is complex.

Example 1

Define a function $H\colon R^2 \times R^2 \to R$ by

$$H\left(\begin{pmatrix} a_1 \\ a_2 \end{pmatrix}, \begin{pmatrix} b_1 \\ b_2 \end{pmatrix}\right) = 2a_1b_1 + 3a_1b_2 + 4a_2b_1 - a_2b_2 \quad \text{for} \quad \begin{pmatrix} a_1 \\ a_2 \end{pmatrix}, \begin{pmatrix} b_1 \\ b_2 \end{pmatrix} \in R^2.$$

We could verify directly that H is a bilinear form on R^2. However, it is more
enlightening and less tedious to observe that if

$$A = \begin{pmatrix} 2 & 3 \\ 4 & -1 \end{pmatrix}, \quad x = \begin{pmatrix} a_1 \\ a_2 \end{pmatrix}, \quad \text{and} \quad y = \begin{pmatrix} b_1 \\ b_2 \end{pmatrix},$$

then

$$H(x, y) = x^t A y.$$

The bilinearity of H now follows directly from the distributive property of
matrix multiplication over matrix addition.   ◆

The preceding bilinear form is a special case of the next example.

Example 2

Let $V = F^n$, where the vectors are considered as column vectors. For any $A \in M_{n \times n}(F)$, define $H \colon V \times V \to F$ by

$$H(x, y) = x^t A y \quad \text{for } x, y \in V.$$

Notice that since x and y are $n \times 1$ matrices and A is an $n \times n$ matrix, $H(x, y)$ is a 1×1 matrix. We identify this matrix with its single entry. The bilinearity of H follows as in Example 1. For example, for $a \in F$ and $x_1, x_2, y \in V$, we have

$$
\begin{aligned}
H(ax_1 + x_2, y) &= (ax_1 + x_2)^t A y = (ax_1^t + x_2^t) A y \\
&= ax_1^t A y + x_2^t A y \\
&= aH(x_1, y) + H(x_2, y). \quad \blacklozenge
\end{aligned}
$$

We list several properties possessed by all bilinear forms. Their proofs are left to the reader (see Exercise 2).

For any bilinear form H on a vector space V over a field F, the following properties hold.

1. If, for any $x \in V$, the functions $\mathsf{L}_x, \mathsf{R}_x \colon V \to F$ are defined by

   $$\mathsf{L}_x(y) = H(x, y) \quad \text{and} \quad \mathsf{R}_x(y) = H(y, x) \quad \text{for all } y \in V,$$

   then L_x and R_x are linear.
2. $H(0, x) = H(x, 0) = 0$ for all $x \in V$.
3. For all $x, y, z, w \in V$,

   $$H(x + y, z + w) = H(x, z) + H(x, w) + H(y, z) + H(y, w).$$

4. If $J \colon V \times V \to F$ is defined by $J(x, y) = H(y, x)$, then J is a bilinear form.

Definitions. Let V be a vector space, let H_1 and H_2 be bilinear forms on V, and let a be a scalar. We define the **sum** $H_1 + H_2$ and the **scalar product** aH_1 by the equations

$$(H_1 + H_2)(x, y) = H_1(x, y) + H_2(x, y)$$

and

$$(aH_1)(x, y) = a(H_1(x, y)) \quad \text{for all } x, y \in V.$$

The following theorem is an immediate consequence of the definitions.

Theorem 6.31. *For any vector space* V, *the sum of two bilinear forms and the product of a scalar and a bilinear form on* V *are again bilinear forms on* V. *Furthermore,* $\mathcal{B}(V)$ *is a vector space with respect to these operations.*

Proof. Exercise. ∎

Let $\beta = \{v_1, v_2, \dots, v_n\}$ be an ordered basis for an n-dimensional vector space V, and let $H \in \mathcal{B}(V)$. We can associate with H an $n \times n$ matrix A whose entry in row i and column j is defined by

$$A_{ij} = H(v_i, v_j) \qquad \text{for } i, j = 1, 2, \dots, n.$$

Definition. *The matrix A above is called the **matrix representation** of H with respect to the ordered basis β and is denoted by $\psi_\beta(H)$.*

We can therefore regard ψ_β as a mapping from $\mathcal{B}(V)$ to $M_{n \times n}(F)$, where F is the field of scalars for V, that takes a bilinear form H into its matrix representation $\psi_\beta(H)$. We first consider an example and then show that ψ_β is an isomorphism.

Example 3

Consider the bilinear form H of Example 1, and let

$$\beta = \left\{ \begin{pmatrix} 1 \\ 1 \end{pmatrix}, \begin{pmatrix} 1 \\ -1 \end{pmatrix} \right\} \quad \text{and} \quad B = \psi_\beta(H).$$

Then

$$B_{11} = H\left(\begin{pmatrix} 1 \\ 1 \end{pmatrix}, \begin{pmatrix} 1 \\ 1 \end{pmatrix} \right) = 2 + 3 + 4 - 1 = 8,$$

$$B_{12} = H\left(\begin{pmatrix} 1 \\ 1 \end{pmatrix}, \begin{pmatrix} 1 \\ -1 \end{pmatrix} \right) = 2 - 3 + 4 + 1 = 4,$$

$$B_{21} = H\left(\begin{pmatrix} 1 \\ -1 \end{pmatrix}, \begin{pmatrix} 1 \\ 1 \end{pmatrix} \right) = 2 + 3 - 4 + 1 = 2,$$

and

$$B_{22} = H\left(\begin{pmatrix} 1 \\ -1 \end{pmatrix}, \begin{pmatrix} 1 \\ -1 \end{pmatrix} \right) = 2 - 3 - 4 - 1 = -6.$$

So

$$\psi_\beta(H) = \begin{pmatrix} 8 & 4 \\ 2 & -6 \end{pmatrix}.$$

If γ is the standard ordered basis for R^2, the reader can verify that

$$\psi_\gamma(H) = \begin{pmatrix} 2 & 3 \\ 4 & -1 \end{pmatrix}. \quad \blacklozenge$$

Theorem 6.32.   *For any n-dimensional vector space* V *over* F *and any ordered basis* β *for* V, $\psi_\beta \colon \mathcal{B}(\mathsf{V}) \to \mathsf{M}_{n \times n}(F)$ *is an isomorphism.*

Proof. We leave the proof that ψ_β is linear to the reader.

To show that ψ_β is one-to-one, suppose that $\psi_\beta(H) = O$ for some $H \in \mathcal{B}(\mathsf{V})$. Fix $v_i \in \beta$, and recall the mapping $\mathsf{L}_{v_i} \colon \mathsf{V} \to F$, which is linear by property 1 on page 423. By hypothesis, $\mathsf{L}_{v_i}(v_j) = H(v_i, v_j) = 0$ for all $v_j \in \beta$. Hence L_{v_i} is the zero transformation from V to F. So

$$H(v_i, x) = \mathsf{L}_{v_i}(x) = 0 \quad \text{for all } x \in \mathsf{V} \text{ and } v_i \in \beta. \tag{7}$$

Next fix an arbitrary $y \in \mathsf{V}$, and recall the linear mapping $\mathsf{R}_y \colon \mathsf{V} \to F$ defined in property 1 on page 423. By (7), $\mathsf{R}_y(v_i) = H(v_i, y) = 0$ for all $v_i \in \beta$, and hence R_y is the zero transformation. So $H(x, y) = \mathsf{R}_y(x) = 0$ for all $x, y \in \mathsf{V}$. Thus H is the zero bilinear form, and therefore ψ_β is one-to-one.

To show that ψ_β is onto, consider any $A \in \mathsf{M}_{n \times n}(F)$. Recall the isomorphism $\phi_\beta \colon \mathsf{V} \to F^n$ defined in Section 2.4. For $x \in \mathsf{V}$, we view $\phi_\beta(x) \in F^n$ as a column vector. Let $H \colon \mathsf{V} \times \mathsf{V} \to F$ be the mapping defined by

$$H(x, y) = [\phi_\beta(x)]^t A [\phi_\beta(y)] \quad \text{for all } x, y \in \mathsf{V}.$$

A slight embellishment of the method of Example 2 can be used to prove that $H \in \mathcal{B}(\mathsf{V})$. We show that $\psi_\beta(H) = A$. Let $v_i, v_j \in \beta$. Then $\phi_\beta(v_i) = e_i$ and $\phi_\beta(v_j) = e_j$; hence, for any i and j,

$$H(v_i, v_j) = [\phi_\beta(v_i)]^t A [\phi_\beta(v_j)] = e_i^t A e_j = A_{ij}.$$

We conclude that $\psi_\beta(H) = A$ and ψ_β is onto. ∎

Corollary 1.   *For any n-dimensional vector space* V, $\mathcal{B}(\mathsf{V})$ *has dimension* n^2.

Proof. Exercise. ∎

The following corollary is easily established by reviewing the proof of Theorem 6.32.

Corollary 2.   *Let* V *be an n-dimensional vector space over* F *with ordered basis* β. *If* $H \in \mathcal{B}(\mathsf{V})$ *and* $A \in \mathsf{M}_{n \times n}(F)$, *then* $\psi_\beta(H) = A$ *if and only if* $H(x, y) = [\phi_\beta(x)]^t A [\phi_\beta(y)]$ *for all* $x, y \in \mathsf{V}$.

The following result is now an immediate consequence of Corollary 2.

Corollary 3.   *Let* F *be a field, n a positive integer, and* β *be the standard ordered basis for* F^n. *Then for any* $H \in \mathcal{B}(F^n)$, *there exists a unique matrix* $A \in \mathsf{M}_{n \times n}(F)$, *namely,* $A = \psi_\beta(H)$, *such that*

$$H(x, y) = x^t A y \quad \text{for all } x, y \in F^n.$$

Example 4

Define a function $H \colon \mathsf{R}^2 \times \mathsf{R}^2 \to R$ by

$$H\left(\begin{pmatrix} a_1 \\ a_2 \end{pmatrix}, \begin{pmatrix} b_1 \\ b_2 \end{pmatrix}\right) = \det \begin{pmatrix} a_1 & b_1 \\ a_2 & b_2 \end{pmatrix} = a_1 b_2 - a_2 b_1 \quad \text{for} \quad \begin{pmatrix} a_1 \\ a_2 \end{pmatrix}, \begin{pmatrix} b_1 \\ b_2 \end{pmatrix} \in \mathsf{R}^2.$$

It can be shown that H is a bilinear form. We find the matrix A in Corollary 3 such that $H(x, y) = x^t A y$ for all $x, y \in \mathsf{R}^2$.

Since $A_{ij} = H(e_i, e_j)$ for all i and j, we have

$$A_{11} = \det \begin{pmatrix} 1 & 1 \\ 0 & 0 \end{pmatrix} = 0 \qquad\qquad A_{12} = \det \begin{pmatrix} 1 & 0 \\ 0 & 1 \end{pmatrix} = 1,$$

$$A_{21} = \det \begin{pmatrix} 0 & 1 \\ 1 & 0 \end{pmatrix} = -1 \quad \text{and} \quad A_{22} = \det \begin{pmatrix} 0 & 0 \\ 1 & 1 \end{pmatrix} = 0.$$

Therefore $A = \begin{pmatrix} 0 & 1 \\ -1 & 0 \end{pmatrix}$. ◆

There is an analogy between bilinear forms and linear operators on finite-dimensional vector spaces in that both are associated with unique square matrices and the correspondences depend on the choice of an ordered basis for the vector space. As in the case of linear operators, one can pose the following question: How does the matrix corresponding to a fixed bilinear form change when the ordered basis is changed? As we have seen, the corresponding question for matrix representations of linear operators leads to the definition of the similarity relation on square matrices. In the case of bilinear forms, the corresponding question leads to another relation on square matrices, the *congruence* relation.

Definition. *Let $A, B \in \mathsf{M}_{n \times n}(F)$. Then B is said to be **congruent** to A if there exists an invertible matrix $Q \in \mathsf{M}_{n \times n}(F)$ such that $B = Q^t A Q$.*

Observe that the relation of congruence is an equivalence relation (see Exercise 12).

The next theorem relates congruence to the matrix representation of a bilinear form.

Theorem 6.33. *Let V be a finite-dimensional vector space with ordered bases $\beta = \{v_1, v_2, \ldots, v_n\}$ and $\gamma = \{w_1, w_2, \ldots, w_n\}$, and let Q be the change of coordinate matrix changing γ-coordinates into β-coordinates. Then, for any $H \in \mathcal{B}(\mathsf{V})$, we have $\psi_\gamma(H) = Q^t \psi_\beta(H) Q$. Therefore $\psi_\gamma(H)$ is congruent to $\psi_\beta(H)$.*

Proof. There are essentially two proofs of this theorem. One involves a direct computation, while the other follows immediately from a clever observation. We give the more direct proof here, leaving the other proof for the exercises (see Exercise 13).

Suppose that $A = \psi_\beta(H)$ and $B = \psi_\gamma(H)$. Then for $1 \le i, j \le n$,

$$w_i = \sum_{k=1}^{n} Q_{ki} v_k \quad \text{and} \quad w_j = \sum_{r=1}^{n} Q_{rj} v_r.$$

Thus

$$B_{ij} = H(w_i, w_j) = H\left(\sum_{k=1}^{n} Q_{ki} v_k, w_j\right)$$

$$= \sum_{k=1}^{n} Q_{ki} H(v_k, w_j)$$

$$= \sum_{k=1}^{n} Q_{ki} H\left(v_k, \sum_{r=1}^{n} Q_{rj} v_r\right)$$

$$= \sum_{k=1}^{n} Q_{ki} \sum_{r=1}^{n} Q_{rj} H(v_k, v_r)$$

$$= \sum_{k=1}^{n} Q_{ki} \sum_{r=1}^{n} Q_{rj} A_{kr}$$

$$= \sum_{k=1}^{n} Q_{ki} \sum_{r=1}^{n} A_{kr} Q_{rj}$$

$$= \sum_{k=1}^{n} Q_{ki} (AQ)_{kj}$$

$$= \sum_{k=1}^{n} Q^t_{ik} (AQ)_{kj} = (Q^t A Q)_{ij}.$$

Hence $B = Q^t A Q$. ∎

The following result is the converse of Theorem 6.33.

Corollary. Let V be an n-dimensional vector space with ordered basis β, and let H be a bilinear form on V. For any $n \times n$ matrix B, if B is congruent to $\psi_\beta(H)$, then there exists an ordered basis γ for V such that $\psi_\gamma(H) = B$. Furthermore, if $B = Q^t \psi_\beta(H) Q$ for some invertible matrix Q, then Q changes γ-coordinates into β-coordinates.

Proof. Suppose that $B = Q^t \psi_\beta(H) Q$ for some invertible matrix Q and that $\beta = \{v_1, v_2, \ldots, v_n\}$. Let $\gamma = \{w_1, w_2, \ldots, w_n\}$, where

$$w_j = \sum_{i=1}^{n} Q_{ij} v_i \quad \text{for } 1 \le j \le n.$$

Since Q is invertible, γ is an ordered basis for V, and Q is the change of coordinate matrix that changes γ-coordinates into β-coordinates. Therefore, by Theorem 6.32,

$$B = Q^t \psi_\beta(H) Q = \psi_\gamma(H). \qquad\blacksquare$$

Symmetric Bilinear Forms

Like the diagonalization problem for linear operators, there is an analogous *diagonalization* problem for bilinear forms, namely, the problem of determining those bilinear forms for which there are diagonal matrix representations. As we will see, there is a close relationship between *diagonalizable* bilinear forms and those that are called *symmetric*.

Definition. *A bilinear form H on a vector space V is **symmetric** if $H(x,y) = H(y,x)$ for all $x, y \in V$.*

As the name suggests, symmetric bilinear forms correspond to symmetric matrices.

Theorem 6.34. *Let H be a bilinear form on a finite-dimensional vector space V, and let β be an ordered basis for V. Then H is symmetric if and only if $\psi_\beta(H)$ is symmetric.*

Proof. Let $\beta = \{v_1, v_2, \ldots, v_n\}$ and $B = \psi_\beta(H)$.
First assume that H is symmetric. Then for $1 \le i, j \le n$,

$$B_{ij} = H(v_i, v_j) = H(v_j, v_i) = B_{ji},$$

and it follows that B is symmetric.

Conversely, suppose that B is symmetric. Let $J\colon V \times V \to F$, where F is the field of scalars for V, be the mapping defined by $J(x,y) = H(y,x)$ for all $x, y \in V$. By property 4 on page 423, J is a bilinear form. Let $C = \psi_\beta(J)$. Then, for $1 \le i, j \le n$,

$$C_{ij} = J(v_i, v_j) = H(v_j, v_i) = B_{ji} = B_{ij}.$$

Thus $C = B$. Since ψ_β is one-to-one, we have $J = H$. Hence $H(y,x) = J(x,y) = H(x,y)$ for all $x, y \in V$, and therefore H is symmetric. $\blacksquare$

Definition. *A bilinear form H on a finite-dimensional vector space V is called **diagonalizable** if there is an ordered basis β for V such that $\psi_\beta(H)$ is a diagonal matrix.*

Corollary. *Let H be a diagonalizable bilinear form on a finite-dimensional vector space V. Then H is symmetric.*

Proof. Suppose that H is diagonalizable. Then there is an ordered basis β for V such that $\psi_\beta(H) = D$ is a diagonal matrix. Trivially, D is a symmetric matrix, and hence, by Theorem 6.34, H is symmetric. ∎

Unfortunately, the converse is not true, as is illustrated by the following example.

Example 5

Let $F = Z_2$, $V = F^2$, and $H : V \times V \to F$ be the bilinear form defined by

$$H\left(\begin{pmatrix} a_1 \\ a_2 \end{pmatrix}, \begin{pmatrix} b_1 \\ b_2 \end{pmatrix} \right) = a_1 b_2 + a_2 b_1.$$

Clearly H is symmetric. In fact, if β is the standard ordered basis for V, then

$$A = \psi_\beta(H) = \begin{pmatrix} 0 & 1 \\ 1 & 0 \end{pmatrix},$$

a symmetric matrix. We show that H is not diagonalizable.

By way of contradiction, suppose that H is diagonalizable. Then there is an ordered basis γ for V such that $B = \psi_\gamma(H)$ is a diagonal matrix. So by Theorem 6.33, there exists an invertible matrix Q such that $B = Q^t A Q$. Since Q is invertible, it follows that $\text{rank}(B) = \text{rank}(A) = 2$, and consequently the diagonal entries of B are nonzero. Since the only nonzero scalar of F is 1,

$$B = \begin{pmatrix} 1 & 0 \\ 0 & 1 \end{pmatrix}.$$

Suppose that

$$Q = \begin{pmatrix} a & b \\ c & d \end{pmatrix}.$$

Then

$$\begin{pmatrix} 1 & 0 \\ 0 & 1 \end{pmatrix} = B = Q^t A Q$$

$$= \begin{pmatrix} a & c \\ b & d \end{pmatrix} \begin{pmatrix} 0 & 1 \\ 1 & 0 \end{pmatrix} \begin{pmatrix} a & b \\ c & d \end{pmatrix} = \begin{pmatrix} ac + ac & bc + ad \\ bc + ad & bd + bd \end{pmatrix}.$$

But $p + p = 0$ for all $p \in F$; hence $ac + ac = 0$. Thus, comparing the row 1, column 1 entries of the matrices in the equation above, we conclude that $1 = 0$, a contradiction. Therefore H is not diagonalizable. ◆

The bilinear form of Example 5 is an anomaly. Its failure to be diagonalizable is due to the fact that the scalar field Z_2 is of characteristic two. Recall from Appendix C that a field F is of **characteristic two** if $1 + 1 = 0$ in F. If F is not of characteristic two, then $1 + 1 = 2$ has a multiplicative inverse, which we denote by $1/2$.

Before proving the converse of the corollary to Theorem 6.34 for scalar fields that are not of characteristic two, we establish the following lemma.

Lemma. *Let H be a nonzero symmetric bilinear form on a vector space V over a field F not of characteristic two. Then there is a vector x in V such that $H(x, x) \neq 0$.*

Proof. Since H is nonzero, we can choose vectors $u, v \in V$ such that $H(u, v) \neq 0$. If $H(u, u) \neq 0$ or $H(v, v) \neq 0$, there is nothing to prove. Otherwise, set $x = u + v$. Then

$$H(x, x) = H(u, u) + H(u, v) + H(v, u) + H(v, v) = 2H(u, v) \neq 0$$

because $2 \neq 0$ and $H(u, v) \neq 0$.   ∎

Theorem 6.35. *Let V be a finite-dimensional vector space over a field F not of characteristic two. Then every symmetric bilinear form on V is diagonalizable.*

Proof. We use mathematical induction on $n = \dim(V)$. If $n = 1$, then every element of $\mathcal{B}(V)$ is diagonalizable. Now suppose that the theorem is valid for vector spaces of dimension less than n for some fixed integer $n > 1$, and suppose that $\dim(V) = n$. If H is the zero bilinear form on V, then trivially H is diagonalizable; so suppose that H is a nonzero symmetric bilinear form on V. By the lemma, there exists a nonzero vector x in V such that $H(x, x) \neq 0$. Recall the function $L_x \colon V \to F$ defined by $L_x(y) = H(x, y)$ for all $y \in V$. By property 1 on page 423, L_x is linear. Furthermore, since $L_x(x) = H(x, x) \neq 0$, L_x is nonzero. Consequently, $\operatorname{rank}(L_x) = 1$, and hence $\dim(N(L_x)) = n - 1$.

The restriction of H to $N(L_x)$ is obviously a symmetric bilinear form on a vector space of dimension $n - 1$. Thus, by the induction hypothesis, there exists an ordered basis $\{v_1, v_2, \ldots, v_{n-1}\}$ for $N(L_x)$ such that $H(v_i, v_j) = 0$ for $i \neq j$ $(1 \leq i, j \leq n - 1)$. Set $v_n = x$. Then $v_n \notin N(L_x)$, and so $\beta = \{v_1, v_2, \ldots, v_n\}$ is an ordered basis for V. In addition, $H(v_i, v_n) = H(v_n, v_i) = 0$ for $i = 1, 2, \ldots, n - 1$. We conclude that $\psi_\beta(H)$ is a diagonal matrix, and therefore H is diagonalizable.   ∎

Corollary. *Let F be a field that is not of characteristic two. If $A \in M_{n \times n}(F)$ is a symmetric matrix, then A is congruent to a diagonal matrix.*

Proof. Exercise.   ∎

Diagonalization of Symmetric Matrices

Let A be a symmetric $n \times n$ matrix with entries from a field F not of characteristic two. By the corollary to Theorem 6.35, there are matrices $Q, D \in \mathsf{M}_{n \times n}(F)$ such that Q is invertible, D is diagonal, and $Q^t A Q = D$. We now give a method for computing Q and D. This method requires familiarity with elementary matrices and their properties, which the reader may wish to review in Section 3.1.

If E is an elementary $n \times n$ matrix, then AE can be obtained by performing an elementary column operation on A. By Exercise 21, $E^t A$ can be obtained by performing the same operation on the rows of A rather than on its columns. Thus $E^t A E$ can be obtained from A by performing an elementary operation on the columns of A and then performing the same operation on the rows of AE. (Note that the order of the operations can be reversed because of the associative property of matrix multiplication.) Suppose that Q is an invertible matrix and D is a diagonal matrix such that $Q^t A Q = D$. By Corollary 3 to Theorem 3.6 (p. 159), Q is a product of elementary matrices, say $Q = E_1 E_2 \cdots E_k$. Thus

$$D = Q^t A Q = E_k^t E_{k-1}^t \cdots E_1^t A E_1 E_2 \cdots E_k.$$

From the preceding equation, we conclude that *by means of several elementary column operations and the corresponding row operations, A can be transformed into a diagonal matrix D. Furthermore, if $E_1, E_2, \ldots, E_k$ are the elementary matrices corresponding to these elementary column operations indexed in the order performed, and if $Q = E_1 E_2 \cdots E_k$, then $Q^t A Q = D$.*

Example 6

Let A be the symmetric matrix in $\mathsf{M}_{3 \times 3}(R)$ defined by

$$A = \begin{pmatrix} 1 & -1 & 3 \\ -1 & 2 & 1 \\ 3 & 1 & 1 \end{pmatrix}.$$

We use the procedure just described to find an invertible matrix Q and a diagonal matrix D such that $Q^t A Q = D$.

We begin by eliminating all of the nonzero entries in the first row and first column except for the entry in column 1 and row 1. To this end, we add the first column of A to the second column to produce a zero in row 1 and column 2. The elementary matrix that corresponds to this elementary column operation is

$$E_1 = \begin{pmatrix} 1 & 1 & 0 \\ 0 & 1 & 0 \\ 0 & 0 & 1 \end{pmatrix}.$$

We perform the corresponding elementary operation on the rows of AE_1 to obtain

$$E_1^t AE_1 = \begin{pmatrix} 1 & 0 & 3 \\ 0 & 1 & 4 \\ 3 & 4 & 1 \end{pmatrix}.$$

We now use the first column of $E_1^t AE_1$ to eliminate the 3 in row 1 column 3, and follow this operation with the corresponding row operation. The corresponding elementary matrix E_2 and the result of the elementary operations $E_2^t E_1^t AE_1 E_2$ are, respectively,

$$E_2 = \begin{pmatrix} 1 & 0 & -3 \\ 0 & 1 & 0 \\ 0 & 0 & 1 \end{pmatrix} \quad \text{and} \quad E_2^t E_1^t AE_1 E_2 = \begin{pmatrix} 1 & 0 & 0 \\ 0 & 1 & 4 \\ 0 & 4 & -8 \end{pmatrix}.$$

Finally, we subtract 4 times the second column of $E_2^t E_1^t AE_1 E_2$ from the third column and follow this with the corresponding row operation. The corresponding elementary matrix E_3 and the result of the elementary operations $E_3^t E_2^t E_1^t AE_1 E_2 E_3$ are, respectively,

$$E_3 = \begin{pmatrix} 1 & 0 & 0 \\ 0 & 1 & -4 \\ 0 & 0 & 1 \end{pmatrix} \quad \text{and} \quad E_3^t E_2^t E_1^t AE_1 E_2 E_3 = \begin{pmatrix} 1 & 0 & 0 \\ 0 & 1 & 0 \\ 0 & 0 & -24 \end{pmatrix}.$$

Since we have obtained a diagonal matrix, the process is complete. So we let

$$Q = E_1 E_2 E_3 = \begin{pmatrix} 1 & 1 & -7 \\ 0 & 1 & -4 \\ 0 & 0 & 1 \end{pmatrix} \quad \text{and} \quad D = \begin{pmatrix} 1 & 0 & 0 \\ 0 & 1 & 0 \\ 0 & 0 & -24 \end{pmatrix}$$

to obtain the desired diagonalization $Q^t AQ = D$. ◆

The reader should justify the following method for computing Q without recording each elementary matrix separately. The method is inspired by the algorithm for computing the inverse of a matrix developed in Section 3.2. We use a sequence of elementary column operations and corresponding row operations to change the $n \times 2n$ matrix $(A|I)$ into the form $(D|B)$, where D is a diagonal matrix and $B = Q^t$. It then follows that $D = Q^t AQ$.

Starting with the matrix A of the preceding example, this method produces the following sequence of matrices:

$$(A|I) = \begin{pmatrix} 1 & -1 & 3 & | & 1 & 0 & 0 \\ -1 & 2 & 1 & | & 0 & 1 & 0 \\ 3 & 1 & 1 & | & 0 & 0 & 1 \end{pmatrix} \longrightarrow \begin{pmatrix} 1 & 0 & 3 & | & 1 & 0 & 0 \\ -1 & 1 & 1 & | & 0 & 1 & 0 \\ 3 & 4 & 1 & | & 0 & 0 & 1 \end{pmatrix}$$

$$\longrightarrow \left(\begin{array}{ccc|ccc} 1 & 0 & 3 & 1 & 0 & 0 \\ 0 & 1 & 4 & 1 & 1 & 0 \\ 3 & 4 & 1 & 0 & 0 & 1 \end{array}\right) \longrightarrow \left(\begin{array}{ccc|ccc} 1 & 0 & 0 & 1 & 0 & 0 \\ 0 & 1 & 4 & 1 & 1 & 0 \\ 3 & 4 & -8 & 0 & 0 & 1 \end{array}\right)$$

$$\longrightarrow \left(\begin{array}{ccc|ccc} 1 & 0 & 0 & 1 & 0 & 0 \\ 0 & 1 & 4 & 1 & 1 & 0 \\ 0 & 4 & -8 & -3 & 0 & 1 \end{array}\right) \longrightarrow \left(\begin{array}{ccc|ccc} 1 & 0 & 0 & 1 & 0 & 0 \\ 0 & 1 & 0 & 1 & 1 & 0 \\ 0 & 4 & -24 & -3 & 0 & 1 \end{array}\right)$$

$$\longrightarrow \left(\begin{array}{ccc|ccc} 1 & 0 & 0 & 1 & 0 & 0 \\ 0 & 1 & 0 & 1 & 1 & 0 \\ 0 & 0 & -24 & -7 & -4 & 1 \end{array}\right) = (D|Q^t).$$

Therefore

$$D = \begin{pmatrix} 1 & 0 & 0 \\ 0 & 1 & 0 \\ 0 & 0 & -24 \end{pmatrix}, \quad Q^t = \begin{pmatrix} 1 & 0 & 0 \\ 1 & 1 & 0 \\ -7 & -4 & 1 \end{pmatrix}, \quad \text{and} \quad Q = \begin{pmatrix} 1 & 1 & -7 \\ 0 & 1 & -4 \\ 0 & 0 & 1 \end{pmatrix}.$$

Quadratic Forms

Associated with symmetric bilinear forms are functions called *quadratic forms*.

Definition.   Let V be a vector space over F. A function $K: \mathsf{V} \to F$ is called a **quadratic form** if there exists a symmetric bilinear form $H \in \mathcal{B}(\mathsf{V})$ such that

$$K(x) = H(x, x) \quad \text{for all } x \in \mathsf{V}. \tag{8}$$

If the field F is not of characteristic two, there is a one-to-one correspondence between symmetric bilinear forms and quadratic forms given by (8). In fact, if K is a quadratic form on a vector space V over a field F not of characteristic two, and $K(x) = H(x, x)$ for some symmetric bilinear form H on V, then we can recover H from K because

$$H(x, y) = \frac{1}{2}[K(x + y) - K(x) - K(y)] \tag{9}$$

(See Exercise 16.)

Example 7

The classic example of a quadratic form is the homogeneous second-degree polynomial of several variables. Given the variables $t_1, t_2, \ldots, t_n$ that take values in a field F not of characteristic two and given (not necessarily distinct) scalars a_{ij} ($1 \le i \le j \le n$), define the polynomial

$$f(t_1, t_2, \ldots, t_n) = \sum_{i \le j} a_{ij} t_i t_j.$$

Any such polynomial is a quadratic form. In fact, if β is the standard ordered basis for F^n, then the symmetric bilinear form H corresponding to the quadratic form f has the matrix representation $\psi_\beta(H) = A$, where

$$A_{ij} = A_{ji} = \begin{cases} a_{ii} & \text{if } i = j \\ \frac{1}{2}a_{ij} & \text{if } i \neq j. \end{cases}$$

To see this, apply (9) to obtain $H(e_i, e_j) = A_{ij}$ from the quadratic form K, and verify that f is computable from H by (8) using f in place of K.

For example, given the polynomial

$$f(t_1, t_2, t_3) = 2t_1^2 - t_2^2 + 6t_1t_2 - 4t_2t_3$$

with real coefficients, let

$$A = \begin{pmatrix} 2 & 3 & 0 \\ 3 & -1 & -2 \\ 0 & -2 & 0 \end{pmatrix}.$$

Setting $H(x, y) = x^t A y$ for all $x, y \in \mathsf{R}^3$, we see that

$$f(t_1, t_2, t_3) = (t_1, t_2, t_3) A \begin{pmatrix} t_1 \\ t_2 \\ t_3 \end{pmatrix} \quad \text{for} \quad \begin{pmatrix} t_1 \\ t_2 \\ t_3 \end{pmatrix} \in \mathsf{R}^3. \quad \blacklozenge$$

Quadratic Forms Over the Field R

Since symmetric matrices over R are *orthogonally diagonalizable* (see Theorem 6.20 p. 384), the theory of symmetric bilinear forms and quadratic forms on finite-dimensional vector spaces over R is especially nice. The following theorem and its corollary are useful.

Theorem 6.36. *Let* V *be a finite-dimensional real inner product space, and let* H *be a symmetric bilinear form on* V. *Then there exists an orthonormal basis* β *for* V *such that* $\psi_\beta(H)$ *is a diagonal matrix.*

Proof. Choose any orthonormal basis $\gamma = \{v_1, v_2, \ldots, v_n\}$ for V, and let $A = \psi_\gamma(H)$. Since A is symmetric, there exists an orthogonal matrix Q and a diagonal matrix D such that $D = Q^t A Q$ by Theorem 6.20. Let $\beta = \{w_1, w_2, \ldots, w_n\}$ be defined by

$$w_j = \sum_{i=1}^{n} Q_{ij} v_i \quad \text{for } 1 \leq j \leq n.$$

By Theorem 6.33, $\psi_\beta(H) = D$. Furthermore, since Q is orthogonal and γ is orthonormal, β is orthonormal by Exercise 30 of Section 6.5.   ∎

Corollary. *Let K be a quadratic form on a finite-dimensional real inner product space* V. *There exists an orthonormal basis* $\beta = \{v_1, v_2, \ldots, v_n\}$ *for* V *and scalars* $\lambda_1, \lambda_2, \ldots, \lambda_n$ *(not necessarily distinct) such that if* $x \in$ V *and*

$$x = \sum_{i=1}^{n} s_i v_i, \quad s_i \in R,$$

then

$$K(x) = \sum_{i=1}^{n} \lambda_i s_i^2.$$

In fact, if H is the symmetric bilinear form determined by K, then β can be chosen to be any orthonormal basis for V *such that $\psi_\beta(H)$ is a diagonal matrix.*

Proof. Let H be the symmetric bilinear form for which $K(x) = H(x, x)$ for all $x \in$ V. By Theorem 6.36, there exists an orthonormal basis $\beta = \{v_1, v_2, \ldots, v_n\}$ for V such that $\psi_\beta(H)$ is the diagonal matrix

$$D = \begin{pmatrix} \lambda_1 & 0 & \cdots & 0 \\ 0 & \lambda_2 & \cdots & 0 \\ \vdots & \vdots & & \vdots \\ 0 & 0 & \cdots & \lambda_n \end{pmatrix}.$$

Let $x \in$ V, and suppose that $x = \sum_{i=1}^{n} s_i v_i$. Then

$$K(x) = H(x, x) = [\phi_\beta(x)]^t D [\phi_\beta(x)] = (s_1, s_2, \ldots, s_n) D \begin{pmatrix} s_1 \\ s_2 \\ \vdots \\ s_n \end{pmatrix} = \sum_{i=1}^{n} \lambda_i s_i^2. \quad \blacksquare$$

Example 8

For the homogeneous real polynomial of degree 2 defined by

$$f(t_1, t_2) = 5t_1^2 + 2t_2^2 + 4t_1 t_2, \tag{10}$$

we find an orthonormal basis $\gamma = \{x_1, x_2\}$ for R^2 and scalars λ_1 and λ_2 such that if

$$\begin{pmatrix} t_1 \\ t_2 \end{pmatrix} \in R^2 \quad \text{and} \quad \begin{pmatrix} t_1 \\ t_2 \end{pmatrix} = s_1 x_1 + s_2 x_2,$$

then $f(t_1, t_2) = \lambda_1 s_1^2 + \lambda_2 s_2^2$. We can think of s_1 and s_2 as the coordinates of (t_1, t_2) relative to γ. Thus the polynomial $f(t_1, t_2)$, as an expression involving the coordinates of a point with respect to the standard ordered basis for R^2, is transformed into a new polynomial $g(s_1, s_2) = \lambda_1 s_1^2 + \lambda_2 s_2^2$ interpreted as an expression involving the coordinates of a point relative to the new ordered basis γ.

Let H denote the symmetric bilinear form corresponding to the quadratic form defined by (10), let β be the standard ordered basis for R^2, and let $A = \psi_\beta(H)$. Then

$$A = \psi_\beta(H) = \begin{pmatrix} 5 & 2 \\ 2 & 2 \end{pmatrix}.$$

Next, we find an orthogonal matrix Q such that $Q^t A Q$ is a diagonal matrix. For this purpose, observe that $\lambda_1 = 6$ and $\lambda_2 = 1$ are the eigenvalues of A with corresponding orthonormal eigenvectors

$$v_1 = \frac{1}{\sqrt{5}} \begin{pmatrix} 2 \\ 1 \end{pmatrix} \quad \text{and} \quad v_2 = \frac{1}{\sqrt{5}} \begin{pmatrix} 1 \\ -2 \end{pmatrix}.$$

Let $\gamma = \{v_1, v_2\}$. Then γ is an orthonormal basis for R^2 consisting of eigenvectors of A. Hence, setting

$$Q = \frac{1}{\sqrt{5}} \begin{pmatrix} 2 & 1 \\ 1 & -2 \end{pmatrix},$$

we see that Q is an orthogonal matrix and

$$Q^t A Q = \begin{pmatrix} 6 & 0 \\ 0 & 1 \end{pmatrix}.$$

Clearly Q is also a change of coordinate matrix. Consequently,

$$\psi_\gamma(H) = Q^t \psi_\beta(H) Q = Q^t A Q = \begin{pmatrix} 6 & 0 \\ 0 & 1 \end{pmatrix}.$$

Thus by the corollary to Theorem 6.36,

$$K(x) = 6 s_1^2 + s_2^2$$

for any $x = s_1 v_1 + s_2 v_2 \in \mathsf{R}^2$. So $g(s_1, s_2) = 6 s_1^2 + s_2^2$.   ◆

The next example illustrates how the theory of quadratic forms can be applied to the problem of describing quadratic surfaces in R^3.

Example 9

Let S be the surface in R^3 defined by the equation

$$2t_1^2 + 6t_1 t_2 + 5t_2^2 - 2t_2 t_3 + 2t_3^2 + 3t_1 - 2t_2 - t_3 + 14 = 0. \tag{11}$$

Then (11) describes the points of $\mathcal{S}$ in terms of their coordinates relative to β, the standard ordered basis for R^3. We find a new orthonormal basis γ for R^3 so that the equation describing the coordinates of $\mathcal{S}$ relative to γ is simpler than (11).

We begin with the observation that the terms of second degree on the left side of (11) add to form a quadratic form K on R^3:

$$K \begin{pmatrix} t_1 \\ t_2 \\ t_3 \end{pmatrix} = 2t_1^2 + 6t_1t_2 + 5t_2^2 - 2t_2t_3 + 2t_3^2.$$

Next, we diagonalize K. Let H be the symmetric bilinear form corresponding to K, and let $A = \psi_\beta(H)$. Then

$$A = \begin{pmatrix} 2 & 3 & 0 \\ 3 & 5 & -1 \\ 0 & -1 & 2 \end{pmatrix}.$$

The characteristic polynomial of A is $(-1)(t-2)(t-7)t$; hence A has the eigenvalues $\lambda_1 = 2$, $\lambda_2 = 7$, and $\lambda_3 = 0$. Corresponding unit eigenvectors are

$$v_1 = \frac{1}{\sqrt{10}} \begin{pmatrix} 1 \\ 0 \\ 3 \end{pmatrix}, \quad v_2 = \frac{1}{\sqrt{35}} \begin{pmatrix} 3 \\ 5 \\ -1 \end{pmatrix}, \quad \text{and} \quad v_3 = \frac{1}{\sqrt{14}} \begin{pmatrix} -3 \\ 2 \\ 1 \end{pmatrix}.$$

Set $\gamma = \{v_1, v_2, v_3\}$ and

$$Q = \begin{pmatrix} \dfrac{1}{\sqrt{10}} & \dfrac{3}{\sqrt{35}} & \dfrac{-3}{\sqrt{14}} \\[2ex] 0 & \dfrac{5}{\sqrt{35}} & \dfrac{2}{\sqrt{14}} \\[2ex] \dfrac{3}{\sqrt{10}} & \dfrac{-1}{\sqrt{35}} & \dfrac{1}{\sqrt{14}} \end{pmatrix}.$$

As in Example 8, Q is a change of coordinate matrix changing γ-coordinates to β-coordinates, and

$$\psi_\gamma(H) = Q^t\psi_\beta(H)Q = Q^t A Q = \begin{pmatrix} 2 & 0 & 0 \\ 0 & 7 & 0 \\ 0 & 0 & 0 \end{pmatrix}.$$

By the corollary to Theorem 6.36, if $x = s_1v_1 + s_2v_2 + s_3v_3$, then

$$K(x) = 2s_1^2 + 7s_2^2. \tag{12}$$

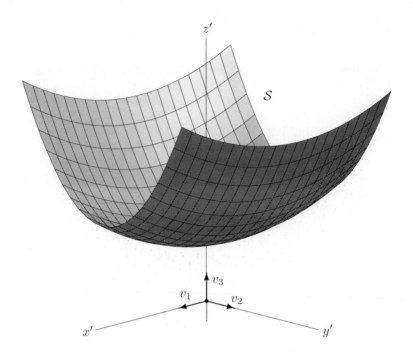

Figure 6.7

We are now ready to transform (11) into an equation involving coordinates relative to γ. Let $x = (t_1, t_2, t_3) \in \mathsf{R}^3$, and suppose that $x = s_1 v_1 + s_2 v_2 + s_3 v_3$. Then, by Theorem 2.22 (p. 111),

$$x = \begin{pmatrix} t_1 \\ t_2 \\ t_3 \end{pmatrix} = Q \begin{pmatrix} s_1 \\ s_2 \\ s_3 \end{pmatrix},$$

and therefore

$$t_1 = \frac{s_1}{\sqrt{10}} + \frac{3s_2}{\sqrt{35}} - \frac{3s_3}{\sqrt{14}},$$

$$t_2 = \frac{5s_2}{\sqrt{35}} + \frac{2s_3}{\sqrt{14}},$$

and

$$t_3 = \frac{3s_1}{\sqrt{10}} - \frac{s_2}{\sqrt{35}} + \frac{s_3}{\sqrt{14}}.$$

Thus

$$3t_1 - 2t_2 - t_3 = -\frac{14s_3}{\sqrt{14}} = -\sqrt{14}s_3.$$

Combining (11), (12), and the preceding equation, we conclude that if $x \in \mathsf{R}^3$ and $x = s_1 v_1 + s_2 v_2 + s_3 v_3$, then $x \in \mathcal{S}$ if and only if

$$2s_1^2 + 7s_2^2 - \sqrt{14}s_3 + 14 = 0 \quad \text{or} \quad s_3 = \frac{\sqrt{14}}{7}s_1^2 + \frac{\sqrt{14}}{2}s_2^2 + \sqrt{14}.$$

Consequently, if we draw new axes x', y', and z' in the directions of v_1, v_2, and v_3, respectively, the graph of the equation, rewritten as

$$z' = \frac{\sqrt{14}}{7}(x')^2 + \frac{\sqrt{14}}{2}(y')^2 + \sqrt{14},$$

coincides with the surface $\mathcal{S}$. We recognize $\mathcal{S}$ to be an elliptic paraboloid.

Figure 6.7 is a sketch of the surface $\mathcal{S}$ drawn so that the vectors v_1, v_2 and v_3 are oriented to lie in the principal directions. For practical purposes, the scale of the z' axis has been adjusted so that the figure fits the page.   ♦

The Second Derivative Test for Functions of Several Variables

We now consider an application of the theory of quadratic forms to multivariable calculus—the derivation of the second derivative test for local extrema of a function of several variables. We assume an acquaintance with the calculus of functions of several variables to the extent of Taylor's theorem. The reader is undoubtedly familiar with the one-variable version of Taylor's theorem. For a statement and proof of the multivariable version, consult, for example, *An Introduction to Analysis* 2d ed, by William R. Wade (Prentice Hall, Upper Saddle River, N.J., 2000).

Let $z = f(t_1, t_2, \ldots, t_n)$ be a fixed real-valued function of n real variables for which all third-order partial derivatives exist and are continuous. The function f is said to have a **local maximum** at a point $p \in \mathsf{R}^n$ if there exists a $\delta > 0$ such that $f(p) \geq f(x)$ whenever $||x - p|| < \delta$. Likewise, f has a **local minimum** at $p \in \mathsf{R}^n$ if there exists a $\delta > 0$ such that $f(p) \leq f(x)$ whenever $||x - p|| < \delta$. If f has either a local minimum or a local maximum at p, we say that f has a **local extremum** at p. A point $p \in \mathsf{R}^n$ is called a **critical point** of f if $\partial f(p)/\partial t_i = 0$ for $i = 1, 2, \ldots, n$. It is a well-known fact that if f has a local extremum at a point $p \in \mathsf{R}^n$, then p is a critical point of f. For, if f has a local extremum at $p = (p_1, p_2, \ldots, p_n)$, then for any $i = 1, 2, \ldots, n$ the function ϕ_i defined by $\phi_i(t) = f(p_1, p_2, \ldots, p_{i-1}, t, p_{i+1}, \ldots, p_n)$ has a local extremum at $t = p_i$. So, by an elementary single-variable argument,

$$\frac{\partial f(p)}{\partial t_i} = \frac{d\phi_i(p_i)}{dt} = 0.$$

Thus p is a critical point of f. But critical points are not necessarily local extrema.

The second-order partial derivatives of f at a critical point p can often be used to test for a local extremum at p. These partials determine a matrix $A(p)$ in which the row i, column j entry is

$$\frac{\partial^2 f(p)}{(\partial t_i)(\partial t_j)}.$$

This matrix is called the **Hessian matrix** of f at p. Note that if the third-order partial derivatives of f are continuous, then the mixed second-order partials of f at p are independent of the order in which they are taken, and hence $A(p)$ is a symmetric matrix. In this case, all of the eigenvalues of $A(p)$ are real.

Theorem 6.37 (The Second Derivative Test). Let $f(t_1, t_2, \ldots, t_n)$ be a real-valued function in n real variables for which all third-order partial derivatives exist and are continuous. Let $p = (p_1, p_2, \ldots, p_n)$ be a critical point of f, and let $A(p)$ be the Hessian of f at p.
  (a) If all eigenvalues of $A(p)$ are positive, then f has a local minimum at p.
  (b) If all eigenvalues of $A(p)$ are negative, then f has a local maximum at p.
  (c) If $A(p)$ has at least one positive and at least one negative eigenvalue, then f has no local extremum at p (p is called a **saddle-point** of f).
  (d) If $\text{rank}(A(p)) < n$ and $A(p)$ does not have both positive and negative eigenvalues, then the second derivative test is inconclusive.

Proof. If $p \neq 0$, we may define a function $g \colon R^n \to R$ by

$$g(t_1, t_2, \ldots, t_n) = f(t_1 + p_1, t_2 + p_2, \ldots, p_n + t_n) - f(p).$$

The following facts are easily verified.

  1. The function f has a local maximum [minimum] at p if and only if g has a local maximum [minimum] at $0 = (0, 0, \ldots, 0)$.
  2. The partial derivatives of g at 0 are equal to the corresponding partial derivatives of f at p.
  3. 0 is a critical point of g.
  4. $A_{ij}(p) = \dfrac{\partial^2 g(0)}{(\partial t_i)(\partial t_j)}$ for all i and j.

In view of these facts, we may assume without loss of generality that $p = 0$ and $f(p) = 0$.

Now we apply Taylor's theorem to f to obtain the first-order approximation of f around 0. We have

$$f(t_1, t_2, \ldots, t_n) = f(0) + \sum_{i=1}^{n} \frac{\partial f(0)}{\partial t_i} t_i + \frac{1}{2} \sum_{i,j=1}^{n} \frac{\partial^2 f(0)}{(\partial t_i)(\partial t_j)} t_i t_j + S(t_1, t_2, \ldots, t_n)$$

$$= \frac{1}{2} \sum_{i,j=1}^{n} \frac{\partial^2 f(0)}{(\partial t_i)(\partial t_j)} t_i t_j + S(t_1, t_2, \ldots, t_n),$$

(13)

where S is a real-valued function on $\mathbf{R}^n$ such that

$$\lim_{x \to 0} \frac{S(x)}{||x||^2} = \lim_{(t_1, t_2, \ldots, t_n) \to 0} \frac{S(t_1, t_2, \ldots, t_n)}{t_1^2 + t_2^2 + \cdots + t_n^2} = 0.$$

(14)

Let $K \colon \mathbf{R}^n \to R$ be the quadratic form defined by

$$K \begin{pmatrix} t_1 \\ t_2 \\ \vdots \\ t_n \end{pmatrix} = \frac{1}{2} \sum_{i,j=1}^{n} \frac{\partial^2 f(0)}{(\partial t_i)(\partial t_j)} t_i t_j,$$

(15)

H be the symmetric bilinear form corresponding to K, and β be the standard ordered basis for $\mathbf{R}^n$. It is easy to verify that $\psi_\beta(H) = \frac{1}{2} A(p)$. Since $A(p)$ is symmetric, Theorem 6.20 (p. 384) implies that there exists an orthogonal matrix Q such that

$$Q^t A(p) Q = \begin{pmatrix} \lambda_1 & 0 & \cdots & 0 \\ 0 & \lambda_2 & \cdots & 0 \\ \vdots & \vdots & & \vdots \\ 0 & 0 & \cdots & \lambda_n \end{pmatrix}$$

is a diagonal matrix whose diagonal entries are the eigenvalues of $A(p)$. Let $\gamma = \{v_1, v_2, \ldots, v_n\}$ be the orthogonal basis for $\mathbf{R}^n$ whose ith vector is the ith column of Q. Then Q is the change of coordinate matrix changing γ-coordinates into β-coordinates, and by Theorem 6.33

$$\psi_\gamma(H) = Q^t \psi_\beta(H) Q = \frac{1}{2} Q^t A(p) Q = \begin{pmatrix} \dfrac{\lambda_1}{2} & 0 & \cdots & 0 \\ 0 & \dfrac{\lambda_2}{2} & \cdots & 0 \\ \vdots & \vdots & & \vdots \\ 0 & 0 & \cdots & \dfrac{\lambda_n}{2} \end{pmatrix}.$$

Suppose that $A(p)$ is not the zero matrix. Then $A(p)$ has nonzero eigenvalues. Choose $\epsilon > 0$ such that $\epsilon < |\lambda_i|/2$ for all $\lambda_i \neq 0$. By (14), there exists $\delta > 0$ such that for any $x \in \mathsf{R}^n$ satisfying $0 < ||x|| < \delta$, we have $|S(x)| < \epsilon||x||^2$. Consider any $x \in \mathsf{R}^n$ such that $0 < ||x|| < \delta$. Then, by (13) and (15),

$$|f(x) - K(x)| = |S(x)| < \epsilon||x||^2,$$

and hence

$$K(x) - \epsilon||x||^2 < f(x) < K(x) + \epsilon||x||^2. \tag{16}$$

Suppose that $x = \sum_{i=1}^{n} s_i v_i$. Then

$$||x||^2 = \sum_{i=1}^{n} s_i^2 \quad \text{and} \quad K(x) = \frac{1}{2} \sum_{i=1}^{n} \lambda_i s_i^2.$$

Combining these equations with (16), we obtain

$$\sum_{i=1}^{n} \left(\frac{1}{2}\lambda_i - \epsilon \right) s_i^2 < f(x) < \sum_{i=1}^{n} \left(\frac{1}{2}\lambda_i + \epsilon \right) s_i^2. \tag{17}$$

Now suppose that all eigenvalues of $A(p)$ are positive. Then $\frac{1}{2}\lambda_i - \epsilon > 0$ for all i, and hence, by the left inequality in (17),

$$f(0) = 0 \leq \sum_{i=1}^{n} \left(\frac{1}{2}\lambda_i - \epsilon \right) s_i^2 < f(x).$$

Thus $f(0) \leq f(x)$ for $||x|| < \delta$, and so f has a local minimum at 0. By a similar argument using the right inequality in (17), we have that if all of the eigenvalues of $A(p)$ are negative, then f has a local maximum at 0. This establishes (a) and (b) of the theorem.

Next, suppose that $A(p)$ has both a positive and a negative eigenvalue, say, $\lambda_i > 0$ and $\lambda_j < 0$ for some i and j. Then $\frac{1}{2}\lambda_i - \epsilon > 0$ and $\frac{1}{2}\lambda_j + \epsilon < 0$. Let s be any real number such that $0 < |s| < \delta$. Substituting $x = sv_i$ and $x = sv_j$ into the left inequality and the right inequality of (17), respectively, we obtain

$$f(0) = 0 < (\tfrac{1}{2}\lambda_i - \epsilon)s^2 < f(sv_i) \quad \text{and} \quad f(sv_j) < (\tfrac{1}{2}\lambda_j + \epsilon)s^2 < 0 = f(0).$$

Thus f attains both positive and negative values arbitrarily close to 0; so f has neither a local maximum nor a local minimum at 0. This establishes (c).

To show that the second-derivative test is inconclusive under the conditions stated in (d), consider the functions

$$f(t_1, t_2) = t_1^2 - t_2^4 \quad \text{and} \quad g(t_1, t_2) = t_1^2 + t_2^4$$

at $p = 0$. In both cases, the function has a critical point at p, and

$$A(p) = \begin{pmatrix} 2 & 0 \\ 0 & 0 \end{pmatrix}.$$

However, f does not have a local extremum at 0, whereas g has a local minimum at 0.   ∎

Sylvester's Law of Inertia

Any two matrix representations of a bilinear form have the same rank because rank is preserved under congruence. We can therefore define the **rank** of a bilinear form to be the rank of any of its matrix representations. If a matrix representation is a diagonal matrix, then the rank is equal to the number of nonzero diagonal entries of the matrix.

We confine our analysis to symmetric bilinear forms on finite-dimensional real vector spaces. Each such form has a diagonal matrix representation in which the diagonal entries may be positive, negative, or zero. Although these entries are not unique, we show that the number of entries that are positive and the number that are negative are unique. That is, they are independent of the choice of diagonal representation. This result is called *Sylvester's law of inertia*. We prove the law and apply it to describe the equivalence classes of congruent symmetric real matrices.

Theorem 6.38 (Sylvester's Law of Inertia).   *Let H be a symmetric bilinear form on a finite-dimensional real vector space* V. *Then the number of positive diagonal entries and the number of negative diagonal entries in any diagonal matrix representation of H are each independent of the diagonal representation.*

Proof. Suppose that β and γ are ordered bases for V that determine diagonal representations of H. Without loss of generality, we may assume that β and γ are ordered so that on each diagonal the entries are in the order of positive, negative, and zero. It suffices to show that both representations have the same number of positive entries because the number of negative entries is equal to the difference between the rank and the number of positive entries. Let p and q be the number of positive diagonal entries in the matrix representations of H with respect to β and γ, respectively. We suppose that $p \neq q$ and arrive at a contradiction. Without loss of generality, assume that $p < q$. Let

$$\beta = \{v_1, v_2, \ldots, v_p, \ldots, v_r, \ldots, v_n\} \text{ and } \gamma = \{w_1, w_2, \ldots, w_q, \ldots, w_r, \ldots, w_n\},$$

where r is the rank of H and $n = \dim(V)$. Let $L\colon V \to R^{p+r-q}$ be the mapping defined by

$$L(x) = (H(x, v_1), H(x, v_2), \ldots, H(x, v_p), H(x, w_{q+1}), \ldots, H(x, w_r)).$$

It is easily verified that L is linear and $\operatorname{rank}(L) \le p + r - q$. Hence

$$\operatorname{nullity}(L) \ge n - (p + r - q) > n - r.$$

So there exists a nonzero vector v_0 such that $v_0 \notin \operatorname{span}(\{v_{r+1}, v_{r+2}, \ldots, v_n\})$, but $v_0 \in N(L)$. Since $v_0 \in N(L)$, it follows that $H(v_0, v_i) = 0$ for $i \le p$ and $H(v_0, w_i) = 0$ for $q < i \le r$. Suppose that

$$v_0 = \sum_{j=1}^n a_j v_j = \sum_{j=1}^n b_j w_j.$$

For any $i \le p$,

$$H(v_0, v_i) = H\left(\sum_{j=1}^n a_j v_j, v_i\right) = \sum_{j=1}^n a_j H(v_j, v_i) = a_i H(v_i, v_i).$$

But for $i \le p$, we have $H(v_i, v_i) > 0$ and $H(v_0, v_i) = 0$, so that $a_i = 0$. Similarly, $b_i = 0$ for $q + 1 \le i \le r$. Since v_0 is not in the span of $\{v_{r+1}, v_{r+2}, \ldots, v_n\}$, it follows that $a_i \ne 0$ for some $p < i \le r$. Thus

$$H(v_0, v_0) = H\left(\sum_{j=1}^n a_j v_j, \sum_{i=1}^n a_i v_i\right) = \sum_{j=1}^n a_j^2 H(v_j, v_j) = \sum_{j=p+1}^r a_j^2 H(v_j, v_j) < 0.$$

Furthermore,

$$H(v_0, v_0) = H\left(\sum_{j=1}^n b_j w_j, \sum_{i=1}^n b_i w_i\right) = \sum_{j=1}^n b_j^2 H(w_j, w_j) = \sum_{j=p+1}^r b_j^2 H(w_j, w_j) \ge 0.$$

So $H(v_0, v_0) < 0$ and $H(v_0, v_0) \ge 0$, which is a contradiction. We conclude that $p = q$. ∎

Definitions. *The number of positive diagonal entries in a diagonal representation of a symmetric bilinear form on a real vector space is called the **index** of the form. The difference between the number of positive and the number of negative diagonal entries in a diagonal representation of a symmetric bilinear form is called the **signature** of the form. The three terms rank, index, and signature are called the **invariants** of the bilinear form because they are invariant with respect to matrix representations. These same terms apply to the associated quadratic form. Notice that the values of any two of these invariants determine the value of the third.*

Example 10

The bilinear form corresponding to the quadratic form K of Example 9 has a 3×3 diagonal matrix representation with diagonal entries of 2, 7, and 0. Therefore the rank, index, and signature of K are each 2.   ◆

Example 11

The matrix representation of the bilinear form corresponding to the quadratic form $K(x, y) = x^2 - y^2$ on R^2 with respect to the standard ordered basis is the diagonal matrix with diagonal entries of 1 and -1. Therefore the rank of K is 2, the index of K is 1, and the signature of K is 0.   ◆

Since the congruence relation is intimately associated with bilinear forms, we can apply Sylvester's law of inertia to study this relation on the set of real symmetric matrices. Let A be an $n \times n$ real symmetric matrix, and suppose that D and E are each diagonal matrices congruent to A. By Corollary 3 to Theorem 6.32, A is the matrix representation of the bilinear form H on R^n defined by $H(x, y) = x^t A y$ with respect to the standard ordered basis for R^n. Therefore Sylvester's law of inertia tells us that D and E have the same number of positive and negative diagonal entries. We can state this result as the matrix version of Sylvester's law.

Corollary 1 (Sylvester's Law of Inertia for Matrices). *Let A be a real symmetric matrix. Then the number of positive diagonal entries and the number of negative diagonal entries in any diagonal matrix congruent to A is independent of the choice of the diagonal matrix.*

Definitions. *Let A be a real symmetric matrix, and let D be a diagonal matrix that is congruent to A. The number of positive diagonal entries of D is called the **index** of A. The difference between the number of positive diagonal entries and the number of negative diagonal entries of D is called the **signature** of A. As before, the rank, index, and signature of a matrix are called the **invariants** of the matrix, and the values of any two of these invariants determine the value of the third.*

Any two of these invariants can be used to determine an equivalence class of congruent real symmetric matrices.

Corollary 2. *Two real symmetric $n \times n$ matrices are congruent if and only if they have the same invariants.*

Proof. If A and B are congruent $n \times n$ symmetric matrices, then they are both congruent to the same diagonal matrix, and it follows that they have the same invariants.

Conversely, suppose that A and B are $n \times n$ symmetric matrices with the same invariants. Let D and E be diagonal matrices congruent to A and B, respectively, chosen so that the diagonal entries are in the order of positive, negative, and zero. (Exercise 23 allows us to do this.) Since A and B have the same invariants, so do D and E. Let p and r denote the index and the rank, respectively, of both D and E. Let d_i denote the ith diagonal entry of D, and let Q be the $n \times n$ diagonal matrix whose ith diagonal entry q_i is given by

$$
q_i = \begin{cases}
\dfrac{1}{\sqrt{d_i}} & \text{if } 1 \le i \le p \\[2ex]
\dfrac{1}{\sqrt{-d_i}} & \text{if } p < i \le r \\[2ex]
1 & \text{if } r < i.
\end{cases}
$$

Then $Q^t DQ = J_{pr}$, where

$$
J_{pr} = \begin{pmatrix} I_p & O & O \\ O & -I_{r-p} & O \\ O & O & O \end{pmatrix}.
$$

It follows that A is congruent to J_{pr}. Similarly, B is congruent to J_{pr}, and hence A is congruent to B. $\blacksquare$

The matrix J_{pr} acts as a canonical form for the theory of real symmetric matrices. The next corollary, whose proof is contained in the proof of Corollary 2, describes the role of J_{pr}.

Corollary 3. *A real symmetric $n \times n$ matrix A has index p and rank r if and only if A is congruent to J_{pr} (as just defined).*

Example 12

Let

$$
A = \begin{pmatrix} 1 & 1 & -3 \\ -1 & 2 & 1 \\ 3 & 1 & 1 \end{pmatrix}, \quad B = \begin{pmatrix} 1 & 2 & 1 \\ 2 & 3 & 2 \\ 1 & 2 & 1 \end{pmatrix}, \quad \text{and} \quad C = \begin{pmatrix} 1 & 0 & 1 \\ 0 & 1 & 2 \\ 1 & 2 & 1 \end{pmatrix}.
$$

We apply Corollary 2 to determine which pairs of the matrices A, B, and C are congruent.

The matrix A is the 3×3 matrix of Example 6, where it is shown that A is congruent to a diagonal matrix with diagonal entries 1, 1, and -24. Therefore, A has rank 3 and index 2. Using the methods of Example 6 (it is not necessary to compute Q), it can be shown that B and C are congruent, respectively, to the diagonal matrices

$$
\begin{pmatrix} 1 & 0 & 0 \\ 0 & -1 & 0 \\ 0 & 0 & -1 \end{pmatrix} \quad \text{and} \quad \begin{pmatrix} 1 & 0 & 0 \\ 0 & 1 & 0 \\ 0 & 0 & -4 \end{pmatrix}.
$$

It follows that both A and C have rank 3 and index 2, while B has rank 3 and index 1. We conclude that A and C are congruent but that B is congruent to neither A nor C.   ◆

EXERCISES

1. Label the following statements as true or false.

   (a) Every quadratic form is a bilinear form.
   (b) If two matrices are congruent, they have the same eigenvalues.
   (c) Symmetric bilinear forms have symmetric matrix representations.
   (d) Any symmetric matrix is congruent to a diagonal matrix.
   (e) The sum of two symmetric bilinear forms is a symmetric bilinear form.
   (f) Two symmetric matrices with the same characteristic polynomial are matrix representations of the same bilinear form.
   (g) There exists a bilinear form H such that $H(x, y) \neq 0$ for all x and y.
   (h) If V is a vector space of dimension n, then $\dim(\mathcal{B}(V)) = 2n$.
   (i) Let H be a bilinear form on a finite-dimensional vector space V with $\dim(V) > 1$. For any $x \in V$, there exists $y \in V$ such that $y \neq 0$, but $H(x, y) = 0$.
   (j) If H is any bilinear form on a finite-dimensional real inner product space V, then there exists an ordered basis β for V such that $\psi_\beta(H)$ is a diagonal matrix.

2. Prove properties 1, 2, 3, and 4 on page 423.

3. (a) Prove that the sum of two bilinear forms is a bilinear form.
   (b) Prove that the product of a scalar and a bilinear form is a bilinear form.
   (c) Prove Theorem 6.31.

4. Determine which of the mappings that follow are bilinear forms. Justify your answers.

   (a) Let $V = C[0, 1]$ be the space of continuous real-valued functions on the closed interval $[0, 1]$. For $f, g \in V$, define
   $$H(f, g) = \int_0^1 f(t)g(t)dt.$$

   (b) Let V be a vector space over F, and let $J \in \mathcal{B}(V)$ be nonzero. Define $H: V \times V \to F$ by
   $$H(x, y) = [J(x, y)]^2 \quad \text{for all } x, y \in V.$$

(c)   Define $H \colon R \times R \to R$ by $H(t_1, t_2) = t_1 + 2t_2$.

(d)   Consider the vectors of R^2 as column vectors, and let $H \colon R^2 \to R$ be the function defined by $H(x, y) = \det(x, y)$, the determinant of the 2×2 matrix with columns x and y.

(e)   Let V be a real inner product space, and let $H \colon V \times V \to R$ be the function defined by $H(x, y) = \langle x, y \rangle$ for $x, y \in V$.

(f)   Let V be a complex inner product space, and let $H \colon V \times V \to C$ be the function defined by $H(x, y) = \langle x, y \rangle$ for $x, y \in V$.

5. Verify that each of the given mappings is a bilinear form. Then compute its matrix representation with respect to the given ordered basis β.

(a) $H \colon R^3 \times R^3 \to R$, where

$$
H\left(\begin{pmatrix} a_1 \\ a_2 \\ a_3 \end{pmatrix}, \begin{pmatrix} b_1 \\ b_2 \\ b_3 \end{pmatrix} \right) = a_1 b_1 - 2a_1 b_2 + a_2 b_1 - a_3 b_3
$$

and

$$
\beta = \left\{ \begin{pmatrix} 1 \\ 0 \\ 1 \end{pmatrix}, \begin{pmatrix} 1 \\ 0 \\ -1 \end{pmatrix}, \begin{pmatrix} 0 \\ 1 \\ 0 \end{pmatrix} \right\}.
$$

(b)   Let $V = M_{2 \times 2}(R)$ and

$$
\beta = \left\{ \begin{pmatrix} 1 & 0 \\ 0 & 0 \end{pmatrix}, \begin{pmatrix} 0 & 1 \\ 0 & 0 \end{pmatrix}, \begin{pmatrix} 0 & 0 \\ 1 & 0 \end{pmatrix}, \begin{pmatrix} 0 & 0 \\ 0 & 1 \end{pmatrix} \right\}.
$$

Define $H \colon V \times V \to R$ by $H(A, B) = \operatorname{tr}(A) \cdot \operatorname{tr}(B)$.

(c)   Let $\beta = \{\cos t, \sin t, \cos 2t, \sin 2t\}$. Then β is an ordered basis for $V = \operatorname{span}(\beta)$, a four-dimensional subspace of the space of all continuous functions on R. Let $H \colon V \times V \to R$ be the function defined by $H(f, g) = f'(0) \cdot g''(0)$.

6. Let $H \colon R^2 \to R$ be the function defined by

$$
H\left(\begin{pmatrix} a_1 \\ a_2 \end{pmatrix}, \begin{pmatrix} b_1 \\ b_2 \end{pmatrix} \right) = a_1 b_2 + a_2 b_1 \quad \text{for} \quad \begin{pmatrix} a_1 \\ a_2 \end{pmatrix}, \begin{pmatrix} b_1 \\ b_2 \end{pmatrix} \in R^2.
$$

(a)   Prove that H is a bilinear form.

(b)   Find the 2×2 matrix A such that $H(x, y) = x^t A y$ for all $x, y \in R^2$.

For a 2×2 matrix M with columns x and y, the bilinear form $H(M) = H(x, y)$ is called the **permanent** of M.

7. Let V and W be vector spaces over the same field, and let $T \colon V \to W$ be a linear transformation. For any $H \in \mathcal{B}(W)$, define $\widehat{T}(H) \colon V \times V \to F$ by $\widehat{T}(H)(x, y) = H(T(x), T(y))$ for all $x, y \in V$. Prove the following results.

(a) If $H \in \mathcal{B}(\mathsf{W})$, then $\widehat{\mathsf{T}}(H) \in \mathcal{B}(\mathsf{V})$.

(b) $\widehat{\mathsf{T}} \colon \mathcal{B}(\mathsf{W}) \to \mathcal{B}(\mathsf{V})$ is a linear transformation.

(c) If T is an isomorphism, then so is $\widehat{\mathsf{T}}$.

8. Assume the notation of Theorem 6.32.

(a) Prove that for any ordered basis β, ψ_β is linear.

(b) Let β be an ordered basis for an n-dimensional space V over F, and let $\phi_\beta \colon \mathsf{V} \to F^n$ be the standard representation of V with respect to β. For $A \in \mathsf{M}_{n \times n}(F)$, define $H \colon \mathsf{V} \times \mathsf{V} \to F$ by $H(x,y) = [\phi_\beta(x)]^t A[\phi_\beta(y)]$. Prove that $H \in \mathcal{B}(\mathsf{V})$. Can you establish this as a corollary to Exercise 7?

(c) Prove the converse of (b): Let H be a bilinear form on V. If $A = \psi_\beta(H)$, then $H(x,y) = [\phi_\beta(x)]^t A[\phi_\beta(y)]$.

9. (a) Prove Corollary 1 to Theorem 6.32.

(b) For a finite-dimensional vector space V, describe a method for finding an ordered basis for $\mathcal{B}(\mathsf{V})$.

10. Prove Corollary 2 to Theorem 6.32.

11. Prove Corollary 3 to Theorem 6.32.

12. Prove that the relation of congruence is an equivalence relation.

13. The following outline provides an alternative proof to Theorem 6.33.

(a) Suppose that β and γ are ordered bases for a finite-dimensional vector space V, and let Q be the change of coordinate matrix changing γ-coordinates to β-coordinates. Prove that $\phi_\beta = L_Q \phi_\gamma$, where ϕ_β and ϕ_γ are the standard representations of V with respect to β and γ, respectively.

(b) Apply Corollary 2 to Theorem 6.32 to (a) to obtain an alternative proof of Theorem 6.33.

14. Let V be a finite-dimensional vector space and $H \in \mathcal{B}(\mathsf{V})$. Prove that, for any ordered bases β and γ of V, $\operatorname{rank}(\psi_\beta(H)) = \operatorname{rank}(\psi_\gamma(H))$.

15. Prove the following results.

(a) Any square diagonal matrix is symmetric.

(b) Any matrix congruent to a diagonal matrix is symmetric.

(c) the corollary to Theorem 6.35

16. Let V be a vector space over a field F not of characteristic two, and let H be a symmetric bilinear form on V. Prove that if $K(x) = H(x,x)$ is the quadratic form associated with H, then, for all $x, y \in \mathsf{V}$,

$$H(x,y) = \frac{1}{2}[K(x+y) - K(x) - K(y)].$$

17. For each of the given quadratic forms K on a real inner product space
    V, find a symmetric bilinear form H such that $K(x) = H(x, x)$ for all
    $x \in V$. Then find an orthonormal basis β for V such that $\psi_\beta(H)$ is a
    diagonal matrix.

    (a) $K \colon \mathsf{R}^2 \to \mathsf{R}$ defined by $K \begin{pmatrix} t_1 \\ t_2 \end{pmatrix} = -2t_1^2 + 4t_1 t_2 + t_2^2$

    (b) $K \colon \mathsf{R}^2 \to \mathsf{R}$ defined by $K \begin{pmatrix} t_1 \\ t_2 \end{pmatrix} = 7t_1^2 - 8t_1 t_2 + t_2^2$

    (c) $K \colon \mathsf{R}^3 \to \mathsf{R}$ defined by $K \begin{pmatrix} t_1 \\ t_2 \\ t_3 \end{pmatrix} = 3t_1^2 + 3t_2^2 + 3t_3^2 - 2t_1 t_3$

18. Let S be the set of all $(t_1, t_2, t_3) \in \mathsf{R}^3$ for which

    $$3t_1^2 + 3t_2^2 + 3t_3^2 - 2t_1 t_3 + 2\sqrt{2}(t_1 + t_3) + 1 = 0.$$

    Find an orthonormal basis β for R^3 for which the equation relating
    the coordinates of points of S relative to β is simpler. Describe S
    geometrically.

19. Prove the following refinement of Theorem 6.37(d).

    (a)   If $0 < \operatorname{rank}(A) < n$ and A has no negative eigenvalues, then f has
          no local maximum at p.
    (b)   If $0 < \operatorname{rank}(A) < n$ and A has no positive eigenvalues, then f has
          no local minimum at p.

20. Prove the following variation of the second-derivative test for the case
    $n = 2$: Define

    $$D = \left[\frac{\partial^2 f(p)}{\partial t_1^2} \right] \left[\frac{\partial^2 f(p)}{\partial t_2^2} \right] - \left[\frac{\partial^2 f(p)}{\partial t_1 \partial t_2} \right]^2 .$$

    (a)   If $D > 0$ and $\partial^2 f(p)/\partial t_1^2 > 0$, then f has a local minimum at p.
    (b)   If $D > 0$ and $\partial^2 f(p)/\partial t_1^2 < 0$, then f has a local maximum at p.
    (c)   If $D < 0$, then f has no local extremum at p.
    (d)   If $D = 0$, then the test is inconclusive.

    Hint: Observe that, as in Theorem 6.37, $D = \det(A) = \lambda_1 \lambda_2$, where λ_1
    and λ_2 are the eigenvalues of A.

21. Let A and E be in $\mathsf{M}_{n \times n}(F)$, with E an elementary matrix. In Sec-
    tion 3.1, it was shown that AE can be obtained from A by means of
    an elementary column operation. Prove that $E^t A$ can be obtained by
    means of the same elementary operation performed on the rows rather
    than on the columns of A. *Hint:* Note that $E^t A = (A^t E)^t$.

22. For each of the following matrices A with entries from R, find a diagonal matrix D and an invertible matrix Q such that $Q^t A Q = D$.

(a) $\begin{pmatrix} 1 & 3 \\ 3 & 2 \end{pmatrix}$ (b) $\begin{pmatrix} 0 & 1 \\ 1 & 0 \end{pmatrix}$ (c) $\begin{pmatrix} 3 & 1 & 2 \\ 1 & 4 & 0 \\ 2 & 0 & -1 \end{pmatrix}$

Hint for (b): Use an elementary operation other than interchanging columns.

23. Prove that if the diagonal entries of a diagonal matrix are permuted, then the resulting diagonal matrix is congruent to the original one.

24. Let T be a linear operator on a real inner product space V, and define $H: V \times V \to R$ by $H(x, y) = \langle x, \mathsf{T}(y) \rangle$ for all $x, y \in V$.

(a) Prove that H is a bilinear form.
(b) Prove that H is symmetric if and only if T is self-adjoint.
(c) What properties must T have for H to be an inner product on V?
(d) Explain why H may fail to be a bilinear form if V is a complex inner product space.

25. Prove the converse to Exercise 24(a): Let V be a finite-dimensional real inner product space, and let H be a bilinear form on V. Then there exists a unique linear operator T on V such that $H(x, y) = \langle x, \mathsf{T}(y) \rangle$ for all $x, y \in V$. *Hint:* Choose an orthonormal basis β for V, let $A = \psi_\beta(H)$, and let T be the linear operator on V such that $[\mathsf{T}]_\beta = A$. Apply Exercise 8(c) of this section and Exercise 15 of Section 6.2 (p. 355).

26. Prove that the number of distinct equivalence classes of congruent $n \times n$ real symmetric matrices is

$$\frac{(n+1)(n+2)}{2}.$$

6.9* EINSTEIN'S SPECIAL THEORY OF RELATIVITY

As a consequence of physical experiments performed in the latter half of the nineteenth century (most notably the Michelson–Morley experiment of 1887), physicists concluded that *the results obtained in measuring the speed of light are independent of the velocity of the instrument used to measure the speed of light.* For example, suppose that while on Earth, an experimenter measures the speed of light emitted from the sun and finds it to be 186,000 miles per second. Now suppose that the experimenter places the measuring equipment in a spaceship that leaves Earth traveling at 100,000 miles per second in a direction away from the sun. A repetition of the same experiment from the spaceship yields the same result: Light is traveling at 186,000 miles per second relative to the spaceship, rather than 86,000 miles per second as one might expect!

This revelation led to a new way of relating coordinate systems used to locate events in space–time. The result was Albert Einstein's *special theory of relativity*. In this section, we develop via a linear algebra viewpoint the essence of Einstein's theory.

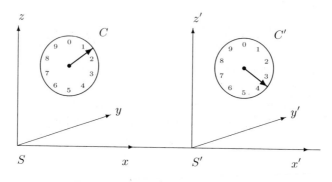

Figure 6.8

The basic problem is to compare two different inertial (nonaccelerating) coordinate systems S and S' in three-space (R^3) that are in motion relative to each other under the assumption that the speed of light is the same when measured in either system. We assume that S' moves at a constant velocity in relation to S as measured from S. (See Figure 6.8.) To simplify matters, let us suppose that the following conditions hold:

1. The corresponding axes of S and S' (x and x', y and y', z and z') are parallel, and the origin of S' moves in the positive direction of the x-axis of S at a constant velocity $v > 0$ relative to S.

2. Two clocks C and C' are placed in space—the first stationary relative to the coordinate system S and the second stationary relative to the coordinate system S'. These clocks are designed to give real numbers in units of seconds as readings. The clocks are calibrated so that at the instant the origins of S and S' coincide, both clocks give the reading zero.

3. The unit of length is the **light second** (the distance light travels in 1 second), and the unit of time is the second. Note that, with respect to these units, the speed of light is 1 light second per second.

Given any event (something whose position and time of occurrence can be described), we may assign a set of *space–time coordinates* to it. For example, if p is an event that occurs at position

$$\begin{pmatrix} x \\ y \\ z \end{pmatrix}$$

relative to S and at time t as read on clock C, we can assign to p the set of coordinates

$$\begin{pmatrix} x \\ y \\ z \\ t \end{pmatrix}.$$

This ordered 4-tuple is called the **space–time coordinates** of p relative to S and C. Likewise, p has a set of space–time coordinates

$$\begin{pmatrix} x' \\ y' \\ z' \\ t' \end{pmatrix}$$

relative to S' and C'.

For a fixed velocity v, let $\mathsf{T}_v \colon \mathsf{R}^4 \to \mathsf{R}^4$ be the mapping defined by

$$\mathsf{T}_v \begin{pmatrix} x \\ y \\ z \\ t \end{pmatrix} = \begin{pmatrix} x' \\ y' \\ z' \\ t' \end{pmatrix},$$

where

$$\begin{pmatrix} x \\ y \\ z \\ t \end{pmatrix} \quad \text{and} \quad \begin{pmatrix} x' \\ y' \\ z' \\ t' \end{pmatrix}$$

are the space–time coordinates of the same event with respect to S and C and with respect to S' and C', respectively.

Einstein made certain assumptions about T_v that led to his special theory of relativity. We formulate an equivalent set of assumptions.

Axioms of the Special Theory of Relativity

(R 1) The speed of any light beam, when measured in either coordinate system using a clock stationary relative to that coordinate system, is 1.

(R 2) The mapping $T_v \colon R^4 \to R^4$ is an isomorphism.

(R 3) If

$$T_v \begin{pmatrix} x \\ y \\ z \\ t \end{pmatrix} = \begin{pmatrix} x' \\ y' \\ z' \\ t' \end{pmatrix},$$

then $y' = y$ and $z' = z$.

(R 4) If

$$T_v \begin{pmatrix} x \\ y_1 \\ z_1 \\ t \end{pmatrix} = \begin{pmatrix} x' \\ y' \\ z' \\ t' \end{pmatrix} \quad \text{and} \quad T_v \begin{pmatrix} x \\ y_2 \\ z_2 \\ t \end{pmatrix} = \begin{pmatrix} x'' \\ y'' \\ z'' \\ t'' \end{pmatrix},$$

then $x'' = x'$ and $t'' = t'$.

(R 5) The origin of S moves in the negative direction of the x'-axis of S' at the constant velocity $-v < 0$ as measured from S'.

Axioms (R 3) and (R 4) tell us that for $p \in R^4$, the second and third coordinates of $T_v(p)$ are unchanged and the first and fourth coordinates of $T_v(p)$ are independent of the second and third coordinates of p.

As we will see, these five axioms completely characterize T_v. The operator T_v is called the **Lorentz transformation** in direction x. We intend to compute T_v and use it to study the curious phenomenon of time contraction.

Theorem 6.39. *On R^4, the following statements are true.*

(a) $T_v(e_i) = e_i$ *for $i = 2, 3$.*

(b) $\mathrm{span}(\{e_2, e_3\})$ *is T_v-invariant.*

(c) $\mathrm{span}(\{e_1, e_4\})$ *is T_v-invariant.*

(d) *Both $\mathrm{span}(\{e_2, e_3\})$ and $\mathrm{span}(\{e_1, e_4\})$ are T_v^*-invariant.*

(e) $T_v^*(e_i) = e_i$ *for $i = 2, 3$.*

Proof. (a) By axiom (R 2),

$$T_v \begin{pmatrix} 0 \\ 0 \\ 0 \\ 0 \end{pmatrix} = \begin{pmatrix} 0 \\ 0 \\ 0 \\ 0 \end{pmatrix},$$

and hence, by axiom (R 4), the first and fourth coordinates of

$$T_v \begin{pmatrix} 0 \\ a \\ b \\ 0 \end{pmatrix}$$

are both zero for any $a, b \in R$. Thus, by axiom (R 3),

$$\mathsf{T}_v \begin{pmatrix} 0 \\ 1 \\ 0 \\ 0 \end{pmatrix} = \begin{pmatrix} 0 \\ 1 \\ 0 \\ 0 \end{pmatrix} \quad \text{and} \quad \mathsf{T}_v \begin{pmatrix} 0 \\ 0 \\ 1 \\ 0 \end{pmatrix} = \begin{pmatrix} 0 \\ 0 \\ 1 \\ 0 \end{pmatrix}.$$

The proofs of (b), (c), and (d) are left as exercises.

(e) For any $j \neq 2$, $\langle \mathsf{T}_v^*(e_2), e_j \rangle = \langle e_2, \mathsf{T}_v(e_j) \rangle = 0$ by (a) and (c); for $j = 2$, $\langle \mathsf{T}_v^*(e_2), e_j \rangle = \langle e_2, \mathsf{T}_v(e_2) \rangle = \langle e_2, e_2 \rangle = 1$ by (a). We conclude that $\mathsf{T}_v^*(e_2)$ is a multiple of e_2 (i.e., that $\mathsf{T}_v^*(e_2) = ke_2$ for some $k \in R$). Thus,

$$1 = \langle e_2, e_2 \rangle = \langle e_2, \mathsf{T}_v(e_2) \rangle = \langle \mathsf{T}_v^*(e_2), e_2 \rangle = \langle ke_2, e_2 \rangle = k,$$

and hence $\mathsf{T}_v^*(e_2) = e_2$. Similarly, $\mathsf{T}_v^*(e_3) = e_3$. ∎

Suppose that, at the instant the origins of S and S' coincide, a light flash is emitted from their common origin. The event of the light flash when measured either relative to S and C or relative to S' and C' has space–time coordinates

$$\begin{pmatrix} 0 \\ 0 \\ 0 \\ 0 \end{pmatrix}.$$

Let P be the set of all events whose space–time coordinates

$$\begin{pmatrix} x \\ y \\ z \\ t \end{pmatrix}$$

relative to S and C are such that the flash is observable from the point with coordinates

$$\begin{pmatrix} x \\ y \\ z \end{pmatrix}$$

(as measured relative to S) at the time t (as measured on C). Let us characterize P in terms of x, y, z, and t. Since the speed of light is 1, at any time $t \geq 0$ the light flash is observable from any point whose distance to the origin of S (as measured on S) is $t \cdot 1 = t$. These are precisely the points that lie on the sphere of radius t with center at the origin. The coordinates (relative to

S) of such points satisfy the equation $x^2 + y^2 + z^2 - t^2 = 0$. Hence an event lies in P if and only if its space–time coordinates

$$\begin{pmatrix} x \\ y \\ z \\ t \end{pmatrix} \quad (t \geq 0)$$

relative to S and C satisfy the equation $x^2 + y^2 + z^2 - t^2 = 0$. By virtue of axiom (R 1), we can characterize P in terms of the space–time coordinates relative to S' and C' similarly: An event lies in P if and only if, relative to S' and C', its space–time coordinates

$$\begin{pmatrix} x' \\ y' \\ z' \\ t' \end{pmatrix} \quad (t \geq 0)$$

satisfy the equation $(x')^2 + (y')^2 + (z')^2 - (t')^2 = 0$.

Let

$$A = \begin{pmatrix} 1 & 0 & 0 & 0 \\ 0 & 1 & 0 & 0 \\ 0 & 0 & 1 & 0 \\ 0 & 0 & 0 & -1 \end{pmatrix}.$$

Theorem 6.40. *If* $\langle \mathsf{L}_A(w), w \rangle = 0$ *for some* $w \in \mathsf{R}^4$, *then*

$$\langle \mathsf{T}_v^* \mathsf{L}_A \mathsf{T}_v(w), w \rangle = 0.$$

Proof. Let

$$w = \begin{pmatrix} x \\ y \\ z \\ t \end{pmatrix} \in \mathsf{R}^4,$$

and suppose that $\langle \mathsf{L}_A(w), w \rangle = 0$.

CASE 1. $t \geq 0$. Since $\langle \mathsf{L}_A(w), w \rangle = x^2 + y^2 + z^2 - t^2$, the vector w gives the coordinates of an event in P relative to S and C. Because

$$\begin{pmatrix} x \\ y \\ z \\ t \end{pmatrix} \quad \text{and} \quad \begin{pmatrix} x' \\ y' \\ z' \\ t' \end{pmatrix}$$

are the space–time coordinates of the same event relative to S' and C', the discussion preceding Theorem 6.40 yields

$$(x')^2 + (y')^2 + (z')^2 - (t')^2 = 0.$$

Thus $\langle T_v^* L_A T_v(w), w \rangle = \langle L_A T_v(w), T_v(w) \rangle = (x')^2 + (y')^2 + (z')^2 - (t')^2 = 0,$ and the conclusion follows.

CASE 2. $t < 0$. The proof follows by applying case 1 to $-w$.   ∎

We now proceed to deduce information about T_v. Let

$$w_1 = \begin{pmatrix} 1 \\ 0 \\ 0 \\ 1 \end{pmatrix} \quad \text{and} \quad w_2 = \begin{pmatrix} 1 \\ 0 \\ 0 \\ -1 \end{pmatrix}.$$

By Exercise 3, $\{w_1, w_2\}$ is an orthogonal basis for span($\{e_1, e_4\}$), and span($\{e_1, e_4\}$) is $T_v^* L_A T_v$-invariant. The next result tells us even more.

Theorem 6.41. *There exist nonzero scalars a and b such that*
(a) $T_v^* L_A T_v(w_1) = aw_2.$
(b) $T_v^* L_A T_v(w_2) = bw_1.$

Proof. (a) Because $\langle L_A(w_1), w_1 \rangle = 0$, $\langle T_v^* L_A T_v(w_1), w_1 \rangle = 0$ by Theorem 6.40. Thus $T_v^* L_A T_v(w_1)$ is orthogonal to w_1. Since span($\{e_1, e_4\}$) = span($\{w_1, w_2\}$) is $T_v^* L_A T_v$-invariant, $T_v^* L_A T_v(w_1)$ must lie in this set. But $\{w_1, w_2\}$ is an orthogonal basis for this subspace, and so $T_v^* L_A T_v(w_1)$ must be a multiple of w_2. Thus $T_v^* L_A T_v(w_1) = aw_2$ for some scalar a. Since T_v and A are invertible, so is $T_v^* L_A T_v$. Thus $a \neq 0$, proving (a).

The proof of (b) is similar to (a).   ∎

Corollary. *Let $B_v = [T_v]_\beta$, where β is the standard ordered basis for R^4. Then*
(a) $B_v^* A B_v = A.$
(b) $T_v^* L_A T_v = L_A.$

We leave the proof of the corollary as an exercise. For hints, see Exercise 4.

Now consider the situation 1 second after the origins of S and S' have coincided as measured by the clock C. Since the origin of S' is moving along the x-axis at a velocity v as measured in S, its space–time coordinates relative to S and C are

$$\begin{pmatrix} v \\ 0 \\ 0 \\ 1 \end{pmatrix}.$$

Similarly, the space–time coordinates for the origin of S' relative to S' and C' must be

$$\begin{pmatrix} 0 \\ 0 \\ 0 \\ t' \end{pmatrix}$$

for some $t' > 0$. Thus we have

$$\mathsf{T}_v \begin{pmatrix} v \\ 0 \\ 0 \\ 1 \end{pmatrix} = \begin{pmatrix} 0 \\ 0 \\ 0 \\ t' \end{pmatrix} \qquad \text{for some } t' > 0. \tag{18}$$

By the corollary to Theorem 6.41,

$$\left\langle \mathsf{T}_v^* \mathsf{L}_A \mathsf{T}_v \begin{pmatrix} v \\ 0 \\ 0 \\ 1 \end{pmatrix}, \begin{pmatrix} v \\ 0 \\ 0 \\ 1 \end{pmatrix} \right\rangle = \left\langle \mathsf{L}_A \begin{pmatrix} v \\ 0 \\ 0 \\ 1 \end{pmatrix}, \begin{pmatrix} v \\ 0 \\ 0 \\ 1 \end{pmatrix} \right\rangle = v^2 - 1. \tag{19}$$

But also

$$\left\langle \mathsf{T}_v^* \mathsf{L}_A \mathsf{T}_v \begin{pmatrix} v \\ 0 \\ 0 \\ 1 \end{pmatrix}, \begin{pmatrix} v \\ 0 \\ 0 \\ 1 \end{pmatrix} \right\rangle = \left\langle \mathsf{L}_A \mathsf{T}_v \begin{pmatrix} v \\ 0 \\ 0 \\ 1 \end{pmatrix}, \mathsf{T}_v \begin{pmatrix} v \\ 0 \\ 0 \\ 1 \end{pmatrix} \right\rangle$$

$$= \left\langle \mathsf{L}_A \begin{pmatrix} 0 \\ 0 \\ 0 \\ t' \end{pmatrix}, \begin{pmatrix} 0 \\ 0 \\ 0 \\ t' \end{pmatrix} \right\rangle = -(t')^2. \tag{20}$$

Combining (19) and (20), we conclude that $v^2 - 1 = -(t')^2$, or

$$t' = \sqrt{1 - v^2}. \tag{21}$$

Thus, from (18) and (21), we obtain

$$\mathsf{T}_v \begin{pmatrix} v \\ 0 \\ 0 \\ 1 \end{pmatrix} = \begin{pmatrix} 0 \\ 0 \\ 0 \\ \sqrt{1 - v^2} \end{pmatrix}. \tag{22}$$

Next recall that the origin of S moves in the negative direction of the x'-axis of S' at the constant velocity $-v < 0$ as measured from S'. [This fact is axiom (R 5).] Consequently, 1 second after the origins of S and S' have coincided as measured on clock C, there exists a time $t'' > 0$ as measured on clock C' such that

$$
\mathsf{T}_v \begin{pmatrix} 0 \\ 0 \\ 0 \\ 1 \end{pmatrix} = \begin{pmatrix} -vt'' \\ 0 \\ 0 \\ t'' \end{pmatrix}.
\tag{23}
$$

From (23), it follows in a manner similar to the derivation of (22) that

$$
t'' = \frac{1}{\sqrt{1 - v^2}};
\tag{24}
$$

hence, from (23) and (24),

$$
\mathsf{T}_v \begin{pmatrix} 0 \\ 0 \\ 0 \\ 1 \end{pmatrix} = \begin{pmatrix} \dfrac{-v}{\sqrt{1 - v^2}} \\ 0 \\ 0 \\ \dfrac{1}{\sqrt{1 - v^2}} \end{pmatrix}.
\tag{25}
$$

The following result is now easily proved using (22), (25), and Theorem 6.39.

Theorem 6.42.   *Let β be the standard ordered basis for R^4. Then*

$$
[\mathsf{T}_v]_\beta = B_v = \begin{pmatrix} \dfrac{1}{\sqrt{1 - v^2}} & 0 & 0 & \dfrac{-v}{\sqrt{1 - v^2}} \\ 0 & 1 & 0 & 0 \\ 0 & 0 & 1 & 0 \\ \dfrac{-v}{\sqrt{1 - v^2}} & 0 & 0 & \dfrac{1}{\sqrt{1 - v^2}} \end{pmatrix}.
$$

Time Contraction

A most curious and paradoxical conclusion follows if we accept Einstein's theory. Suppose that an astronaut leaves our solar system in a space vehicle traveling at a fixed velocity v as measured relative to our solar system. It follows from Einstein's theory that, at the end of time t as measured on Earth, the time that passes on the space vehicle is only $t\sqrt{1 - v^2}$. To establish this result, consider the coordinate systems S and S' and clocks C and C' that we have been studying. Suppose that the origin of S' coincides with the space vehicle and the origin of S coincides with a point in the solar system (stationary relative to the sun) so that the origins of S and S' coincide and clocks C and C' read zero at the moment the astronaut embarks on the trip.

As viewed from S, the space–time coordinates of the vehicle at any time $t > 0$ as measured by C are

$$\begin{pmatrix} vt \\ 0 \\ 0 \\ t \end{pmatrix},$$

whereas, as viewed from S', the space–time coordinates of the vehicle at any time $t' > 0$ as measured by C' are

$$\begin{pmatrix} 0 \\ 0 \\ 0 \\ t' \end{pmatrix}.$$

But if two sets of space–time coordinates

$$\begin{pmatrix} vt \\ 0 \\ 0 \\ t \end{pmatrix} \quad \text{and} \quad \begin{pmatrix} 0 \\ 0 \\ 0 \\ t' \end{pmatrix}$$

are to describe the same event, it must follow that

$$\mathsf{T}_v \begin{pmatrix} vt \\ 0 \\ 0 \\ t \end{pmatrix} = \begin{pmatrix} 0 \\ 0 \\ 0 \\ t' \end{pmatrix}.$$

Thus

$$[\mathsf{T}_v]_\beta = B_v = \begin{pmatrix} \dfrac{1}{\sqrt{1-v^2}} & 0 & 0 & \dfrac{-v}{\sqrt{1-v^2}} \\ 0 & 1 & 0 & 0 \\ 0 & 0 & 1 & 0 \\ \dfrac{-v}{\sqrt{1-v^2}} & 0 & 0 & \dfrac{1}{\sqrt{1-v^2}} \end{pmatrix} \begin{pmatrix} vt \\ 0 \\ 0 \\ t \end{pmatrix} = \begin{pmatrix} 0 \\ 0 \\ 0 \\ t' \end{pmatrix}.$$

From the preceding equation, we obtain $\dfrac{-v^2 t}{\sqrt{1-v^2}} + \dfrac{t}{\sqrt{1-v^2}} = t'$, or

$$t' = t\sqrt{1-v^2}. \tag{26}$$

This is the desired result.

A dramatic consequence of time contraction is that distances are contracted along the line of motion (see Exercise 9).

Let us make one additional point. Suppose that we consider units of distance and time more commonly used than the light second and second, such as the mile and hour, or the kilometer and second. Let c denote the speed of light relative to our chosen units of distance. It is easily seen that if an object travels at a velocity v relative to a set of units, then it is traveling at a velocity v/c in units of light seconds per second. Thus, for an arbitrary set of units of distance and time, (26) becomes

$$t' = t\sqrt{1 - \frac{v^2}{c^2}}.$$

EXERCISES

1. Prove (b), (c), and (d) of Theorem 6.39.

2. Complete the proof of Theorem 6.40 for the case $t < 0$.

3. For

$$w_1 = \begin{pmatrix} 1 \\ 0 \\ 0 \\ 1 \end{pmatrix} \quad \text{and} \quad w_2 = \begin{pmatrix} 1 \\ 0 \\ 0 \\ -1 \end{pmatrix},$$

   show that

   (a) $\{w_1, w_2\}$ is an orthogonal basis for span($\{e_1, e_4\}$);
   (b) span($\{e_1, e_4\}$) is $T_v^* L_A T_v$-invariant.

4. Prove the corollary to Theorem 6.41.

   Hints:

   (a) Prove that

   $$B_v^* A B_v = \begin{pmatrix} p & 0 & 0 & q \\ 0 & 1 & 0 & 0 \\ 0 & 0 & 1 & 0 \\ -q & 0 & 0 & -p \end{pmatrix},$$

   where

   $$p = \frac{a+b}{2} \quad \text{and} \quad q = \frac{a-b}{2}.$$

(b) Show that $q = 0$ by using the fact that $B_v^* A B_v$ is self-adjoint.

(c) Apply Theorem 6.40 to

$$
w = \begin{pmatrix} 0 \\ 1 \\ 0 \\ 1 \end{pmatrix}
$$

to show that $p = 1$.

5. Derive (24), and prove that

$$
\mathsf{T}_v \begin{pmatrix} 0 \\ 0 \\ 0 \\ 1 \end{pmatrix} = \begin{pmatrix} \dfrac{-v}{\sqrt{1-v^2}} \\ 0 \\ 0 \\ 1 \\ \dfrac{1}{\sqrt{1-v^2}} \end{pmatrix}. \tag{25}
$$

Hint: Use a technique similar to the derivation of (22).

6. Consider three coordinate systems S, S', and S'' with the corresponding axes $(x, x', x''; \; y, y', y''; \;$ and $z, z', z'')$ parallel and such that the x-, x'-, and x''-axes coincide. Suppose that S' is moving past S at a velocity $v_1 > 0$ (as measured on S), S'' is moving past S' at a velocity $v_2 > 0$ (as measured on S'), and S'' is moving past S at a velocity $v_3 > 0$ (as measured on S), and that there are three clocks C, C', and C'' such that C is stationary relative to S, C' is stationary relative to S', and C'' is stationary relative to S''. Suppose that when measured on any of the three clocks, all the origins of S, S', and S'' coincide at time 0. Assuming that $\mathsf{T}_{v_3} = \mathsf{T}_{v_2} \mathsf{T}_{v_1}$ (i.e., $B_{v_3} = B_{v_2} B_{v_1}$), prove that

$$
v_3 = \frac{v_1 + v_2}{1 + v_1 v_2}.
$$

Note that substituting $v_2 = 1$ in this equation yields $v_3 = 1$. This tells us that the speed of light as measured in S or S' is the same. Why would we be surprised if this were not the case?

7. Compute $(B_v)^{-1}$. Show $(B_v)^{-1} = B_{(-v)}$. Conclude that if S' moves at a negative velocity v relative to S, then $[\mathsf{T}_v]_\beta = B_v$, where B_v is of the form given in Theorem 6.42.

8. Suppose that an astronaut left Earth in the year 2000 and traveled to a star 99 light years away from Earth at 99% of the speed of light and that upon reaching the star immediately turned around and returned to Earth at the same speed. Assuming Einstein's special theory of

relativity, show that if the astronaut was 20 years old at the time of departure, then he or she would return to Earth at age 48.2 in the year 2200. Explain the use of Exercise 7 in solving this problem.

9. Recall the moving space vehicle considered in the study of time contraction. Suppose that the vehicle is moving toward a fixed star located on the x-axis of S at a distance b units from the origin of S. If the space vehicle moves toward the star at velocity v, Earthlings (who remain "almost" stationary relative to S) compute the time it takes for the vehicle to reach the star as $t = b/v$. Due to the phenomenon of time contraction, the astronaut perceives a time span of $t' = t\sqrt{1 - v^2} = (b/v)\sqrt{1 - v^2}$. A paradox appears in that the astronaut perceives a time span inconsistent with a distance of b and a velocity of v. The paradox is resolved by observing that the distance from the solar system to the star as measured by the astronaut is less than b.

Assuming that the coordinate systems S and S' and clocks C and C' are as in the discussion of time contraction, prove the following results.

(a) At time t (as measured on C), the space–time coordinates of star relative to S and C are

$$\begin{pmatrix} b \\ 0 \\ 0 \\ t \end{pmatrix}.$$

(b) At time t (as measured on C), the space–time coordinates of the star relative to S' and C' are

$$\begin{pmatrix} \dfrac{b - vt}{\sqrt{1 - v^2}} \\ 0 \\ 0 \\ \dfrac{t - bv}{\sqrt{1 - v^2}} \end{pmatrix}.$$

(c) For

$$x' = \frac{b - tv}{\sqrt{1 - v^2}} \quad \text{and} \quad t' = \frac{t - bv}{\sqrt{1 - v^2}},$$

we have $x' = b\sqrt{1 - v^2} - t'v$.

This result may be interpreted to mean that at time t' as measured by the astronaut, the distance from the astronaut to the star as measured by the astronaut (see Figure 6.9) is

$$b\sqrt{1 - v^2} - t'v.$$

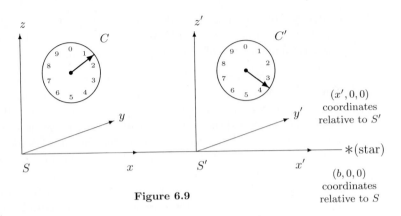

Figure 6.9

(d) Conclude from the preceding equation that

(1) the speed of the space vehicle relative to the star, as measured by the astronaut, is v;

(2) the distance from Earth to the star, as measured by the astronaut, is $b\sqrt{1 - v^2}$.

Thus distances along the line of motion of the space vehicle appear to be contracted by a factor of $\sqrt{1 - v^2}$.

6.10* CONDITIONING AND THE RAYLEIGH QUOTIENT

In Section 3.4, we studied specific techniques that allow us to solve systems of linear equations in the form $Ax = b$, where A is an $m \times n$ matrix and b is an $m \times 1$ vector. Such systems often arise in applications to the real world. The coefficients in the system are frequently obtained from experimental data, and, in many cases, both m and n are so large that a computer must be used in the calculation of the solution. Thus two types of errors must be considered. First, experimental errors arise in the collection of data since no instruments can provide completely accurate measurements. Second, computers introduce roundoff errors. One might intuitively feel that small relative changes in the coefficients of the system cause small relative errors in the solution. A system that has this property is called **well-conditioned**; otherwise, the system is called **ill-conditioned**.

We now consider several examples of these types of errors, concentrating primarily on changes in b rather than on changes in the entries of A. In addition, we assume that A is a square, complex (or real), invertible matrix since this is the case most frequently encountered in applications.

Example 1

Consider the system

$$x_1 + x_2 = 5$$
$$x_1 - x_2 = 1.$$

The solution to this system is

$$\begin{pmatrix} 3 \\ 2 \end{pmatrix}.$$

Now suppose that we change the system somewhat and consider the new system

$$x_1 + x_2 = 5$$
$$x_1 - x_2 = 1.0001.$$

This modified system has the solution

$$\begin{pmatrix} 3.00005 \\ 1.99995 \end{pmatrix}.$$

We see that a change of 10^{-4} in one coefficient has caused a change of less than 10^{-4} in each coordinate of the new solution. More generally, the system

$$x_1 + x_2 = 5$$
$$x_1 - x_2 = 1 + \delta$$

has the solution

$$\begin{pmatrix} 3 + \delta/2 \\ 2 - \delta/2 \end{pmatrix}.$$

Hence small changes in b introduce small changes in the solution. Of course, we are really interested in *relative changes* since a change in the solution of, say, 10, is considered large if the original solution is of the order 10^{-2}, but small if the original solution is of the order 10^6.

We use the notation δb to denote the vector $b' - b$, where b is the vector in the original system and b' is the vector in the modified system. Thus we have

$$\delta b = \begin{pmatrix} 5 \\ 1+h \end{pmatrix} - \begin{pmatrix} 5 \\ 1 \end{pmatrix} = \begin{pmatrix} 0 \\ h \end{pmatrix}.$$

We now define the **relative change** in b to be the scalar $\|\delta b\|/\|b\|$, where $\|\cdot\|$ denotes the standard norm on $\mathbf{C}^n$ (or $\mathbf{R}^n$); that is, $\|b\| = \sqrt{\langle b, b \rangle}$. Most of what follows, however, is true for any norm. Similar definitions hold for the **relative change** in x. In this example,

$$\frac{\|\delta b\|}{\|b\|} = \frac{|h|}{\sqrt{26}} \quad \text{and} \quad \frac{\|\delta x\|}{\|x\|} = \frac{\left\| \begin{pmatrix} 3 + (h/2) \\ 2 - (h/2) \end{pmatrix} - \begin{pmatrix} 3 \\ 2 \end{pmatrix} \right\|}{\left\| \begin{pmatrix} 3 \\ 2 \end{pmatrix} \right\|} = \frac{|h|}{\sqrt{26}}.$$

Thus the relative change in x equals, coincidentally, the relative change in b; so the system is well-conditioned.   ◆

Example 2

Consider the system

$$\begin{array}{rcl} x_1 + x_2 & = & 3 \\ x_1 + 1.00001x_2 & = & 3.00001, \end{array}$$

which has

$$\begin{pmatrix} 2 \\ 1 \end{pmatrix}$$

as its solution. The solution to the related system

$$\begin{array}{l} x_1 + x_2 = 3 \\ x_1 + 1.00001x_2 = 3.00001 + \delta \end{array}$$

is

$$\begin{pmatrix} 2 - (10^5)h \\ 1 + (10^5)h \end{pmatrix}.$$

Hence,

$$\frac{\|\delta x\|}{\|x\|} = 10^5 \sqrt{2/5}\, |h| \geq 10^4 |h|,$$

while

$$\frac{\|\delta b\|}{\|b\|} \approx \frac{|h|}{3\sqrt{2}}.$$

Thus the relative change in x is at least 10^4 times the relative change in b! This system is very ill-conditioned. Observe that the lines defined by the two equations are nearly coincident. So a small change in either line could greatly alter the point of intersection, that is, the solution to the system.   ◆

To apply the full strength of the theory of self-adjoint matrices to the study of conditioning, we need the notion of the norm of a matrix. (See Exercise 24 of Section 6.1 for further results about norms.)

Definition.   *Let A be a complex (or real) $n \times n$ matrix. Define the* **(Euclidean) norm** *of A by*

$$\|A\| = \max_{x \neq 0} \frac{\|Ax\|}{\|x\|},$$

where $x \in C^n$ or $x \in R^n$.

Intuitively, $\|A\|$ represents the maximum *magnification* of a vector by the matrix A. The question of whether or not this maximum exists, as well as the problem of how to compute it, can be answered by the use of the so-called *Rayleigh quotient.*

Definition.   *Let B be an $n \times n$ self-adjoint matrix. The* **Rayleigh quotient** *for $x \neq 0$ is defined to be the scalar $R(x) = \langle Bx, x \rangle / \|x\|^2$.*

The following result characterizes the extreme values of the Rayleigh quotient of a self-adjoint matrix.

Theorem 6.43. *For a self-adjoint matrix $B \in \mathsf{M}_{n \times n}(F)$, we have that $\max_{x \neq 0} R(x)$ is the largest eigenvalue of B and $\min_{x \neq 0} R(x)$ is the smallest eigenvalue of B.*

Proof. By Theorems 6.19 (p. 384) and 6.20 (p. 384), we may choose an orthonormal basis $\{v_1, v_2, \ldots, v_n\}$ of eigenvectors of B such that $Bv_i = \lambda_i v_i$ $(1 \leq i \leq n)$, where $\lambda_1 \geq \lambda_2 \geq \cdots \geq \lambda_n$. (Recall that by the lemma to Theorem 6.17, p. 373, the eigenvalues of B are real.) Now, for $x \in F^n$, there exist scalars $a_1, a_2, \ldots, a_n$ such that

$$x = \sum_{i=1}^{n} a_i v_i.$$

Hence

$$
\begin{aligned}
R(x) &= \frac{\langle Bx, x \rangle}{\|x\|^2} = \frac{\left\langle \sum_{i=1}^{n} a_i \lambda_i v_i, \sum_{j=1}^{n} a_j v_j \right\rangle}{\|x\|^2} \\
&= \frac{\sum_{i=1}^{n} \lambda_i |a_i|^2}{\|x\|^2} \leq \frac{\lambda_1 \sum_{i=1}^{n} |a_i|^2}{\|x\|^2} = \frac{\lambda_1 \|x\|^2}{\|x\|^2} = \lambda_1.
\end{aligned}
$$

It is easy to see that $R(v_1) = \lambda_1$, so we have demonstrated the first half of the theorem. The second half is proved similarly. ∎

Corollary 1. *For any square matrix A, $\|A\|$ is finite and, in fact, equals $\sqrt{\lambda}$, where λ is the largest eigenvalue of A^*A.*

Proof. Let B be the self-adjoint matrix A^*A, and let λ be the largest eigenvalue of B. Since, for $x \neq 0$,

$$0 \leq \frac{\|Ax\|^2}{\|x\|^2} = \frac{\langle Ax, Ax \rangle}{\|x\|^2} = \frac{\langle A^*Ax, x \rangle}{\|x\|^2} = \frac{\langle Bx, x \rangle}{\|x\|^2} = R(x),$$

it follows from Theorem 6.43 that $\|A\|^2 = \lambda$.   ∎

Observe that the proof of Corollary 1 shows that all the eigenvalues of A^*A are nonnegative. For our next result, we need the following lemma.

Lemma. *For any square matrix A, λ is an eigenvalue of A^*A if and only if λ is an eigenvalue of AA^*.*

Proof. Let λ be an eigenvalue of A^*A. If $\lambda = 0$, then A^*A is not invertible. Hence A and A^* are not invertible, so that λ is also an eigenvalue of AA^*. The proof of the converse is similar.

Suppose now that $\lambda \neq 0$. Then there exists $x \neq 0$ such that $A^*Ax = \lambda x$. Apply A to both sides to obtain $(AA^*)(Ax) = \lambda(Ax)$. Since $Ax \neq 0$ (lest $\lambda x = 0$), we have that λ is an eigenvalue of AA^*. The proof of the converse is left as an exercise.   ∎

Corollary 2. *Let A be an invertible matrix. Then $\|A^{-1}\| = 1/\sqrt{\lambda}$, where λ is the smallest eigenvalue of A^*A.*

Proof. Recall that λ is an eigenvalue of an invertible matrix if and only if λ^{-1} is an eigenvalue of its inverse.

Now let $\lambda_1 \geq \lambda_2 \geq \cdots \geq \lambda_n$ be the eigenvalues of A^*A, which by the lemma are the eigenvalues of AA^*. Then $\|A^{-1}\|^2$ equals the largest eigenvalue of $(A^{-1})^*A^{-1} = (AA^*)^{-1}$, which equals $1/\lambda_n$.   ∎

For many applications, it is only the largest and smallest eigenvalues that are of interest. For example, in the case of vibration problems, the smallest eigenvalue represents the lowest frequency at which vibrations can occur.

We see the role of both of these eigenvalues in our study of conditioning.

Example 3

Let

$$A = \begin{pmatrix} 1 & 0 & 1 \\ -1 & 1 & 0 \\ 0 & 1 & 1 \end{pmatrix}.$$

Then

$$B = A^*A = \begin{pmatrix} 2 & -1 & 1 \\ -1 & 2 & 1 \\ 1 & 1 & 2 \end{pmatrix}.$$

The eigenvalues of B are 3, 3, and 0. Therefore, $\|A\| = \sqrt{3}$. For any

$$x = \begin{pmatrix} a \\ b \\ c \end{pmatrix} \neq 0,$$

we may compute $R(x)$ for the matrix B as

$$3 \geq R(x) = \frac{\langle Bx, x \rangle}{\|x\|^2} = \frac{2(a^2 + b^2 + c^2 - ab + ac + bc)}{a^2 + b^2 + c^2}. \quad \blacklozenge$$

Now that we know $\|A\|$ exists for every square matrix A, we can make use of the inequality $\|Ax\| \leq \|A\| \cdot \|x\|$, which holds for every x.

Assume in what follows that A is invertible, $b \neq 0$, and $Ax = b$. For a given δb, let δx be the vector that satisfies $A(x + \delta x) = b + \delta b$. Then $A(\delta x) = \delta b$, and so $\delta x = A^{-1}(\delta b)$. Hence

$$\|b\| = \|Ax\| \leq \|A\| \cdot \|x\| \quad \text{and} \quad \|\delta x\| = \|A^{-1}(\delta b)\| \leq \|A^{-1}\| \cdot \|\delta b\|.$$

Thus

$$\frac{\|\delta x\|}{\|x\|} \leq \frac{\|x\|^2}{\|b\|/\|A\|} \leq \frac{\|A^{-1}\| \cdot \|\delta b\| \cdot \|A\|}{\|b\|} = \|A\| \cdot \|A^{-1}\| \cdot \left(\frac{\|\delta b\|}{\|b\|}\right).$$

Similarly (see Exercise 9),

$$\frac{1}{\|A\| \cdot \|A^{-1}\|} \left(\frac{\|\delta b\|}{\|b\|}\right) \leq \frac{\|\delta x\|}{\|x\|}.$$

The number $\|A\| \cdot \|A^{-1}\|$ is called the **condition number** of A and is denoted cond(A). It should be noted that the definition of cond(A) depends on how the norm of A is defined. There are many reasonable ways of defining the norm of a matrix. In fact, the only property needed to establish the inequalities above is that $\|Ax\| \leq \|A\| \cdot \|x\|$ for all x. We summarize these results in the following theorem.

Theorem 6.44. For the system $Ax = b$, where A is invertible and $b \neq 0$, the following statements are true.

(a) For any norm $\|\cdot\|$, we have $\dfrac{1}{\text{cond}(A)} \dfrac{\|\delta b\|}{\|b\|} \leq \dfrac{\|\delta x\|}{\|x\|} \leq \text{cond}(A) \dfrac{\|\delta b\|}{\|b\|}.$

(b) *If $\| \cdot \|$ is the Euclidean norm, then $\mathrm{cond}(A) = \sqrt{\lambda_1/\lambda_n}$, where λ_1 and λ_n are the largest and smallest eigenvalues, respectively, of A^*A.*

Proof. Statement (a) follows from the previous inequalities, and (b) follows from Corollaries 1 and 2 to Theorem 6.43. ∎

It is clear from Theorem 6.44 that $\mathrm{cond}(A) \geq 1$. It is left as an exercise to prove that $\mathrm{cond}(A) = 1$ if and only if A is a scalar multiple of a unitary or orthogonal matrix. Moreover, it can be shown with some work that equality can be obtained in (a) by an appropriate choice of b and δb.

We can see immediately from (a) that if $\mathrm{cond}(A)$ is close to 1, then a small relative error in b forces a small relative error in x. If $\mathrm{cond}(A)$ is large, however, then the relative error in x may be small even though the relative error in b is large, or the relative error in x may be large even though the relative error in b is small! In short, $\mathrm{cond}(A)$ merely indicates the *potential* for large relative errors.

We have so far considered only errors in the vector b. If there is an error δA in the coefficient matrix of the system $Ax = b$, the situation is more complicated. For example, $A + \delta A$ may fail to be invertible. But under the appropriate assumptions, it can be shown that a bound for the relative error in x can be given in terms of $\mathrm{cond}(A)$. For example, Charles Cullen (Charles G. Cullen, *An Introduction to Numerical Linear Algebra*, PWS Publishing Co., Boston 1994, p. 60) shows that if $A + \delta A$ is invertible, then

$$\frac{\|\delta x\|}{\|x + \delta x\|} \leq \mathrm{cond}(A)\frac{\|\delta A\|}{\|A\|}.$$

It should be mentioned that, in practice, one never computes $\mathrm{cond}(A)$ from its definition, for it would be an unnecessary waste of time to compute A^{-1} merely to determine its norm. In fact, if a computer is used to find A^{-1}, the computed inverse of A in all likelihood only approximates A^{-1}, and the error in the computed inverse is affected by the size of $\mathrm{cond}(A)$. So we are caught in a vicious circle! There are, however, some situations in which a usable approximation of $\mathrm{cond}(A)$ can be found. Thus, in most cases, the estimate of the relative error in x is based on an estimate of $\mathrm{cond}(A)$.

EXERCISES

1. Label the following statements as true or false.
   (a) If $Ax = b$ is well-conditioned, then $\mathrm{cond}(A)$ is small.
   (b) If $\mathrm{cond}(A)$ is large, then $Ax = b$ is ill-conditioned.
   (c) If $\mathrm{cond}(A)$ is small, then $Ax = b$ is well-conditioned.
   (d) The norm of A equals the Rayleigh quotient.
   (e) The norm of A always equals the largest eigenvalue of A.

2. Compute the norms of the following matrices.

(a) $\begin{pmatrix} 4 & 0 \\ 1 & 3 \end{pmatrix}$ (b) $\begin{pmatrix} 5 & 3 \\ -3 & 3 \end{pmatrix}$ (c) $\begin{pmatrix} 1 & \frac{-2}{\sqrt{3}} & 0 \\ 0 & \frac{-2}{\sqrt{3}} & 1 \\ 0 & \frac{2}{\sqrt{3}} & 1 \end{pmatrix}$

3. Prove that if B is symmetric, then $\|B\|$ is the largest eigenvalue of B.

4. Let A and A^{-1} be as follows:

$$A = \begin{pmatrix} 6 & 13 & -17 \\ 13 & 29 & -38 \\ -17 & -38 & 50 \end{pmatrix} \quad \text{and} \quad A^{-1} = \begin{pmatrix} 6 & -4 & 1 \\ -4 & 11 & 7 \\ -1 & 7 & 5 \end{pmatrix}.$$

The eigenvalues of A are approximately 84.74, 0.2007, and 0.0588.

(a) Approximate $\|A\|$, $\|A^{-1}\|$, and cond(A). (Note Exercise 3.)
(b) Suppose that we have vectors x and $\tilde{x}$ such that $Ax = b$ and $\|b - A\tilde{x}\| \le 0.001$. Use (a) to determine upper bounds for $\|\tilde{x} - A^{-1}b\|$ (the absolute error) and $\|\tilde{x} - A^{-1}b\|/\|A^{-1}b\|$ (the relative error).

5. Suppose that x is the actual solution of $Ax = b$ and that a computer arrives at an approximate solution $\tilde{x}$. If cond(A) = 100, $\|b\| = 1$, and $\|b - A\tilde{x}\| = 0.1$, obtain upper and lower bounds for $\|x - \tilde{x}\|/\|x\|$.

6. Let

$$B = \begin{pmatrix} 2 & 1 & 1 \\ 1 & 2 & 1 \\ 1 & 1 & 2 \end{pmatrix}.$$

Compute

$$R\begin{pmatrix} 1 \\ -2 \\ 3 \end{pmatrix}, \quad \|B\|, \quad \text{and} \quad \text{cond}(B).$$

7. Let B be a symmetric matrix. Prove that $\min_{x \ne 0} R(x)$ equals the smallest eigenvalue of B.

8. Prove that if λ is an eigenvalue of AA^*, then λ is an eigenvalue of A^*A. This completes the proof of the lemma to Corollary 2 to Theorem 6.43.

9. Prove that if A is an invertible matrix and $Ax = b$, then

$$\frac{1}{\|A\| \cdot \|A^{-1}\|} \left(\frac{\|\delta b\|}{\|b\|} \right) \le \frac{\|\delta x\|}{\|x\|}.$$

10. Prove the left inequality of (a) in Theorem 6.44.

11. Prove that $\text{cond}(A) = 1$ if and only if A is a scalar multiple of a unitary or orthogonal matrix.

12. (a) Let A and B be square matrices that are unitarily equivalent. Prove that $\|A\| = \|B\|$.

   (b) Let T be a linear operator on a finite-dimensional inner product space V. Define

$$\|\mathsf{T}\| = \max_{x \neq 0} \frac{\|\mathsf{T}(x)\|}{\|x\|}.$$

   Prove that $\|\mathsf{T}\| = \|[\mathsf{T}]_\beta\|$, where β is any orthonormal basis for V.

   (c) Let V be an infinite-dimensional inner product space with an orthonormal basis $\{v_1, v_2, \ldots\}$. Let T be the linear operator on V such that $\mathsf{T}(v_k) = k v_k$. Prove that $\|\mathsf{T}\|$ (defined in (b)) does not exist.

The next exercise assumes the definitions of *singular value* and *pseudoinverse* and the results of Section 6.7.

13. Let A be an $n \times n$ matrix of rank r with the nonzero singular values $\sigma_1 \geq \sigma_2 \geq \cdots \geq \sigma_r$. Prove each of the following results.

   (a) $\|A\| = \sigma_1$.

   (b) $\|A^\dagger\| = \dfrac{1}{\sigma_r}$.

   (c) If A is invertible (and hence $r = n$), then $\text{cond}(A) = \dfrac{\sigma_1}{\sigma_n}$.

6.11* THE GEOMETRY OF ORTHOGONAL OPERATORS

By Theorem 6.22 (p. 386), any rigid motion on a finite-dimensional real inner product space is the composite of an orthogonal operator and a translation. Thus, to understand the geometry of rigid motions thoroughly, we must analyze the structure of orthogonal operators. Such is the aim of this section. We show that any orthogonal operator on a finite-dimensional real inner product space is the composite of rotations and reflections.

   This material assumes familiarity with the results about direct sums developed at the end of Section 5.2, and familiarity with the definition and elementary properties of the determinant of a linear operator defined in Exercise 7 of Section 5.1.

   Definitions. *Let* T *be a linear operator on a finite-dimensional real inner product space* V. *The operator* T *is called a* **rotation** *if* T *is the identity on*

V *or if there exists a two-dimensional subspace* W *of* V, *an orthonormal basis* $\beta = \{x_1, x_2\}$ *for* W, *and a real number* θ *such that*

$$T(x_1) = (\cos\theta)x_1 + (\sin\theta)x_2, \quad T(x_2) = (-\sin\theta)x_1 + (\cos\theta)x_2,$$

and $T(y) = y$ *for all* $y \in W^{\perp}$. *In this context,* T *is called a* **rotation of** W **about** $W^{\perp}$. *The subspace* $W^{\perp}$ *is called the* **axis of rotation**.

Rotations are defined in Section 2.1 for the special case that $V = R^2$.

Definitions.   *Let* T *be a linear operator on a finite-dimensional real inner product space* V. *The operator* T *is called a* **reflection** *if there exists a one-dimensional subspace* W *of* V *such that* $T(x) = -x$ *for all* $x \in W$ *and* $T(y) = y$ *for all* $y \in W^{\perp}$. *In this context,* T *is called a* **reflection of** V **about** $W^{\perp}$.

It should be noted that rotations and reflections (or composites of these) are orthogonal operators (see Exercise 2). The principal aim of this section is to establish that the converse is also true, that is, any orthogonal operator on a finite-dimensional real inner product space is the composite of rotations and reflections.

Example 1

A Characterization of Orthogonal Operators on a One-Dimensional Real Inner Product Space

Let T be an orthogonal operator on a one-dimensional inner product space V. Choose any nonzero vector x in V. Then $V = \text{span}(\{x\})$, and so $T(x) = \lambda x$ for some $\lambda \in R$. Since T is orthogonal and λ is an eigenvalue of T, $\lambda = \pm 1$. If $\lambda = 1$, then T is the identity on V, and hence T is a rotation. If $\lambda = -1$, then $T(x) = -x$ for all $x \in V$; so T is a reflection of V about $V^{\perp} = \{0\}$. Thus T is either a rotation or a reflection. Note that in the first case, $\det(T) = 1$, and in the second case, $\det(T) = -1$. $\blacklozenge$

Example 2

Some Typical Reflections

(a) Define $T\colon R^2 \to R^2$ by $T(a,b) = (-a,b)$, and let $W = \text{span}(\{e_1\})$. Then $T(x) = -x$ for all $x \in W$, and $T(y) = y$ for all $y \in W^{\perp}$. Thus T is a reflection of R^2 about $W^{\perp} = \text{span}(\{e_2\})$, the y-axis.

(b) Let $T\colon R^3 \to R^3$ be defined by $T(a,b,c) = (a,b,-c)$, and let $W = \text{span}(\{e_3\})$. Then $T(x) = -x$ for all $x \in W$, and $T(y) = y$ for all $y \in W^{\perp} = \text{span}(\{e_1, e_2\})$, the xy-plane. Hence T is a reflection of R^3 about $W^{\perp}$. $\blacklozenge$

Example 1 characterizes all orthogonal operators on a one-dimensional real inner product space. The following theorem characterizes all orthogonal operators on a two-dimensional real inner product space V. The proof follows from Theorem 6.23 (p. 387) since all two-dimensional real inner product spaces are structurally identical. For a rigorous justification, apply Theorem 2.21 (p. 104), where β is an orthonormal basis for V. By Exercise 15 of Section 6.2, the resulting isomorphism $\phi_\beta \colon V \to R^2$ preserves inner products. (See Exercise 8.)

Theorem 6.45. *Let* T *be an orthogonal operator on a two-dimensional real inner product space* V. *Then* T *is either a rotation or a reflection. Furthermore,* T *is a rotation if and only if* $\det(T) = 1$, *and* T *is a reflection if and only if* $\det(T) = -1$.

A complete description of the reflections of R^2 is given in Section 6.5.

Corollary. *Let* V *be a two-dimensional real inner product space. The composite of a reflection and a rotation on* V *is a reflection on* V.

Proof. If T_1 is a reflection on V and T_2 is a rotation on V, then by Theorem 6.45, $\det(T_1) = 1$ and $\det(T_2) = -1$. Let $T = T_2 T_1$ be the composite. Since T_2 and T_1 are orthogonal, so is T. Moreover, $\det(T) = \det(T_2) \cdot \det(T_1) = -1$. Thus, by Theorem 6.45, T is a reflection. The proof for $T_1 T_2$ is similar. ∎

We now study orthogonal operators on spaces of higher dimension.

Lemma. *If* T *is a linear operator on a nonzero finite-dimensional real vector space* V, *then there exists a* T*-invariant subspace* W *of* V *such that* $1 \le \dim(W) \le 2$.

Proof. Fix an ordered basis $\beta = \{y_1, y_2, \ldots, y_n\}$ for V, and let $A = [T]_\beta$. Let $\phi_\beta \colon V \to R^n$ be the linear transformation defined by $\phi_\beta(y_i) = e_i$ for $i = 1, 2, \ldots, n$. Then ϕ_β is an isomorphism, and, as we have seen in Section 2.4, the diagram in Figure 6.10 commutes, that is, $L_A \phi_\beta = \phi_\beta T$. As a consequence, it suffices to show that there exists an L_A-invariant subspace Z of R^n such that $1 \le \dim(Z) \le 2$. If we then define $W = \phi_\beta^{-1}(Z)$, it follows that W satisfies the conclusions of the lemma (see Exercise 13).

$$
\begin{array}{ccc}
V & \xrightarrow{\ T\ } & V \\
\downarrow{\scriptstyle \phi_\beta} & & \downarrow{\scriptstyle \phi_\beta} \\
R^n & \xrightarrow{\ L_A\ } & R^n
\end{array}
$$

Figure 6.10

The matrix A can be considered as an $n \times n$ matrix over C and, as such, can be used to define a linear operator U on C^n by $\mathsf{U}(v) = Av$. Since U is a linear operator on a finite-dimensional vector space over C, it has an eigenvalue $\lambda \in C$. Let $x \in C^n$ be an eigenvector corresponding to λ. We may write $\lambda = \lambda_1 + i\lambda_2$, where λ_1 and λ_2 are real, and

$$x = \begin{pmatrix} a_1 + ib_1 \\ a_2 + ib_2 \\ \vdots \\ a_n + ib_n \end{pmatrix},$$

where the a_i's and b_i's are real. Thus, setting

$$x_1 = \begin{pmatrix} a_1 \\ a_2 \\ \vdots \\ a_n \end{pmatrix} \quad \text{and} \quad x_2 = \begin{pmatrix} b_1 \\ b_2 \\ \vdots \\ b_n \end{pmatrix},$$

we have $x = x_1 + ix_2$, where x_1 and x_2 have real entries. Note that at least one of x_1 or x_2 is nonzero since $x \neq 0$. Hence

$$\mathsf{U}(x) = \lambda x = (\lambda_1 + i\lambda_2)(x_1 + ix_2) = (\lambda_1 x_1 - \lambda_2 x_2) + i(\lambda_1 x_2 + \lambda_2 x_1).$$

Similarly,

$$\mathsf{U}(x) = A(x_1 + ix_2) = Ax_1 + iAx_2.$$

Comparing the real and imaginary parts of these two expressions for $\mathsf{U}(x)$, we conclude that

$$Ax_1 = \lambda_1 x_1 - \lambda_2 x_2 \quad \text{and} \quad Ax_2 = \lambda_1 x_2 + \lambda_2 x_1.$$

Finally, let $Z = \text{span}(\{x_1, x_2\})$, the span being taken as a subspace of R^n. Since $x_1 \neq 0$ or $x_2 \neq 0$, Z is a nonzero subspace. Thus $1 \leq \dim(Z) \leq 2$, and the preceding pair of equations shows that Z is L_A-invariant. ∎

Theorem 6.46. *Let* T *be an orthogonal operator on a nonzero finite-dimensional real inner product space* V. *Then there exists a collection of pairwise orthogonal* T*-invariant subspaces* $\{\mathsf{W}_1, \mathsf{W}_2, \ldots, \mathsf{W}_m\}$ *of* V *such that*

(a) $1 \leq \dim(\mathsf{W}_i) \leq 2$ *for* $i = 1, 2, \ldots, m$.

(b) $\mathsf{V} = \mathsf{W}_1 \oplus \mathsf{W}_2 \oplus \cdots \oplus \mathsf{W}_m$.

Proof. The proof is by mathematical induction on $\dim(\mathsf{V})$. If $\dim(\mathsf{V}) = 1$, the result is obvious. So assume that the result is true whenever $\dim(\mathsf{V}) < n$ for some fixed integer $n > 1$.

Suppose $\dim(V) = n$. By the lemma, there exists a T-invariant subspace W_1 of V such that $1 \leq \dim(W) \leq 2$. If $W_1 = V$, the result is established. Otherwise, $W_1^\perp \neq \{0\}$. By Exercise 14, $W_1^\perp$ is T-invariant and the restriction of T to $W_1^\perp$ is orthogonal. Since $\dim(W_1^\perp) < n$, we may apply the induction hypothesis to $T_{W_1^\perp}$ and conclude that there exists a collection of pairwise orthogonal T-invariant subspaces $\{W_1, W_2, \ldots, W_m\}$ of $W_1^\perp$ such that $1 \leq \dim(W_i) \leq 2$ for $i = 2, 3, \ldots, m$ and $W_1^\perp = W_2 \oplus W_3 \oplus \cdots \oplus W_m$. Thus $\{W_1, W_2, \ldots, W_m\}$ is pairwise orthogonal, and by Exercise 13(d) of Section 6.2,

$$V = W_1 \oplus W_1^\perp = W_1 \oplus W_2 \oplus \cdots \oplus W_m. \qquad \blacksquare$$

Applying Example 1 and Theorem 6.45 in the context of Theorem 6.46, we conclude that the restriction of T to W_i is either a rotation or a reflection for each $i = 2, 3, \ldots, m$. Thus, in some sense, T is composed of rotations and reflections. Unfortunately, very little can be said about the uniqueness of the decomposition of V in Theorem 6.46. For example, the W_i's, the number m of W_i's, and the number of W_i's for which T_{W_i} is a reflection are not unique. Although the number of W_i's for which T_{W_i} is a reflection is not unique, whether this number is even or odd is an intrinsic property of T. Moreover, we can always decompose V so that T_{W_i} is a reflection for at most one W_i. These facts are established in the following result.

Theorem 6.47. Let T, V, $W_1, \ldots, W_m$ be as in Theorem 6.46.
(a) The number of W_i's for which T_{W_i} is a reflection is even or odd according to whether $\det(T) = 1$ or $\det(T) = -1$.
(b) It is always possible to decompose V as in Theorem 6.46 so that the number of W_i's for which T_{W_i} is a reflection is zero or one according to whether $\det(T) = 1$ or $\det(T) = -1$. Furthermore, if T_{W_i} is a reflection, then $\dim(W_i) = 1$.

Proof. (a) Let r denote the number of W_i's in the decomposition for which T_{W_i} is a reflection. Then, by Exercise 15,

$$\det(T) = \det(T_{W_1}) \cdot \det(T_{W_2}) \cdot \cdots \cdot \det(T_{W_m}) = (-1)^r,$$

proving (a).

(b) Let $E = \{x \in V: T(x) = -x\}$; then E is a T-invariant subspace of V. If $W = E^\perp$, then W is T-invariant. So by applying Theorem 6.46 to T_W, we obtain a collection of pairwise orthogonal T-invariant subspaces $\{W_1, W_2, \ldots, W_k\}$ of W such that $W = W_1 \oplus W_2 \oplus \cdots \oplus W_k$ and for $1 \leq i \leq k$, the dimension of each W_i is either 1 or 2. Observe that, for each $i = 1, 2, \ldots, k$, T_{W_i} is a rotation. For otherwise, if T_{W_i} is a reflection, there exists a nonzero $x \in W_i$ for which $T(x) = -x$. But then, $x \in W_i \cap E \subseteq E^\perp \cap E = \{0\}$, a contradiction. If $E = \{0\}$, the result follows. Otherwise, choose an orthonormal basis β for E containing p vectors $(p > 0)$. It is possible to decompose β into a pairwise disjoint union $\beta = \beta_1 \cup \beta_2 \cup \cdots \cup \beta_r$ such that each β_i contains exactly two vectors for $i < r$, and β_r contains two vectors if p is even and one vector if p is odd. For each $i = 1, 2, \ldots, r$, let $\mathsf{W}_{k+i} = \operatorname{span}(\beta_i)$. Then, clearly, $\{\mathsf{W}_1, \mathsf{W}_2, \ldots, \mathsf{W}_k, \ldots, \mathsf{W}_{k+r}\}$ is pairwise orthogonal, and

$$\mathsf{V} = \mathsf{W}_1 \oplus \mathsf{W}_2 \oplus \cdots \oplus \mathsf{W}_k \oplus \cdots \oplus \mathsf{W}_{k+r}. \tag{27}$$

Moreover, if any β_i contains two vectors, then

$$\det(\mathsf{T}_{\mathsf{W}_{k+i}}) = \det([\mathsf{T}_{\mathsf{W}_{k+i}}]_{\beta_i}) = \det \begin{pmatrix} -1 & 0 \\ 0 & -1 \end{pmatrix} = 1.$$

So $\mathsf{T}_{\mathsf{W}_{k+i}}$ is a rotation, and hence $\mathsf{T}_{\mathsf{W}_j}$ is a rotation for $j < k + r$. If β_r consists of one vector, then $\dim(\mathsf{W}_{k+r}) = 1$ and

$$\det(\mathsf{T}_{\mathsf{W}_{k+r}}) = \det([\mathsf{T}_{\mathsf{W}_{k+r}}]_{\beta_r}) = \det(-1) = -1.$$

Thus $\mathsf{T}_{\mathsf{W}_{k+r}}$ is a reflection by Theorem 6.46, and we conclude that the decomposition in (27) satisfies the condition of (b). ∎

As a consequence of the preceding theorem, an orthogonal operator can be factored as a product of rotations and reflections.

Corollary. *Let T be an orthogonal operator on a finite-dimensional real inner product space V. Then there exists a collection $\{\mathsf{T}_1, \mathsf{T}_2, \ldots, \mathsf{T}_m\}$ of orthogonal operators on V such that the following statements are true.*
(a) *For each i, T_i is either a reflection or a rotation.*
(b) *For at most one i, T_i is a reflection.*
(c) *$\mathsf{T}_i \mathsf{T}_j = \mathsf{T}_j \mathsf{T}_i$ for all i and j.*
(d) *$\mathsf{T} = \mathsf{T}_1 \mathsf{T}_2 \cdots \mathsf{T}_m$.*

(e) *$\det(\mathsf{T}) = \begin{cases} 1 & \text{if } \mathsf{T}_i \text{ is a rotation for each } i \\ -1 & \text{otherwise.} \end{cases}$*

Proof. As in the proof of Theorem 6.47(b), we can write

$$\mathsf{V} = \mathsf{W}_1 \oplus \mathsf{W}_2 \oplus \cdots \oplus \mathsf{W}_m,$$

where $\mathsf{T}_{\mathsf{W}_i}$ is a rotation for $i < m$. For each $i = 1, 2, \ldots, m$, define $\mathsf{T}_i \colon \mathsf{V} \to \mathsf{V}$ by

$$\mathsf{T}_i(x_1 + x_2 + \cdots + x_m) = x_1 + x_2 + \cdots + x_{i-1} + \mathsf{T}(x_i) + x_{i+1} + \cdots + x_m,$$

where $x_j \in \mathsf{W}_j$ for all j. It is easily shown that each T_i is an orthogonal operator on V. In fact, T_i is a rotation or a reflection according to whether $\mathsf{T}_{\mathsf{W}_i}$ is a rotation or a reflection. This establishes (a) and (b). The proofs of (c), (d), and (e) are left as exercises. (See Exercise 16.) ∎

Example 3

Orthogonal Operators on a Three-Dimensional Real Inner Product Space

Let T be an orthogonal operator on a three-dimensional real inner product space V. We show that T can be decomposed into the composite of a rotation and at most one reflection. Let

$$\mathsf{V} = \mathsf{W}_1 \oplus \mathsf{W}_2 \oplus \cdots \oplus \mathsf{W}_m$$

be a decomposition as in Theorem 6.47(b). Clearly, $m = 2$ or $m = 3$.

If $m = 2$, then $\mathsf{V} = \mathsf{W}_1 \oplus \mathsf{W}_2$. Without loss of generality, suppose that $\dim(\mathsf{W}_1) = 1$ and $\dim(\mathsf{W}_2) = 2$. Thus $\mathsf{T}_{\mathsf{W}_1}$ is a reflection or the identity on W_1, and $\mathsf{T}_{\mathsf{W}_2}$ is a rotation. Defining T_1 and T_2 as in the proof of the corollary to Theorem 6.47, we have that $\mathsf{T} = \mathsf{T}_1 \mathsf{T}_2$ is the composite of a rotation and at most one reflection. (Note that if $\mathsf{T}_{\mathsf{W}_1}$ is not a reflection, then T_1 is the identity on V and $\mathsf{T} = \mathsf{T}_2$.)

If $m = 3$, then $\mathsf{V} = \mathsf{W}_1 \oplus \mathsf{W}_2 \oplus \mathsf{W}_3$ and $\dim(\mathsf{W}_i) = 1$ for all i. For each i, let T_i be as in the proof of the corollary to Theorem 6.47. If $\mathsf{T}_{\mathsf{W}_i}$ is not a reflection, then T_i is the identity on W_i. Otherwise, T_i is a reflection. Since $\mathsf{T}_{\mathsf{W}_i}$ is a reflection for at most one i, we conclude that T is either a single reflection or the identity (a rotation).   ◆

EXERCISES

1. Label the following statements as true or false. Assume that the underlying vector spaces are finite-dimensional real inner product spaces.

   (a) Any orthogonal operator is either a rotation or a reflection.

   (b) The composite of any two rotations on a two-dimensional space is a rotation.

   (c) The composite of any two rotations on a three-dimensional space is a rotation.

   (d) The composite of any two rotations on a four-dimensional space is a rotation.

   (e) The identity operator is a rotation.

   (f) The composite of two reflections is a reflection.

   (g) Any orthogonal operator is a composite of rotations.

   (h) For any orthogonal operator T, if $\det(\mathsf{T}) = -1$, then T is a reflection.

   (i) Reflections always have eigenvalues.

   (j) Rotations always have eigenvalues.

2. Prove that rotations, reflections, and composites of rotations and reflections are orthogonal operators.

3. Let

$$A = \begin{pmatrix} \dfrac{1}{2} & \dfrac{\sqrt{3}}{2} \\ \dfrac{\sqrt{3}}{2} & -\dfrac{1}{2} \end{pmatrix} \quad \text{and} \quad B = \begin{pmatrix} 1 & 0 \\ 0 & -1 \end{pmatrix}.$$

(a) Prove that L_A is a reflection.
(b) Find the axis in R^2 about which L_A reflects, that is, the subspace of R^2 on which L_A acts as the identity.
(c) Prove that L_{AB} and L_{BA} are rotations.

4. For any real number ϕ, let

$$A = \begin{pmatrix} \cos\phi & \sin\phi \\ \sin\phi & -\cos\phi \end{pmatrix}.$$

(a) Prove that L_A is a reflection.
(b) Find the axis in R^2 about which L_A reflects.

5. For any real number ϕ, define $T_\phi = L_A$, where

$$A = \begin{pmatrix} \cos\phi & -\sin\phi \\ \sin\phi & \cos\phi \end{pmatrix}.$$

(a) Prove that any rotation on R^2 is of the form T_ϕ for some ϕ.
(b) Prove that $T_\phi T_\psi = T_{(\phi+\psi)}$ for any $\phi, \psi \in R$.
(c) Deduce that any two rotations on R^2 commute.

6. Prove that the composite of any two rotations on R^3 is a rotation on R^3.

7. Given real numbers ϕ and ψ, define matrices

$$A = \begin{pmatrix} 1 & 0 & 0 \\ 0 & \cos\phi & -\sin\phi \\ 0 & \sin\phi & \cos\phi \end{pmatrix} \quad \text{and} \quad B = \begin{pmatrix} \cos\psi & -\sin\psi & 0 \\ \sin\psi & \cos\psi & 0 \\ 0 & 0 & 1 \end{pmatrix}.$$

(a) Prove that L_A and L_B are rotations.
(b) Prove that L_{AB} is a rotation.
(c) Find the axis of rotation for L_{AB}.

8. Prove Theorem 6.45 using the hints preceding the statement of the theorem.

9. Prove that no orthogonal operator can be both a rotation and a reflection.

10. Prove that if V is a two- or three-dimensional real inner product space, then the composite of two reflections on V is a rotation of V.

11. Give an example of an orthogonal operator that is neither a reflection nor a rotation.

12. Let V be a finite-dimensional real inner product space. Define T: V → V by T(x) = $-x$. Prove that T is a product of rotations if and only if dim(V) is even.

13. Complete the proof of the lemma to Theorem 6.46 by showing that W = $\phi_\beta^{-1}(Z)$ satisfies the required conditions.

14. Let T be an orthogonal [unitary] operator on a finite-dimensional real [complex] inner product space V. If W is a T-invariant subspace of V, prove the following results.

    (a) T$_W$ is an orthogonal [unitary] operator on W.
    (b) W$^\perp$ is a T-invariant subspace of V. *Hint:* Use the fact that T$_W$ is one-to-one and onto to conclude that, for any $y \in$ W, T$^*(y)$ = T$^{-1}(y) \in$ W.
    (c) T$_{W^\perp}$ is an orthogonal [unitary] operator on W.

15. Let T be a linear operator on a finite-dimensional vector space V, where V is a direct sum of T-invariant subspaces, say, V = $W_1 \oplus W_2 \oplus \cdots \oplus W_k$. Prove that det(T) = det(T$_{W_1}$)· det(T$_{W_2}$)· $\cdots$ · det(T$_{W_k}$).

16. Complete the proof of the corollary to Theorem 6.47.

17. Let T be a linear operator on an n-dimensional real inner product space V. Suppose that T is not the identity. Prove the following results.

    (a) If n is odd, then T can be expressed as the composite of at most one reflection and at most $\frac{1}{2}(n-1)$ rotations.
    (b) If n is even, then T can be expressed as the composite of at most $\frac{1}{2}n$ rotations or as the composite of one reflection and at most $\frac{1}{2}(n-2)$ rotations.

18. Let V be a real inner product space of dimension 2. For any $x, y \in$ V such that $x \neq y$ and $\|x\| = \|y\| = 1$, show that there exists a unique rotation T on V such that T(x) = y.

INDEX OF DEFINITIONS FOR CHAPTER 6

7

Canonical Forms

$\boldsymbol{A}$s we learned in Chapter 5, the advantage of a diagonalizable linear operator lies in the simplicity of its description. Such an operator has a diagonal matrix representation, or, equivalently, there is an ordered basis for the underlying vector space consisting of eigenvectors of the operator. However, not every linear operator is diagonalizable, even if its characteristic polynomial splits. Example 3 of Section 5.2 describes such an operator.

It is the purpose of this chapter to consider alternative matrix representations for nondiagonalizable operators. These representations are called *canonical forms*. There are different kinds of canonical forms, and their advantages and disadvantages depend on how they are applied. The choice of a canonical form is determined by the appropriate choice of an ordered basis. Naturally, the canonical forms of a linear operator are not diagonal matrices if the linear operator is not diagonalizable.

In this chapter, we treat two common canonical forms. The first of these, the *Jordan canonical form*, requires that the characteristic polynomial of the operator splits. This form is always available if the underlying field is algebraically closed, that is, if every polynomial with coefficients from the field splits. For example, the field of complex numbers is algebraically closed by the fundamental theorem of algebra (see Appendix D). The first two sections deal with this form. The *rational canonical form*, treated in Section 7.4, does not require such a factorization.

7.1 THE JORDAN CANONICAL FORM I

Let T be a linear operator on a finite-dimensional vector space V, and suppose that the characteristic polynomial of T splits. Recall from Section 5.2 that the diagonalizability of T depends on whether the union of ordered bases for the distinct eigenspaces of T is an ordered basis for V. So a lack of diagonalizability means that at least one eigenspace of T is too "small."

In this section, we extend the definition of eigenspace to *generalized eigenspace*. From these subspaces, we select ordered bases whose union is an ordered basis β for V such that

$$[\mathsf{T}]_\beta = \begin{pmatrix} A_1 & O & \cdots & O \\ O & A_2 & \cdots & O \\ \vdots & \vdots & & \vdots \\ O & O & \cdots & A_k \end{pmatrix},$$

where each O is a zero matrix, and each A_i is a square matrix of the form (λ) or

$$\begin{pmatrix} \lambda & 1 & 0 & \cdots & 0 & 0 \\ 0 & \lambda & 1 & \cdots & 0 & 0 \\ \vdots & \vdots & \vdots & & \vdots & \vdots \\ 0 & 0 & 0 & \cdots & \lambda & 1 \\ 0 & 0 & 0 & \cdots & 0 & \lambda \end{pmatrix}$$

for some eigenvalue λ of T. Such a matrix A_i is called a **Jordan block** corresponding to λ, and the matrix $[\mathsf{T}]_\beta$ is called a **Jordan canonical form** of T. We also say that the ordered basis β is a **Jordan canonical basis** for T. Observe that each Jordan block A_i is "almost" a diagonal matrix—in fact, $[\mathsf{T}]_\beta$ is a diagonal matrix if and only if each A_i is of the form (λ).

Example 1

Suppose that T is a linear operator on C^8, and $\beta = \{v_1, v_2, \ldots, v_8\}$ is an ordered basis for C^8 such that

$$J = [\mathsf{T}]_\beta = \left(\begin{array}{ccc|c|cc|cc} 2 & 1 & 0 & 0 & 0 & 0 & 0 & 0 \\ 0 & 2 & 1 & 0 & 0 & 0 & 0 & 0 \\ 0 & 0 & 2 & 0 & 0 & 0 & 0 & 0 \\ \hline 0 & 0 & 0 & 2 & 0 & 0 & 0 & 0 \\ \hline 0 & 0 & 0 & 0 & 3 & 1 & 0 & 0 \\ 0 & 0 & 0 & 0 & 0 & 3 & 0 & 0 \\ \hline 0 & 0 & 0 & 0 & 0 & 0 & 0 & 1 \\ 0 & 0 & 0 & 0 & 0 & 0 & 0 & 0 \end{array} \right)$$

is a Jordan canonical form of T. Notice that the characteristic polynomial of T is $\det(J - tI) = (t-2)^4(t-3)^2 t^2$, and hence the multiplicity of each eigenvalue is the number of times that the eigenvalue appears on the diagonal of J. Also observe that v_1, v_4, v_5, and v_7 are the only vectors in β that are eigenvectors of T. These are the vectors corresponding to the columns of J with no 1 above the diagonal entry. ◆

In Sections 7.1 and 7.2, we prove that every linear operator whose characteristic polynomial splits has a Jordan canonical form that is unique up to the order of the Jordan blocks. Nevertheless, it is not the case that the Jordan canonical form is completely determined by the characteristic polynomial of the operator. For example, let T' be the linear operator on C^8 such that $[\mathsf{T}']_\beta = J'$, where β is the ordered basis in Example 1 and

$$
J' = \begin{pmatrix}
2 & 0 & 0 & 0 & 0 & 0 & 0 & 0 \\
0 & 2 & 0 & 0 & 0 & 0 & 0 & 0 \\
0 & 0 & 2 & 0 & 0 & 0 & 0 & 0 \\
0 & 0 & 0 & 2 & 0 & 0 & 0 & 0 \\
0 & 0 & 0 & 0 & 3 & 0 & 0 & 0 \\
0 & 0 & 0 & 0 & 0 & 3 & 0 & 0 \\
0 & 0 & 0 & 0 & 0 & 0 & 0 & 0 \\
0 & 0 & 0 & 0 & 0 & 0 & 0 & 0
\end{pmatrix}.
$$

Then the characteristic polynomial of T' is also $(t-2)^4(t-3)^2t^2$. But the operator T' has the Jordan canonical form J', which is different from J, the Jordan canonical form of the linear operator T of Example 1.

Consider again the matrix J and the ordered basis β of Example 1. Notice that $\mathsf{T}(v_2) = v_1 + 2v_2$ and therefore, $(\mathsf{T} - 2\mathsf{I})(v_2) = v_1$. Similarly, $(\mathsf{T} - 2\mathsf{I})(v_3) = v_2$. Since v_1 and v_4 are eigenvectors of T corresponding to $\lambda = 2$, it follows that $(\mathsf{T} - 2\mathsf{I})^3(v_i) = 0$ for $i = 1, 2, 3$, and 4. Similarly $(\mathsf{T} - 3\mathsf{I})^2(v_i) = 0$ for $i = 5, 6$, and $(\mathsf{T} - 0\mathsf{I})^2(v_i) = 0$ for $i = 7, 8$.

Because of the structure of each Jordan block in a Jordan canonical form, we can generalize these observations: *If v lies in a Jordan canonical basis for a linear operator T and is associated with a Jordan block with diagonal entry λ, then $(\mathsf{T} - \lambda\mathsf{I})^p(v) = 0$ for sufficiently large p.* Eigenvectors satisfy this condition for $p = 1$.

Definition. *Let T be a linear operator on a vector space V, and let λ be a scalar. A nonzero vector x in V is called a **generalized eigenvector of T corresponding to** λ if $(\mathsf{T} - \lambda\mathsf{I})^p(x) = 0$ for some positive integer p.*

Notice that if x is a generalized eigenvector of T corresponding to λ, and p is the smallest positive integer for which $(\mathsf{T} - \lambda\mathsf{I})^p(x) = 0$, then $(\mathsf{T} - \lambda\mathsf{I})^{p-1}(x)$ is an eigenvector of T corresponding to λ. Therefore λ is an eigenvalue of T.

In the context of Example 1, each vector in β is a generalized eigenvector of T. In fact, v_1, v_2, v_3 and v_4 correspond to the scalar 2, v_5 and v_6 correspond to the scalar 3, and v_7 and v_8 correspond to the scalar 0.

Just as eigenvectors lie in eigenspaces, generalized eigenvectors lie in "generalized eigenspaces."

Definition. *Let T be a linear operator on a vector space V, and let λ be an eigenvalue of T. The **generalized eigenspace of T corresponding to***

λ, denoted K_λ, is the subset of V defined by

$$K_\lambda = \{x \in V : (T - \lambda I)^p(x) = 0 \text{ for some positive integer } p\}.$$

Note that K_λ consists of the zero vector and all generalized eigenvectors corresponding to λ.

Recall that a subspace W of V is T-invariant for a linear operator T if $T(W) \subseteq W$. In the development that follows, we assume the results of Exercises 3 and 4 of Section 5.4. In particular, for any polynomial $g(x)$, if W is T-invariant, then it is also $g(T)$-invariant. Furthermore, the range of a linear operator T is T-invariant.

Theorem 7.1. *Let* T *be a linear operator on a vector space* V, *and let* λ *be an eigenvalue of* T. *Then*

   (a) K_λ *is a* T-*invariant subspace of* V *containing* E_λ *(the eigenspace of* T *corresponding to* λ).
   (b) *For any scalar* $\mu \neq \lambda$, *the restriction of* $T - \mu I$ *to* K_λ *is one-to-one.*

Proof. (a) Clearly, $0 \in K_\lambda$. Suppose that x and y are in K_λ. Then there exist positive integers p and q such that

$$(T - \lambda I)^p(x) = (T - \lambda I)^q(y) = 0.$$

Therefore

$$\begin{aligned}(T - \lambda I)^{p+q}(x + y) &= (T - \lambda I)^{p+q}(x) + (T - \lambda I)^{p+q}(y) \\ &= (T - \lambda I)^q(0) + (T - \lambda I)^p(0) \\ &= 0,\end{aligned}$$

and hence $x + y \in K_\lambda$. The proof that K_λ is closed under scalar multiplication is straightforward.

To show that K_λ is T-invariant, consider any $x \in K_\lambda$. Choose a positive integer p such that $(T - \lambda I)^p(x) = 0$. Then

$$(T - \lambda I)^p T(x) = T(T - \lambda I)^p(x) = T(0) = 0.$$

Therefore $T(x) \in K_\lambda$.

Finally, it is a simple observation that E_λ is contained in K_λ.

(b) Let $x \in K_\lambda$ and $(T - \mu I)(x) = 0$. By way of contradiction, suppose that $x \neq 0$. Let p be the smallest integer for which $(T - \lambda I)^p(x) = 0$, and let $y = (T - \lambda I)^{p-1}(x)$. Then

$$(T - \lambda I)(y) = (T - \lambda I)^p(x) = 0,$$

and hence $y \in E_\lambda$. Furthermore,

$$(T - \mu I)(y) = (T - \mu I)(T - \lambda I)^{p-1}(x) = (T - \lambda I)^{p-1}(T - \mu I)(x) = 0,$$

so that $y \in E_\mu$. But $E_\lambda \cap E_\mu = \{0\}$, and thus $y = 0$, contrary to the hypothesis. So $x = 0$, and the restriction of $T - \mu I$ to K_λ is one-to-one. ∎

Theorem 7.2. *Let* T *be a linear operator on a finite-dimensional vector space* V *such that the characteristic polynomial of* T *splits. Suppose that* λ *is an eigenvalue of* T *with multiplicity* m. *Then*

(a) $\dim(\mathsf{K}_\lambda) \le m$.

(b) $\mathsf{K}_\lambda = \mathsf{N}((\mathsf{T} - \lambda\mathsf{I})^m)$.

Proof. (a) Let $\mathsf{W} = \mathsf{K}_\lambda$, and let $h(t)$ be the characteristic polynomial of T_W. By Theorem 5.21 (p. 314), $h(t)$ divides the characteristic polynomial of T, and by Theorem 7.1(b), λ is the only eigenvalue of T_W. Hence $h(t) = (-1)^d(t-\lambda)^d$, where $d = \dim(\mathsf{W})$, and $d \le m$.

(b) Clearly $\mathsf{N}((\mathsf{T} - \lambda\mathsf{I})^m) \subseteq \mathsf{K}_\lambda$. Now let W and $h(t)$ be as in (a). Then $h(\mathsf{T}_\mathsf{W})$ is identically zero by the Cayley–Hamilton theorem (p. 317); therefore $(\mathsf{T} - \lambda\mathsf{I})^d(x) = 0$ for all $x \in \mathsf{W}$. Since $d \le m$, we have $\mathsf{K}_\lambda \subseteq \mathsf{N}((\mathsf{T} - \lambda\mathsf{I})^m)$. ∎

Theorem 7.3. *Let* T *be a linear operator on a finite-dimensional vector space* V *such that the characteristic polynomial of* T *splits, and let* $\lambda_1, \lambda_2, \ldots, \lambda_k$ *be the distinct eigenvalues of* T. *Then, for every* $x \in \mathsf{V}$, *there exist vectors* $v_i \in \mathsf{K}_{\lambda_i}$, $1 \le i \le k$, *such that*

$$x = v_1 + v_2 + \cdots + v_k.$$

Proof. The proof is by mathematical induction on the number k of distinct eigenvalues of T. First suppose that $k = 1$, and let m be the multiplicity of λ_1. Then $(\lambda_1 - t)^m$ is the characteristic polynomial of T, and hence $(\lambda_1\mathsf{I} - \mathsf{T})^m = \mathsf{T}_0$ by the Cayley-Hamilton theorem (p.317). Thus $\mathsf{V} = \mathsf{K}_{\lambda_1}$, and the result follows.

Now suppose that for some integer $k > 1$, the result is established whenever T has fewer than k distinct eigenvalues, and suppose that T has k distinct eigenvalues. Let m be the multiplicity of λ_k, and let $f(t)$ be the characteristic polynomial of T. Then $f(t) = (t - \lambda_k)^m g(t)$ for some polynomial $g(t)$ not divisible by $(t - \lambda_k)$. Let $\mathsf{W} = \mathsf{R}((\mathsf{T} - \lambda_k\mathsf{I})^m)$. Clearly W is T-invariant. Observe that $(\mathsf{T} - \lambda_k\mathsf{I})^m$ maps K_{λ_i} onto itself for $i < k$. For suppose that $i < k$. Since $(\mathsf{T} - \lambda_k\mathsf{I})^m$ maps K_{λ_i} into itself and $\lambda_k \ne \lambda_i$, the restriction of $\mathsf{T} - \lambda_k\mathsf{I}$ to K_{λ_i} is one-to-one (by Theorem 7.1(b)) and hence is onto. One consequence of this is that for $i < k$, K_{λ_i} is contained in W; hence λ_i is an eigenvalue of T_W for $i < k$.

Next, observe that λ_k is not an eigenvalue of T_W. For suppose that $\mathsf{T}(v) = \lambda_k v$ for some $v \in \mathsf{W}$. Then $v = (\mathsf{T} - \lambda_k\mathsf{I})^m(y)$ for some $y \in \mathsf{V}$, and it follows that

$$0 = (\mathsf{T} - \lambda_k\mathsf{I})(v) = (\mathsf{T} - \lambda_k\mathsf{I})^{m+1}(y).$$

Therefore $y \in \mathsf{K}_{\lambda_k}$. So by Theorem 7.2, $v = (\mathsf{T} - \lambda_k\mathsf{I})^m(y) = 0$.

Since every eigenvalue of T_W is an eigenvalue of T, the distinct eigenvalues of T_W are $\lambda_1, \lambda_2, \ldots, \lambda_{k-1}$.

Now let $x \in V$. Then $(T - \lambda_k I)^m(x) \in W$. Since T_W has the $k - 1$ distinct eigenvalues $\lambda_1, \lambda_2, \ldots, \lambda_{k-1}$, the induction hypothesis applies. Hence there are vectors $w_i \in K'_{\lambda_i}$, $1 \le i \le k - 1$, such that

$$(T - \lambda_k I)^m(x) = w_1 + w_2 + \cdots + w_{k-1}.$$

Since $K'_{\lambda_i} \subseteq K_{\lambda_i}$ for $i < k$ and $(T - \lambda_k I)^m$ maps K_{λ_i} onto itself for $i < k$, there exist vectors $v_i \in K_{\lambda_i}$ such that $(T - \lambda_k I)^m(v_i) = w_i$ for $i < k$. Thus we have

$$(T - \lambda_k I)^m(x) = (T - \lambda_k I)^m(v_1) + (T - \lambda_k I)^m(v_2) + \cdots + (T - \lambda_k I)^m(v_{k-1}),$$

and it follows that $x - (v_1 + v_2 + \cdots + v_{k-1}) \in K_{\lambda_k}$. Therefore there exists a vector $v_k \in K_{\lambda_k}$ such that

$$x = v_1 + v_2 + \cdots + v_k. \qquad \blacksquare$$

The next result extends Theorem 5.9(b) (p. 268) to all linear operators whose characteristic polynomials split. In this case, the eigenspaces are replaced by generalized eigenspaces.

Theorem 7.4. *Let T be a linear operator on a finite-dimensional vector space V such that the characteristic polynomial of T splits, and let $\lambda_1, \lambda_2, \ldots, \lambda_k$ be the distinct eigenvalues of T with corresponding multiplicities $m_1, m_2, \ldots, m_k$. For $1 \le i \le k$, let β_i be an ordered basis for K_{λ_i}. Then the following statements are true.*
(a) *$\beta_i \cap \beta_j = \varnothing$ for $i \ne j$.*
(b) *$\beta = \beta_1 \cup \beta_2 \cup \cdots \cup \beta_k$ is an ordered basis for V.*
(c) *$\dim(K_{\lambda_i}) = m_i$ for all i.*

Proof. (a) Suppose that $x \in \beta_i \cap \beta_j \subseteq K_{\lambda_i} \cap K_{\lambda_j}$, where $i \ne j$. By Theorem 7.1(b), $T - \lambda_i I$ is one-to-one on K_{λ_j}, and therefore $(T - \lambda_i I)^p(x) \ne 0$ for any positive integer p. But this contradicts the fact that $x \in K_{\lambda_i}$, and the result follows.

(b) Let $x \in V$. By Theorem 7.3, for $1 \le i \le k$, there exist vectors $v_i \in K_{\lambda_i}$ such that $x = v_1 + v_2 + \cdots + v_k$. Since each v_i is a linear combination of the vectors of β_i, it follows that x is a linear combination of the vectors of β. Therefore β spans V. Let q be the number of vectors in β. Then $\dim V \le q$. For each i, let $d_i = \dim(K_{\lambda_i})$. Then, by Theorem 7.2(a),

$$q = \sum_{i=1}^{k} d_i \le \sum_{i=1}^{k} m_i = \dim(V).$$

Hence $q = \dim(V)$. Consequently β is a basis for V by Corollary 2 to the replacement theorem (p. 47).

(c) Using the notation and result of (b), we see that $\displaystyle\sum_{i=1}^{k} d_i = \sum_{i=1}^{k} m_i$. But $d_i \leq m_i$ by Theorem 7.2(a), and therefore $d_i = m_i$ for all i. ∎

Corollary. *Let* T *be a linear operator on a finite-dimensional vector space* V *such that the characteristic polynomial of* T *splits. Then* T *is diagonalizable if and only if* $E_\lambda = K_\lambda$ *for every eigenvalue* λ *of* T.

Proof. Combining Theorems 7.4 and 5.9(a) (p. 268), we see that T is diagonalizable if and only if $\dim(E_\lambda) = \dim(K_\lambda)$ for each eigenvalue λ of T. But $E_\lambda \subseteq K_\lambda$, and hence these subspaces have the same dimension if and only if they are equal. ∎

We now focus our attention on the problem of selecting suitable bases for the generalized eigenspaces of a linear operator so that we may use Theorem 7.4 to obtain a Jordan canonical basis for the operator. For this purpose, we consider again the basis β of Example 1. We have seen that the first four vectors of β lie in the generalized eigenspace K_2. Observe that the vectors in β that determine the first Jordan block of J are of the form

$$\{v_1, v_2, v_3\} = \{(\mathsf{T} - 2\mathsf{I})^2(v_3), (\mathsf{T} - 2\mathsf{I})(v_3), v_3\}.$$

Furthermore, observe that $(\mathsf{T} - 2\mathsf{I})^3(v_3) = 0$. The relation between these vectors is the key to finding Jordan canonical bases. This leads to the following definitions.

Definitions. *Let* T *be a linear operator on a vector space* V, *and let* x *be a generalized eigenvector of* T *corresponding to the eigenvalue* λ. *Suppose that* p *is the smallest positive integer for which* $(\mathsf{T} - \lambda\mathsf{I})^p(x) = 0$. *Then the ordered set*

$$\{(\mathsf{T} - \lambda\mathsf{I})^{p-1}(x), (\mathsf{T} - \lambda\mathsf{I})^{p-2}(x), \ldots, (\mathsf{T} - \lambda\mathsf{I})(x), x\}$$

is called a **cycle of generalized eigenvectors** *of* T *corresponding to* λ. *The vectors* $(\mathsf{T} - \lambda\mathsf{I})^{p-1}(x)$ *and* x *are called the* **initial vector** *and the* **end vector** *of the cycle, respectively. We say that the* **length** *of the cycle is* p.

Notice that the initial vector of a cycle of generalized eigenvectors of a linear operator T is the only eigenvector of T in the cycle. Also observe that if x is an eigenvector of T corresponding to the eigenvalue λ, then the set $\{x\}$ is a cycle of generalized eigenvectors of T corresponding to λ of length 1.

In Example 1, the subsets $\beta_1 = \{v_1, v_2, v_3\}$, $\beta_2 = \{v_4\}$, $\beta_3 = \{v_5, v_6\}$, and $\beta_4 = \{v_7, v_8\}$ are the cycles of generalized eigenvectors of T that occur in β. Notice that β is a disjoint union of these cycles. Furthermore, setting $W_i = \text{span}(\beta_i)$ for $1 \leq i \leq 4$, we see that β_i is a basis for W_i and $[\mathsf{T}_{W_i}]_{\beta_i}$ is the ith Jordan block of the Jordan canonical form of T. This is precisely the condition that is required for a Jordan canonical basis.

Theorem 7.5. *Let* T *be a linear operator on a finite-dimensional vector space* V *whose characteristic polynomial splits, and suppose that* β *is a basis for* V *such that* β *is a disjoint union of cycles of generalized eigenvectors of* T. *Then the following statements are true.*

(a) *For each cycle* γ *of generalized eigenvectors contained in* β, W $= \text{span}(\gamma)$ *is* T*-invariant, and* $[\mathsf{T}_\mathsf{W}]_\gamma$ *is a Jordan block.*

(b) β *is a Jordan canonical basis for* V.

Proof. (a) Suppose that γ corresponds to λ, γ has length p, and x is the end vector of γ. Then $\gamma = \{v_1, v_2, \ldots, v_p\}$, where

$$v_i = (\mathsf{T} - \lambda\mathsf{I})^{p-i}(x) \text{ for } i < p \quad \text{and} \quad v_p = x.$$

So

$$(\mathsf{T} - \lambda\mathsf{I})(v_1) = (\mathsf{T} - \lambda\mathsf{I})^p(x) = 0,$$

and hence $\mathsf{T}(v_1) = \lambda v_1$. For $i > 1$,

$$(\mathsf{T} - \lambda\mathsf{I})(v_i) = (\mathsf{T} - \lambda\mathsf{I})^{p-(i-1)}(x) = v_{i-1}.$$

Therefore T maps W into itself, and, by the preceding equations, we see that $[\mathsf{T}_\mathsf{W}]_\gamma$ is a Jordan block.

For (b), simply repeat the arguments of (a) for each cycle in β in order to obtain $[\mathsf{T}]_\beta$. We leave the details as an exercise. ∎

In view of this result, we must show that, under appropriate conditions, there exist bases that are disjoint unions of cycles of generalized eigenvectors. Since the characteristic polynomial of a Jordan canonical form splits, this is a necessary condition. We will soon see that it is also sufficient. The next result moves us toward the desired existence theorem.

Theorem 7.6. *Let* T *be a linear operator on a vector space* V, *and let* λ *be an eigenvalue of* T. *Suppose that* $\gamma_1, \gamma_2, \ldots, \gamma_q$ *are cycles of generalized eigenvectors of* T *corresponding to* λ *such that the initial vectors of the* γ_i's *are distinct and form a linearly independent set. Then the* γ_i's *are disjoint, and their union* $\gamma = \bigcup_{i=1}^{q} \gamma_i$ *is linearly independent.*

Proof. Exercise 5 shows that the γ_i's are disjoint.

The proof that γ is linearly independent is by mathematical induction on the number of vectors in γ. If this number is less than 2, then the result is clear. So assume that, for some integer $n > 1$, the result is valid whenever γ has fewer than n vectors, and suppose that γ has exactly n vectors. Let W be the subspace of V generated by γ. Clearly W is $(\mathsf{T} - \lambda\mathsf{I})$-invariant, and $\dim(\mathsf{W}) \leq n$. Let U denote the restriction of $\mathsf{T} - \lambda\mathsf{I}$ to W.

For each i, let γ_i' denote the cycle obtained from γ_i by deleting the end vector. Note that if γ_i has length one, then $\gamma_i' = \varnothing$. In the case that $\gamma_i' \neq \varnothing$, each vector of γ_i' is the image under U of a vector in γ_i, and conversely, every nonzero image under U of a vector of γ_i is contained in γ_i'. Let $\gamma' = \bigcup_i \gamma_i'$.

Then by the last statement, γ' generates $\mathsf{R}(\mathsf{U})$. Furthermore, γ' consists of $n - q$ vectors, and the initial vectors of the γ_i''s are also initial vectors of the γ_i's. Thus we may apply the induction hypothesis to conclude that γ' is linearly independent. Therefore γ' is a basis for $\mathsf{R}(\mathsf{U})$. Hence $\dim(\mathsf{R}(\mathsf{U})) = n - q$. Since the q initial vectors of the γ_i's form a linearly independent set and lie in $\mathsf{N}(\mathsf{U})$, we have $\dim(\mathsf{N}(\mathsf{U})) \geq q$. From these inequalities and the dimension theorem, we obtain

$$
\begin{aligned}
n &\geq \dim(\mathsf{W}) \\
&= \dim(\mathsf{R}(\mathsf{U})) + \dim(\mathsf{N}(\mathsf{U})) \\
&\geq (n - q) + q \\
&= n.
\end{aligned}
$$

We conclude that $\dim(\mathsf{W}) = n$. Since γ generates W and consists of n vectors, it must be a basis for W. Hence γ is linearly independent. $\blacksquare$

Corollary. *Every cycle of generalized eigenvectors of a linear operator is linearly independent.*

Theorem 7.7. *Let* T *be a linear operator on a finite-dimensional vector space* V, *and let* λ *be an eigenvalue of* T. *Then* K_λ *has an ordered basis consisting of a union of disjoint cycles of generalized eigenvectors corresponding to* λ.

Proof. The proof is by mathematical induction on $n = \dim(\mathsf{K}_\lambda)$. The result is clear for $n = 1$. So suppose that for some integer $n > 1$ the result is valid whenever $\dim(\mathsf{K}_\lambda) < n$, and assume that $\dim(\mathsf{K}_\lambda) = n$. Let U denote the restriction of $\mathsf{T} - \lambda\mathsf{I}$ to K_λ. Then $\mathsf{R}(\mathsf{U})$ is a subspace of K_λ of lesser dimension, and $\mathsf{R}(\mathsf{U})$ is the space of generalized eigenvectors corresponding to λ for the restriction of T to $\mathsf{R}(\mathsf{U})$. Therefore, by the induction hypothesis, there exist disjoint cycles $\gamma_1, \gamma_2, \ldots, \gamma_q$ of generalized eigenvectors of this restriction, and hence of T itself, corresponding to λ for which $\gamma = \bigcup_{i=1}^{q} \gamma_i$ is a basis for $\mathsf{R}(\mathsf{U})$. For $1 \leq i \leq q$, the end vector of γ_i is the image under U of a vector $v_i \in \mathsf{K}_\lambda$, and so we can extend each γ_i to a larger cycle $\tilde{\gamma}_i = \gamma_i \cup \{v_i\}$ of generalized eigenvectors of T corresponding to λ. For $1 \leq i \leq q$, let w_i be the initial vector of $\tilde{\gamma}_i$ (and hence of γ_i). Since $\{w_1, w_2, \ldots, w_q\}$ is a linearly independent subset of E_λ, this set can be extended to a basis $\{w_1, w_2, \ldots, w_q, u_1, u_2, \ldots, u_s\}$

for E_λ. Then $\tilde{\gamma}_1, \tilde{\gamma}_2, \ldots, \tilde{\gamma}_q, \{u_1\}, \{u_2\}, \ldots, \{u_s\}$ are disjoint cycles of generalized eigenvectors of T corresponding to λ such that the initial vectors of these cycles are linearly independent. Therefore their union $\tilde{\gamma}$ is a linearly independent subset of K_λ by Theorem 7.6.

We show that $\tilde{\gamma}$ is a basis for K_λ. Suppose that γ consists of $r = \operatorname{rank}(\mathsf{U})$ vectors. Then $\tilde{\gamma}$ consists of $r + q + s$ vectors. Furthermore, since $\{w_1, w_2, \ldots, w_q, u_1, u_2, \ldots, u_s\}$ is a basis for $\mathsf{E}_\lambda = \mathsf{N}(\mathsf{U})$, it follows that $\operatorname{nullity}(\mathsf{U}) = q + s$. Therefore

$$\dim(\mathsf{K}_\lambda) = \operatorname{rank}(\mathsf{U}) + \operatorname{nullity}(\mathsf{U}) = r + q + s.$$

So $\tilde{\gamma}$ is a linearly independent subset of K_λ containing $\dim(\mathsf{K}_\lambda)$ vectors. It follows that $\tilde{\gamma}$ is a basis for K_λ. ∎

The following corollary is immediate.

Corollary 1. *Let* T *be a linear operator on a finite-dimensional vector space* V *whose characteristic polynomial splits. Then* T *has a Jordan canonical form.*

Proof. Let $\lambda_1, \lambda_2, \ldots, \lambda_k$ be the distinct eigenvalues of T. By Theorem 7.7, for each i there is an ordered basis β_i consisting of a disjoint union of cycles of generalized eigenvectors corresponding to λ_i. Let $\beta = \beta_1 \cup \beta_2 \cup \cdots \cup \beta_k$. Then, by Theorem 7.4(b), β is an ordered basis for V. ∎

The Jordan canonical form also can be studied from the viewpoint of matrices.

Definition. *Let* $A \in \mathsf{M}_{n \times n}(F)$ *be such that the characteristic polynomial of* A *(and hence of* L_A*) splits. Then the* **Jordan canonical form** *of* A *is defined to be the Jordan canonical form of the linear operator* L_A *on* F^n.

The next result is an immediate consequence of this definition and Corollary 1.

Corollary 2. *Let* A *be an* $n \times n$ *matrix whose characteristic polynomial splits. Then* A *has a Jordan canonical form* J, *and* A *is similar to* J.

Proof. Exercise. ∎

We can now compute the Jordan canonical forms of matrices and linear operators in some simple cases, as is illustrated in the next two examples. The tools necessary for computing the Jordan canonical forms in general are developed in the next section.

Example 2

Let

$$A = \begin{pmatrix} 3 & 1 & -2 \\ -1 & 0 & 5 \\ -1 & -1 & 4 \end{pmatrix} \in \mathsf{M}_{3\times 3}(R).$$

To find the Jordan canonical form for A, we need to find a Jordan canonical basis for $\mathsf{T} = \mathsf{L}_A$.

The characteristic polynomial of A is

$$f(t) = \det(A - tI) = -(t - 3)(t - 2)^2.$$

Hence $\lambda_1 = 3$ and $\lambda_2 = 2$ are the eigenvalues of A with multiplicities 1 and 2, respectively. By Theorem 7.4, $\dim(\mathsf{K}_{\lambda_1}) = 1$, and $\dim(\mathsf{K}_{\lambda_2}) = 2$. By Theorem 7.2, $\mathsf{K}_{\lambda_1} = \mathsf{N}(\mathsf{T} - 3\mathsf{I})$, and $\mathsf{K}_{\lambda_2} = \mathsf{N}((\mathsf{T} - 2\mathsf{I})^2)$. Since $\mathsf{E}_{\lambda_1} = \mathsf{N}(\mathsf{T} - 3\mathsf{I})$, we have that $\mathsf{E}_{\lambda_1} = \mathsf{K}_{\lambda_1}$. Observe that $(-1, 2, 1)$ is an eigenvector of T corresponding to $\lambda_1 = 3$; therefore

$$\beta_1 = \left\{ \begin{pmatrix} -1 \\ 2 \\ 1 \end{pmatrix} \right\}$$

is a basis for K_{λ_1}.

Since $\dim(\mathsf{K}_{\lambda_2}) = 2$ and a generalized eigenspace has a basis consisting of a union of cycles, this basis is either a union of two cycles of length 1 or a single cycle of length 2. The former case is impossible because the vectors in the basis would be eigenvectors—contradicting the fact that $\dim(\mathsf{E}_{\lambda_2}) = 1$. Therefore the desired basis is a single cycle of length 2. A vector v is the end vector of such a cycle if and only if $(A - 2I)v \neq 0$, but $(A - 2I)^2 v = 0$. It can easily be shown that

$$\left\{ \begin{pmatrix} 1 \\ -3 \\ -1 \end{pmatrix}, \begin{pmatrix} -1 \\ 2 \\ 0 \end{pmatrix} \right\}$$

is a basis for the solution space of the homogeneous system $(A - 2I)^2 x = 0$. Now choose a vector v in this set so that $(A - 2I)v \neq 0$. The vector $v = (-1, 2, 0)$ is an acceptable candidate for v. Since $(A - 2I)v = (1, -3, -1)$, we obtain the cycle of generalized eigenvectors

$$\beta_2 = \{(A - 2I)v, v\} = \left\{ \begin{pmatrix} 1 \\ -3 \\ -1 \end{pmatrix}, \begin{pmatrix} -1 \\ 2 \\ 0 \end{pmatrix} \right\}$$

as a basis for K_{λ_2}. Finally, we take the union of these two bases to obtain

$$\beta = \beta_1 \cup \beta_2 = \left\{ \begin{pmatrix} -1 \\ 2 \\ 1 \end{pmatrix}, \begin{pmatrix} 1 \\ -3 \\ -1 \end{pmatrix}, \begin{pmatrix} -1 \\ 2 \\ 0 \end{pmatrix} \right\},$$

which is a Jordan canonical basis for A. Therefore,

$$J = [T]_\beta = \left(\begin{array}{c|cc} 3 & 0 & 0 \\ \hline 0 & 2 & 1 \\ 0 & 0 & 2 \end{array} \right)$$

is a Jordan canonical form for A. Notice that A is similar to J. In fact, $J = Q^{-1}AQ$, where Q is the matrix whose columns are the vectors in β.

$\blacklozenge$

Example 3

Let T be the linear operator on $P_2(R)$ defined by $T(g(x)) = -g(x) - g'(x)$. We find a Jordan canonical form of T and a Jordan canonical basis for T.

Let β be the standard ordered basis for $P_2(R)$. Then

$$[T]_\beta = \begin{pmatrix} -1 & -1 & 0 \\ 0 & -1 & -2 \\ 0 & 0 & -1 \end{pmatrix},$$

which has the characteristic polynomial $f(t) = -(t+1)^3$. Thus $\lambda = -1$ is the only eigenvalue of T, and hence $K_\lambda = P_2(R)$ by Theorem 7.4. So β is a basis for K_λ. Now

$$\dim(E_\lambda) = 3 - \text{rank}(A + I) = 3 - \text{rank} \begin{pmatrix} 0 & -1 & 0 \\ 0 & 0 & -2 \\ 0 & 0 & 0 \end{pmatrix} = 3 - 2 = 1.$$

Therefore a basis for K_λ cannot be a union of two or three cycles because the initial vector of each cycle is an eigenvector, and there do not exist two or more linearly independent eigenvectors. So the desired basis must consist of a single cycle of length 3. If γ is such a cycle, then γ determines a single Jordan block

$$[T]_\gamma = \begin{pmatrix} -1 & 1 & 0 \\ 0 & -1 & 1 \\ 0 & 0 & -1 \end{pmatrix},$$

which is a Jordan canonical form of T.

The end vector $h(x)$ of such a cycle must satisfy $(T + I)^2(h(x)) \neq 0$. In any basis for K_λ, there must be a vector that satisfies this condition, or else no vector in K_λ satisfies this condition, contrary to our reasoning. Testing the vectors in β, we see that $h(x) = x^2$ is acceptable. Therefore

$$\gamma = \{(T+I)^2(x^2), (T+I)(x^2), x^2\} = \{2, -2x, x^2\}$$

is a Jordan canonical basis for T.   ◆

In the next section, we develop a computational approach for finding a Jordan canonical form and a Jordan canonical basis. In the process, we prove that Jordan canonical forms are unique up to the order of the Jordan blocks.

Let T be a linear operator on a finite-dimensional vector space V, and suppose that the characteristic polynomial of T splits. By Theorem 5.11 (p. 278), T is diagonalizable if and only if V is the direct sum of the eigenspaces of T. If T is diagonalizable, then the eigenspaces and the generalized eigenspaces coincide. The next result, which is optional, extends Theorem 5.11 to the nondiagonalizable case.

Theorem 7.8. *Let T be a linear operator on a finite-dimensional vector space V whose characteristic polynomial splits. Then V is the direct sum of the generalized eigenspaces of T.*

Proof. Exercise.   ∎

EXERCISES

1. Label the following statements as true or false.

    (a) Eigenvectors of a linear operator T are also generalized eigenvectors of T.

    (b) It is possible for a generalized eigenvector of a linear operator T to correspond to a scalar that is not an eigenvalue of T.

    (c) Any linear operator on a finite-dimensional vector space has a Jordan canonical form.

    (d) A cycle of generalized eigenvectors is linearly independent.

    (e) There is exactly one cycle of generalized eigenvectors corresponding to each eigenvalue of a linear operator on a finite-dimensional vector space.

    (f) Let T be a linear operator on a finite-dimensional vector space whose characteristic polynomial splits, and let $\lambda_1, \lambda_2, \ldots, \lambda_k$ be the distinct eigenvalues of T. If, for each i, β_i is a basis for K_{λ_i}, then $\beta_1 \cup \beta_2 \cup \cdots \cup \beta_k$ is a Jordan canonical basis for T.

    (g) For any Jordan block J, the operator L_J has Jordan canonical form J.

    (h) Let T be a linear operator on an n-dimensional vector space whose characteristic polynomial splits. Then, for any eigenvalue λ of T, $K_\lambda = N((T - \lambda I)^n)$.

2. For each matrix A, find a basis for each generalized eigenspace of L_A consisting of a union of disjoint cycles of generalized eigenvectors. Then find a Jordan canonical form J of A.

(a) $A = \begin{pmatrix} 1 & 1 \\ -1 & 3 \end{pmatrix}$

(b) $A = \begin{pmatrix} 1 & 2 \\ 3 & 2 \end{pmatrix}$

(c) $A = \begin{pmatrix} 11 & -4 & -5 \\ 21 & -8 & -11 \\ 3 & -1 & 0 \end{pmatrix}$

(d) $A = \begin{pmatrix} 2 & 1 & 0 & 0 \\ 0 & 2 & 1 & 0 \\ 0 & 0 & 3 & 0 \\ 0 & 1 & -1 & 3 \end{pmatrix}$

3. For each linear operator T, find a basis for each generalized eigenspace of T consisting of a union of disjoint cycles of generalized eigenvectors. Then find a Jordan canonical form J of T.

(a) T is the linear operator on $P_2(R)$ defined by $T(f(x)) = 2f(x) - f'(x)$

(b) V is the real vector space of functions spanned by the set of real valued functions $\{1, t, t^2, e^t, te^t\}$, and T is the linear operator on V defined by $T(f) = f'$.

(c) T is the linear operator on $M_{2\times2}(R)$ defined by $T(A) = \begin{pmatrix} 1 & 1 \\ 0 & 1 \end{pmatrix} \cdot A$

for all $A \in M_{2\times2}(R)$.

(d) $T(A) = 2A + A^t$ for all $A \in M_{2\times2}(R)$.

4.† Let T be a linear operator on a vector space V, and let γ be a cycle of generalized eigenvectors that corresponds to the eigenvalue λ. Prove that span(γ) is a T-invariant subspace of V.

5. Let $\gamma_1, \gamma_2, \ldots, \gamma_p$ be cycles of generalized eigenvectors of a linear operator T corresponding to an eigenvalue λ. Prove that if the initial eigenvectors are distinct, then the cycles are disjoint.

6. Let $T: V \to W$ be a linear transformation. Prove the following results.
    (a) $N(T) = N(-T)$.
    (b) $N(T^k) = N((-T)^k)$.
    (c) If $V = W$ (so that T is a linear operator on V) and λ is an eigenvalue of T, then for any positive integer k

$$N((T - \lambda I_V)^k) = N((\lambda I_V - T)^k).$$

7. Let U be a linear operator on a finite-dimensional vector space V. Prove the following results.
    (a) $N(U) \subseteq N(U^2) \subseteq \cdots \subseteq N(U^k) \subseteq N(U^{k+1}) \subseteq \cdots$.

(b) If $\text{rank}(\mathsf{U}^m) = \text{rank}(\mathsf{U}^{m+1})$ for some positive integer m, then $\text{rank}(\mathsf{U}^m) = \text{rank}(\mathsf{U}^k)$ for any positive integer $k \geq m$.

(c) If $\text{rank}(\mathsf{U}^m) = \text{rank}(\mathsf{U}^{m+1})$ for some positive integer m, then $\mathsf{N}(\mathsf{U}^m) = \mathsf{N}(\mathsf{U}^k)$ for any positive integer $k \geq m$.

(d) Let T be a linear operator on V, and let λ be an eigenvalue of T. Prove that if $\text{rank}((\mathsf{T}-\lambda\mathsf{I})^m) = \text{rank}((\mathsf{T}-\lambda\mathsf{I})^{m+1})$ for some integer m, then $\mathsf{K}_\lambda = \mathsf{N}((\mathsf{T} - \lambda\mathsf{I})^m)$.

(e) *Second Test for Diagonalizability.* Let T be a linear operator on V whose characteristic polynomial splits, and let $\lambda_1, \lambda_2, \ldots, \lambda_k$ be the distinct eigenvalues of T. Then T is diagonalizable if and only if $\text{rank}(\mathsf{T} - \lambda_i\mathsf{I}) = \text{rank}((\mathsf{T} - \lambda_i\mathsf{I})^2)$ for $1 \leq i \leq k$.

(f) Use (e) to obtain a simpler proof of Exercise 24 of Section 5.4: If T is a diagonalizable linear operator on a finite-dimensional vector space V and W is a T-invariant subspace of V, then T_W is diagonalizable.

8. Use Theorem 7.4 to prove that the vectors $v_1, v_2, \ldots, v_k$ in the statement of Theorem 7.3 are unique.

9. Let T be a linear operator on a finite-dimensional vector space V whose characteristic polynomial splits.

(a) Prove Theorem 7.5(b).

(b) Suppose that β is a Jordan canonical basis for T, and let λ be an eigenvalue of T. Let $\beta' = \beta \cap \mathsf{K}_\lambda$. Prove that β' is a basis for K_λ.

10. Let T be a linear operator on a finite-dimensional vector space whose characteristic polynomial splits, and let λ be an eigenvalue of T.

(a) Suppose that γ is a basis for K_λ consisting of the union of q disjoint cycles of generalized eigenvectors. Prove that $q \leq \dim(\mathsf{E}_\lambda)$.

(b) Let β be a Jordan canonical basis for T, and suppose that $J = [\mathsf{T}]_\beta$ has q Jordan blocks with λ in the diagonal positions. Prove that $q \leq \dim(\mathsf{E}_\lambda)$.

11. Prove Corollary 2 to Theorem 7.7.

Exercises 12 and 13 are concerned with direct sums of matrices, defined in Section 5.4 on page 320.

12. Prove Theorem 7.8.

13. Let T be a linear operator on a finite-dimensional vector space V such that the characteristic polynomial of T splits, and let $\lambda_1, \lambda_2, \ldots, \lambda_k$ be the distinct eigenvalues of T. For each i, let J_i be the Jordan canonical form of the restriction of T to K_{λ_i}. Prove that

$$J = J_1 \oplus J_2 \oplus \cdots \oplus J_k$$

is the Jordan canonical form of J.

7.2   THE JORDAN CANONICAL FORM II

For the purposes of this section, we fix a linear operator T on an n-dimensional vector space V such that the characteristic polynomial of T splits. Let $\lambda_1, \lambda_2, \ldots, \lambda_k$ be the distinct eigenvalues of T.

By Theorem 7.7 (p. 490), each generalized eigenspace K_{λ_i} contains an ordered basis β_i consisting of a union of disjoint cycles of generalized eigenvectors corresponding to λ_i. So by Theorems 7.4(b) (p. 487) and 7.5 (p. 489), the union $\beta = \bigcup\limits_{i=1}^{k} \beta_i$ is a Jordan canonical basis for T. For each i, let T_i be the restriction of T to K_{λ_i}, and let $A_i = [T_i]_{\beta_i}$. Then A_i is the Jordan canonical form of T_i, and

$$
J = [T]_\beta = \begin{pmatrix} A_1 & O & \cdots & O \\ O & A_2 & \cdots & O \\ \vdots & \vdots & & \vdots \\ O & O & \cdots & A_k \end{pmatrix}
$$

is the Jordan canonical form of T. In this matrix, each O is a zero matrix of appropriate size.

In this section, we compute the matrices A_i and the bases β_i, thereby computing J and β as well. While developing a method for finding J, it becomes evident that in some sense the matrices A_i are unique.

To aid in formulating the uniqueness theorem for J, we adopt the following convention: The basis β_i for K_{λ_i} will henceforth be ordered in such a way that the cycles appear in order of decreasing length. That is, if β_i is a disjoint union of cycles $\gamma_1, \gamma_2, \ldots, \gamma_{n_i}$ and if the length of the cycle γ_j is p_j, we index the cycles so that $p_1 \geq p_2 \geq \cdots \geq p_{n_i}$. This ordering of the cycles limits the possible orderings of vectors in β_i, which in turn determines the matrix A_i. It is in this sense that A_i is unique. It then follows that the Jordan canonical form for T is unique up to an ordering of the eigenvalues of T. As we will see, there is no uniqueness theorem for the bases β_i or for β. Specifically, we show that for each i, the number n_i of cycles that form β_i, and the length p_j ($j = 1, 2, \ldots, n_i$) of each cycle, is completely determined by T.

Example 1

To illustrate the discussion above, suppose that, for some i, the ordered basis β_i for K_{λ_i} is the union of four cycles $\beta_i = \gamma_1 \cup \gamma_2 \cup \gamma_3 \cup \gamma_4$ with respective lengths $p_1 = 3$, $p_2 = 3$, $p_3 = 2$, and $p_4 = 1$. Then

$$
A_i = \begin{pmatrix}
\lambda_i & 1 & 0 & 0 & 0 & 0 & 0 & 0 & 0 \\
0 & \lambda_i & 1 & 0 & 0 & 0 & 0 & 0 & 0 \\
0 & 0 & \lambda_i & 0 & 0 & 0 & 0 & 0 & 0 \\
0 & 0 & 0 & \lambda_i & 1 & 0 & 0 & 0 & 0 \\
0 & 0 & 0 & 0 & \lambda_i & 1 & 0 & 0 & 0 \\
0 & 0 & 0 & 0 & 0 & \lambda_i & 0 & 0 & 0 \\
0 & 0 & 0 & 0 & 0 & 0 & \lambda_i & 1 & 0 \\
0 & 0 & 0 & 0 & 0 & 0 & 0 & \lambda_i & 0 \\
0 & 0 & 0 & 0 & 0 & 0 & 0 & 0 & \lambda_i
\end{pmatrix}. \quad \blacklozenge
$$

To help us visualize each of the matrices A_i and ordered bases β_i, we use an array of dots called a **dot diagram** of T_i, where T_i is the restriction of T to K_{λ_i}. Suppose that β_i is a disjoint union of cycles of generalized eigenvectors $\gamma_1, \gamma_2, \ldots, \gamma_{n_i}$ with lengths $p_1 \geq p_2 \geq \cdots \geq p_{n_i}$, respectively. The dot diagram of T_i contains one dot for each vector in β_i, and the dots are configured according to the following rules.

1. The array consists of n_i columns (one column for each cycle).
2. Counting from left to right, the jth column consists of the p_j dots that correspond to the vectors of γ_j starting with the initial vector at the top and continuing down to the end vector.

Denote the end vectors of the cycles by $v_1, v_2, \ldots, v_{n_i}$. In the following dot diagram of T_i, each dot is labeled with the name of the vector in β_i to which it corresponds.

$\bullet\, (\mathsf{T} - \lambda_i \mathsf{I})^{p_1 - 1}(v_1)$ $\bullet\, (\mathsf{T} - \lambda_i \mathsf{I})^{p_2 - 1}(v_2)$ $\cdots$ $\bullet\, (\mathsf{T} - \lambda_i \mathsf{I})^{p_{n_i} - 1}(v_{n_i})$
$\bullet\, (\mathsf{T} - \lambda_i \mathsf{I})^{p_1 - 2}(v_1)$ $\bullet\, (\mathsf{T} - \lambda_i \mathsf{I})^{p_2 - 2}(v_2)$ $\cdots$ $\bullet\, (\mathsf{T} - \lambda_i \mathsf{I})^{p_{n_i} - 2}(v_{n_i})$

$\vdots$ $\vdots$ $\vdots$

                        $\bullet\, (\mathsf{T} - \lambda_i \mathsf{I})(v_{n_i})$
                        $\bullet\, v_{n_i}$

         $\bullet\, (\mathsf{T} - \lambda_i \mathsf{I})(v_2)$
         $\bullet\, v_2$

$\bullet\, (\mathsf{T} - \lambda_i \mathsf{I})(v_1)$
$\bullet\, v_1$

Notice that the dot diagram of T_i has n_i columns (one for each cycle) and p_1 rows. Since $p_1 \geq p_2 \geq \cdots \geq p_{n_i}$, the columns of the dot diagram become shorter (or at least not longer) as we move from left to right.

Now let r_j denote the number of dots in the jth row of the dot diagram. Observe that $r_1 \geq r_2 \geq \cdots \geq r_{p_1}$. Furthermore, the diagram can be re-constructed from the values of the r_i's. The proofs of these facts, which are combinatorial in nature, are treated in Exercise 9.

In Example 1, with $n_i = 4$, $p_1 = p_2 = 3$, $p_3 = 2$, and $p_4 = 1$, the dot diagram of T_i is as follows:

$$
\begin{matrix}
\bullet & \bullet & \bullet & \bullet \\
\bullet & \bullet & \bullet & \\
\bullet & \bullet & &
\end{matrix}
$$

Here $r_1 = 4$, $r_2 = 3$, and $r_3 = 2$.

We now devise a method for computing the dot diagram of T_i using the ranks of linear operators determined by T and λ_i. Hence the dot diagram is completely determined by T, from which it follows that it is unique. On the other hand, β_i is not unique. For example, see Exercise 8. (It is for this reason that we associate the dot diagram with T_i rather than with β_i.)

To determine the dot diagram of T_i, we devise a method for computing each r_j, the number of dots in the jth row of the dot diagram, using only T and λ_i. The next three results give us the required method. To facilitate our arguments, we fix a basis β_i for K_{λ_i} so that β_i is a disjoint union of n_i cycles of generalized eigenvectors with lengths $p_1 \geq p_2 \geq \cdots \geq p_{n_i}$.

Theorem 7.9. *For any positive integer r, the vectors in β_i that are associated with the dots in the first r rows of the dot diagram of T_i constitute a basis for $N((T - \lambda_i I)^r)$. Hence the number of dots in the first r rows of the dot diagram equals* nullity$((T - \lambda_i I)^r)$.

Proof. Clearly, $N((T - \lambda_i I)^r) \subseteq K_{\lambda_i}$, and K_{λ_i} is invariant under $(T - \lambda_i I)^r$. Let U denote the restriction of $(T - \lambda_i I)^r$ to K_{λ_i}. By the preceding remarks, $N((T - \lambda_i I)^r) = N(U)$, and hence it suffices to establish the theorem for U. Now define

$$S_1 = \{x \in \beta_i \colon U(x) = 0\} \quad \text{and} \quad S_2 = \{x \in \beta_i \colon U(x) \neq 0\}.$$

Let a and b denote the number of vectors in S_1 and S_2, respectively, and let $m_i = \dim(K_{\lambda_i})$. Then $a + b = m_i$. For any $x \in \beta_i$, $x \in S_1$ if and only if x is one of the first r vectors of a cycle, and this is true if and only if x corresponds to a dot in the first r rows of the dot diagram. Hence a is the number of dots in the first r rows of the dot diagram. For any $x \in S_2$, the effect of applying U to x is to move the dot corresponding to x exactly r places up its column to another dot. It follows that U maps S_2 in a one-to-one fashion into β_i. Thus $\{U(x) \colon x \in S_2\}$ is a basis for $R(U)$ consisting of b vectors. Hence rank$(U) = b$, and so nullity$(U) = m_i - b = a$. But S_1 is a linearly independent subset of $N(U)$ consisting of a vectors; therefore S_1 is a basis for $N(U)$. ∎

In the case that $r = 1$, Theorem 7.9 yields the following corollary.

Corollary. *The dimension of E_{λ_i} is n_i. Hence in a Jordan canonical form of T, the number of Jordan blocks corresponding to λ_i equals the dimension of E_{λ_i}.*

Proof. Exercise. ∎

We are now able to devise a method for describing the dot diagram in terms of the ranks of operators.

Theorem 7.10. *Let r_j denote the number of dots in the jth row of the dot diagram of T_i, the restriction of T to K_{λ_i}. Then the following statements are true.*

(a) $r_1 = \dim(V) - \text{rank}(T - \lambda_i I)$.

(b) $r_j = \text{rank}((T - \lambda_i I)^{j-1}) - \text{rank}((T - \lambda_i I)^j)$ *if $j > 1$.*

Proof. By Theorem 7.9, for $1 \leq j \leq p_1$, we have

$$r_1 + r_2 + \cdots + r_j = \text{nullity}((T - \lambda_i I)^j)$$
$$= \dim(V) - \text{rank}((T - \lambda_i I)^j).$$

Hence

$$r_1 = \dim(V) - \text{rank}(T - \lambda_i I),$$

and for $j > 1$,

$$r_j = (r_1 + r_2 + \cdots + r_j) - (r_1 + r_2 + \cdots + r_{j-1})$$
$$= [\dim(V) - \text{rank}((T - \lambda_i I)^j)] - [\dim(V) - \text{rank}((T - \lambda_i I)^{j-1})]$$
$$= \text{rank}((T - \lambda_i I)^{j-1}) - \text{rank}((T - \lambda_i I)^j). \quad ∎$$

Theorem 7.10 shows that the dot diagram of T_i is completely determined by T and λ_i. Hence we have proved the following result.

Corollary. *For any eigenvalue λ_i of T, the dot diagram of T_i is unique. Thus, subject to the convention that the cycles of generalized eigenvectors for the bases of each generalized eigenspace are listed in order of decreasing length, the Jordan canonical form of a linear operator or a matrix is unique up to the ordering of the eigenvalues.*

We apply these results to find the Jordan canonical forms of two matrices and a linear operator.

Example 2

Let

$$A = \begin{pmatrix} 2 & -1 & 0 & 1 \\ 0 & 3 & -1 & 0 \\ 0 & 1 & 1 & 0 \\ 0 & -1 & 0 & 3 \end{pmatrix}.$$

We find the Jordan canonical form of A and a Jordan canonical basis for the linear operator $T = L_A$. The characteristic polynomial of A is

$$\det(A - tI) = (t-2)^3(t-3).$$

Thus A has two distinct eigenvalues, $\lambda_1 = 2$ and $\lambda_2 = 3$, with multiplicities 3 and 1, respectively. Let T_1 and T_2 be the restrictions of L_A to the generalized eigenspaces K_{λ_1} and K_{λ_2}, respectively.

Suppose that β_1 is a Jordan canonical basis for T_1. Since λ_1 has multiplicity 3, it follows that $\dim(K_{\lambda_1}) = 3$ by Theorem 7.4(c) (p. 487); hence the dot diagram of T_1 has three dots. As we did earlier, let r_j denote the number of dots in the jth row of this dot diagram. Then, by Theorem 7.10,

$$r_1 = 4 - \text{rank}(A - 2I) = 4 - \text{rank} \begin{pmatrix} 0 & -1 & 0 & 1 \\ 0 & 1 & -1 & 0 \\ 0 & 1 & -1 & 0 \\ 0 & -1 & 0 & 1 \end{pmatrix} = 4 - 2 = 2,$$

and

$$r_2 = \text{rank}(A - 2I) - \text{rank}((A - 2I)^2) = 2 - 1 = 1.$$

(Actually, the computation of r_2 is unnecessary in this case because $r_1 = 2$ and the dot diagram only contains three dots.) Hence the dot diagram associated with β_1 is

So

$$A_1 = [T_1]_{\beta_1} = \begin{pmatrix} 2 & 1 & 0 \\ 0 & 2 & 0 \\ 0 & 0 & 2 \end{pmatrix}.$$

Since $\lambda_2 = 3$ has multiplicity 1, it follows that $\dim(K_{\lambda_2}) = 1$, and consequently any basis β_2 for K_{λ_2} consists of a single eigenvector corresponding to $\lambda_2 = 3$. Therefore

$$A_2 = [T_2]_{\beta_2} = (3).$$

Setting $\beta = \beta_1 \cup \beta_2$, we have

$$J = [L_A]_\beta = \begin{pmatrix} 2 & 1 & 0 & 0 \\ 0 & 2 & 0 & 0 \\ 0 & 0 & 2 & 0 \\ 0 & 0 & 0 & 3 \end{pmatrix},$$

and so J is the Jordan canonical form of A.

We now find a Jordan canonical basis for $T = L_A$. We begin by determining a Jordan canonical basis β_1 for T_1. Since the dot diagram of T_1 has two columns, each corresponding to a cycle of generalized eigenvectors, there are two such cycles. Let v_1 and v_2 denote the end vectors of the first and second cycles, respectively. We reprint below the dot diagram with the dots labeled with the names of the vectors to which they correspond.

$$\bullet\,(T - 2I)(v_1) \quad \bullet\, v_2$$
$$\bullet\, v_1$$

From this diagram we see that $v_1 \in N((T - 2I)^2)$ but $v_1 \notin N(T - 2I)$. Now

$$A - 2I = \begin{pmatrix} 0 & -1 & 0 & 1 \\ 0 & 1 & -1 & 0 \\ 0 & 1 & -1 & 0 \\ 0 & -1 & 0 & 1 \end{pmatrix} \quad \text{and} \quad (A - 2I)^2 = \begin{pmatrix} 0 & -2 & 1 & 1 \\ 0 & 0 & 0 & 0 \\ 0 & 0 & 0 & 0 \\ 0 & -2 & 1 & 1 \end{pmatrix}.$$

It is easily seen that

$$\left\{ \begin{pmatrix} 1 \\ 0 \\ 0 \\ 0 \end{pmatrix}, \begin{pmatrix} 0 \\ 1 \\ 2 \\ 0 \end{pmatrix}, \begin{pmatrix} 0 \\ 1 \\ 0 \\ 2 \end{pmatrix} \right\}$$

is a basis for $N((T - 2I)^2) = K_{\lambda_1}$. Of these three basis vectors, the last two do not belong to $N(T - 2I)$, and hence we select one of these for v_1. Suppose that we choose

$$v_1 = \begin{pmatrix} 0 \\ 1 \\ 2 \\ 0 \end{pmatrix}.$$

Then

$$(T - 2I)(v_1) = (A - 2I)(v_1) = \begin{pmatrix} 0 & -1 & 0 & 1 \\ 0 & 1 & -1 & 0 \\ 0 & 1 & -1 & 0 \\ 0 & -1 & 0 & 1 \end{pmatrix} \begin{pmatrix} 0 \\ 1 \\ 2 \\ 0 \end{pmatrix} = \begin{pmatrix} -1 \\ -1 \\ -1 \\ -1 \end{pmatrix}.$$

Now simply choose v_2 to be a vector in E_{λ_1} that is linearly independent of $(T - 2I)(v_1)$; for example, select

$$v_2 = \begin{pmatrix} 1 \\ 0 \\ 0 \\ 0 \end{pmatrix}.$$

Thus we have associated the Jordan canonical basis

$$\beta_1 = \left\{ \begin{pmatrix} -1 \\ -1 \\ -1 \\ -1 \end{pmatrix}, \begin{pmatrix} 0 \\ 1 \\ 2 \\ 0 \end{pmatrix}, \begin{pmatrix} 1 \\ 0 \\ 0 \\ 0 \end{pmatrix} \right\}$$

with the dot diagram in the following manner.

$$\bullet \begin{pmatrix} -1 \\ -1 \\ -1 \\ -1 \end{pmatrix} \quad \bullet \begin{pmatrix} 1 \\ 0 \\ 0 \\ 0 \end{pmatrix}$$

$$\bullet \begin{pmatrix} 0 \\ 1 \\ 2 \\ 0 \end{pmatrix}$$

By Theorem 7.6 (p. 489), the linear independence of β_1 is guaranteed since v_2 was chosen to be linearly independent of $(\mathsf{T} - 2\mathsf{I})(v_1)$.

Since $\lambda_2 = 3$ has multiplicity 1, $\dim(\mathsf{K}_{\lambda_2}) = \dim(\mathsf{E}_{\lambda_2}) = 1$. Hence any eigenvector of L_A corresponding to $\lambda_2 = 3$ constitutes an appropriate basis β_2. For example,

$$\beta_2 = \left\{ \begin{pmatrix} 1 \\ 0 \\ 0 \\ 1 \end{pmatrix} \right\}.$$

Thus

$$\beta = \beta_1 \cup \beta_2 = \left\{ \begin{pmatrix} -1 \\ -1 \\ -1 \\ -1 \end{pmatrix}, \begin{pmatrix} 0 \\ 1 \\ 2 \\ 0 \end{pmatrix}, \begin{pmatrix} 1 \\ 0 \\ 0 \\ 0 \end{pmatrix}, \begin{pmatrix} 1 \\ 0 \\ 0 \\ 1 \end{pmatrix} \right\}$$

is a Jordan canonical basis for L_A.

Notice that if

$$Q = \begin{pmatrix} -1 & 0 & 1 & 1 \\ -1 & 1 & 0 & 0 \\ -1 & 2 & 0 & 0 \\ -1 & 0 & 0 & 1 \end{pmatrix},$$

then $J = Q^{-1}AQ$.   ◆

Example 3

Let

$$A = \begin{pmatrix} 2 & -4 & 2 & 2 \\ -2 & 0 & 1 & 3 \\ -2 & -2 & 3 & 3 \\ -2 & -6 & 3 & 7 \end{pmatrix}.$$

We find the Jordan canonical form J of A, a Jordan canonical basis for L_A, and a matrix Q such that $J = Q^{-1}AQ$.

The characteristic polynomial of A is $\det(A - tI) = (t - 2)^2(t - 4)^2$. Let $\mathsf{T} = \mathsf{L}_A$, $\lambda_1 = 2$, and $\lambda_2 = 4$, and let T_i be the restriction of L_A to K_{λ_i} for $i = 1, 2$.

We begin by computing the dot diagram of T_1. Let r_1 denote the number of dots in the first row of this diagram. Then

$$r_1 = 4 - \text{rank}(A - 2I) = 4 - 2 = 2;$$

hence the dot diagram of T_1 is as follows.

$$\bullet \quad \bullet$$

Therefore

$$A_1 = [\mathsf{T}_1]_{\beta_1} = \begin{pmatrix} 2 & 0 \\ 0 & 2 \end{pmatrix},$$

where β_1 is any basis corresponding to the dots. In this case, β_1 is an arbitrary basis for $\mathsf{E}_{\lambda_1} = \mathsf{N}(\mathsf{T} - 2\mathsf{I})$, for example,

$$\beta_1 = \left\{ \begin{pmatrix} 2 \\ 1 \\ 0 \\ 2 \end{pmatrix}, \begin{pmatrix} 0 \\ 1 \\ 2 \\ 0 \end{pmatrix} \right\}.$$

Next we compute the dot diagram of T_2. Since $\text{rank}(A - 4I) = 3$, there is only $4 - 3 = 1$ dot in the first row of the diagram. Since $\lambda_2 = 4$ has multiplicity 2, we have $\dim(\mathsf{K}_{\lambda_2}) = 2$, and hence this dot diagram has the following form:

$$\bullet$$
$$\bullet$$

Thus

$$A_2 = [\mathsf{T}_2]_{\beta_2} = \begin{pmatrix} 4 & 1 \\ 0 & 4 \end{pmatrix},$$

where β_2 is any basis for K_{λ_2} corresponding to the dots. In this case, β_2 is a cycle of length 2. The end vector of this cycle is a vector $v \in K_{\lambda_2} = N((T - 4I)^2)$ such that $v \notin N(T - 4I)$. One way of finding such a vector was used to select the vector v_1 in Example 2. In this example, we illustrate another method. A simple calculation shows that a basis for the null space of $L_A - 4I$ is

$$\left\{ \begin{pmatrix} 0 \\ 1 \\ 1 \\ 1 \end{pmatrix} \right\}.$$

Choose v to be any solution to the system of linear equations

$$(A - 4I)x = \begin{pmatrix} 0 \\ 1 \\ 1 \\ 1 \end{pmatrix},$$

for example,

$$v = \begin{pmatrix} 1 \\ -1 \\ -1 \\ 0 \end{pmatrix}.$$

Thus

$$\beta_2 = \{(L_A - 4I)(v), v\} = \left\{ \begin{pmatrix} 0 \\ 1 \\ 1 \\ 1 \end{pmatrix}, \begin{pmatrix} 1 \\ -1 \\ -1 \\ 0 \end{pmatrix} \right\}.$$

Therefore

$$\beta = \beta_1 \cup \beta_2 = \left\{ \begin{pmatrix} 2 \\ 1 \\ 0 \\ 2 \end{pmatrix}, \begin{pmatrix} 0 \\ 1 \\ 2 \\ 0 \end{pmatrix}, \begin{pmatrix} 0 \\ 1 \\ 1 \\ 1 \end{pmatrix}, \begin{pmatrix} 1 \\ -1 \\ -1 \\ 0 \end{pmatrix} \right\}$$

is a Jordan canonical basis for L_A. The corresponding Jordan canonical form is given by

$$J = [L_A]_\beta = \begin{pmatrix} A_1 & O \\ O & A_2 \end{pmatrix} = \left(\begin{array}{cc|cc} 2 & 0 & 0 & 0 \\ 0 & 2 & 0 & 0 \\ \hline 0 & 0 & 4 & 1 \\ 0 & 0 & 0 & 4 \end{array} \right).$$

Finally, we define Q to be the matrix whose columns are the vectors of β listed in the same order, namely,

$$Q = \begin{pmatrix} 2 & 0 & 0 & 1 \\ 1 & 1 & 1 & -1 \\ 0 & 2 & 1 & -1 \\ 2 & 0 & 1 & 0 \end{pmatrix}.$$

Then $J = Q^{-1}AQ$.   ◆

Example 4

Let V be the vector space of polynomial functions in two real variables x and y of degree at most 2. Then V is a vector space over R and $\alpha = \{1, x, y, x^2, y^2, xy\}$ is an ordered basis for V. Let T be the linear operator on V defined by

$$\mathsf{T}(f(x,y)) = \frac{\partial}{\partial x} f(x,y).$$

For example, if $f(x,y) = x + 2x^2 - 3xy + y$, then

$$\mathsf{T}(f(x,y)) = \frac{\partial}{\partial x}(x + 2x^2 - 3xy + y) = 1 + 4x - 3y.$$

We find the Jordan canonical form and a Jordan canonical basis for T.

Let $A = [\mathsf{T}]_\alpha$. Then

$$A = \begin{pmatrix} 0 & 1 & 0 & 0 & 0 & 0 \\ 0 & 0 & 0 & 2 & 0 & 0 \\ 0 & 0 & 0 & 0 & 0 & 1 \\ 0 & 0 & 0 & 0 & 0 & 0 \\ 0 & 0 & 0 & 0 & 0 & 0 \\ 0 & 0 & 0 & 0 & 0 & 0 \end{pmatrix},$$

and hence the characteristic polynomial of T is

$$\det(A - tI) = \det \begin{pmatrix} -t & 1 & 0 & 0 & 0 & 0 \\ 0 & -t & 0 & 2 & 0 & 0 \\ 0 & 0 & -t & 0 & 0 & 1 \\ 0 & 0 & 0 & -t & 0 & 0 \\ 0 & 0 & 0 & 0 & -t & 0 \\ 0 & 0 & 0 & 0 & 0 & -t \end{pmatrix} = t^6.$$

Thus $\lambda = 0$ is the only eigenvalue of T, and $\mathsf{K}_\lambda = \mathsf{V}$. For each j, let r_j denote the number of dots in the jth row of the dot diagram of T. By Theorem 7.10,

$$r_1 = 6 - \operatorname{rank}(A) = 6 - 3 = 3,$$

and since

$$A^2 = \begin{pmatrix} 0 & 0 & 0 & 2 & 0 & 0 \\ 0 & 0 & 0 & 0 & 0 & 0 \\ 0 & 0 & 0 & 0 & 0 & 0 \\ 0 & 0 & 0 & 0 & 0 & 0 \\ 0 & 0 & 0 & 0 & 0 & 0 \\ 0 & 0 & 0 & 0 & 0 & 0 \end{pmatrix},$$

$r_2 = \text{rank}(A) - \text{rank}(A^2) = 3 - 1 = 2.$

Because there are a total of six dots in the dot diagram and $r_1 = 3$ and $r_2 = 2$, it follows that $r_3 = 1$. So the dot diagram of T is

We conclude that the Jordan canonical form of T is

$$J = \left(\begin{array}{ccc|ccc} 0 & 1 & 0 & 0 & 0 & 0 \\ 0 & 0 & 1 & 0 & 0 & 0 \\ 0 & 0 & 0 & 0 & 0 & 0 \\ \hline 0 & 0 & 0 & 0 & 1 & 0 \\ 0 & 0 & 0 & 0 & 0 & 0 \\ 0 & 0 & 0 & 0 & 0 & 0 \end{array} \right).$$

We now find a Jordan canonical basis for T. Since the first column of the dot diagram of T consists of three dots, we must find a polynomial $f_1(x, y)$ such that $\dfrac{\partial^2}{\partial x^2} f_1(x, y) \neq 0$. Examining the basis $\alpha = \{1, x, y, x^2, y^2, xy\}$ for $K_\lambda = V$, we see that x^2 is a suitable candidate. Setting $f_1(x, y) = x^2$, we see that

$$(\mathsf{T} - \lambda\mathsf{I})(f_1(x, y)) = \mathsf{T}(f_1(x, y)) = \frac{\partial}{\partial x}(x^2) = 2x$$

and

$$(\mathsf{T} - \lambda\mathsf{I})^2(f_1(x, y)) = \mathsf{T}^2(f_1(x, y)) = \frac{\partial^2}{\partial x^2}(x^2) = 2.$$

Likewise, since the second column of the dot diagram consists of two dots, we must find a polynomial $f_2(x, y)$ such that

$$\frac{\partial}{\partial x}(f_2(x, y)) \neq 0, \quad \text{but} \quad \frac{\partial^2}{\partial x^2}(f_2(x, y)) = 0.$$

Since our choice must be linearly independent of the polynomials already chosen for the first cycle, the only choice in α that satisfies these constraints is xy. So we set $f_2(x, y) = xy$. Thus

$$(\mathsf{T} - \lambda\mathsf{I})(f_2(x, y)) = \mathsf{T}(f_2(x, y)) = \frac{\partial}{\partial x}(xy) = y.$$

Finally, the third column of the dot diagram consists of a single polynomial that lies in the null space of T. The only remaining polynomial in α is y^2, and it is suitable here. So set $f_3(x, y) = y^2$. Therefore we have identified polynomials with the dots in the dot diagram as follows.

$$
\begin{array}{lll}
\bullet\, 2 & \bullet\, y & \bullet\, y^2 \\
\bullet\, 2x & \bullet\, xy & \\
\bullet\, x^2 & &
\end{array}
$$

Thus $\beta = \{2, 2x, x^2, y, xy, y^2\}$ is a Jordan canonical basis for T. ◆

In the three preceding examples, we relied on our ingenuity and the context of the problem to find Jordan canonical bases. The reader can do the same in the exercises. We are successful in these cases because the dimensions of the generalized eigenspaces under consideration are small. We do not attempt, however, to develop a general algorithm for computing Jordan canonical bases, although one could be devised by following the steps in the proof of the existence of such a basis (Theorem 7.7 p. 490).

The following result may be thought of as a corollary to Theorem 7.10.

Theorem 7.11. Let A and B be $n \times n$ matrices, each having Jordan canonical forms computed according to the conventions of this section. Then A and B are similar if and only if they have (up to an ordering of their eigenvalues) the same Jordan canonical form.

Proof. If A and B have the same Jordan canonical form J, then A and B are each similar to J and hence are similar to each other.

Conversely, suppose that A and B are similar. Then A and B have the same eigenvalues. Let J_A and J_B denote the Jordan canonical forms of A and B, respectively, with the same ordering of their eigenvalues. Then A is similar to both J_A and J_B, and therefore, by the corollary to Theorem 2.23 (p. 115), J_A and J_B are matrix representations of L_A. Hence J_A and J_B are Jordan canonical forms of L_A. Thus $J_A = J_B$ by the corollary to Theorem 7.10. ∎

Example 5

We determine which of the matrices

$$
A = \begin{pmatrix} -3 & 3 & -2 \\ -7 & 6 & -3 \\ 1 & -1 & 2 \end{pmatrix}, \qquad
B = \begin{pmatrix} 0 & 1 & -1 \\ -4 & 4 & -2 \\ -2 & 1 & 1 \end{pmatrix},
$$

$$C = \begin{pmatrix} 0 & -1 & -1 \\ -3 & -1 & -2 \\ 7 & 5 & 6 \end{pmatrix}, \quad \text{and} \quad D = \begin{pmatrix} 0 & 1 & 2 \\ 0 & 1 & 1 \\ 0 & 0 & 2 \end{pmatrix}$$

are similar. Observe that A, B, and C have the same characteristic polynomial $-(t-1)(t-2)^2$, whereas D has $-t(t-1)(t-2)$ as its characteristic polynomial. Because similar matrices have the same characteristic polynomials, D cannot be similar to A, B, or C. Let J_A, J_B, and J_C be the Jordan canonical forms of A, B, and C, respectively, using the ordering 1, 2 for their common eigenvalues. Then (see Exercise 4)

$$J_A = \begin{pmatrix} 1 & 0 & 0 \\ 0 & 2 & 1 \\ 0 & 0 & 2 \end{pmatrix}, \quad J_B = \begin{pmatrix} 1 & 0 & 0 \\ 0 & 2 & 0 \\ 0 & 0 & 2 \end{pmatrix}, \quad \text{and} \quad J_C = \begin{pmatrix} 1 & 0 & 0 \\ 0 & 2 & 1 \\ 0 & 0 & 2 \end{pmatrix}.$$

Since $J_A = J_C$, A is similar to C. Since J_B is different from J_A and J_C, B is similar to neither A nor C. ◆

The reader should observe that any diagonal matrix is a Jordan canonical form. Thus a linear operator T on a finite-dimensional vector space V *is diagonalizable if and only if its Jordan canonical form is a diagonal matrix. Hence* T *is diagonalizable if and only if the Jordan canonical basis for* T *consists of eigenvectors of* T. Similar statements can be made about matrices. Thus, of the matrices A, B, and C in Example 5, A and C are not diagonalizable because their Jordan canonical forms are not diagonal matrices.

EXERCISES

1. Label the following statements as true or false. Assume that the characteristic polynomial of the matrix or linear operator splits.

    (a) The Jordan canonical form of a diagonal matrix is the matrix itself.
    (b) Let T be a linear operator on a finite-dimensional vector space V that has a Jordan canonical form J. If β is any basis for V, then the Jordan canonical form of $[T]_\beta$ is J.
    (c) Linear operators having the same characteristic polynomial are similar.
    (d) Matrices having the same Jordan canonical form are similar.
    (e) Every matrix is similar to its Jordan canonical form.
    (f) Every linear operator with the characteristic polynomial $(-1)^n(t-\lambda)^n$ has the same Jordan canonical form.
    (g) Every linear operator on a finite-dimensional vector space has a unique Jordan canonical basis.
    (h) The dot diagrams of a linear operator on a finite-dimensional vector space are unique.

2. Let T be a linear operator on a finite-dimensional vector space V such that the characteristic polynomial of T splits. Suppose that $\lambda_1 = 2$, $\lambda_2 = 4$, and $\lambda_3 = -3$ are the distinct eigenvalues of T and that the dot diagrams for the restriction of T to K_{λ_i} ($i = 1, 2, 3$) are as follows:

$$\lambda_1 = 2 \qquad\qquad \lambda_2 = 4 \qquad\qquad \lambda_3 = -3$$

Find the Jordan canonical form J of T.

3. Let T be a linear operator on a finite-dimensional vector space V with Jordan canonical form

$$\begin{pmatrix} 2 & 1 & 0 & 0 & 0 & 0 & 0 \\ 0 & 2 & 1 & 0 & 0 & 0 & 0 \\ 0 & 0 & 2 & 0 & 0 & 0 & 0 \\ 0 & 0 & 0 & 2 & 1 & 0 & 0 \\ 0 & 0 & 0 & 0 & 2 & 0 & 0 \\ 0 & 0 & 0 & 0 & 0 & 3 & 0 \\ 0 & 0 & 0 & 0 & 0 & 0 & 3 \end{pmatrix}.$$

(a) Find the characteristic polynomial of T.
(b) Find the dot diagram corresponding to each eigenvalue of T.
(c) For which eigenvalues λ_i, if any, does $E_{\lambda_i} = K_{\lambda_i}$?
(d) For each eigenvalue λ_i, find the smallest positive integer p_i for which $K_{\lambda_i} = N((T - \lambda_i I)^{p_i})$.
(e) Compute the following numbers for each i, where U_i denotes the restriction of $T - \lambda_i I$ to K_{λ_i}.
    (i) rank(U_i)
    (ii) rank(U_i^2)
    (iii) nullity(U_i)
    (iv) nullity(U_i^2)

4. For each of the matrices A that follow, find a Jordan canonical form J and an invertible matrix Q such that $J = Q^{-1}AQ$. Notice that the matrices in (a), (b), and (c) are those used in Example 5.

(a) $A = \begin{pmatrix} -3 & 3 & -2 \\ -7 & 6 & -3 \\ 1 & -1 & 2 \end{pmatrix}$ (b) $A = \begin{pmatrix} 0 & 1 & -1 \\ -4 & 4 & -2 \\ -2 & 1 & 1 \end{pmatrix}$

(c) $A = \begin{pmatrix} 0 & -1 & -1 \\ -3 & -1 & -2 \\ 7 & 5 & 6 \end{pmatrix}$ (d) $A = \begin{pmatrix} 0 & -3 & 1 & 2 \\ -2 & 1 & -1 & 2 \\ -2 & 1 & -1 & 2 \\ -2 & -3 & 1 & 4 \end{pmatrix}$

5. For each linear operator T, find a Jordan canonical form J of T and a Jordan canonical basis β for T.

   (a) V is the real vector space of functions spanned by the set of real-valued functions $\{e^t, te^t, t^2e^t, e^{2t}\}$, and T is the linear operator on V defined by $T(f) = f'$.

   (b) T is the linear operator on $P_3(R)$ defined by $T(f(x)) = xf''(x)$.

   (c) T is the linear operator on $P_3(R)$ defined by
$T(f(x)) = f''(x) + 2f(x)$.

   (d) T is the linear operator on $M_{2\times2}(R)$ defined by

$$T(A) = \begin{pmatrix} 3 & 1 \\ 0 & 3 \end{pmatrix} \cdot A - A^t.$$

   (e) T is the linear operator on $M_{2\times2}(R)$ defined by

$$T(A) = \begin{pmatrix} 3 & 1 \\ 0 & 3 \end{pmatrix} \cdot (A - A^t).$$

   (f) V is the vector space of polynomial functions in two real variables x and y of degree at most 2, as defined in Example 4, and T is the linear operator on V defined by

$$T(f(x,y)) = \frac{\partial}{\partial x} f(x,y) + \frac{\partial}{\partial y} f(x,y).$$

6. Let A be an $n \times n$ matrix whose characteristic polynomial splits. Prove that A and A^t have the same Jordan canonical form, and conclude that A and A^t are similar. *Hint:* For any eigenvalue λ of A and A^t and any positive integer r, show that $\operatorname{rank}((A - \lambda I)^r) = \operatorname{rank}((A^t - \lambda I)^r)$.

7. Let A be an $n \times n$ matrix whose characteristic polynomial splits, γ be a cycle of generalized eigenvectors corresponding to an eigenvalue λ, and W be the subspace spanned by γ. Define γ' to be the ordered set obtained from γ by reversing the order of the vectors in γ.

   (a) Prove that $[T_W]_{\gamma'} = ([T_W]_\gamma)^t$.

   (b) Let J be the Jordan canonical form of A. Use (a) to prove that J and J^t are similar.

   (c) Use (b) to prove that A and A^t are similar.

8. Let T be a linear operator on a finite-dimensional vector space, and suppose that the characteristic polynomial of T splits. Let β be a Jordan canonical basis for T.

   (a) Prove that for any nonzero scalar c, $\{cx : x \in \beta\}$ is a Jordan canonical basis for T.

(b) Suppose that γ is one of the cycles of generalized eigenvectors that forms β, and suppose that γ corresponds to the eigenvalue λ and has length greater than 1. Let x be the end vector of γ, and let y be a nonzero vector in E_λ. Let γ' be the ordered set obtained from γ by replacing x by $x + y$. Prove that γ' is a cycle of generalized eigenvectors corresponding to λ, and that if γ' replaces γ in the union that defines β, then the new union is also a Jordan canonical basis for T.

(c) Apply (b) to obtain a Jordan canonical basis for L_A, where A is the matrix given in Example 2, that is different from the basis given in the example.

9. Suppose that a dot diagram has k columns and m rows with p_j dots in column j and r_i dots in row i. Prove the following results.

(a) $m = p_1$ and $k = r_1$.

(b) $p_j = \max\{i\colon r_i \geq j\}$ for $1 \leq j \leq k$ and $r_i = \max\{j\colon p_j \geq i\}$ for $1 \leq i \leq m$. *Hint:* Use mathematical induction on m.

(c) $r_1 \geq r_2 \geq \cdots \geq r_m$.

(d) Deduce that the number of dots in each column of a dot diagram is completely determined by the number of dots in the rows.

10. Let T be a linear operator whose characteristic polynomial splits, and let λ be an eigenvalue of T.

(a) Prove that $\dim(K_\lambda)$ is the sum of the lengths of all the blocks corresponding to λ in the Jordan canonical form of T.

(b) Deduce that $E_\lambda = K_\lambda$ if and only if all the Jordan blocks corresponding to λ are 1×1 matrices.

The following definitions are used in Exercises 11–19.

Definitions. *A linear operator* T *on a vector space* V *is called* **nilpotent** *if* $T^p = T_0$ *for some positive integer* p. *An* $n \times n$ *matrix* A *is called* **nilpotent** *if* $A^p = O$ *for some positive integer* p.

11. Let T be a linear operator on a finite-dimensional vector space V, and let β be an ordered basis for V. Prove that T is nilpotent if and only if $[T]_\beta$ is nilpotent.

12. Prove that any square upper triangular matrix with each diagonal entry equal to zero is nilpotent.

13. Let T be a nilpotent operator on an n-dimensional vector space V, and suppose that p is the smallest positive integer for which $T^p = T_0$. Prove the following results.

(a) $N(T^i) \subseteq N(T^{i+1})$ for every positive integer i.

(b) There is a sequence of ordered bases $\beta_1, \beta_2, \ldots, \beta_p$ such that β_i is a basis for $N(T^i)$ and β_{i+1} contains β_i for $1 \leq i \leq p - 1$.

(c) Let $\beta = \beta_p$ be the ordered basis for $N(T^p) = V$ in (b). Then $[T]_\beta$ is an upper triangular matrix with each diagonal entry equal to zero.

(d) The characteristic polynomial of T is $(-1)^n t^n$. Hence the characteristic polynomial of T splits, and 0 is the only eigenvalue of T.

14. Prove the converse of Exercise 13(d): If T is a linear operator on an n-dimensional vector space V and $(-1)^n t^n$ is the characteristic polynomial of T, then T is nilpotent.

15. Give an example of a linear operator T on a finite-dimensional vector space such that T is not nilpotent, but zero is the only eigenvalue of T. Characterize all such operators.

16. Let T be a nilpotent linear operator on a finite-dimensional vector space V. Recall from Exercise 13 that $\lambda = 0$ is the only eigenvalue of T, and hence $V = K_\lambda$. Let β be a Jordan canonical basis for T. Prove that for any positive integer i, if we delete from β the vectors corresponding to the last i dots in each column of a dot diagram of β, the resulting set is a basis for $R(T^i)$. (If a column of the dot diagram contains fewer than i dots, all the vectors associated with that column are removed from β.)

17. Let T be a linear operator on a finite-dimensional vector space V such that the characteristic polynomial of T splits, and let $\lambda_1, \lambda_2, \ldots, \lambda_k$ be the distinct eigenvalues of T. Let $S: V \to V$ be the mapping defined by

$$S(x) = \lambda_1 v_1 + \lambda_2 v_2 + \cdots + \lambda_k v_k,$$

where, for each i, v_i is the unique vector in K_{λ_i} such that $x = v_1 + v_2 + \cdots + v_k$. (This unique representation is guaranteed by Theorem 7.3 (p. 486) and Exercise 8 of Section 7.1.)

(a) Prove that S is a diagonalizable linear operator on V.

(b) Let $U = T - S$. Prove that U is nilpotent and commutes with S, that is, $SU = US$.

18. Let T be a linear operator on a finite-dimensional vector space V, and let J be the Jordan canonical form of T. Let D be the diagonal matrix whose diagonal entries are the diagonal entries of J, and let $M = J - D$. Prove the following results.

(a) M is nilpotent.

(b) $MD = DM$.

(c) If p is the smallest positive integer for which $M^p = O$, then, for any positive integer $r < p$,

$$J^r = D^r + rD^{r-1}M + \frac{r(r-1)}{2!}D^{r-2}M^2 + \cdots + rDM^{r-1} + M^r,$$

and, for any positive integer $r \geq p$,

$$J^r = D^r + rD^{r-1}M + \frac{r(r-1)}{2!}D^{r-2}M^2 + \cdots$$

$$+ \frac{r!}{(r-p+1)!(p-1)!}D^{r-p+1}M^{p-1}.$$

19. Let

$$J = \begin{pmatrix} \lambda & 1 & 0 & \cdots & 0 \\ 0 & \lambda & 1 & \cdots & 0 \\ 0 & 0 & \lambda & \cdots & 0 \\ \vdots & \vdots & \vdots & & \vdots \\ 0 & 0 & 0 & \cdots & 1 \\ 0 & 0 & 0 & \cdots & \lambda \end{pmatrix}$$

be the $m \times m$ Jordan block corresponding to λ, and let $N = J - \lambda I_m$. Prove the following results:

(a) $N^m = O$, and for $1 \leq r < m$,

$$N_{ij}^r = \begin{cases} 1 & \text{if } j = i + r \\ 0 & \text{otherwise.} \end{cases}$$

(b) For any integer $r \geq m$,

$$J^r = \begin{pmatrix} \lambda^r & r\lambda^{r-1} & \frac{r(r-1)}{2!}\lambda^{r-2} & \cdots & \frac{r(r-1)\cdots(r-m+2)}{(m-1)!}\lambda^{r-m+1} \\ 0 & \lambda^r & r\lambda^{r-1} & \cdots & \frac{r(r-1)\cdots(r-m+3)}{(m-2)!}\lambda^{r-m+2} \\ \vdots & \vdots & \vdots & & \vdots \\ 0 & 0 & 0 & \cdots & \lambda^r \end{pmatrix}.$$

(c) $\lim_{r \to \infty} J^r$ exists if and only if one of the following holds:

(i) $|\lambda| < 1$.

(ii) $\lambda = 1$ and $m = 1$.

(Note that $\lim\limits_{r\to\infty} \lambda^r$ exists under these conditions. See the discussion preceding Theorem 5.13 on page 285.) Furthermore, $\lim\limits_{r\to\infty} J^r$ is the zero matrix if condition (i) holds and is the 1×1 matrix (1) if condition (ii) holds.

(d) Prove Theorem 5.13 on page 285.

The following definition is used in Exercises 20 and 21.

Definition. *For any $A \in \mathsf{M}_{n\times n}(C)$, define the* norm *of A by*

$$\|A\| = \max\{|A_{ij}| : 1 \leq i, j \leq n\}.$$

20. Let $A, B \in \mathsf{M}_{n\times n}(C)$. Prove the following results.
    (a) $\|A\| \geq 0$ and $\|A\| = 0$ if and only if $A = O$.
    (b) $\|cA\| = |c| \cdot \|A\|$ for any scalar c.
    (c) $\|A + B\| \leq \|A\| + \|B\|$.
    (d) $\|AB\| \leq n\|A\|\|B\|$.

21. Let $A \in \mathsf{M}_{n\times n}(C)$ be a transition matrix. (See Section 5.3.) Since C is an algebraically closed field, A has a Jordan canonical form J to which A is similar. Let P be an invertible matrix such that $P^{-1}AP = J$. Prove the following results.
    (a) $\|A^m\| \leq 1$ for every positive integer m.
    (b) There exists a positive number c such that $\|J^m\| \leq c$ for every positive integer m.
    (c) Each Jordan block of J corresponding to the eigenvalue $\lambda = 1$ is a 1×1 matrix.
    (d) $\lim\limits_{m\to\infty} A^m$ exists if and only if 1 is the only eigenvalue of A with absolute value 1.
    (e) Theorem 5.20(a) using (c) and Theorem 5.19.

The next exercise requires knowledge of absolutely convergent series as well as the definition of e^A for a matrix A. (See page 312.)

22. Use Exercise 20(d) to prove that e^A exists for every $A \in \mathsf{M}_{n\times n}(C)$.

23. Let $x' = Ax$ be a system of n linear differential equations, where x is an n-tuple of differentiable functions $x_1(t), x_2(t), \ldots, x_n(t)$ of the real variable t, and A is an $n \times n$ coefficient matrix as in Exercise 15 of Section 5.2. In contrast to that exercise, however, do not assume that A is diagonalizable, but assume that the characteristic polynomial of A splits. Let $\lambda_1, \lambda_2, \ldots, \lambda_k$ be the distinct eigenvalues of A.

(a) Prove that if u is the end vector of a cycle of generalized eigenvectors of L_A of length p and u corresponds to the eigenvalue λ_i, then for any polynomial $f(t)$ of degree less than p, the function

$$e^{\lambda_i t}[f(t)(A - \lambda_i I)^{p-1} + f'(t)(A - \lambda_i I)^{p-2} + \cdots + f^{(p-1)}(t)]u$$

is a solution to the system $x' = Ax$.

(b) Prove that the general solution to $x' = Ax$ is a sum of the functions of the form given in (a), where the vectors u are the end vectors of the distinct cycles that constitute a fixed Jordan canonical basis for L_A.

24. Use Exercise 23 to find the general solution to each of the following systems of linear equations, where x, y, and z are real-valued differentiable functions of the real variable t.

$$\text{(a)}\quad
\begin{aligned}
x' &= 2x + y \\
y' &= \quad\;\; 2y - z \\
z' &= \qquad\qquad 3z
\end{aligned}
\qquad
\text{(b)}\quad
\begin{aligned}
x' &= 2x + y \\
y' &= \quad\;\; 2y + z \\
z' &= \qquad\qquad 2z
\end{aligned}$$

7.3   THE MINIMAL POLYNOMIAL

The Cayley-Hamilton theorem (Theorem 5.23 p. 317) tells us that for any linear operator T on an n-dimensional vector space, there is a polynomial $f(t)$ of degree n such that $f(\mathsf{T}) = \mathsf{T}_0$, namely, the characteristic polynomial of T. Hence there is a polynomial of least degree with this property, and this degree is at most n. If $g(t)$ is such a polynomial, we can divide $g(t)$ by its leading coefficient to obtain another polynomial $p(t)$ of the same degree with leading coefficient 1, that is, $p(t)$ is a *monic* polynomial. (See Appendix E.)

Definition.   *Let* T *be a linear operator on a finite-dimensional vector space. A polynomial* $p(t)$ *is called a **minimal polynomial** of* T *if* $p(t)$ *is a monic polynomial of least positive degree for which* $p(\mathsf{T}) = \mathsf{T}_0$.

The preceding discussion shows that every linear operator on a finite-dimensional vector space has a minimal polynomial. The next result shows that it is unique.

Theorem 7.12. *Let* $p(t)$ *be a minimal polynomial of a linear operator* T *on a finite-dimensional vector space* V.

(a) *For any polynomial* $g(t)$, *if* $g(\mathsf{T}) = \mathsf{T}_0$, *then* $p(t)$ *divides* $g(t)$. *In particular,* $p(t)$ *divides the characteristic polynomial of* T.

(b) *The minimal polynomial of* T *is unique.*

Proof. (a) Let $g(t)$ be a polynomial for which $g(\mathsf{T}) = \mathsf{T}_0$. By the division algorithm for polynomials (Theorem E.1 of Appendix E, p. 562), there exist polynomials $q(t)$ and $r(t)$ such that

$$g(t) = q(t)p(t) + r(t), \tag{1}$$

where $r(t)$ has degree less than the degree of $p(t)$. Substituting T into (1) and using that $g(\mathsf{T}) = p(\mathsf{T}) = \mathsf{T}_0$, we have $r(\mathsf{T}) = \mathsf{T}_0$. Since $r(t)$ has degree less than $p(t)$ and $p(t)$ is the minimal polynomial of T, $r(t)$ must be the zero polynomial. Thus (1) simplifies to $g(t) = q(t)p(t)$, proving (a).

(b) Suppose that $p_1(t)$ and $p_2(t)$ are each minimal polynomials of T. Then $p_1(t)$ divides $p_2(t)$ by (a). Since $p_1(t)$ and $p_2(t)$ have the same degree, we have that $p_2(t) = cp_1(t)$ for some nonzero scalar c. Because $p_1(t)$ and $p_2(t)$ are monic, $c = 1$; hence $p_1(t) = p_2(t)$. ∎

The minimal polynomial of a linear operator has an obvious analog for a matrix.

Definition. Let $A \in \mathsf{M}_{n \times n}(F)$. The **minimal polynomial** $p(t)$ of A is the monic polynomial of least positive degree for which $p(A) = O$.

The following results are now immediate.

Theorem 7.13. Let T be a linear operator on a finite-dimensional vector space V, and let β be an ordered basis for V. Then the minimal polynomial of T is the same as the minimal polynomial of $[\mathsf{T}]_\beta$.

Proof. Exercise. ∎

Corollary. For any $A \in \mathsf{M}_{n \times n}(F)$, the minimal polynomial of A is the same as the minimal polynomial of L_A.

Proof. Exercise. ∎

In view of the preceding theorem and corollary, Theorem 7.12 and all subsequent theorems in this section that are stated for operators are also valid for matrices.

For the remainder of this section, we study primarily minimal polynomials of operators (and hence matrices) whose characteristic polynomials split. A more general treatment of minimal polynomials is given in Section 7.4.

Theorem 7.14. Let T be a linear operator on a finite-dimensional vector space V, and let $p(t)$ be the minimal polynomial of T. A scalar λ is an eigenvalue of T if and only if $p(\lambda) = 0$. Hence the characteristic polynomial and the minimal polynomial of T have the same zeros.

Proof. Let $f(t)$ be the characteristic polynomial of T. Since $p(t)$ divides $f(t)$, there exists a polynomial $q(t)$ such that $f(t) = q(t)p(t)$. If λ is a zero of $p(t)$, then

$$f(\lambda) = q(\lambda)p(\lambda) = q(\lambda)\cdot 0 = 0.$$

So λ is a zero of $f(t)$; that is, λ is an eigenvalue of T.

Conversely, suppose that λ is an eigenvalue of T, and let $x \in \mathsf{V}$ be an eigenvector corresponding to λ. By Exercise 22 of Section 5.1, we have

$$0 = \mathsf{T}_0(x) = p(\mathsf{T})(x) = p(\lambda)x.$$

Since $x \neq 0$, it follows that $p(\lambda) = 0$, and so λ is a zero of $p(t)$.   ∎

The following corollary is immediate.

Corollary. *Let* T *be a linear operator on a finite-dimensional vector space* V *with minimal polynomial* $p(t)$ *and characteristic polynomial* $f(t)$. *Suppose that* $f(t)$ *factors as*

$$f(t) = (\lambda_1 - t)^{n_1}(\lambda_2 - t)^{n_2} \cdots (\lambda_k - t)^{n_k},$$

where $\lambda_1, \lambda_2, \ldots, \lambda_k$ *are the distinct eigenvalues of* T. *Then there exist integers* $m_1, m_2, \ldots, m_k$ *such that* $1 \leq m_i \leq n_i$ *for all* i *and*

$$p(t) = (t - \lambda_1)^{m_1}(t - \lambda_2)^{m_2} \cdots (t - \lambda_k)^{m_k}.$$

Example 1

We compute the minimal polynomial of the matrix

$$A = \begin{pmatrix} 3 & -1 & 0 \\ 0 & 2 & 0 \\ 1 & -1 & 2 \end{pmatrix}.$$

Since A has the characteristic polynomial

$$f(t) = \det \begin{pmatrix} 3-t & -1 & 0 \\ 0 & 2-t & 0 \\ 1 & -1 & 2-t \end{pmatrix} = -(t-2)^2(t-3),$$

the minimal polynomial of A must be either $(t-2)(t-3)$ or $(t-2)^2(t-3)$ by the corollary to Theorem 7.14. Substituting A into $p(t) = (t-2)(t-3)$, we find that $p(A) = O$; hence $p(t)$ is the minimal polynomial of A.   ◆

Example 2

Let T be the linear operator on R^2 defined by

$$\mathsf{T}(a, b) = (2a + 5b, 6a + b)$$

and β be the standard ordered basis for R^2. Then

$$[\mathsf{T}]_\beta = \begin{pmatrix} 2 & 5 \\ 6 & 1 \end{pmatrix},$$

and hence the characteristic polynomial of T is

$$f(t) = \det \begin{pmatrix} 2-t & 5 \\ 6 & 1-t \end{pmatrix} = (t-7)(t+4).$$

Thus the minimal polynomial of T is also $(t-7)(t+4)$.   ◆

Example 3

Let D be the linear operator on $P_2(R)$ defined by $D(g(x)) = g'(x)$, the derivative of $g(x)$. We compute the minimal polynomial of T. Let β be the standard ordered basis for $P_2(R)$. Then

$$[D]_\beta = \begin{pmatrix} 0 & 1 & 0 \\ 0 & 0 & 2 \\ 0 & 0 & 0 \end{pmatrix},$$

and it follows that the characteristic polynomial of D is $-t^3$. So by the corollary to Theorem 7.14, the minimal polynomial of D is t, t^2, or t^3. Since $D^2(x^2) = 2 \neq 0$, it follows that $D^2 \neq T_0$; hence the minimal polynomial of D must be t^3.   ◆

In Example 3, it is easily verified that $P_2(R)$ is a D-cyclic subspace (of itself). Here the minimal and characteristic polynomials are of the same degree. This is no coincidence.

Theorem 7.15. *Let T be a linear operator on an n-dimensional vector space V such that V is a T-cyclic subspace of itself. Then the characteristic polynomial $f(t)$ and the minimal polynomial $p(t)$ have the same degree, and hence $f(t) = (-1)^n p(t)$.*

Proof. Since V is a T-cyclic space, there exists an $x \in V$ such that

$$\beta = \{x, T(x), \dots, T^{n-1}(x)\}$$

is a basis for V (Theorem 5.22 p. 315). Let

$$g(t) = a_0 + a_1 t + \cdots + a_k t^k,$$

be a polynomial of degree $k < n$. Then $a_k \neq 0$ and

$$g(T)(x) = a_0 x + a_1 T(x) + \cdots + a_k T^k(x),$$

and so $g(T)(x)$ is a linear combination of the vectors of β having at least one nonzero coefficient, namely, a_k. Since β is linearly independent, it follows that $g(T)(x) \neq 0$; hence $g(T) \neq T_0$. Therefore the minimal polynomial of T has degree n, which is also the degree of the characteristic polynomial of T.   ∎

Theorem 7.15 gives a condition under which the degree of the minimal polynomial of an operator is as large as possible. We now investigate the other extreme. By Theorem 7.14, the degree of the minimal polynomial of an operator must be greater than or equal to the number of distinct eigenvalues of the operator. The next result shows that the operators for which the degree of the minimal polynomial is as small as possible are precisely the diagonalizable operators.

Theorem 7.16. *Let* T *be a linear operator on a finite-dimensional vector space* V. *Then* T *is diagonalizable if and only if the minimal polynomial of* T *is of the form*

$$p(t) = (t - \lambda_1)(t - \lambda_2) \cdots (t - \lambda_k),$$

where $\lambda_1, \lambda_2, \ldots, \lambda_k$ *are the distinct eigenvalues of* T.

Proof. Suppose that T is diagonalizable. Let $\lambda_1, \lambda_2, \ldots, \lambda_k$ be the distinct eigenvalues of T, and define

$$p(t) = (t - \lambda_1)(t - \lambda_2) \cdots (t - \lambda_k).$$

By Theorem 7.14, $p(t)$ divides the minimal polynomial of T. Let $\beta = \{v_1, v_2, \ldots, v_n\}$ be a basis for V consisting of eigenvectors of T, and consider any $v_i \in \beta$. Then $(T - \lambda_j I)(v_i) = 0$ for some eigenvalue λ_j. Since $t - \lambda_j$ divides $p(t)$, there is a polynomial $q_j(t)$ such that $p(t) = q_j(t)(t - \lambda_j)$. Hence

$$p(T)(v_i) = q_j(T)(T - \lambda_j I)(v_i) = 0.$$

It follows that $p(T) = T_0$, since $p(T)$ takes each vector in a basis for V into 0. Therefore $p(t)$ is the minimal polynomial of T.

Conversely, suppose that there are distinct scalars $\lambda_1, \lambda_2, \ldots, \lambda_k$ such that the minimal polynomial $p(t)$ of T factors as

$$p(t) = (t - \lambda_1)(t - \lambda_2) \cdots (t - \lambda_k).$$

By Theorem 7.14, the λ_i's are eigenvalues of T. We apply mathematical induction on $n = \dim(V)$. Clearly T is diagonalizable for $n = 1$. Now assume that T is diagonalizable whenever $\dim(V) < n$ for some $n > 1$, and let $\dim(V) = n$ and $W = R(T - \lambda_k I)$. Obviously $W \neq V$, because λ_k is an eigenvalue of T. If $W = \{0\}$, then $T = \lambda_k I$, which is clearly diagonalizable. So suppose that $0 < \dim(W) < n$. Then W is T-invariant, and for any $x \in W$,

$$(T - \lambda_1 I)(T - \lambda_2 I) \cdots (T - \lambda_{k-1} I)(x) = 0.$$

It follows that the minimal polynomial of T_W divides the polynomial $(t - \lambda_1)(t - \lambda_2) \cdots (t - \lambda_{k-1})$. Hence by the induction hypothesis, T_W is diagonalizable. Furthermore, λ_k is not an eigenvalue of T_W by Theorem 7.14. Therefore $W \cap N(T - \lambda_k I) = \{0\}$. Now let $\beta_1 = \{v_1, v_2, \ldots, v_m\}$ be a basis for W consisting of eigenvectors of T_W (and hence of T), and let $\beta_2 = \{w_1, w_2, \ldots, w_p\}$ be a basis for $N(T - \lambda_k I)$, the eigenspace of T corresponding to λ_k. Then β_1 and β_2 are disjoint by the previous comment. Moreover, $m + p = n$ by the dimension theorem applied to $T - \lambda_k I$. We show that $\beta = \beta_1 \cup \beta_2$ is linearly independent. Consider scalars $a_1, a_2, \ldots, a_m$ and $b_1, b_2, \ldots, b_p$ such that

$$a_1 v_1 + a_2 v_2 + \cdots + a_m v_m + b_1 w_1 + b_2 w_2 + \cdots + b_p w_p = 0.$$

Let

$$x = \sum_{i=1}^{m} a_i v_i \qquad \text{and} \qquad y = \sum_{i=1}^{p} b_i w_i.$$

Then $x \in \mathsf{W}$, $y \in \mathsf{N}(\mathsf{T} - \lambda_k \mathsf{I})$, and $x + y = 0$. It follows that $x = -y \in \mathsf{W} \cap \mathsf{N}(\mathsf{T} - \lambda_k \mathsf{I})$, and therefore $x = 0$. Since β_1 is linearly independent, we have that $a_1 = a_2 = \cdots = a_m = 0$. Similarly, $b_1 = b_2 = \cdots = b_p = 0$, and we conclude that β is a linearly independent subset of V consisting of n eigenvectors. It follows that β is a basis for V consisting of eigenvectors of T, and consequently T is diagonalizable. ∎

In addition to diagonalizable operators, there are methods for determining the minimal polynomial of any linear operator on a finite-dimensional vector space. In the case that the characteristic polynomial of the operator splits, the minimal polynomial can be described using the Jordan canonical form of the operator. (See Exercise 13.) In the case that the characteristic polynomial does not split, the minimal polynomial can be described using the *rational canonical form*, which we study in the next section. (See Exercise 7 of Section 7.4.)

Example 4

We determine all matrices $A \in \mathsf{M}_{2 \times 2}(R)$ for which $A^2 - 3A + 2I = O$. Let $g(t) = t^2 - 3t + 2 = (t-1)(t-2)$. Since $g(A) = O$, the minimal polynomial $p(t)$ of A divides $g(t)$. Hence the only possible candidates for $p(t)$ are $t - 1$, $t - 2$, and $(t-1)(t-2)$. If $p(t) = t - 1$ or $p(t) = t - 2$, then $A = I$ or $A = 2I$, respectively. If $p(t) = (t-1)(t-2)$, then A is diagonalizable with eigenvalues 1 and 2, and hence A is similar to

$$\begin{pmatrix} 1 & 0 \\ 0 & 2 \end{pmatrix}. \quad \blacklozenge$$

Example 5

Let $A \in \mathsf{M}_{n \times n}(R)$ satisfy $A^3 = A$. We show that A is diagonalizable. Let $g(t) = t^3 - t = t(t+1)(t-1)$. Then $g(A) = O$, and hence the minimal polynomial $p(t)$ of A divides $g(t)$. Since $g(t)$ has no repeated factors, neither does $p(t)$. Thus A is diagonalizable by Theorem 7.16. $\blacklozenge$

Example 6

In Example 3, we saw that the minimal polynomial of the differential operator D on $\mathsf{P}_2(R)$ is t^3. Hence, by Theorem 7.16, D is not diagonalizable. $\blacklozenge$

EXERCISES

1. Label the following statements as true or false. Assume that all vector spaces are finite-dimensional.

   (a) Every linear operator T has a polynomial $p(t)$ of largest degree for which $p(\mathsf{T}) = \mathsf{T}_0$.
   (b) Every linear operator has a unique minimal polynomial.
   (c) The characteristic polynomial of a linear operator divides the minimal polynomial of that operator.
   (d) The minimal and the characteristic polynomials of any diagonalizable operator are equal.
   (e) Let T be a linear operator on an n-dimensional vector space V, $p(t)$ be the minimal polynomial of T, and $f(t)$ be the characteristic polynomial of T. Suppose that $f(t)$ splits. Then $f(t)$ divides $[p(t)]^n$.
   (f) The minimal polynomial of a linear operator always has the same degree as the characteristic polynomial of the operator.
   (g) A linear operator is diagonalizable if its minimal polynomial splits.
   (h) Let T be a linear operator on a vector space V such that V is a T-cyclic subspace of itself. Then the degree of the minimal polynomial of T equals $\dim(\mathsf{V})$.
   (i) Let T be a linear operator on a vector space V such that T has n distinct eigenvalues, where $n = \dim(\mathsf{V})$. Then the degree of the minimal polynomial of T equals n.

2. Find the minimal polynomial of each of the following matrices.

   (a) $\begin{pmatrix} 2 & 1 \\ 1 & 2 \end{pmatrix}$ (b) $\begin{pmatrix} 1 & 1 \\ 0 & 1 \end{pmatrix}$

   (c) $\begin{pmatrix} 4 & -14 & 5 \\ 1 & -4 & 2 \\ 1 & -6 & 4 \end{pmatrix}$ (d) $\begin{pmatrix} 3 & 0 & 1 \\ 2 & 2 & 2 \\ -1 & 0 & 1 \end{pmatrix}$

3. For each linear operator T on V, find the minimal polynomial of T.

   (a) $\mathsf{V} = \mathsf{R}^2$ and $\mathsf{T}(a, b) = (a + b, a - b)$
   (b) $\mathsf{V} = \mathsf{P}_2(R)$ and $\mathsf{T}(g(x)) = g'(x) + 2g(x)$
   (c) $\mathsf{V} = \mathsf{P}_2(R)$ and $\mathsf{T}(f(x)) = -xf''(x) + f'(x) + 2f(x)$
   (d) $\mathsf{V} = \mathsf{M}_{n \times n}(R)$ and $\mathsf{T}(A) = A^t$. *Hint:* Note that $\mathsf{T}^2 = \mathsf{I}$.

4. Determine which of the matrices and operators in Exercises 2 and 3 are diagonalizable.

5. Describe all linear operators T on R^2 such that T is diagonalizable and $\mathsf{T}^3 - 2\mathsf{T}^2 + \mathsf{T} = \mathsf{T}_0$.

6. Prove Theorem 7.13 and its corollary.

7. Prove the corollary to Theorem 7.14.

8. Let T be a linear operator on a finite-dimensional vector space, and let $p(t)$ be the minimal polynomial of T. Prove the following results.

   (a) T is invertible if and only if $p(0) \neq 0$.
   (b) If T is invertible and $p(t) = t^n + a_{n-1}t^{n-1} + \cdots + a_1 t + a_0$, then

   $$\mathsf{T}^{-1} = -\frac{1}{a_0}\left(\mathsf{T}^{n-1} + a_{n-1}\mathsf{T}^{n-2} + \cdots + a_2\mathsf{T} + a_1\mathsf{I}\right).$$

9. Let T be a diagonalizable linear operator on a finite-dimensional vector space V. Prove that V is a T-cyclic subspace if and only if each of the eigenspaces of T is one-dimensional.

10. Let T be a linear operator on a finite-dimensional vector space V, and suppose that W is a T-invariant subspace of V. Prove that the minimal polynomial of T_W divides the minimal polynomial of T.

11. Let $g(t)$ be the auxiliary polynomial associated with a homogeneous linear differential equation with constant coefficients (as defined in Section 2.7), and let V denote the solution space of this differential equation. Prove the following results.

    (a) V is a D-invariant subspace, where D is the differentiation operator on C^∞.
    (b) The minimal polynomial of D_V (the restriction of D to V) is $g(t)$.
    (c) If the degree of $g(t)$ is n, then the characteristic polynomial of D_V is $(-1)^n g(t)$.

    Hint: Use Theorem 2.32 (p. 135) for (b) and (c).

12. Let D be the differentiation operator on $P(R)$, the space of polynomials over R. Prove that there exists no polynomial $g(t)$ for which $g(\mathsf{D}) = \mathsf{T}_0$. Hence D has no minimal polynomial.

13. Let T be a linear operator on a finite-dimensional vector space, and suppose that the characteristic polynomial of T splits. Let $\lambda_1, \lambda_2, \ldots, \lambda_k$ be the distinct eigenvalues of T, and for each i let p_i be the order of the largest Jordan block corresponding to λ_i in a Jordan canonical form of T. Prove that the minimal polynomial of T is

$$(t - \lambda_1)^{p_1}(t - \lambda_2)^{p_2} \cdots (t - \lambda_k)^{p_k}.$$

The following exercise requires knowledge of direct sums (see Section 5.2).

14. Let T be linear operator on a finite-dimensional vector space V, and let W_1 and W_2 be T-invariant subspaces of V such that $V = W_1 \oplus W_2$. Suppose that $p_1(t)$ and $p_2(t)$ are the minimal polynomials of T_{W_1} and T_{W_2}, respectively. Prove or disprove that $p_1(t)p_2(t)$ is the minimal polynomial of T.

Exercise 15 uses the following definition.

Definition. *Let T be a linear operator on a finite-dimensional vector space V, and let x be a nonzero vector in V. The polynomial $p(t)$ is called a **T-annihilator** of x if $p(t)$ is a monic polynomial of least degree for which $p(T)(x) = 0$.*

15.[†] Let T be a linear operator on a finite-dimensional vector space V, and let x be a nonzero vector in V. Prove the following results.

(a) The vector x has a unique T-annihilator.

(b) The T-annihilator of x divides any polynomial $g(t)$ for which $g(T) = T_0$.

(c) If $p(t)$ is the T-annihilator of x and W is the T-cyclic subspace generated by x, then $p(t)$ is the minimal polynomial of T_W, and $\dim(W)$ equals the degree of $p(t)$.

(d) The degree of the T-annihilator of x is 1 if and only if x is an eigenvector of T.

16. T be a linear operator on a finite-dimensional vector space V, and let W_1 be a T-invariant subspace of V. Let $x \in V$ such that $x \notin W_1$. Prove the following results.

(a) There exists a unique monic polynomial $g_1(t)$ of least positive degree such that $g_1(T)(x) \in W_1$.

(b) If $h(t)$ is a polynomial for which $h(T)(x) \in W_1$, then $g_1(t)$ divides $h(t)$.

(c) $g_1(t)$ divides the minimal and the characteristic polynomials of T.

(d) Let W_2 be a T-invariant subspace of V such that $W_2 \subseteq W_1$, and let $g_2(t)$ be the unique monic polynomial of least degree such that $g_2(T)(x) \in W_2$. Then $g_1(t)$ divides $g_2(t)$.

7.4* THE RATIONAL CANONICAL FORM

Until now we have used eigenvalues, eigenvectors, and generalized eigenvectors in our analysis of linear operators with characteristic polynomials that split. In general, characteristic polynomials need not split, and indeed, operators need not have eigenvalues! However, the unique factorization theorem for polynomials (see Appendix E) guarantees that the characteristic polynomial $f(t)$ of any linear operator T on an n-dimensional vector space factors uniacely as

$$f(t) = (-1)^n (\phi_1(t))^{n_1} (\phi_2(t))^{n_2} \cdots (\phi_k(t))^{n_k},$$

where the $\phi_i(t)$'s $(1 \leq i \leq k)$ are distinct irreducible monic polynomials and the n_i's are positive integers. In the case that $f(t)$ splits, each irreducible monic polynomial factor is of the form $\phi_i(t) = t - \lambda_i$, where λ_i is an eigenvalue of T, and there is a one-to-one correspondence between eigenvalues of T and the irreducible monic factors of the characteristic polynomial. In general, eigenvalues need not exist, but the irreducible monic factors always exist. In this section, we establish structure theorems based on the irreducible monic factors of the characteristic polynomial instead of eigenvalues.

In this context, the following definition is the appropriate replacement for eigenspace and generalized eigenspace.

Definition. *Let T be a linear operator on a finite-dimensional vector space V with characteristic polynomial*

$$f(t) = (-1)^n (\phi_1(t))^{n_1} (\phi_2(t))^{n_2} \cdots (\phi_k(t))^{n_k},$$

where the $\phi_i(t)$'s $(1 \leq i \leq k)$ are distinct irreducible monic polynomials and the n_i's are positive integers. For $1 \leq i \leq k$, we define the subset K_{ϕ_i} of V by

$$\mathsf{K}_{\phi_i} = \{x \in \mathsf{V} : (\phi_i(\mathsf{T}))^p (x) = 0 \text{ for some positive integer } p\}.$$

We show that each K_{ϕ_i} is a nonzero T-invariant subspace of V. Note that if $\phi_i(t) = t - \lambda$ is of degree one, then K_{ϕ_i} is the generalized eigenspace of T corresponding to the eigenvalue λ.

Having obtained suitable generalizations of the related concepts of eigenvalue and eigenspace, our next task is to describe a canonical form of a linear operator suitable to this context. The one that we study is called the *rational canonical form*. Since a canonical form is a description of a matrix representation of a linear operator, it can be defined by specifying the form of the ordered bases allowed for these representations.

Here the bases of interest naturally arise from the generators of certain cyclic subspaces. For this reason, the reader should recall the definition of a T-cyclic subspace generated by a vector and Theorem 5.22 (p. 315). We briefly review this concept and introduce some new notation and terminology.

Let T be a linear operator on a finite-dimensional vector space V, and let x be a nonzero vector in V. We use the notation C_x for the T-cyclic subspace generated by x. Recall (Theorem 5.22) that if $\dim(\mathsf{C}_x) = k$, then the set

$$\{x, \mathsf{T}(x), \mathsf{T}^2(x), \dots, \mathsf{T}^{k-1}(x)\}$$

is an ordered basis for C_x. To distinguish this basis from all other ordered bases for C_x, we call it the **T-cyclic basis generated by x** and denote it by

β_x. Let A be the matrix representation of the restriction of T to C_x relative to the ordered basis β_x. Recall from the proof of Theorem 5.22 that

$$A = \begin{pmatrix} 0 & 0 & \cdots & 0 & -a_0 \\ 1 & 0 & \cdots & 0 & -a_1 \\ 0 & 1 & \cdots & 0 & -a_2 \\ \vdots & \vdots & & \vdots & \vdots \\ 0 & 0 & \cdots & 1 & -a_{k-1} \end{pmatrix},$$

where

$$a_0 x + a_1 \mathsf{T}(x) + \cdots + a_{k-1} \mathsf{T}^{k-1}(x) + \mathsf{T}^k(x) = 0.$$

Furthermore, the characteristic polynomial of A is given by

$$\det(A - tI) = (-1)^k (a_0 + a_1 t + \cdots + a_{k-1} t^{k-1} + t^k).$$

The matrix A is called the **companion matrix** of the monic polynomial $h(t) = a_0 + a_1 t + \cdots + a_{k-1} t^{k-1} + t^k$. Every monic polynomial has a companion matrix, and the characteristic polynomial of the companion matrix of a monic polynomial $g(t)$ of degree k is equal to $(-1)^k g(t)$. (See Exercise 19 of Section 5.4.) By Theorem 7.15 (p. 519), the monic polynomial $h(t)$ is also the minimal polynomial of A. Since A is the matrix representation of the restriction of T to C_x, $h(t)$ is also the minimal polynomial of this restriction. By Exercise 15 of Section 7.3, $h(t)$ is also the T-annihilator of x.

It is the object of this section to prove that for every linear operator T on a finite-dimensional vector space V, there exists an ordered basis β for V such that the matrix representation $[\mathsf{T}]_\beta$ is of the form

$$\begin{pmatrix} C_1 & O & \cdots & O \\ O & C_2 & \cdots & O \\ \vdots & \vdots & & \vdots \\ O & O & \cdots & C_r \end{pmatrix},$$

where each C_i is the companion matrix of a polynomial $(\phi(t))^m$ such that $\phi(t)$ is a monic irreducible divisor of the characteristic polynomial of T and m is a positive integer. A matrix representation of this kind is called a **rational canonical form** of T. We call the accompanying basis a **rational canonical basis** for T.

The next theorem is a simple consequence of the following lemma, which relies on the concept of T-annihilator, introduced in the Exercises of Section 7.3.

Lemma. *Let T be a linear operator on a finite-dimensional vector space V, let x be a nonzero vector in V, and suppose that the T-annihilator of x is of the form $(\phi(t))^p$ for some irreducible monic polynomial $\phi(t)$. Then $\phi(t)$ divides the minimal polynomial of T, and $x \in \mathsf{K}_\phi$.*

Proof. By Exercise 15(b) of Section 7.3, $(\phi(t))^p$ divides the minimal polynomial of T. Therefore $\phi(t)$ divides the minimal polynomial of T. Furthermore, $x \in K_\phi$ by the definition of K_ϕ. ∎

Theorem 7.17. *Let* T *be a linear operator on a finite-dimensional vector space* V, *and let* β *be an ordered basis for* V. *Then* β *is a rational canonical basis for* T *if and only if* β *is the disjoint union of* T-*cyclic bases* β_{v_i}, *where each* v_i *lies in* K_ϕ *for some irreducible monic divisor* $\phi(t)$ *of the characteristic polynomial of* T.

Proof. Exercise. ∎

Example 1

Suppose that T is a linear operator on R^8 and

$$\beta = \{v_1, v_2, v_3, v_4, v_5, v_6, v_7, v_8\}$$

is a rational canonical basis for T such that

$$C = [T]_\beta = \begin{pmatrix} 0 & -3 & 0 & 0 & 0 & 0 & 0 & 0 \\ 1 & 1 & 0 & 0 & 0 & 0 & 0 & 0 \\ 0 & 0 & 0 & 0 & 0 & -1 & 0 & 0 \\ 0 & 0 & 1 & 0 & 0 & 0 & 0 & 0 \\ 0 & 0 & 0 & 1 & 0 & -2 & 0 & 0 \\ 0 & 0 & 0 & 0 & 1 & 0 & 0 & 0 \\ 0 & 0 & 0 & 0 & 0 & 0 & 0 & -1 \\ 0 & 0 & 0 & 0 & 0 & 0 & 1 & 0 \end{pmatrix}$$

is a rational canonical form of T. In this case, the submatrices C_1, C_2, and C_3 are the companion matrices of the polynomials $\phi_1(t)$, $(\phi_2(t))^2$, and $\phi_2(t)$, respectively, where

$$\phi_1(t) = t^2 - t + 3 \quad \text{and} \quad \phi_2(t) = t^2 + 1.$$

In the context of Theorem 7.17, β is the disjoint union of the T-cyclic bases; that is,

$$\beta = \beta_{v_1} \cup \beta_{v_3} \cup \beta_{v_7}$$
$$= \{v_1, v_2\} \cup \{v_3, v_4, v_5, v_6\} \cup \{v_7, v_8\}.$$

By Exercise 40 of Section 5.4, the characteristic polynomial $f(t)$ of T is the product of the characteristic polynomials of the companion matrices:

$$f(t) = \phi_1(t)(\phi_2(t))^2\phi_2(t) = \phi_1(t)(\phi_2(t))^3. \quad \blacklozenge$$

The rational canonical form C of the operator T in Example 1 is constructed from matrices of the form C_i, each of which is the companion matrix of some power of a monic irreducible divisor of the characteristic polynomial of T. Furthermore, each such divisor is used in this way at least once.

In the course of showing that every linear operator T on a finite dimensional vector space has a rational canonical form C, we show that the companion matrices C_i that constitute C are always constructed from powers of the monic irreducible divisors of the characteristic polynomial of T. A key role in our analysis is played by the subspaces K_ϕ, where $\phi(t)$ is an irreducible monic divisor of the minimal polynomial of T. Since the minimal polynomial of an operator divides the characteristic polynomial of the operator, every irreducible divisor of the former is also an irreducible divisor of the latter. We eventually show that the converse is also true; that is, the minimal polynomial and the characteristic polynomial have the same irreducible divisors.

We begin with a result that lists several properties of irreducible divisors of the minimal polynomial. The reader is advised to review the definition of T-annihilator and the accompanying Exercise 15 of Section 7.3.

Theorem 7.18. *Let* T *be a linear operator on a finite-dimensional vector space* V, *and suppose that*

$$p(t) = (\phi_1(t))^{m_1}(\phi_2(t))^{m_2}\cdots(\phi_k(t))^{m_k}$$

is the minimal polynomial of T, *where the* $\phi_i(t)$*'s* $(1 \le i \le k)$ *are the distinct irreducible monic factors of* $p(t)$ *and the* m_i*'s are positive integers. Then the following statements are true.*

(a) K_{ϕ_i} *is a nonzero* T*-invariant subspace of* V *for each* i.

(b) *If* x *is a nonzero vector in some* K_{ϕ_i}, *then the* T*-annihilator of* x *is of the form* $(\phi_i(t))^p$ *for some integer* p.

(c) $\mathsf{K}_{\phi_i} \cap \mathsf{K}_{\phi_j} = \{0\}$ *for* $i \ne j$. .

(d) K_{ϕ_i} *is invariant under* $\phi_j(\mathsf{T})$ *for* $i \ne j$, *and the restriction of* $\phi_j(\mathsf{T})$ *to* K_{ϕ_i} *is one-to-one and onto.*

(e) $\mathsf{K}_{\phi_i} = \mathsf{N}((\phi_i(\mathsf{T}))^{m_i})$ *for each* i.

Proof. If $k = 1$, then (a), (b), and (e) are obvious, while (c) and (d) are vacuously true. Now suppose that $k > 1$.

(a) The proof that K_{ϕ_i} is a T-invariant subspace of V is left as an exercise. Let $f_i(t)$ be the polynomial obtained from $p(t)$ by omitting the factor $(\phi_i(t))^{m_i}$. To prove that K_{ϕ_i} is nonzero, first observe that $f_i(t)$ is a proper divisor of $p(t)$; therefore there exists a vector $z \in \mathsf{V}$ such that $x = f_i(\mathsf{T})(z) \ne 0$. Then $x \in \mathsf{K}_{\phi_i}$ because

$$(\phi_i(\mathsf{T}))^{m_i}(x) = (\phi_i(\mathsf{T}))^{m_i}f_i(\mathsf{T})(z) = p(\mathsf{T})(z) = 0.$$

(b) Assume the hypothesis. Then $(\phi_i(\mathsf{T}))^q(x) = 0$ for some positive integer q. Hence the T-annihilator of x divides $(\phi_i(t))^q$ by Exercise 15(b) of Section 7.3, and the result follows.

(c) Assume $i \neq j$. Let $x \in \mathsf{K}_{\phi_i} \cap \mathsf{K}_{\phi_j}$, and suppose that $x \neq 0$. By (b), the T-annihilator of x is a power of both $\phi_i(t)$ and $\phi_j(t)$. But this is impossible because $\phi_i(t)$ and $\phi_j(t)$ are relatively prime (see Appendix E). We conclude that $x = 0$.

(d) Assume $i \neq j$. Since K_{ϕ_i} is T-invariant, it is also $\phi_j(\mathsf{T})$-invariant. Suppose that $\phi_j(\mathsf{T})(x) = 0$ for some $x \in \mathsf{K}_{\phi_i}$. Then $x \in \mathsf{K}_{\phi_i} \cap \mathsf{K}_{\phi_j} = \{0\}$ by (c). Therefore the restriction of $\phi_j(\mathsf{T})$ to K_{ϕ_i} is one-to-one. Since V is finite-dimensional, this restriction is also onto.

(e) Suppose that $1 \leq i \leq k$. Clearly, $\mathsf{N}((\phi_i(\mathsf{T}))^{m_i}) \subseteq \mathsf{K}_{\phi_i}$. Let $f_i(t)$ be the polynomial defined in (a). Since $f_i(t)$ is a product of polynomials of the form $\phi_j(t)$ for $j \neq i$, we have by (d) that the restriction of $f_i(\mathsf{T})$ to K_{ϕ_i} is onto. Let $x \in \mathsf{K}_{\phi_i}$. Then there exists $y \in \mathsf{K}_{\phi_i}$ such that $f_i(\mathsf{T})(y) = x$. Therefore

$$((\phi_i(\mathsf{T}))^{m_i})(x) = ((\phi_i(\mathsf{T}))^{m_i})f_i(\mathsf{T})(y) = p(\mathsf{T})(y) = 0,$$

and hence $x \in \mathsf{N}((\phi_i(\mathsf{T}))^{m_i})$. Thus $\mathsf{K}_{\phi_i} = \mathsf{N}((\phi_i(\mathsf{T}))^{m_i})$.   ∎

Since a rational canonical basis for an operator T is obtained from a union of T-cyclic bases, we need to know when such a union is linearly independent. The next major result, Theorem 7.19, reduces this problem to the study of T-cyclic bases within K_ϕ, where $\phi(t)$ is an irreducible monic divisor of the minimal polynomial of T. We begin with the following lemma.

Lemma. *Let* T *be a linear operator on a finite-dimensional vector space* V, *and suppose that*

$$p(t) = (\phi_1(t))^{m_1}(\phi_2(t))^{m_2} \cdots (\phi_k(t))^{m_k}$$

is the minimal polynomial of T, *where the* ϕ_i's $(1 \leq i \leq k)$ *are the distinct irreducible monic factors of* $p(t)$ *and the* m_i's *are positive integers. For* $1 \leq i \leq k$, *let* $v_i \in \mathsf{K}_{\phi_i}$ *be such that*

$$v_1 + v_2 + \cdots + v_k = 0. \tag{2}$$

Then $v_i = 0$ *for all* i.

Proof. The result is trivial if $k = 1$, so suppose that $k > 1$. Consider any i. Let $f_i(t)$ be the polynomial obtained from $p(t)$ by omitting the factor $(\phi_i(t))^{m_i}$. As a consequence of Theorem 7.18, $f_i(\mathsf{T})$ is one-to-one on K_{ϕ_i}, and $f_i(\mathsf{T})(v_j) = 0$ for $i \neq j$. Thus, applying $f_i(\mathsf{T})$ to (2), we obtain $f_i(\mathsf{T})(v_i) = 0$, from which it follows that $v_i = 0$.   ∎

Theorem 7.19. *Let* T *be a linear operator on a finite-dimensional vector space* V, *and suppose that*

$$p(t) = (\phi_1(t))^{m_1}(\phi_2(t))^{m_2} \cdots (\phi_k(t))^{m_k}$$

is the minimal polynomial of T, where the ϕ_i's $(1 \leq i \leq k)$ are the distinct irreducible monic factors of $p(t)$ and the m_i's are positive integers. For $1 \leq i \leq k$, let S_i be a linearly independent subset of K_{ϕ_i}. Then
(a) $S_i \cap S_j = \varnothing$ for $i \neq j$
(b) $S_1 \cup S_2 \cup \cdots \cup S_k$ is linearly independent.

Proof. If $k = 1$, then (a) is vacuously true and (b) is obvious. Now suppose that $k > 1$. Then (a) follows immediately from Theorem 7.18(c). Furthermore, the proof of (b) is identical to the proof of Theorem 5.8 (p. 267) with the eigenspaces replaced by the subspaces K_{ϕ_i}. ∎

In view of Theorem 7.19, we can focus on bases of individual spaces of the form $\mathsf{K}_\phi(t)$, where $\phi(t)$ is an irreducible monic divisor of the minimal polynomial of T. The next several results give us ways to construct bases for these spaces that are unions of T-cyclic bases. These results serve the dual purposes of leading to the existence theorem for the rational canonical form and of providing methods for constructing rational canonical bases.

For Theorems 7.20 and 7.21 and the latter's corollary, we fix a linear operator T on a finite-dimensional vector space V and an irreducible monic divisor $\phi(t)$ of the minimal polynomial of T.

Theorem 7.20. Let $v_1, v_2, \ldots, v_k$ be distinct vectors in K_ϕ such that

$$S_1 = \beta_{v_1} \cup \beta_{v_2} \cup \cdots \cup \beta_{v_k}$$

is linearly independent. For each i, choose $w_i \in \mathsf{V}$ such that $\phi(\mathsf{T})(w_i) = v_i$. Then

$$S_2 = \beta_{w_1} \cup \beta_{w_2} \cup \cdots \cup \beta_{w_k}$$

is also linearly independent.

Proof. Consider any linear combination of vectors in S_2 that sums to zero, say,

$$\sum_{i=1}^{k} \sum_{j=0}^{n_i} a_{ij} \mathsf{T}^j(w_i) = 0. \tag{3}$$

For each i, let $f_i(t)$ be the polynomial defined by

$$f_i(t) = \sum_{j=0}^{n_i} a_{ij} t^j.$$

Then (3) can be rewritten as

$$\sum_{i=1}^{k} f_i(\mathsf{T})(w_i) = 0. \tag{4}$$

Apply $\phi(\mathsf{T})$ to both sides of (4) to obtain

$$\sum_{i=1}^{k} \phi(\mathsf{T}) f_i(\mathsf{T})(w_i) = \sum_{i=1}^{k} f_i(\mathsf{T}) \phi(\mathsf{T})(w_i) = \sum_{i=1}^{k} f_i(\mathsf{T})(v_i) = 0.$$

This last sum can be rewritten as a linear combination of the vectors in S_1 so that each $f_i(\mathsf{T})(v_i)$ is a linear combination of the vectors in β_{v_i}. Since S_1 is linearly independent, it follows that

$$f_i(\mathsf{T})(v_i) = 0 \quad \text{for all } i.$$

Therefore the T-annihilator of v_i divides $f_i(t)$ for all i. (See Exercise 15 of Section 7.3.) By Theorem 7.18(b), $\phi(t)$ divides the T-annihilator of v_i, and hence $\phi(t)$ divides $f_i(t)$ for all i. Thus, for each i, there exists a polynomial $g_i(t)$ such that $f_i(t) = g_i(t)\phi(t)$. So (4) becomes

$$\sum_{i=1}^{k} g_i(\mathsf{T}) \phi(\mathsf{T})(w_i) = \sum_{i=1}^{k} g_i(\mathsf{T})(v_i) = 0.$$

Again, linear independence of S_1 requires that

$$f_i(\mathsf{T})(w_i) = g_i(\mathsf{T})(v_i) = 0 \quad \text{for all } i.$$

But $f_i(\mathsf{T})(w_i)$ is the result of grouping the terms of the linear combination in (3) that arise from the linearly independent set β_{w_i}. We conclude that for each i, $a_{ij} = 0$ for all j. Therefore S_2 is linearly independent. ∎

We now show that K_ϕ has a basis consisting of a union of T-cycles.

Lemma. *Let W be a T-invariant subspace of K_ϕ, and let β be a basis for W. Then the following statements are true.*
(a) *Suppose that $x \in \mathsf{N}(\phi(\mathsf{T}))$, but $x \notin \mathsf{W}$. Then $\beta \cup \beta_x$ is linearly independent.*
(b) *For some $w_1, w_2, \ldots, w_s$ in $\mathsf{N}(\phi(\mathsf{T}))$, β can be extended to the linearly independent set*

$$\beta' = \beta \cup \beta_{w_1} \cup \beta_{w_2} \cup \cdots \cup \beta_{w_s},$$

whose span contains $\mathsf{N}(\phi(\mathsf{T}))$.

Proof. (a) Let $\beta = \{v_1, v_2, \ldots, v_k\}$, and suppose that

$$\sum_{i=1}^{k} a_i v_i + z = 0 \quad \text{and} \quad z = \sum_{j=0}^{d-1} b_j \mathsf{T}^j(x),$$

where d is the degree of $\phi(t)$. Then $z \in C_x \cap W$, and hence $C_z \subseteq C_x \cap W$. Suppose that $z \neq 0$. Then z has $\phi(t)$ as its T-annihilator, and therefore

$$d = \dim(C_z) \leq \dim(C_x \cap W) \leq \dim(C_x) = d.$$

It follows that $C_x \cap W = C_x$, and consequently $x \in W$, contrary to hypothesis. Therefore $z = 0$, from which it follows that $b_j = 0$ for all j. Since β is linearly independent, it follows that $a_i = 0$ for all i. Thus $\beta \cup \beta_x$ is linearly independent.

(b) Suppose that W does not contain $N(\phi(T))$. Choose a vector $w_1 \in N(\phi(t))$ that is not in W. By (a), $\beta_1 = \beta \cup \beta_{w_1}$ is linearly independent. Let $W_1 = \mathrm{span}(\beta_1)$. If W_1 does not contain $N(\phi(t))$, choose a vector w_2 in $N(\phi(t))$, but not in W_1, so that $\beta_2 = \beta_1 \cup \beta_{w_2} = \beta \cup \beta_{w_1} \cup \beta_{w_2}$ is linearly independent. Continuing this process, we eventually obtain vectors $w_1, w_2, \ldots, w_s$ in $N(\phi(T))$ such that the union

$$\beta' = \beta \cup \beta_{w_1} \cup \beta_{w_2} \cup \cdots \cup \beta_{w_s}$$

is a linearly independent set whose span contains $N(\phi(T))$. ∎

Theorem 7.21. *If the minimal polynomial of* T *is of the form* $p(t) = (\phi(t))^m$, *then there exists a rational canonical basis for* T.

Proof. The proof is by mathematical induction on m. Suppose that $m = 1$. Apply (b) of the lemma to $W = \{0\}$ to obtain a linearly independent subset of V of the form $\beta_{v_1} \cup \beta_{v_2} \cup \cdots \cup \beta_{v_k}$, whose span contains $N(\phi(T))$. Since $V = N(\phi(T))$, this set is a rational canonical basis for V.

Now suppose that, for some integer $m > 1$, the result is valid whenever the minimal polynomial of T is of the form $(\phi(T))^k$, where $k < m$, and assume that the minimal polynomial of T is $p(t) = (\phi(t))^m$. Let $r = \mathrm{rank}(\phi(T))$. Then $R(\phi(T))$ is a T-invariant subspace of V, and the restriction of T to this subspace has $(\phi(t))^{m-1}$ as its minimal polynomial. Therefore we may apply the induction hypothesis to obtain a rational canonical basis for the restriction of T to $R(T)$. Suppose that $v_1, v_2, \ldots, v_k$ are the generating vectors of the T-cyclic bases that constitute this rational canonical basis. For each i, choose w_i in V such that $v_i = \phi(T)(w_i)$. By Theorem 7.20, the union β of the sets β_{w_i} is linearly independent. Let $W = \mathrm{span}(\beta)$. Then W contains $R(\phi(T))$. Apply (b) of the lemma and adjoin additional T-cyclic bases $\beta_{w_{k+1}}, \beta_{w_{k+2}}, \ldots, \beta_{w_s}$ to β, if necessary, where w_i is in $N(\phi(T))$ for $i \geq k$, to obtain a linearly independent set

$$\beta' = \beta_{w_1} \cup \beta_{w_2} \cup \cdots \cup \beta_{w_k} \cup \cdots \cup \beta_{w_s}$$

whose span W' contains both W and $N(\phi(T))$.

We show that $W' = V$. Let U denote the restriction of $\phi(T)$ to W', which is $\phi(T)$-invariant. By the way in which W' was obtained from $R(\phi(T))$, it follows that $R(U) = R(\phi(T))$ and $N(U) = N(\phi(T))$. Therefore

$$\begin{aligned}
\dim(W') &= \operatorname{rank}(U) + \operatorname{nullity}(U) \\
&= \operatorname{rank}(\phi(T)) + \operatorname{nullity}(\phi(T)) \\
&= \dim(V).
\end{aligned}$$

Thus $W' = V$, and β' is a rational canonical basis for T. ∎

Corollary. K_ϕ *has a basis consisting of the union of* T-*cyclic bases.*

Proof. Apply Theorem 7.21 to the restriction of T to K_ϕ. ∎

We are now ready to study the general case.

Theorem 7.22. *Every linear operator on a finite-dimensional vector space has a rational canonical basis and, hence, a rational canonical form.*

Proof. Let T be a linear operator on a finite-dimensional vector space V, and let $p(t) = (\phi_1(t))^{m_1}(\phi_2(t))^{m_2} \cdots (\phi_k(t))^{m_k}$ be the minimal polynomial of T, where the $\phi_i(t)$'s are the distinct irreducible monic factors of $p(t)$ and $m_i > 0$ for all i. The proof is by mathematical induction on k. The case $k = 1$ is proved in Theorem 7.21.

Suppose that the result is valid whenever the minimal polynomial contains fewer than k distinct irreducible factors for some $k > 1$, and suppose that $p(t)$ contains k distinct factors. Let U be the restriction of T to the T-invariant subspace $W = R((\phi_k(T)^{m_k})$, and let $q(t)$ be the minimal polynomial of U. Then $q(t)$ divides $p(t)$ by Exercise 10 of Section 7.3. Furthermore, $\phi_k(t)$ does not divide $q(t)$. For otherwise, there would exist a nonzero vector $x \in W$ such that $\phi_k(U)(x) = 0$ and a vector $y \in V$ such that $x = (\phi_k(T))^{m_k}(y)$. It follows that $(\phi_k(T))^{m_k+1}(y) = 0$, and hence $y \in K_{\phi_k}$ and $x = (\phi_k(T))^{m_k}(y) = 0$ by Theorem 7.18(e), a contradiction. Thus $q(t)$ contains fewer than k distinct irreducible divisors. So by the induction hypothesis, U has a rational canonical basis β_1 consisting of a union of U-cyclic bases (and hence T-cyclic bases) of vectors from some of the subspaces K_{ϕ_i}, $1 \le i \le k - 1$. By the corollary to Theorem 7.21, K_{ϕ_k} has a basis β_2 consisting of a union of T-cyclic bases. By Theorem 7.19, β_1 and β_2 are disjoint, and $\beta = \beta_1 \cup \beta_2$ is linearly independent. Let s denote the number of vectors in β. Then

$$\begin{aligned}
s &= \dim(R((\phi_k(T))^{m_k})) + \dim(K_{\phi_k}) \\
&= \operatorname{rank}((\phi_k(T))^{m_k}) + \operatorname{nullity}((\phi_k(T))^{m_k}) \\
&= n.
\end{aligned}$$

We conclude that β is a basis for V. Therefore β is a rational canonical basis, and T has a rational canonical form. ∎

In our study of the rational canonical form, we relied on the minimal polynomial. We are now able to relate the rational canonical form to the characteristic polynomial.

Theorem 7.23. *Let* T *be a linear operator on an n-dimensional vector space* V *with characteristic polynomial*

$$f(t) = (-1)^n (\phi_1(t))^{n_1} (\phi_2(t))^{n_2} \cdots (\phi_k(t))^{n_k},$$

where the $\phi_i(t)$'s $(1 \le i \le k)$ *are distinct irreducible monic polynomials and the n_i's are positive integers. Then the following statements are true.*

(a) $\phi_1(t), \phi_2(t), \dots, \phi_k(t)$ *are the irreducible monic factors of the minimal polynomial.*

(b) *For each i,* $\dim(\mathsf{K}_{\phi_i}) = d_i n_i$, *where d_i is the degree of $\phi_i(t)$.*

(c) *If β is a rational canonical basis for T, then $\beta_i = \beta \cap \mathsf{K}_{\phi_i}$ is a basis for K_{ϕ_i} for each i.*

(d) *If γ_i is a basis for K_{ϕ_i} for each i, then $\gamma = \gamma_1 \cup \gamma_2 \cup \cdots \cup \gamma_k$ is a basis for V. In particular, if each γ_i is a disjoint union of T-cyclic bases, then γ is a rational canonical basis for T.*

Proof. (a) By Theorem 7.22, T has a rational canonical form C. By Exercise 40 of Section 5.4, the characteristic polynomial of C, and hence of T, is the product of the characteristic polynomials of the companion matrices that compose C. Therefore each irreducible monic divisor $\phi_i(t)$ of $f(t)$ divides the characteristic polynomial of at least one of the companion matrices, and hence for some integer p, $(\phi_i(t))^p$ is the T-annihilator of a nonzero vector of V. We conclude that $(\phi_i(t))^p$, and so $\phi_i(t)$, divides the minimal polynomial of T. Conversely, if $\phi(t)$ is an irreducible monic polynomial that divides the minimal polynomial of T, then $\phi(t)$ divides the characteristic polynomial of T because the minimal polynomial divides the characteristic polynomial.

(b), (c), and (d) Let $C = [\mathsf{T}]_\beta$, which is a rational canonical form of T. Consider any i, $(1 \le i \le k)$. Since $f(t)$ is the product of the characteristic polynomials of the companion matrices that compose C, we may multiply those characteristic polynomials that arise from the T-cyclic bases in β_i to obtain the factor $(\phi_i(t))^{n_i}$ of $f(t)$. Since this polynomial has degree $n_i d_i$, and the union of these bases is a linearly independent subset β_i of K_{ϕ_i}, we have

$$n_i d_i \le \dim(\mathsf{K}_{\phi_i}).$$

Furthermore, $n = \displaystyle\sum_{i=1}^{k} d_i n_i$, because this sum is equal to the degree of $f(t)$. Now let s denote the number of vectors in γ. By Theorem 7.19, γ is linearly independent, and therefore

$$n = \sum_{i=1}^{k} d_i n_i \le \sum_{i=1}^{k} \dim(\mathsf{K}_{\phi_i}) = s \le n.$$

Hence $n = s$, and $d_i n_i = \dim(\mathsf{K}_{\phi_i})$ for all i. It follows that γ is a basis for V and β_i is a basis for K_{ϕ_i} for each i.   ∎

Uniqueness of the Rational Canonical Form

Having shown that a rational canonical form exists, we are now in a position to ask about the extent to which it is unique. Certainly, the rational canonical form of a linear operator T can be modified by permuting the T-cyclic bases that constitute the corresponding rational canonical basis. This has the effect of permuting the companion matrices that make up the rational canonical form. As in the case of the Jordan canonical form, we show that except for these permutations, the rational canonical form is unique, although the rational canonical bases are not.

To simplify this task, we adopt the convention of ordering every rational canonical basis so that all the T-cyclic bases associated with the same irreducible monic divisor of the characteristic polynomial are grouped together. Furthermore, within each such grouping, we arrange the T-cyclic bases in decreasing order of size. Our task is to show that, subject to this order, the rational canonical form of a linear operator is unique up to the arrangement of the irreducible monic divisors.

As in the case of the Jordan canonical form, we introduce arrays of dots from which we can reconstruct the rational canonical form. For the Jordan canonical form, we devised a dot diagram for each eigenvalue of the given operator. In the case of the rational canonical form, we define a dot diagram for each irreducible monic divisor of the characteristic polynomial of the given operator. A proof that the resulting dot diagrams are completely determined by the operator is also a proof that the rational canonical form is unique.

In what follows, T is a linear operator on a finite-dimensional vector space with rational canonical basis β; $\phi(t)$ is an irreducible monic divisor of the characteristic polynomial of T; $\beta_{v_1}, \beta_{v_2}, \ldots, \beta_{v_k}$ are the T-cyclic bases of β that are contained in K_{ϕ}; and d is the degree of $\phi(t)$. For each j, let $(\phi(t))^{p_j}$ be the annihilator of v_j. This polynomial has degree dp_j; therefore, by Exercise 15 of Section 7.3, β_{v_j} contains dp_j vectors. Furthermore, $p_1 \geq p_2 \geq \cdots \geq p_k$ since the T-cyclic bases are arranged in decreasing order of size. We define the **dot diagram** of $\phi(t)$ to be the array consisting of k columns of dots with p_j dots in the jth column, arranged so that the jth column begins at the top and terminates after p_j dots. For example, if $k = 3$, $p_1 = 4$, $p_2 = 2$, and $p_3 = 2$, then the dot diagram is

$$
\begin{array}{ccc}
\bullet & \bullet & \bullet \\
\bullet & \bullet & \bullet \\
\bullet & & \\
\bullet & &
\end{array}
$$

Although each column of a dot diagram corresponds to a T-cyclic basis

536 Chap. 7 Canonical Forms

β_{v_i} in K_ϕ, there are fewer dots in the column than there are vectors in the basis.

Example 2

Recall the linear operator T of Example 1 with the rational canonical basis β and the rational canonical form $C = [\mathsf{T}]_\beta$. Since there are two irreducible monic divisors of the characteristic polynomial of T, $\phi_1(t) = t^2 - t + 3$ and $\phi_2(t) = t^2 + 1$, there are two dot diagrams to consider. Because $\phi_1(t)$ is the T-annihilator of v_1 and β_{v_1} is a basis for K_{ϕ_1}, the dot diagram for $\phi_1(t)$ consists of a single dot. The other two T cyclic bases, β_{v_3} and β_{v_7}, lie in K_{ϕ_2}. Since v_3 has T-annihilator $(\phi_2(t))^2$ and v_7 has T-annihilator $\phi_2(t)$, in the dot diagram of $\phi_2(t)$ we have $p_1 = 2$ and $p_2 = 1$. These diagrams are as follows:

$$\bullet \qquad\qquad\qquad \begin{matrix} \bullet & \bullet \\ \bullet & \end{matrix}$$

$$\text{Dot diagram for } \phi_1(t) \qquad \text{Dot diagram for } \phi_2(t) \qquad \blacklozenge$$

In practice, we obtain the rational canonical form of a linear operator from the information provided by dot diagrams. This is illustrated in the next example.

Example 3

Let T be a linear operator on a finite-dimensional vector space over R, and suppose that the irreducible monic divisors of the characteristic polynomial of T are

$$\phi_1(t) = t - 1, \quad \phi_2(t) = t^2 + 2, \quad \text{and} \quad \phi_3(t) = t^2 + t + 1.$$

Suppose, furthermore, that the dot diagrams associated with these divisors are as follows:

$$\begin{matrix} \bullet & \bullet \\ \bullet & \end{matrix} \qquad\qquad \begin{matrix} \bullet & \bullet \end{matrix} \qquad\qquad \bullet$$

$$\text{Diagram for } \phi_1(t) \qquad \text{Diagram for } \phi_2(t) \qquad \text{Diagram for } \phi_3(t)$$

Since the dot diagram for $\phi_1(t)$ has two columns, it contributes two companion matrices to the rational canonical form. The first column has two dots, and therefore corresponds to the 2×2 companion matrix of $(\phi_1(t))^2 = (t-1)^2$. The second column, with only one dot, corresponds to the 1×1 companion matrix of $\phi_1(t) = t - 1$. These two companion matrices are given by

$$C_1 = \begin{pmatrix} 0 & -1 \\ 1 & 2 \end{pmatrix} \quad \text{and} \quad C_2 = \begin{pmatrix} 1 \end{pmatrix}.$$

The dot diagram for $\phi_2(t) = t^2 + 2$ consists of two columns. each containing a single dot; hence this diagram contributes two copies of the 2×2 companion matrix for $\phi_2(t)$, namely,

$$C_3 = C_4 = \begin{pmatrix} 0 & -2 \\ 1 & 0 \end{pmatrix}.$$

The dot diagram for $\phi_3(t) = t^2 + t + 1$ consists of a single column with a single dot contributing the single 2×2 companion matrix

$$C_5 = \begin{pmatrix} 0 & -1 \\ 1 & -1 \end{pmatrix}.$$

Therefore the rational canonical form of T is the 9×9 matrix

$$C = \begin{pmatrix} C_1 & O & O & O & O \\ O & C_2 & O & O & O \\ O & O & C_3 & O & O \\ O & O & O & C_4 & O \\ O & O & O & O & C_5 \end{pmatrix}$$

$$= \begin{pmatrix} 0 & -1 & 0 & 0 & 0 & 0 & 0 & 0 & 0 \\ 1 & 2 & 0 & 0 & 0 & 0 & 0 & 0 & 0 \\ 0 & 0 & 1 & 0 & 0 & 0 & 0 & 0 & 0 \\ 0 & 0 & 0 & 0 & -2 & 0 & 0 & 0 & 0 \\ 0 & 0 & 0 & 1 & 0 & 0 & 0 & 0 & 0 \\ 0 & 0 & 0 & 0 & 0 & 0 & -2 & 0 & 0 \\ 0 & 0 & 0 & 0 & 0 & 1 & 0 & 0 & 0 \\ 0 & 0 & 0 & 0 & 0 & 0 & 0 & 0 & -1 \\ 0 & 0 & 0 & 0 & 0 & 0 & 0 & 1 & -1 \end{pmatrix}. \quad \blacklozenge$$

We return to the general problem of finding dot diagrams. As we did before, we fix a linear operator T on a finite-dimensional vector space and an irreducible monic divisor $\phi(t)$ of the characteristic polynomial of T. Let U denote the restriction of the linear operator $\phi(T)$ to K_ϕ. By Theorem 7.18(d), $U^q = T_0$ for some positive integer q. Consequently, by Exercise 12 of Section 7.2, the characteristic polynomial of U is $(-1)^m t^m$, where $m = \dim(K_\phi)$. Therefore K_ϕ is the generalized eigenspace of U corresponding to $\lambda = 0$, and U has a Jordan canonical form. The dot diagram associated with the Jordan canonical form of U gives us a key to understanding the dot diagram of T that is associated with $\phi(t)$. We now relate the two diagrams.

Let β be a rational canonical basis for T, and $\beta_{v_1}, \beta_{v_2}, \ldots, \beta_{v_k}$ be the T-cyclic bases of β that are contained in K_ϕ. Consider one of these T-cyclic bases β_{v_j}, and suppose again that the T-annihilator of v_j is $(\phi(t))^{p_j}$. Then β_{v_j} consists of dp_j vectors in β. For $0 \le i < d$, let γ_i be the cycle of generalized eigenvectors of U corresponding to $\lambda = 0$ with end vector $T^i(v_j)$, where $\mathsf{T}^0(v_j) = b_j$. Then

$$\gamma_i = \{(\phi(\mathsf{T}))^{p_j-1}\mathsf{T}^i(v_j), (\phi(\mathsf{T}))^{p_j-2}\mathsf{T}^i(v_j), \ldots, (\phi(\mathsf{T}))\mathsf{T}^i(v_j), \mathsf{T}^i(v_j)\}.$$

By Theorem 7.1 (p. 485), γ_i is a linearly independent subset of C_{v_i}. Now let

$$\alpha_j = \gamma_0 \cup \gamma_1 \cup \cdots \cup \gamma_{d-1}.$$

Notice that α_j contains $p_j d$ vectors.

Lemma 1. α_j *is an ordered basis for* C_{v_j}.

Proof. The key to this proof is Theorem 7.4 (p. 487). Since α_j is the union of cycles of generalized eigenvectors of U corresponding to $\lambda = 0$, it suffices to show that the set of initial vectors of these cycles

$$\{(\phi(\mathsf{T}))^{p_j-1}(v_j), (\phi(\mathsf{T}))^{p_j-1}\mathsf{T}(v_j), \ldots, (\phi(\mathsf{T}))^{p_j-1}\mathsf{T}^{d-1}(v_j)\}$$

is linearly independent. Consider any linear combination of these vectors

$$a_0(\phi(\mathsf{T}))^{p_j-1}(v_j) + a_1(\phi(\mathsf{T}))^{p_j-1}\mathsf{T}(v_j) + \cdots + a_{d-1}(\phi(\mathsf{T}))^{p_j-1}\mathsf{T}^{d-1}(v_j),$$

where not all of the coefficients are zero. Let $g(t)$ be the polynomial defined by $g(t) = a_0 + a_1 t + \cdots + a_{d-1}t^{d-1}$. Then $g(t)$ is a nonzero polynomial of degree less than d, and hence $(\phi(t))^{p_j-1}g(t)$ is a nonzero polynomial with degree less than $p_j d$. Since $(\phi(t))^{p_j}$ is the T-annihilator of v_j, it follows that $(\phi(\mathsf{T}))^{p_j-1}g(\mathsf{T})(v_j) \neq 0$. Therefore the set of initial vectors is linearly independent. So by Theorem 7.4, α_j is linearly independent, and the γ_i's are disjoint. Consequently, α_j consists of $p_j d$ linearly independent vectors in C_{v_j}, which has dimension $p_j d$. We conclude that α_j is a basis for C_{v_j}. ∎

Thus we may replace β_{v_j} by α_j as a basis for C_{v_j}. We do this for each j to obtain a subset $\alpha = \alpha_1 \cup \alpha_2 \cdots \cup \alpha_k$ of K_ϕ.

Lemma 2. α *is a Jordan canonical basis for* K_ϕ.

Proof. Since $\beta_{v_1} \cup \beta_{v_2} \cup \cdots \cup \beta_{v_k}$ is a basis for K_ϕ, and since $\text{span}(\alpha_i) = \text{span}(\beta_{v_i}) = \mathsf{C}_{v_i}$, Exercise 9 implies that α is a basis for K_ϕ. Because α is a union of cycles of generalized eigenvectors of U, we conclude that α is a Jordan canonical basis. ∎

We are now in a position to relate the dot diagram of T corresponding to $\phi(t)$ to the dot diagram of U, bearing in mind that in the first case we are considering a rational canonical form and in the second case we are considering a Jordan canonical form. For convenience, we designate the first diagram D_1, and the second diagram D_2. For each j, the presence of the T-cyclic basis β_{x_j} results in a column of p_j dots in D_1. By Lemma 1, this basis is replaced by the union α_j of d cycles of generalized eigenvectors of U, each of length p_j, which becomes part of the Jordan canonical basis for U. In effect, α_j determines d columns each containing p_j dots in D_2. So each column in D_1 determines d columns in D_2 of the same length, and all columns in D_2 are obtained in this way. Alternatively, each row in D_2 has d times as many dots as the corresponding row in D_1. Since Theorem 7.10 (p. 500) gives us the number of dots in any row of D_2, we may divide the appropriate expression in this theorem by d to obtain the number of dots in the corresponding row of D_1. Thus we have the following result.

Theorem 7.24. *Let* T *be a linear operator on a finite-dimensional vector space* V, *let* $\phi(t)$ *be an irreducible monic divisor of the characteristic polynomial of* T *of degree* d, *and let* r_i *denote the number of dots in the ith row of the dot diagram for* $\phi(t)$ *with respect to a rational canonical basis for* T. *Then*

(a) $r_1 = \dfrac{1}{d}[\dim(\mathsf{V}) - \operatorname{rank}(\phi(\mathsf{T}))]$

(b) $r_i = \dfrac{1}{d}[\operatorname{rank}((\phi(\mathsf{T}))^{i-1}) - \operatorname{rank}((\phi(\mathsf{T}))^{i})]$ *for* $i > 1$.

Thus the dot diagrams associated with a rational canonical form of an operator are completely determined by the operator. Since the rational canonical form is completely determined by its dot diagrams, we have the following uniqueness condition.

Corollary. *Under the conventions described earlier, the rational canonical form of a linear operator is unique up to the arrangement of the irreducible monic divisors of the characteristic polynomial.*

Since the rational canonical form of a linear operator is unique, the polynomials corresponding to the companion matrices that determine this form are also unique. These polynomials, which are powers of the irreducible monic divisors, are called the **elementary divisors** of the linear operator. Since a companion matrix may occur more than once in a rational canonical form, the same is true for the elementary divisors. We call the number of such occurrences the **multiplicity** of the elementary divisor.

Conversely, the elementary divisors and their multiplicities determine the companion matrices and, therefore, the rational canonical form of a linear operator.

Example 4

Let

$$\beta = \{e^x \cos 2x, e^x \sin 2x, xe^x \cos 2x, xe^x \sin 2x\}$$

be viewed as a subset of $\mathcal{F}(R, R)$, the space of all real-valued functions defined on R, and let $V = \text{span}(\beta)$. Then V is a four-dimensional subspace of $\mathcal{F}(R, R)$, and β is an ordered basis for V. Let D be the linear operator on V defined by $D(y) = y'$, the derivative of y, and let $A = [D]_\beta$. Then

$$A = \begin{pmatrix} 1 & 2 & 1 & 0 \\ -2 & 1 & 0 & 1 \\ 0 & 0 & 1 & 2 \\ 0 & 0 & -2 & 1 \end{pmatrix},$$

and the characteristic polynomial of D, and hence of A, is

$$f(t) = (t^2 - 2t + 5)^2.$$

Thus $\phi(t) = t^2 - 2t + 5$ is the only irreducible monic divisor of $f(t)$. Since $\phi(t)$ has degree 2 and V is four-dimensional, the dot diagram for $\phi(t)$ contains only two dots. Therefore the dot diagram is determined by r_1, the number of dots in the first row. Because ranks are preserved under matrix representations, we can use A in place of D in the formula given in Theorem 7.24. Now

$$\phi(A) = \begin{pmatrix} 0 & 0 & 0 & 4 \\ 0 & 0 & -4 & 0 \\ 0 & 0 & 0 & 0 \\ 0 & 0 & 0 & 0 \end{pmatrix},$$

and so

$$r_1 = \tfrac{1}{2}[4 - \text{rank}(\phi(A))] = \tfrac{1}{2}[4 - 2] = 1.$$

It follows that the second dot lies in the second row, and the dot diagram is as follows:

$$\bullet$$
$$\bullet$$

Hence V is a D-cyclic space generated by a single function with D-annihilator $(\phi(t))^2$. Furthermore, its rational canonical form is given by the companion matrix of $(\phi(t))^2 = t^4 - 4t^3 + 14t^2 - 20t + 25$, which is

$$\begin{pmatrix} 0 & 0 & 0 & -25 \\ 1 & 0 & 0 & 20 \\ 0 & 1 & 0 & -14 \\ 0 & 0 & 1 & 4 \end{pmatrix}.$$

Thus $(\phi(t))^2$ is the only elementary divisor of D, and it has multiplicity 1. For the cyclic generator, it suffices to find a function g in V for which $\phi(D)(g) \neq 0$.

Since $\phi(A)(e_3) \neq 0$, it follows that $\phi(\mathsf{D})(xe^x \cos 2x) \neq 0$; therefore $g(x) = xe^x \cos 2x$ can be chosen as the cyclic generator. Hence

$$\beta_g = \{xe^x \cos 2x, \mathsf{D}(xe^x \cos 2x), \mathsf{D}^2(xe^x \cos 2x), \mathsf{D}^3(xe^x \cos 2x)\}$$

is a rational canonical basis for D. Notice that the function h defined by $h(x) = xe^x \sin 2x$ can be chosen in place of g. This shows that the rational canonical basis is not unique. ◆

It is convenient to refer to the rational canonical form and elementary divisors of a matrix, which are defined in the obvious way.

Definitions.   Let $A \in \mathsf{M}_{n \times n}(F)$. The **rational canonical form** of A is defined to be the rational canonical form of L_A. Likewise, for A, the **elementary divisors** and their **multiplicities** are the same as those of L_A.

Let A be an $n \times n$ matrix, let C be a rational canonical form of A, and let β be the appropriate rational canonical basis for L_A. Then $C = [\mathsf{L}_A]_\beta$, and therefore A is similar to C. In fact, if Q is the matrix whose columns are the vectors of β in the same order, then $Q^{-1}AQ = C$.

Example 5

For the following real matrix A, we find the rational canonical form C of A and a matrix Q such that $Q^{-1}AQ = C$.

$$A = \begin{pmatrix} 0 & 2 & 0 & -6 & 2 \\ 1 & -2 & 0 & 0 & 2 \\ 1 & 0 & 1 & -3 & 2 \\ 1 & -2 & 1 & -1 & 2 \\ 1 & -4 & 3 & -3 & 4 \end{pmatrix}$$

The characteristic polynomial of A is $f(t) = -(t^2 + 2)^2(t - 2)$; therefore $\phi_1(t) = t^2 + 2$ and $\phi_2(t) = t - 2$ are the distinct irreducible monic divisors of $f(t)$. By Theorem 7.23, $\dim(K_{\phi_1}) = 4$ and $\dim(K_{\phi_2}) = 1$. Since the degree of $\phi_1(t)$ is 2, the total number of dots in the dot diagram of $\phi_1(t)$ is $4/2 = 2$, and the number of dots r_1 in the first row is given by

$$r_1 = \tfrac{1}{2}[\dim(\mathsf{R}^5) - \operatorname{rank}(\phi_1(A))]$$
$$= \tfrac{1}{2}[5 - \operatorname{rank}(A^2 + 2I)]$$
$$= \tfrac{1}{2}[5 - 1] = 2.$$

Thus the dot diagram of $\phi_1(t)$ is

● ●

and each column contributes the companion matrix

$$\begin{pmatrix} 0 & -2 \\ 1 & 0 \end{pmatrix}$$

for $\phi_1(t) = t^2 + 2$ to the rational canonical form C. Consequently $\phi_1(t)$ is an elementary divisor with multiplicity 2. Since $\dim(\mathsf{K}_{\phi_2}) = 1$, the dot diagram of $\phi_2(t) = t - 2$ consists of a single dot, which contributes the 1×1 matrix (2). Hence $\phi_2(t)$ is an elementary divisor with multiplicity 1. Therefore the rational canonical form C is

$$C = \left(\begin{array}{cc|cc|c} 0 & -2 & 0 & 0 & 0 \\ 1 & 0 & 0 & 0 & 0 \\ \hline 0 & 0 & 0 & -2 & 0 \\ 0 & 0 & 1 & 0 & 0 \\ \hline 0 & 0 & 0 & 0 & 2 \end{array} \right).$$

We can infer from the dot diagram of $\phi_1(t)$ that if β is a rational canonical basis for L_A, then $\beta \cap \mathsf{K}_{\phi_1}$ is the union of two cyclic bases β_{v_1} and β_{v_2}, where v_1 and v_2 each have annihilator $\phi_1(t)$. It follows that both v_1 and v_2 lie in $\mathsf{N}(\phi_1(\mathsf{L}_A))$. It can be shown that

$$\left\{ \begin{pmatrix} 1 \\ 0 \\ 0 \\ 0 \\ 0 \end{pmatrix}, \begin{pmatrix} 0 \\ 1 \\ 0 \\ 0 \\ 0 \end{pmatrix}, \begin{pmatrix} 0 \\ 0 \\ 2 \\ 1 \\ 0 \end{pmatrix}, \begin{pmatrix} 0 \\ 0 \\ -1 \\ 0 \\ 1 \end{pmatrix} \right\}$$

is a basis for $\mathsf{N}(\phi_1(\mathsf{L}_A))$. Setting $v_1 = e_1$, we see that

$$Av_1 = \begin{pmatrix} 0 \\ 1 \\ 1 \\ 1 \\ 1 \end{pmatrix}.$$

Next choose v_2 in $\mathsf{K}_{\phi_1} = \mathsf{N}(\phi(\mathsf{L}_A))$, but not in the span of $\beta_{v_1} = \{v_1, Av_1\}$. For example, $v_2 = e_2$. Then it can be seen that

$$Av_2 = \begin{pmatrix} 2 \\ -2 \\ 0 \\ -2 \\ -4 \end{pmatrix},$$

and $\beta_{v_1} \cup \beta_{v_2}$ is a basis for K_{ϕ_1}.

Since the dot diagram of $\phi_2(t) = t - 2$ consists of a single dot, any nonzero vector in K_{ϕ_2} is an eigenvector of A corresponding to the eigenvalue $\lambda = 2$. For example, choose

$$v_3 = \begin{pmatrix} 0 \\ 1 \\ 1 \\ 1 \\ 2 \end{pmatrix}.$$

By Theorem 7.23, $\beta = \{v_1, Av_1, v_2, Av_2, v_3\}$ is a rational canonical basis for L_A. So setting

$$Q = \begin{pmatrix} 1 & 0 & 0 & 2 & 0 \\ 0 & 1 & 1 & -2 & 1 \\ 0 & 1 & 0 & 0 & 1 \\ 0 & 1 & 0 & -2 & 1 \\ 0 & 1 & 0 & -4 & 2 \end{pmatrix},$$

we have $Q^{-1}AQ = C$.   ◆

Example 6

For the following matrix A, we find the rational canonical form C and a matrix Q such that $Q^{-1}AQ = C$:

$$A = \begin{pmatrix} 2 & 1 & 0 & 0 \\ 0 & 2 & 1 & 0 \\ 0 & 0 & 2 & 0 \\ 0 & 0 & 0 & 2 \end{pmatrix}.$$

Since the characteristic polynomial of A is $f(t) = (t-2)^4$, the only irreducible monic divisor of $f(t)$ is $\phi(t) = t - 2$, and so $K_\phi = R^4$. In this case, $\phi(t)$ has degree 1; hence in applying Theorem 7.24 to compute the dot diagram for $\phi(t)$, we obtain

$$r_1 = 4 - \operatorname{rank}(\phi(A)) = 4 - 2 = 2,$$
$$r_2 = \operatorname{rank}(\phi(A)) - \operatorname{rank}((\phi(A))^2) = 2 - 1 = 1,$$

and

$$r_3 = \operatorname{rank}((\phi(A))^2) - \operatorname{rank}((\phi(A))^3) = 1 - 0 = 1,$$

where r_i is the number of dots in the ith row of the dot diagram. Since there are $\dim(R^4) = 4$ dots in the diagram, we may terminate these computations with r_3. Thus the dot diagram for A is

$$\begin{matrix} \bullet & \bullet \\ \bullet & \\ \bullet & \end{matrix}$$

Since $(t-2)^3$ has the companion matrix

$$\begin{pmatrix} 0 & 0 & 8 \\ 1 & 0 & -12 \\ 0 & 1 & 6 \end{pmatrix}$$

and $(t-2)$ has the companion matrix (2), the rational canonical form of A is given by

$$C = \left(\begin{array}{ccc|c} 0 & 0 & 8 & 0 \\ 1 & 0 & -12 & 0 \\ 0 & 1 & 6 & 0 \\ \hline 0 & 0 & 0 & 2 \end{array} \right).$$

Next we find a rational canonical basis for L_A. The preceding dot diagram indicates that there are two vectors v_1 and v_2 in R^4 with annihilators $(\phi(t))^3$ and $\phi(t)$, respectively, and such that

$$\beta = \{\beta_{v_1} \cup \beta_{v_1}\} = \{v_1, Av_1, A^2 v_1, v_2\}$$

is a rational canonical basis for L_A. Furthermore, $v_1 \notin \mathsf{N}((\mathsf{L}_A - 2\mathsf{I})^2)$, and $v_2 \in \mathsf{N}(\mathsf{L}_A - 2\mathsf{I})$. It can easily be shown that

$$\mathsf{N}(\mathsf{L}_A - 2\mathsf{I}) = \mathrm{span}(\{e_1, e_4\})$$

and

$$\mathsf{N}((\mathsf{L}_A - 2\mathsf{I})^2) = \mathrm{span}(\{e_1, e_2, e_4\}).$$

The standard vector e_3 meets the criteria for v_1; so we set $v_1 = e_3$. It follows that

$$Av_1 = \begin{pmatrix} 0 \\ 1 \\ 2 \\ 0 \end{pmatrix} \quad \text{and} \quad A^2 v_1 = \begin{pmatrix} 1 \\ 4 \\ 4 \\ 0 \end{pmatrix}.$$

Next we choose a vector $v_2 \in \mathsf{N}(\mathsf{L}_A - 2\mathsf{I})$ that is not in the span of β_{v_1}. Clearly, $v_2 = e_4$ satisfies this condition. Thus

$$\left\{ \begin{pmatrix} 0 \\ 0 \\ 1 \\ 0 \end{pmatrix}, \begin{pmatrix} 0 \\ 1 \\ 2 \\ 0 \end{pmatrix}, \begin{pmatrix} 1 \\ 4 \\ 4 \\ 0 \end{pmatrix}, \begin{pmatrix} 0 \\ 0 \\ 0 \\ 1 \end{pmatrix} \right\}$$

is a rational canonical basis for L_A.

Finally, let Q be the matrix whose columns are the vectors of β in the same order:

$$Q = \begin{pmatrix} 0 & 0 & 1 & 0 \\ 0 & 1 & 4 & 0 \\ 1 & 2 & 4 & 0 \\ 0 & 0 & 0 & 1 \end{pmatrix}.$$

Then $C = Q^{-1}AQ$.   ◆

Direct Sums*

The next theorem is a simple consequence of Theorem 7.23.

Theorem 7.25 (Primary Decomposition Theorem).   *Let* T *be a linear operator on an* n-*dimensional vector space* V *with characteristic polynomial*

$$f(t) = (-1)^n (\phi_1(t))^{n_1} (\phi_2(t))^{n_2} \cdots (\phi_k(t))^{n_k},$$

where the $\phi_i(t)$'s $(1 \le i \le k)$ *are distinct irreducible monic polynomials and the* n_i's *are positive integers. Then the following statements are true.*
  (a) $\mathsf{V} = \mathsf{K}_{\phi_1} \oplus \mathsf{K}_{\phi_2} \oplus \cdots \oplus \mathsf{K}_{\phi_k}$.
  (b) *If* T_i $(1 \le i \le k)$ *is the restriction of* T *to* K_{ϕ_i} *and* C_i *is the rational canonical form of* T_i, *then* $C_1 \oplus C_2 \oplus \cdots \oplus C_k$ *is the rational canonical form of* T.

Proof. Exercise.   ∎

The next theorem is a simple consequence of Theorem 7.17.

Theorem 7.26. *Let* T *be a linear operator on a finite-dimensional vector space* V. *Then* V *is a direct sum of* T-*cyclic subspaces* C_{v_i}, *where each* v_i *lies in* K_ϕ *for some irreducible monic divisor* $\phi(t)$ *of the characteristic polynomial of* T.

Proof. Exercise.   ∎

EXERCISES

1. Label the following statements as true or false.

    (a) Every rational canonical basis for a linear operator T is the union of T-cyclic bases.

(b) If a basis is the union of T-cyclic bases for a linear operator T, then it is a rational canonical basis for T.

(c) There exist square matrices having no rational canonical form.

(d) A square matrix is similar to its rational canonical form.

(e) For any linear operator T on a finite-dimensional vector space, any irreducible factor of the characteristic polynomial of T divides the minimal polynomial of T.

(f) Let $\phi(t)$ be an irreducible monic divisor of the characteristic polynomial of a linear operator T. The dots in the diagram used to compute the rational canonical form of the restriction of T to K_ϕ are in one-to-one correspondence with the vectors in a basis for K_ϕ.

(g) If a matrix has a Jordan canonical form, then its Jordan canonical form and rational canonical form are similar.

2. For each of the following matrices $A \in \mathsf{M}_{n \times n}(F)$, find the rational canonical form C of A and a matrix $Q \in \mathsf{M}_{n \times n}(F)$ such that $Q^{-1}AQ = C$.

(a) $A = \begin{pmatrix} 3 & 1 & 0 \\ 0 & 3 & 1 \\ 0 & 0 & 3 \end{pmatrix}$ $F = R$

(b) $A = \begin{pmatrix} 0 & -1 \\ 1 & -1 \end{pmatrix}$ $F = R$

(c) $A = \begin{pmatrix} 0 & -1 \\ 1 & -1 \end{pmatrix}$ $F = C$

(d) $A = \begin{pmatrix} 0 & -7 & 14 & -6 \\ 1 & -4 & 6 & -3 \\ 0 & -4 & 9 & -4 \\ 0 & -4 & 11 & -5 \end{pmatrix}$ $F = R$

(e) $A = \begin{pmatrix} 0 & -4 & 12 & -7 \\ 1 & -1 & 3 & -3 \\ 0 & -1 & 6 & -4 \\ 0 & -1 & 8 & -5 \end{pmatrix}$ $F = R$

3. For each of the following linear operators T, find the elementary divisors, the rational canonical form C, and a rational canonical basis β.

(a) T is the linear operator on $\mathsf{P}_3(R)$ defined by
$$T(f(x)) = f(0)x - f'(1).$$

(b) Let $S = \{\sin x, \cos x, x \sin x, x \cos x\}$, a subset of $\mathcal{F}(R, R)$, and let $V = \text{span}(S)$. Define T to be the linear operator on V such that
$$T(f) = f'.$$

(c) T is the linear operator on $\mathsf{M}_{2 \times 2}(R)$ defined by

$$T(A) = \begin{pmatrix} 0 & 1 \\ -1 & 1 \end{pmatrix} \cdot A.$$

(d) Let $S = \{\sin x \sin y, \sin x \cos y, \cos x \sin y, \cos x \cos y\}$, a subset of $\mathcal{F}(R \times R, R)$, and let $V = \text{span}(S)$. Define T to be the linear operator on V such that

$$T(f)(x,y) = \frac{\partial f(x,y)}{\partial x} + \frac{\partial f(x,y)}{\partial y}.$$

4. Let T be a linear operator on a finite-dimensional vector space V with minimal polynomial $(\phi(t))^m$ for some positive integer m.

   (a) Prove that $R(\phi(T)) \subseteq N((\phi(T))^{m-1})$.
   (b) Give an example to show that the subspaces in (a) need not be equal.
   (c) Prove that the minimal polynomial of the restriction of T to $R(\phi(T))$ equals $(\phi(t))^{m-1}$.

5. Let T be a linear operator on a finite-dimensional vector space. Prove that the rational canonical form of T is a diagonal matrix if and only if T is diagonalizable.

6. Let T be a linear operator on a finite-dimensional vector space V with characteristic polynomial $f(t) = (-1)^n \phi_1(t)\phi_2(t)$, where $\phi_1(t)$ and $\phi_2(t)$ are distinct irreducible monic polynomials and $n = \dim(V)$.

   (a) Prove that there exist $v_1, v_2 \in V$ such that v_1 has T-annihilator $\phi_1(t)$, v_2 has T-annihilator $\phi_2(t)$, and $\beta_{v_1} \cup \beta_{v_2}$ is a basis for V.
   (b) Prove that there is a vector $v_3 \in V$ with T-annihilator $\phi_1(t)\phi_2(t)$ such that β_{v_3} is a basis for V.
   (c) Describe the difference between the matrix representation of T with respect to $\beta_{v_1} \cup \beta_{v_2}$ and the matrix representation of T with respect to β_{v_3}.

   Thus, to assure the uniqueness of the rational canonical form, we require that the generators of the T-cyclic bases that constitute a rational canonical basis have T-annihilators equal to powers of irreducible monic factors of the characteristic polynomial of T.

7. Let T be a linear operator on a finite-dimensional vector space with minimal polynomial

$$f(t) = (\phi_1(t))^{m_1}(\phi_2(t))^{m_2} \cdots (\phi_k(t))^{m_k},$$

   where the $\phi_i(t)$'s are distinct irreducible monic factors of $f(t)$. Prove that for each i, m_i is the number of entries in the first column of the dot diagram for $\phi_i(t)$.

8. Let T be a linear operator on a finite-dimensional vector space V. Prove that for any irreducible polynomial $\phi(t)$, if $\phi(\mathsf{T})$ is not one-to-one, then $\phi(t)$ divides the characteristic polynomial of T. *Hint:* Apply Exercise 15 of Section 7.3.

9. Let V be a vector space and $\beta_1, \beta_2, \ldots, \beta_k$ be disjoint subsets of V whose union is a basis for V. Now suppose that $\gamma_1, \gamma_2, \ldots, \gamma_k$ are linearly independent subsets of V such that $\text{span}(\gamma_i) = \text{span}(\beta_i)$ for all i. Prove that $\gamma_1 \cup \gamma_2 \cup \cdots \cup \gamma_k$ is also a basis for V.

10. Let T be a linear operator on a finite-dimensional vector space, and suppose that $\phi(t)$ is an irreducible monic factor of the characteristic polynomial of T. Prove that if $\phi(t)$ is the T-annihilator of vectors x and y, then $x \in \mathsf{C}_y$ if and only if $\mathsf{C}_x = \mathsf{C}_y$.

Exercises 11 and 12 are concerned with direct sums.

11. Prove Theorem 7.25.

12. Prove Theorem 7.26.

INDEX OF DEFINITIONS FOR CHAPTER 7

Appendices

APPENDIX A   SETS

A **set** is a collection of objects, called **elements** of the set. If x is an element of the set A, then we write $x \in A$; otherwise, we write $x \notin A$. For example, if Z is the set of integers, then $3 \in Z$ and $\frac{1}{2} \notin Z$.

One set that appears frequently is the set of real numbers, which we denote by R throughout this text.

Two sets A and B are called **equal**, written $A = B$, if they contain exactly the same elements. Sets may be described in one of two ways:

1. By listing the elements of the set between set braces { }.
2. By describing the elements of the set in terms of some characteristic property.

For example, the set consisting of the elements 1, 2, 3, and 4 can be written as $\{1, 2, 3, 4\}$ or as

$$\{x \colon x \text{ is a positive integer less than } 5\}.$$

Note that the order in which the elements of a set are listed is immaterial; hence

$$\{1, 2, 3, 4\} = \{3, 1, 2, 4\} = \{1, 3, 1, 4, 2\}.$$

Example 1

Let A denote the set of real numbers between 1 and 2. Then A may be written as

$$A = \{x \in R \colon 1 < x < 2\}. \quad \blacklozenge$$

A set B is called a **subset** of a set A, written $B \subseteq A$ or $A \supseteq B$, if every element of B is an element of A. For example, $\{1, 2, 6\} \subseteq \{2, 8, 7, 6, 1\}$. If $B \subseteq A$, and $B \neq A$, then B is called a **proper subset** of A. Observe that $A = B$ if and only if $A \subseteq B$ and $B \subseteq A$, a fact that is often used to prove that two sets are equal.

The **empty set**, denoted by $\varnothing$, is the set containing no elements. The empty set is a subset of every set.

Sets may be combined to form other sets in two basic ways. The **union** of two sets A and B, denoted $A \cup B$, is the set of elements that are in A, or B, or both; that is,

$$A \cup B = \{x \colon x \in A \text{ or } x \in B\}.$$

The **intersection** of two sets A and B, denoted $A \cap B$, is the set of elements that are in both A and B; that is,

$$A \cap B = \{x \colon x \in A \text{ and } x \in B\}.$$

Two sets are called **disjoint** if their intersection equals the empty set.

Example 2

Let $A = \{1, 3, 5\}$ and $B = \{1, 5, 7, 8\}$. Then

$$A \cup B = \{1, 3, 5, 7, 8\} \quad \text{and} \quad A \cap B = \{1, 5\}.$$

Likewise, if $X = \{1, 2, 8\}$ and $Y = \{3, 4, 5\}$, then

$$X \cup Y = \{1, 2, 3, 4, 5, 8\} \quad \text{and} \quad X \cap Y = \varnothing.$$

Thus X and Y are disjoint sets. ◆

The union and intersection of more than two sets can be defined analogously. Specifically, if $A_1, A_2, \ldots, A_n$ are sets, then the union and intersections of these sets are defined, respectively, by

$$\bigcup_{i=1}^{n} A_i = \{x \colon x \in A_i \text{ for some } i = 1, 2, \ldots, n\}$$

and

$$\bigcap_{i=1}^{n} A_i = \{x \colon x \in A_i \text{ for all } i = 1, 2, \ldots, n\}.$$

Similarly, if Λ is an index set and $\{A_\alpha \colon \alpha \in \Lambda\}$ is a collection of sets, the union and intersection of these sets are defined, respectively, by

$$\bigcup_{\alpha \in \Lambda} A_\alpha = \{x \colon x \in A_\alpha \text{ for some } \alpha \in \Lambda\}$$

and

$$\bigcap_{\alpha \in \Lambda} A_\alpha = \{x \colon x \in A_\alpha \text{ for all } \alpha \in \Lambda\}.$$

Example 3

Let $\Lambda = \{\alpha \in R \colon \alpha > 1\}$, and let

$$A_\alpha = \left\{ x \in R \colon \frac{-1}{\alpha} \le x \le 1 + \alpha \right\}$$

for each $\alpha \in \Lambda$. Then

$$\bigcup_{\alpha \in \Lambda} A_\alpha = \{x \in R \colon x > -1\} \quad \text{and} \quad \bigcap_{\alpha \in \Lambda} A_\alpha = \{x \in R \colon 0 \le x \le 2\}. \quad ◆$$

By a relation on a set A, we mean a rule for determining whether or not, for any elements x and y in A, x stands in a given relationship to y. More precisely, a **relation** on A is a set S of ordered pairs of elements of A such that $(x, y) \in S$ if and only if x stands in the given relationship to y. On the set of real numbers, for instance, "is equal to," "is less than," and "is greater than or equal to" are familiar relations. If S is a relation on a set A, we often write $x \sim y$ in place of $(x, y) \in S$.

A relation S on a set A is called an **equivalence relation** on A if the following three conditions hold:

1. For each $x \in A$, $x \sim x$ *(reflexivity)*.
2. If $x \sim y$, then $y \sim x$ *(symmetry)*.
3. If $x \sim y$ and $y \sim z$, then $x \sim z$ *(transitivity)*.

For example, if we define $x \sim y$ to mean that $x - y$ is divisible by a fixed integer n, then $\sim$ is an equivalence relation on the set of integers.

APPENDIX B   FUNCTIONS

If A and B are sets, then a **function** f from A to B, written $f\colon A \to B$, is a rule that associates to each element x in A a unique element denoted $f(x)$ in B. The element $f(x)$ is called the **image** of x (under f), and x is called a **preimage** of $f(x)$ (under f). If $f\colon A \to B$, then A is called the **domain** of f, B is called the **codomain** of f, and the set $\{f(x)\colon x \in A\}$ is called the **range** of f. Note that the range of f is a subset of B. If $S \subseteq A$, we denote by $f(S)$ the set $\{f(x)\colon x \in S\}$ of all images of elements of S. Likewise, if $T \subseteq B$, we denote by $f^{-1}(T)$ the set $\{x \in A\colon f(x) \in T\}$ of all preimages of elements in T. Finally, two functions $f\colon A \to B$ and $g\colon A \to B$ are **equal**, written $f = g$, if $f(x) = g(x)$ for all $x \in A$.

Example 1

Suppose that $A = [-10, 10]$. Let $f\colon A \to R$ be the function that assigns to each element x in A the element $x^2 + 1$ in R; that is, f is defined by $f(x) = x^2 + 1$. Then A is the domain of f, R is the codomain of f, and $[1, 101]$ is the range of f. Since $f(2) = 5$, the image of 2 is 5, and 2 is a preimage of 5. Notice that -2 is another preimage of 5. Moreover, if $S = [1, 2]$ and $T = [82, 101]$, then $f(S) = [2, 5]$ and $f^{-1}(T) = [-10, -9] \cup [9, 10]$.   ◆

As Example 1 shows, the preimage of an element in the range need not be unique. Functions such that each element of the range has a unique preimage are called **one-to-one**; that is $f\colon A \to B$ is one-to-one if $f(x) = f(y)$ implies $x = y$ or, equivalently, if $x \neq y$ implies $f(x) \neq f(y)$.

If $f\colon A \to B$ is a function with range B, that is, if $f(A) = B$, then f is called **onto**. So f is onto if and only if the range of f equals the codomain of f.

Let $f: A \rightarrow B$ be a function and $S \subseteq A$. Then a function $f_S: S \rightarrow B$, called the **restriction** of f to S, can be formed by defining $f_S(x) = f(x)$ for each $x \in S$.

The next example illustrates these concepts.

Example 2

Let $f: [-1, 1] \rightarrow [0, 1]$ be defined by $f(x) = x^2$. This function is onto, but not one-to-one since $f(-1) = f(1) = 1$. Note that if $S = [0, 1]$, then f_S is both onto and one-to-one. Finally, if $T = [\frac{1}{2}, 1]$, then f_T is one-to-one, but not onto. ◆

Let A, B, and C be sets and $f: A \rightarrow B$ and $g: B \rightarrow C$ be functions. By following f with g, we obtain a function $g \circ f: A \rightarrow C$ called the **composite** of g and f. Thus $(g \circ f)(x) = g(f(x))$ for all $x \in A$. For example, let $A = B = C = R$, $f(x) = \sin x$, and $g(x) = x^2 + 3$. Then $(g \circ f)(x) = (g(f(x)) = \sin^2 x + 3$, whereas $(f \circ g)(x) = f(g(x)) = \sin(x^2 + 3)$. Hence, $g \circ f \neq f \circ g$. Functional composition is associative, however; that is, if $h: C \rightarrow D$ is another function, then $h \circ (g \circ f) = (h \circ g) \circ f$.

A function $f: A \rightarrow B$ is said to be **invertible** if there exists a function $g: B \rightarrow A$ such that $(f \circ g)(y) = y$ for all $y \in B$ and $(g \circ f)(x) = x$ for all $x \in A$. If such a function g exists, then it is unique and is called the **inverse** of f. We denote the inverse of f (when it exists) by f^{-1}. It can be shown that f is invertible if and only if f is both one-to-one and onto.

Example 3

The function $f: R \rightarrow R$ defined by $f(x) = 3x + 1$ is one-to-one and onto; hence f is invertible. The inverse of f is the function $f^{-1}: R \rightarrow R$ defined by $f^{-1}(x) = (x - 1)/3$. ◆

The following facts about invertible functions are easily proved.

1. If $f: A \rightarrow B$ is invertible, then f^{-1} is invertible, and $(f^{-1})^{-1} = f$.
2. If $f: A \rightarrow B$ and $g: B \rightarrow C$ are invertible, then $g \circ f$ is invertible, and $(g \circ f)^{-1} = f^{-1} \circ g^{-1}$.

APPENDIX C   FIELDS

The set of real numbers is an example of an algebraic structure called a *field*. Basically, a field is a set in which four operations (called addition, multiplication, subtraction, and division) can be defined so that, with the exception of division by zero, the sum, product, difference, and quotient of any two elements in the set is an element of the set. More precisely, a field is defined as follows.

Definitions. *A field F is a set on which two operations $+$ and $\cdot$ (called* **addition** *and* **multiplication**, *respectively) are defined so that, for each pair of elements x, y in F, there are unique elements $x + y$ and $x \cdot y$ in F for which the following conditions hold for all elements a, b, c in F.*

(F 1) $a + b = b + a$ *and* $a \cdot b = b \cdot a$
   (commutativity of addition and multiplication)

(F 2) $(a + b) + c = a + (b + c)$ *and* $(a \cdot b) \cdot c = a \cdot (b \cdot c)$
   (associativity of addition and multiplication)

(F 3) *There exist distinct elements 0 and 1 in F such that*

$$0 + a = a \quad \text{and} \quad 1 \cdot a = a$$

   (existence of identity elements for addition and multiplication)

(F 4) *For each element a in F and each nonzero element b in F, there exist elements c and d in F such that*

$$a + c = 0 \quad \text{and} \quad b \cdot d = 1$$

   (existence of inverses for addition and multiplication)

(F 5) $a \cdot (b + c) = a \cdot b + a \cdot c$
   (distributivity of multiplication over addition)

The elements $x + y$ and $x \cdot y$ are called the **sum** *and* **product**, *respectively, of x and y. The elements 0 (read "***zero***") and 1 (read "***one***") mentioned in (F 3) are called* **identity elements** *for addition and multiplication, respectively, and the elements c and d referred to in (F 4) are called an* **additive inverse** *for a and a* **multiplicative inverse** *for b, respectively.*

Example 1

The set of real numbers R with the usual definitions of addition and multiplication is a field.   ◆

Example 2

The set of rational numbers with the usual definitions of addition and multiplication is a field.   ◆

Example 3

The set of all real numbers of the form $a + b\sqrt{2}$, where a and b are rational numbers, with addition and multiplication as in R is a field.   ◆

Example 4

The field Z_2 consists of two elements 0 and 1 with the operations of addition and multiplication defined by the equations

$$0 + 0 = 0, \quad 0 + 1 = 1 + 0 = 1, \quad 1 + 1 = 0,$$
$$0 \cdot 0 = 0, \quad 0 \cdot 1 = 1 \cdot 0 = 0, \quad \text{and} \quad 1 \cdot 1 = 1. \quad ◆$$

Example 5

Neither the set of positive integers nor the set of integers with the usual definitions of addition and multiplication is a field, for in either case (F 4) does not hold. ◆

The identity and inverse elements guaranteed by (F 3) and (F 4) are unique; this is a consequence of the following theorem.

Theorem C.1 (Cancellation Laws). *For arbitrary elements a, b, and c in a field, the following statements are true.*
(a) *If $a + b = c + b$, then $a = c$.*
(b) *If $a \cdot b = c \cdot b$ and $b \neq 0$, then $a = c$.*

Proof. (a) The proof of (a) is left as an exercise.
(b) If $b \neq 0$, then (F 4) guarantees the existence of an element d in the field such that $b \cdot d = 1$. Multiply both sides of the equality $a \cdot b = c \cdot b$ by d to obtain $(a \cdot b) \cdot d = (c \cdot b) \cdot d$. Consider the left side of this equality: By (F 2) and (F 3), we have

$$(a \cdot b) \cdot d = a \cdot (b \cdot d) = a \cdot 1 = a.$$

Similarly, the right side of the equality reduces to c. Thus $a = c$. ∎

Corollary. *The elements 0 and 1 mentioned in (F 3), and the elements c and d mentioned in (F 4), are unique.*

Proof. Suppose that $0' \in F$ satisfies $0' + a = a$ for each $a \in F$. Since $0 + a = a$ for each $a \in F$, we have $0' + a = 0 + a$ for each $a \in F$. Thus $0' = 0$ by Theorem C.1.
The proofs of the remaining parts are similar. ∎

Thus each element b in a field has a unique additive inverse and, if $b \neq 0$, a unique multiplicative inverse. (It is shown in the corollary to Theorem C.2 that 0 has no multiplicative inverse.) The additive inverse and the multiplicative inverse of b are denoted by $-b$ and b^{-1}, respectively. Note that $-(-b) = b$ and $(b^{-1})^{-1} = b$.

Subtraction and *division* can be defined in terms of addition and multiplication by using the additive and multiplicative inverses. Specifically, subtraction of b is defined to be addition of $-b$ and division by $b \neq 0$ is defined to be multiplication by b^{-1}; that is,

$$a - b = a + (-b) \quad \text{and} \quad \frac{a}{b} = a \cdot b^{-1}.$$

In particular, the symbol $\frac{1}{b}$ denotes b^{-1}. Division by zero is undefined, but, with this exception, the sum, product, difference, and quotient of any two elements of a field are defined.

Many of the familiar properties of multiplication of real numbers are true in any field, as the next theorem shows.

Theorem C.2. *Let a and b be arbitrary elements of a field. Then each of the following statements are true.*

(a) $a \cdot 0 = 0$.
(b) $(-a) \cdot b = a \cdot (-b) = -(a \cdot b)$.
(c) $(-a) \cdot (-b) = a \cdot b$.

Proof. (a) Since $0 + 0 = 0$, (F 5) shows that

$$0 + a \cdot 0 = a \cdot 0 = a \cdot (0 + 0) = a \cdot 0 + a \cdot 0.$$

Thus $0 = a \cdot 0$ by Theorem C.1.

(b) By definition, $-(a \cdot b)$ is the unique element of F with the property $a \cdot b + [-(a \cdot b)] = 0$. So in order to prove that $(-a) \cdot b = -(a \cdot b)$, it suffices to show that $a \cdot b + (-a) \cdot b = 0$. But $-a$ is the element of F such that $a + (-a) = 0$; so

$$a \cdot b + (-a) \cdot b = [a + (-a)] \cdot b = 0 \cdot b = b \cdot 0 = 0$$

by (F 5) and (a). Thus $(-a) \cdot b = -(a \cdot b)$. The proof that $a \cdot (-b) = -(a \cdot b)$ is similar.

(c) By applying (b) twice, we find that

$$(-a) \cdot (-b) = -[a \cdot (-b)] = -[-(a \cdot b)] = a \cdot b. \qquad \blacksquare$$

Corollary. *The additive identity of a field has no multiplicative inverse.*

In an arbitrary field F, it may happen that a sum $1 + 1 + \cdots + 1$ (p summands) equals 0 for some positive integer p. For example, in the field Z_2 (defined in Example 4), $1 + 1 = 0$. In this case, the smallest positive integer p for which a sum of p 1's equals 0 is called the **characteristic** of F; if no such positive integer exists, then F is said to have **characteristic zero**. Thus Z_2 has characteristic two, and R has characteristic zero. Observe that if F is a field of characteristic $p \neq 0$, then $x + x + \cdots + x$ (p summands) equals 0 for all $x \in F$. In a field having nonzero characteristic (especially characteristic two), many unnatural problems arise. For this reason, some of the results about vector spaces stated in this book require that the field over which the vector space is defined be of characteristic zero (or, at least, of some characteristic other than two).

Finally, note that in other sections of this book, the product of two elements a and b in a field is usually denoted ab rather than $a \cdot b$.

APPENDIX D   COMPLEX NUMBERS

For the purposes of algebra, the field of real numbers is not sufficient, for there are polynomials of nonzero degree with real coefficients that have no zeros in the field of real numbers (for example, $x^2 + 1$). It is often desirable to have a field in which any polynomial of nonzero degree with coefficients from that field has a zero in that field. It is possible to "enlarge" the field of real numbers to obtain such a field.

   Definitions. *A **complex number** is an expression of the form $z = a+bi$, where a and b are real numbers called the **real part** and the **imaginary part** of z, respectively.*
   *The **sum** and **product** of two complex numbers $z = a+bi$ and $w = c+di$ (where a, b, c, and d are real numbers) are defined, respectively, as follows:*

$$z + w = (a + bi) + (c + di) = (a + c) + (b + d)i$$

and

$$zw = (a + bi)(c + di) = (ac - bd) + (bc + ad)i.$$

Example 1

The sum and product of $z = 3 - 5i$ and $w = 9 + 7i$ are, respectively,

$$z + w = (3 - 5i) + (9 + 7i) = (3 + 9) + [(-5) + 7]i = 12 + 2i$$

and

$$zw = (3 - 5i)(9 + 7i) = [3 \cdot 9 - (-5) \cdot 7] + [(-5) \cdot 9 + 3 \cdot 7]i = 62 - 24i. \quad \blacklozenge$$

   Any real number c may be regarded as a complex number by identifying c with the complex number $c + 0i$. Observe that this correspondence preserves sums and products; that is,

$$(c + 0i) + (d + 0i) = (c + d) + 0i \quad \text{and} \quad (c + 0i)(d + 0i) = cd + 0i.$$

   Any complex number of the form $bi = 0 + bi$, where b is a nonzero real number, is called **imaginary**. The product of two imaginary numbers is real since

$$(bi)(di) = (0 + bi)(0 + di) = (0 - bd) + (b \cdot 0 + 0 \cdot d)i = -bd.$$

In particular, for $i = 0 + 1i$, we have $i \cdot i = -1$.
   The observation that $i^2 = i \cdot i = -1$ provides an easy way to remember the definition of multiplication of complex numbers: simply multiply two complex numbers as you would any two algebraic expressions, and replace i^2 by -1. Example 2 illustrates this technique.

Example 2

The product of $-5 + 2i$ and $1 - 3i$ is

$$
\begin{aligned}
(-5 + 2i)(1 - 3i) &= -5(1 - 3i) + 2i(1 - 3i) \\
&= -5 + 15i + 2i - 6i^2 \\
&= -5 + 15i + 2i - 6(-1) \\
&= 1 + 17i. \quad \blacklozenge
\end{aligned}
$$

The real number 0, regarded as a complex number, is an additive identity element for the complex numbers since

$$(a + bi) + 0 = (a + bi) + (0 + 0i) = (a + 0) + (b + 0)i = a + bi.$$

Likewise the real number 1, regarded as a complex number, is a multiplicative identity element for the set of complex numbers since

$$(a + bi) \cdot 1 = (a + bi)(1 + 0i) = (a \cdot 1 - b \cdot 0) + (b \cdot 1 + a \cdot 0)i = a + bi.$$

Every complex number $a + bi$ has an additive inverse, namely $(-a) + (-b)i$. But also each complex number except 0 has a multiplicative inverse. In fact,

$$(a + bi)^{-1} = \left(\frac{a}{a^2 + b^2} \right) - \left(\frac{b}{a^2 + b^2} \right)i.$$

In view of the preceding statements, the following result is not surprising.

Theorem D.1. *The set of complex numbers with the operations of addition and multiplication previously defined is a field.*

Proof. Exercise. ∎

Definition. The **(complex) conjugate** of a complex number $a + bi$ is the complex number $a - bi$. We denote the conjugate of the complex number z by $\bar{z}$.

Example 3

The conjugates of $-3 + 2i$, $4 - 7i$, and 6 are, respectively,

$$\overline{-3 + 2i} = -3 - 2i, \quad \overline{4 - 7i} = 4 + 7i, \quad \text{and} \quad \bar{6} = \overline{6 + 0i} = 6 - 0i = 6. \quad \blacklozenge$$

The next theorem contains some important properties of the conjugate of a complex number.

Theorem D.2. *Let z and w be complex numbers. Then the following statements are true.*

(a) $\overline{\overline{z}} = z$.

(b) $\overline{(z + w)} = \overline{z} + \overline{w}$.

(c) $\overline{zw} = \overline{z} \cdot \overline{w}$.

(d) $\overline{\left(\dfrac{z}{w}\right)} = \dfrac{\overline{z}}{\overline{w}}$ if $w \neq 0$.

(e) z is a real number if and only if $\overline{z} = z$.

Proof. We leave the proofs of (a), (d), and (e) to the reader.

(b) Let $z = a + bi$ and $w = c + di$, where $a, b, c, d \in R$. Then

$$\overline{(z + w)} = \overline{(a + c) + (b + d)i} = (a + c) - (b + d)i$$
$$= (a - bi) + (c - di) = \overline{z} + \overline{w}.$$

(c) For z and w, we have

$$\overline{zw} = \overline{(a + bi)(c + di)} = \overline{(ac - bd) + (ad + bc)i}$$
$$= (ac - bd) - (ad + bc)i = (a - bi)(c - di) = \overline{z} \cdot \overline{w}. \qquad \blacksquare$$

For any complex number $z = a + bi$, $z\overline{z}$ is real and nonnegative, for

$$z\overline{z} = (a + bi)(a - bi) = a^2 + b^2.$$

This fact can be used to define the absolute value of a complex number.

Definition.   Let $z = a + bi$, where $a, b \in R$. The **absolute value** (or **modulus**) of z is the real number $\sqrt{a^2 + b^2}$. We denote the absolute value of z by $|z|$.

Observe that $z\overline{z} = |z|^2$. The fact that the product of a complex number and its conjugate is real provides an easy method for determining the quotient of two complex numbers; for if $c + di \neq 0$, then

$$\frac{a + bi}{c + di} = \frac{a + bi}{c + di} \cdot \frac{c - di}{c - di} = \frac{(ac + bd) + (bc - ad)i}{c^2 + d^2} = \frac{ac + bd}{c^2 + d^2} + \frac{bc - ad}{c^2 + d^2}i.$$

Example 4

To illustrate this procedure, we compute the quotient $(1 + 4i)/(3 - 2i)$:

$$\frac{1 + 4i}{3 - 2i} = \frac{1 + 4i}{3 - 2i} \cdot \frac{3 + 2i}{3 + 2i} = \frac{-5 + 14i}{9 + 4} = -\frac{5}{13} + \frac{14}{13}i. \qquad \blacklozenge$$

The absolute value of a complex number has the familiar properties of the absolute value of a real number, as the following result shows.

Theorem D.3. *Let z and w denote any two complex numbers. Then the following statements are true.*

(a) $|zw| = |z| \cdot |w|$.

(b) $\left|\dfrac{z}{w}\right| = \dfrac{|z|}{|w|}$ if $w \neq 0$.

(c) $|z + w| \leq |z| + |w|$.

(d) $|z| - |w| \leq |z + w|$.

Proof. (a) By Theorem D.2, we have

$$|zw|^2 = (zw)\overline{(zw)} = (zw)(\overline{z} \cdot \overline{w}) = (z\overline{z})(w\overline{w}) = |z|^2|w|^2,$$

proving (a).

(b) For the proof of (b), apply (a) to the product $\left(\dfrac{z}{w}\right) w$.

(c) For any complex number $x = a + bi$, where $a, b \in R$, observe that

$$x + \overline{x} = (a + bi) + (a - bi) = 2a \leq 2\sqrt{a^2 + b^2} = 2|x|.$$

Thus $x + \overline{x}$ is real and satisfies the inequality $x + \overline{x} \leq 2|x|$. Taking $x = w\overline{z}$, we have, by Theorem D.2 and (a),

$$w\overline{z} + \overline{w}z \leq 2|w\overline{z}| = 2|w||\overline{z}| = 2|z||w|.$$

Using Theorem D.2 again gives

$$|z + w|^2 = (z + w)\overline{(z + w)} = (z + w)(\overline{z} + \overline{w}) = z\overline{z} + w\overline{z} + z\overline{w} + w\overline{w}$$
$$\leq |z|^2 + 2|z||w| + |w|^2 = (|z| + |w|)^2.$$

By taking square roots, we obtain (c).

(d) From (a) and (c), it follows that

$$|z| = |(z + w) - w| \leq |z + w| + |-w| = |z + w| + |w|.$$

So

$$|z| - |w| \leq |z + w|,$$

proving (d). ∎

It is interesting as well as useful that complex numbers have both a geometric and an algebraic representation. Suppose that $z = a + bi$, where a and b are real numbers. We may represent z as a vector in the complex plane (see Figure D.1(a)). Notice that, as in R^2, there are two axes, the **real axis** and the **imaginary axis**. The real and imaginary parts of z are the first and second coordinates, and the absolute value of z gives the length of the vector z. It is clear that addition of complex numbers may be represented as in R^2 using the parallelogram law.

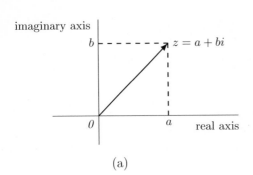

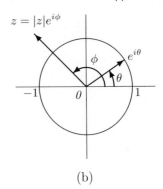

(a)

(b)

Figure D.1

In Section 2.7 (p.132), we introduce Euler's formula. The special case $e^{i\theta} = \cos\theta + i\sin\theta$ is of particular interest. Because of the geometry we have introduced, we may represent the vector $e^{i\theta}$ as in Figure D.1(b); that is, $e^{i\theta}$ is the unit vector that makes an angle θ with the positive real axis. From this figure, we see that any nonzero complex number z may be depicted as a multiple of a unit vector, namely, $z = |z|e^{i\phi}$, where ϕ is the angle that the vector z makes with the positive real axis. Thus multiplication, as well as addition, has a simple geometric interpretation: If $z = |z|e^{i\theta}$ and $w = |w|e^{i\omega}$ are two nonzero complex numbers, then from the properties established in Section 2.7 and Theorem D.3, we have

$$zw = |z|e^{i\theta} \cdot |w|e^{i\omega} = |zw|e^{i(\theta+\omega)}.$$

So zw is the vector whose length is the product of the lengths of z and w, and makes the angle $\theta + \omega$ with the positive real axis.

Our motivation for enlarging the set of real numbers to the set of complex numbers is to obtain a field such that every polynomial with nonzero degree having coefficients in that field has a zero. Our next result guarantees that the field of complex numbers has this property.

Theorem D.4 (The Fundamental Theorem of Algebra). *Suppose that $p(z) = a_n z^n + a_{n-1}z^{n-1} + \cdots + a_1 z + a_0$ is a polynomial in $\mathsf{P}(C)$ of degree $n \geq 1$. Then $p(z)$ has a zero.*

The following proof is based on one in the book *Principles of Mathematical Analysis* 3d., by Walter Rudin (McGraw-Hill Higher Education, New York, 1976).

Proof. We want to find z_0 in C such that $p(z_0) = 0$. Let m be the greatest lower bound of $\{|p(z)| : z \in C\}$. For $|z| = s > 0$, we have

$$|p(z)| = |a_n z^n + a_{n-1}z^{n-1} + \cdots + a_0|$$

$$\geq |a_n||z^n| - |a_{n-1}||z|^{n-1} - \cdots - |a_0|$$
$$= |a_n|s^n - |a_{n-1}|s^{n-1} - \cdots - |a_0|$$
$$= s^n[|a_n| - |a_{n-1}|s^{-1} - \cdots - |a_0|s^{-n}].$$

Because the last expression approaches infinity as s approaches infinity, we may choose a closed disk D about the origin such that $|p(z)| > m + 1$ if z is not in D. It follows that m is the greatest lower bound of $\{|p(z)|: z \in D\}$. Because D is closed and bounded and $p(z)$ is continuous, there exists z_0 in D such that $|p(z_0)| = m$. We want to show that $m = 0$. We argue by contradiction.

Assume that $m \neq 0$. Let $q(z) = \dfrac{p(z + z_0)}{p(z_0)}$. Then $q(z)$ is a polynomial of degree n, $q(0) = 1$, and $|q(z)| \geq 1$ for all z in C. So we may write

$$q(z) = 1 + b_k z^k + b_{k+1} z^{k+1} + \cdots + b_n z^n,$$

where $b_k \neq 0$. Because $-\dfrac{|b_k|}{b_k}$ has modulus one, we may pick a real number θ such that $e^{ik\theta} = -\dfrac{|b_k|}{b_k}$, or $e^{ik\theta}b_k = -|b_k|$. For any $r > 0$, we have

$$q(re^{i\theta}) = 1 + b_k r^k e^{ik\theta} + b_{k+1} r^{k+1} e^{i(k+1)\theta} + \cdots + b_n r^n e^{in\theta}$$
$$= 1 - |b_k|r^k + b_{k+1} r^{k+1} e^{i(k+1)\theta} + \cdots + b_n r^n e^{in\theta}.$$

Choose r small enough so that $1 - |b_k|r^k > 0$. Then

$$|q(re^{i\theta})| \leq 1 - |b_k|r^k + |b_{k+1}|r^{k+1} + \cdots + |b_n|r^n$$
$$= 1 - r^k[|b_k| - |b_{k+1}|r - \cdots - |b_n|r^{n-k}].$$

Now choose r even smaller, if necessary, so that the expression within the brackets is positive. We obtain that $|q(re^{i\theta})| < 1$. But this is a contradiction. ∎

The following important corollary is a consequence of Theorem D.4 and the division algorithm for polynomials (Theorem E.1).

Corollary. If $p(z) = a_n z^n + a_{n-1} z^{n-1} + \cdots + a_1 z + a_0$ is a polynomial of degree $n \geq 1$ with complex coefficients, then there exist complex numbers $c_1, c_2, \cdots, c_n$ (not necessarily distinct) such that

$$p(z) = a_n(z - c_1)(z - c_2) \cdots (z - c_n).$$

Proof. Exercise. ∎

A field is called **algebraically closed** if it has the property that every polynomial of positive degree with coefficients from that field factors as a product of polynomials of degree 1. Thus the preceding corollary asserts that the field of complex numbers is algebraically closed.

APPENDIX E   POLYNOMIALS

In this appendix, we discuss some useful properties of the polynomials with coefficients from a field. For the definition of a polynomial, refer to Section 1.2. Throughout this appendix, we assume that all polynomials have coefficients from a fixed field F.

Definition. *A polynomial $f(x)$ **divides** a polynomial $g(x)$ if there exists a polynomial $q(x)$ such that $g(x) = f(x)q(x)$.*

Our first result shows that the familiar long division process for polynomials with real coefficients is valid for polynomials with coefficients from an arbitrary field.

Theorem E.1 (The Division Algorithm for Polynomials). *Let $f(x)$ be a polynomial of degree n, and let $g(x)$ be a polynomial of degree $m \geq 0$. Then there exist unique polynomials $q(x)$ and $r(x)$ such that*

$$f(x) = q(x)g(x) + r(x), \tag{1}$$

where the degree of $r(x)$ is less than m.

Proof. We begin by establishing the existence of $q(x)$ and $r(x)$ that satisfy (1).

CASE 1.   If $n < m$, take $q(x) = 0$ and $r(x) = f(x)$ to satisfy (1).

CASE 2.   When $0 \leq m \leq n$, we apply mathematical induction on n. First suppose that $n = 0$. Then $m = 0$, and it follows that $f(x)$ and $g(x)$ are nonzero constants. Hence we may take $q(x) = f(x)/g(x)$ and $r(x) = 0$ to satisfy (1).

Now suppose that the result is valid for all polynomials with degree less than n for some fixed $n > 0$, and assume that $f(x)$ has degree n. Suppose that

$$f(x) = a_n x^n + a_{n-1} x^{n-1} + \cdots + a_1 x + a_0$$

and

$$g(x) = b_m x^m + b_{m-1} x^{m-1} + \cdots + b_1 x + b_0,$$

and let $h(x)$ be the polynomial defined by

$$h(x) = f(x) - a_n b_m^{-1} x^{n-m} g(x). \tag{2}$$

Then $h(x)$ is a polynomial of degree less than n, and therefore we may apply the induction hypothesis or CASE 1 (whichever is relevant) to obtain polynomials $q_1(x)$ and $r(x)$ such that $r(x)$ has degree less than m and

$$h(x) = q_1(x)g(x) + r(x). \tag{3}$$

Combining (2) and (3) and solving for $f(x)$ gives us $f(x) = q(x)g(x) + r(x)$ with $q(x) = a_n b_m^{-1} x^{n-m} + q_1(x)$, which establishes (a) and (b) for any $n \geq 0$ by mathematical induction. This establishes the existence of $q(x)$ and $r(x)$.

We now show the uniqueness of $q(x)$ and $r(x)$. Suppose that $q_1(x)$, $q_2(x)$, $r_1(x)$, and $r_2(x)$ exist such that $r_1(x)$ and $r_2(x)$ each has degree less than m and

$$f(x) = q_1(x)g(x) + r_1(x) = q_2(x)g(x) + r_2(x).$$

Then

$$[q_1(x) - q_2(x)] \, g(x) = r_2(x) - r_1(x). \tag{4}$$

The right side of (4) is a polynomial of degree less than m. Since $g(x)$ has degree m, it must follow that $q_1(x) - q_2(x)$ is the zero polynomial. Hence $q_1(x) = q_2(x)$; thus $r_1(x) = r_2(x)$ by (4). ∎

In the context of Theorem E.1, we call $q(x)$ and $r(x)$ the **quotient** and **remainder**, respectively, for the division of $f(x)$ by $g(x)$. For example, suppose that F is the field of complex numbers. Then the quotient and remainder for the division of

$$f(x) = (3+i)x^5 - (1-i)x^4 + 6x^3 + (-6+2i)x^2 + (2+i)x + 1$$

by

$$g(x) = (3+i)x^2 - 2ix + 4$$

are, respectively,

$$q(x) = x^3 + ix^2 - 2 \quad \text{and} \quad r(x) = (2-3i)x + 9.$$

Corollary 1. Let $f(x)$ be a polynomial of positive degree, and let $a \in F$. Then $f(a) = 0$ if and only if $x - a$ divides $f(x)$.

Proof. Suppose that $x - a$ divides $f(x)$. Then there exists a polynomial $q(x)$ such that $f(x) = (x-a)q(x)$. Thus $f(a) = (a-a)q(a) = 0 \cdot q(a) = 0$.

Conversely, suppose that $f(a) = 0$. By the division algorithm, there exist polynomials $q(x)$ and $r(x)$ such that $r(x)$ has degree less than one and

$$f(x) = q(x)(x-a) + r(x).$$

Substituting a for x in the equation above, we obtain $r(a) = 0$. Since $r(x)$ has degree less than 1, it must be the constant polynomial $r(x) = 0$. Thus $f(x) = q(x)(x-a)$. ∎

For any polynomial $f(x)$ with coefficients from a field F, an element $a \in F$ is called a **zero** of $f(x)$ if $f(a) = 0$. With this terminology, the preceding corollary states that a is a zero of $f(x)$ if and only if $x - a$ divides $f(x)$.

Corollary 2. *Any polynomial of degree $n \geq 1$ has at most n distinct zeros.*

Proof. The proof is by mathematical induction on n. The result is obvious if $n = 1$. Now suppose that the result is true for some positive integer n, and let $f(x)$ be a polynomial of degree $n + 1$. If $f(x)$ has no zeros, then there is nothing to prove. Otherwise, if a is a zero of $f(x)$, then by Corollary 1 we may write $f(x) = (x - a)q(x)$ for some polynomial $q(x)$. Note that $q(x)$ must be of degree n; therefore, by the induction hypothesis, $q(x)$ can have at most n distinct zeros. Since any zero of $f(x)$ distinct from a is also a zero of $q(x)$, it follows that $f(x)$ can have at most $n + 1$ distinct zeros. ∎

Polynomials having no common divisors arise naturally in the study of *canonical forms.* (See Chapter 7.)

Definition. *Two nonzero polynomials are called **relatively prime** if no polynomial of positive degree divides each of them.*

For example, the polynomials with real coefficients $f(x) = x^2(x - 1)$ and $h(x) = (x - 1)(x - 2)$ are not relatively prime because $x - 1$ divides each of them. On the other hand, consider $f(x)$ and $g(x) = (x - 2)(x - 3)$, which do not appear to have common factors. Could other factorizations of $f(x)$ and $g(x)$ reveal a hidden common factor? We will soon see (Theorem E.9) that the preceding factors are the only ones. Thus $f(x)$ and $g(x)$ are relatively prime because they have no common factors of positive degree.

Theorem E.2. *If $f_1(x)$ and $f_2(x)$ are relatively prime polynomials, there exist polynomials $q_1(x)$ and $q_2(x)$ such that*

$$q_1(x)f_1(x) + q_2(x)f_2(x) = 1,$$

where 1 denotes the constant polynomial with value 1.

Proof. Without loss of generality, assume that the degree of $f_1(x)$ is greater than or equal to the degree of $f_2(x)$. The proof is by mathematical induction on the degree of $f_2(x)$. If $f_2(x)$ has degree 0, then $f_2(x)$ is a nonzero constant c. In this case, we can take $q_1(x) = 0$ and $q_2(x) = 1/c$.

Now suppose that the theorem holds whenever the polynomial of lesser degree has degree less than n for some positive integer n, and suppose that $f_2(x)$ has degree n. By the division algorithm, there exist polynomials $q(x)$ and $r(x)$ such that $r(x)$ has degree less than n and

$$f_1(x) = q(x)f_2(x) + r(x). \tag{5}$$

Since $f_1(x)$ and $f_2(x)$ are relatively prime, $r(x)$ is not the zero polynomial. We claim that $f_2(x)$ and $r(x)$ are relatively prime. Suppose otherwise; then there exists a polynomial $g(x)$ of positive degree that divides both $f_2(x)$ and $r(x)$. Hence, by (5), $g(x)$ also divides $f_1(x)$, contradicting the fact that $f_1(x)$ and $f_2(x)$ are relatively prime. Since $r(x)$ has degree less than n, we may apply the induction hypothesis to $f_2(x)$ and $r(x)$. Thus there exist polynomials $g_1(x)$ and $g_2(x)$ such that

$$g_1(x)f_2(x) + g_2(x)r(x) = 1. \tag{6}$$

Combining (5) and (6), we have

$$\begin{aligned} 1 &= g_1(x)f_2(x) + g_2(x)\left[f_1(x) - q(x)f_2(x)\right] \\ &= g_2(x)f_1(x) + \left[g_1(x) - g_2(x)q(x)\right]f_2(x). \end{aligned}$$

Thus, setting $q_1(x) = g_2(x)$ and $q_2(x) = g_1(x) - g_2(x)q(x)$, we obtain the desired result.   ∎

Example 1

Let $f_1(x) = x^3 - x^2 + 1$ and $f_2(x) = (x - 1)^2$. As polynomials with real coefficients, $f_1(x)$ and $f_2(x)$ are relatively prime. It is easily verified that the polynomials $q_1(x) = -x + 2$ and $q_2(x) = x^2 - x - 1$ satisfy

$$q_1(x)f_1(x) + q_2(x)f_2(x) = 1,$$

and hence these polynomials satisfy the conclusion of Theorem E.2.   ♦

Throughout Chapters 5, 6, and 7, we consider linear operators that are polynomials in a particular operator T and matrices that are polynomials in a particular matrix A. For these operators and matrices, the following notation is convenient.

Definitions.   *Let*

$$f(x) = a_0 + a_1(x) + \cdots + a_n x^n$$

be a polynomial with coefficients from a field F. If T is a linear operator on a vector space V over F, we define

$$f(\mathsf{T}) = a_0\mathsf{I} + a_1\mathsf{T} + \cdots + a_n\mathsf{T}^n.$$

Similarly, if A is a $n \times n$ matrix with entries from F, we define

$$f(A) = a_0 I + a_1 A + \cdots + a_n A^n.$$

Example 2

Let T be the linear operator on R^2 defined by $\mathsf{T}(a,b) = (2a + b, a - b)$, and let $f(x) = x^2 + 2x - 3$. It is easily checked that $\mathsf{T}^2(a,b) = (5a + b, a + 2b)$; so

$$
\begin{aligned}
f(\mathsf{T})(a,b) &= (\mathsf{T}^2 + 2\mathsf{T} - 3\mathsf{I})(a,b) \\
&= (5a + b, a + 2b) + (4a + 2b, 2a - 2b) - 3(a,b) \\
&= (6a + 3b, 3a - 3b).
\end{aligned}
$$

Similarly, if

$$
A = \begin{pmatrix} 2 & 1 \\ 1 & -1 \end{pmatrix},
$$

then

$$
f(A) = A^2 + 2A - 3I = \begin{pmatrix} 5 & 1 \\ 1 & 2 \end{pmatrix} + 2\begin{pmatrix} 2 & 1 \\ 1 & -1 \end{pmatrix} - 3\begin{pmatrix} 1 & 0 \\ 0 & 1 \end{pmatrix} = \begin{pmatrix} 6 & 3 \\ 3 & -3 \end{pmatrix}. \quad \blacklozenge
$$

The next three results use this notation.

Theorem E.3. *Let $f(x)$ be a polynomial with coefficients from a field F, and let T be a linear operator on a vector space V over F. Then the following statements are true.*
(a) *$f(\mathsf{T})$ is a linear operator on V.*
(b) *If β is a finite ordered basis for V and $A = [\mathsf{T}]_\beta$, then $[f(\mathsf{T})]_\beta = f(A)$.*

Proof. Exercise.   ∎

Theorem E.4. *Let T be a linear operator on a vector space V over a field F, and let A be a square matrix with entries from F. Then, for any polynomials $f_1(x)$ and $f_2(x)$ with coefficients from F,*
(a) *$f_1(\mathsf{T})f_2(\mathsf{T}) = f_2(\mathsf{T})f_1(\mathsf{T})$*
(b) *$f_1(A)f_2(A) = f_2(A)f_1(A)$.*

Proof. Exercise.   ∎

Theorem E.5. *Let T be a linear operator on a vector space V over a field F, and let A be an $n \times n$ matrix with entries from F. If $f_1(x)$ and $f_2(x)$ are relatively prime polynomials with entries from F, then there exist polynomials $q_1(x)$ and $q_2(x)$ with entries from F such that*
(a) *$q_1(\mathsf{T})f_1(\mathsf{T}) + q_2(\mathsf{T})f_2(\mathsf{T}) = \mathsf{I}$*
(b) *$q_1(A)f_1(A) + q_2(A)f_2(A) = I$.*

Proof. Exercise.   ∎

In Chapters 5 and 7, we are concerned with determining when a linear operator T on a finite-dimensional vector space can be *diagonalized* and with finding a simple (canonical) representation of T. Both of these problems are affected by the factorization of a certain polynomial determined by T (the *characteristic polynomial* of T). In this setting, particular types of polynomials play an important role.

Definitions. *A polynomial $f(x)$ with coefficients from a field F is called* **monic** *if its leading coefficient is 1. If $f(x)$ has positive degree and cannot be expressed as a product of polynomials with coefficients from F each having positive degree, then $f(x)$ is called* **irreducible**.

Observe that whether a polynomial is irreducible depends on the field F from which its coefficients come. For example, $f(x) = x^2 + 1$ is irreducible over the field of real numbers, but it is not irreducible over the field of complex numbers since $x^2 + 1 = (x + i)(x - i)$.

Clearly any polynomial of degree 1 is irreducible. Moreover, for polynomials with coefficients from an algebraically closed field, the polynomials of degree 1 are the only irreducible polynomials.

The following facts are easily established.

Theorem E.6. *Let $\phi(x)$ and $f(x)$ be polynomials. If $\phi(x)$ is irreducible and $\phi(x)$ does not divide $f(x)$, then $\phi(x)$ and $f(x)$ are relatively prime.*

Proof. Exercise. ∎

Theorem E.7. *Any two distinct irreducible monic polynomials are relatively prime.*

Proof. Exercise. ∎

Theorem E.8. *Let $f(x)$, $g(x)$, and $\phi(x)$ be polynomials. If $\phi(x)$ is irreducible and divides the product $f(x)g(x)$, then $\phi(x)$ divides $f(x)$ or $\phi(x)$ divides $g(x)$.*

Proof. Suppose that $\phi(x)$ does not divide $f(x)$. Then $\phi(x)$ and $f(x)$ are relatively prime by Theorem E.6, and so there exist polynomials $q_1(x)$ and $q_2(x)$ such that

$$1 = q_1(x)\phi(x) + q_2(x)f(x).$$

Multiplying both sides of this equation by $g(x)$ yields

$$g(x) = q_1(x)\phi(x)g(x) + q_2(x)f(x)g(x). \qquad (7)$$

Since $\phi(x)$ divides $f(x)g(x)$, there is a polynomial $h(x)$ such that $f(x)g(x) = \phi(x)h(x)$. Thus (7) becomes

$$g(x) = q_1(x)\phi(x)g(x) + q_2(x)\phi(x)h(x) = \phi(x)\left[q_1(x)g(x) + q_2(x)h(x)\right].$$

So $\phi(x)$ divides $g(x)$. ∎

Corollary. Let $\phi(x), \phi_1(x), \phi_2(x), \ldots, \phi_n(x)$ be irreducible monic polynomials. If $\phi(x)$ divides the product $\phi_1(x)\phi_2(x)\cdots\phi_n(x)$, then $\phi(x) = \phi_i(x)$ for some i $(i = 1, 2, \ldots, n)$.

Proof. We prove the corollary by mathematical induction on n. For $n = 1$, the result is an immediate consequence of Theorem E.7. Suppose then that for some $n > 1$, the corollary is true for any $n-1$ irreducible monic polynomials, and let $\phi_1(x), \phi_2(x), \ldots, \phi_n(x)$ be n irreducible polynomials. If $\phi(x)$ divides

$$\phi_1(x)\phi_2(x)\cdots\phi_n(x) = [\phi_1(x)\phi_2(x)\cdots\phi_{n-1}(x)]\,\phi_n(x),$$

then $\phi(x)$ divides the product $\phi_1(x)\phi_2(x)\cdots\phi_{n-1}(x)$ or $\phi(x)$ divides $\phi_n(x)$ by Theorem E.8. In the first case, $\phi(x) = \phi_i(x)$ for some i $(i = 1, 2, \ldots, n-1)$ by the induction hypothesis; in the second case, $\phi(x) = \phi_n(x)$ by Theorem E.7. ∎

We are now able to establish the unique factorization theorem, which is used throughout Chapters 5 and 7. This result states that every polynomial of positive degree is uniquely expressible as a constant times a product of irreducible monic polynomials.

Theorem E.9 (Unique Factorization Theorem for Polynomials). *For any polynomial $f(x)$ of positive degree, there exist a unique constant c; unique distinct irreducible monic polynomials $\phi_1(x), \phi_2(x), \ldots, \phi_k(x)$; and unique positive integers $n_1, n_2, \ldots, n_k$ such that*

$$f(x) = c[\phi_1(x)]^{n_1}[\phi_2(x)]^{n_2}\cdots[\phi_k(x)]^{n_k}.$$

Proof. We begin by showing the existence of such a factorization using mathematical induction on the degree of $f(x)$. If $f(x)$ is of degree 1, then $f(x) = ax + b$ for some constants a and b with $a \neq 0$. Setting $\phi(x) = x + b/a$, we have $f(x) = a\phi(x)$. Since $\phi(x)$ is an irreducible monic polynomial, the result is proved in this case. Now suppose that the conclusion is true for any polynomial with positive degree less than some integer $n > 1$, and let $f(x)$ be a polynomial of degree n. Then

$$f(x) = a_n x^n + \cdots + a_1 x + a_0$$

for some constants a_i with $a_n \neq 0$. If $f(x)$ is irreducible, then

$$f(x) = a_n\left(x^n + \frac{a_{n-1}}{a_n}x^{n-1} + \cdots + \frac{a_1}{a_n} + \frac{a_0}{a_n}\right)$$

is a representation of $f(x)$ as a product of a_n and an irreducible monic polynomial. If $f(x)$ is not irreducible, then $f(x) = g(x)h(x)$ for some polynomials $g(x)$ and $h(x)$, each of positive degree less than n. The induction hypothesis guarantees that both $g(x)$ and $h(x)$ factor as products of a constant and powers of distinct irreducible monic polynomials. Consequently $f(x) = g(x)h(x)$ also factors in this way. Thus, in either case, $f(x)$ can be factored as a product of a constant and powers of distinct irreducible monic polynomials.

It remains to establish the uniqueness of such a factorization. Suppose that

$$
\begin{aligned}
f(x) &= c[\phi_1(x)]^{n_1}[\phi_2(x)]^{n_2}\cdots[\phi_k(x)]^{n_k} \\
&= d[\psi_1(x)]^{m_1}[\psi_2(x)]^{m_2}\cdots[\psi_r(x)]^{m_r},
\end{aligned}
\tag{8}
$$

where c and d are constants, $\phi_i(x)$ and $\psi_j(x)$ are irreducible monic polynomials, and n_i and m_j are positive integers for $i = 1, 2, \ldots, k$ and $j = 1, 2, \ldots, r$. Clearly both c and d must be the leading coefficient of $f(x)$; hence $c = d$. Dividing by c, we find that (8) becomes

$$
[\phi_1(x)]^{n_1}[\phi_2(x)]^{n_2}\cdots[\phi_k(x)]^{n_k} = [\psi_1(x)]^{m_1}[\psi_2(x)]^{m_2}\cdots[\psi_r(x)]^{m_r}.
\tag{9}
$$

So $\phi_i(x)$ divides the right side of (9) for $i = 1, 2, \ldots, k$. Consequently, by the corollary to Theorem E.8, each $\phi_i(x)$ equals some $\psi_j(x)$, and similarly, each $\psi_j(x)$ equals some $\phi_i(x)$. We conclude that $r = k$ and that, by renumbering if necessary, $\phi_i(x) = \psi_i(x)$ for $i = 1, 2, \ldots, k$. Suppose that $n_i \neq m_i$ for some i. Without loss of generality, we may suppose that $i = 1$ and $n_1 > m_1$. Then by canceling $[\phi_1(x)]^{m_1}$ from both sides of (9), we obtain

$$
[\phi_1(x)]^{n_1 - m_1}[\phi_2(x)]^{n_2}\cdots[\phi_k(x)]^{n_k} = [\phi_2(x)]^{m_2}\cdots[\phi_k(x)]^{m_k}.
\tag{10}
$$

Since $n_1 - m_1 > 0$, $\phi_1(x)$ divides the left side of (10) and hence divides the right side also. So $\phi_1(x) = \phi_i(x)$ for some $i = 2, \ldots, k$ by the corollary to Theorem E.8. But this contradicts that $\phi_1(x), \phi_2(x), \ldots, \phi_k(x)$ are distinct. Hence the factorizations of $f(x)$ in (8) are the same. ∎

It is often useful to regard a polynomial $f(x) = a_n x^n + \cdots + a_1 x + a_0$ with coefficients from a field F as a function $f\colon F \to F$. In this case, the value of f at $c \in F$ is $f(c) = a_n c^n + \cdots + a_1 c + a_0$. Unfortunately, for arbitrary fields there is not a one-to-one correspondence between polynomials and polynomial functions. For example, if $f(x) = x^2$ and $g(x) = x$ are two polynomials over the field Z_2 (defined in Example 4 of Appendix C), then $f(x)$ and $g(x)$ have different degrees and hence are not equal as polynomials. But $f(a) = g(a)$ for all $a \in Z_2$, so that f and g are equal polynomial functions. Our final result shows that this anomaly cannot occur over an infinite field.

Theorem E.10. *Let $f(x)$ and $g(x)$ be polynomials with coefficients from an infinite field F. If $f(a) = g(a)$ for all $a \in F$, then $f(x)$ and $g(x)$ are equal.*

Proof. Suppose that $f(a) = g(a)$ for all $a \in F$. Define $h(x) = f(x) - g(x)$, and suppose that $h(x)$ is of degree $n \geq 1$. It follows from Corollary 2 to

Theorem E.1 that $h(x)$ can have at most n zeroes. But

$$h(a) = f(a) - g(a) = 0$$

for every $a \in F$, contradicting the assumption that $h(x)$ has positive degree. Thus $h(x)$ is a constant polynomial, and since $h(a) = 0$ for each $a \in F$, it follows that $h(x)$ is the zero polynomial. Hence $f(x) = g(x)$. ∎

Answers
to Selected Exercises

CHAPTER 1

SECTION 1.1

1. Only the pairs in (b) and (c) are parallel.

2. (a) $x = (3, -2, 4) + t(-8, 9, -3)$ (c) $x = (3, 7, 2) + t(0, 0, -10)$

3. (a) $x = (2, -5, -1) + s(-2, 9, 7) + t(-5, 12, 2)$
   (c) $x = (-8, 2, 0) + s(9, 1, 0) + t(14, -7, 0)$

SECTION 1.2

1. (a) T   (b) F   (c) F   (d) F   (e) T   (f) F
   (g) F   (h) F   (i) T   (j) T   (k) T

3. $M_{13} = 3$, $M_{21} = 4$, and $M_{22} = 5$

4. (a) $\begin{pmatrix} 6 & 3 & 2 \\ -4 & 3 & 9 \end{pmatrix}$ (c) $\begin{pmatrix} 8 & 20 & -12 \\ 4 & 0 & 28 \end{pmatrix}$

   (e) $2x^4 + x^3 + 2x^2 - 2x + 10$ (g) $10x^7 - 30x^4 + 40x^2 - 15x$

13. No, (VS 4) fails.

14. Yes

15. No

17. No, (VS 5) fails.

22. 2^{mn}

SECTION 1.3

1. (a) F   (b) F   (c) T   (d) F   (e) T   (f) F   (g) F

2. (a) $\begin{pmatrix} -4 & 5 \\ 2 & -1 \end{pmatrix}$; the trace is -5 (c) $\begin{pmatrix} -3 & 0 & 6 \\ 9 & -2 & 1 \end{pmatrix}$

   (e) $\begin{pmatrix} 1 \\ -1 \\ 3 \\ 5 \end{pmatrix}$ (g) $\begin{pmatrix} 5 & 6 & 7 \end{pmatrix}$

8. (a) Yes   (c) Yes   (e) No

11. No, the set is not closed under addition.

15. Yes

SECTION 1.4

1. (a) T   (b) F   (c) T   (d) F   (e) T   (f) F
2. (a) $\{r(1,1,0,0) + s(-3,0,-2,1) + (5,0,4,0) : r,s \in R\}$
   (c) There are no solutions.
   (e) $\{r(10,-3,1,0,0) + s(-3,2,0,1,0) + (-4,3,0,0,5) : r,s \in R\}$
3. (a) Yes   (c) No   (e) No
4. (a) Yes   (c) Yes   (e) No
5. (a) Yes   (c) No   (e) Yes   (g) Yes

SECTION 1.5

1. (a) F   (b) T   (c) F   (d) F   (e) T   (f) T
2. (a) linearly dependent   (c) linearly independent   (e) linearly dependent
   (g) linearly dependent   (i) linearly independent
7. $\left\{\begin{pmatrix} 1 & 0 \\ 0 & 0 \end{pmatrix}, \begin{pmatrix} 0 & 0 \\ 0 & 1 \end{pmatrix}\right\}$
11. 2^n

SECTION 1.6

1. (a) F   (b) T   (c) F   (d) F   (e) T   (f) F
   (g) F   (h) T   (i) F   (j) T   (k) T   (l) T
2. (a) Yes   (c) Yes   (e) No
3. (a) No   (c) Yes   (e) No
4. No
5. No
7. $\{u_1, u_2, u_5\}$
9. $(a_1, a_2, a_3, a_4) = a_1 u_1 + (a_2 - a_1)u_2 + (a_3 - a_2)u_3 + (a_4 - a_3)u_4$
10. (a) $-4x^2 - x + 8$ (c) $-x^3 + 2x^2 + 4x - 5$
13. $\{(1,1,1)\}$
15. $n^2 - 1$
17. $\frac{1}{2}n(n-1)$
26. n
30. $\dim(W_1) = 3$, $\dim(W_2) = 2$, $\dim(W_1 + W_2) = 4$, and $\dim(W_1 \cap W_2) = 1$

SECTION 1.7

1. (a) F   (b) F   (c) F   (d) T   (e) T   (f) T

CHAPTER 2

SECTION 2.1

1. (a) T   (b) F   (c) F   (d) T   (e) F   (f) F   (g) T   (h) F

2. The nullity is 1, and the rank is 2. T is not one-to-one but is onto.

4. The nullity is 4, and the rank is 2. T is neither one-to-one nor onto.

5. The nullity is 0, and the rank is 3. T is one-to-one but not onto.

10. $T(2,3) = (5,11)$. T is one-to-one. **12.** No.

SECTION 2.2

1. (a) T **(b)** T **(c)** F **(d)** T **(e)** T **(f)** F

2. (a) $\begin{pmatrix} 2 & -1 \\ 3 & 4 \\ 1 & 0 \end{pmatrix}$ **(c)** $(2 \quad 1 \quad -3)$ **(d)** $\begin{pmatrix} 0 & 2 & 1 \\ -1 & 4 & 5 \\ 1 & 0 & 1 \end{pmatrix}$

(f) $\begin{pmatrix} 0 & 0 & \cdots & 0 & 1 \\ 0 & 0 & \cdots & 1 & 0 \\ \vdots & \vdots & & \vdots & \vdots \\ 0 & 1 & \cdots & 0 & 0 \\ 1 & 0 & \cdots & 0 & 0 \end{pmatrix}$ **(g)** $(1 \quad 0 \quad \cdots \quad 0 \quad 1)$

3. $[\mathsf{T}]_\beta^\gamma = \begin{pmatrix} -\frac{1}{3} & -1 \\ 0 & 1 \\ \frac{2}{3} & 0 \end{pmatrix}$ and $[\mathsf{T}]_\alpha^\gamma = \begin{pmatrix} -\frac{7}{3} & -\frac{11}{3} \\ 2 & 3 \\ \frac{2}{3} & \frac{4}{3} \end{pmatrix}$

5. (a) $\begin{pmatrix} 1 & 0 & 0 & 0 \\ 0 & 0 & 1 & 0 \\ 0 & 1 & 0 & 0 \\ 0 & 0 & 0 & 1 \end{pmatrix}$ **(b)** $\begin{pmatrix} 0 & 1 & 0 \\ 2 & 2 & 2 \\ 0 & 0 & 0 \\ 0 & 0 & 2 \end{pmatrix}$ **(e)** $\begin{pmatrix} 1 \\ -2 \\ 0 \\ 4 \end{pmatrix}$

10. $\begin{pmatrix} 1 & 1 & 0 & \cdots & 0 \\ 0 & 1 & 1 & \cdots & 0 \\ 0 & 0 & 1 & \cdots & 0 \\ \vdots & \vdots & \vdots & & \vdots \\ 0 & 0 & 0 & \cdots & 1 \\ 0 & 0 & 0 & \cdots & 1 \end{pmatrix}$

SECTION 2.3

1. (a) F **(b)** T **(c)** F **(d)** T **(e)** F **(f)** F
(g) F **(h)** F **(i)** T **(j)** T

2. (a) $A(2B + 3C) = \begin{pmatrix} 20 & -9 & 18 \\ 5 & 10 & 8 \end{pmatrix}$ and $A(BD) = \begin{pmatrix} 29 \\ -26 \end{pmatrix}$

(b) $A^t B = \begin{pmatrix} 23 & 19 & 0 \\ 26 & -1 & 10 \end{pmatrix}$ and $CB = (27 \quad 7 \quad 9)$

3. (a) $[\mathsf{T}]_\beta = \begin{pmatrix} 2 & 3 & 0 \\ 0 & 3 & 6 \\ 0 & 0 & 4 \end{pmatrix}$, $[\mathsf{U}]_\beta^\gamma = \begin{pmatrix} 1 & 1 & 0 \\ 0 & 0 & 1 \\ 1 & -1 & 0 \end{pmatrix}$, and $[\mathsf{UT}]_\beta^\gamma = \begin{pmatrix} 2 & 6 & 6 \\ 0 & 0 & 4 \\ 2 & 0 & -6 \end{pmatrix}$

4. (a) $\begin{pmatrix} 1 \\ -1 \\ 4 \\ 6 \end{pmatrix}$ **(c)** (5)

12. (a) No.   **(b)** No.

SECTION 2.4

1. (a) F   **(b)** T   **(c)** F   **(d)** F   **(e)** T   **(f)** F
    (g) T   **(h)** T   **(i)** T

2. (a) No   **(b)** No   **(c)** Yes   **(d)** No   **(e)** No   **(f)** Yes

3. (a) No   **(b)** Yes   **(c)** Yes   **(d)** No

19. (b) $[T]_\beta = \begin{pmatrix} 1 & 0 & 0 & 0 \\ 0 & 0 & 1 & 0 \\ 0 & 1 & 0 & 0 \\ 0 & 0 & 0 & 1 \end{pmatrix}$

SECTION 2.5

1. (a) F   **(b)** T   **(c)** T   **(d)** F   **(e)** T

2. (a) $\begin{pmatrix} a_1 & b_1 \\ a_2 & b_2 \end{pmatrix}$ **(c)** $\begin{pmatrix} 3 & -1 \\ 5 & -2 \end{pmatrix}$

3. (a) $\begin{pmatrix} a_2 & b_2 & c_2 \\ a_1 & b_1 & c_1 \\ a_0 & b_0 & c_0 \end{pmatrix}$ **(c)** $\begin{pmatrix} 0 & -1 & 0 \\ 1 & 0 & 0 \\ -3 & 2 & 1 \end{pmatrix}$ **(e)** $\begin{pmatrix} 5 & -6 & 3 \\ 0 & 4 & -1 \\ 3 & -1 & 2 \end{pmatrix}$

4. $[T]_{\beta'} = \begin{pmatrix} 2 & -1 \\ -1 & 1 \end{pmatrix} \begin{pmatrix} 2 & 1 \\ 1 & -3 \end{pmatrix} \begin{pmatrix} 1 & 1 \\ 1 & 2 \end{pmatrix} = \begin{pmatrix} 8 & 13 \\ -5 & -9 \end{pmatrix}$

5. $[T]_{\beta'} = \begin{pmatrix} \frac{1}{2} & \frac{1}{2} \\ \frac{1}{2} & -\frac{1}{2} \end{pmatrix} \begin{pmatrix} 0 & 1 \\ 0 & 0 \end{pmatrix} \begin{pmatrix} 1 & 1 \\ 1 & -1 \end{pmatrix} = \begin{pmatrix} \frac{1}{2} & -\frac{1}{2} \\ \frac{1}{2} & -\frac{1}{2} \end{pmatrix}$

6. (a) $Q = \begin{pmatrix} 1 & 1 \\ 1 & 2 \end{pmatrix}$, $[L_A]_\beta = \begin{pmatrix} 6 & 11 \\ -2 & -4 \end{pmatrix}$

   (c) $Q = \begin{pmatrix} 1 & 1 & 1 \\ 1 & 0 & 1 \\ 1 & 1 & 2 \end{pmatrix}$, $[L_A]_\beta = \begin{pmatrix} 2 & 2 & 2 \\ -2 & -3 & -4 \\ 1 & 1 & 2 \end{pmatrix}$

7. (a) $T(x, y) = \dfrac{1}{1 + m^2}((1 - m^2)x + 2my, 2mx + (m^2 - 1)y)$

SECTION 2.6

1. (a) F   **(b)** T   **(c)** T   **(d)** T   **(e)** F   **(f)** T   **(g)** T   **(h)** F

2. The functions in (a), (c), (e), and (f) are linear functionals.

3. (a) $f_1(x, y, z) = x - \frac{1}{2}y$, $f_2(x, y, z) = \frac{1}{2}y$, and $f_3(x, y, z) = -x + z$

5. The basis for V is $\{p_1(x), p_2(x)\}$, where $p_1(x) = 2 - 2x$ and $p_2(x) = -\frac{1}{2} + x$.

7. (a) $T^t(f) = g$, where $g(a + bx) = -3a - 4b$

   (b) $[T^t]_{\gamma^*}^{\beta^*} = \begin{pmatrix} -1 & 1 \\ -2 & 1 \end{pmatrix}$ **(c)** $[T]_\beta^\gamma = \begin{pmatrix} -1 & -2 \\ 1 & 1 \end{pmatrix}$

SECTION 2.7

1. (a) T   (b) T   (c) F   (d) F   (e) T   (f) F   (g) T
2. (a) F   (b) F   (c) T   (d) T   (e) F
3. (a) $\{e^{-t}, te^{-t}\}$ (c) $\{e^{-t}, te^{-t}, e^t, te^t\}$ (e) $\{e^{-t}, e^t \cos 2t, e^t \sin 2t\}$
4. (a) $\{e^{(1+\sqrt{5})t/2}, e^{(1-\sqrt{5})t/2}\}$ (c) $\{1, e^{-4t}, e^{-2t}\}$

CHAPTER 3

SECTION 3.1

1. (a) T   (b) F   (c) T   (d) F   (e) T   (f) F
   (g) T   (h) F   (i) T
2. Adding -2 times column 1 to column 2 transforms A into B.

3. (a) $\begin{pmatrix} 0 & 0 & 1 \\ 0 & 1 & 0 \\ 1 & 0 & 0 \end{pmatrix}$ (c) $\begin{pmatrix} 1 & 0 & 0 \\ 0 & 1 & 0 \\ 2 & 0 & 1 \end{pmatrix}$

SECTION 3.2

1. (a) F   (b) F   (c) T   (d) T   (e) F   (f) T
   (g) T   (h) T   (i) T
2. (a) 2   (c) 2   (e) 3   (g) 1

4. (a) $\begin{pmatrix} 1 & 0 & 0 & 0 \\ 0 & 1 & 0 & 0 \\ 0 & 0 & 0 & 0 \end{pmatrix}$; the rank is 2.

5. (a) The rank is 2, and the inverse is $\begin{pmatrix} -1 & 2 \\ 1 & -1 \end{pmatrix}$.

   (c) The rank is 2, and so no inverse exists.

   (e) The rank is 3, and the inverse is $\begin{pmatrix} \frac{1}{6} & -\frac{1}{3} & \frac{1}{2} \\ \frac{1}{2} & 0 & -\frac{1}{2} \\ -\frac{1}{6} & \frac{1}{3} & \frac{1}{2} \end{pmatrix}$.

   (g) The rank if 4, and the inverse is $\begin{pmatrix} -51 & 15 & 7 & 12 \\ 31 & -9 & -4 & -7 \\ -10 & 3 & 1 & 2 \\ -3 & 1 & 1 & 1 \end{pmatrix}$.

6. (a) $\mathsf{T}^{-1}(ax^2 + bx + c) = -ax^2 - (4a+b)x - (10a + 2b + c)$
   (c) $\mathsf{T}^{-1}(a, b, c) = \left(\frac{1}{6}a - \frac{1}{3}b + \frac{1}{2}c, \frac{1}{2}a - \frac{1}{2}c, -\frac{1}{6} + \frac{1}{3}b + \frac{1}{2}c \right)$
   (e) $\mathsf{T}^{-1}(a, b, c) = \left(\frac{1}{2}a - b + \frac{1}{2}c \right) x^2 + \left(-\frac{1}{2}a + \frac{1}{2}c \right) x + b$

7. $\begin{pmatrix} 1 & 0 & 0 \\ 0 & 1 & 0 \\ 1 & 0 & 1 \end{pmatrix} \begin{pmatrix} 1 & 0 & 0 \\ 1 & 1 & 0 \\ 0 & 0 & 1 \end{pmatrix} \begin{pmatrix} 1 & 0 & 0 \\ 0 & -2 & 0 \\ 0 & 0 & 1 \end{pmatrix} \begin{pmatrix} 1 & 2 & 0 \\ 0 & 1 & 0 \\ 0 & 0 & 1 \end{pmatrix} \begin{pmatrix} 1 & 0 & 0 \\ 0 & 1 & 0 \\ 0 & -1 & 1 \end{pmatrix} \begin{pmatrix} 1 & 0 & 1 \\ 0 & 1 & 0 \\ 0 & 0 & 1 \end{pmatrix}$

20. (a) $\begin{pmatrix} 1 & 3 & 0 & 0 & 0 \\ -2 & 1 & 0 & 0 & 0 \\ 1 & 0 & 0 & 0 & 0 \\ 0 & -2 & 0 & 0 & 0 \\ 0 & 1 & 0 & 0 & 0 \end{pmatrix}$

SECTION 3.3

1. (a) F   (b) F   (c) T   (d) F   (e) F   (f) F   (g) T   (h) F

2. (a) $\left\{ \begin{pmatrix} -3 \\ 1 \end{pmatrix} \right\}$ (c) $\left\{ \begin{pmatrix} -1 \\ 1 \\ 1 \end{pmatrix} \right\}$

(e) $\left\{ \begin{pmatrix} -2 \\ 1 \\ 0 \\ 0 \end{pmatrix}, \begin{pmatrix} 3 \\ 0 \\ 1 \\ 0 \end{pmatrix}, \begin{pmatrix} -1 \\ 0 \\ 0 \\ 1 \end{pmatrix} \right\}$ (g) $\left\{ \begin{pmatrix} -3 \\ 1 \\ 1 \\ 0 \end{pmatrix}, \begin{pmatrix} 1 \\ -1 \\ 0 \\ 1 \end{pmatrix} \right\}$

3. (a) $\left\{ \begin{pmatrix} 5 \\ 0 \end{pmatrix} + t \begin{pmatrix} -3 \\ 1 \end{pmatrix} : t \in R \right\}$ (c) $\left\{ \begin{pmatrix} 2 \\ 1 \\ 1 \end{pmatrix} + t \begin{pmatrix} -1 \\ 1 \\ 1 \end{pmatrix} : t \in R \right\}$

(e) $\left\{ \begin{pmatrix} 1 \\ 0 \\ 0 \\ 0 \end{pmatrix} + r \begin{pmatrix} -2 \\ 1 \\ 0 \\ 0 \end{pmatrix} + s \begin{pmatrix} 3 \\ 0 \\ 1 \\ 0 \end{pmatrix} + t \begin{pmatrix} -1 \\ 0 \\ 0 \\ 1 \end{pmatrix} : r, s, t \in R \right\}$

(g) $\left\{ \begin{pmatrix} 0 \\ 0 \\ 0 \\ 1 \end{pmatrix} + r \begin{pmatrix} -3 \\ 1 \\ 1 \\ 0 \end{pmatrix} + s \begin{pmatrix} 1 \\ -1 \\ 0 \\ 1 \end{pmatrix} : r, s, \in R \right\}$

4. (b) (1) $A^{-1} = \begin{pmatrix} \frac{1}{3} & 0 & \frac{1}{3} \\ \frac{1}{9} & \frac{1}{3} & -\frac{2}{9} \\ -\frac{4}{9} & \frac{2}{3} & -\frac{1}{9} \end{pmatrix}$ (2) $\begin{pmatrix} x_1 \\ x_2 \\ x_3 \end{pmatrix} = \begin{pmatrix} 3 \\ 0 \\ -2 \end{pmatrix}$

6. $T^{-1}\{(1,11)\} = \left\{ \begin{pmatrix} \frac{11}{2} \\ -\frac{9}{2} \\ 0 \end{pmatrix} + t \begin{pmatrix} 1 \\ -1 \\ 2 \end{pmatrix} : t \in R \right\}$

7. The systems in parts (b), (c), and (d) have solutions.

11. The farmer, tailor, and carpenter must have incomes in the proportions 4 : 3 : 4.

13. There must be 7.8 units of the first commodity and 9.5 units of the second.

SECTION 3.4

1. (a) F   (b) T   (c) T   (d) T   (e) F   (f) T   (g) T

2. (a) $\begin{pmatrix} 4 \\ -3 \\ -1 \end{pmatrix}$ **(c)** $\begin{pmatrix} 2 \\ 3 \\ -2 \\ -1 \end{pmatrix}$ **(e)** $\left\{ \begin{pmatrix} 4 \\ 0 \\ 1 \\ 0 \end{pmatrix} + r \begin{pmatrix} 4 \\ 1 \\ 0 \\ 0 \end{pmatrix} + s \begin{pmatrix} 1 \\ 0 \\ 2 \\ 1 \end{pmatrix} : r, s \in R \right\}$

(g) $\left\{ \begin{pmatrix} -23 \\ 0 \\ 7 \\ 9 \\ 0 \end{pmatrix} + r \begin{pmatrix} 1 \\ 1 \\ 0 \\ 0 \\ 0 \end{pmatrix} + s \begin{pmatrix} -23 \\ 0 \\ 6 \\ 9 \\ 1 \end{pmatrix} : r, s \in R \right\}$

(i) $\left\{ \begin{pmatrix} 2 \\ 0 \\ 0 \\ -1 \\ 0 \end{pmatrix} + r \begin{pmatrix} 0 \\ 2 \\ 1 \\ 0 \\ 0 \end{pmatrix} + s \begin{pmatrix} 1 \\ -4 \\ 0 \\ -2 \\ 1 \end{pmatrix} : r, s \in R \right\}$

4. (a) $\left\{ \begin{pmatrix} \frac{4}{3} \\ \frac{1}{3} \\ 0 \\ 0 \end{pmatrix} + t \begin{pmatrix} 1 \\ -1 \\ 1 \\ 2 \end{pmatrix} : t \in R \right\}$ **(c)** There are no solutions.

5. $\begin{pmatrix} 1 & 0 & 2 & 1 & 4 \\ -1 & -1 & 3 & -2 & -7 \\ 3 & 1 & 1 & 0 & -9 \end{pmatrix}$

7. $\{u_1, u_2, u_5\}$

11. (b) $\{(1,2,1,0,0), (2,1,0,0,0), (1,0,0,1,0), (-2,0,0,0,1)\}$

13. (b) $\{(1,0,1,1,1,0), (0,2,1,1,0,0), (1,1,1,0,0,0), (-3,-2,0,0,0,1)\}$

CHAPTER 4

SECTION 4.1

1. (a) F   **(b)** T   **(c)** F   **(d)** F   **(e)** T

2. (a) 30   **(c)** -8

3. (a) $-10 + 15i$ **(c)** -24

4. (a) 19   **(c)** 14

SECTION 4.2

1. (a) F   **(b)** T   **(c)** T   **(d)** T   **(e)** F   **(f)** F   **(g)** F   **(h)** T

3. 42   **5.** -12 **7.** -12 **9.** 22   **11.** -3

13. -8 **15.** 0   **17.** -49 **19.** $-28 - i$ **21.** 95

SECTION 4.3

1. (a) F **(b)** T **(c)** F **(d)** T **(e)** F **(f)** T **(g)** F **(h)** F

3. $(4, -3, 0)$ **5.** $(-20, -48, -8)$ **7.** $(0, -12, 16)$

24. $t^n + a_{n-1}t^{n-1} + \cdots + a_1 t + a_0$

26. (a) $\begin{pmatrix} A_{22} & -A_{12} \\ -A_{21} & A_{11} \end{pmatrix}$ **(c)** $\begin{pmatrix} 10 & 0 & 0 \\ 0 & -20 & 0 \\ 0 & 0 & -8 \end{pmatrix}$

(e) $\begin{pmatrix} -3i & 0 & 0 \\ 4 & -1+i & 0 \\ 10+16i & -5-3i & 3+3i \end{pmatrix}$ **(g)** $\begin{pmatrix} 18 & 28 & -6 \\ -20 & -21 & 37 \\ 48 & 14 & -16 \end{pmatrix}$

SECTION 4.4

1. (a) T **(b)** T **(c)** T **(d)** F **(e)** F **(f)** T
(g) T **(h)** F **(i)** T **(j)** T **(k)** T

2. (a) 22 **(c)** $2 - 4i$

3. (a) -12 **(c)** -12 **(e)** 22 **(g)** -3

4. (a) 0 **(c)** -49 **(e)** $-28 - i$ **(g)** 95

SECTION 4.5

1. (a) F **(b)** T **(c)** T **(d)** F **(e)** F **(f)** T

3. No **5.** Yes **7.** Yes **9.** No

CHAPTER 5

SECTION 5.1

1. (a) F **(b)** T **(c)** T **(d)** F **(e)** F **(f)** F
(g) F **(h)** T **(i)** T **(j)** F **(k)** F

2. (a) $[\mathsf{T}]_\beta = \begin{pmatrix} 0 & 2 \\ -1 & 0 \end{pmatrix}$, no **(c)** $[\mathsf{T}]_\beta = \begin{pmatrix} -1 & 0 & 0 \\ 0 & 1 & 0 \\ 0 & 0 & -1 \end{pmatrix}$, yes

(e) $[\mathsf{T}]_\beta = \begin{pmatrix} -1 & 1 & 0 & 0 \\ 0 & -1 & 1 & 0 \\ 0 & 0 & -1 & 0 \\ 0 & 0 & 0 & -1 \end{pmatrix}$, no

3. (a) The eigenvalues are 4 and -1, a basis of eigenvectors is
$$\left\{ \begin{pmatrix} 2 \\ 3 \end{pmatrix}, \begin{pmatrix} 1 \\ -1 \end{pmatrix} \right\}, \quad Q = \begin{pmatrix} 2 & 1 \\ 3 & -1 \end{pmatrix}, \quad \text{and} \quad D = \begin{pmatrix} 4 & 0 \\ 0 & -1 \end{pmatrix}.$$
(c) The eigenvalues are 1 and -1, a basis of eigenvectors is
$$\left\{ \begin{pmatrix} 1 \\ 1-i \end{pmatrix}, \begin{pmatrix} 1 \\ -1-i \end{pmatrix} \right\}, \quad Q = \begin{pmatrix} 1 & 1 \\ 1-i & -1-i \end{pmatrix}, \quad \text{and} \quad D = \begin{pmatrix} 1 & 0 \\ 0 & -1 \end{pmatrix}.$$

4. (a) $\lambda = 3, 4$ $\qquad\qquad$ $\beta = \{(3, 5), (1, 2)\}$
$\quad$ **(b)** $\lambda = -1, 1, 2$ $\qquad$ $\beta = \{(1, 2, 0), (1, -1, -1), (2, 0, -1)\}$
$\quad$ **(f)** $\lambda = 1, 3$ $\qquad\quad$ $\beta = \{-2 + x, -4 + x^2, -8 + x^3, x\}$

$\quad$ **(h)** $\lambda = -1, 1, 1, 1$ $\quad$ $\beta = \left\{ \begin{pmatrix} -1 & 0 \\ 0 & 1 \end{pmatrix}, \begin{pmatrix} 0 & 1 \\ 0 & 0 \end{pmatrix}, \begin{pmatrix} 1 & 0 \\ 0 & 1 \end{pmatrix}, \begin{pmatrix} 0 & 0 \\ 1 & 0 \end{pmatrix} \right\}$

$\quad$ **(i)** $\lambda = 1, 1, -1, -1$ $\quad$ $\beta = \left\{ \begin{pmatrix} 1 & 0 \\ 1 & 0 \end{pmatrix}, \begin{pmatrix} 0 & 1 \\ 0 & 1 \end{pmatrix}, \begin{pmatrix} -1 & 0 \\ 1 & 0 \end{pmatrix}, \begin{pmatrix} 0 & -1 \\ 0 & 1 \end{pmatrix} \right\}$

$\quad$ **(j)** $\lambda = -1, 1, 5$ $\qquad$ $\beta = \left\{ \begin{pmatrix} 0 & 1 \\ -1 & 0 \end{pmatrix}, \begin{pmatrix} 1 & 0 \\ 0 & -1 \end{pmatrix}, \begin{pmatrix} 0 & 1 \\ 1 & 0 \end{pmatrix}, \begin{pmatrix} 1 & 0 \\ 0 & 1 \end{pmatrix} \right\}$

26. 4

SECTION 5.2

1. (a) F $\quad$ **(b)** F $\quad$ **(c)** F $\quad$ **(d)** T $\quad$ **(e)** T $\quad$ **(f)** F
$\quad$ **(g)** T $\quad$ **(h)** T $\quad$ **(i)** F

2. (a) Not diagonalizable $\qquad$ **(c)** $Q = \begin{pmatrix} 1 & 4 \\ 1 & -3 \end{pmatrix}$

$\quad$ **(e)** Not diagonalizable $\qquad$ **(g)** $Q = \begin{pmatrix} 1 & 1 & 1 \\ 2 & -1 & 0 \\ -1 & 0 & -1 \end{pmatrix}$

3. (a) Not diagonalizable $\qquad$ **(c)** Not diagonalizable

$\quad$ **(d)** $\beta = \{x - x^2, 1 - x - x^2, x + x^2\}$ $\qquad$ **(e)** $\beta = \{(1, 1), (1, -1)\}$

7. $A^n = \dfrac{1}{3} \begin{pmatrix} 5^n + 2(-1)^n & 2(5^n) - 2(-1)^n \\ 5^n - (-1)^n & 2(5)^n + (-1)^n \end{pmatrix}$

14. (b) $x(t) = c_1 e^{3t} \begin{pmatrix} -2 \\ 1 \end{pmatrix} + c_2 e^{-2t} \begin{pmatrix} 1 \\ -1 \end{pmatrix}$

$\quad$ **(c)** $x(t) = e^t \left[c_1 \begin{pmatrix} 1 \\ 0 \\ 0 \end{pmatrix} + c_2 \begin{pmatrix} 0 \\ 1 \\ 0 \end{pmatrix} \right] + c_3 e^{2t} \begin{pmatrix} 1 \\ 1 \\ 1 \end{pmatrix}$

SECTION 5.3

1. (a) T $\quad$ **(b)** T $\quad$ **(c)** F $\quad$ **(d)** F $\quad$ **(e)** T $\quad$ **(f)** T
$\quad$ **(g)** T $\quad$ **(h)** F $\quad$ **(i)** F $\quad$ **(j)** T

2. (a) $\begin{pmatrix} 0 & 0 \\ 0 & 0 \end{pmatrix}$ $\quad$ **(c)** $\begin{pmatrix} \frac{7}{13} & \frac{7}{13} \\ \frac{6}{13} & \frac{6}{13} \end{pmatrix}$ $\quad$ **(e)** No limit exists.

$\quad$ **(g)** $\begin{pmatrix} -1 & 0 & -1 \\ -4 & 1 & -2 \\ 2 & 0 & 2 \end{pmatrix}$ $\quad$ **(i)** No limit exists.

6. One month after arrival, 25% of the patients have recovered, 20% are ambulatory, 41% are bedridden, and 14% have died. Eventually $\frac{59}{90}$ recover and $\frac{31}{90}$ die.

7. $\frac{3}{7}$.

8. Only the matrices in (a) and (b) are regular transition matrices.

9. (a) $\begin{pmatrix} \frac{1}{3} & \frac{1}{3} & \frac{1}{3} \\ \frac{1}{3} & \frac{1}{3} & \frac{1}{3} \\ \frac{1}{3} & \frac{1}{3} & \frac{1}{3} \end{pmatrix}$ **(c)** No limit exists.

(e) $\begin{pmatrix} 0 & 0 & 0 \\ \frac{1}{2} & 1 & 0 \\ \frac{1}{2} & 0 & 1 \end{pmatrix}$ **(g)** $\begin{pmatrix} 0 & 0 & 0 & 0 \\ 0 & 0 & 0 & 0 \\ \frac{1}{2} & \frac{1}{2} & 1 & 0 \\ \frac{1}{2} & \frac{1}{2} & 0 & 1 \end{pmatrix}$

10. (a) $\begin{pmatrix} 0.225 \\ 0.441 \\ 0.334 \end{pmatrix}$ after two stages and $\begin{pmatrix} 0.20 \\ 0.60 \\ 0.20 \end{pmatrix}$ eventually

(c) $\begin{pmatrix} 0.372 \\ 0.225 \\ 0.403 \end{pmatrix}$ after two stages and $\begin{pmatrix} 0.50 \\ 0.20 \\ 0.30 \end{pmatrix}$ eventually

(e) $\begin{pmatrix} 0.329 \\ 0.334 \\ 0.337 \end{pmatrix}$ after two stages and $\begin{pmatrix} \frac{1}{3} \\ \frac{1}{3} \\ \frac{1}{3} \end{pmatrix}$ eventually

12. $\frac{9}{19}$ new, $\frac{6}{19}$ once-used, and $\frac{4}{19}$ twice-used

13. In 1995, 24% will own large cars, 34% will own intermediate-sized cars, and 42% will own small cars; the corresponding eventual percentages are 10%, 30%, and 60%.

20. $e^O = I$ and $e^I = eI$.

SECTION 5.4

1. (a) F   **(b)** T   **(c)** F   **(d)** F   **(e)** T   **(f)** T   **(g)** T

2. The subspaces in (a), (c), and (d) are T-invariant.

6. (a) $\left\{ \begin{pmatrix} 1 \\ 0 \\ 0 \\ 0 \end{pmatrix}, \begin{pmatrix} 1 \\ 0 \\ 1 \\ 1 \end{pmatrix}, \begin{pmatrix} 1 \\ -1 \\ 2 \\ 2 \end{pmatrix} \right\}$ **(c)** $\left\{ \begin{pmatrix} 0 & 1 \\ 1 & 0 \end{pmatrix} \right\}$

9. (a) $-t(t^2 - 3t + 3)$ **(c)** $1 - t$

10. (a) $t(t-1)(t^2 - 3t + 3)$ **(c)** $(t-1)^3(t+1)$

18. (c) $A^{-1} = \frac{1}{2} \begin{pmatrix} 2 & -2 & -4 \\ 0 & 1 & 3 \\ 0 & 0 & -2 \end{pmatrix}$

31. (a) $t^2 - 6t + 6$ **(c)** $-(t+1)(t^2 - 6t + 6)$

CHAPTER 6

SECTION 6.1

1. (a) T   **(b)** T   **(c)** F   **(d)** F   **(e)** F   **(f)** F   **(g)** F   **(h)** T

2. $\langle x, y \rangle = 8 + 5i$, $\|x\| = \sqrt{7}$, $\|y\| = \sqrt{14}$, and $\|x + y\| = \sqrt{37}$.

3. $\langle f, g \rangle = 1$, $\|f\| = \frac{\sqrt{3}}{3}$, $\|g\| = \sqrt{\dfrac{e^2 - 1}{2}}$, and $\|f + g\| = \sqrt{\dfrac{11 + 3e^2}{6}}$.

16. (b) No

SECTION 6.2

1. (a) F   **(b)** T   **(c)** T   **(d)** F   **(e)** T   **(f)** F   **(g)** T

2. For each part the orthonormal basis and the Fourier coefficients are given.

(b) $\left\{ \frac{\sqrt{3}}{3}(1,1,1), \frac{\sqrt{6}}{6}(-2,1,1), \frac{\sqrt{2}}{2}(0,-1,1) \right\}$; $\frac{2\sqrt{3}}{3}, -\frac{\sqrt{6}}{6}, \frac{\sqrt{2}}{2}$.

(c) $\{1, 2\sqrt{3}(x - \frac{1}{2}), 6\sqrt{5}(x^2 - x + \frac{1}{6})\}$; $\frac{3}{2}, \frac{\sqrt{3}}{6}, 0$.

(e) $\left\{ \frac{1}{5}(2,-1,-2,4), \frac{1}{\sqrt{30}}(-4,2,-3,1), \frac{1}{\sqrt{155}}(-3,4,9,7) \right\}$; $10, 3\sqrt{30}, \sqrt{155}$

(g) $\left\{ \frac{1}{6}\begin{pmatrix} 3 & 5 \\ -1 & 1 \end{pmatrix}, \frac{1}{6\sqrt{2}}\begin{pmatrix} -4 & 4 \\ 6 & -2 \end{pmatrix}, \frac{1}{9\sqrt{2}}\begin{pmatrix} 9 & -3 \\ 6 & -6 \end{pmatrix} \right\}$; $24, 6\sqrt{2}, -9\sqrt{2}$

(i) $\left\{ \sqrt{\frac{2}{\pi}}\sin t, \sqrt{\frac{2}{\pi}}\cos t, \sqrt{\frac{\pi}{\pi^2 - 8}}(1 - \frac{4}{\pi}\sin t), \sqrt{\frac{12\pi}{\pi^4 - 96}}(t + \frac{4}{\pi}\cos t - \frac{\pi}{2}) \right\}$;

   $\sqrt{\frac{2}{\pi}}(2\pi + 2), -4\sqrt{\frac{2}{\pi}}, \sqrt{\frac{\pi^2 - 8}{\pi}}(1 + \pi), \sqrt{\frac{\pi^4 - 96}{3\pi}}$

(k) $\left\{ \frac{1}{\sqrt{47}}(-4, 3 - 2i, i, 1 - 4i), \frac{1}{\sqrt{60}}(3 - i, -5i, -2 + 4i, 2 + i), \right.$

     $\left. \frac{1}{\sqrt{1160}}(-17 - i, -9 + 8i, -18 + 6i, -9 + 8i) \right\}$;

   $\sqrt{47}(-1 - i), \sqrt{60}(-1 + 2i), \sqrt{1160}(1 + i)$

(m) $\left\{ \frac{1}{\sqrt{18}}\begin{pmatrix} -1 + i & -i \\ 2 - i & 1 + 3i \end{pmatrix}, \frac{1}{\sqrt{246}}\begin{pmatrix} -4i & -11 - 9i \\ 1 + 5i & 1 - i \end{pmatrix}, \right.$

     $\left. \frac{1}{\sqrt{39063}}\begin{pmatrix} -5 - 118i & -7 - 26i \\ -145i & -58 \end{pmatrix} \right\}$; $\sqrt{18}(2 + i), \sqrt{246}(-1 - i), 0$

4. $S^\perp = \text{span}(\{(i, -\frac{1}{2}(1 + i), 1)\})$

5. $S_0^\perp$ is the plane through the origin that is perpendicular to x_0; $S^\perp$ is the line through the origin that is perpendicular to the plane containing x_1 and x_2.

19. (a) $\frac{1}{17}\begin{pmatrix} 26 \\ 104 \end{pmatrix}$ **(b)** $\frac{1}{14}\begin{pmatrix} 29 \\ 17 \\ 40 \end{pmatrix}$

20. (b) $\dfrac{1}{\sqrt{14}}$

SECTION 6.3

1. (a) T   (b) F   (c) F   (d) T   (e) F   (f) T   (g) T
2. (a) $y = (1, -2, 4)$ (c) $y = 210x^2 - 204x + 33$
3. (a) $\mathsf{T}^*(x) = (11, -12)$ (c) $\mathsf{T}^*(f(t)) = 12 + 6t$
14. $\mathsf{T}^*(x) = \langle x, z \rangle y$
20. (a) The linear function is $y = -2t + 5/2$ with $E = 1$, and the quadratic function is $y = t^2/3 - 4t/3 + 2$ with $E = 0$.
    (b) The linear function is $y = 1.25t + 0.55$ with $E = 0.3$, and the quadratic function is $t^2/56 + 15t/14 + 239/280$ with $E = 0.22857$ (approximation).
21. The spring constant is approximately 2.1.
22. (a) $x = \frac{2}{7}$, $y = \frac{3}{7}$, $z = \frac{1}{7}$ (d) $x = \frac{7}{12}$, $y = \frac{1}{12}$, $z = \frac{1}{4}$, $w = -\frac{1}{12}$

SECTION 6.4

1. (a) T   (b) F   (c) F   (d) T   (e) T   (f) T   (g) F   (h) T
2. (a) T is self-adjoint. An orthonormal basis of eigenvectors is
$$\left\{ \frac{1}{\sqrt{5}}(1, -2), \frac{1}{\sqrt{5}}(2, 1) \right\},$$ with corresponding eigenvalues 6 and 1.
   (c) T is normal, but not self-adjoint. An orthonormal basis of eigenvectors is
$$\left\{ \frac{1}{2}(1 + i, \sqrt{2}), \frac{1}{2}(1 + i, -\sqrt{2}) \right\}$$ with corresponding eigenvalues
$$2 + \frac{1+i}{\sqrt{2}} \text{ and } 2 - \frac{1+i}{\sqrt{2}}.$$
   (e) T is self-adjoint. An orthonormal basis of eigenvectors is
$$\left\{ \frac{1}{\sqrt{2}} \begin{pmatrix} 0 & 1 \\ 1 & 0 \end{pmatrix}, \frac{1}{\sqrt{2}} \begin{pmatrix} 1 & 0 \\ 0 & 1 \end{pmatrix}, \frac{1}{\sqrt{2}} \begin{pmatrix} 0 & -1 \\ 1 & 0 \end{pmatrix}, \frac{1}{\sqrt{2}} \begin{pmatrix} -1 & 0 \\ 0 & 1 \end{pmatrix} \right\}$$
   with corresponding eigenvalues $1, 1, -1, -1$.

SECTION 6.5

1. (a) T   (b) F   (c) F   (d) T   (e) F   (f) T
   (g) F   (h) F   (i) F
2. (a) $P = \dfrac{1}{\sqrt{2}} \begin{pmatrix} 1 & 1 \\ 1 & -1 \end{pmatrix}$ and $D = \begin{pmatrix} 3 & 0 \\ 0 & -1 \end{pmatrix}$
   (d) $P = \begin{pmatrix} \frac{1}{\sqrt{2}} & \frac{1}{\sqrt{6}} & \frac{1}{\sqrt{3}} \\ -\frac{1}{\sqrt{2}} & \frac{1}{\sqrt{6}} & \frac{1}{\sqrt{3}} \\ 0 & -\frac{2}{\sqrt{6}} & \frac{1}{\sqrt{3}} \end{pmatrix}$ and $D = \begin{pmatrix} -2 & 0 & 0 \\ 0 & -2 & 0 \\ 0 & 0 & 4 \end{pmatrix}$
4. T_z is normal for all $z \in C$, T_z is self-adjoint if and only if $z \in R$, and T_z is unitary if and only if $|z| = 1$.
5. Only the pair of matrices in (d) are unitarily equivalent.

25. $2(\psi - \phi)$

26. (a) $\psi - \dfrac{\phi}{2}$ (b) $\psi + \dfrac{\phi}{2}$

27. (a) $x = \dfrac{1}{\sqrt{2}}x' + \dfrac{1}{\sqrt{2}}y'$ and $y = \dfrac{1}{\sqrt{2}}x' - \dfrac{1}{\sqrt{2}}y'$

   The new quadratic form is $3(x')^2 - (y')^2$.

   (c) $x = \dfrac{3}{\sqrt{13}}x' + \dfrac{2}{\sqrt{13}}y'$ and $y = \dfrac{-2}{\sqrt{13}}x' + \dfrac{2}{\sqrt{13}}y'$

   The new quadratic form is $5(x')^2 - 8(y')^2$.

29. (c) $P = \begin{pmatrix} \frac{1}{\sqrt{2}} & \frac{1}{\sqrt{3}} & -\frac{6}{\sqrt{6}} \\ \frac{1}{\sqrt{2}} & -\frac{1}{\sqrt{3}} & \frac{\sqrt{6}}{6} \\ 0 & \frac{1}{\sqrt{3}} & \frac{\sqrt{6}}{3} \end{pmatrix}$ and $R = \begin{pmatrix} \sqrt{2} & \sqrt{2} & 2\sqrt{2} \\ 0 & \sqrt{3} & \frac{\sqrt{3}}{3} \\ 0 & 0 & \frac{\sqrt{6}}{3} \end{pmatrix}$

   (e) $x_1 = 3,\ x_2 = -5,\ x_3 = 4$

SECTION 6.6

1. (a) F   (b) T   (c) T   (d) F   (e) F

2. For $W = \text{span}(\{(1,2)\})$, $[T]_\beta = \begin{pmatrix} \frac{1}{5} & \frac{2}{5} \\ \frac{2}{5} & \frac{4}{5} \end{pmatrix}$.

3. (2) (a) $T_1(a, b) = \frac{1}{2}(a + b, a + b)$ and $T_2(a, b) = \frac{1}{2}(a - b, -a + b)$

   (d) $T_1(a, b, c) = \frac{1}{3}(2a - b - c, -a + 2b - c, -a - b + 2c)$ and

   $T_2(a, b, c) = \frac{1}{3}(a + b + c, a + b + c, a + b + c)$

SECTION 6.7

1. (a) F   (b) F   (c) T   (d) T   (e) F   (f) F   (g) T

2. (a) $v_1 = \begin{pmatrix} 1 \\ 0 \end{pmatrix}$, $v_2 = \begin{pmatrix} 0 \\ 1 \end{pmatrix}$, $u_1 = \dfrac{1}{\sqrt{3}} \begin{pmatrix} 1 \\ 1 \\ 1 \end{pmatrix}$, $u_2 = \dfrac{1}{\sqrt{2}} \begin{pmatrix} 0 \\ 1 \\ -1 \end{pmatrix}$, $u_3 = \dfrac{1}{\sqrt{6}} \begin{pmatrix} 2 \\ -1 \\ -1 \end{pmatrix}$

   $\sigma_1 = \sqrt{3},\ \sigma_2 = \sqrt{2}$

   (c) $v_1 = \dfrac{1}{\sqrt{\pi}} \sin x,\ v_2 = \dfrac{1}{\sqrt{\pi}} \cos x,\ v_3 = \dfrac{1}{\sqrt{2\pi}}$

   $u_1 = \dfrac{\cos x + 2\sin x}{\sqrt{5\pi}},\ u_2 = \dfrac{2\cos x - \sin x}{\sqrt{5\pi}},\ u_3 = \dfrac{1}{\sqrt{2\pi}},$

   $\sigma_1 = \sqrt{5},\ \sigma_2 = \sqrt{5},\ \sigma_3 = 2$

3. (a) $\begin{pmatrix} \frac{1}{\sqrt{3}} & \frac{1}{\sqrt{2}} & \frac{1}{\sqrt{6}} \\ \frac{1}{\sqrt{3}} & -\frac{1}{\sqrt{2}} & \frac{1}{\sqrt{6}} \\ -\frac{1}{\sqrt{3}} & 0 & \frac{2}{\sqrt{6}} \end{pmatrix} \begin{pmatrix} \sqrt{6} & 0 \\ 0 & 0 \\ 0 & 0 \end{pmatrix} \begin{pmatrix} \frac{1}{\sqrt{2}} & \frac{1}{\sqrt{2}} \\ \frac{1}{\sqrt{2}} & -\frac{1}{\sqrt{2}} \end{pmatrix}^*$

(c) $\begin{pmatrix} \frac{2}{\sqrt{10}} & 0 & \frac{1}{\sqrt{2}} & \frac{1}{\sqrt{10}} \\ \frac{1}{\sqrt{10}} & -\frac{1}{\sqrt{2}} & 0 & -\frac{2}{\sqrt{10}} \\ \frac{1}{\sqrt{10}} & \frac{1}{\sqrt{2}} & 0 & -\frac{2}{\sqrt{10}} \\ \frac{2}{\sqrt{10}} & 0 & -\frac{1}{\sqrt{2}} & \frac{1}{\sqrt{10}} \end{pmatrix} \begin{pmatrix} \sqrt{5} & 0 \\ 0 & 1 \\ 0 & 0 \\ 0 & 0 \end{pmatrix} \begin{pmatrix} \frac{1}{\sqrt{2}} & \frac{1}{\sqrt{2}} \\ \frac{1}{\sqrt{2}} & -\frac{1}{\sqrt{2}} \end{pmatrix}^*$

(e) $\begin{pmatrix} \frac{1+i}{2} & \frac{1+i}{2} \\ \frac{1-i}{2} & \frac{-1+i}{2} \end{pmatrix} \begin{pmatrix} \sqrt{6} & 0 \\ 0 & 0 \end{pmatrix} \begin{pmatrix} \frac{2}{\sqrt{6}} & \frac{1-i}{\sqrt{6}} \\ \frac{1+i}{\sqrt{6}} & -\frac{2}{\sqrt{6}} \end{pmatrix}^*$

4. (a) $WP = \begin{pmatrix} \frac{1}{\sqrt{2}} & \frac{1}{\sqrt{2}} \\ \frac{1}{\sqrt{2}} & -\frac{1}{\sqrt{2}} \end{pmatrix} \begin{pmatrix} \frac{\sqrt{8}+\sqrt{2}}{2} & \frac{-\sqrt{8}+\sqrt{2}}{2} \\ \frac{-\sqrt{8}+\sqrt{2}}{2} & \frac{\sqrt{8}+\sqrt{2}}{2} \end{pmatrix}$

5. (a) $T^\dagger(x,y,z) = \left(\dfrac{x+y+z}{3}, \dfrac{y-z}{2} \right)$

(c) $T^\dagger(a + b\sin x + c\cos x) = T^{-1}(a + b\sin x + c\cos x) =$

$\dfrac{a}{2} + \dfrac{(2b+c)\sin x + (-b+2c)\cos x}{5}$

6. (a) $\dfrac{1}{6}\begin{pmatrix} 1 & 1 & -1 \\ 1 & 1 & -1 \end{pmatrix}$ **(c)** $\dfrac{1}{5}\begin{pmatrix} 1 & -2 & 3 & 1 \\ 1 & 3 & -2 & 1 \end{pmatrix}$ **(e)** $\dfrac{1}{6}\begin{pmatrix} 1-i & 1+i \\ 1 & i \end{pmatrix}$

7. (a) $Z_1 = N(T)^\perp = R^2$ and $Z_2 = R(T) = \text{span}\{(1,1,1),(0,1,-1)\}$
(c) $Z_1 = N(T)^\perp = V$ and $Z_2 = R(T) = V$

8. (a) No solution $\dfrac{1}{2}\begin{pmatrix} 1 \\ 1 \end{pmatrix}$

SECTION 6.8

1. (a) F   **(b)** F   **(c)** T   **(d)** F   **(e)** T   **(f)** F
(g) F   **(h)** F   **(i)** T   **(j)** F
4. (a) Yes   **(b)** No   **(c)** No   **(d)** Yes   **(e)** Yes   **(f)** No

5. (a) $\begin{pmatrix} 0 & 2 & -2 \\ 2 & 0 & -2 \\ 1 & 1 & 0 \end{pmatrix}$ **(b)** $\begin{pmatrix} 1 & 0 & 0 & 1 \\ 0 & 0 & 0 & 0 \\ 0 & 0 & 0 & 0 \\ 1 & 0 & 0 & 1 \end{pmatrix}$ **(c)** $\begin{pmatrix} 0 & 0 & 0 & 0 \\ -1 & 0 & -4 & 0 \\ 0 & 0 & 0 & 0 \\ -2 & 0 & -8 & 0 \end{pmatrix}$

17. (a) and (b) $\left\{ \begin{pmatrix} \frac{2}{\sqrt{5}} \\ -\frac{1}{\sqrt{5}} \end{pmatrix}, \begin{pmatrix} \frac{1}{\sqrt{5}} \\ \frac{2}{\sqrt{5}} \end{pmatrix} \right\}$ **(c)** $\left\{ \begin{pmatrix} \frac{1}{\sqrt{2}} \\ 0 \\ \frac{1}{\sqrt{2}} \end{pmatrix}, \begin{pmatrix} 0 \\ 1 \\ 0 \end{pmatrix}, \begin{pmatrix} \frac{1}{\sqrt{2}} \\ 0 \\ -\frac{1}{\sqrt{2}} \end{pmatrix} \right\}$

18. Same as Exercise 17(c)

22. (a) $Q = \begin{pmatrix} 1 & -3 \\ 0 & 1 \end{pmatrix}$ and $D = \begin{pmatrix} 1 & 0 \\ 0 & -7 \end{pmatrix}$

(b) $Q = \begin{pmatrix} 1 & -\frac{1}{2} \\ 1 & \frac{1}{2} \end{pmatrix}$ and $D = \begin{pmatrix} 2 & 0 \\ 0 & -\frac{1}{2} \end{pmatrix}$

(c) $Q = \begin{pmatrix} 0 & 0 & 1 \\ 0 & 1 & -0.25 \\ 1 & 0 & 2 \end{pmatrix}$ and $D = \begin{pmatrix} -1 & 0 & 0 \\ 0 & 4 & 0 \\ 0 & 0 & 6.75 \end{pmatrix}$

SECTION 6.9

7. $(B_v)^{-1} = \begin{pmatrix} \dfrac{1}{\sqrt{1-v^2}} & 0 & 0 & \dfrac{v}{\sqrt{1-v^2}} \\ 0 & 1 & 0 & 0 \\ 0 & 0 & 1 & 0 \\ \dfrac{v}{\sqrt{1-v^2}} & 0 & 0 & \dfrac{1}{\sqrt{1-v^2}} \end{pmatrix}$

SECTION 6.10

1. (a) F (b) T (c) T (d) F (e) F

2. (a) $\sqrt{18}$ (c) approximately 2.34

4. (a) $\|A\| \approx 84.74$, $\|A^{-1}\| \approx 17.01$, and $\text{cond}(A) \approx 1441$
   (b) $\|\tilde{x} - A^{-1}b\| \le \|A^{-1}\| \cdot \|A\tilde{x} - b\| \approx 0.17$ and

$$\frac{\|\tilde{x} - A^{-1}b\|}{\|A^{-1}b\|} \le \text{cond}(A)\frac{\|b - A\tilde{x}\|}{\|b\|} \approx \frac{14.41}{\|b\|}$$

5. $0.001 \le \dfrac{\|x - \tilde{x}\|}{\|x\|} \le 10$

6. $R\begin{pmatrix} 1 \\ -2 \\ 3 \end{pmatrix} = \dfrac{9}{7}$, $\|B\| = 2$, and $\text{cond}(B) = 2$.

SECTION 6.11

1. (a) F   (b) T   (c) T   (d) F   (e) T   (f) F
   (g) F   (h) F   (i) T   (j) F

3. (b) $\left\{ t\begin{pmatrix} \sqrt{3} \\ 1 \end{pmatrix} : t \in R \right\}$

4. (b) $\left\{ t\begin{pmatrix} 1 \\ 0 \end{pmatrix} : t \in R \right\}$ if $\phi = 0$ and $\left\{ t\begin{pmatrix} \cos\phi + 1 \\ \sin\phi \end{pmatrix} : t \in R \right\}$ if $\phi \ne 0$

7. (c) There are six possibilities:

   (1) Any line through the origin if $\phi = \psi = 0$

   (2) $\left\{ t\begin{pmatrix} 0 \\ 0 \\ 1 \end{pmatrix} : t \in R \right\}$ if $\phi = 0$ and $\psi = \pi$

   (3) $\left\{ t\begin{pmatrix} \cos\psi + 1 \\ -\sin\psi \\ 0 \end{pmatrix} : t \in R \right\}$ if $\phi = \pi$ and $\psi \ne \pi$

   (4) $\left\{ t\begin{pmatrix} 0 \\ \cos\phi - 1 \\ \sin\phi \end{pmatrix} : t \in R \right\}$ if $\psi = \pi$ and $\phi \ne \pi$

   (5) $\left\{ t\begin{pmatrix} 0 \\ 1 \\ 0 \end{pmatrix} : t \in R \right\}$ if $\phi = \psi = \pi$

$$(6) \quad \left\{ t \begin{pmatrix} \sin\phi(\cos\psi + 1) \\ -\sin\phi\sin\psi \\ \sin\psi(\cos\phi + 1) \end{pmatrix} : t \in R \right\} \quad \text{otherwise}$$

CHAPTER 7

SECTION 7.1

1. (a) T   (b) F   (c) F   (d) T   (e) F   (f) F   (g) T   (h) T

2. (a) For $\lambda = 2$, $\left\{ \begin{pmatrix} -1 \\ -1 \end{pmatrix}, \begin{pmatrix} 1 \\ 0 \end{pmatrix} \right\}$ $J = \begin{pmatrix} 2 & 1 \\ 0 & 2 \end{pmatrix}$

   (c) For $\lambda = -1$, $\left\{ \begin{pmatrix} 1 \\ 3 \\ 0 \end{pmatrix} \right\}$ For $\lambda = 2$, $\left\{ \begin{pmatrix} 1 \\ 1 \\ 1 \end{pmatrix}, \begin{pmatrix} 1 \\ 2 \\ 0 \end{pmatrix} \right\}$ $J = \begin{pmatrix} -1 & 0 & 0 \\ 0 & 2 & 1 \\ 0 & 0 & 2 \end{pmatrix}$

3. (a) For $\lambda = 2$, $\{2, -2x, x^2\}$ $J = \begin{pmatrix} 2 & 1 & 0 \\ 0 & 2 & 1 \\ 0 & 0 & 2 \end{pmatrix}$

   (c) For $\lambda = 1$, $\left\{ \begin{pmatrix} 1 & 0 \\ 0 & 0 \end{pmatrix}, \begin{pmatrix} 0 & 0 \\ 1 & 0 \end{pmatrix}, \begin{pmatrix} 0 & 1 \\ 0 & 0 \end{pmatrix}, \begin{pmatrix} 0 & 0 \\ 0 & 1 \end{pmatrix} \right\}$ $J = \begin{pmatrix} 1 & 1 & 0 & 0 \\ 0 & 1 & 0 & 0 \\ 0 & 0 & 1 & 1 \\ 0 & 0 & 0 & 1 \end{pmatrix}$

SECTION 7.2

1. (a) T   (b) T   (c) F   (d) T   (e) T   (f) F   (g) F   (h) T

2. $J = \begin{pmatrix} A_1 & O & O \\ O & A_2 & O \\ O & O & A_3 \end{pmatrix}$ where $A_1 = \begin{pmatrix} 2 & 1 & 0 & 0 & 0 & 0 \\ 0 & 2 & 1 & 0 & 0 & 0 \\ 0 & 0 & 2 & 0 & 0 & 0 \\ 0 & 0 & 0 & 2 & 1 & 0 \\ 0 & 0 & 0 & 0 & 2 & 0 \\ 0 & 0 & 0 & 0 & 0 & 2 \end{pmatrix}$,

   $A_2 = \begin{pmatrix} 4 & 1 & 0 & 0 \\ 0 & 4 & 1 & 0 \\ 0 & 0 & 4 & 0 \\ 0 & 0 & 0 & 4 \end{pmatrix}$ and $A_3 = \begin{pmatrix} -3 & 0 \\ 0 & -3 \end{pmatrix}$

3. (a) $-(t-2)^5(t-3)^2$ (b)

$$\lambda_1 = 2 \qquad\qquad \lambda_2 = 3$$

   (c) $\lambda_2 = 3$ (d) $p_1 = 3$ and $p_2 = 1$
   (e) (i) rank$(U_1) = 3$ and rank$(U_2) = 0$
       (ii) rank$(U_1^2) = 1$ and rank$(U_2^2) = 0$
       (iii) nullity$(U_1) = 2$ and nullity$(U_2) = 2$
       (iv) nullity$(U_1^2) = 4$ and nullity$(U_2^2) = 2$

4. (a) $J = \begin{pmatrix} 1 & 0 & 0 \\ 0 & 2 & 1 \\ 0 & 0 & 2 \end{pmatrix}$ and $Q = \begin{pmatrix} 1 & 1 & 1 \\ 2 & 1 & 2 \\ 1 & -1 & 0 \end{pmatrix}$

(d) $J = \begin{pmatrix} 0 & 1 & 0 & 0 \\ 0 & 0 & 0 & 0 \\ 0 & 0 & 2 & 0 \\ 0 & 0 & 0 & 2 \end{pmatrix}$ and $Q = \begin{pmatrix} 1 & 0 & 1 & -1 \\ 1 & -1 & 0 & 1 \\ 1 & -2 & 0 & 1 \\ 1 & 0 & 1 & 0 \end{pmatrix}$

5. (a) $J = \begin{pmatrix} 1 & 1 & 0 & 0 \\ 0 & 1 & 1 & 0 \\ 0 & 0 & 1 & 0 \\ 0 & 0 & 0 & 2 \end{pmatrix}$ and $\beta = \{2e^t, 2te^t, t^2e^t, e^{2t}\}$

(c) $J = \begin{pmatrix} 2 & 1 & 0 & 0 \\ 0 & 2 & 0 & 0 \\ 0 & 0 & 2 & 1 \\ 0 & 0 & 0 & 2 \end{pmatrix}$ and $\beta = \{6x, x^3, 2, x^2\}$

(d) $J = \begin{pmatrix} 2 & 1 & 0 & 0 \\ 0 & 2 & 1 & 0 \\ 0 & 0 & 2 & 0 \\ 0 & 0 & 0 & 4 \end{pmatrix}$ and

$$\beta = \left\{ \begin{pmatrix} 1 & 0 \\ 0 & 0 \end{pmatrix}, \begin{pmatrix} 0 & 1 \\ 1 & 0 \end{pmatrix}, \begin{pmatrix} 0 & -1 \\ 0 & 2 \end{pmatrix}, \begin{pmatrix} 1 & -2 \\ 2 & 0 \end{pmatrix} \right\}$$

24. (a) $\begin{pmatrix} x \\ y \\ z \end{pmatrix} = e^{2t} \left[(c_1 + c_2 t) \begin{pmatrix} 1 \\ 0 \\ 0 \end{pmatrix} + c_2 \begin{pmatrix} 0 \\ 1 \\ 0 \end{pmatrix} \right] + c_3 e^{3t} \begin{pmatrix} 1 \\ 1 \\ -1 \end{pmatrix}$

(b) $\begin{pmatrix} x \\ y \\ z \end{pmatrix} = e^{2t} \left[(c_1 + c_2 t + c_3 t^2) \begin{pmatrix} 1 \\ 0 \\ 0 \end{pmatrix} + (c_2 + 2c_3 t) \begin{pmatrix} 0 \\ 1 \\ 0 \end{pmatrix} + 2c_3 \begin{pmatrix} 0 \\ 0 \\ 1 \end{pmatrix} \right]$

SECTION 7.3

1. (a) F **(b)** T **(c)** F **(d)** F **(e)** T **(f)** F
(g) F **(h)** T **(i)** T
2. (a) $(t-1)(t-3)$ **(c)** $(t-1)^2(t-2)$ **(d)** $(t-2)^2$
3. (a) $t^2 - 2$ **(c)** $(t-2)^2$ **(d)** $(t-1)(t+1)$
4. For (2), (a); for (3), (a) and (d)
5. The operators are T_0, I, and all operators having both 0 and 1 as eigenvalues.

SECTION 7.4

1. (a) T **(b)** F **(c)** F **(d)** T **(e)** T **(f)** F **(g)** T

2. (a) $\begin{pmatrix} 0 & 0 & 27 \\ 1 & 0 & -27 \\ 0 & 1 & 9 \end{pmatrix}$ **(b)** $\begin{pmatrix} 0 & -1 \\ 1 & -1 \end{pmatrix}$

(c) $\begin{pmatrix} \frac{1}{2}(-1+i\sqrt{3}) & 0 \\ 0 & \frac{1}{2}(-1-i\sqrt{3}) \end{pmatrix}$ **(e)** $\begin{pmatrix} 0 & -2 & 0 & 0 \\ 1 & 0 & 0 & 0 \\ 0 & 0 & 0 & -3 \\ 0 & 0 & 1 & 0 \end{pmatrix}$

3. (a) $t^2 + 1$ and t^2 $\quad C = \begin{pmatrix} 0 & -1 & 0 & 0 \\ 1 & 0 & 0 & 0 \\ 0 & 0 & 0 & 0 \\ 0 & 0 & 0 & 0 \end{pmatrix}$; $\quad \beta = \{1, x, -2x + x^2, -3x + x^3\}$

(c) $t^2 - t + 1$ $\quad C = \begin{pmatrix} 0 & -1 & 0 & 0 \\ 1 & 1 & 0 & 0 \\ 0 & 0 & 0 & -1 \\ 0 & 0 & 1 & 1 \end{pmatrix}$

$\beta = \left\{ \begin{pmatrix} 1 & 0 \\ 0 & 0 \end{pmatrix}, \begin{pmatrix} 0 & 0 \\ -1 & 0 \end{pmatrix}, \begin{pmatrix} 0 & 1 \\ 0 & 0 \end{pmatrix}, \begin{pmatrix} 0 & 0 \\ 0 & -1 \end{pmatrix} \right\}$

Index

List of Symbols (continued)